The Mycota

Edited by
K. Esser

The Mycota

The Mycota

A Comprehensive Treatise
on Fungi as Experimental Systems
for Basic and Applied Research

Edited by K. Esser

XV Physiology and Genetics
Selected Basic and
Applied Aspects

Volume Editors:
T. Anke and D. Weber

 Springer

Series Editor

Professor Dr. Dr. h.c. mult. Karl Esser
Allgemeine Botanik
Ruhr-Universität
44780 Bochum, Germany

Tel.: +49 (234)32-22211
Fax.: +49 (234)32-14211
e-mail: Karl.Esser@rub.de

Volume Editors

Professor Dr. Timm Anke
Institute for Biotechnology and Drug Research,
IBWF e.V., Erwin-Schroedinger-Strasse 56,
67663 Kaiserslautern, Germany

Tel.: +49 (631) 205-2697
Fax: +49 (631) 205-2699
e-mail: timm.anke@ibwf.de

Dr. Daniela Weber
MIP International Pharma Research GmbH
Mühlstr. 50
66386 St. Ingbert, Germany

Tel.: +49 (163) 3111278
e-mail: weber-daniela@gmx.de

Library of Congress Control Number: 2009926066

ISBN 978-3-642-00285-4   e-ISBN 978-3-642-00286-1

Cover design: Erich Kirchner and WMXDesign GmbH, Heidelberg, Germany

Printed on acid-free paper   5 4 3 2 1 0

Karl Esser

(born 1924) is retired Professor of General Botany and Director of the Botanical Garden at the Ruhr-Universität Bochum (Germany). His scientific work focused on basic research in classical and molecular genetics in relation to practical application. His studies were carried out mostly on fungi. Together with his collaborators he was the first to detect plasmids in higher fungi. This has led to the integration of fungal genetics in biotechnology. His scientific work was distinguished by many national and international honors, especially three honorary doctoral degrees.

Timm Anke

(born 1944) studied Biochemistry at the University of Tuebingen, where he completed his PhD degree with a dissertation on the biosynthesis of fungal siderophores. In 1973 he joined Fritz Lipmann's group of at the Rockefeller University in New York City, where he investigated the biosynthesis of valinomycin, a streptomycete ionophore. After his return to Tuebingen in 1975 he started organizing a group searching for new antibiotics from basidiomycetes within the framework of Hans Zaehner's Collaboratorive Research Center (SFB 76) focusing on the chemistry and biology of microorganisms. In 1981 he became full Professor of Biotechnology at the University of Kaiserslautern and since 1998 he has headed the Institute of Biotechnology and Drug Research (IBWF e. V.) in Kaiserslautern. One of his outstanding achievements in the field of antibiotic research is the discovery of the strobilurins, a major class of agricultural fungicides, for which he was awarded the Karl-Heinz-Beckurts Prize in 1996.

Daniela Weber

(born 1978) studied natural products from endophytic fungi during her graduate work at the University of Kaiserslautern's Department of Biotechnology. She received a Chemiefonds fellowship for Ph.D. students. Her Ph.D., completed in 2006, focused on the isolation of endophytic fungi from medicinal plants and the investigation of the fungal metabolites. She later continued her work at the Institute of Biotechnology and Drug Research (IBWF, Kasiserslautern). In 2008 she accepted a position at MIP International Pharma Research GmbH (St. Ingbert, Germany). Her fields of activities are pharmaco-toxicological and clinical expert reports, pharmacovigilance, and regulatory affairs.

Series Preface

Mycology, the study of fungi, originated as a sub discipline of botany and was a descriptive discipline, largely neglected as an experimental science until the early years of this century. A seminal paper by Blakeslee in 1904 provided evidence for self incompatibility, termed "heterothallism", and stimulated interest in studies related to the control of sexual reproduction in fungi by mating-type specificities. Soon to follow was the demonstration that sexually reproducing fungi exhibit Mendelian inheritance and that it was possible to conduct formal genetic analysis with fungi. The names Burgeff, Kniep and Lindegren are all associated with this early period of fungal genetics research.

These studies and the discovery of penicillin by Fleming, who shared a Nobel Prize in 1945, provided further impetus for experimental research with fungi. Thus began a period of interest in mutation induction and analysis of mutants for biochemical traits. Such fundamental research, conducted largely with Neurospora crassa, led to the one gene: one enzyme hypothesis and to a second Nobel Prize for fungal research awarded to Beadle and Tatum in 1958. Fundamental research in biochemical genetics was extended to other fungi, especially to Saccharomyces cerevisiae, and by the mid-1960s fungal systems were much favored for studies in eukaryotic molecular biology and were soon able to compete with bacterial systems in the molecular arena.

The experimental achievements in research on the genetics and molecular biology of fungi have benefited more generally studies in the related fields of fungal biochemistry, plant pathology, medical mycology, and systematics. Today, there is much interest in the genetic manipulation of fungi for applied research. This current interest in biotechnical genetics has been augmented by the development of DNA-mediated transformation systems in fungi and by an understanding of gene expression and regulation at the molecular level. Applied research initiatives involving fungi extend broadly to areas of interest not only to industry but to agricultural and environmental sciences as well.

It is this burgeoning interest in fungi as experimental systems for applied as well as basic research that has prompted publication of this series of books under the title The Mycota. This title knowingly relegates fungi into a separate realm, distinct from that of either plants, animals, or protozoa. For consistency throughout this Series of Volumes the names adopted for major groups of fungi (representative genera in parentheses) areas follows:

Pseudomycota

Division:	Oomycota (Achlya, Phytophthora, Pythium)
Division:	Hyphochytriomycota

Eumycota

Division:	Chytridiomycota (Allomyces)
Division:	Zygomycota (Mucor, Phycomyces, Blakeslea)

Division:	Dikaryomycota
Subdivision:	Ascomycotina
Class:	Saccharomycetes (Saccharomyces, Schizosaccharomyces)
Class:	Ascomycetes (Neurospora, Podospora, Aspergillus)
Subdivision:	Basidiomycotina
Class:	Heterobasidiomycetes (Ustilago, Tremella)
Class:	Homobasidiomycetes (Schizophyllum, Coprinus)

We have made the decision to exclude from The Mycota the slime molds which, although they have traditional and strong ties to mycology, truly represent nonfungal forms insofar as they ingest nutrients by phagocytosis, lack a cell wall during the assimilative phase, and clearly show affinities with certain protozoan taxa.

The Series throughout will address three basic questions: what are the fungi, what dothey do, and what is their relevance to human affairs? Such a focused and comprehensive treatment of the fungi is long overdue in the opinion of the editors.

A volume devoted to systematics would ordinarily have been the first to appear in this Series. However, the scope of such a volume, coupled with the need to give serious and sustained consideration to any reclassification of major fungal groups, has delayed early publication. We wish, however, to provide a preamble on the nature off ungi, to acquaint readers who are unfamiliar with fungi with certain characteristics that are representative of these organisms and which make them attractive subjects for experimentation.

The fungi represent a heterogeneous assemblage of eukaryotic microorganisms. Fungal metabolism is characteristically heterotrophic or assimilative for organic carbon and some nonelemental source of nitrogen. Fungal cells characteristically imbibe or absorb, rather thaningest, nutrients and they have rigid cellwalls. The vast majority of fungi are haploid organisms reproducing either sexually or asexually through spores. The spore forms and details on their method of production have been used to delineate most fungal taxa. Although there is amultitude of spore forms, fungal spores are basically only of two types: (i) asexual spores are formed following mitosis (mitospores) and culminate vegetative growth, and (ii) sexual spores are formed following meiosis (meiospores) and are borne in or upon specialized generative structures, the latter frequently clustered in a fruit body. The vegetative forms of fungi are either unicellular, yeasts are an example, or hyphal; the latter may be branched to form an extensive mycelium.

Regardless of these details, it is the accessibility of spores, especially the direct recovery of meiospores coupled with extended vegetative haploidy, that have made fungi especially attractive as objects for experimental research.

The ability of fungi, especially the saprobic fungi, to absorb and grow on rather simple and defined substrates and to convert these substances, not only into essential metabolites but into important secondary metabolites, is also noteworthy. The metabolic capacities of fungi have attracted much interest in natural products chemistry and in the production of antibiotics and other bioactive compounds. Fungi, especially yeasts, are important in fermentation processes. Other fungi are important in the production of enzymes, citric acid and other organic compounds as well as in the fermentation of foods.

Fungi have invaded every conceivable ecological niche. Saprobic forms abound, especially in the decay of organic debris. Pathogenic forms exist with both plant and animal hosts. Fungi even grow on other fungi. They are found in aquatic as well as soil environments, and their spores may pollute the air. Some are edible; others are poisonous. Many are variously associated with plants as copartners in the formation of lichens and mycorrhizae, as symbiotic endophytes or as overt pathogens. Association with animal systems varies; examples include the predaceous fungi that trap nematodes, the micro fungi that grow in the anaerobic environment of the rumen, the many insect associated fungi and the medically important pathogens afflicting humans. Yes, fungi are ubiquitous and important.

There are many fungi, conservative estimates are in the order of 100,000 species, and there are many ways to study them, from descriptive accounts of organisms found in nature to laboratory experimentation at the cellular and molecular level. All such studies expand our knowledge of fungi and of fungal processes and improve our ability to utilize and to control fungi for the benefit of humankind.

We have invited leading research specialists in the field of mycology to contributeto this Series. We are especially indebted and grateful for the initiative and leadership shown by the Volume Editors in selecting topics and assembling the experts. We have all been a bit ambitious in producing these Volumes on a timely basis and there in lies the possibility of mistakes and oversights in this first edition. We encourage the readership to draw our attention to any error, omission or inconsistency in this Series in order that improvements can be made in any subsequent edition.

Finally, we wish to acknowledge the willingness of Springer-Verlag to host this project, which is envisioned to require more than 5 years of effort and the publication of at least nine Volumes.

Bochum, Germany
Auburn, AL, USA
April 1994

KARL ESSER
PAUL A. LEMKE
Series Editors

It is Time to Retire

During the Fourth International Mycological Congress in Regensburg (1989) while relaxing in a beer garden with Paul Lemke (USA), Dr. Czeschlik (Springer Verlag) discussed with us the possibility to publish a series about Fungi. We both were at first somewhat reserved, but after a comprehensive discussion this idea looked promising. In analogy to another series by Springer Verlag entitled *The Prokaryota* we decided to name the new series *The Mycota*.

Then Paul Lemke and I created a program involving seven volumes covering a wide area of Mycology. The first volume was presented in 1994 at the Fifth International Mycological Congress in Vancover (Canada). The other volumes followed step by step. After the early death of Paul Lemke (1995) I proceeded alone as Series Editor. Since evidently the series was well accepted by the scientific community and since the broad area of Fungi was not completely covered, it was decided to proceed with eight more volumes. In addition, in the following years second editions of eight volumes were published.

Now we present *Volume XV*. This will be the last volume of this series. As its title *"Physiology and Genetics: Selected Basic and Applied Aspects"* expresses, it contains special papers of various fields of Mycology which have been missing in the previous volumes.

Now, after 20 years of editing this series and at the age of 85 years, I guess it is the right time to terminate my editorship. I would like to express my sincerest thanks to all the volume editors, to the numerous authors for their successful cooperation and last but not least to Joan Bennett who supported me in editing three volumes.

I would also like to thank Springer Verlag, represented by Drs. Czeschlick and Schlitzberger for their support and cooperation.

I hope that *The Mycota* will also in the future find the interest of mycologists and other scholars interrested in Fungi and maybe some more second editions.

Bochum, Germany   KARL ESSER
May 2009

Volume Preface

More than 120 000 different fungal species have been described and it is estimated that there exist more than 1.5×10^6 species. Fungi have adopted many different ways of living in very diverse habitats as saprophytes, pathogens, symbionts or endophytes. Fungi and their products are used for the fermentation and processing of food and feeds, for biological control and for the production of vitamins and amino acids. Some of their secondary metabolites are used in medicine, e.g. as antibiotics, immuno-supressants, cholesterol-lowering drugs or as agrochemical fungicides. Recently, progress in the field of mycology has been substantial due to new methodological approaches and technologies, many of them DNA-based, strongly adding to the motivation to compile a new volume of "The Mycota".

Mycota XV "*Physiology and Genetics: Selected Basic and Applied Aspects*" provides a selection of state of the art reviews in traditional and new fields of mycology. As addressed in Chap. 1, DNA sequencing of functionally conserved homologue genes has resulted in new approaches to taxonomy, which hopefully will result in phylogenetic trees and taxonomic groups which can be understood and accepted by the majority of mycologists. As a homothallic fungus *Sordaria macrospora*, producing only sexual meiospores and no asexual conidia, is an ideal model for studying all stages of fruiting body formation and development. The application of several genetic tools allows advanced studies of genetic interactions controlling developmental plasticity (Chap. 2). Chapter 3 summarizes the current knowledge of *inteins*, internal protein sequences, which are "selfish elements" in fungal genomes. They are transcribed and translated together with their host protein and excised at the protein level. Chapter 4 gives an overview of fungal apoptosis in yeasts and filamentous fungi. Similarities and differences between fungal and mammalian apoptosis are discussed and the role of apoptosis in development and ageing are described and evaluated. A broad view on ways and means of the vegetative and sexual interaction of fungal colonies as well as the communication of fungi with bacteria, plants and animals is offered in Chap. 5. In the interaction of yeasts killer toxins play an important role. Their structures, modes of action and resistance as well as possible applications are discussed in Chap. 6. Chapter 7 deals with aspects of evolutionary and ecological interactions of fungi and insects; and Chap. 8 offers an insight into the occurrence and metabolites of endophytic fungi. Not all compounds isolated from plants are genuine plant metabolites but are produced by fungi. Thus the ergoline alkaloids present in *Convolvulaceae* are produced by fungi living in close association with secretory glands on the leaf surface (Chap. 9). Basidiomycetes are a rich source of unique secondary metabolites in most cases not found in other fungi. Chapter 10 offers a survey of new compounds isolated within the past decade, with special emphasis on bioactive metabolites. Genome-wide approaches to identify genes or gene products essential for the establishment of pathogenic interactions between plant host and fungal pathogen are discussed in Chap. 11. In addition, the authors stress the important role of small molecules in identifying and validating new targets for fungicides. Helminths can pose serious problems to animal and human health. It is therefore quite remarkable that fungi produce low molecular weight compounds specifically interfering with reactions not present in the mammalian hosts, paving the way to non-toxic medications or agrochemicals (Chap. 12). Chapter 13 describes the occurrence, structures and biological activities of peptides and depsipeptides produced by fungi and discusses the importance of these compounds as lead compounds for agricultural and pharmaceutical applications. The sequencing of whole fungal genomes has revealed that there are many more genes supposedly coding for secondary metabolites than there are compounds already isolated and characterized. As can be seen in Chap. 14, homologue overexpression of a regulatory gene construct can indeed lead to silent gene expression and production of the corresponding metabolite. Chapters 15–17 deal with genes, enzymes and products of important biogenetic pathways. Nonribosomal peptide synthetases of fungi differ in some major aspects from the corresponding bacterial enzymes. The same is true for fungal polyketides which constitute a large part of the fungal secondary metabolome. The importance of the detrimental mycotoxin ochratoxin A for human health and the alimentary industry was recognized only recently. The investigation of its biosynthesis and regulation is important for developing strategies for its avoidance in food and feed. Chapter 18 is devoted to genetic and metabolic engineering in fungi, key areas of research aiming at the improved application of these organisms in biotechnology.

We do hope that readers enjoy reading this volume of *The Mycota*. We are very grateful to the contributing authors, whose expertise and efforts have made this project possible. We thank Dr. Andrea Schlitzberger of Springer Verlag for her support and engagement during the preparation of this volume.

Kaiserslautern and St. Ingbert, Germany   TIMM ANKE
May 2009   DANIELA WEBER
                                                                   Volume Editors

Contents

List of Contributors

HEIDRUN ANKE
(e-mail: anke@ibwf.de, Tel.: +49-631-31672-10, Fax: +49-631-31672-15)
Institute for Biotechnology and Drug Research, IBWF e.V., Erwin-Schroedinger-Strasse 56, 67663 Kaiserslautern, Germany

TIMM ANKE
(e-mail: anke@rhrk.uni-kl.de, Tel.: +49-631-205-3046, Fax: +49-631-205-2999)
Institute for Biotechnology and Drug Research, IBWF e.V., Erwin-Schroedinger-Strasse 56, 67663 Kaiserslautern, Germany

LUIS ANTELO
(e-mail: antelo@ibwf.de, Fax: +49-631-31672-15)
Institute for Biotechnology and Drug Research, IBWF e.V., Erwin-Schroedinger-Strasse 56, 67663 Kaiserslautern, Germany

SEBASTIAN BERGMANN
Molecular and Applied Microbiology, Leibniz Institute for Natural Product Research and Infection Biology (HKI), and Friedrich Schiller University, Beutenbergstrasse 11a, 07745 Jena, Germany

AXEL A. BRAKHAGE
(e-mail: axel.brakhage@hki-jena.de, Tel.: +49-3641-65-6601, Fax: +49-3641-65-6825)
Molecular and Applied Microbiology, Leibniz Institute for Natural Product Research and Infection Biology (HKI), and Friedrich Schiller University, Beutenbergstrasse 11a, 07745 Jena, Germany

DIANA BRUST
(e-mail: brust@bio.uni-frankfurt.de)
Institute for Molecular Biosciences, Department of Biosciences and Cluster of Excellence Macromolecular Complexes, Goethe-University, Max-von-Laue-Strasse 9, 60438 Frankfurt, Germany

KATRIN EISFELD
(e-mail: eisfeld@ibwf.de, Tel.: +49-631-31672-0, Fax: +49-631-31672-15)
Institute of Biotechnology and Drug Research, Erwin-Schrödinger-Strasse 56, 67663 Kaiserslautern, Germany

SKANDER ELLEUCHE
Abteilung Genetik eukaryotischer Mikroorganismen, Institut für Mikrobiologie und Genetik, Georg-August-Universität Göttingen, Grisebachstrasse 8, 37077 Göttingen, Germany

INES ENGH
(e-mail: ines.engh@rub.de, Tel.: +49-234-3224974, Fax: +49-234-3214184)
Lehrstuhl für Allgemeine und Molekulare Botanik, Fakultät für Biologie, Ruhr-Universität Bochum, Universitätsstrasse 150, 44780 Bochum, Germany

ULRIKE FOHGRUB
Botanical Institute, Department of Plant Genetics and Molecular Biology, Christian-Albrechts-University of Kiel, Olshausenstrasse 40, 24098 Kiel, Germany

ANDREW J. FOSTER
(e-mail: foster@ibwf.de, Tel.: +49-631-2054063, Fax: +49-631-3167215)
Institut für Biotechnologie und Wirkstoff-Forschung e.V./Institute for Biotechnology and Drug Research, Erwin-Schrödinger-Strasse 56, 67663 Kaiserslautern, Germany

ROLF GEISEN
(e-mail: Rolf.Geisen@mri.bund.de, Tel.: +49-721-6625-450, Fax: +49-721-6625-453)
Max Rubner Institute, Federal Research Institute of Nutrition and Food, Haid-und-Neu-Strasse 9, 76131 Karlsruhe, Germany

ANDREA HAMANN
(e-mail: a.hamann@bio.uni-frankfurt.de, Tel.: +49-69-798-29548, Fax: +49-69-798-29435)
Institute for Molecular Biosciences, Department of Biosciences and Cluster of Excellence Macromolecular Complexes, Goethe-University, Max-von-Laue-Strasse 9, 60438 Frankfurt, Germany

CHRISTIAN HERTWECK
(e-mail: christian.hertweck@hki-jena.de, Tel.: +49-3641-5321100/01, Fax: +49-3641-5320408)
Leibniz Institute for Natural Product Research and Infection Biology, HKI, and Friedrich Schiller University, Beutenbergstrasse 11a, 07745 Jena, Germany

FRANK KEMPKEN
(e-mail: fkempken@bot.uni-kiel.de, Tel.: +49-431-880-4274, Fax: +49-431-880-4248)
Botanical Institute, Department of Plant Genetics and Molecular Biology, Christian-Albrechts-University of Kiel, Olshausenstrasse 40, 24098 Kiel, Germany

ROLAND KLASSEN
(e-mail: roland.klassen@uni-muenster.de, Fax: +49-251-83-38388)
Institut für Molekulare Mikrobiologie und Biotechnologie, Westfälische Wilhelms-Universität Münster, Corrensstrasse 3, 48149 Münster, Germany

ULRICH KÜCK
(e-mail: ulrich.kueck@ruhr-uni-bochum.de, Tel.: +49-0234-322-6212, Fax: +49-234-32-14184)
Lehrstuhl für Allgemeine und Molekulare Botanik, Fakultät für Biologie, Ruhr-Universität Bochum, Universitätsstrasse 150, 44780 Bochum, Germany

URSULA KÜES
(e-mail: ukuees@gwdg.de, Tel.: +49-551-397024, Fax: +49-551-392705)
Division of Molecular Wood Biotechnology and Technical Mycology, Büsgen-Institute, Georg-August-University Göttingen, Büsgenweg 2, 37077 Göttingen, Germany

ECKHARD LEISTNER
(e-mail: eleistner@uni-bonn.de, Tel.: +49-228-73-3199, Fax: +49-228-73-3250)
Institut für Pharmazeutische Biologie, Rheinische Friedrich Wilhelm-Universität Bonn, Nussallee 6, 53115 Bonn, Germany

ROKURO MASUMA
(e-mail: masuma@lisci.kitasato-u.ac.jp)
The Kitasato Institute for Life Sciences and Graduate School of Infection Control Sciences, Kitasato University, 5-9-1 Shirokane, Minato-ku, Tokyo 108-8641, Japan

FRIEDHELM MEINHARDT
(e-mail: meinhar@uni-muenster.de, Tel.: +49-251-83-39825, Fax: +49-251-83-38388)
Institut für Molekulare Mikrobiologie und Biotechnologie, Westfälische Wilhelms-Universität Münster, Corrensstrasse 3, 48149 Münster, Germany

VERA MEYER
University of Technology Berlin, Department of Microbiology and Genetics, Gustav-Meyer-Allee 25, 13355 Berlin, Germany; and Leiden University, Institute of Biology, Research Group for Molecular Microbiology, Wassenaarseweg 64, 2333 AL Leiden, The Netherlands

MÓNICA NAVARRO-GONZÁLEZ
(e-mail: mnavarr@gwdg.de, Fax: +49-551-392705)
Division of Molecular Wood Biotechnology and Technical Mycology, Büsgen-Institute, Georg-August-University Göttingen, Büsgenweg 2, 37077 Göttingen, Germany

NICOLE NOLTING
Abteilung Genetik eukaryotischer Mikroorganismen, Institut für Mikrobiologie und Genetik, Georg-August-Universität Göttingen, Grisebachstrasse 8, 37077 Göttingen, Germany

MINOU NOWROUSIAN
Lehrstuhl für Allgemeine und Molekulare Botanik, Fakultät für Biologie, Ruhr-Universität Bochum, Universitätsstrasse 150, 44780 Bochum, Germany

SATOSHI ŌMURA
(e-mail: omura-s@kitasato.or.jp, Tel.: +81-3-5791-6101, Fax: +81-3444-8360)
The Kitasato Institute for Life Sciences and Graduate School of Infection Control Sciences, Kitasato University, 5-9-1 Shirokane, Minato-ku, Tokyo 108-8641, Japan

Heinz D. Osiewacz
(e-mail: osiewacz@em.uni-frankfurt.de, Tel.: +49-69-798-24827, Fax: +49-69-798-24822)
Institute for Molecular Biosciences, Department of Biosciences and Cluster of
Excellence Macromolecular Complexes, Goethe-University, Max-von-Laue-Strasse 9,
60438 Frankfurt, Germany

Stefanie Pöggeler
(e-mail: spoegge@gwdg.de, Tel.: +49-551-3913930, Fax: +49-551-3910123)
Abteilung Genetik eukaryotischer Mikroorganismen, Institut für Mikrobiologie und
Genetik, Georg-August-Universität Göttingen, Grisebachstrasse 8, 37077 Göttingen,
Germany

Marko Rohlfs
Zoological Institute, Department of Animal Ecology, Christian-Albrechts-University
of Kiel, Olshausenstrasse 40, 24098 Kiel, Germany

Kirstin Scherlach
Biomolecular Chemistry, Leibniz Institute for Natural Product Research and Infection
Biology (HKI), and Friedrich Schiller University, Beutenbergstrasse 11a, 07745 Jena,
Germany

Jochen Schmid
(e-mail: j.schmid@lb.tu-berlin.de, Tel.: +49-30-314-72750, Fax: +49-30-314-72922)
University of Technology Berlin, Department of Microbiology and Genetics, Gustav-
Meyer-Allee 25, 13355 Berlin, Germany

Markus Schmidt-Heydt
(e-mail: Markus.Schmidt-Heydt@MRI.Bund.de, Tel.: +49-721-6625-450, Fax: +49-721-
6625-453)
Max Rubner Institute, Federal Research Institute of Nutrition and Food, Haid-und-
Neu-Strasse 9, 76131 Karlsruhe, Germany

Julia Schuemann
Biomolecular Chemistry, Leibniz Institute for Natural Product Research and Infection
Biology (HKI), and Friedrich Schiller University, Beutenbergstrasse 11a, 07745 Jena,
Germany

Anja Schüffler
Institut für Biotechnologie und Wirkstoff-Forschung e.V./Institute for Biotechnology
and Drug Research, Erwin-Schroedinger-Strasse 56, 67663 Kaiserslautern, Germany

Kazuro Shiomi
(e-mail: shiomi@lisci.kitasato-u.ac.jp)
The Kitasato Institute for Life Sciences and Graduate School of Infection Control
Sciences, Kitasato University, 5-9-1 Shirokane, Minato-ku, Tokyo 108-8641, Japan

Ulf Stahl
(e-mail: Ulf.Stahl@lb.tu-berlin.de, Tel.: +49-30-314-72750, Fax: +49-30-314-72922)
University of Technology Berlin, Department of Microbiology and Genetics, Gustav-
Meyer-Allee 25, 13355 Berlin, Germany

ULRIKE STEINER
(e-mail: u-steiner@uni-bonn.de)
Institut für Nutzpflanzenwissenschaften und Ressourcenschutz (INRES), Rheinische Friedrich Wilhelm-Universität Bonn, Nussallee 9, 53115 Bonn, Germany

ECKHARD THINES
(e-mail: thines@ibwf.de, Tel.: +49-631-316720, Fax: +49-631-3167215)
Institut für Biotechnologie und Wirkstoff-Forschung e.V./Institute for Biotechnology and Drug Research, Erwin-Schrödinger-Strasse 56, 67663 Kaiserslautern, Germany

MONIKA TRIENENS
Zoological Institute, Department of Animal Ecology, Christian-Albrechts-University of Kiel, Olshausenstrasse 40, 24098 Kiel, Germany

DANIELA WEBER
(e-mail: weber-daniela@gmx.de, Tel.: +49-68-94-590-950, Fax: +49-68-94-590-95-275)
Institute of Biotechnology and Drug Research, Erwin-Schrödinger-Strasse 56, 67663 Kaiserslautern, Germany; *present address*: MIP International Pharma Research GmbH, Mühlstrasse 50, 66386 St. Ingbert, Germany

ROLAND W.S. WEBER
(e-mail: Roland.Weber@lwk-niedersachsen.de, Tel.: +49-4162-6016-133, Fax: +49-4162-6016-99133)
Obstbau Versuchs- und Beratungszentrum/Fruit Research and Advisory Centre (OVB) Jork, Moorende 53, 21635 Jork, Germany

1 Recent Developments in the Molecular Taxonomy of Fungi

Roland W.S. Weber[1]

CONTENTS

I. Introduction

Taxonomy is the science of classification, i.e. the grouping of organisms into defined categories (taxa). Ideally, a classification scheme should be 'natural' in the sense that all organisms in a given taxon should be related to each other by common ancestry. Traditionally, however, this was not necessarily the case with fungi because taxonomic approaches based on morphological and microscopic characters, later augmented by ultrastructural and biochemical features, did not always distinguish between homologies and analogies. Traditional taxonomies were arbitrary also in the sense that different taxonomists proposed widely differing classification schemes, depending on which features they regarded as relevant and at which taxonomic level.

The past two decades have witnessed a revolution in the taxonomy of fungi because it became possible to generate and analyse homologous DNA sequences of functionally conserved genes (initially mainly ribosomal DNA sequences) on a routine basis, leading to a more 'objective' comparison of taxa. Initial attempts at molecular phylogeny created considerable complications because each phylogenetic tree was usually based on the analysis of a single gene, and widely different phylogenies could result if different genes were used. The situation around the turn of the millennium and in the following years was therefore comparable to a football game with shifting goal-posts, as especially those involved in teaching fungal taxonomy will recall with horror. Several publications also reflect the fluid state of the discipline at that time, especially *The Mycota VIIA* and *VIIB* (McLaughlin et al. 2001) and the ninth edition of *Ainsworth and Bisby's Dictionary of the Fungi* (Kirk et al. 2001). These works have become outdated quite rapidly in terms of taxonomic concepts, though not, of course, in their valuable information on general biological features of the fungi. The third edition of *Introduction to Fungi* (Webster and Weber 2007), although principally intended to be used as a textbook pursuing a holistic approach to the fungi rather than as a taxonomic work, also had to be based on the taxonomy of the period. By the time of manuscript submission in March 2006, the outlines of certain 'natural' groups of fungi had become apparent, but these could neither be delimited clearly, nor was it possible to assign formal names to many of them for lack of existence, validation or general acceptance.

This state of transition led a large consortium of fungal taxonomists to join forces and develop a classification scheme based on multi-gene phylogenies in the hope that this would provide a

---

[1]Obstbau Versuchs- und Beratungszentrum/Fruit Research and Advisory Centre (OVB) Jork, Moorende 53, 21635 Jork, Germany; e-mail: roland.weber@lwk-niedersachsen.de

Physiology and Genetics, 1st Edition
The Mycota XV
T. Anke and D. Weber (Eds.)
© Springer-Verlag Berlin Heidelberg 2009

sound, permanent framework. Following the establishment of a communications platform ('deep hypha'), the 'AFTOL' (assembling the fungal tree of life) project supported most of the immense amount of DNA sequencing work and data analysis (Blackwell et al. 2006). Both undertakings were funded by the United States National Science Foundation (NSF). The entire November/December 2006 issue of *Mycologia* was dedicated to reports of the results of these efforts, and a classification scheme synthesising the main findings was published by Hibbett et al. (2007). These authors emphasised the broad support and input which their scheme had received from numerous taxonomists, and called on the worldwide mycological community to adopt their unified system with generally accepted taxon names in future. To this end, many of the names of higher taxa proposed by Hibbett et al. (2007) are based on concepts readily recognised by most mycologists, which greatly facilitates the usage of their scheme. An arguable exception is the re-naming of higher taxa after key species, which would replace e.g. Urediniomycetes by Pucciniomycotina or Hymenomycetes by Agaricomycotina. It is reassuring to see that the introduction of these terms is an ongoing process with occasional inconsistencies even in publications by leading 'AFTOL' authors.

Therefore, the main purpose of the present chapter is to give a brief overview of the current taxonomic concept, highlighting changes from equivalent taxonomic groupings delimited in earlier schemes represented by Webster and Weber (2007), McLaughlin et al. (2001) and Kirk et al. (2001). At the same time, the likely stability of the 'AFTOL' system is evaluated. Since work by this consortium did not cover organisms unrelated to Fungi but nonetheless studied by mycologists, their current phylogenetic and taxonomic placement is also summarised. Finally, there is a brief consideration of problems and potential of current fungal taxonomy in mycology teaching.

II. Non-Fungal Organisms Studied by Mycologists

Genera such as *Phytophthora*, *Pythium*, *Peronospora*, *Plasmopara* or *Plasmodiophora* are so intrinsically linked with the history of mycology and fungal plant pathology that they continue to be studied in mycological laboratories and to be taught in mycology lectures and practical classes. Slime moulds and other non-fungal protists will be encountered by mycologists during forays. Protists are subjected to two mutually incompatible classification systems: a zoological one governed by the International Code of Zoological Nomenclature and a mycological one according to the International Code of Botanical Nomenclature. The resulting confusion is multiplied by the rapid discovery of new protist species and new relationships between existing taxa by molecular phylogenetic approaches. Adl et al. (2005, 2007), recognising that neither of the two Codes provides a stable classification system for these organisms, broke free of both and established a scheme which retains as many of the commonly known names as possible. This move is supported by many protozoologists.

Some five or six kingdom-sized super-groups of eukaryotes can be resolved by current phylogenetic studies based on multi-gene analyses and modern analytical methods (Simpson and Roger 2004; Keeling et al. 2005). All but one of these harbour organisms studied by mycologists (Fig. 1.1). An overview of formal and informal names of relevant higher taxa is given in Fig. 1.2.

A. Slime Moulds

Slime moulds may exist vegetatively as amoebae (cellular slime moulds) or as multinucleate plasmodia (plasmodial slime moulds). Both ingest particulate food by phagocytosis. Reproduction is by means of single-spored or multi-spored structures termed sporocarps or sorocarps, respectively. Flagellate stages (swarmers) bearing two whiplash-type flagella may be present in the life cycle of certain groups (myxogastrids, dictyostelids).

The separation of a small group, the acrasids, from the bulk of slime moulds has long been recognised on the basis of phylogenetic studies (Baldauf et al. 2000) and finds its morphological manifestation in the shape of the amoeba which produces a single lobed (lobose) anterior pseudopodium in acrasids, as opposed to the pointed (filose) pseudopodia emitted by amoebae of all other slime moulds. Acrasids are now grouped among the Excavata. All other slime moulds, including both cellular and plasmodial forms, belong to three phylogenetically related groups

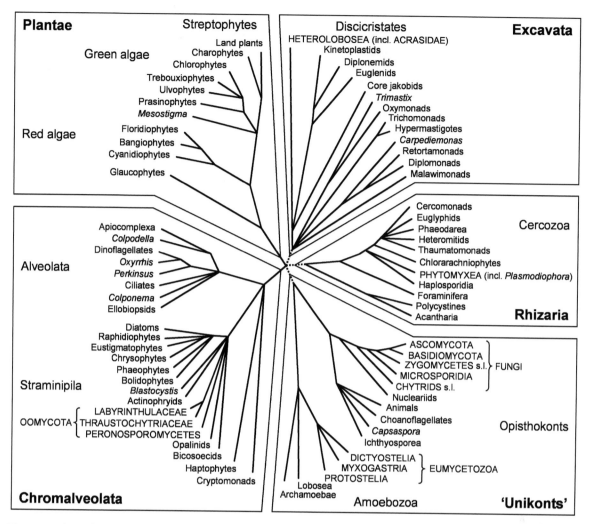

Fig. 1.1. A hypothetical phylogenetic tree of eukaryotes showing five 'supergroups', based on molecular phylogenies, other molecular characters and morphological and biochemical evidence. *Dotted lines* indicate relationships resolved on preliminary evidence only, whereas relationships were left unresolved where there was no evidence of branching order. Groups traditionally studied by mycologists are printed in *capital letters*. Re-drawn from Keeling et al. (2005), with permission from Elsevier

within the Amoebozoa (Fig. 1.1). Dictyostelid slime moulds are cellular forms, represented by the well-known *Dictyostelium discoideum* with its cAMP-mediated aggregation of thousands of amoebae into a pseudoplasmodium, which goes on to form a slug and ultimately a sorocarp (see Kessin 2001). Protostelid slime moulds may form filose amoebae or small plasmodia, whereas in myxogastrid forms such as *Fuligo septica* or *Physarum polycephalum*, the plasmodium is dominant, amoebae being reduced in the life cycle to a brief period following spore germination (see Webster and Weber 2007).

B. *Plasmodiophora* and Related Species

Plasmodiophora brassicae is the cause of club root of brassicas. Infection is initiated when a zoospore with two whiplash-type flagella encysts on a root hair. The penetration mechanism is highly characteristic in that a bullet-shaped stylet is forced by turgor pressure from the spore cyst through the root cell wall, injecting a small wall-less amoeba into the host cytoplasm. Plasmodia capable of phagocytosis develop from this amoeba and migrate deeper into the root tissue, breaking through plant cell walls along their way.

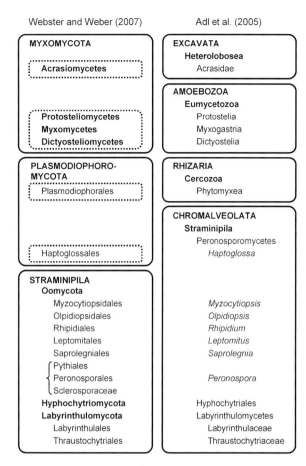

Webster and Weber (2007)   Adl et al. (2005)

Fig. 1.2. Organisms not belonging to the Fungi yet traditionally studied by mycologists, in a mycological system (Webster and Weber 2007) and in the protistological classification scheme of Adl et al. (2005). Organisms grouped within a phylum but already suspected at the time of being phylogenetically unrelated to each other are separated by *dotted boxes*. Orders united by a *bracket* are currently grouped together in Peronosporales by many authors

Archibald and Keeling (2004) and others have provided evidence that *Plasmodiophora* and allied genera (*Spongospora*, *Polymyxa*) are related to Cercozoan protists. Adl et al. (2005) have also placed plasmodiophorids among the Cercozoa within the 'kingdom' Rhizaria (see Fig. 1.1).

C. Straminipila

Organisms possessing cellulose-containing cell walls, an inner mitochondrial membrane folded into tubular rather than plate-like cristae, morphologically recognisable Golgi stacks, biflagellate heterokont zoospores (i.e. with one flagellum of the tinsel and the other of the whiplash type), and a

lysine biosynthetic pathway via α,ε-diaminopimelic acid (DAP) rather than α-aminoadipic acid (AAA), are accommodated in a kingdom named Chromista by workers emphasising the lack of chlorophyll b in its photosynthetic members, or Straminipila by others highlighting the tinsel-type flagellum with its uniquely complex architecture. Straminipila/Chromista are now grouped together with Alveolata (dinoflagellates, ciliates, and other taxa; Fast et al. 2002) in the 'kingdom' Chromalveolata (Simpson and Roger 2004; Keeling et al. 2005).

Several phyla within the Straminipila are of relevance to mycology. The phylum Labyrinthulomycota contains marine organisms in which individual vegetative cells produce a network of slime tracks. In addition, each cell is surrounded by a wall comprising a Golgi-derived ʟ-galactose polymer. In Labyrinthulales, the cells move within their slime tracks, whereas in Thraustochytriales the slime track forms a rhizoid-like network at the base of a thallus-like structure (see Webster and Weber 2007). Both orders produce heterokont zoospores.

Hyphochytriomycota resemble chytrid fungi in forming thalli comprised of one or more sporangia linked by rhizoids. The zoospore contains only a forward-directed tinsel flagellum, the backward-pointing whiplash flagellum having been lost during the course of evolution.

By far the most important group of straminipilous organisms from a mycological perspective is the phylum Oomycota. Included here are organisms with heterokont zoospores and oogamous sexual reproduction, e.g. filamentous water moulds such as the Saprolegniales (*Saprolegnia*, *Achlya*), thallic aquatic saprotrophs or parasites of algae and other organisms, and filamentous species adapted to the terrestrial environment. Terrestrial straminipilous fungi have traditionally been separated into Pythiales (saprotrophic forms and necrotrophic pathogens especially of plants), Peronosporales and Sclerosporaceae (obligately biotrophic downy mildews of dicotyledonous plants and grasses, respectively). However, several recent phylogenetic analyses point towards intercalations between these taxa (Riethmüller et al. 2002; Villa et al. 2006; Thines et al. 2008) so that they may all eventually be united in one group (e.g. Peronosporales), at the possible exclusion of *Albugo* (Hudspeth et al. 2003). The collateral abandonment or revision of well-known genera such as *Pythium*, *Phytophthora* or *Peronospora* has not yet been completed. In view of the ongoing trend to name higher taxa on the basis of exemplar genera, it may be deemed necessary to replace the term Oomycota/Oomycetes by Peronosporomycetes (Dick 2001).

D. *Haptoglossa*

Haptoglossa spp. infect rotifers or nematodes by a forceful injection mechanism very similar to that of *Plasmodiophora* (see above), although a walled sporidium instead of a wall-less amoeba enters the living host. Further, zoospores of *Haptoglossa*, where formed, appear to be anisokont, i.e. with two whiplash-type flagella of unequal length, like those of *Plasmodiophora*. Nonetheless, molecular phylogenetic studies have shown *Haptoglossa* spp. to be members of Straminipila, and indeed to belong to the Peronosporomycetes (Hakariya et al. 2007). *Haptoglossa* had also been grouped there by Adl et al. (2005) and, more intuitively, by earlier workers.

III. The 'Basal Fungi'

James et al. (2006a) published a 70-author paper about the trunk of the fungal phylogenetic tree based on analyses of six functionally conserved genes, i.e. the nuclear ribosomal DNA operon (18S-, 28S-, 5.8S-rDNA), the ribosomal elongation factor gene *EF1α*, and the RNA polymerase II subunit genes *RPB1* and *RPB2*. As a result of this fundamental work and numerous other contributions, the taxonomy of the basal fungi was modified extensively by Hibbett et al. (2007), as summarised in Fig. 1.3. The two established phyla, Chytridiomycota and Zygomycota, were broken into a total of eight groups, to which a further phylum, the Microsporidia, was added.

A. Microsporidia

The inclusion of the Microsporidia within the Fungi has been highly controversial until very recently (e.g. Keeling et al. 2000; Tanabe et al. 2002), and the continued use of the zoological term 'Microsporidia' is a reminder of that debate. Included in this phylum are the most basal fungi according to current phylogenetic knowledge.

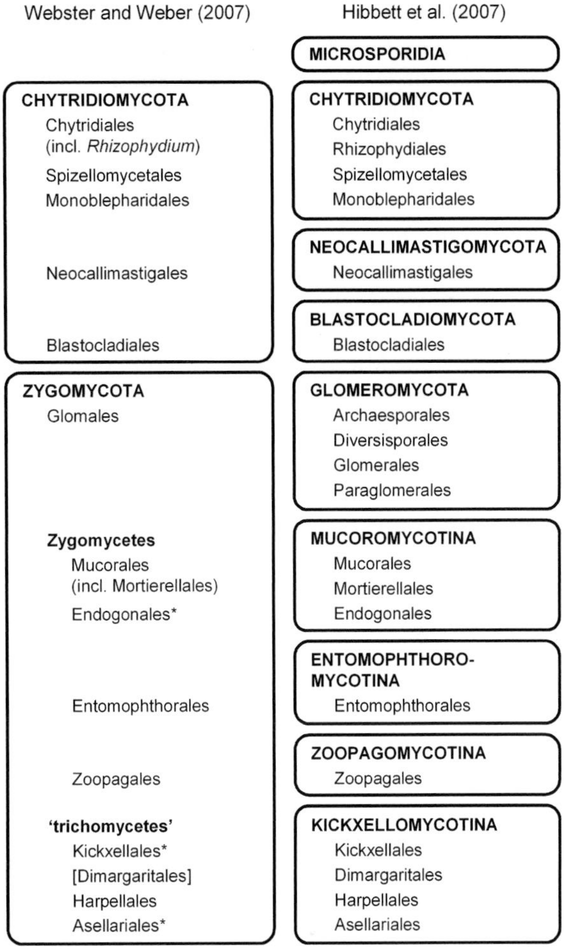

Webster and Weber (2007)	Hibbett et al. (2007)
	MICROSPORIDIA
CHYTRIDIOMYCOTA Chytridiales (incl. *Rhizophydium*) Spizellomycetales Monoblepharidales	**CHYTRIDIOMYCOTA** Chytridiales Rhizophydiales Spizellomycetales Monoblepharidales
Neocallimastigales	**NEOCALLIMASTIGOMYCOTA** Neocallimastigales
Blastocladiales	**BLASTOCLADIOMYCOTA** Blastocladiales
ZYGOMYCOTA Glomales	**GLOMEROMYCOTA** Archaesporales Diversisporales Glomerales Paraglomerales
Zygomycetes Mucorales (incl. Mortierellales) Endogonales*	**MUCOROMYCOTINA** Mucorales Mortierellales Endogonales
Entomophthorales	**ENTOMOPHTHORO-MYCOTINA** Entomophthorales
Zoopagales	**ZOOPAGOMYCOTINA** Zoopagales
'trichomycetes' Kickxellales* [Dimargaritales] Harpellales Asellariales*	**KICKXELLOMYCOTINA** Kickxellales Dimargaritales Harpellales Asellariales

Fig. 1.3. Classification of the lower fungi according to the current 'AFTOL' scheme (Hibbett et al. 2007) in comparison with Webster and Weber (2007) summarising earlier approaches. Groups highlighted by an *asterisk* were mentioned by these authors but not discussed in detail. Supplementary data from Kirk et al. (2001) are indicated in *square brackets*

These organisms are obligate parasites of animals, infecting their host cells by injection of a small protoplast from a spore through a tube initially coiled up inside the spore, then extruded by osmotic pressure (Keeling and Fast 2002). Such an infection mechanism is reminiscent of that found in *Plasmodiophora* (Cercozoa) or *Haptoglossa* (Chromalveolata, Straminipila) with which Microsporidia are not related. Instead, their closest relative in the analysis by James et al. (2006a) was the chytrid fungus *Rozella allomycis*, an endoparasite of *Allomyces*. Whereas *R. allomycis* possesses a posteriorly uniflagellated zoospore typical of the chytrids, Microsporidia do not, indicating that the flagellum must have been lost during the course of evolution. Microsporidia show an accelerated rate of DNA sequence evolution as compared to other fungi, and this may be explained by the loss of function of many genes due to an increased reliance upon the host's metabolic machinery (Inagaki et al. 2004). Such trends may cause 'long branch attractions', i.e. the false grouping-together of unrelated organisms at the base of phylogenetic schemes. Therefore, the trunk of the fungal tree and the circumscription of the phylum Microsporidia are still somewhat diffuse at present.

B. Chytrids

The original phylum Chytridiomycota contained all true Fungi with zoospores bearing a single posterior whiplash-type flagellum, or several such flagella (as in the anaerobic rumen fungi). The thallus is either holocarpic, consisting of a sac-like structure which later converts to a zoosporangium, or eucarpic, i.e. sporangia are formed from rhizoids or a rhizomycelium. Monophyly of the Chytridiomycota could not be upheld on the basis of the studies by James et al. (2006a, b) and Hibbett et al. (2007). Instead, this taxon was retained as a more narrowly defined group, and two additional phyla were created: Neocallimastigomycota and Blastocladiomycota (Fig. 1.3). The Blastocladiomycota were identified as a sister group to the basal non-flagellated fungi formerly called Zygomycota, and to higher fungi (Ascomycota, Basidiomycota).

Nonetheless, there are still some escapees from a unified chytrid taxonomy, e.g. *R. allomycis* with an affinity for Microsporidia as mentioned above, and *Olpidium brassicae* which groups with *Basidiobolus ranarum* close to the Entomophthorales (Zygomycota). Placement of *Basidiobolus* in the vicinity of Entomophthorales is a relief to teaching mycologists because this fungus shares so many key biological features with the insect-pathogenic Entomophthorales that its inclusion in the Chytridiomycota as suggested by preliminary phylogenetic studies (e.g. Berbee and Taylor 2001) did not make any biological sense (Webster and Weber 2007). All the more curious is the current association of *O. brassicae*, a seemingly typical chytrid, with the Entomophthorales; it will be interesting to see whether this remains its final resting place.

Rhizophlyctis and *Caulochytrium* also remain to be accommodated at ordinal level (Hibbett et al. 2007), and the establishment of further chytrid orders therefore cannot be ruled out.

James et al. (2006a) discussed the number of occasions on which the flagellum may have been lost in the course of fungal evolution, and counted at least four; once in *Rozella*, once in *Hyaloraphidium curvatum*, formerly thought to be derived from a marine alga (Ustinova et al. 2000), and at least twice en route to the zygomycetous fungi.

C. Zygomycete-Type Fungi

Zygomycetous fungi reproduce sexually by zygospores arising from suspensors which are usually isogamous. Asexual reproduction is typically by sporangiospores arising from internal cleavage of protoplasm, or by single-spored sporangia or even true conidia (Webster and Weber 2007). Fungi formerly included in the phylum Zygomycota present one of the greatest challenges to current phylogeny-based classification because several recent studies have given conflicting results of monophyly or polyphyly (Tanabe et al. 2004; Liu et al. 2006; White et al. 2006). Mindful of these uncertainties, Hibbett et al. (2007) broke up the Zygomycota into four subphyla, but held open the possibility that a revised Zygomycota of more restricted circumscription might be re-established and validated in future. This would be highly desirable, especially from the perspective of mycology teaching.

The subphylum Mucoromycotina includes the largest order, Mucorales, together with Mortierellales and Endogonales. The former two contain mainly saprotrophic members, the latter facultatively ectomycorrhizal species forming 'pea truffles' in which zygospores are clustered together in subterranean fruit-bodies. Webster and Weber (2007) did not describe Endogonales in detail and included *Mortierella* and its allies in Mucorales on the grounds of unifying biological themes while being aware of their possible status as a separate order. Two other subphyla, the mostly entomopathogenic Entomophthoromycotina and the Zoopagomycotina (pathogens of animals and fungi), were treated as orders by Webster and Weber (2007); that change to sub-phylum is easily accommodated.

A difficult case is presented by the Kickxellomycotina, a group formerly regarded as 'trichomycetes'.

Included here are biologically heterogeneous saprotrophs and mycoparasites (Kickxellales, Dimargaritales) as well as commensals or weak parasites of insect guts (Harpellales and Asellariales, i.e. trichomycetes sensu stricto). Other orders formerly regarded as trichomycetes, especially Eccrinales and Amoebidiales, were relegated to protozoa (Cafaro 2005), where they were classified among Ichthyosporea by Adl et al. (2005; see Fig. 1.1). Nonetheless, the Kickxellomycotina represent a likely candidate for substantial future taxonomic rearrangements among zygomycete-type fungi.

D. Glomeromycota

Formerly included in the order Glomales among the Zygomycota (Morton and Benny 1990), this group of obligately symbiotic fungi associated with vesicular-arbuscular mycorrhiza was given the status of a phylum, Glomeromycota (Schüssler et al. 2001). This placement has won widespread support as being monophyletic. The only known non-mycorrhizal member is *Geosiphon pyriformis*, which contains intracellular cyanobacteria as symbionts. Because glomeralean fungi cannot be grown in pure culture, species are difficult to define, and considerable re-arrangements below the phylum level may well have to be performed in future (Redecker and Raab 2006). Among all basal fungi, Glomeromycota are the closest known relatives of Basidiomycota and Ascomycota (James et al. 2006a).

IV. Ascomycota

The subkingdom Dikarya, referring to the presence of at least a short dikaryotic phase in the life cycle, was erected by Hibbett et al. (2007) to accommodate Ascomycota and Basidiomycota, thereby reflecting their close phylogenetic relationship. It is debatable whether this subkingdom is necessary from a taxonomic point of view if there is no counterpart to group together the 'lower fungi'. The retention of Ascomycota and Basidiomycota as phyla is welcome as these terms are in universal use (e.g. Kirk et al. 2001). Within the phylum Ascomycota, three subphyla are currently recognised and are discussed in their turn. A summary of current classes and orders of Ascomycota is given in Fig. 1.4.

A. Taphrinomycotina

Nishida and Sugiyama (1994) gave the provisional name 'Archiascomycetes' to this most basal group of Ascomycota, but Hibbett et al. (2007) accepted the term Taphrinomycotina, originally coined by Eriksson and Winka (1997), to describe the same taxonomic grouping. Included are the yeasts *Schizosaccharomyces* and *Saitoella*, the plant pathogens *Taphrina* and *Protomyces*, the mammal pathogen *Pneumocystis*, and the unusual apothecial fungus *Neolecta* (Sugiyama et al. 2006). There is still some controversy over the monophyly of this group even after numerous phylogenetic studies, but also because there are so few obvious morphological characters uniting its members. One of these is that ascogenous hyphae are lacking, even in the filamentous *Neolecta*, and that ascospores are capable of germinating by budding. *Taphrina* is believed to be close to the evolutionary origin of both Ascomycota and Basidiomycota, and in this context it is tantalising that both yeasts and filamentous fungi are found in the Taphrinomycotina, thereby leaving the question open to future discussion as to which state is ancestral in the Dikarya.

B. Saccharomycotina

This subphylum, equivalent to the 'Hemiascomycetes' (Kurtzman and Sugiyama 2001; Webster and Weber 2007), contains most of the ascomycete yeasts. It is phylogenetically well-defined with only one order, Saccharomycetales (James et al. 2006a; Suh et al. 2006). Considerable rearrangements at the genus level are certain as more DNA sequences become available and the unreliable nature of physiological tests becomes more obvious. *Pichia* and *Candida* are extreme examples of polyphyletic genera within the Saccharomycotina.

C. Pezizomycotina

With over 30 000 species, this is by far the largest group of Ascomycota, and it is regarded as monophyletic (Spatafora et al. 2006). It is equivalent in its circumscription to the 'filamentous ascomycetes' (Alexopoulos et al. 1996) or 'Euascomycetes' (Kurtzman and Sugiyama 2001). Inevitably, there have been major rearrangements within the subphylum, and these will continue for years to come.

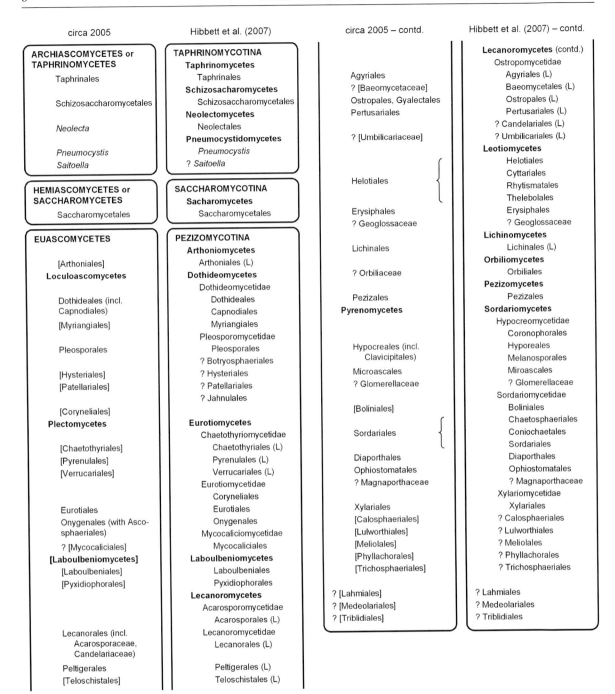

Fig. 1.4. Classification of the Ascomycota according to the current 'AFTOL' scheme (Hibbett et al. 2007) in comparison with the state of knowledge in 2005 as used by Webster and Weber (2007). Supplementary data from Kirk et al. (2001) are indicated in *square brackets*. *Question marks* indicate taxonomic positions *incertae sedes*, the letter *L* the inclusion of lichenised forms in the order

Most groups within the Pezizomycotina have now been accommodated in classes, although there are still some nomadic units, among them families including important or widely known fungi such as Magnaporthaceae (*Magnaporthe grisea*), Glomerellaceae (*Glomerella*, *Colletotrichum*), and Geoglossaceae (*Geoglossum*, *Trichoglossum*). Further, several established orders cannot yet be anchored securely in higher taxa and must remain *incertae sedes* at present. A recent phylogenetic analysis of classes within the Pezizomycotina may be found in Spatafora et al. (2006). Clearly, there is ample scope for further rearrangements which questions the stability of the system proposed by Hibbett et al. (2007).

The most basal classes of Pezizomycotina are the small group Orbiliomycetes and the larger Pezizomycetes, each with a single order. These groups are well-separated from a 'crown' of other pezizomycotinoid ascomycetes (James et al. 2006a). Complex fruit-bodies (stromata) are absent, and asci are arranged in apothecia in most cases, only rarely in hypogeous ascocarps (truffles) or in cleistothecium- or perithecium-like forms. Some progress is being made with the classification of Pezizales at the sub-ordinal level (Hansen and Pfister 2006).

A notoriously difficult group is the class Dothideomycetes, formerly known as Loculoascomycetes. The multi-gene analysis by Schoch et al. (2006) confirmed the existence of two previously suspected groups, now called Dothideomycetidae and Pleosporomycetidae, but left several orders *incertae sedes*. Lichenised fungi were not included in that analysis, and these are likely to complicate the matter further.

A group with a similarly high degree of morphological and biological variation is the class Eurotiomycetes (formerly Plectomycetes; Geiser and LoBuglio 2001). 'Core groups' are Onygenales (which form gymnothecia or cleistothecia and comprise important human and animal pathogens) and Eurotiales (containing *Aspergillus* and *Penicillium*); a second clade identified by Geiser et al. (2006) contains lichen-forming as well as non-lichenised orders (Verrucariales, Chaetothyriales, Pyrenulales).

Another phylogenetically defined but biologically heterogeneous group is the class Leotiomycetes containing 'inoperculate discomycetes' (Helotiales) as their core group (Wang et al. 2006). Although indicated by previous studies (Tsuneda and Currah 2004), the confirmation of typical gymnothecial fungi such as *Myxotrichum* and *Pseudogymnoascus* as members of this class will come as a surprise to many, as will the inclusion of Erysiphales (powdery mildews) and the exclusion of Geoglossaceae (earth-tongues). It will be a challenge to teach Leotiomycetes, or give a written account of them, in a taxonomic context!.

Sordariomycetes is the name now applied to a large group formerly called Pyrenomycetes (Samuels and Blackwell 2001). However, while the name may have changed, most of the important orders have been confirmed in an excellent phylogenetic account by Zhang et al. (2006), indicating areas of further work such as the need to accommodate Magnaporthaceae and Glomerellaceae as well as several orders of uncertain position.

Probably the most difficult phylogenetic problem within the Ascomycota is presented by lichenised fungi which make up almost all the species within Lecanoromycetes, the largest class of Ascomycota, but are also found in Dothideomycetes, Eurotiomycetes and in the smaller classes Lichinomycetes and Arthoniomycetes. The comprehensive study by Miadlikowska et al. (2006) on Lecanoromycetes is remarkable in being the first major phylogenetic work to take account of photobiont identity as a much-needed additional taxonomic marker. Nonetheless, extensive future rearrangements may be expected.

V. Basidiomycota

Almost all Basidiomycota can be assigned to one of three broad subphyla (i.e. Pucciniomycotina, Ustilaginomycotina, Agaricomycotina) erected by Bauer et al. (2006) to replace the former classes Urediniomycetes, Ustilaginomycetes and Hymenomycetes, respectively. Currently homeless taxa are Entorrhizales, a small group of smut-like pathogens of monocot roots formerly included among the Ustilaginomycotina or a synonym, and *Wallemia*, a mitosporic mould of relevance as a food spoilage fungus and in indoor situations (Hibbett 2006). A summary of the current taxonomic concept by Hibbett et al. (2007) is presented in Fig. 1.5.

A. Pucciniomycotina

This subphylum, equivalent in its circumscription to the Urediniomycetes (Kirk et al. 2001), has been subjected to numerous taxonomic and phylogenetic studies. Earlier schemes were summarised by Swann et al. (2001) who recognised that the simple, ascomycete-like architecture of hyphal septa distinguished the Pucciniomycotina from that of the other two major subphyla of Basidiomycota, i.e. Ustilaginomycotina (membrane-bounded septal caps) and Agaricomycotina (dolipore septa). These and many other ultrastructural

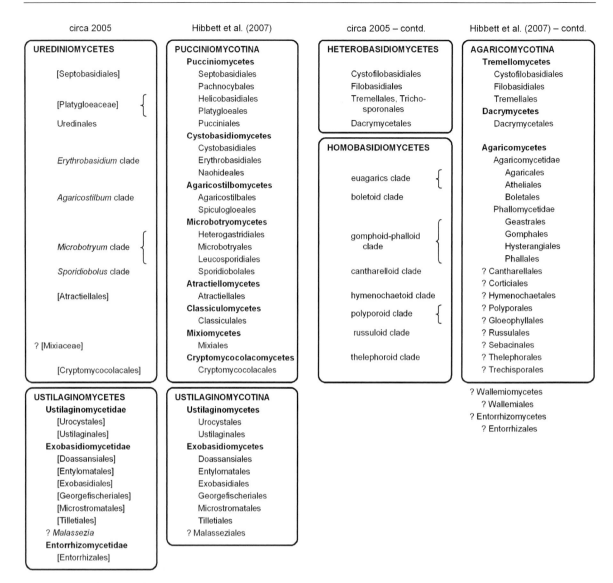

Fig. 1.5. Classification of the Basidiomycota according to the current 'AFTOL' scheme (Hibbett et al. 2007) in comparison with data available in 2005 (Webster and Weber 2007). Supplementary data from Kirk et al. (2001) and McLaughlin et al. (2001) are indicated in *square brackets*. Positions *incertae sedes* are indicated by *question marks*

and biochemical features are consistent with the basal position within the Basidiomycota which Pucciniomycotina occupy in many phylogenetic analyses (Bauer et al. 2006). However, whilst the range of genera to be included in this subphylum had become evident, their grouping into higher taxa was still tentative. Bauer et al. (2006) evaluated over 30 years of research into ultrastructural features, backed up by phylogenetic studies of rDNA loci. Their classification scheme was supported by the comprehensive phylogenetic study of Aime et al. (2006) and accepted by Hibbett et al. (2007).

The class Pucciniomycetes is by far the most important one in this subphylum. The order Pucciniales (formerly Uredinales) includes all known rust fungi, obligately biotrophic plant pathogens of which about 7000 species are known. These have distinct monokaryotic and dikaryotic phases which may alternate between two taxonomically unrelated host plants, producing up to five different spore stages in the process (Webster and Weber 2007). An even more extreme case of host alternation is presented by the Helicobasidiales (ca. 6 spp.) which parasitise rust fungi in their haploid conidial phase (*Tuberculina*), whereas their dikaryotic phase (*Rhizoctonia*-like as a sterile mycelium, *Helicobasidium* when producing basidia) causes violet root rot on a wide range of terrestrial plants (Lutz et al. 2004). The Septobasidiales (170 spp.) are dimorphic; yeast cells produced from basidiospores can infect scale insects where a mycelium is formed which ultimately produces resupinate fruit-bodies outside the dead insect host.

The remaining seven classes contain fungi with an astonishing diversity of life styles (plant-pathogenic, mycoparasitic, aquatic, saprotrophic) and diverse growth forms, as summarised in the insightful discussion of biological and ultrastructural features by Bauer et al. (2006). Yeasts of the *Rhodotorula* and *Sporobolomyces* types occur in the classes Cystobasidiomycetes (Cystobasidiales) and in Microbotryomycetes (Sporidiales). Other yeast genera are also found in these two classes. Macroscopic fruit-bodies, where present, are stilboid, i.e. small, jelly-like and stalked with an enlarged head. The class Microbotryomycetes includes the Microbotryales, an order of fungi formerly called 'dicot *Ustilago* spp.' and included in the Ustilaginales because of their striking biological and morphological similarities to the true smut fungi (discussed by Webster and Weber 2007).

Because most of the remaining classes and orders are mono- or oligotypic, we may expect the discovery of many more species especially from tropical habitats, and this could destabilise the current taxonomic system (Aime et al. 2006).

B. Ustilaginomycotina

This group of approximately 1500 species contains the true smut fungi and their allies with unifying biological features of saprotrophic haploid yeast cells which copulate and establish a dikaryotic mycelium as prerequisite for infection of host plants (Webster and Weber 2007). Septal pores are commonly surrounded by simple membrane caps. Despite a relatively limited number of groups, the taxonomy of the Ustilaginomycotina is still far from settled. Recent advances in our understanding both of ultrastructural features and rDNA-based phylogenetics have led to a revised classification of Ustilaginomycotina (Bauer et al. 1997; Begerow et al. 1997) which has repeatedly been

modified (see Begerow et al. 2006). There are two firmly established classes, Ustilaginomycetes and Exobasidiomycetes (Hibbett et al. 2007). The Entorrhizales were included by Begerow et al. (2006) but excluded by Hibbett et al. (2007), whereas the Malasseziales, a group of anamorphic yeasts associated with warm-blooded animals, are *incertae sedes* within the Ustilaginomycotina.

Members of the Ustilaginomycetes (Urocystales, Ustilaginales) almost exclusively infect monocots (grasses, sedges, rushes). No true haustoria are formed, but intracellular hyphae may be present, surrounded by a thick sheath (Bauer et al. 1997). Infection culminates in the production of thick-walled teliospores from which basidia arise. The scorched appearance of infected, teliospore-releasing plant tissue has given the smut fungi their name. The class Exobasidiomycetes contains a more diverse assemblage of fungi. Members of the order Exobasidiales parasitise dicot trees and shrubs, forming basidia without the teliospore stage on the infected and often swollen host surface. The electron-dense sheath around intracellular hyphae of Ustilaginomycetes is reduced to a few plaques in *Exobasidium* (Begerow et al. 2002). Confusingly, typical smut fungi are also included in the Exobasidiomycetes, e.g. Tilletiales, Entylomatales, and Georgefischeriales, in addition to aquatic smuts (Doassansiales) and *Tilletiopsis*-type yeasts (Entylomatales).

The conclusion we must draw is that typical 'smut fungi' are found in several phylogenetically distant groups, including Microbotryales (Pucciniomycotina), Entorrhizomycetes (Basidiomycota *incertae sedes*), and Ustilaginomycotina (Ustilaginomycetes and Exobasidiomycetes). Further, the taxonomy of Ustilaginomycetes is still in a state of flux which even the comprehensive recent multigene study by Begerow et al. (2006) has not altogether consolidated.

C. Agaricomycotina

This subphylum, approximately synonymous with the Hymenomycetes (Swann and Taylor 1993) or Basidiomycetes (Kirk et al. 2001), provides the most striking fruit-bodies of all fungi, and these have been studied for centuries in a taxonomic context. Unfortunately, fruit-body morphology has been found in many cases to be a poor indicator of true phylogenetic relationships, explaining the unusually high rate of rearrangements within the Agaricomycotina in recent years. Hibbett (2006) and Hibbett et al. (2007), summarising the current viewpoint of AFTOL project members, subdivided the Agaricomycotina into three classes, i.e. Tremellomycetes, Dacrymycetes, and Agaricomycetes, but this is unlikely to remain the last word on the matter as numerous orders remain *incertae sedes*.

In the past, fungi have been called heterobasidiomycetes or phragmobasidiomycetes if: (1) their basidia possess highly differentiated sterigmata (epibasidia) or are even divided by transverse or longitudinal septa, (2) basidiospores germinate by budding in a yeast-like manner or by producing ballistosporic conidia, and (3) fruit-bodies, if present, are jelly-like and capable of discharging basidiospores even after several drying/rehydration cycles (Kirk et al. 2001; Wells and Bandoni 2001; Webster and Weber 2007). Unfortunately, these features, like many others, are now known to be phylogenetically irrelevant, and 'heterobasidiomycetes' may be found in Tremellomycetes (Tremellales), Dacrymycetes (*Dacrymyces, Calocera*) and certain orders of the Agaricomycetes, such as Auriculariales (*Auricularia, Exidia*), Cantharellales (*Tulasnella*, possibly

Ceratobasidium and other teleomorphs of *Rhizoctonia*), and Sebacinales (Weiss et al. 2004; Binder et al. 2005; Hibbett 2006; Moncalvo et al. 2006). The class Tremellomycetes as currently circumscribed contains the well-delimited Tremellales and two orders of yeasts, Cystofilobasidiales and Filobasidiales. This placement of most if not all agaricomycetous yeasts in the vicinity of *Tremella* confirms earlier results (Fell et al. 2001) and correlates with the germination of *Tremella* basidiospores as yeast cells.

Many phylogenetic studies have attempted to subdivide fungi currently grouped in the class Agaricomycetes (Hibbett et al. 2007), approximately equivalent to Homobasidiomycetes. Major consternation arose when it became apparent just how frequently the various forms of basidiocarps and hymenophore – chiefly lamellate, cantharelloid, poroid, clavate, hydnoid, resupinate, and gasteroid – appeared in different phylogenetic groups (Fig. 1.6). Key biological features have also evolved repeatedly. To give one example, ectomycorrhizal associations may have arisen independently 11 times in the Agaricales (Matheny et al. 2006), twice in Boletales (Binder and Hibbett 2006), and on several occasions elsewhere within

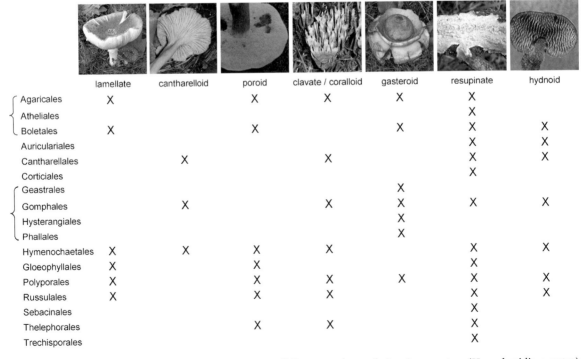

	lamellate	cantharelloid	poroid	clavate / coralloid	gasteroid	resupinate	hydnoid
Agaricales	X		X	X	X	X	
Atheliales						X	
Boletales	X		X		X	X	X
Auriculariales						X	X
Cantharellales		X		X		X	X
Corticiales						X	
Geastrales					X		
Gomphales		X		X	X	X	X
Hysterangiales					X		
Phallales					X		
Hymenochaetales	X	X	X	X		X	X
Gloeophyllales	X		X			X	
Polyporales	X		X	X	X	X	X
Russulales	X		X	X		X	X
Sebacinales						X	
Thelephorales			X	X		X	
Trechisporales						X	

Fig. 1.6. Distribution of hymenial surfaces among different orders of Agaricomycetes (Homobasidiomycetes). Phylogenetically closely related orders are united by *brackets*

Agaricomycetes (Hibbett and Thorn 2001). In the Sebacinales, a small order of heterobasidiomycete-like fungi now placed at the base of Agaricomycetes, different types such as ectomycorrhiza, ericoid, and orchid mycorrhiza may all be found (Weiss et al. 2004). Many groups therefore lack any obvious character by which their members can be recognised.

Initial attempts at re-classifying agaricomycetoid fungi defined phylogenetic clades instead of formal order names (Hibbett et al. 1997; Hibbett and Thorn 2001; Moncalvo et al. 2002; Webster and Weber 2007). Orders have now been formally re-described for the eight clades of Hibbett and Thorn (2001), i.e. cantharelloid (Cantharellales; Moncalvo et al. 2006), hymenochaetoid (Hymenochaetales; Larsson et al. 2006), gomphoid-phalloid (Geastrales, Gomphales, Hysterangiales, Phallales; Hosaka et al. 2006), russuloid (Russulales; Miller et al. 2006), boletoid (Boletales; Binder and Hibbett 2006), euagarics (Agaricales; Matheny et al. 2006), polyporoid (Polyporales; see Hibbett and Thorn 2001; Matheny et al. 2007), and thelephoroid (Thelephorales; see Hibbett et al. 2007). Additional (mostly smaller) clades were resolved subsequent to the summary by Hibbett and Thorn (2001), and these were described as Trechisporales, Sebacinales, Gloeophyllales, Corticiales, and Atheliales (see Weiss et al. 2004; Binder et al. 2005; Hibbett et al. 2007). However, a complete sub-class system is not yet in place, only two groups (Agaricomycetidae and Phallomycetidae) having been formalised in the new taxonomic context as yet (Hibbett et al. 2007). Indeed, a new research discipline, fungal phylogenomics, may have to be established before the Agaricomycetes can be tidied up (Hibbett 2006), and even then a working taxonomic system based on well-known species may be destabilised by the future discovery of new species from the immense unknown biodiversity thought to exist among Agaricomycotina.

VI. Conclusions

Modern multi-gene phylogenetic approaches are carrying forward the revolution of fungal taxonomy begun by single-gene studies some two decades ago and are beginning to place the resulting taxonomy on a firmer, more 'natural' base. However, as we have seen, many areas of ambiguity remain. In addressing them, every effort should be made to reduce name changes to the bare minimum, i.e.

to situations in which there is sound evidence of their necessity, general acceptance, and future durability. Further, a hierarchical scheme with a limited number of phyla should be maintained from the perspective of mycology teaching. Each phylum should contain an equivalent structure of lower ranks. In this context a re-erection of the Zygomycota or an equivalent in rank is desirable. The 'AFTOL' project has a model character in addressing most of the above requirements, and it is particularly promising in that so many influential taxonomists have united behind it. It is certainly possible to adopt this classification for use with existing textbooks such as Webster and Weber (2007), new texts or editions, or as a framework for mycology classes. The AFTOL scheme presents the first realistic opportunity to introduce a unified taxonomic scheme into global mycological use.

However, taxonomy is merely a foundation on which the ultimate goal of mycology – an understanding of the *biology* of fungi – can be approached. In many institutions, the teaching of mycology at undergraduate level is currently being carried out by non-mycologists. The perception of fungal taxonomy as an obsession with name changes of ever shorter shelf-life must be avoided, as this would simply lead to the elimination of fungi from biology textbooks and curricula, as has arguably happened with protozoa (Adl et al. 2007). In this context as in many others, one must hope for a high acceptance and durability of the AFTOL scheme.

References

Adl S, et al (2005) The new higher level classification of eukaryotes with emphasis on the taxonomy of protists. J Eukaryot Microbiol 52:399–451

Adl S, et al (2007) Diversity, nomenclature, and taxonomy of protists. Syst Biol 56:684–689

Aime MC, Matheny PB, Henk DA, Frieders EM, Nilsson RH, Piepenbring M, McLaughlin DJ, Szabo LJ, Begerow D, Sampaio JP, Bauer R, Weiss M, Oberwinkler F, Hibbett D (2006) An overview of the higher-level classification of Pucciniomycotina based on combined analyses of nuclear large and small subunit rDNA sequences. Mycologia 98:869–905

Alexopoulos CJ, Mims CW, Blackwell M (1996) Introductory mycology, fourth edn. Wiley, New York

Archibald JM, Keeling PJ (2004) Actin and ubiquitin protein sequences support a cercozoan/foraminiferan ancestry for the plasmodiophorid plant pathogens. J Eukaryot Microbiol 51:113–118

Baldauf SL, Roger AJ, Wenk-Siefert I, Doolittle WF (2000) A kingdom-level phylogeny of eukaryotes based on combined protein data. Science 290:972–977

Bauer R, Oberwinkler F, Vánky K (1997) Ultrastructural markers and systematics in smut fungi and allied taxa. Can J Bot 75:1273–1314

Bauer R, Begerow D, Sampaio JP, Weiss M, Oberwinkler F (2006) The simple-septate basidiomycetes: a synopsis. Mycol Progr 5:41–66

Begerow D, Bauer R, Oberwinkler F (1997) Phylogenetic studies on the nuclear large subunit ribosomal DNA of smut fungi and related taxa. Can J Bot 75:2045–2056

Begerow M, Bauer R, Oberwinkler F (2002) The Exobasidiales: an evolutionary hypothesis. Mycol Prog 1:187–199

Begerow D, Stoll M, Bauer R (2006) A phylogenetic hypothesis of Ustilaginomycotina based on multiple gene analyses and morphological data. Mycologia 98:906–916

Berbee ML, Taylor JW (2001) Fungal molecular evolution: gene trees and geologic time. In: McLaughlin DJ, McLaughlin EG, Lemke PA (eds) The Mycota VIIB: systematics and evolution. Springer, Heidelberg, pp 229–245

Binder M, Hibbett DS (2006) Molecular systematics and biological diversification of Boletales. Mycologia 98:971–981

Binder M, Hibbett DS, Larsson K-H, Larsson E, Langer E, Langer G (2005) The phylogenetic distribution of resupinate forms across the major clades of mushroom-forming fungi (Homobasidiomycetes). Syst Biodiv 3:113–157

Blackwell M, Hibbett DS, Taylor JW, Spatafora JW (2006) Research coordination networks: a phylogeny for kingdom Fungi (Deep Hypha). Mycologia 98:829–837

Cafaro M (2005) Eccrinales (Trichomycetes) are not fungi, but a clade of protists at the early divergence of animals and fungi. Mol Phylogen Evol 35:21–34

Dick MW (2001) Straminipilous fungi. Kluwer, Dordrecht

Eriksson OE, Winka K (1997) Supraordinal taxa of the Ascomycota. Myconet 1:1–16

Fast NM, Xue L, Bingham S, Keeling PJ (2002) Re-examining alveolate evolution using multiple protein molecular phylogenies. J Eukaryot Microbiol 49:30–37

Fell JW, Boekhout T, Fonseca A, Sampaio JP (2001) Basidiomycetous yeasts. In: McLaughlin DJ, McLaughlin EG, Lemke PA (eds) The Mycota VIIB: systematics and evolution. Springer, Heidelberg, pp 3–35

Geiser DM, LoBuglio KF (2001) The monophyletic Plectomycetes: Ascosphaerales, Onygenales, Eurotiales. In: McLaughlin DJ, McLaughlin EG, Lemke PA (eds) The Mycota VIIA: systematics and evolution. Springer, Heidelberg, pp 201–219

Geiser DM, Gueidan C, Miadlikowska J, Lutzoni F, Kauff F, Hofstetter V, Fraker E, Schoch CL, Tibell L, Untereiner WA, Aptroot A (2006) Eurotiomycetes: Eurotiomycetidae and Chaetothyriomycetidae. Mycologia 98:1953–1064

Hakariya M, Hirose D, Tokumasu S (2007) A molecular phylogeny of Haptoglossa species, terrestrial peronosporomycetes (oomycetes) endoparasitic on nematodes. Mycoscience 48:169–175

Hansen K, Pfister DH (2006) Systematics of the Pezizomycetes – the operculate discomycetes. Mycologia 98:1029–1040

Hibbett DS (2006) A phylogenetic overview of the Agaricomycotina. Mycologia 98:917–925

Hibbett DS, Thorn RG (2001) Basidiomycota: Homobasidiomycetes. In: McLaughlin DJ, McLaughlin EG, Lemke PA (eds) The Mycota VIIB: systematics and evolution. Springer, Heidelberg, pp 121–168

Hibbett DS, Pine EM, Langer E, Langer G, Donoghue MJ (1997) Evolution of gilled mushrooms and puffballs inferred from ribosomal DNA sequences. Proc Natl Acad Sci USA 94:12002–12006

Hibbett DS, et al (2007) A higher-level phylogenetic classification of the Fungi. Mycol Res 111:509–547

Hosaka K, Bates ST, Beever RE, Castellano MA, Colgan W, Domínguez LS, Nouhra ER, Geml J, Giachini AJ, Kenney SR, Simpson NB, Spatafora JW, Trappe J (2006) Molecular phylogenetics of the gomphoid–phalloid fungi with an establishment of the new subclass Phallomycetidae and two new orders. Mycologia 98:949–959

Hudspeth DSS, Stenger D, Hudspeth MES (2003) A cox2 phylogenetic hypothesis for the downy mildews and white rusts. Fungal Divers 13:47–57

Inagaki Y, Susko E, Fast NM, Roger AJ (2004) Covarion shifts cause a long-branch attraction artifact that unites Microsporidia and Archaebacteria in EF-1α phylogenies. Mol Biol Evol 21:1340–1349

James TY, et al (2006a) Reconstructing the early evolution of fungi using a six-gene phylogeny. Nature 443:818–822

James TY, Letcher PM, Longcore JE, Mozley-Standridge SE, Powell MJ, Griffith GW, Vilgalys R (2006b) A molecular phylogeny of the flagellated fungi (Chytridiomycota) and description of a new phylum (Blastocladiomycota). Mycologia 98:860–871

Keeling PJ, Fast NM (2002) Microsporidia: biology and evolution of highly reduced intracellular parasites. Annu Rev Microbiol 56:93–116

Keeling PJ, Luker MA, Palmer JD (2000) Evidence from beta-tubulin phylogeny that Microsporidia evolved from within the Fungi. Mol Biol Evol 17:23–31

Keeling PJ, Burger G, Durnford DG, Lang BF, Lee RW, Pearlman RE, Roger AJ, Gray MW (2005) The tree of eukaryotes. Trends Ecol Evol 20:670–676

Kessin RH (2001) Dictyostelium. Evolution, cell biology, and the evolution of multicellularity. Cambridge University Press, Cambridge

Kirk PM, Cannon PF, David JC, Stalpers JA (eds) (2001) Ainsworth and Bisby's dictionary of the fungi, 9th edn. CABI, Wallingford

Kurtzman CP, Sugiyama J (2001) Ascomycetous yeasts and yeastlike taxa. In: McLaughlin DJ, McLaughlin EG, Lemke PA (eds) The mycota VIIA: systematics and evolution. Springer, Heidelberg, pp 179–200

Larsson K-H, Parmasto E, Fischer M, Langer E, Nakasone KK, Redhead SA (2006) Hymenochaetales: a molecular phylogeny for the hymenochaetoid clade. Mycologia 98:926–936

Liu YJ, Hodson MC, Hall BD (2006) Loss of the flagellum happened only once in the fungal lineage: phylogenetic structure of Kingdom Fungi inferred from RNA polymerase II subunit genes. BMC Evol Biol 6:74

Lutz M, Bauer R, Begerow D, Oberwinkler F (2004) *Tuberculina – Thanatophytum/Rhizoctonia crocorum – Helicobasidium*: a unique mycoparasitic–phytoparasitic life strategy. Mycol Res 108:227–238

Matheny PB, et al (2006) Major clades of Agaricales: a multilocus phylogenetic overview. Mycologia 98:982–995

Matheny PB, et al (2007) Contributions of *rpb1* and *tef1* to the phylogeny of mushrooms and allies (Basidiomycota, Fungi). Mol Phylogen Evol 43:430–451

McLaughlin DJ, McLaughlin EG, Lemke PA (eds) (2001) The mycota VIIA, VIIB: systematics and evolution. Springer, Heidelberg

Miadlikowska J, et al (2006) New insights into classification and evolution od the Lecanoromycetes (Pezizomycotina, Ascomycota) from phylogenetic analyses of three ribosomal RNA- and two protein-encoding genes. Mycologia 98:1088–1103

Miller SL, Larsson E, Larsson K-H, Verbeken A, Nuytinck J (2006) Perspectives in the new Russulales. Mycologia 98:960–970

Moncalvo J-M, Vilgalys R, Redhead SA, Johnson JE, James TY, Aime MC, Hofstetter V, Verduin SJW, Larsson E, Baroni TJ, Thorn RG, Jacobsson S, Clémençon H, Miller OK (2002) One hundred and seventeen clades of euagarics. Mol Phylogen Evol 23:357–400

Moncalvo J-M, Nilsson RH, Koster B, Dunham SM, Bernauer T, Matheny PB, Porter TM, Margaritescu S, Weiss M, Garnica S, Danell E, Langer G, Langer E, Larsson E, Larsson K-H, Vilgalys R (2006) The cantharelloid clade: dealing with incongruent gene trees and phylogenetic reconstruction methods. Mycologia 98:937–948

Morton JB, Benny GL (1990) Revised classification of arbuscular mycorrhizal fungi (Zygomycetes): a new order, Glomales, two new suborders, Glomineae and Gigasporineae, and two new families, Acaulosporaceae and Gigasporaceae, with an emendation of Glomaceae. Mycotaxon 37:471–491

Nishida H, Sugiyama J (1994) Archiascomycetes: detection of a major new lineage within the Ascomycota. Mycoscience 35:361–366

Redecker D, Raab P (2006) Phylogeny of the Glomeromycota (arbuscular mycorrhizal fungi): recent developments and new gene markers. Mycologia 98:885–895

Riethmüller A, Voglmayr H, Göker M, Weiss M, Oberwinkler F (2002) Phylogenetic relationships of the downy mildews (Peronosporales) and related groups based on nuclear large subunit ribosomal DNA sequences. Mycologia 94:834–849

Samuels GJ, Blackwell M (2001) Pyrenomycetes – fungi with perithecia. In: McLaughlin DJ, McLaughlin EG, Lemke PA (eds) The mycota VIIA: systematics and evolution. Springer, Heidelberg, pp 221–255

Schoch CL, Shoemaker RA, Seifert KA, Hambleton S, Spatafora JW, Crous PW (2006) A multigene phylogeny of the Dothideomycetes using four nuclear loci. Mycologia 98:1041–1052

Schüssler A, Schwarzott D, Walker C (2001) A new fungal phylum, the *Glomeromycota*: phylogeny and evolution. Mycol Res 105:1413–1421

Simpson AGB, Roger AJ (2004) The real 'kingdoms' of eukaryotes. Curr Biol 14:R693–R696

Spatafora JW, et al (2006) A five-gene phylogeny of Pezizomycotina. Mycologia 98:1018–1028

Sugiyama J, Hosaka K, Suh S-U (2006) Early diverging Ascomycota: phylogenetic divergence and related evolutionary origins. Mycologia 98:996–1005

Suh S-O, Blackwell M, Kurtzman CP, Lachance M-A (2006) Phylogenetics of Saccharomycetales, the ascomycete yeasts. Mycologia 98:1006–1017

Swann EC, Taylor JW (1993) Higher taxa of basidiomycetes: an 18S rRNA perspective. Mycologia 85:923–936

Swann EC, Frieders EM, McLaughlin DJ (2001) Urediniomycetes. In: McLaughlin DJ, McLaughlin EG, Lemke PA (eds) The mycota VIIB: systematics and evolution. Springer, Heidelberg, pp 37–56

Tanabe Y, Watanabe MM, Sugiyama J (2002) Are *Microsporidia* really related to *Fungi*? A reappraisal based on additional gene sequences from basal fungi. Mycol Res 106:1380–1391

Tanabe Y, Saikawa M, Watanabe MM, Sugiyama J (2004) Molecular phylogeny of Zygomycota based on EF-1α and RPB1 sequences: limitations and utility of alternative markers to rDNA. Mol Phylogen Evol 30:438–449

Thines M, Göker M, Telle S, Ryley M, Mathur K, Narayana YD, Spring O, Thakur RP (2008) Phylogenetic relationships of graminicolous downy mildews based on *cox2* sequence data. Mycol Res 112:345–351

Tsuneda A, Currah RS (2004) Ascomatal morphogenesis in *Myxotrichum arcticum* supports the derivation of Myxotrichaceae from a discomycetous ancestor. Mycologia 96:627–635

Ustinova I, Krienitz L, Huss VAR (2000) *Hyaloraphidium curvatum* is not a green alga, but a lower fungus; *Amoebidium parasiticum* is not a fungus, but a member of the DRIPs. Protist 151:253–262

Villa NO, Kageyama K, Asano T, Suga H (2006) Phylogenetic relationships of *Pythium* and *Phytophthora* species based on ITS rDNA, cytochrome oxidase II and β-tubulin gene sequences. Mycologia 98:410–422

Wang Z, Johnston PR, Takamatsu S, Spatafora JW, Hibbett DS (2006) Toward a phylogenetic classification of the Leotiomycetes based on rDNA data. Mycologia 98:1065–1075

Webster J, Weber RWS (2007) Introduction to fungi, 3rd edn. Cambridge University Press, Cambridge

Weiss M, Selosse M-A, Rexer K-H, Urban A, Oberwinkler F (2004) *Sebacinales*: a hitherto overlooked cosm of heterobasidiomycetes with a broad mycorrhizal potential. Mycol Res 108:1003–1010

Wells K, Bandoni RJ (2001) Heterobasidiomycetes. In: McLaughlin DJ, McLaughlin EG, Lemke PA (eds) The mycota VIIB: systematics and evolution. Springer, Heidelberg, pp 86–120

White MW, James TY, O'Donnell K, Cafaro MJ, Tanabe Y, Sugiyama J (2006) Phylogeny of the Zygomycota based on nuclear ribosomal sequence data. Mycologia 98:872–884

Zhang N, Castlebury LA, Miller AN, Huhndorf SM, Schoch CL, Seifert KA, Rossman AY, Rogers JD, Kohlmeyer J, Volkmann-Kohlmeyer B, Sung G-H (2006) An overview of the systematics of the Sordariomycetes based on a four-gene phylogeny. Mycologia 98:1076–1087

2 *Sordaria macrospora*, a Model System for Fungal Development

Ulrich Kück[1], Stefanie Pöggeler[2], Minou Nowrousian[1], Nicole Nolting[2,*], Ines Engh[1]

CONTENTS

I. Introduction

The application of forward or reverse genetic approaches in model organisms has made great contributions to our fundamental biological knowledge. Associated with these studies is the technical progress in understanding and manipulating the genomes of the relevant organisms. Basic biological studies with model organisms have led, for example, to the identification of signal transduction pathways, transcription control circuits, and cell cycle checkpoints. Moreover, studies using model systems continue to uncover the details of genetic interactions controlling developmental plasticity. The acquired knowledge usually can be applied to organisms of higher order having more complex morphologies that are difficult to manipulate experimentally. A further advantage of model systems compared to complex highly developed organisms (e.g. humans) is their potential for a rapid and inexpensive genetic analysis.

In this review, we focus on the filamentous fungus *Sordaria macrospora*, which for long has been used as a model for conventional genetic analysis, fruiting body development, and the analysis of meiosis (Esser and Straub 1958; Heslot 1958; Zickler 1977; van Heemst et al. 1999; Pöggeler et al. 2006a). Since the first combination of classical tetrad analysis and modern molecular genetics, this fungus has become an important eukaryotic model system to study developmental biology.

While this review was in preparation, the genome of *S. macrospora* was sequenced, using next generation sequencing techniques. De novo assembly of Illumina/Solexa paired end reads and subsequent comparative assembly with genome data from *Neurospora* species yielded a first draft version with about 10 000 predicted genes. This genome sequence will greatly enhance the value of *S. macrospora* as a eukaryotic model system (Nowrousian M, Kamerewerd J, Engh I, Pöggeler S, Stajich J, Read N, Kück U, Freitag M, personal communication).

A. Fungal Organisms as Model Systems for Developmental Biology

So far about 74 000 fungal species have been identified and classified, but even conservative estimates suspect that about 1.5×10^6 species exist in nature

[1]Lehrstuhl für Allgemeine und Molekulare Botanik, Fakultät für Biologie, Ruhr-Universität Bochum, Universitätsstrasse 150, 44780 Bochum Germany; e-mail: ulrich.kueck@ruhr-uni-bochum.de

[2]Abteilung Genetik eukaryotischer Mikroorganismen, Institut für Mikrobiologie und Genetik, Georg-August-Universität Göttingen, Grisebachstrasse 8, 37077 Göttingen, Germany; e-mail: spoegge@gwdg.de

*Present address: Heinrich Petk Institute for Experimental Virology and Immunology at the University of Hambury, Germany

Physiology and Genetics, 1st Edition
The Mycota XV
T. Anke and D. Weber (Eds.)
© Springer-Verlag Berlin Heidelberg 2009

(Hawksworth 2001). Fungi are true eukaryotes, with a nucleus and (in most cases) mitochondria. Exceptions are those species having mitosomes or mitochondria-related hydrogenosomes instead of mitochondria. However, recent evolutionary studies indicate that both mitosomes and hydrogenosomes might be derived from mitochondrial ancestors (Embley and Martin 2006). Fungi can be either uni- or multicellular. Irrespective of their taxonomy, unicellular fungi are named yeasts, while the terms filamentous or mycelial fungi indicate multicelluar fungi with complex, multicellular propagation structures. Depending on their environmental growth conditions, some fungi are able to undergo a transition from uni- to multicellular growth and vice versa. This dimorphic development is a characteristic of certain species that are considered to be either yeasts or filamentous fungi.

Baker's yeast, *Saccharomyces cerevisiae*, is the most prominent fungal model system and has contributed enormously to the basic understanding for example of eukaryotic signalling pathways or the biogenesis of eukaryotic organelles (e.g. mitochondria, peroxisomes, endoplasmic reticulum). Thus, yeast is the main model system for eukaryotic cell biology (Botstein and Fink 1988).

However, multicellular models are essential since many genes play different roles in defined cell types at different developmental stages. These complex cell and tissue interactions cannot be mimicked in unicellular organisms like yeasts. Therefore other model organisms, such as filamentous fungi, have proven their value in specialized research areas.

Filamentous fungi are characterized by the extreme polar growth of their hyphae, which is reminiscent of the growth behaviour in neurons or plant pollen tubes. This polarized growth in fungi is characterized by vesicle transport towards the apex. In contrast to yeasts, filamentous fungi can undergo major morphogenetic changes that are accompanied by a switch from polarized to isotropic growth (Harris and Momany 2004; Fischer et al. 2008). These changes in cellular development occur regularly during asexual or sexual propagation.

During asexual propagation, the typical formation of conidiospores is observed (e.g. *Aspergillus* spp., *Penicillium* spp.). Another, often overlooked, asexual differentiation process is the fragmentation of hyphae into arthrospores. Both these processes are dependent on environmental conditions, such as light.

Similarly, sexual propagation structures occur in true filamentous fungi due to environmental changes such as nutrient starvation. In almost all filamentous fungi, fruiting body formation is the major developmental step during the sexual reproduction cycle. Fruiting bodies vary in size between 200 μm and several centimetres as found for example in the morel. Their major function is the protection and further distribution of the sexual spores that are generated after the meiotic divisions that follow karyogamy (Pöggeler et al. 2006a).

B. Why Choose *Sordaria macrospora*?

Sordaria macrospora is a coprophilic homothallic pyrenomycete first described in 1866 (Auerswald 1866), which is closely related to *Podospora anserina* and *Neurospora crassa*. In contrast to these heterothallic species, single strains of *S. macrospora* produce fruiting bodies (perithecia) without the presence of a mating partner. As a consequence of self-fertility, fruiting body development is an apandrous process, allowing the direct testing of recessive mutations that lead to defects in fruiting body formation. The life cycle of *S. macrospora*, as depicted in Fig. 2.1, can be completed in the laboratory within seven days. During vegetative growth, chemical and physical stimuli such as biotin or light induce branching of hyphal tips, which is followed by adhesion of several hyphae to each other. This entry into the sexual phase results in the formation of a network of interconnecting hyphae, leading to the first of two consecutive morphological stages: first, the development of primordia (protoperithecia), and second, the transition of protoperithecia into mature fruiting bodies, called perithecia. In the homothallic fungus *S. macrospora*, transition between these two stages takes 72 h. In heterothallic fungi such as *N. crassa*, this transition requires crossing between two strains of opposite mating types.

Ascus development, as detailed in the next section, starts with the formation of female gametangia called ascogonia. The ascogenous cells are enveloped by sterile hyphae to form fruiting body precursors. Subsequent tissue differentiation gives rise to an outer pigmented peridial tissue and, following karyogamy, inner ascus initials embedded in sterile paraphyses are formed. Mature fruiting bodies from *S. macrospora* harbor 50–300 asci, which after meiosis and postmeiotic

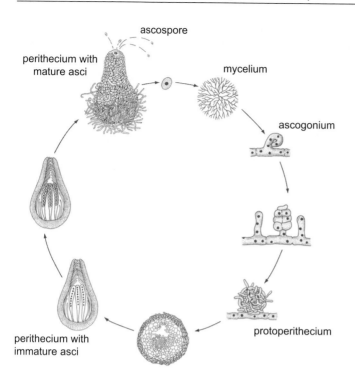

ascospore

perithecium with
mature asci

mycelium

ascogonium

protoperithecium

perithecium with
immature asci

Fig. 2.1. Life cycle of *Sordaria macrospora*: The cycle starts with a germinating ascospore and is completed in the laboratory within seven days by the formation of mature fruiting bodies (perithecia) that discharge the mature and melanized ascospores

divisions contain eight ascospores each. Mature black ascospores are discharged through an apical pore (ostiole) at the neck of the fruiting body. In conclusion, fruiting body development requires the differentiation of the mycelia into several specialized tissues; and regulation of these morphological and physiological changes requires a number of different genes (Esser and Straub 1958).

There is another developmental feature that makes *S. macrospora* a favourable object for fruiting body development. *S. macrospora* produces only meiotically derived ascospores, whereas asexual spores, such as conidia, are absent. Thus, there is no interference between two different developmental programmes, which makes it easier, for example, to analyse differentially expressed genes involved in ascocarp development.

II. Biology

In order to manipulate the development of a fungal model system, detailed knowledge of its life cycle is an essential prerequisite for any experimental study. The sexual system of fungi can be either heterothallic (self-incompatible), pseudo-homothallic (secondary homothallic, pseudo-compatible), or homothallic (compatible) (Lin and Heitman 2007).

A. Life Cycle

Similar to heterothallic filamentous ascomycetes, the homothallic *S. macrospora* has a dikaryophase in its life cycle. During this phase, pairs of nuclei synchronously and repeatedly divide inside the ascogonium, thereby producing a great number of nuclei. The nuclei then migrate in pairs to the developing dikaryotic ascogenous hyphae emerging from the ascogonium. The ascogenous hyphal tips develop a U-shaped crozier containing two nuclei. These nuclei divide mitotically in synchrony and, subsequently, two septa appear which divide the hook cell into three sections: a lateral and a basal cell with one nucleus each, and an apical cell with two nuclei. The two nuclei in the apical cell then fuse. Immediately after karyogamy within the young ascus, the diploid nucleus undergoes meiosis, which is followed by a post-meiotic mitosis, resulting in the formation of eight nuclei. Each of the eight nuclei is incorporated into its own ascospore (Esser and Straub 1958; Esser 1982). The self-fertilization mode of *S. macrospora* is termed autogamous fertilization. This means that a pair-wise fusion of nuclei present within the ascus initial occurs, without cell fusion having taken place before (Esser and Straub 1958).

B. Homothallism

In homothallic species, like *S. macrospora*, a mycelium derived from a uninucleate ascospore is self-fertile and able to perform all steps of meiosis. In contrast, heterothallic species, like *N. crassa*, require another compatible individual for sexual reproduction. The mating partners of heterothallic fungi are morphologically identical and are only distinguished by their mating type (*MAT*) loci. In *N. crassa*, these are termed *MAT1-1* (*matA*) and *MAT1-2* (*mata*). The alternative versions of the mating-type locus on homologous chromosomes of the mating partners are called idiomorphs because they are completely dissimilar in the genes they carry (Metzenberg and Glass 1990). The *MAT1-1* idiomorph of *N. crassa* contains three genes: *MAT1-1-1* (*mat A-1*), *MAT1-1-2* (*mat A-2*), and *MAT1-1-3* (*mat A-3*; Glass et al. 1990a; Ferreira et al. 1996, 1998; Fig. 2.2). The MAT1-1 polypeptide has a DNA-binding motif that shows similarity to the *S. cerevisiae* MATα1p mating-type protein. *MAT1-1-2* encodes a protein without a characteristic DNA-binding motif, but with a conserved region encompassing 25 amino acids with three invariant *h*istidine, *p*roline, and *g*lycine residues (the HPG domain; Debuchy and Turgeon 2006). The *MAT1-1-3* gene encodes a protein with a *h*igh *m*obility *g*roup (HMG) DNA-binding motif (Ferreira et al. 1996). The *N. crassa MAT1-2* idiomorph contains two genes,

MAT1-2-1 (*mat a-1*) and MAT1-2-3 (*mat a-2*). While the *MAT1-2-1* gene encodes a HMG-domain protein that is the major regulator of mating in *MAT1-2* strains, the function of the *MAT1-2-2* encoded protein is unknown (Staben and Yanofsky 1990; Pöggeler and Kück 2000).

Pseudo-homothallic members of the Sordariaceae (e.g. *Neurospora tetrasperma*) or the Lasiosphaeriaceae (e.g. *P. anserina*) develop four-spored asci in which most ascospores carry two nuclei, one of each mating type. Another form of pseudo-homothallism is accomplished by mating-type switching, but this has not been identified in members of the Sordariales (Lin and Heitman 2007).

There are three ways fungi can be homothallic: (1) they can harbour a fused *MAT* locus of both idiomorphs, (2) they can harbour both *MAT* alleles at different loci in the genome, and (3) they can sexually reproduce but carry only one *MAT* locus (Lin and Heitman 2007). In Sordariales all three forms of homothallism can be found in the same order.

S. macrospora as well as *Neurospora pannonica* and *Neurospora terricola* belong to the first group and contain sequences similar to both the *MAT1-1* and the *MAT1-2* idiomorphs. In the homothallic *Chaetomium globosum* in addition to a *MAT1-1* locus, a *MAT1-2* locus was identified at another genomic locus (Glass et al. 1990a; Pöggeler et al. 1997a; Debuchy and Turgeon 2006). However, according to data from Southern hybridization, the homothallic Neurospora species *N. africana*, *N. dodgei*, *N. galapagonensis*, and *N. lineolata* possess only the *MAT1-1* locus (Glass et al. 1990a; Pöggeler 1999). The fused mating-type-loci of *S. macrospora*, *N. pannonica*, and *N. terricola* appear to be derived from a heterothallic ancestor. The molecular mechanisms responsible for the fusion of *MAT* regions are

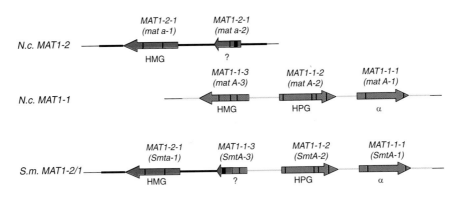

Fig. 2.2. Comparative maps of mating type loci from the two *N. crassa* (*N. c.*) mating type idiomorphs and from the *S. macrospora* (*S. m.*) mating type locus. For further details of designations see text

so far unknown, but it is likely that a recombination event led to the close linkage of *MAT1-1-* and *MAT1-2*-specific sequences on a single chromosome.

Cloning and sequencing of the entire mating-type locus of *S. macrospora* revealed that it encodes four different ORFs: *MAT1-2-1* (*Smt a-1*), *MAT1-1-3* (*Smt A-3*), *MAT1-1-2* (*Smt A-2*), and *MAT1-1-1* (*SmtA-1*). Except for *MAT1-1-3*, proteins encoded by the *S. macrospora* mating type genes show strong sequence similarities to the corresponding *N. crassa* mating-type ORFs. The *S. macrospora* *MAT1-1-3* gene has a chimeric character and exhibits sequence similarity to the *MAT1-1-3* and the *MAT1-2-2* ORF of *N. crassa*. In contrast to the *N. crassa* MAT1-1-3, the *S. macrospora* MAT1-1-3 protein lacks an HMG domain (Fig. 2.2).

Sequence analysis of the cDNA from the *S. macrospora* *MAT1-2-1*, *MAT1-1-3*, *MAT1-1-2*, and *MAT1-1-1* revealed that introns in each gene are spliced, indicating that all of the mating-type genes identified in the mating-type locus of *S. macrospora* are transcriptionally active (Pöggeler et al. 1997b). Surprisingly, RT-PCR experiments demonstrated co-transcription of the *MAT1-1-3* gene and *MAT1-2-1* as well as optional splicing of two introns within the *MAT1-1-2* gene (Pöggeler and Kück 2000).

To investigate the functional conservation of mating-type genes of *S. macrospora*, cosmid clones containing the entire mating-type locus from *S. macrospora* were transformed into both *P. anserina MAT1-1* and *MAT1-2* strains. Similar to *N. crassa MAT1-1* strains, the *P. anserina MAT1-1* contains three genes (*FMR1* = *MAT1-1-1*, *FMR2* = *MAT1-1-2*, *SMR3* = *Mat1-1-3*), while the *P. anserina MAT1-2* locus contains only a single gene (*FPR1* = *MAT1-2-1*) encoding an HMG-domain protein (Debuchy and Coppin 1992; Debuchy et al. 1993). After introduction of the *S. macrospora MAT* locus into the homokaryotic strains of *P. anserina*, transformants were capable of forming fruiting bodies without crossing with a mating partner of the opposite mating type. However, instead of asci with four ascospores, the perithecia of the transformants contain only a gelatinous mass and have no structures such as hook cells, croziers, asci or spores (Pöggeler et al. 1997b). Interestingly, transformation of the entire *S. macrospora MAT* locus into a *P. anserina* Δ*MAT* strain without an endogenous *MAT* locus led to fertile perithecia containing

rosettes of asci (Pöggeler et al. 1997b). Thus, the mating-type genes from *S. macrospora* confer self-fertility to the heterothallic *P. anserina* when they do not interfere with a resident mating-type locus. This result suggests that the *S. macrospora* mating-type genes are functional and most probably are involved in fruiting body development and ascosporogenesis.

Due to the presence of *MAT1-1*-type homothallic species in the genus *Neurospora* (Glass et al. 1990b; Metzenberg and Glass 1990) and due to the fact that *MAT1-1* strains of the heterothallic *S. brevicollis* were shown to produce apandrous fruiting bodies with ascospores under certain culture conditions (Robertson et al. 1998), it was unclear whether *MAT1-2* specific genes are essential for sexual reproduction in homothallic members of the family Sordariaceae. Deletion of *MAT1-2-1* converts the self-fertile *S. macrospora* to a self-sterile fungus, no longer able to produce fruiting bodies and ascospores. Thus, the Δ*MAT1-2-1* mutant phenotypically resembles the *S. macrospora* *pro* mutants (see Sect. IV) which also do not produce any perithecia. However, no irregularities in vegetative growth and mycelial morphology were observed when the mutant strain was compared to *S. macrospora* wild type This finding demonstrates that, at least in the homothallic *S. macrospora*, the *MAT1-2-1* gene is required for fruiting-body development and sexual reproduction (Pöggeler et al. 2006b).

Cross-species microarray technology (see Sect. V.D) showed that the *S. macrospora* MAT1-2-1 affects the expression of at least 107 genes, including a common set of ten putative developmental genes deregulated in sterile *pro* mutants. At least 74 and 33 genes are two-fold up- or down-regulated in the *S. macrospora* Δ*MAT1-2-1* mutant strain. Of these 107 genes, 80 have homologues with known or putative function and were sorted into ten putative functional categories (Pöggeler et al. 2006b).

Transcription of cell type-specific genes in the ascomycetous yeast *Saccharomyces cerevisiae* has been shown to rely on the interaction of mating-type proteins (Herskowitz 1989). A two-hybrid approach in conjunction with protein cross-linking analysis demonstrated that in *Sordaria macrospora* the MAT1-1-1 and MAT1-2-1 proteins, homologous to polypeptides encoded by opposite mating partners of the heterothallic *N. crassa*, are able to form a heterodimer (Jacobsen et al. 2002).

III. Phylogeny

S. macrospora belongs to the family Sordariaceae (Sordariales, Ascomycetes) which comprises taxa characterized by dark, ostiolate fruiting bodies, and unitunicate, cylindrical asci (Kirk et al. 2001). The Sordariaceae family presently contains 7–10 genera (Kirk et al. 2001, Eriksson et al. 2004) and is represented by well known genera such as Gelasinospora, Neurospora, and Sordaria. It is also morphologically related to the Lasiosphaeriaceae, another family within the Sordariales, which contains the important genus *Podospora* (Huhndorf et al. 2004). Phylogenetic studies have revealed that species of the genus *Sordaria* are closely related to the genus *Neurospora* (Pöggeler 1999; Cai et al. 2006) which includes the fungal model organism *N. crassa* (Davis and Perkins 2002; Galagan et al. 2003; Borkovich et al. 2004; Fig. 2.3).

Phylogenetic trees based on sequences of the conserved glyceraldehyde-3-phosphate dehydrogenase gene (*gpd*) and on gene fragments of *MAT1-1-1* and *MAT1-2-1* have shown that *Neurospora* and *Sordaria* are monophyletic units (Pöggeler 1999). Moreover, there is a strict separation of heterothallic and homothallic species within both genera and a separation of homothallic strains with fused mating-type loci and *MAT1-1*-type homothallic species within *Neurospora*. The phylogenetic analyses suggest that changes in the reproductive strategy may represent a single event in each genus (Pöggeler 1999).

The *MAT* genes of members of the Sordariaceae appear to evolve more quickly than other regions of the genome; a phenomenon also observed in other ascomycetes (Turgeon 1998; Pöggeler 1999; Voigt et al. 2005). Interestingly, analysis of non-synonymous and synonymous substitution rates within *MAT* genes of the genus *Neurospora* revealed that the rapid divergence of *MAT* genes is due to an adaptive evolution within heterothallic members of this genus, whereas it is due to the lack of selective constraints within homothallic members of *Neurospora* (Wik et al. 2008).

IV. Mutants and Morphology

A major step in the molecular genetic analysis of components controlling cellular development is the functional analysis of mutants having a defect or block in defined morphogenetic steps. Such morphological mutants can be generated either by conventional mutagenesis or by the application of homologous recombination procedures (see Sect. V.B). The detailed molecular and cellular analysis of mutants will decipher molecular determinants involved in cellular differentiation processes. Using a conventional genetic approach, more than hundred developmental mutants with defects in fruiting body formation were generated. Mutants were obtained after UV mutagenesis of a protoplast suspension derived from the wild-type strain. Protoplasts were exposed to UV light (254 nm) for 15 min, with a survival rate of 0.1%. The protoplasts were regenerated on complete medium, supplemented with 10.8% sucrose (Masloff et al. 1999). After 24 h, individual clones were transferred to corn-meal medium to investigate clones with phenotypic variations in fruiting body development. In total, more than 23 000 clones were screened in this way. Presumptive developmental mutants were further tested for mitotic stability, before single-spore isolates were generated for a detailed molecular analysis. The mutants were grouped into four different phenotypic classes (see Fig. 2.4A):

1. Asc mutants. These mutants show a very early developmental block in the life cycle. They form ascogonia that sometimes have a non-wild-type structure. In more than 100 developmental mutants, we detected only five having this phenotype. They are sterile and never form protoperithecia.

2. Pro-mutants (Fig. 2.4B). Some 45 mutants have been isolated and they show a developmental block just after protoperithecia formation. Therefore, all pro-mutants are sterile. Protoperithecia have a round-shaped structure and a diameter of 30–55 m. The average size can vary in the different mutants but has been found to be constant in a defined mutant when grown on corn meal agar. As shown in Sect. VI. E, these mutants have contributed greatly to our current understanding of the molecular genetic network controlling fruiting body development. The first mutant used for a complementation analysis was pro1, where the *pro1* gene encodes a development-specific GAL4-like C_2H_6 transcription factor (Masloff et al. 1999).

3. Per-mutants. The 44 mutants of this type are characterized by the presence of a more or less

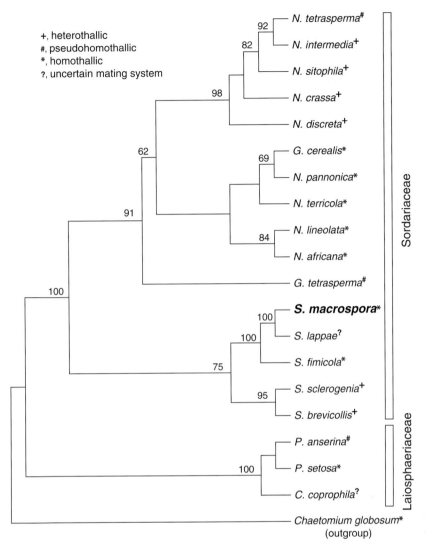

Fig. 2.3. Phylogenetic tree of partial *gpd* sequences generated by neighbor-joining analysis. 436 nucleotide positions were included in the analysis. The tree shown is based on a consensus tree calculated with the Neighbor program (Phylip). The *numbers* (percentages) indicate the bootstrap support based on 1000 replications. Abbreviations and accession numbers are as follows: *N. tetrasperma* (*Neurospora tetrasperma*, AJ133018.1); *N. intermedia* (*Neurospora intermedia*, AJ133019.1); *N. sitophila* (*Neurospora sitophila*, AJ133020.1); *N. crassa* (*Neurospora crassa*, U67457.1); *N. discreta* (*Neurospora discreta*, AJ133021.1); *G. cerealis* (*Gelasinospora cerealis*, AF388934.1); *G. tetrasperma* (*Gelasinospora tetrasperma*, AF388935.1); *N. pan-* *nonica-*. (*Neurospora pannonica*, AJ133016.1); *N. terricola* (*Neurospora terricola*, AJ133017.1); *N. lineolata* (*Neurospora lineolata*, AJ133015.1); *N. africana* (*Neurospora africana*, AJ133012.1); *S. macrospora* (*Sordaria macrospora*, AJ133007.1); *S. lappae* (*Sordaria lappae*, EF197111.1); *S. fimicola* (*Sordaria fimicola*, AJ133009.1); *S. sclerogenia* (*Sordaria sclerogenia*, AJ133011.1); *S. brevicollis* (*Sordaria brevicollis*, AJ133010.1); *P. anserina* (*Podospora anserina*, EF197096.1); *P. setosa* (*Podospora setosa*, EF197110.1); *C. coprophila* (*Cercophora coprophila*, EF197091.1). *Chaetomium globosum-* (CHGG_07980.1; http://www.broad.mit.edu/annotation/genome/chaetomium_globosum) was used as the outgroup

developed perithecium, but otherwise they are sterile. This is due to the fact that they are unable to generate mature ascospores, although in some cases, they are able to form immature asci. A good example is mutant per5 (see Sect. VI.E), which was one of two mutants that provided the first example for the applicability of the *S. macrospora* system for a molecular genetic analysis (Nowrousian et al. 1999).

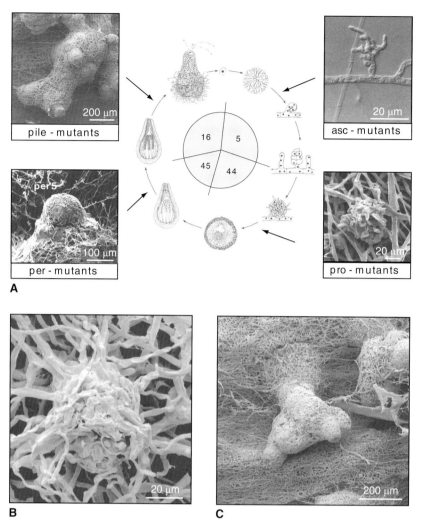

Fig. 2.4. Phenotypes of developmental mutants from *Sordaria macrospora*. **A** Life cycle and number of mutants that can be grouped in four different morphological classes having distinct developmental blocks. **B, C** Scanning electron micrographs of a pro (**B**) and pile mutant (**C**)

4. Pile mutants (Fig. 2.4C). These mutants show a rather complex phenotype. The spatial order of perithecium formation is disturbed. One perithecium can sit on top of another, thus forming a pile of many perithecia. Initially, we obtained 16 isolates of this mutant type which seem to be genetically instable. Pile mutants are often fertile, but show a defect in the formation of the pigment melanin (Engh et al. 2007a).

As described in the sections to come, different developmental mutants were used for the further molecular investigation of the developmental process. Restoration of a fertile phenotype in these sterile mutants allowed isolation of a wild-type gene copy, which is defective in the recipient strains.

V. Molecular and Genetic Tools

Development of an organism as a genetic model system is greatly dependent on the application of modern molecular genetic techniques to a hitherto only conventionally manipulated genetic system. In *S. macrospora* the use of molecular genetics started with DNA-mediated transformations and recently applied techniques such as microarray hybridization and novel cellular markers for fluorescent microscopic investigations (Nowrousian et al. 2005; Engh et al. 2007b; Nowrousian et al. 2007a).

A. Tetrad Analysis

Tetrad analysis investigates the four products that are generated from a diploid cell during two meiotic divisions. In fungi, meiosis occurs during sexual spore formation, resulting in each of the spores in the meiosporangium (the ascus in the case of ascomycetes) resembling one product of meiosis. In ascomycetes ordered tetrads as can be found in *S. macrospora* can be distinguished from non-ordered tetrads present in yeast. If alleles are segregated during the first meiotic division, this leads to ascus halves in ordered tetrads with spores of the same genotype. However, cross-over may occur during prophase of the first meiotic division between the centromere and the gene of interest. Different alleles are then present on sister chromatids and do not segregate until the second meiotic division. This leads to ascus halves with spores of different genotypes. For further analysis, spores can be isolated and germinated on agar medium.

    S. macrospora is an excellent organism for tetrad analysis for several reasons (Esser and Straub 1958; Esser 1982), some of which are outlined here.

    The tetrad is ordered and the ascospores are linearly arranged. A post-meiotic mitosis leads to eight instead of four spores. Sister products remain adjacent to one another; thus, the position of each ascospore reflects the preceding nuclear division.

    The ascospores of *S. macrospora* are about 28 μm long, allowing for easy manual isolation.

    Several spore colour mutants exist that can be used in crosses. This is important, because *S. macrospora* is self-fertile and one cannot distinguish selfed and non-selfed fruiting bodies. However, when using a strain with a spore colour defect in sexual crosses, recombinant perithecia are identified by the segregation of spore colour alleles within the asci (Fig. 2.5).

    Experiments have shown that not only spore colour alleles but also heterologous marker genes like *egfp* segregate in a Mendelian manner (Pöggeler et al. 2003).

    S. macrospora has a very short sexual cycle that can be completed within seven days under laboratory conditions.

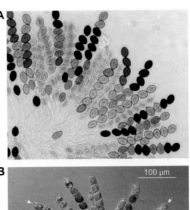

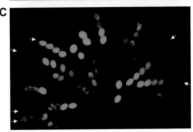

Fig. 2.5. Ascus rosettes from a cross between two different *S. macrospora* strains. **A** Tetrad analysis from a cross between a wild-type strain having black ascospores and a mutant strain with yellow ascospores. Light (**B**) and fluorescence (**C**) micrographs from ascus rosettes that were derived from a cross between a strain carrying two different ascospore colour mutations and a transformant carrying the gene for the green fluorescence protein (*gfp*). *Arrows* indicate asci in which the alleles for the ascospore colours or green fluorescent protein were separated by first and second division segregation, respectively

What is the advantage of a tetrad analysis? Traditionally, it is used to examine gene distribution and gene–centromere distances by analysing the frequency of second division segregation. This was made mostly obsolete by genome sequencing. However, tetrad analysis is still a powerful tool for genetic analysis. For example, recently a Δku70 mutant was generated that facilitates the generation of knockout strains (Pöggeler and Kück 2006), which still carry the *ku70* deletion (see next chapter). In order to obtain a knockout strain in the wild-type background, crosses against a spore colour mutant are performed for further tetrad analysis. This is a fast and easy method leading to homokaryotic mutant strains

(Pöggeler and Kück 2006). Another important advantage of a tetrad analysis is the possibility to generate double mutants by crossing the two mutants carrying distinct single mutations. The double mutant can be isolated even if: (1) both single mutants were generated using the same selection marker, or (2) the double mutant has the same phenotype as one of the single mutants. This is feasible, because segregation into parental and recombinant spores is taken into account and the ordered tetrad allows for the identification of both. Using this method, numerous *S. macrospora* double mutants have been generated (Mayrhofer et al. 2006; Kamerewerd et al. 2008). As a consequence, conventional genetic analysis strongly increases the potential of *S. macrospora* as a model organism for studying diverse biological phenomena.

B. DNA-Mediated Transformations and Gene Libraries

A variety of options recently became available to introduce and control genetic elements, as well as to help the construction of different molecular libraries, displaying the importance of *S. macrospora* as model organism for basic research (Kück and Pöggeler 2005).

At the beginning, the development of a reproducible DNA-mediated transformation protocol was of vital importance. Transformation of *S. macrospora* was first reported by Le Chevanton and co-workers. They used an auxotrophic *ura5* strain as the recipient. Appropriate selection was performed by complementation of the uracil auxotrophy, by transforming with a homologous *ura5* gene (Le Chevanton et al. 1989).

Selection of *S. macrospora* transformants with a dominant marker became feasible using the *Escherichia coli* hygromycin B phosphotransferase gene (*hph*) under control of fungal promoter and terminator elements (Walz and Kück 1995). The advantage of such a dominant selectable marker is that basically any recipient strain can be used in transformation experiments. Expression of the *hph* gene was initially directed by the upstream region of the isopenicillin N synthetase gene (*pcbC*) from the cephalosporin C producing fungus *Acremonium chrysogenum*. Recently, the strong constitutive promoters of the *A. nidulans gpd* and *trpC* genes were shown to be useful for

driving *hph* expression (Walz and Kück 1995; Pöggeler et al. 2003), with transformation frequencies of about 5–15 transformants per 10 µg of DNA (Walz and Kück 1995).

In order to generate an alternative dominant marker system that does not exhibit cross-resistance to hygromycin B, the *nat1* gene conferring resistance to nourseothricin was used to transform *S. macrospora* (Hoff and Kück 2006). The *nat1* gene product is the nourseothricin acetyltransferase from *Streptomyces noursei* (Krügel et al. 1993) and was expressed under the control of the *A. nidulans trpC* promoter. Transformation frequencies of 10–40 transformants per 10 µg of plasmid DNA were obtained with this dominant selection system (Hoff and Kück 2006).

Alternative selection systems were developed based on novel auxotrophic recipient strains. Using conventional mutagenesis, a leucine auxotrophic strain was generated, which lacks the complete wild-type copy of the *leu1* gene, encoding β-isopropylmalate dehydrogenase (Kück 2005). A uracil auxotrophic mutant strain was constructed by isolating the *ura3* (syn. *pyr4*) gene, encoding orotidine-5′-phosphate decarboxylase from the *S. macrospora* genomic DNA (Nowrousian and Kück 1998) and subsequently disrupting the *ura3* open reading frame in vitro by insertion of a *hph* resistance cassette. The resulting recombinant plasmid was used in *S. macrospora* transformations to isolate a fungal derivative, in which the wild-type *ura3* copy was substituted by the chimeric *ura3-hph* gene through site specific homologous recombination (Kück and Pöggeler 2004). The *leu1* and *ura3* strains can be transformed with the corresponding wild-type genes from *S. macrospora*; and the transformation frequency is comparable to that described for the *hph* gene.

Transformed DNA was shown to be ectopically integrated into the genomic DNA of *S. macrospora*. The copy number varied between transformants. Interestingly, the analysis of *S. macrospora* transformants by pulse field gelelectrophoresis revealed that all vector molecules usually integrate into the same chromosome (Walz and Kück 1995).

Targeted gene replacement via homologous recombination is a routinely used approach to elucidate the function of unknown genes. Integration of exogenous DNA in the genomic DNA requires the action of double-strand repair mechanisms. However, similar to most filamentous fungi, plants, and animals, transformed DNA is predominately ectopically integrated into the genome of *S. macrospora*.

Most probably this occurs by non-homologous end joining (NHEJ), a mechanism that involves the binding of the Ku heterodimer (Ku70/Ku80) at the ends of a DNA double-strand break (DSB; Pastwa and Blasiak 2003).

Hence, the low rate of homologous recombination exhibited by the *S. macrospora* wild-type has hindered highly efficient gene knockout protocols.

In *N. crassa*, Ninomiya and co-workers demonstrated that deletion of the genes *ku70* and *ku80* results in a dramatic increase in the frequency of homologous recombination resulting in 100% of transformants exhibiting integration only at homologous sites (Ninomiya et al. 2004). This strategy for targeted gene integration was extended to *S. macrospora* and a ku70 deletion mutant was generated by replacing the *ku70* gene with a *nat1* resistance cassette. No phenotypic defects regarding vegetative or sexual development were observed in the ku70 strain, making it an attractive recipient strain for transformation and a valuable tool for gene disruption of developmental genes. Compared to the wild type, which shows homologous integration efficiency between 0.1% and 5.0%, a Δku70 strain showed up to 100% of transformants with homologous integration of exogenous DNA (Pöggeler and Kück 2006). As described in the previous section, the Δ*ku70*::nat1 mutation can easily be eliminated from a disruption strain by crossing to a spore color mutant with an otherwise wild-type genetic background. Tetrad analyses allow the isolation of ascospore isolates that lack the Δ*ku70* mutation, but carry the substituted gene copy.

Based on procedures described for *N. crassa* (Colot et al. 2006), the *S. macrospora* transformation procedure was expanded by substituting conventional cloning with recombinational cloning in yeast (Oldenburg et al. 1997; Raymond et al. 1999). Using this method, traditional restriction digestion and ligation for the generation of knockout fragments are unnecessary and components of the final construct are synthesized individually by PCR. Amplified fragments contain short overlapping sequences and are cotransformed into yeast for assembly by the recombination machinery. This "two-step" transformation procedure including homologous recombination in yeast and transformation of the knockout fragment into

S. macrospora Δ*ku70* strain presents a valuable tool for gene disruption and a crucial improvement for the molecular research on *S. macrospora* (Mayrhofer et al. 2006; Nolting and Pöggeler 2006a).

Genetic transformation protocols contribute greatly to the molecular dissection of the *S. macrospora* sexual development. Another important development for genetic engineering of *S. macrospora* comes from the construction of various gene libraries of *S. macrospora* DNA. An indexed cosmid library and a cDNA library, which is suitable for yeast two-hybrid screens, were used to identify important components involved in sexual reproduction and fruiting-body development (Pöggeler et al. 1997a; Mayrhofer and Pöggeler 2005). Clones of indexed cosmid libraries are kept individually in microtitre dishes to ensure equal representation of each clone. Several methods for the rapid screening of pooled cosmid DNA have been published, such as isolation of conserved fungal genes by means of colony filter hybridization of cosmid clones or rapid screening by means of PCR using pooled cosmid DNA (Pöggeler et al. 1997a). In other approaches, this library was used to clone novel developmental genes by restoring the fertile wild-type phenotypes in otherwise sterile mutants of *S. macrospora* (see Sects. IV, VI.E; Masloff et al. 1999; Nowrousian et al. 1999, 2007a; Pöggeler and Kück 2004; Engh et al. 2007b). The second genomic library was generated for yeast two-hybrid studies and consists of yeast vectors containing *S. macrospora* cDNA (Nolting and Pöggeler 2006a). This library facilitated the screening of interaction partners of developmental proteins and was shown to be useful for the identification of transcriptional activators and other developmental genes (Nolting and Pöggeler 2006a).

C. Tools for Fluorescence Microscopy

Fluorescence microscopy has become a powerful tool for studying the biology of filamentous fungi in detail, e.g. gene regulation, signal transduction, and protein localization. It relies on the fluorescent labelling of proteins or cellular components. This can be achieved by different methods, i.e. fluorescent dyes, fluorescent proteins or a combination of specific primary antibodies with conjugated secondary antibodies. While fluorescent dyes are mostly used to label cell organelles or membranes, fluorescent proteins and antibodies are often employed to visualize proteins. All three applications were conducted with *S. macrospora* as an experimental system.

Using immunofluorescence, *S. macrospora* was employed as a model organism for cytoskeleton interactions during ascus development (Thompson-Coffe and Zickler 1992, 1993), using antibodies against actin and tubulins in combination with secondary antibodies conjugated with fluorescent probes. In addition, antibodies against meiotic proteins like e.g. RAD-51 were generated for the analysis of meiosis (Tessé et al. 2003; Storlazzi et al. 2008). A similar approach was used to determine the subcellular localization of the developmental protein ACL1 (Nowrousian et al. 1999).

The advantage of fluorescent proteins over immunofluorescence is that no antibodies are necessary and in vivo time-lapse imaging is possible. For studying the localization of developmental proteins, a number of plasmids with genes for different fluorescent proteins were generated (Table 2.1, Fig. 2.6). These plasmids encode N- or C-terminal fusion proteins between the gene of interest and the fluorescent protein. These fluorescent markers segregate during meiotic division (Pöggeler et al. 2003). Using EGFP and DsRed, the localization of the developmental proteins PRO40, PRO41, and MCM1 were identified as Woronin bodies, the ER, and the nucleus, respectively (Nolting and Pöggeler 2006b; Engh et al. 2007b; Nowrousian et al. 2007a). For PPG1, the pheromone precursor protein, and PRO41, a novel component of the ER, putative signal sequences for translational insertion into the ER were verified by Western analysis with an anti-EGFP-antibody that detected the secreted PPG1-EGFP and PRO41-EGFP fusion proteins (Mayrhofer and Pöggeler 2005; Nowrousian et al. 2007a).

However, fluorescent proteins are not only powerful tools for characterizing protein localization. In fungi, they have been used to study the dynamics of organelles (Mouriño-Pérez et al. 2006) and to screen for targeting mutants (Ohneda et al. 2005). For *S. macrospora*, several plasmids encoding organelle-targeted fluorescent proteins were constructed (Table 2.1, Fig. 6A, C, E–H) by fusing targeting signals to genes encoding fluorescent proteins. One example is plasmid pDsRed-SKL, encoding the *DsRed* gene fused to a short sequence, encoding the SKL motif. This C-terminal peroxisomal targeting sequence (PTS1) targets the fusion protein to the peroxisome (Elleuche and Pöggeler 2008). Likewise, ER markers pEGFP-KDEL and pDsRed-KDEL were constructed using EGFP and DsRed together with the C-terminal retention signal KDEL and the N-terminal secretion signal sequence of the *ppg1* pheromone precursor gene (Mayrhofer and Pöggeler 2005; Nowrousian et al. 2007a). The ER marker proteins were successfully used to identify the localization of fluorescent signals from labelled developmental proteins (Engh et al. 2007b, Nowrousian et al. 2007a). Using organelle-targeted GFP proteins, we have shown in two mutants (pro40, pro41) that the Woronin body

Table 2.1. Plasmids encoding fluorescent proteins for microscopic investigations. Plasmids encode either solely the fluorescent protein and are designed for C- or N-terminal fusion with genes of interest, or they encode an organelle-targeted fluorescent protein

Plasmid	Description	Localization	Colour	References
pEH3	hph::Pgpd-egfp-TtrpC, C-terminal	Cytoplasm	Green	Nowrousian et al. (2007a)
pYHN3	hph::Pgpd-eyfp-TtrpC, C-terminal	Cytoplasm	Yellow	Rech et al. (2007)
pCHN3	hph::Pgpd-ecfp-TtrpC, C-terminal	Cytoplasm	Cyan	Rech et al. (2007)
pRHN3	hph::Pgpd-DsRed-TtrpC, C-terminal	Cytoplasm	Red	Engh et al. (2007b)
pMHN2	hph::Pgpd-mRFP1-TtrpC, C-terminal	Cytoplasm	Red	Unpublished data
pTomato	hph::Pgpd-tdTomato-TtrpC, C-terminal	Cytoplasm	Red	Unpublished data
pmKalama1	hph::Pgpd-mKalama1-TtrpC, C-terminal	Cytoplasm	Blue	Unpublished data
pN-EGFP	hph::Pgpd-egfp-TtrpC, N-terminal	Cytoplasm	Green	Unpublished data
pN-Tomato	hph::Pgpd-tdTomato-TtrpC, N-terminal	Cytoplasm	Red	Unpublished data
pYH2A	hph::Pgpd-hh2a-eyfp-TtrpC	Nucleus	Yellow	Rech et al. (2007)
pCH2B	hph::Pgpd-hh2b-ecfp-TtrpC	Nucleus	Cyan	Rech et al. (2007)
pDsRed-SKL	nat::Pgpd-DsRed-SKL-TtrpC	Peroxisome	Red	Elleuche and Pöggeler (2008)
pGFP-SKL	phleo::Pgpd-egfp-SKL-TtrpC	Peroxisome	Green	Ruprich-Robert et al. (2002)
pCW15	his3::Pccg1-sgfp-hex-1	Woronin body	Green	Engh et al. (2007b)
pEGFP-KDEL	nat::Pgpd-Sppg1-egfp-KDEL-TtrpC	ER	Green	Nowrousian et al. (2007a)
pDsRed-KDEL	nat::Pgpd-pro41-DsRed-KDEL-TtrpC	ER	Red	Nowrousian et al. (2007a)

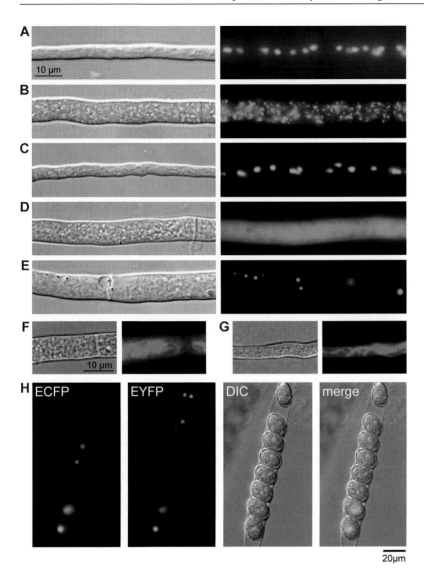

Fig. 2.6. DIC and fluorescence microscopic images of *S. macrospora*. Strains were transformed with plasmids pCH2B (**A**), pGFP-SKL (**B**), pYH2A (**C**), pTomato (**D**), pCW15 (**E**) and pEGFP-KDEL (**F**, **G**). **H** Ascus from a sexual cross of different strains transformed with pYH2A and pCH2B, respectively. Note the Mendelian segregation of the fluorescent markers. Bar size is the same for A–E and F–G, respectively

and ER appear to have a wild-type morphology, although the proteins defective in these mutants localize to these organelles (Engh et al. 2007b; Nowrousian et al. 2007a).

A particular application of fluorescence microscopy was developed for the quantification of hyphal fusion events between different *S. macrospora* strains (Rech et al. 2007). During hyphal fusion, cytoplasmic mixing occurs, which also enables mixing of genetic information by exchange of nuclei between hyphal compartments. Different strains were labelled with histone fusion proteins H2A-YFP and H2B-CFP. Following successful fusion, differently labelled nuclei are present in one hyphal compartment, and differently labelled histones are targeted to the same nucleus. Subsequently, fluorescence microscopy shows both cyan and yellow fluorescence in one nucleus as a result of hyphal fusion. Besides immunofluorescence and fluorescent proteins, vital fluorescent dyes have been used for *S. macrospora*.

An example is Calcofluor white, an exceptional dye for staining chitin in fungal cell walls. As such, it can be applied for studying morphological changes, and its use uncovered the hyperbranching phenotype of the mcm1 mutant (Nolting and Pöggeler 2006b). The DNA-staining dye 4′,6-diamidino-2-phenylindol (DAPI) was used for studying meiotic and post-meiotic events (Storlazzi et al. 2003, 2008) as well as ascus development, which is closely coupled to meiosis. For example, DAPI staining showed abnormal ascus development in several *S. macrospora* mutants like Δste12 and the pheromone receptor/phero-mone precursor double mutant Δpre2/Δppg2 (Mayrhofer et al. 2006; Nolting and Pöggeler 2006b). DAPI staining was also used to show that the mouse striatin gene is able to complement the developmental defects of the sterile pro11 mutant (Pöggeler and Kück 2004).

In summary, a great variety of tools for fluores-cence microscopy has been developed for *S. macrospora* that allow not only protein local-ization but also specialized experiments for further studying the cell biology of this model organism.

D. Functional Genomics

In recent years, high-throughput methods for the analysis of gene expression have been established for a number of filamentous fungi. Most prominent among these new techniques are microarrays that allow the parallel analysis of gene expression for (nearly) all genes within a genome (Nowrousian et al. 2004a). Microarrays can be established from cDNA sequences or, if the genome of an organism has been sequenced, oligonucleotides as array probes can be derived from genomic sequences; and both types of arrays have been used for fila-mentous fungi (Breakspear and Momany 2007). In most cases, expression analyses with microarrays are performed with the organisms for which the array was developed (Nowrousian 2007). However, if two species are closely related, it is possible to do cross-species microarray hybridizations where the targets that are used for hybridization are derived from another organism than the one for which the array was originally developed. This saves the effort of setting up microarrays, especially for organisms where the genome has not yet been sequenced. *S. macrospora* was the first filamentous fungus for which cross-species array hybridizations were established (Nowrousian et al. 2005). Cross-species hybridizations were performed first using cDNA microarrays for *N. crassa*, a close relative of *S. macrospora* with ~90% nucleic acid sequence identity within coding regions (Nowrousian et al. 2004b). It was subsequently shown that these cross-species hybridizations could also be done with oligonucleotide microarrays from *N. crassa* that carry 70mer oligonucleotides as probes (Nowrou-sian et al. 2007a).

Gene expression analysis with microarrays can serve several purposes, for example: (1) the identi-fication of differentially expressed genes for further studies, and (2) the analysis of genetic networks by making use of expression data as a molecular phe-notype (Nowrousian 2007). Both aspects were stud-ied in *S. macrospora* by comparing gene expression in developmental mutants and a wild-type strain (Nowrousian et al. 2005; Pöggeler et al. 2006b; Nowrousian et al. 2007a). Microarray analysis was performed using mutants pro1, pro11, pro22, pro41, and ΔSmta-1, all of which produce only protoperithecia (see Sects. IV, VI); and a number of genes were identified that are up- or down-regu-lated in the mutants compared to the wild type. Among these are the genes for the pheromone precursors *ppg1* and *ppg2*, the putative lectin-encoding gene *tap1*, a gene that was subsequently shown to encode the fruiting body-specific protein APP and the melanin biosynthesis genes *pks* and *sdh*. These genes were further studied to decipher the details of their expression patterns and their roles in fruiting body development (Nowrousian and Cebula 2005; Mayrhofer et al. 2006; Engh et al. 2007a; Nowrousian et al. 2007b). Additionally, overall expression patterns in the mutants were used as molecular phenotypes to establish a genetic network of *pro* genes (see also Sect. VI.E): expres-sion patterns in mutants pro1, pro11 and pro22 are more similar to each other than to the ΔSmta-1 mutant, indicating that *Smta-1* may act in parallel to these *pro* genes (Nowrousian et al. 2005; Pöggeler et al. 2006b). Furthermore, it was shown that *pro41* acts genetically downstream of the *pro1* gene (Nowrousian et al. 2007a).

Taken together, *S. macrospora* was among the first filamentous fungi for which microarray experiments were performed; and, in recent years, the results of these high throughput analyses have identified developmentally regu-lated genes and aided analysis of genetic networks that control fruiting body formation.

VI. Developmental Biology and Components of Signalling Pathways

There is increasing evidence that both external and internal signals are essential for fungal fruiting body development. In several filamentous ascomy-cetes these signals have been shown to be trans-duced by conserved signal transduction pathways. The components involved can be subdivided into receptors that receive the signal, components that transmit the signal into the cell, and nuclear transcription factors that regulate gene expression (Pöggeler et al. 2006a). The molecular mechanisms underlying sexual developmental processes in *S. macrospora* are only poorly understood. However, some of the key components of the signalling cascades from the cell surface to the nucleus have been genetically characterized in *S. macrospora*. Moreover, novel developmental proteins were discovered that might function further downstream of conserved signalling pathways, ensuring cell-specific differentiation.

A. Pheromones and Pheromone Receptors

Male and female fertility of heterothallic filamentous ascomycetes depends on interactions of pheromones with their specific receptors. Pheromone precursor genes encoding two different types of pheromones have been isolated from heterothallic filamentous ascomycetes. One of the precursor genes encodes a polypeptide containing multiple repeats of a putative pheromone sequence bordered by protease processing sites and resembles the α-factor precursor gene of *Saccharomyces cerevisiae*. The other gene encodes a short polypeptide similar to the *S. cerevisiae* a-factor precursor. The short precursor has a C-terminal CaaX motif (C = cysteine, a = aliphatic, X = any amino acid residue) expected to produce a lipopeptide pheromone with a C-terminal carboxy-methyl-isoprenylated cysteine. The two types of pheromone precursor genes are present in the same genome and expression of pheromone genes seems to occur in a mating type-specific manner (Bobrowicz et al. 2002; Coppin et al. 2005).

When pheromone genes were deleted in heterothallic ascomycetes, male spermatia were no longer able to fertilize the female partner, proving that pheromones are crucial for the fertility of male spermatia (Turina et al. 2003; Coppin et al. 2005; Kim and Borkovich 2006). Sensing of the pheromone signal depends on (G) protein-coupled seven transmembrane-spanning receptors (GPCRs). Two genes, designated *pre-1* and *pre-2*, encode pheromone receptors similar to *S. cerevisiae* a-factor receptor (Ste3p) and α-factor receptor (Ste2p), respectively, and have been shown to be essential for mating type-specific directional growth and fusion of trichogynes and female fertility in *Neurospora crassa* (Kim and Borkovich

2004, 2006). Thus, in heterothallic filamentous ascomycetes the function of pheromones and their cognate receptor seems to be restricted to the fertilization process.

Interestingly, genes encoding two different pheromone precursors and two pheromone receptors have also been identified in the homothallic *Sordaria macrospora* (Pöggeler 2000; Pöggeler and Kück 2001). The pheromone precursor gene *ppg1* is predicted to encode a α-factor-like peptide pheromone and the *ppg2* gene an a-factor-like lipopeptide pheromone (Pöggeler 2000). The deduced gene products of the *S. macrospora pre1* and *pre2* genes are predicted to be seven transmembrane domain proteins (Pöggeler and Kück 2001). For *S. macrospora*, it has been demonstrated that disruption of the pheromone precursor *ppg1* gene prevents the production of the peptide pheromone, but does not affect vegetative growth, fruiting-body, or ascospore development (Mayrhofer and Pöggeler 2005). Similarly, no effect on vegetative growth, fruiting-body, and ascospore development was observed in the single pheromone-mutant Δ*ppg2* and single receptor-mutants Δ*pre1* and Δ*pre2*. However, double-knockout strains lacking any compatible pheromone/receptor pair (Δ*pre2*/Δ*ppg2*, Δ*pre1*/Δ*ppg1*) and the double-pheromone mutant strain (Δ*ppg1*/Δ*ppg2*) have a drastically reduced number of perithecia and sexual spores, whereas deletion of both receptor genes (Δ*pre1*/Δ*pre2*) almost completely eliminates fruiting body and ascospore formation (Mayrhofer et al. 2006). Taken together, these results suggest that pheromones and pheromone receptors are involved in post-fertilization events and are required for optimal sexual reproduction of the homothallic *S. macrospora*. Moreover, by heterologously expressing the *S. macrospora pre2* in *Saccharomycs cerevisiae MATa* cells, lacking the Ste2p receptor, PRE2 was shown to facilitate all aspects of the *S. cerevisiae* pheromone pathway when activated by the *Sordaria macrospora* peptide pheromone (Mayrhofer and Pöggeler 2005). Therefore, one may conclude that the receptors encoded by *pre2* and *pre1* also function as G protein-coupled receptors (GPCRs) in *S. macrospora*.

B. Heterotrimeric G Proteins

Upon activation by GPRC receptors, heterotrimeric G proteins catalyse the exchange of GDP for GTP on the G protein α subunit. This leads to dissociation of the βγ subunits. Either Gα, or Gβγ, or both are then free to activate downstream effectors. Signalling persists until GTP is hydrolysed to GDP, and the subunits re-associate (Dohlman 2002). The genome of *S. macrospora* contains three genes (*gsa1*, *gsa2* and *gsa3*) encoding Gα subunits (Kamerewerd et al. 2008). In addition genes encoding the β-subunit and the γ-subunits of the G protein are present in *S. macrospora* (Mayrhofer and Pöggeler 2005).

To explore the functional role of G protein α subunits (GSA) in the sexual life cycle of this fungus, knockout strains for all three *gsa*-genes (Δgsa1, Δgsa2, Δgsa3) and double mutants were generated. Phenotypic analysis of single and double mutants showed that the genes for Gα subunits contribute differently to sexual development. While the Δgsa2 mutant revealed wild-type-like fertility, Δgsa3 developed fruiting bodies, but the ascospores had a drastically reduced germination rate. Δgsa1 showed a delay in sexual development and a 50% reduction in the number of fruiting bodies. A more impaired phenotype can be observed in Δgsa1Δgsa2 and Δgsa1Δgsa3 double mutants. These mutants produce only protoperithecia, indicating that all Gα subunits contribute significantly to sexual development. To test whether the pheromone receptors PRE1 and PRE2 mediate signalling via distinct Gα subunits, Δpre strains were crossed with all Δgsa strains. The data obtained with double mutants carrying a disrupted *gsa* gene together with a disrupted receptor gene indicated that the pheromone receptors interact differently with GSA1 or GSA2 and that GSA1 is the main component of a signal transduction cascade downstream of the pheromone receptors. As shown in Fig. 2.7, the third Gα protein, GSA3, seems to be involved in a parallel signalling pathway but also contributes to fruiting body development and fertility (Kamerewerd et al. 2008).

C. Adenylyl Cyclase

Activated Gα or Gβγ subunits can regulate downstream effectors such as adenylyl cyclase and

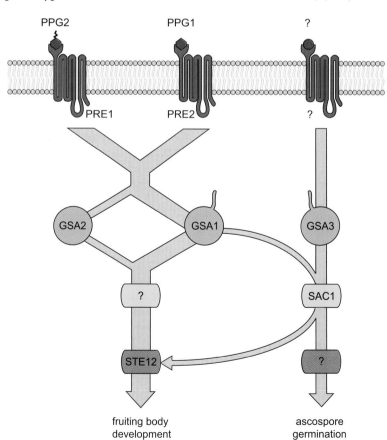

Fig. 2.7. Pheromone signalling pathway showing the function of pheromones, pheromone receptors and G protein alpha subunits, as well as downstream signalling components in the development of perithecia (modified from Kamerewerd et al. 2008)

mitogen-activated protein kinase (MAPK) cascades. The fungal adenylyl cyclase is a soluble enzyme that produces the second messenger cyclic AMP (cAMP) from ATP. The cAMP signalling pathway of fungi is involved in several important processes including nutrient sensing, stress response, metabolism, pathogenicity, and sexual development (Lengeler et al. 2000; D'Souza and Heitman 2001). Several lines of evidence indicate that many of the known fungal G proteins influence the intracellular level of cAMP by either stimulating or inhibiting adenylyl cyclase (Bölker 1998). In order to investigate a component that acts supposedly downstream of the GSAs, the *S. macrospora* gene encoding adenylyl cyclase (*sac1*) was isolated. Deletion of *sac1* prevents cAMP synthesis. Moreover, a Δsac1 mutant exhibits reduced fertility with a significant number of the fruiting bodies that are embedded in the solid media. A similar phenotype was observed in a Δgna3 mutant of *N. crassa*. In addition, ascospores derived from the Δsac1 mutant have a highly reduced germination rate and this phenotype also resembles that of Gα subunit 3 disruption mutants from *N. crassa* (Kays et al. 2000) and *S. macrospora* (Kamerewerd et al. 2008). Analysis of the three Δgsa/Δsac1 double mutants indicates that SAC1 acts downstream of GSA3, parallel to a GSA1-GSA2 mediated signalling pathway (Kamerewerd et al. 2008).

D. Transcription Factors

Signal transduction initiates morphogenesis by finally activating transcription factors which in turn activate or repress cell- or tissue-specific expression of developmental genes. To elucidate whether conserved fungal transcription factors are involved in sexual development of *S. macrospora*, *mcm1* and *ste12* were functionally analysed. The *mcm1* gene encodes a putative homologue of the *Saccharomyces cerevisiae* MADS box protein Mcm1p and *ste12* encodes a homeodomain similar to the *S. cerevisiae* Ste12p, respectively (Nolting and Pöggeler 2006a, b). Deletion of the *mcm1* gene led to a pleiotropic phenotype, including reduced biomass, increased hyphal branching, and reduced hyphal compartment length during vegetative growth, as well as sexual sterility. The *mcm1* mutant was only capable of producing protoperithecia, but was unable to form either ascospores or

perithecia and, thus, phenotypically resembled pro-mutants of *Sordaria macrospora* (see Sect. IV). A two-hybrid analysis demonstrated that MCM1 may physically interact with itself and with the α-domain mating-type protein MAT1-1-1 (Nolting and Pöggeler 2006b). The pleiotropic phenotype of *S. macrospora* Δmcm1 suggests that the *S. macrospora* ΔMCM1 protein might be involved in a wide range of functions via interaction with diverse transcriptional regulators.

From a yeast two-hybrid screen, among several other proteins, a putative homologue of the *Saccharomyces cerevisiae* homeodomain protein Ste12p was identified as an additional MCM1 interaction partner (Nolting and Pöggeler 2006a). In *S. cerevisiae*, Ste12p is known to be a transcription factor acting downstream of Fus3p/Kss1 MAP kinases. In haploid yeast cells, Ste12p is required for the response to the mating pheromone produced by the opposite mating type and for invasive growth in response to limited nutrients. In diploid yeast cells, Ste12p regulates pseudohyphal development in response to nitrogen starvation (Madhani and Fink 1997; Gustin et al. 1998; Roberts et al. 2000). Transcriptional regulation of different classes of genes is thereby triggered through interactions with different transcriptional regulators. Ste12p forms a homodimer and binds either Mcm1p or Matα1p to regulate pheromone-responsive genes and cell-type-specific genes, respectively, and with Tec1p to activate genes required for filamentous growth (Dolan et al. 1989; Yuan et al. 1993; Bruhn and Sprague 1994; Madhani and Fink 1997).

In *Sordaria macrospora*, two-hybrid and biochemical studies showed that STE12 in addition to MCM1 is also able to interact with the mating-type protein MAT1-1-1. Analysis of a *S. macrospora* Δste12 knockout mutant demonstrated that STE12 is not needed for fruiting body formation and vegetative growth but is involved in ascosporogenesis. The *S. macrospora* Δste12 mutant is able to form protoperithecia and perithecia, but the latter contain a drastically reduced number of asci with predominantly non-viable ascospores. Particularly, cell walls of asci and ascospores appear to be fragile (Nolting and Pöggeler 2006a). This implies that STE12, in addition to its role in pheromone signal transduction, might be involved in cell wall integrity of asci and ascospores. To study the functional connections between the GSA subunits and the STE12 transcription factor in *S. macrospora*, the pheno-types of three ΔgsaΔste12 double mutants were analysed (Kamerewerd et al. 2008). While Δgsa2Δste12 double mutants show a wild-type-like formation of perithecia containing fragile asci and ascospores, the Δgsa1Δste12 mutant develops only a few fruiting bodies. Furthermore, perithecia of the Δgsa1Δste12 mutant contain only a few asci compared to the Δste12 single or the Δgsa2Δste12 double mutant. The Δgsa3Δste12 mutant exhibits the most severe phenotype. Protoperithecia are produced, but no mature perithecia can be detected. The sterility of the Δgsa3Δste12 mutant is a strong evidence for the STE12 transcription factor being one of the key regulators downstream of the pheromone receptors.

E. Novel Developmental Proteins

As shown in Fig. 2.4 (see Sect. IV), more than 100 *S. macrospora* mutants have been described with blocks at various stages of the developmental cycle. For seven of these mutants, the affected genes have already been identified: The wild-type phenotype was restored in the mutants by trans-formation with an indexed cosmid library (Pög-geler et al. 1997a). Thus it was possible to isolate the wild-type alleles of the mutated genes. Using this experimental approach, novel developmental genes were identified, most of which had not been implicated in fungal development before, thus validating a forward genetic approach.

The mutants that were complemented are pro1, pro4, pro11, pro22, pro40, pro41, and per5, all of which have a block at the stage of protoper-ithecium formation (pro mutants) or produce perithecia but no mature ascospores (per5). The corresponding genes can be grouped according to the functions of their gene products (Table 2.2). In mutants pro4 and per5, the *leu1* and *acl1* genes encoding β-isopropylmalate dehydrogenase and ATP citrate lyase, respectively, are defective (Nowrousian et al. 1999; Kück 2005). They are involved in amino acid and fatty acid biosynthe-sis, respectively, and their requirement for fruit-ing body formation, but not for vegetative growth may indicate an increased energy demand during sexual development. Expression analyses showed that both genes are under transcriptional control during the developmental cycle of *S. macrospora*,

indicating a tight integration of metabolic activity and developmental processes.

The first *pro* gene to be identified was *pro1*, a gene encoding a zinc finger transcription factor (Masloff et al. 1999, 2002). The gene product belongs to the fungal-specific class of zinc cluster proteins, the best known member of which is the yeast transcriptional regulator Gal4 (MacPherson et al. 2006). *pro1* was the first gene encoding a transcription factor of this class that was shown to be essential for sexual development in filamentous fungi. Since then, *pro1* orthologues from several other filamentous ascomycetes have been identified that also have a role in development; however, there seems to be a great diversity in the function of *pro1* orthologues in that their roles can include functions in vegetative sporulation or growth (Colot et al. 2006; Vienken and Fischer 2006).

The last four *pro* genes that were cloned en-code proteins that are either membrane or mem-brane-associated proteins and/or localize to organelles. Among these is PRO11, a WD40 repeat protein that associates with membranes (Pöggeler and Kück 2004). It is a functional homologue of the mammalian protein striatin as was demon-strated by the ability of the mouse orthologue to restore fertility in the pro11 mutant. This indi-cates that PRO11 is a member of a protein family with functionally conserved members from lower to higher eukaryotes. Another putative membrane or membrane-associated protein that is encoded by a *pro* gene is PRO22, an orthologue of the *N. crassa* protein HAM-2. The latter was shown to be necessary for hyphal fusion in *N. crassa* and, like the corresponding *N. crassa* mutant, the pro22 mutant is defective in hyphal fusion (Xiang et al. 2002; Rech et al. 2007).

Two identified organellar proteins are PRO40 and PRO41, that localize to the Woronin body and the ER, respectively (Engh et al. 2007b; Nowrou-sian et al. 2007a). PRO40 is a WW domain protein and the first of its class that was shown to be essential for fruiting body development. Addition-ally, it is the first Woronin body protein for which a role in development could be demonstrated. The Woronin body is an organelle that is specific to filamentous ascomycetes where it is necessary for septal plugging to prevent loss of cytoplasm after hyphal injury (Markham and Collinge 1987). Woronin bodies have not been previously impli-cated in fruiting body formation; however, detec-tion of PRO40 (and its corresponding *N. crassa*

Table 2.2. Developmental genes from *S. macrospora*. Localizations given in brackets are derived from sequence homologies, the others were verified experimentally. The genes *acl1* and *leu1* complement mutants per5 and pro4, respectively, all other genes carry the same name as the corresponding mutant

Gene	Gene product/conserved domains	Localization	References
Primary metabolism			
acl1	Subunit of the ATP citrate lyase	Cytoplasm	Nowrousian et al. (1999)
leu1	β-Isopropylmalate dehydrogenase	(Cytoplasm)	Kück (2005)
Secondary metabolism			
fbm1	Dehydrogenase (polyketide biosynthesis)	(Extracellular)	Nowrousian (2009)
pks	Polyketide synthase (melanin biosynthesis)	(Cytoplasm)	Engh et al. (2007b)
sdh	scytalon dehydratase (melanin biosynthesis)	(Cytoplasm)	Engh et al. (2007b)
Transcription factors			
mcm1	MADS-box transcription factor	Nucleus	Nolting and Pöggeler (2006b)
pro1	C_6 zinc finger transcription factor	(Nucleus)	Masloff et al. (1999)
Smta-1	HMG domain transcription factor	(Nucleus)	Pöggeler et al. (1997b)
ste12	Homeodomain/zinc finger transcription factor	(Nucleus)	Nolting and Pöggeler (2006a)
Pheromones and pheromone receptors			
ppg1	Peptide pheromone	Extracellular	Mayrhofer and Pöggeler (2005)
ppg2	Lipopeptide pheromone	(Extracellular)	Mayrhofer et al. (2006)
pre1	Pheromone receptor	(Plasma membrane)	Mayrhofer et al. (2006)
pre2	Pheromone receptor	(Plasma membrane)	Mayrhofer et al. (2006)
Signal transduction			
gsa1	G protein α-subunit	(Membrane-associated)	Kamerewerd et al. (2008)
gsa2	G protein α-subunit	(Cytoplasm)	Kamerewerd et al. (2008)
gsa3	G protein α-subunit	(Membrane-associated)	Kamerewerd et al. (2008)
sac1	Adenylate cyclase	(Cytoplasm)	Kamerewerd et al. (2008)
Membrane proteins and membrane-associated proteins			
pro11	WD40-repeat protein	Membrane-associated	Pöggeler and Kück (2004)
pro22	Membrane protein	(Membrane)	Rech et al. (2007)
Organellar proteins			
pro40	WW domain protein	Woronin body	Engh et al. (2007a)
pro41	Membrane protein	ER membrane	Nowrousian et al. (2007)

orthologue SO that also localizes to septal plugs and is essential for sexual development) makes it tempting to speculate about as yet unsuspected roles for this fungal-specific organelle (Engh et al. 2007b; Fleißner and Glass 2007).

The last *pro* gene to be isolated was *pro41* that encodes a small protein of the ER membrane (Nowrousian et al. 2007a). PRO41 orthologues exist only in filamentous ascomycetes and the *S. macrospora* gene was the first of these to be functionally characterized. Similar to the other *pro* genes, *pro41* is dispensable for overall vegetative growth but is necessary for the formation of mature fruiting bodies. Expression studies showed that it is transcriptionally upregulated during fruiting body formation; and microarray analyses of single and double mutants indicated that it acts genetically downstream of the transcription factor gene *pro1*. The identification of the development-specific proteins PRO40 and PRO41 throws a spotlight on the role of organelles in fungal development. Future studies should include the identification of interaction partners of the proteins as well as further analysis of the genetic interactions between the genes to unravel metabolic and signalling networks that integrate subcellular functions with multicellular development.

Another interesting aspect that came to light during the analysis of the pro-mutants is the fact that there might be a connection between hyphal fusion events and the ability to form sexual structures: Mutants pro22 and pro40 are greatly impaired in their ability to form anastomoses between vegetative hyphae, and the same is true for the corresponding *N. crassa* mutants ham-2 and soft, both of which were initially identified in screens for hyphal fusion mutants (Xiang et al. 2002; Fleißner and Glass 2007; Rech et al. 2007). There are a number of hyphal fusion events necessary for sexual development. These include fusion of sub-apical cells in crozier formation as well as (putative) fusion of hyphae that form the outer perithecial shell. In heterothallic ascomycetes like *N. crassa*, an additional fusion event has to occur to allow fertilization; however, the SO protein is not necessary for this specific type of fusion (Fleissner et al. 2005). Thus, the connection between hyphal fusion events and fruiting body development has yet to be discovered and poses one of the most challenging future questions in fungal biology.

VII Conclusions

Sordaria macrospora has a long-standing history as a classic genetic model system for conventional tetrad analysis. The application of several molecular tools, such as DNA mediated transformation, site-specific recombination or functional genomics to this filamentous fungus makes it an ideal experimental system to uncover the details of genetic interactions controlling developmental plasticity. The rapid and inexpensive genetic analysis of developmental mutants with distinct and defined morphological defects will increase our knowledge of multicellular differentiation processes in eukaryotes.

Acknowledgements. We thank Ms. Gabriele Frenßen-Schenkel for artwork, Ms. Melanie Mees for typing the manuscript and Dr. Sandra Masloff for preparing scanning electron microscopic images from developmental mutants. The work of the authors is substantially supported by funding from the Deutsche Forschungsgemeinschaft (Bonn-Bad Godesberg, Germany).

References

Auerswald B (1866) *Sordaria macrospora*. Hedwigia 5:192

Bobrowicz P, Pawlak R, Correa A, Bell-Pedersen D, Ebbole DJ (2002) The *Neurospora crassa* pheromone precursor genes are regulated by the mating type locus and the circadian clock. Mol Microbiol 45:795–804

Bölker M (1998) Sex and crime: heterotrimeric G proteins in fungal mating and pathogenesis. Fungal Genet Biol 25:143–156

Borkovich KA, Alex LA, Yarden O, Freitag M, Turner GE, Read ND, Seiler S, Bell-Pedersen D, Paietta J, Plesofsky N, Plamann M, Goodrich-Tanrikulu M, Schulte U, Mannhaupt G, Nargang FE, Radford A, Selitrennikoff C, Galagan JE, Dunlap JC, Loros JJ, Catcheside D, Inoue H, Aramayo R, Polymenis M, Selker EU, Sachs MS, Marzluf GA, Paulsen I, Davis

R, Ebbole DJ, Zelter A, Kalkman ER, O'Rourke R, Bowring F, Yeadon J, Ishii C, Suzuki K, Sakai W, Pratt R (2004) Lessons from the genome sequence of *Neurospora crassa*: tracing the path from genomic blueprint to multicellular organism. Microbiol Mol Biol Rev 68:1–108

Botstein D, Fink GR (1988) Yeast: an experimental organism for modern biology. Science 240:1439–1443

Breakspear A, Momany M (2007) The first fifty microarray studies in filamentous fungi. Microbiology 153:7–15

Bruhn L, Sprague GFJ (1994) MCM1 point mutants deficient in expression of alpha-specific genes: residues important for interaction with alpha 1. Mol Cell Biol 14:2534–2544

Cai L, Jeewon R, Hyde KD (2006) Phylogenetic investigations of Sordariaceae based on multiple gene sequences and morphology. Mycol Res 110:137–150

Colot HV, Park G, Turner GE, Ringelberg C, Crew CM, Litvinkova L, Weiss RL, Borkovich KA, Dunlap JC (2006) A high-throughput gene knockout procedure for Neurospora reveals functions for multiple transcription factors. Proc Nat Acad Sci USA 103:10352–10357

Coppin E, de Renty C, Debuchy R (2005) The function of the coding sequences for the putative pheromone precursors in *Podospora anserina* is restricted to fertilization. Eukaryot Cell 4:407–420

D'Souza CA, Heitman J (2001) Conserved cAMP signaling cascades regulate fungal development and virulence. FEMS Microbiol Rev 25:349–364

Davis RH, Perkins DD (2002) Timeline: *Neurospora*: a model of model microbes. Nat Rev Genet 3:397–403

Debuchy R, Coppin E (1992) The mating types of *Podospora anserina*: functional analysis and sequence of the fertilization domains. Mol Gen Genet 233:113–121

Debuchy R, Turgeon BG (2006) Mating-type structure, evolution, and function in euascomycetes. In: Kües U, Fischer R (eds.) Growth, differentiation and sexuality. Springer, Heidelberg pp 293–323

Debuchy R, Arnaise S, Lecellier G (1993) The mat- allele of *Podospora anserina* contains three regulatory genes required for the development of fertilized female organs. Mol Gen Genet 241:667–673

Dohlman HG (2002) G proteins and pheromone signaling. Annu Rev Physiol 64:129–152

Dolan JW, Kirkman C, Fields S (1989) The Yeast STE12 protein binds to the DNA sequence mediating pheromone induction. Proc Natl Acad Sci USA 86:5703–5707

Elleuche S, Pöggeler S (2008) Visualization of peroxisomes via SKL-tagged DsRed protein in *Sordaria macrospora*. Fungal Genet Rep 55:9–12

Embley TM, Martin W (2006) Eukaryotic evolution, changes and challenges. Nature 440:623–630

Engh I, Nowrousian M, Kück U (2007a) Regulation of melain biosynthesis via the dihydroxynaphtalene pathway is dependent on sexual development in the ascomycete *Sordaria macrospora*. FEMS Microbiol Lett 275:62–70

Engh I, Würtz C, Witzel-Schlömp K, Zhang HY, Hoff B, Nowrousian M, Rottensteiner H, Kück U (2007b) The WW domain protein PRO40 is required for fungal fertility and associates with Woronin bodies. Eukaryot Cell 6:831–843

Eriksson OE, Baral HO, Currah RS, Hansen K, Kurtzmann CP, Rambold G, Klassoe T (2004) Outline of ascomycota – 2004. Myconet 10:1–99

Esser K, Straub J (1958) Genetische Untersuchungen an *Sordaria macrospora* Auersw.: Kompensation und Induktion bei genbedingten Entwicklungsdefekten. Z Vererbungsl 89:729–746

Esser K (1982) Cryptogams – cyanobacteria, algae, fungi, lichens. Cambridge University Press, Cambridge

Ferreira AV, Saupe S, Glass NL (1996) Transcriptional analysis of the mtA idiomorph of *Neurospora crassa* identifies two genes in addition to mtA-1. Mol Gen Genet 250:767–774

Ferreira AV, An Z, Metzenberg RL, Glass NL (1998) Characterization of mat A-2, mat A-3 and deltamatA mating-type mutants of *Neurospora crassa*. Genetics 148:1069–1079

Fischer R, Zekert N, Takeshita N (2008) Polarized growth in fungi – interplay between the cytoskeleton, positional markers and membrane domains. Mol Microbiol 68:813–826

Fleißner A, Glass NL (2007) SO, a protein involved in hyphal fusion in *Neurospora crassa*, localizes to septal plugs. Eukaryot Cell 6:94–94

Fleissner A, Sarkar S, Jacobson DJ, Roca MG, Read ND, Glass NL (2005) The *so* locus is required for vegetative cell fusion and postfertilization events in *Neurospora crassa*. Eukaryot Cell 4:920–930

Galagan JE, Calvo SE, Borkovich KA, Selker EU, Read ND, Jaffe D, FitzHugh W, Ma LJ, Smirnov S, Purcell S, Rehman B, Elkins T, Engels R, Wang S, Nielsen CB, Butler J, Endrizzi M, Qui D, Ianakiev P, Bell-Pedersen D, Nelson MA, Werner-Washburne M, Selitrennikoff CP, Kinsey JA, Braun EL, Zelter A, Schulte U, Kothe GO, Jedd G, Mewes W, Staben C, Marcotte E, Greenberg D, Roy A, Foley K, Naylor J, Stange-Thomann N, Barrett R, Gnerre S, Kamal M, Kamvysselis M, Mauceli E, Bielke C, Rudd S, Frishman D, Krystofova S, Rasmussen C, Metzenberg RL, Perkins DD, Kroken S, Cogoni C, Macino G, Catcheside D, Li W, Pratt RJ, Osmani SA, DeSouza CP, Glass L, Orbach MJ, Berglund JA, Voelker R, Yarden O, Plamann M, Seiler S, Dunlap J, Radford A, Aramayo R, Natvig DO, Alex LA, Mannhaupt G, Ebbole DJ, Freitag M, Paulsen I, Sachs MS, Lander ES, Nusbaum C, Birren B (2003) The genome sequence of the filamentous fungus *Neurospora crassa*. Nature 422:859–868

Glass NL, Grotelueschen J, Metzenberg RL (1990a) *Neurospora crassa* A mating-type region. Proc Natl Acad Sci USA 87:4912–4916

Glass NL, Metzenberg RL, Raju NB (1990b) Homothallic Sordariaceae from nature: The absence of strains containing only the *a* mating type sequence. Exp Mycol 14:274–289

Gustin MC, Albertyn J, Alexander M, Davenport K (1998) MAP kinase pathways in the yeast *Saccharomyces cerevisiae*. Microbiol Mol Biol Rev 62:1264–1300

Harris SD, Momany M (2004) Polarity in filamentous fungi: moving beyond the yeast paradigm. Fungal Genet Biol 41:391–400

Hawksworth DL (2001) The magnitude of fungal diversity: the 1.5 million species estimate revisited. Mycol Res 105:1422–1432

Herskowitz I (1989) A regulatory hierarchy for cell specialization in yeast. Nature 342:749–757

Heslot H (1958) Contribution à l'étude cytogénétique des Sordariacées. Rev Cytol Biol Veg 19:1–235

Hoff B, Kück U (2006) Application of the nourseothricin acetyltransferase gene (*nat1*) as dominant marker for the transformation of filamentous fungi. Fungal Genet Newsl 53:9–11

Huhndorf SM, Miller AN, Fernandez F (2004) Molecular systematics of the Sordariales: the order and the family Lasiosphaeriaceae redefined. Mycologia 96:368–387

Jacobsen S, Wittig M, Pöggeler S (2002) Interaction between mating-type proteins from the homothallic fungus *Sordaria macrospora*. Curr Genet 41:150–158

Kamerewerd J, Jansson M, Nowrousian M, Pöggeler S, Kück U (2008) Three alpha subunits of heterotrimeric G proteins and an adenylyl cyclase have distinct roles in fruiting body development in a homothallic fungus. Genetics 180:191–206

Kays AM, Rowley PS, Baasiri RA, Borkovich KA (2000) Regulation of conidiation and adenylyl cyclase levels by the Galpha protein GNA-3 in *Neurospora crassa*. Mol Cell Biol 20:7693–7705

Kim H, Borkovich KA (2004) A pheromone receptor gene, *pre-1*, is essential for mating type-specific directional growth and fusion of trichogynes and female fertility in *Neurospora crassa*. Mol Microbiol 52:1781–1798

Kim H, Borkovich KA (2006) Pheromones are essential for male fertility and sufficient to direct chemotropic polarized growth of trichogynes during mating in *Neurospora crassa*. Eukaryot Cell 5:544–554

Kirk PM, Cannon PF, David JC, Staplers JA (2001) Ainsworth & Bisby's dictionary of the fungi. CABI International, Wallingford

Krügel HG, Fiedler G, Smith C, Baumberg S (1993) Sequence and transcriptional analysis of the nourseothricin acetyltransferase-encoding gene *nat1* from *Streptomyces noursei*. Gene 127:127–131

Kück U (2005) A *Sordaria macrospora* mutant lacking the *leu1* gene shows a developmental arrest during fruiting body formation. Mol Genet Genomics 274:307–315

Kück U, Pöggeler S (2004) pZHK2, a bi-functional transformation vector, suitable for two step gene targeting. Fungal Genet Newsl 51:4–6

Kück U, Pöggeler S (2005) *Sordaria macrospora*. In: Gellissen G (ed) Production of recombinant proteins. Wiley-VCH, Weinheim, pp 215–231

Le Chevanton L, Leblon G, Lebilcot S (1989) Duplications created by transformation in *Sordaria macrospora* are not inactivated during meiosis. Mol Gen Genet 218:390–396

Lengeler KB, Davidson RC, D'Souza C, Harashima T, Shen WC, Wang P, Pan X, Waugh M, Heitman J (2000) Signal transduction cascades regulating fungal development and virulence. Microbiol Mol Biol Rev 64:746–785

Lin X, Heitman J (2007) Mechanisms of homothallism in fungi and transitions between heterothallism and homothallism. In: Heitman J, Kronstad JW, Taylor JW, Casselton LA (eds) Sex in fungi. ASM, Washington, D.C., pp 35–57

MacPherson S, Larochelle M, Turcotte B (2006) A fungal family of transcriptional regulators: the zinc cluster proteins. Microbiol Mol Biol Rev 70:583–604

Madhani HD, Fink GR (1997) Combinatorial control required for the specificity of yeast MAPK signaling. Science 275:1314–1317

Markham P, Collinge AJ (1987) Woronin bodies of filamentous fungi. FEMS Microbiol Lett 46:1–11

Masloff S, Pöggeler S, Kück U (1999) The *pro1+* gene from *Sordaria macrospora* encodes a C6 zinc finger transcription factor required for fruiting body development. Genetics 152:191–199

Masloff S, Jacobsen S, Pöggeler S, Kück U (2002) Functional analysis of the C_6 zinc finger gene *pro1* involved in fungal sexual development. Fungal Genet Biol 36:107–116

Mayrhofer S, Pöggeler S (2005) Functional characterization of an α-factor-like *Sordaria macrospora* peptide pheromone and analysis of its interaction with its cognate receptor in *Saccharomyces cerevisiae*. Eukaryot Cell 4:661–672

Mayrhofer S, Weber JM, Pöggeler S (2006) Pheromones and pheromone receptors are required for proper sexual development in the homothallic ascomycete *Sordaria macrospora*. Genetics 172:1521–1533

Metzenberg RL, Glass NL (1990) Mating type and mating strategies in Neurospora. Bioessays 12:53–59

Mouriño-Pérez RR, Roberson RW, Bartnicki-García S (2006) Microtubule dynamics and organization during hyphal growth and branching in *Neurospora crassa*. Fungal Genet Biol 43:389–400

Ninomiya Y, Suzuki K, Ishii C, Inoue H (2004) Highly efficient gene replacements in *Neurospora* strains deficient for nonhomologous end-joining. Proc Natl Acad Sci USA 101:12248–12253

Nolting N, Pöggeler S (2006a) A STE12 homologue of the homothallic ascomycete *Sordaria macrospora* interacts with the MADS box protein MCM1 and is required for ascosporogenesis. Mol Microbiol 62:853–868

Nolting N, Pöggeler S (2006b) A MADS box protein interacts with a mating-type protein and is required for fruiting body development in the homothallic ascomycete *Sordaria macrospora*. Eukaryot Cell 5:1043–1056

Nowrousian M (2007) Of patterns and pathways: microarray technologies for the analysis of filamentous fungi. Fungal Biol Rev 21:171–178

Nowrousian M (2009) A novel polyketide biosynthesis gene cluster is involved in fruiting body morphogenesis in the filamentous fungi *Sordaria macrospora and Neurospora crassa*. Curr Genet (in press)

Nowrousian M, Cebula P (2005) The gene for a lectin-like protein is transcriptionally activated during sexual development, but is not essential for fruiting body formation in the filamentous fungus *Sordaria macrospora*. BMC Microbiol 5:64

Nowrousian M, Kück U (1998) Isolation and cloning of the *Sordaria macrospora ura3* gene and its heterologous expression in *Aspergillus niger*. Fungal Genet Newsl 45:34–37

Nowrousian M, Masloff S, Pöggeler S, Kück U (1999) Cell differentiation during sexual development of the fungus *Sordaria macrospora* requires ATP citrate lyase activity. Mol Cell Biol 19:450–460

Nowrousian M, Dunlap JC, Nelson MA (2004a) Functional genomics in fungi. In: Kück U (ed) The Mycota II. Springer, Heidelberg, pp 115–128

Nowrousian M, Würtz C, Pöggeler S, Kück U (2004b) Comparative sequence analysis of *Sordaria macrospora* and *Neurospora crassa* as a means to improve genome annotation. Fungal Genet Biol 41:285–292

Nowrousian M, Ringelberg C, Dunlap JC, Loros JJ, Kück U (2005) Cross-species microarray hybridization to identify developmentally regulated genes in the filamentous fungus *Sordaria macrospora*. Mol Genet Genomics 273:137–149

Nowrousian M, Frank S, Koers S, Strauch P, Weitner T, Ringelberg C, Dunlap JC, Loros JJ, Kück U (2007a) The novel ER membrane protein PRO41 is essential for sexual development in the filamentous fungus *Sordaria macrospora*. Mol Microbiol 64:923–937

Nowrousian M, Piotrowski M, Kück U (2007b) Multiple layers of temporal and spatial control regulate accumulation of the fruiting body-specific protein APP in *Sordaria macrospora* and *Neurospora crassa*. Fungal Genet Biol 44:602–614

Ohneda M, Arioka M, Kitamoto K (2005) Isolation and characterization of *Aspergillus oryzae* vacuolar protein sorting mutants. Appl Environ Microbiol 71:4856–4861

Oldenburg KR, Vo KT, Michaelis S, Paddon C (1997) Recombination-mediated PCR-directed plasmid construction in vivo in yeast. Nucleic Acids Res 25:451–452

Pastwa E, Blasiak J (2003) Non-homologous DNA end joining. Acta Biochim Pol 50:891–908

Pöggeler S (1999) Phylogenetic relationships between mating-type sequences from homothallic and heterothallic ascomycetes. Curr Genet 36:222–231

Pöggeler S (2000) Two pheromone precursor genes are transcriptionally expressed in the homothallic ascomycete *Sordaria macrospora*. Curr Genet 37:403–411

Pöggeler S, Kück U (2000) Comparative analysis of the mating-type loci from *Neurospora crassa* and *Sordaria macrospora*: identification of novel transcribed ORFs. Mol Gen Genet 236:292–301

Pöggeler S, Kück U (2001) Identification of transcriptionally expressed pheromone receptor genes in filamentous ascomycetes. Gene 280:9–17

Pöggeler S, Kück U (2004) A WD40 repeat protein regulates fungal cell differentiation and can be replaced functionally by the mammalian homologue striatin. Eukaryot Cell 3:232–240

Pöggeler S, Kück U (2006) Highly efficient generation of signal transduction knockout mutants using a fungal strain deficient in the mammalian ku70 ortholog. Gene 378:1–10

Pöggeler S, Nowrousian M, Jacobsen S, Kück U (1997a) An efficient procedure to isolate fungal genes from an indexed cosmid library. J Microbiol Meth 29:49–61

Pöggeler S, Risch S, Kück U, Osiewacz HD (1997b) Mating-type genes from the homothallic fungus *Sordaria macrospora* are functionally expressed in a hetero-thallic ascomycete. Genetics 147:567–580

Pöggeler S, Masloff S, Hoff B, Mayrhofer S, Kück U (2003) Versatile EGFP reporter plasmids for cellular locali-zation of recombinant gene products in filamentous fungi. Curr Genet 43:54–61

Pöggeler S, Nowrousian M, Kück U (2006a) Fruiting body development in ascomycetes. In: Kües U, Fischer R (eds) Growth, differentiation and sexuality. Springer, Heidelberg, pp 325–355

Pöggeler S, Nowrousian M, Ringelberg C, Loros JJ, Dunlap JC, Kück U (2006b) Microarray and real time PCR analyses reveal mating type-dependent gene expres-sion in a homothallic fungus. Mol Genet Genomics 275:492–503

Raymond CK, Pownder TA, Sexson SL (1999) General method for plasmid construction using homologous recombination. Biotechniques 26:140–141

Rech C, Engh I, Kück U (2007) Detection of hyphal fusion in filamentous fungi using differently fluorescene-la-beled histones. Curr Genet 52:259–266

Roberts CJ, Nelson B, Marton MJ, Stoughton R, Meyer MR, Bennett HA, He YD, Dai H, Walker WL, Hughes TR, Tyers M, Boone C, Friend SH (2000) Signaling and circuitry of multiple MAPK pathways revealed by a matrix of global gene expression profiles. Science 287:873–880

Robertson SJ, Bond DJ, Read ND (1998) Homothallism and heterothallism in *Sordaria brevicollis*. Mycol Res 102:1215–1223

Ruprich-Robert G, Berteaux-Lecellier V, Zickler D, Pan-vier-Adoutte A, Picard M (2002) Identification of six loci in which mutations partially restore peroxisome biogenesis and/or alleviate the metabolic defect of pex2 mutants in *Podospora*. Genetics 161:1089–1099

Staben C, Yanofsky C (1990) *Neurospora crassa* a mating-type region. Proc Natl Acad Sci USA 87:4917–4921

Storlazzi A, Tessé S, Gargano S, James F, Kleckner NDZ (2003) Meiotic double-strand breaks at the interface of chromosome movement, chromosome remodeling, and reductional division. Genes Dev 17:2675–2687

Storlazzi A, Tesse S, Ruprich-Robert G, Gargano S, Pogge-ler S, Kleckner N, Zickler D (2008) Coupling meiotic chromosome axis integrity to recombination. Genes Dev 22:796–809

Tessé S, Storlazzi A, Kleckner N, Gargano SDZ (2003) Localization and roles of Ski8p protein in Sordaria

meiosis and delineation of three mechanistically dis-tinct steps of meiotic homolog juxtaposition. Proc Nat Acad Sci USA 100:12865–12870

Thompson-Coffe C, Zickler D (1992) Three microtubule-organizing centers are required for ascus growth and sporulation in the fungus *Sordaria macrospora*. Cell Motil Cytoskeleton 22:257–273

Thompson-Coffe C, Zickler D (1993) Cytoskeleton interactions in the ascus development and sporulation of *Sordaria macrospora*. J Cell Sci 104:883–898

Turgeon BG (1998) Application of mating type gene tech-nology to problems in fungal biology. Annu Rev Phy-topathol 36:115–137

Turina M, Prodi A, Alfen NK (2003) Role of the *Mf1-1* pheromone precursor gene of the filamentous asco-mycete *Cryphonectria parasitica*. Fungal Genet Biol 40:242–251

van Heemst D, James F, Pöggeler S, Berteaux-Lecellier V, Zickler D (1999) Spo76p is a conserved chromosome morphogenesis protein that links the mitotic and meiotic programs. Cell 98:261–271

Vienken K, Fischer R (2006) The Zn(II)2Cys6 putative transcription factor NosA controls fruiting body for-mation in *Aspergillus nidulans*. Mol Microbiol 61:544–554

Voigt K, Cozijnsen AJ, Kroymann J, Pöggeler S, Howlett BJ (2005) Phylogenetic relationships between members of the crucifer pathogenic *Leptosphaeria maculans* species complex as shown by mating type (MAT1-2), actin, and beta-tubulin sequences. Mol Phylogenet Evol 37:541–557

Walz M, Kück U (1995) Transformation of *Sordaria macrospora* to hygromycin B resistance: characteriza-tion of transformants by electrophoretic karyotyping and tetrad analysis. Curr Genet 29:88–95

Wik L, Karlsson M, Johannesson H (2008) The evolution-ary trajectory of the mating-type (mat) genes in *Neu-rospora relates* to reproductive behavior of taxa. BMC Evol Biol 8:109

Xiang Q, Rasmussen C, Glass NL (2002) The *ham-2* locus, encoding a putative transmembrane protein, is required for hyphal fusion in *Neurospora crassa*. Genetics 160:169–180

Yuan YO, Stroke IL, Fields S (1993) Coupling of cell iden-tity to signal response in yeast: interaction between the alpha 1 and STE12 proteins. Genes Dev 1584–1597

Zickler D (1977) Development of the synaptonemal com-plex and the "recombination nodules" during meiotic prophase in the seven bivalents of the fungus *Sordaria macrospora* Auersw. Chromosoma 61: 289–316

3 Inteins – Selfish Elements in Fungal Genomes

Skander Elleuche[1], Stefanie Pöggeler[1]

CONTENTS

I. Introduction

Since the late 1970s it has become clear that the coding sequence of many eukaryotic genes is disrupted by genetic elements called intervening sequences, which must be removed prior to host gene function. Intervening sequences can be classified into intron and intein. Introns are excised from the primary RNA transcript by a process termed splicing. In contrast to introns, inteins are internal in-frame insertions transcribed and translated together with their host protein and

excised at the protein level (Fig. 3.1). By analogy to introns and exons, the terms *intein* and *extein* have been proposed, for *intein*al *prot*ein sequence and *extein*al *prot*ein sequence of the precursor, respectively. Upstream exteins are termed the N-extein and downstream exteins the C-extein. The post-translational process that excises the internal region from a precursor protein with subsequent ligation of the N- and C-extein is termed protein splicing (Perler et al. 1994). Thus, protein splicing is a processing event in which the expression of a single gene results in production of two stable proteins, the mature protein and the intein (Fig. 3.1).

Protein splicing elements were first described in fungi. In 1990, two groups reported an in-frame insertion in the *Saccharomyces cerevisiae VMA1* gene encoding a vacuolar membrane H^+-ATPase (Hirata et al. 1990; Kane et al. 1990). The nucleotide sequence of the *S. cerevisiae VMA1* gene predicts a polypeptide of 1071 amino acids (aa) with a calculated molecular mass of 118 kDa, which was much larger than the 67 kDa value estimated from sodium dodecyl sulfate–polyacrylamide (SDS-PAGE) gels. The N- and C-terminal regions of the deduced sequence were shown to be very similar to those of the catalytic subunits of vacuolar membrane H^+-ATPases of other organisms, while the internal region containing 454 amino acid residues displayed no detectable sequence similarities to any known ATPase subunits; instead it exhibited similarity to a yeast endonuclease encoded by the *HO* gene. Subsequently, it was demonstrated that the insertion was still present in the mRNA and translated together with the VMA1 protein, and that the insertion spliced itself out of the protein (Kane et al. 1990).

By now, more than 350 inteins have been recognized in the genomes of bacteria, archaea, and eukaryotes, as well as in viruses. Inteins have been identified predominantly in prokaryotes, but are also found in greater than 70 eukaryotic taxa. Eukaryotic inteins are encoded within the nuclear genomes of fungi, as well as within the nuclear genomes and plastomes of some unicellular algae. For current information, see the intein

[1]Abteilung Genetik eukaryotischer Mikroorganismen, Institut für Mikrobiologie und Genetik, Georg-August-Universität Göttingen, Grisebachstrasse 8, 37077 Göttingen, Germany; e-mail: spoegge@gwdg.de

Physiology and Genetics, 1st Edition
The Mycota XV
T. Anke and D. Weber (Eds.)
© Springer-Verlag Berlin Heidelberg 2009

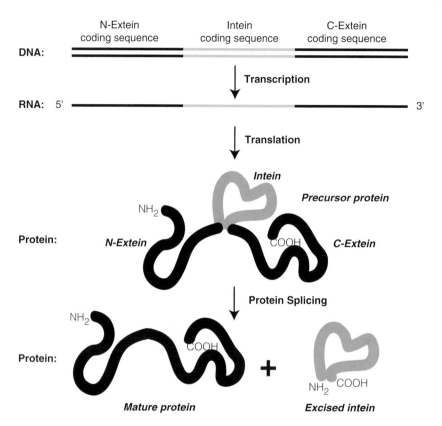

Fig. 3.1. Protein splicing of inteins. The intein coding sequence is transcribed into mRNA and translated to a non-functional protein precursor, which then undergoes a self-catalyzed rearrangement in which the intein is excised and the exteins are joined to yield the mature protein

registry InBase at http://www.neb.com/neb/inteins.html (Perler 2002). According to accepted nomenclature, intein names include a genus and species designation (abbreviated to three letters) and a host gene designation. For example the *S. cerevisiae* VMA1 intein is called *Sce* VMA1. When a protein contains more than one intein, they are given Roman numerals (Perler et al. 1994).

This review focuses on fungal inteins and includes the characteristics, as well as the distribution of fungal inteins. It summarizes recent progress in understanding the protein splicing mechanism, how inteins evolve, and how they can be used for technical applications.

A. General Characteristics of Inteins

Inteins are classified into two groups: large and minimal (mini-) inteins, depending on whether or not they contain a homing endonuclease domain (Liu 2000; Fig. 3.2). Homing endonucleases are

encoded by open reading frames that are embedded within introns and inteins (Chevalier and Stoddard 2001; Stoddard 2005). They are site-specific, double-strand DNA endonucleases that promote the lateral transfer of their own coding region and flanking sequences between genomes by a recombination-dependent process known as 'homing' (Fig. 3.3). Genomic DNA encoding an intein-free allele of the potentially intein-carrying gene is cut by the homing endonuclease, and during DNA repair, the intein is subsequently copied into the empty allele (Gimble and Thorner 1993; Gimble 2000). The nomenclature of homing endonucleases is patterned after that of restriction enzymes. A three-letter, genus- and species-specific designation is followed by a Roman numeral to distinguish multiple enzymes originating from a single organism. Intron-encoded endonucleases are characterized by the prefix I (for intron), whereas intein-derived endonucleases are characterized by the prefix PI (for protein insert; i.e. the endonuclease of *Sce* VMA1 is termed PI *Sce*I; Belfort and Roberts 1997).

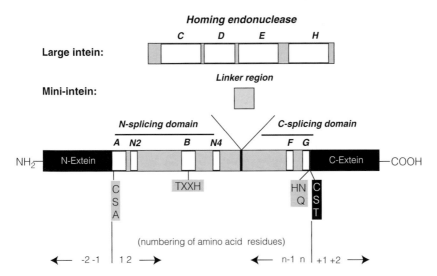

Fig. 3.2. Conserved elements in a large intein and mini-intein. The *white areas* A, N2, B, N4, C, D, E, F and G are conserved intein motifs identified by Pietrokovski (Pietrokovski 1994, 1998b) and Perler et al. (1997). The exteins are illustrated in *black* and the intein sequence in *grey*. The site of insertion of the homing endonuclease and the linker region in large and mini-inteins, respectively, is indicated by the *dark vertical line*. Conserved amino acid residues of the intein and the C-extein are indicated below

Four families of homing endonucleases containing highly conserved amino acid sequences have been identified; including the LAGLIDADG, the GIY-YIG, the His-Cys box, and the H-N-H endonucleases (Chevalier and Stoddard 2001). Fungal intein endonucleases belong to the LAGLIDADG family and are characterized by two copies of the conserved LAGLIDADG motif (Belfort and Roberts 1997).

Mini-inteins are not bifunctional and contain only a splicing domain without endonuclease function. Because several splicing-efficient mini-inteins have been engineered from large inteins by deleting the central endonuclease domain, it is clear that the endonuclease domain is not involved in protein splicing (Chong and Xu 1997; Derbyshire et al. 1997; Shingledecker et al. 1998). The splicing domain of large inteins seems to be split by the endonuclease domain into N-terminal and C-terminal subdomains. Both intein subdomains contain conserved blocks of amino acids: blocks A, N2, B, and N4 for the N-terminal subdomains, and blocks G and F for the C-terminal subdomains (Pietrokovski 1994, 1998b; Perler et al. 1997; Fig. 3.2). These domains can also be identified in mini-inteins. However, motifs C, D, E, and H of the endonuclease domain are missing; instead they have a small linker region (Fig. 3.2). In addition, all known inteins have a Ser, Cys, or Ala at the N terminus and end in His-Asn, or in His-Gln. There is no conserved residue preceding the intein

at the C-terminus of the N-extein, but all inteins have an invariant Ser, Thr, or Cys at the N-terminus of the C-extein (Perler 2002). According to the most common amino acid numbering system in intein precursors, minus and plus signs are used to designate N-extein and C-extein residues, respectively. Numbering begins at the N-terminus of the intein and C-extein, and proceeds to the C-terminus. Numbering of the N-extein begins at its C-terminus and proceeds to its N-terminus (Saleh and Perler 2006; Fig. 3.2).

Three-dimensional structures of inteins reveal that the N- and C-terminal splicing domains form a common horseshoe-like 12-β-strand scaffold termed the *H*edgehog/(Hint) module (Koonin 1995; Hall et al. 1997; Klabunde et al. 1998; Perler 1998; Ding et al. 2003; Sun et al. 2005).

Hedgehog proteins are essential signaling molecules for animal embryonic development and possess an autoprocessing activity that results in an intramolecular cleavage of the full-length Hedgehog protein and covalent attachment of a cholesterol moiety to the newly generated amino-terminal fragment. Cholesterol anchors the signaling domain to the cell surface (Beachy et al. 1997; Ingham 2001). Until now, Hedgehog proteins have been described only in metazoa.

Interestingly, the developmental protein GmGIN1 was identified in the arbuscular mycorrhizal fungus *Glomus mossae* as having an N-terminal domain

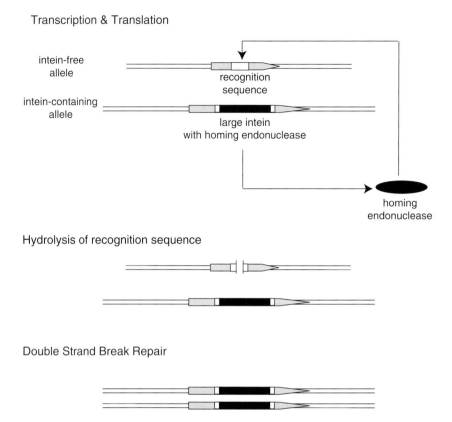

Fig. 3.3. Intein homing. The large intein encodes an endo-nuclease that in heterozygous cells recognizes a recognition sequence in the intein-less allele on the homologous chromosome. The recognition site is hydrolyzed by the intein-encoded endonuclease, while the chromosome carrying the intein is protected from being hydrolyzed. After cleavage, the recominational repair system of the cell is stimulated and the intein-containing chromosome is used as a template for repair. The result of intein homing is a homozygote with two copies of the intein-containing gene

that shares sequence similarity with a novel family of GTP-binding proteins, while the C-terminus has a striking homology to the C-terminal part of the Hedgehog protein family from metazoa. It is therefore hypothesized that the N-terminal part of GmGIN1 might be activated through a similar autoprocessing mechanism, as was demonstrated for Hedgehog proteins (Requena et al. 2002). In addition to Hedgehog proteins, fungal inteins also share sequence and structure properties with bacterial intein-like sequence (BIL) Amitai et al. 2003.

Combining all these characteristics, four main criteria help to differentiate true inteins from other in-frame insertions: (i) an intein-containing protein has a sequence lacking in homologs of other organisms, (ii) the intervening sequence is over 100 aa, (iii) it contains the conserved intein motifs A, B, F, and G, and (iv) the observed protein product is the same size as the predicted open reading frame without the intein (Perler et al. 1997).

B. Protein Splicing Mechanism

The post-translational removal of the intein is an autocatalytic process that depends only on the information of the intein plus the first downstream extein residue (Fig. 3.1). Protein splicing is extremely rapid, and to date, precursors have not been identified in native systems. The first direct evidence for protein splicing was demonstrated for an intein derived from the thermostable DNA polymerase of *Pyrococcus* species GB-D. The intein from *Pyrococcus* was cloned into a foreign gene, and the purified precursor was shown to be capable of in vitro protein splicing in the absence of other proteins or cofactors (Xu et al. 1993).

Protein splicing typically involves four steps: (i) an N-O or N-S acyl shift at the N-extein/intein junction, (ii) transesterification, which transfers the N-extein to the side-chain of the first residue (+1) of the C-extein, (iii) cyclization of a conserved asparagine residue at the C-terminus of the intein and cleavage of the peptide bond, resulting in the release of the intein, and (iv) rearrangement to a peptide bond of the ester/thioester bond linking the N- and C-exteins. The details of the chemical process involved in protein splicing have been comprehensively described and reviewed (Liu 2000; Noren et al. 2000; Paulus 2000; Gogarten et al. 2002; Saleh and Perler 2006; Starokadomskyy 2007).

Several inteins exist as two fragments and are encoded by two separately transcribed and translated genes. These so-called split inteins self-associate and catalyze protein-splicing activity in *trans*. The first native split intein capable of protein *trans*-splicing was identified in the catalytic subunit alpha of DNA polymerase III DnaE of the cyanobacterium *Synechocystis* sp. strain PCC6803 (Wu et al. 1998b).

The N- and C-terminal halves are encoded by two separate genes, *dnaE-n* and *dnaE-c*, respectively. These two genes are located 745 226 bp apart in the genome and are on opposite DNA strands (Wu et al. 1998b).

Currently, native split inteins have been identified in diverse cyanobacteria and the archaeon *Nanoarchaeum equitans* (Wu et al. 1998b; Caspi et al. 2003; Liu and Yang 2003; Choi et al. 2006; Dassa et al. 2007), but so far they have not been found in fungi. *Trans*-splicing inteins have, however, been artificially engineered from *cis*-splicing bacterial and fungal inteins (Mills et al. 1998; Southworth et al. 1998; Mootz and Muir 2002; Elleuche and Pöggeler 2007). In the majority of cases, naturally or artificially split inteins are separated between motifs B and F resulting in N-terminal intein fragments (I_N) of 25–40 aa and C-terminal intein fragments (I_C) of ~100 aa (Fig. 3.4). In addition, it was demonstrated that inteins can be artificially split at other sites. The *Ssp* DnaB mini-intein can be split at different loop regions between β-strands and still maintain the ability to splice in *trans*. Even a three-piece split intein can function in protein *trans*-splicing (Sun et al. 2004). Naturally split inteins, as well as engineered split inteins can be used for various applications in protein chemistry, e.g. protein purification,

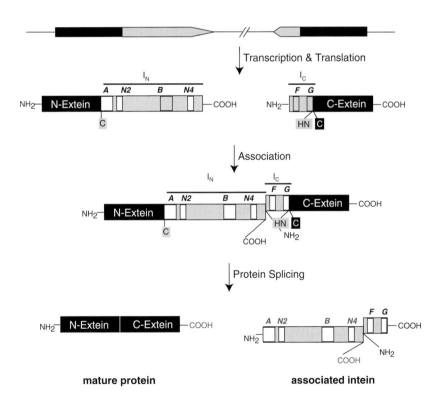

Fig. 3.4. Protein splicing in *trans*. The precursor is encoded by two genes: one gene encodes the N-extein and the N-terminal intein (I_N), the other gene encodes the C-terminal part of the intein (I_C) and the C-extein. Both genes are separately transcribed and translated, the intein fragments associate and form a functional intein. N- and C-exteins become ligated to the mature protein and the intein fragments usually remain associated in solution

regulation of protein activity, and analysis of protein–protein interaction (see Sect. IV).

II. Inteins in Fungi

In fungal organisms, only *cis*-splicing inteins have been detected, and all are encoded by nuclear genes (Poulter et al. 2007). According to InBase (Perler 2002), to date 73 inteins have been detected in fungi. They have been identified in chytrids, zygomycetes, ascomycetes, and basidiomycetes. Most of them are embedded in homologs of the *S. cerevisiae VMA1* gene or within the prp8 gene, but they can also be found in genes encoding glutamate synthases, chitin synthases, threonyl-tRNA synthetases, and subunits of DNA-directed RNA polymerases (Poulter et al. 2007).

A. VMA1 Inteins of Saccharomycete Yeasts

The *VMA1* gene of many saccharomycete yeasts consists of a single open reading frame that comprises two independent regions of genetic information for VMA1, the catalytic 67-kDa subunit of the vacuolar H$^+$-ATPase, and for an intein varying in length from 394 to 517 residues (Table 3.1). Alleles of the *VMA1* gene lacking the intein element are present in the fission yeast *Schizosaccharomyces pombe* and in filamentous ascomycetes, including *Neurospora crassa* (Bowman et al. 1988; Ghislain and Bowman 1992). In addition several saccharomycete yeast species do not contain a VMA1 intein. Koufopanou et al. (2002) surveyed 24 species of saccharomycete yeasts and detected VMA1 inteins in 14 species that diverged in amino acid sequence. All inteins found in *VMA1* genes of saccharomycete yeasts are large inteins and contain two separate domains: a splicing domain and an endonuclease domain (Table 3.1). The catalytic subunit of the vacuolar proton pump is responsible for the ATP hydrolysis that is coupled to the pumping of protons into a variety of intracellular acidic compartments, including the fungal vacuole (Anraku et al. 1992). Under laboratory conditions, *VMA1* mutants of *Saccharomyces cerevisiae* are viable but lack the vacuolar H$^+$-ATPase and are defective in vacuolar acidification. Moreover, they are calcium-sensitive and exhibit low spore viability as well as slow growth (Hirata et al. 1990). Thus, under natural

conditions the *VMA1* gene seems to be essential for yeast cell growth.

Until now, all identified VMA1 inteins are inserted at an identical insertion site. The best-characterized VMA1 intein is *Sce* VMA1 from *S. cerevisiae*. As shown by crystal structure, the protein splicing domain of *Sce* VMA1 is structurally related to bacterial mini-inteins and to the *Drosophila* Hedgehog protein autoprocessing domain (Koonin 1995; Duan et al. 1997; Hall et al. 1997; Pietrokovski 1998b; Moure et al. 2002). Site-directed mutagenesis experiments and deletion of the endonuclease domain show that endonuclease activity is not required for protein splicing (Gimble and Stephens 1995; Chong and Xu 1997). The two-domain structure (self-splicing and endonuclease domain) of PI-*Sce*I was confirmed by X-ray crystallography. However, despite the apparent structural autonomy of the protein splicing and endonuclease domains, the two domains appear to collaborate by interacting with the homing site DNA (Duan et al. 1997; Moure et al. 2002; Werner et al. 2002). Furthermore, it was shown that the two intein domains are functionally independent and have separate evolutionary origins (Dalgaard et al. 1997; Derbyshire et al. 1997; Okuda et al. 2003; see Sect. III).

B. PRP8 Inteins in Fungi

After the discovery of the VMA1 intein, a second, non-allelic intein in fungi was identified within the PRP8 protein of the basidiomycete *Cryptococcus neoformans*. Many fungal inteins have also been found at an allelic insertion site in various fungi; some of these have been characterized experimentally in detail.

1. Distribution of PRP8 Inteins in Fungi

A nuclear-encoded intein has been identified inside the *prp8* gene in two strains (Cn 3511 and JEC 21) of the human pathogen *C. neoformans* (Butler et al. 2001). The PRP8 protein is a major component of the spliceosome and is highly conserved among eukaryotes (Grainger and Beggs 2005). Therefore, the PRP8 protein is essential for cell viability (Luo et al. 1999) and PRP8 inteins must be active, because inactivity leads to host lethality. The *Cne* PRP8 intein (Table 3.2) is a mini-intein composed of only 172 aa and lacks the endonuclease domain, present in the

Table 3.1. Characteristics of VMA1 inteins inserted in the vacuolar ATPase, subunit A

Species/intein	Intein size, endonuclease	Activity	References
Saccharomyces cerevisiae/Sce *VMA1*	454 aa, +	Experimental	Hirata et al. (1990), Kane et al. (1990)
S. cerevisiae DH1-1A/Sce-*DH1-1A VMA1*	454 aa, +	Theoretical	Gimble (2001)
S. cerevisiae OUT7091/Sce-*OUT7091 VMA1*	454 aa, +	Theoretical	Okuda et al. (2003)
S. cerevisiae OUT7112/Sce-*OUT7112 VMA1*	454 aa, +	Theoretical	Okuda et al. (2003)
S. castellii CBS4309/*Sca*-CBS4309 VMA1	517 aa, +	Theoretical	Koufopanou et al. (2002), Okuda et al. (2003)
S. castellii IFO1992/*Sca*-IFO1992 VMA1	517 aa, +	Theoretical	Koufopanou et al. (2002), Okuda et al. (2003)
S. cariocanus/*Scar* VMA1	454 aa, +	Theoretical	Perler (2002)
S. dairenensis/*Sda* VMA1	501 aa, +	Theoretical	Koufopanou et al. (2002), Okuda et al. (2003)
S. exiguous/*Sex* VMA1	499 aa, +	Theoretical	Koufopanou et al. (2002), Okuda et al. (2003)
S. pastorianus/*Spa* VMA1	454 aa, +	Theoretical	Okuda et al. (2003)
S. unisporus/*Sun* VMA1	414 aa, +	Theoretical	Koufopanou et al. (2002), Okuda et al. (2003)
Kazachstania exiguus/*Kex* VMA1	502 aa, +	Theoretical	Koufopanou et al. (2002), Okuda et al. (2003)
Candida tropicalis/*Ctr* VMA1	471 aa, +	Experimental	Gu et al. (1993), Steuer et al. (2004)
C. glabrata/*Cgl* VMA1	415 aa, +	Theoretical	Okuda et al. (2003)
Debaryomyces hansenii/*Dha* VMA1	394 aa, +	Theoretical	Dujon et al. (2004)
Pichia stipitis/*Pst* VMA1	449 aa, +	Theoretical	Bakhrat et al. (2006), Jeffries et al. (2007)
Kluyveromyces lactis CBS683/*Kla*-CBS683 VMA1	410 aa, +	Theoretical	Koufopanou et al. (2002), Okuda et al. (2003)
K. lactis IFO1267/*Kla*-IFO1267 VMA1	410 aa, +	Theoretical	Okuda et al. (2003)
K. lactis NRRLY1140/*Kla*-NRRLY1140 VMA1	410 aa, +	Theoretical	Dujon et al. (2004)
Vanderwaltozyma polyspora/*Vpo* VMA1	433 aa, +	Theoretical	Koufopanou et al. (2002)
Lodderomyces elongisporus/*Lel* VMA1	421 aa, +	Theoretical	Perler (2002)
Torulaspora globosa/*Tgl* VMA1	456 aa, +	Theoretical	Koufopanou et al. (2002)
T. pretoriensis/*Tpr* VMA1	455 aa, +	Theoretical	Koufopanou et al. (2002)
Zygosaccharomyces bailii/*Zba* VMA1	456 aa, +	Theoretical	Koufopanou et al. (2002)
Z. bisporus/*Zbi* VMA1	450 aa, +	Theoretical	Koufopanou et al. (2002)
Z. rouxii/*Zro*	450 aa, +	Theoretical	Koufopanou et al. (2002)
Schizosaccharomyces japonicus/*Sja* VMA1	476 aa, +	Theoretical	Perler (2002)

experimentally characterized VMA1 intein of *S. cerevisiae* (Butler et al. 2001; see Sect. II.A).

Isolates of *C. neoformans* have been classified into five different serotypes and three different varieties. They are named *C. neoformans* var. *grubii* (serotype A), var. *gattii* (serotypes B, C), var. *neoformans* (serotype D), and a hybrid of serotypes A and D (serotype AD; Nelson and Lodes 2006; Ito-Kuwa et al. 2007).

Varieties *grubii* and *neoformans* are distributed all over the world, while serotypes B and C are essentially restricted to tropical and subtropical regions (Granados and Castaneda 2005). Due to the evolutionary distance of *C. neoformans* var. *gattii*, Kwon-Chung et al. (2002) proposed separating the fungus from *C. neoformans* and naming it *C. gattii*.

Liu and Yang (2004) detected highly conserved PRP8 inteins with a size range of 170–172 aa in the genomes of different *C. neoformans* serotypes. In addition to sequence variations within the intein, PRP8 inteins from different serotypes also differ in their sizes. While PRP8 inteins of serotype A are 172 aa in length, PRP8 inteins of serotype B/C are 170 aa and inteins of serotype D are 171 aa in length (Butler and Poulter 2005). *C. neoformans* serotype AD carries two distinct PRP8 inteins, indicating that strains of this serotype may be diploid (Poulter et al. 2007).

Besides the PRP8 mini-inteins identified in *C. neoformans* and *C. gattii*, a large intein was found within the PRP8 protein of *Cryptococcus laurentii* (Butler and Poulter 2005). To date, no

Table 3.2. Characteristics of PRP8 inteins inserted in a subunit of the spliceosome

Species/intein	Intein size, endonuclease	Activity	References
Aspergillus brevipes FRR2439/*Abr* PRP8	165, −	Theoretical	Butler et al. (2006)
A. fumigatus FRR0163/*Afu*-FRR0163 PRP8	820, +	Experimental	Liu and Yang (2004)
A. fumigatus var. ellipticus Af293/*Afu*-Af293 PRP8	819, +	Theoretical	Poulter et al. (2007)
A. giganteus NRRL6136/*Agi* PRP8	167, −	Theoretical	Butler et al. (2006)
A. nidulans FGSC A4/*Ani*-FGSCA4-PRP8	605, +	Experimental	Liu and Yang (2004)
A. viridinutans FRR0577/*Avi* PRP8	169, −	Theoretical	Butler et al. (2006)
Batrachochytrium dendrobatidis JEL423/*Bde*-JEL423 PRP8-1	465, +	Theoretical	Perler (2002)
B. dendrobatidis JEL423/*Bde*-JEL423 PRP8-2	465, +	Theoretical	Perler (2002)
Botrytis cinerea B05.10/Bci PRP8	838, +	Theoretical	Butler et al. (2006)
Cryptococcus gattii (serotype C/B)/*Cga* PRP8	170, −	Theoretical	Liu and Yang (2004), Butler and Poulter (2005)
C. laurentii CBS139/*Cla* PRP8	522, +	Theoretical	Butler and Poulter (2005)
C. neoformans grubii (serotype A)/*Cne* PRP8	171, −	Theoretical	Liu and Yang (2004)
C. neoformans neoformans (serotype D)/*Cne* PRP8	172, −	Experimental	Butler et al. (2001), Liu and Yang (2004)
Eupenicillium baarnense CBS 134.41/*Eba* PRP8	168, −	Experimental	Elleuche et al. (2009)
E. crustaceum CBS 244.32/*Ecr* PRP8	158, −	Experimental	Elleuche et al. (2009)
Histoplasma capsulatum/*Hca* PRP8	534, 535, +	Experimental	Liu and Yang (2004), Butler et al. (2006)
Neosartorya aurata NRRL 4378/*Nau* PRP8	164, −	Theoretical	Butler et al. (2006)
N. fenelliae NRRL 5534/*Nfe* PRP8	155, −	Theoretical	Butler et al. (2006)
N. fischeri FRR0181/*Nfi* PRP8	517, +	Theoretical	Butler et al. (2006)
N. glabra FRR 2163/*Ngl*-2163 PRP8	155, −	Theoretical	Butler et al. (2006)
N. glabra FRR 1833/*Ngl*-FRR1833 PRP8	169, −	Theoretical	Perler (2002)
N. pseudofischeri FRR 0186/*Nps* PRP8	169, −	Theoretical	Butler et al. (2006)
N. quadricincta NRRL 4175/*Nqu* PRP8	169, −	Theoretical	Butler et al. (2006)
N. spinosa FRR4595/*Nsp* PRP8	169, −	Theoretical	Butler et al. (2006)
Phycomyces blakesleeanus NRRL 155/*Pbl* PRP8	198, −	Theoretical	Perler (2002)
Paracoccidioides brasiliensis Pb18/*Pbr* PRP8	573, +	Theoretical	Butler and Poulter (2005), Butler et al. (2006)
Penicillium chrysogenum/*Pch* PRP8	157, −	Experimental	Elleuche et al. (2006)
P. expansum/*Pex* PRP8	162, −	Experimental	Elleuche et al. (2006)
P. vulpinum/*Pvu* PRP8	161, −	Experimental	Elleuche et al. (2006)
Pyrenophora tritici-repentis Pt-1C-BF/*Ptr* PRP8	844, +	Theoretical	Perler (2002)
Spizellomyces punctatus/*Spu* PRP8	473, +	Theoretical	Perler (2002)
Uncinocarpus reesii/*Ure* PRP8	180, −	Theoretical	Butler et al. (2006)

inteins have been identified in any other species of the genus *Cryptococcus*, nor have they been found in the order Tremellales, nor in the phylum Basidiomycota, including *Ustilago maydis*, *Coprinopsis cinerea*, and *Phanerochaete chrysosporium*, species for which the whole genome sequence is available (Butler et al. 2006).

Allelic PRP8 inteins that occupy the same insertion site as the *Cne* PRP8 have, however, been identified in various species of the phylum Ascomycota. Particularly, PRP8 inteins were found in species of the genera *Neosartorya*, *Aspergillus*,

and *Penicillium* of the order Eurotiales. PRP8 inteins of *A. fumigatus* and *A. nidulans* are large inteins and contain an endonuclease domain (Liu and Yang 2004). The *Afu* PRP8 intein is an exceptionally long PRP8 intein and is comprised of 819 aa (*Afu*-FRR0163 PRP8) and 820 aa (*Afu*-Af293 PRP8; Table 3.2). In addition to a 455-aa endonuclease domain, a 222-aa sequence is integrated after block D of the *Afu*-FRR0163 PRP8 intein. At the same position, a so-called tongs subdomain of 69 aa is integrated in the *Sce* VMA1 intein. On this basis, the 222-aa domain of *Afu* PRP8 was defined as a putative tongs domain, although there is no apparent sequence homology between the *Afu* PRP8 and the *Sce* VMA1 domains (Liu and Yang 2004). The only other putative tongs domain was identified in the large PRP8 intein from *Botrytis cinerea*. The domain of *Bci* PRP8 is at exactly the same position as the putative tongs domain of *A. fumigatus* and shares a high degree of similarity (Poulter et al. 2007). Overall the tongs domain seems not to be an essential feature of the intein function, because the domain is located in a region of the intein where mutations frequently occur (Poulter et al. 2007). Also, deletions of selected amino acids within this region in a homologous mini-intein from *Penicillium chrysogenum* did not result in the complete loss of intein function (Elleuche et al. 2008).

PRP8 inteins were also identified in species of the genus *Neosartorya*, a genus closely related to the genus *Aspergillus*, in the genera *Eupenicillium* and *Penicillium*, as well in the pathogens *Histoplasma capsulatum*, *Paracoccidioides brasiliensis*, and in its non-pathogenic relative *Unicarpus reesii* (Butler et al. 2006; Elleuche et al. 2006; Poulter et al. 2007; Theodoro et al. 2008; Table 3.2).

Phylogenetic analysis based on the internal transcribed spacer sequences (ITS) of *Penicillium* and *Eupenicillium* species revealed that species containing a PRP8 intein are distinct from species lacking the intein (Elleuche and Pöggeler, unpublished data). Moreover, we and others observed that intron are often inserted close to the empty insertion sites in intein-free alleles (Butler et al. 2006; Elleuche and Pöggeler, unpublished data). In the genus *Eupenicillium*, all introns are identified in intein-free *prp8* alleles located between 13 base pairs (bp) and 15 bp downstream of the PRP8 intein insertion site (Elleuche et al. 2009). To date, however, introns at the same insertion site as the PRP8 intein have not been found in any fungus. In total, PRP8 inteins were detected in 32 different fungal species of the phyla ascomycota, basidiomycota, chytridiomycota and zygomycota. The 844-aa PRP8 intein of *Pyrenophora tritici-repentis* is the largest PRP8 intein described so far. This intein appears to possess a large insertion after block C within the endonuclease domain (see Note at InBase; Perler 2002)). The smallest PRP8 intein of 155 aa was identified in *Neosartorya fenelliae* (Perler 2002). Only 15 out of the 32 fungal PRP8 inteins (approx. 47%), contain an endonuclease domain, in contrast to the VMA1 inteins of

yeasts, where all known inteins contain an endonuclease domain (Table 3.1). Of the known inteins of all three domains of life, only about 15% lack the endonuclease domain (Perler 2002).

Nearly all PRP8 inteins have clear homology, based on a high degree of sequence identity and the position of their insertion within the *prp8* gene. The vast majority of fungal PRP8 inteins occupy the same insertion site as the prototype allele of *Cne* PRP8. The PRP8 inteins of only two chytrid fungi are not homologous and were identified at two different non-allelic positions: the PRP8 intein of *Spizellomyces punctatus Spu* PRP8 and *Bde*-JEL423 PRP8-1, one of the two PRP8 inteins of *Batrachochytrium dendrobatidis* (Perler 2002; Table 3.2). Nonetheless, PRP8 inteins are not universal. Various fungal species have been analyzed to date that do not contain an intein in the *prp8* gene (Elleuche et al. 2006; Poulter et al. 2007).

2. Activity of Fungal PRP8 Inteins

One of the most interesting questions concerning inteins is their splicing activity. Since most inteins are embedded in essential proteins that are often involved in nucleic acid metabolism, they must be removed from the precursor protein for the mature protein to perform its function.

Most experiments on the functionality of fungal inteins were done by expression in heterologous systems, mainly using *Escherichia coli* as a host. To track the splicing reaction, spliced products were usually visualized by SDS-PAGE and Western blot after overexpression of plasmid-borne fusion constructs. When inteins are embedded in foreign proteins, they can splice themselves out, so heterologous extein sequences such as maltose binding protein, GST- or His-tags, and thioredoxin often serve as tags. All of these can be easily detected by Western blot (Wu et al. 1998a; Liu et al. 2003; Liu and Yang 2004; Elleuche et al. 2006). The yield of spliced products varies with the extein sequence used, although activity is often higher when the splicing reaction occurs with native extein flanks (Xu et al. 1993; Perler 2005).

For the PRP8 inteins, best-characterized is the splicing of PRP8 mini-inteins from *C. neoformans* and from species of *Penicillium* (Liu and Yang 2004; Elleuche et al. 2006; Pearl et al. 2007a, b). To characterize the splicing activity of the

Cne PRP8, as well as large PRP8 inteins from *A. nidulans*, *A. fumigatus*, and *H. capsulatum*, each intein was cloned into a model gene encoding 5 aa of its native extein sequences on each site of the intein and an N-terminal maltose-binding protein and a C-terminal thioredoxin tag (Liu and Yang 2004). After overexpression, SDS-PAGE, and Western blot analysis with an anti-thioredoxin antibody, splicing products were identified. All inteins investigated by Liu and Yang (2004) were demonstrated to be highly active when heterologously expressed in *E. coli*. The *Hca* PRP8 intein exhibited a 75% decrease in splicing activity when shifted from 25°C to 37°C. This temperature shift is an important physiological stimulus for the morphotypic transition from mold to yeast of *H. capsulatum* (Woods 2002). Temperature-independent splicing activity was reported for three *Penicillium* PRP8 inteins (Elleuche et al. 2006).

Another system to characterize PRP8 intein activity was applied by Pearl et al. (2007a), based on an assay previously developed for the *Mycobacterium tuberculosis* RecA large intein (Davis et al. 1992; Buskirk et al. 2004; Skretas and Wood 2005). This method interrupted the α-complementation peptide of *E. coli* β-galactosidase, allowing assessment of intein function in vivo in *E. coli*, in combination with immunoblot analysis of the tagged intein sequences. *E. coli* loses its ability to grow on lactose, if the inserted intein is inactive. The authors examined in detail the influence of adjacent extein residues by replacing amino acids −1, −2 and +1, +2 with various other residues. The results confirmed that splicing efficiency depends on the composition of extein sequences (Pearl et al. 2007b). A similar result was obtained for the *Pex* PRP8 intein, which was able to undergo protein splicing within the green fluorescent protein (GFP). Only the +1 residue (a Serine) was conserved in the GFP and as expected, the yield of spliced product was reduced compared to results obtained with native flanking extein regions (Elleuche et al. 2006).

To gain further insights into the splicing mechanism of fungal inteins, alanine-scanning mutagenesis was applied to investigate the influence of single amino acid residues in *Cne* PRP8 on protein splicing (Pearl et al. 2007a). Similar to bacterial inteins and the large *Sce* VMA1 intein (Southworth et al. 1999; Shingledecker et al. 2000), mutation of the first, penultimate, and last residues, as well as

mutation in a highly conserved TxxH motif in block B, strongly inhibited or completely blocked the splicing reaction. Furthermore, five new residues were found to be crucial for splicing activity in this study. All of these residues are located in the well-conserved block F and have not been previously determined to be involved in the protein splicing of bacterial inteins (Pearl et al. 2007a).

As mentioned above, large inteins derived from bacteria or the VMA1 intein of *S. cerevisiae* remain active after the endonuclease domain has been deleted, indicating that inteins are highly robust genetic elements. To define the catalytic and structural elements involved in protein splicing of naturally occurring mini-inteins, an analysis of the structural requirements of protein splicing has been conducted for the 157-aa *Pch* PRP8 intein, which is among the smallest known nuclear-encoded active splicing protein elements (Table 3.2). Amino acid sequences of *Pch* PRP8 can be deleted at two different sites without affecting splicing activity. One site corresponds to the insertion site of the endonuclease domain in large allelic PRP8 inteins. The other site was detected at a new position corresponding to the insertion site of a putative tongs domain of a large fungal PRP8 intein (see above). The smallest functional intein found after deleting eight and six amino acids at two different sites comprises only 143 residues and is the smallest functional eukaryotic intein engineered so far (Elleuche et al. 2008). Moreover, it was demonstrated that the *Pch* PRP8 intein is capable of protein splicing in *trans* (Elleuche and Pöggeler 2007). After artificially splitting the *Pch* PRP8 intein into an N- and a C-terminal domain at three different sites, protein *trans*-splicing has been shown to occur at two sites. One of the functional split sites corresponds to the insertion site of the endonuclease domain in allelic large PRP8 inteins, and the other was detected at a new position, located N-terminal to the endonuclease insertion site (Elleuche and Pöggeler 2007).

C. Other Fungal Inteins

To date, 12 non-allelic sites of intein gene insertion in nine different nuclear genes have been described in fungi (Perler 2002). In the yeasts *Debaroyomyces hansenii* and *Candida (Pichia) guilliermondii*, and in the filamentous ascomycete *Podospora anserina*, inteins have also been

discovered within the *glt1* gene encoding glutamate synthase. These belong to the group of large
inteins and are all embedded at exactly the same
site of the GLT1 protein. Recently, a further GLT1
intein was identified in the GLT1-protein in the
plant pathogen *Phaeosphaeria nodorum* (Poulter
et al. 2007). In addition to the GLT1 intein, *P.
anserina* contains a non-allelic large intein in the
chitin synthase gene *chs2* (Butler et al. 2005). Both
GLT1 and CHS2 inteins contain domains that
indicate the presence of a homing endonuclease
of the LAGLIDADG type. Interestingly, the *glt1*
gene of *A. nidulans* contains a currently unanno-
tated intron that occupies the same insertion site
as the *Dha* GLT1 (Butler et al. 2005).

Phylogenetic analysis revealed that Pan CHS2
is closely related to the GLT1 inteins. Therefore,
Poulter et al. (2007) proposed that Pan CHS2 may
have been derived from an ectopic movement of a
GLT1 intein into a non-allelic CHS2 site. No CHS2
intein has been found in the genomes of any other
fungus: a possible explanation for this might be a
conserved valine residue at the putative +1 extein
position. The CHS2 protein of *P. anserina* seems to
be the sole known exception with a cysteine at this
position (Butler et al. 2005). A large intein and a
mini-intein have also been detected in the threonyl
tRNA synthetase gene *thrRS* of the yeast *Candida
tropicalis* and *C. parapsilosis*, respectively. The
large *Ctr* ThrRS intein is 345 aa in length with
some endonuclease-resembling domains, but the
region seems to have accumulated multiple of
mutations that might result in an inactive endonuclease. (Poulter et al. 2007). The allelic 182-aa
mini-intein of *C. parapsilosis* (*Cpa* ThrRS) has
lost the endonuclease domain (Poulter et al.
2007). Except for mini-inteins of PRP8 proteins,
Cpa ThrRS is the sole example of a eukaryotic
mini-intein.

In prokaryotes, inteins are often embedded in
proteins involved in replication, transcription, or
in related processes such as the metabolism of
nucleotides (Liu 2000).

In general, eukaryotes encode three RNA
polymerases involved in the synthesis of messenger, ribosomal, and transfer RNAs (Cramer et al.
2008). Detailed investigation of eukaryotic RNA
polymerase subunits resulted in the identification
of inteins within this functional category of genes.
The inteins are found in the second largest subunit of either RNA polymerase I, or RNA polymerase II, or RNA polymerase III. The insertion

sites of inteins in RNA polymerases are among the
most highly conserved regions of these subunits
(Goodwin et al. 2006).

A 456-aa intein is embedded within a subunit of the RNA
polymerase I (RPA2) of the plant pathogen *P. nodorum*. *Pno*
RPA2 contains degenerated motifs of an endonuclease domain that no longer seems to be active (Goodwin et al.
2006). Three large inteins were also identified within the
gene encoding the second largest subunit of RNA polymerase II (RPB2). Two of them are inserted at the same site in
the zygomycete *Spiromyces aspiralis* (*Sas* RPB2-b) and the
chytrid *Coelomomyces stegomyiae* (*Cst* RPB2-b). The third
intein, *Bde* RPB2-c, was identified in another chytridiomy-
cete, *B. dendrobatidis*, at a non-allelic position. Another
non-allelic RPB2 intein, namely at the a-site, was described
in the RNA polymerase II of the green alga *Chlamydomonas
reinhardtii* (Goodwin et al. 2006). According to InBase, an
endonuclease-containing intein was recently identified in
the second largest subunit of RNA polymerase III in the
chytrid *B. dendrobatidis* (Perler 2002). The sequences of the
Bde RPB2 and *Bde* RPC2, as well as the regions of the
insertion sites, are very similar at the amino acid level.
However, the remaining parts of the exteins are not conserved, which suggests an intein transfer that recently occurred (InBase; Poulter, Butler and Goodwin, unpublished
data). In contrast to this, the three non-allelic inteins in the
RNA polymerase II are not closely related to each other
(Poulter et al. 2007).

Inteins have been described in four fungal phyla:
Ascomycota, Basidiomycota, Chytridiomycota,
and Zygomycota. Only the PRP8 intein has been
discovered in all fungal phyla. In fungi, inteins are
embedded within nine different genes and with
few exceptions they are inserted at allelic sites
(Tables 3.1–3.3). In eukaryotes, only a few species
are known that contain more than one intein in
their genome. As can be seen from Table 3.1
and Table 3.3, two inteins have been identified in
D. hansenii (*Dha* VMA1, *Dha* GLT1), *P. anserina*
(*Pan* GLT1, PanCHS2), *C. tropicalis* (*Ctr* VMA1,
Ctr ThRS), and *P. nodorum* (*Pno* GLT1, *Pno*
RPA2), and five inteins can be found in
the genome of the chytrid *B. dendrobaditis*. Prokaryotes often contain a variety of inteins in their
genomes, for example, 19 inteins were identified
in the genome of the methanogenic archaeon
Methanococcus jannaschii (Pietrokovski 1998b).

Nucleus-encoded inteins outside the fungi
were recently identified in the green alga *C. reinhardtii* and the slime mold *Dictyostelium discoideum* (Goodwin et al. 2006). In addition, other
eukaryotic inteins have been described in
plastids from algae and in eukaryotic viruses

Table 3.3. Characteristics of other fungal inteins

Species/intein	Host protein	Intein size, endonuclease	Activity	References
Podospora anserina/Pan CHS2	CHS2	649, +	Theoretical	Butler et al. (2005)
Batrachochytrium dendrobatidis JEL423/*Bde*-JEL423 eIF-5B	eIF-5B	419, +	Theoretical	Perler (2002)
Debaryomyces hansenii CBS767/*Dha* GLT1	GLT1	607, +	Theoretical	Butler et al. (2005)
Phaeosphaeria nodorum SN15/*Pno* GLT1	GLT1	635, +	Theoretical	Poulter et al. (2007)
Pichia (Candida) guilliermondii/Pgu GLT1	GLT1	553, +	Theoretical	Butler et al. (2005)
Podospora anserina/Pan GLT1	GLT1	682, +	Theoretical	Butler et al. (2005)
Phaeosphaeria nodorum SN15/*Pno* RPA2	RPA2	456, +	Theoretical	Goodwin et al. (2006)
Batrachochytrium dendrobatidis JEL197/*Bde*-JEL197 RPB2-c	RPB2-c	488, +	Theoretical	Goodwin et al. (2006)
Coelomomyces stegomyiae/Cst RPB2	RPB2-b	362, +	Theoretical	Goodwin et al. (2006)
Spiromyces aspiralis NRRL 22631/*Sas* RPB2	RPB2-b	354, +	Theoretical	Goodwin et al. (2006)
Batrachochytrium dendrobatidis JEL423/*Bde*-JEL423 RPC2	RPC2	488, +	Theoretical	Perler (2002)
Candida parapsilosis CLIB214/*Cpa* ThrRS	ThrRS	182, −	Theoretical	Poulter et al. (2007)
Candida tropicalis ATCC750/*Ctr* ThrRS	ThrRS	345, +	Theoretical	Perler (2002)

(Pietrokovski 1998a; Ogata et al. 2005). For example, a non-allelic intein was identified in an ancient eukaryotic viral DNA, encoding an RNA polymerase subunit. The viral intein-containing DNA is embedded in the genome of the oomycete *Phytophthora ramorum* (Goodwin et al. 2006).

III. Mobility, Evolution, and Domestication of Inteins

A variety of inteins can invade DNA sequences by virtue of the endonuclease encoded within them (Chevalier and Stoddard 2001). The lateral transfer of an intervening sequence (either an intron or intein) into a homologous intron-less/intein-less allele is termed 'homing' and must be distinguished from the transposition process to non-allelic sites (Dujon 1989). Phylogenetic analyses have revealed that the evolutionary biology of mobile inteins is highly dynamic. Further, VMA1 inteins have been shown to be the ancestors of the HO endonuclease required for mating type switching in yeast.

A. Mobility of Fungal Inteins

Intein-homing relies on endonucleases that create specific double-strand breaks at cognate alleles that do not contain these mobile elements. Mating of an intein-less and intein-containing strain leads to intein transmission to about 75% of the meiotic progeny. This non-Mendelian inheritance pattern

ensures the persistence of homing endonuclease genes in populations and leads to their invasion of new species by horizontal transmission. Intein-containing genes are not deleterious to their host organisms because they have evolved in association with an intein that is removed by protein splicing (Chevalier and Stoddard 2001; Gogarten et al. 2002). Fungal intein-encoded endonucleases belong to the LAGLIDADG family and are characterized by two copies of this conserved motif. As described in Sect. II.A, the VMA1 intein endonuclease PI-*Sce*I of *S. cerevisiae* is among the best-characterized homing endonucleases. PI-*Sce*I initiates the mobility of the intein by cleaving at intein-less alleles of the *VMA1* gene (Gimble and Thorner 1992). Subsequent to purification of PI-*Sce*I, genetic and biochemical studies demonstrated that the endonuclease makes numerous base-specific and phosphate backbone contacts with its 31-bp asymmetrical recognition site (Gimble and Thorner 1993; Gimble and Wang 1996).

It binds to a 36-bp DNA substrate on intein-free *VMA1* alleles and cleaves in a Mg^{2+}- or Mn^{2+}-dependent reaction to yield 4-bp extensions with 3′ overhangs (Gimble and Wang 1996; Wende et al. 1996; Moure et al. 2002; Noël et al. 2004). The intein-containing allele is immune to hydrolysis by PI-*Sce*I, as the VMA1 intein disrupts the 31-bp recognition sequence.

Homing of the *Sce* VMA1 intein occurs only during meiosis and not during mitosis, even if intein-containing and intein-free *VMA1* alleles coexist in the same cell and the *VMA1* gene is expressed in mitosis (Gimble and Thorner 1992). A functional genomic approach revealed that at least two karyopherins, Srp1p and Kap142p, are required for the nuclear localization of PI-SceI. The nuclear localization of PI-SceI was shown to be induced by inactivation of TOR kinases. Moreover, inactivation of TOR signaling or acquisition of an extra nuclear localization signal in the PI-SceI coding region leads to artificial nuclear localization of the endonuclease and thereby induces homing even during mitosis. Thus, the intein-encoded endonuclease utilizes the host systems of nutrient signal transduction and nucleocytoplasmic transport to ensure the propagation of its coding region (Nagai et al. 2003). Furthermore, it was demonstrated that the repair of PI-SceI-induced double-strand breaks occurs at the same period as that of SPO11-initiated meiotic recombination and that it is dependent on the host homologous recombination system as well as on premeiotic DNA replication (Fukuda et al. 2003, 2004). Several VMA1-derived endonucleases were shown to fail to cleave their own intein-less DNA substrate (Posey et al. 2004). Only two enzymes have been demonstrated to be active: PI-ZbaI from *Zygosaccharomyces bailii* and PI-ScaI from *Saccharomyces cariocanus*. Sequence alignment of active site residues revealed that inactive endonucleases lack one or both of the conserved acidic residues corresponding to D-218 and D-326 in PI-SceI (Posey et al. 2004). It is not known whether any other large fungal intein, besides the VMA1-derived endonucleases has an active homing endonuclease. However, the homing endonuclease domains of all of the PRP8 inteins, except *Pbr* PRP8 and *Cla* PRP8, were demonstrated to have both of the conserved aspartate residues (Poulter et al. 2007; Theodoro et al. 2008). Furthermore, an analysis of the frequency of synonymous versus non-synonymous changes indicated that PRP8-encoded endonucleases have been selectively constrained (Butler et al. 2006). Together, these data suggest that many PRP8-derived endonucleases might be active. Fungal GLT1-derived endonucleases, except for *Cgu* GLT1, also possess the two conserved aspartate residues, while the first aspartate residue has been substituted in the *Pan* CHS2 endonuclease domain. The other fungal inteins (Table 3.3) do not even exhibit clear evidence of the conserved motifs C, D, E, and H (Poulter et al. 2007). Thus it seems that only *Dha* GLT1, *Pan* GTL1, and *Pno* GTL1 might contain an active endonuclease domain (see Sect. II.C).

B. Evolution of Fungal Inteins

Goddard and Burt (1999) formulated a cycle model consisting of recurrent cycles of intron invasion, maintenance, degeneration, and loss for mobile introns with an endonuclease activity. A modified model was applied to large inteins (Gogarten et al. 2002; Koufopanou et al. 2002; Burt and Koufopanou 2004). According to this model, a mobile intein initially invades a genome by horizontal transmission. Once situated in the new genome, it is vertically transmitted to successive generations and spread out in the population by homing. When the intein is fixed in the population, there is little or no selection against point or deletion mutations and the activity of the endonuclease is destroyed. After the loss of activity, the endonuclease domain of the intein may eventually be deleted. The splicing domain is predicted to be under a different selection regime. Since the functionality of the splicing domain is critical for the function of the host protein, which in most cases is an essential protein, a strong purifying selection occurs to preserve function. Only a precise deletion of the intein-splicing domain will ensure host protein function and results in a genome that is devoid of the entire mobile intein. Before degeneration of the endonuclease, mutation of the endonuclease domain may create an enzyme that evolves a new specificity and can be transferred by horizontal transfer to a new host genome with the recognition site for the altered enzyme. At this point, the cycle begins again. Regarding VMA1 inteins of saccharomycete yeasts, several lines of evidences support the model of the homing cycle. First, Koufopanou et al. (2002) and Okuda et al. (2003) reported that horizontal transmission of VMA1 inteins from different strains of *S. cerevisiae*, saccharomycete yeasts, and other yeast species has been a regular occurrence in their evolutionary history. Second, the evolutionary state of VMA1-derived endonucleases from 12 yeast species was addressed by assaying their endonuclease activities (Posey et al. 2004).

As stated above, only two enzymes have been shown to be active. PI-ZbaI cleaves the *Z. bailii* recognition sequence significantly faster than the *S. cerevisiae* site, which differs at six nucleotide positions. A mutational analysis indicated that PI-*ZbaI* cleaves the *S. cerevisiae* substrate poorly due to the absence of a DNA-protein contact that is established by PI-*SceI*. These findings demonstrated that intein homing endonucleases evolve altered specificities as they adapt to recognize alternative target sites (Posey et al. 2004).

Finally, in the tetraploid wild-wine *S. cerevisiae* strain DH1-1A, it was shown that an intein-containing *VMA1* allele and an intein-less *VMA1* allele are encoded in one genome. The molecular reason for the co-existence of both alleles in one genome was shown to be due to a loss of activity in the PI-*SceI* analogue encoded by the DH1-1A *VMA1* intein, and mutations in the 31-bp recognition site of the intein-free allele of DH1-1A (Gimble 2001). In contrast, recent analyses of the PRP8 inteins suggest that the homing cycle model is not generally applicable (Gogarten and Hilario 2006). Compared to the endonucleases of VMA1 inteins, the homing endonucleases of PRP8 inteins show high dS/dN values (Butler et al. 2006).

A high value of the quotient of the frequency of synonymous (dS) versus the frequency of non-synonymous changes (dN) implies that the endonuclease is functional (Butler et al. 2006).

In addition, the level of synonymous change in the PRP8 endonuclease has reached saturation while the *VMA1* endonuclease is not saturated by synonymous substitutions. These results imply that VMA1 inteins are of more recent origin than the PRP8 inteins. Furthermore, phylogeny of the euascomycete PRP8 inteins provides no evidence for horizontal transfer. Only basidiomycetes of the genus *Cryptococcus* may have gained the PRP8 intein by horizontal transmission (Butler and Poulter 2005; Butler et al. 2006). Based on these results, Butler et al. (2006) proposed a modified model for the evolution of inteins present in euascomycetes. They suggested that inteins might be maintained by balancing selection. They may not become fixed in a population due to a decreased fitness of the host organism, and thus, the homing cycle might operate in sub-populations only. This scenario would allow continuous selection for the endonuclease function during extended periods of vertical transmission.

C. Domestication of Fungal Inteins

Per se, it cannot be ruled out that inteins contribute to host fitness rather than being detrimental.

It has been reported that PI-*SceI* binds but does not cut the promoter of the *S. cerevisiae GSH11* gene encoding a high-affinity glutathione transporter. *GSH11* is not expressed in a PI-*SceI*-deleted strain, and the inability to express *GSH11* has been shown to be overcome by the introduction of the coding region of PI-*SceI* or the entire *VMA1* gene (Miyake et al. 2003). Another example for the domestication of an intein is the HO endonuclease, required for mating type switching in various yeast species (Haber 1998). Phylogenetic analyses revealed that HO and the VMA1 intein of *S. cerevisiae* are close relatives (Dalgaard et al. 1997). The *HO* gene is thought to have arisen by an ectopic insertion of the coding region of the *VMA1* intein (Butler et al. 2004). The HO endonuclease is encoded by a free-standing gene showing 50% similarity to the full-length VMA1 intein, including the splicing domain, but is unique among LAGLIDADG endonucleases in having a 120-residue C-terminal putative zinc finger domain. Mutational analysis indicated that in addition to the splicing domain and the zinc finger domain, conserved residues between HO and catalytic, active-site residues in PI-*SceI*, and other related homing endonucleases are essential for HO activity (Bakhrat et al. 2004). Phylogenetic analysis from several yeast species revealed a single origin of the *HO* gene from the *VMA1* intein. In contrast to *VMA1* inteins, *HO* shows no evidence for degeneration or horizontal transmission (Koufopanou and Burt 2004; Bakhrat et al. 2006).

IV. Application of Inteins

In recent years, the utilization of inteins has become a major focus in biotechnological research. This section focus on and briefly reviews various intein-related applications for the analysis of protein–protein interactions, regulation of protein activity, and protein purification.

A. Inteins and Their Application in Protein–Protein Interaction Studies

Protein–protein interactions play key roles in various processes of biological systems, including transcriptional regulation, protein sorting, and receptor-ligand interactions. A number of techniques for the detection of protein–protein interactions, such as yeast two-hybrid, split-ubiquitin, bimolecular fluorescence complementation, or fluorescence detected resonance energy transfer have been invented and established in recent years (Fields and Song 1989; Rossi et al. 1997; Hink et al. 2002; Pasch et al. 2005). Several of these methods are often used in fungi, e.g. to detect or prove the interaction of fungal proteins (Hoff and Kück 2005; Nolting and Pöggeler 2006). All strategies used must result in the detection of a specific protein–protein interaction via conversion to a traceable signal. An intein-based method to study in vivo protein protein interactions in a mammalian system, as well as in bacteria, has been published by Ozawa et al. (2000, 2001). The technique relies on the reconstitution of a split enhanced green fluorescent protein (GFP) by intein-mediated protein splicing. Each part of the GFP is fused as an extein sequence to an intein moiety derived from an artificially split version of the *Sce* VMA1 intein. Neither intein construct can self-assemble, but interacting proteins fused to the other sides of the split intein are able to bring the intein halves in close proximity. This results in protein splicing and the release of a functional green fluorescent protein (GFP). The fluorescence intensity increases proportionally with the number of interacting protein pairs (Ozawa and Umezawa 2001). A great advantage of this system is the feasibility to measure protein–protein interactions anywhere in the cell because the method does not require that the interaction takes place in a special cell compartment, e.g. in the nucleus, which is required in the yeast two-hybrid system.

Another genetic approach, using the reconstitution of a split-enhanced green fluorescent protein (EGFP) by splicing a split DnaE intein derived from *Synechocystis* sp. PCC6803, was developed to investigate cellular localization of proteins. In this approach, a fusion protein comprised of the I_C moiety of the naturally split *Ssp* DnaE intein was fused to one half of the GFP protein and to a mitochondrial targeting signal. After transformation, the fusion protein was demonstrated to be translocated into the mitochondrial matrix. The other half of the GFP was fused with the I_N domain of *Ssp* DnaE and this construct was then fused to cDNA libraries. If a test protein contains a functional mitochondrial targeting signal, it translocates into the mitochondrial matrix, where EGFP is then formed by protein splicing. The cells harboring this reconstituted EGFP can be screened rapidly by fluorescence-activated cell sorting, and the cDNAs can be subsequently isolated and identified from the cells (Ozawa et al. 2003). This system has been also elaborated to identify proteins targeted to the endoplasmic reticulum (Ozawa et al. 2005). Although not all of these methods have been tested in fungal systems, they demonstrate interesting approaches to characterize protein–protein interactions and protein sorting in the living fungal cell.

B. Regulation of Protein-Splicing Activity

A major aim of physiologists is to temporally control protein function in living organisms. To accomplish this aim, the GFP-based system to study protein–protein interactions was further elaborated to induce splicing in living cells at a defined time-point. First, the protein of interest must be divided into two parts and fused to inteins, like GFP in the above-mentioned fluorescent-based systems. To keep the target protein inactive, until it is specifically activated in the cell, both fragments of the protein including the intein halves are fused to other proteins that interact or dimerize only under certain inducible conditions. In one approach, two mammalian proteins (FKBP, FK506-binding protein; FRB, the binding domain of the FKBP-rapamycin-associated protein), which are known to bind to rapamycin were fused to the artificially split *Sce* VMA1 intein (Mootz and Muir 2002). Upon addition of rapamycin, the *trans*-splicing reaction can be triggered in vivo or in vitro, and the intein connects the halves of the desired protein with a peptide bond (Mootz et al. 2003; Schwartz et al. 2003). This impressive system was termed the conditional protein splicing (CPS) system and was used to induce luciferase function in cultured cells and multicellular organisms. *Drosophila melanogaster* strains transformed with CPS constructs were fed with food containing rapamycin, resulting in the detection of luminescence in the living flies (Perler 2007; Schwartz et al. 2007).

Since rapamycin is fairly toxic to yeast, a light-dependent CPS system was recently developed for the yeast *S. cerevisiae* that depends on regulation of the photo-dimerization of phytochrome B and the phytochrome interacting factor 3 of *Arabidopsis thaliana* (Tyszkiewicz and Muir 2008).

Dimerization of the plant phytochrome B and the phytochrome interacting factor 3 occurs in red light (660 nm) and reversal of the dimerization in far-red light (750 nm). Thus, a change of the excitation wavelength to the far-red spectrum (750 nm) results in the dissociation of interaction partners. Fusion of phytochrome B and the phytochrome interacting factor 3 to the artificially split *Sce* VMA1 intein, and using the maltose-binding protein (MBP) and a Flag-tag as exteins, enabled Tyszkiewicz und Muir (2008) to demonstrate protein splicing under light-inducing conditions in living yeast cells.

C. Intein-Mediated Protein Purification

To produce recombinant proteins in large amounts and to purify them from crude extracts has become a major task for biotechnologists in recent years. Proteins can be produced in bacterial expression systems and the purification process can be eased by the utilization of affinity tags (Terpe 2003). Different small peptide tags (e.g. poly-His-, poly-Arg-, CBD-, or S-tags) are commonly used, as well as higher molecular weight tags (e.g. GST- or MBP-tags) to induce the solubility of proteins. However, removal of the affinity tag is often a complicated challenge and accompanied by different problems. Intein-mediated protein purification systems attempt to avoid these problems. In principle, a target protein is fused at either the N- or the C-terminus to a modified intein. The other site of the intein is linked to an affinity tag and is mutated at the tag-linked splice junction to undergo inducible specific cleavage with the site on the protein of interest. Crude protein extracts are loaded on an affinity column, the tag immobilizes the intein fusion construct, and after some washing steps, an induced splicing reaction (cleavage step) releases the desired protein from the column. Normally, the cleavage step is inducible under mild conditions, e.g. by the addition of thiolic agents, changes in pH or temperature. The first intein engineered for protein purification was the VMA1 intein from *S. cerevisiae* (Chong et al. 1997). Later, naturally occurring or artificially engineered mini-inteins were predominantly used for protein purification systems (Southworth et al. 1999; Wood et al. 1999; Ding et al. 2007).

Further development in intein-mediated protein purification was achieved using naturally or artificially split inteins (Mills et al. 1998; Wu et al. 1998b). The use of these *trans*-cleavage systems advanced the purification, because it prevented premature in vivo cleavage, a problem often observed when *cis*-splicing inteins are used. The *trans*-splicing system takes advantage of the ability of the intein halves to self-assemble. In the majority of cases, the *Ssp* DnaE intein is used (Chong and Xu 2005). The target protein is fused to the I_N domain of *Ssp* DnaE, while the I_C domain is mutated at the C-extein cleavage site. Furthermore, both intein halves are tagged to facilitate isolation and purification of the expressed intein fusion proteins. Co-incubation of both parts results in the reconstitution of the intein splicing activity and in the release of the target protein.

D. Screening Systems for Protein-Splicing Inhibitors

Using inteins to identify inhibitory agents to prevent fungal growth is a new field in fungal biology. Because inteins are localized inside essential genes, they have been often confirmed as useful drug targets for preventing fungal infections (e.g. *C. neoformans, A. fumigatus*, or *H. capsulatum*; Liu and Yang 2004). Thus, assays were developed to identify new protein-splicing inhibitors. The above-mentioned split GFP reporter system (Ozawa et al. 2000) was applied for the in vitro screening of antimicrobial agents inhibiting the splicing reaction of the *Mycobacterium tuberculosis* RecA intein (Gangopadhyay et al. 2003a, b). The ORF of GFP was disrupted by integration of *Mtu* RecA. Initially, the GFP-intein fusion protein is overexpressed, but remains inactive because inclusion bodies are formed. After release from the inclusion bodies and renaturating of the fusion product, in the absence of an inhibitor the intein undergoes protein splicing. When an inhibitor is present, protein splicing and the emission of fluorescence is prevented (Gangopadhyay et al. 2003a). The system described has already been used to test the fungal *Pex* PRP8 intein for its splicing activity within a foreign host protein and may be used in the future to design an efficient screening system for protein-splicing inhibitors of other fungal inteins (Elleuche et al. 2006).

V. Conclusions

Intein are internal protein domains found inside the coding region of different proteins. They are transcribed and translated together with their host protein and are removed from the unprocessed protein by protein splicing. Fungi encode large inteins comprising independent protein-splicing and endonuclease domains, and mini-inteins which lack the central endonuclease domain. Our knowledge on the distribution, evolution, and functionality of fungal inteins has expanded enormously during the past ten years. Currently, more than 70 inteins have been identified within the nuclear genomes of fungi. Most of them are embedded in homologs of the *S. cerevisiae VMA1* gene or within the *prp8* gene, but they can be also found in glutamate synthases, chitin synthases, threonyl-tRNA synthetases, and subunits of DNA-directed RNA polymerases. Genomic sequencing projects for several fungal species have been completed and many more are under way (http://www.fgsc.net/). This will accelerate the discovery of new fungal inteins. Biochemical and phylogenetic analysis will contribute further insight into the splicing mechanism and the evolution of fungal inteins. Ultimately, this will lead to the engineering of fungal inteins for a variety of applications.

References

Amitai G, Belenkiy O, Dassa B, Shainskaya A, Pietrokovski S (2003) Distribution and function of new bacterial intein-like protein domains. Mol Microbiol 47:61–73

Anraku Y, Hirata R, Wada Y, Ohya Y (1992) Molecular genetics of the yeast vacuolar H(+)-ATPase. J Exp Biol 172:67–81

Bakhrat A, Jurica MS, Stoddard BL, Raveh D (2004) Homology modeling and mutational analysis of Ho endonuclease of yeast. Genetics 166:721–728

Bakhrat A, Baranes K, Krichevsky O, Rom I, Schlenstedt G, Pietrokovski S, Raveh D (2006) Nuclear import of ho endonuclease utilizes two nuclear localization signals and four importins of the ribosomal import system. J Biol Chem 281:12218–12226

Beachy PA, Cooper MK, Young KE, von Kessler DP, Park WJ, Hall TM, Leahy DJ, Porter JA (1997) Multiple roles of cholesterol in hedgehog protein biogenesis and signaling. Cold Spring Harb Symp Quant Biol 62:191–204

Belfort M, Roberts RJ (1997) Homing endonucleass: keeping the house in order. Nucleic Acids Res 25:3379–3388

Bowman EJ, Tenney K, Bowman BJ (1988) Isolation of genes encoding the Neurospora vacuolar ATPase. Analysis of *vma*-1 encoding the 67-kDa subunit reveals homology to other ATPases. J Biol Chem 263:13994–14001

Burt A, Koufopanou V (2004) Homing endonuclease genes: the rise and fall and rise again of a selfish element. Curr Opin Genet Dev 14:609–615

Buskirk AR, Ong YC, Gartner ZJ, Liu DR (2004) Directed evolution of ligand dependence: small-molecule-activated protein splicing. Proc Natl Acad Sci USA 101:10505–10510

Butler MI, Poulter RT (2005) The PRP8 inteins in Cryptococcus are a source of phylogenetic and epidemiological information. Fungal Genet Biol 42:452–463

Butler MI, Goodwin TJ, Poulter RT (2001) A nuclear-encoded intein in the fungal pathogen *Cryptococcus neoformans*. Yeast 18:1365–1370

Butler G, Kenny C, Fagan A, Kurischko C, Gaillardin C, Wolfe KH (2004) Evolution of the MAT locus and its Ho endonuclease in yeast species. Proc Natl Acad Sci USA 101:1632–1637

Butler MI, Goodwin TJ, Poulter RT (2005) Two new fungal inteins. Yeast 22:493–501

Butler MI, Gray J, Goodwin TJ, Poulter RT (2006) The distribution and evolutionary history of the PRP8 intein. BMC Evol Biol 6:42

Caspi J, Amitai G, Belenkiy O, Pietrokovski S (2003) Distribution of split DnaE inteins in cyanobacteria. Mol Microbiol 50:1569–1577

Chevalier BS, Stoddard BL (2001) Homing endonucleases: structural and functional insight into the catalysts of intron/intein mobility. Nucleic Acids Res 29:3757–3774

Choi JJ, Nam KH, Min B, Kim SJ, Söll D, Kwon ST (2006) Protein trans-splicing and characterization of a split family B-type DNA polymerase from the hyperthermophilic archaeal parasite *Nanoarchaeum equitans*. J Mol Biol 356:1093–1106

Chong S, Xu MQ (1997) Protein splicing of the *Saccharomyces cerevisiae* VMA intein without the endonuclease motifs. J Biol Chem 272:15587–155890

Chong S, Xu MQ (2005) Harnessing inteins for protein purification and characterization. In: Belfort M, Derbyshire V, Stoddard BL, Wood DW (eds.) Homing endonucleases and inteins. Springer, Heidelberg, pp 273–292

Chong S, Mersha FB, Comb DG, Scott ME, Landry D, Vence LM, Perler FB, Benner J, Kucera RB, Hirvonen CA, Pelletier JJ, Paulus H, Xu MQ (1997) Single-column purification of free recombinant proteins using a self-cleavable affinity tag derived from a protein splicing element. Gene 192:271–281

Cramer P, Armache KJ, Baumli S, Benkert S, Brueckner F, Buchen C, Damsma GE, Dengl S, Geiger SR, Jasiak AJ, Jawhari A, Jennebach S, Kamenski T, Kettenberger H, Kuhn CD, Lehmann E, Leike K, Sydow JF, Vannini A (2008) Structure of eukaryotic RNA polymerases. Annu Rev Biophys 37:337–352

Dalgaard JZ, Klar AJ, Moser MJ, Holley WR, Chatterjee A, Mian IS (1997) Statistical modeling and analysis of the LAGLIDADG family of site-specific endonucleases and identification of an intein that encodes a site-specific endonuclease of the HNH family. Nucleic Acids Res 25:4626–4638

Dassa B, Amitai G, Caspi J, Schueler-Furman O, Pietro-kovski S (2007) Trans protein splicing of cyanobacterial split inteins in endogenous and exogenous combinations. Biochemistry 46:322–330

Davis EO, Jenner PJ, Brooks PC, Colston MJ, Sedgwick SG (1992) Protein splicing in the maturation of *M. tuberculosis* recA protein: a mechanism for tolerating a novel class of intervening sequence. Cell 71:201–210

Derbyshire V, Wood DW, Wu W, Dansereau JT, Dalgaard JZ, Belfort M (1997) Genetic definition of a protein-splicing domain: functional mini-inteins support structure predictions and a model for intein evolution. Proc Natl Acad Sci USA 94:11466–11471

Ding FX, Yan HL, Mei Q, Xue G, Wang YZ, Gao YJ, Sun SH (2007) A novel, cheap and effective fusion expression system for the production of recombinant proteins. Appl Microbiol Biotechnol 77:483–488

Ding Y, Xu MQ, Ghosh I, Chen X, Ferrandon S, Lesage G, Rao Z (2003) Crystal structure of a mini-intein reveals a conserved catalytic module involved in side chain cyclization of asparagine during protein splicing. J Biol Chem 278:39133–39142

Duan X, Gimble FS, Quiocho FA (1997) Crystal structure of PI-SceI, a homing endonuclease with protein splicing activity. Cell 89:555–564

Dujon B (1989) Group I introns as mobile genetic elements: facts and mechanistic speculations – a review. Gene 82:91–114

Dujon B, Sherman D, Fischer G, Durrens P, Casaregola S, Lafontaine I, De Montigny J, Marck C, Neuvéglise C, Talla E, Goffard N, Frangeul L, Aigle M, Anthouard V, Babour A, Barbe V, Barnay S, Blanchin S, Beckerich JM, Beyne E, Bleykasten C, Boisramé A, Boyer J, Cattolico L, Confanioleri F, De Daruvar A, Despons L, Fabre E, Fairhead C, Ferry-Dumazet H, Groppi A, Hantraye F, Hennequin C, Jauniaux N, Joyet P, Kachouri R, Kerrest A, Koszul R, Lemaire M, Lesur I, Ma L, Muller H, Nicaud JM, Nikolski M, Oztas S, Ozier-Kalogeropoulos O, Pellenz S, Potier S, Richard GF, Straub ML, Suleau A, Swennen D, Tekaia F, Wésolowski-Louvel M, Westhof E, Wirth B, Zeniou-Meyer M, Zivanovic I, Bolotin-Fukuhara M, Thierry A, Bouchier C, Caudron B, Scarpelli C, Gaillardin C, Weissenbach J, Wincker P, Souciet JL (2004) Genome evolution in yeasts. Nature 430:35–44

Elleuche S, Pöggeler S (2007) Trans-splicing of an artificially split fungal mini-intein. Biochem Biophys Res Commun 355:830–834

Elleuche S, Nolting N, Pöggeler S (2006) Protein splicing of PRP8 mini-inteins from species of the genus Penicillium. Appl Microbiol Biotechnol 72:959–967

Elleuche S, Döring K, Pöggeler S (2008) Minimization of a eukaryotic mini-intein. Biochem Biophys Res Commun 366:239–243

Elleuche S, Pelikan C, Nolting N, Pöggeler S (2009) Inteins and introns within the *prp8*-gene of four Eupenicillium species. J Basic Microbiol 49:52–57

Fields S, Song O (1989) A novel genetic system to detect protein-protein interactions. Nature 340:245–246

Fukuda T, Nogami S, Ohya Y (2003) VDE-initiated intein homing in *Saccharomyces cerevisiae* proceeds in a meiotic recombination-like manner. Genes Cell 8:587–602

Fukuda T, Nagai Y, Ohya Y (2004) Molecular mechanism of VDE-initiated intein homing in yeast nuclear genome. Adv Biophys 38:215–232

Gangopadhyay JP, Jiang SQ, Paulus H (2003a) An in vitro screening system for protein splicing inhibitors based on green fluorescent protein as an indicator. Anal Chem 75:2456–2462

Gangopadhyay JP, Jiang SQ, van Berkel P, Paulus H (2003b) In vitro splicing of erythropoietin by the *Mycobacterium tuberculosis* RecA intein without substituting amino acids at the splice junctions. Biochim Biophys Acta 1619:193–200

Ghislain M, Bowman EJ (1992) Sequence of the genes encoding subunits A and B of the vacuolar H(+)-ATPase of *Schizosaccharomyces pombe*. Yeast 8:791–799

Gimble FS (2000) Invasion of a multitude of genetic niches by mobile endonuclease genes. FEMS Microbiol Lett 185:99–107

Gimble FS (2001) Degeneration of a homing endonuclease and its target sequence in a wild yeast strain. Nucleic Acids Res 29:4215–4223

Gimble FS, Thorner J (1992) Homing of a DNA endonuclease gene by meiotic gene conversion in *Saccharomyces cerevisiae*. Nature 357:301–306

Gimble FS, Thorner J (1993) Purification and characterization of VDE, a site-specific endonuclease from the yeast *Saccharomyces cerevisiae*. J Biol Chem 268:21844–21853

Gimble FS, Stephens BW (1995) Substitutions in conserved dodecapeptide motifs that uncouple the DNA binding and DNA cleavage activities of PI-SceI endonuclease. J Biol Chem 270:5849–5856

Gimble FS, Wang J (1996) Substrate recognition and induced DNA distortion by the PI-*Sce*I endonuclease, an enzyme generated by protein splicing. J Mol Biol 263:163–180

Goddard MR, Burt A (1999) Recurrent invasion and extinction of a selfish gene. Proc Natl Acad Sci USA 96:13880–13885

Gogarten JP, Hilario E (2006) Inteins, introns, and homing endonucleases: recent revelations about the life cycle of parasitic genetic elements. BMC Evol Biol 6:94

Gogarten JP, Senejani AG, Zhaxybayeva O, Olendzenski L, Hilario E (2002) Inteins: structure, function, and evolution. Annu Rev Microbiol 56:263–287

Goodwin TJ, Butler MI, Poulter RT (2006) Multiple, non-allelic, intein-coding sequences in eukaryotic RNA polymerase genes. BMC Biol 4:38

Grainger RJ, Beggs JD (2005) Prp8 protein: at the heart of the spliceosome. RNA 11:533–557

Granados DP, Castaneda E (2005) Isolation and characterization of *Cryptococcus neoformans* varieties recovered from natural sources in Bogota, Colombia, and study of ecological conditions in the area. Microb Ecol 49:282–290

Gu HH, Xu J, Gallagher M, Dean GE (1993) Peptide splicing in the vacuolar ATPase subunit A from *Candida tropicalis*. J Biol Chem 268:7372–7381

Haber JE (1998) Mating-type gene switching in *Saccharomyces cerevisiae*. Annu Rev Genet 32:561–599

Hall TM, Porter JA, Young KE, Koonin EV, Beachy PA, Leahy DJ (1997) Crystal structure of a Hedgehog autoprocessing domain: homology between Hedge-hog and self-splicing proteins. Cell 91:85–97

Hink MA, Bisselin T, Visser AJ (2002) Imaging protein-protein interactions in living cells. Plant Mol Biol 50:871–883

Hirata R, Ohsumk Y, Nakano A, Kawasaki H, Suzuki K, Anraku Y (1990) Molecular structure of a gene, VMA1, encoding the catalytic subunit of H(+)-trans-locating adenosine triphosphatase from vacuolar membranes of Saccharomyces cerevisiae. J Biol Chem 265:6726–6733

Hoff B, Kück U (2005) Use of bimolecular fluorescence complementation to demonstrate transcription factor interaction in nuclei of living cells from the filamen-tous fungus Acremonium chrysogenum. Curr Genet 47:132–138

Ingham PW (2001) Hedgehog signaling: a tale of two lipids. Science 294:1879–1881

Ito-Kuwa S, Nakamura K, Aoki S, Vidotto V (2007) Sero-type identification of Cryptococcus neoformans by multiplex PCR. Mycoses 50:277–281

Jeffries TW, Grigoriev IV, Grimwood J, Laplaza JM, Aerts A, Salamov A, Schmutz J, Lindquist E, Dehal P, Sha-piro H, Jin YS, Passoth V, Richardson PM (2007) Genome sequence of the lignocellulose-bioconverting and xylose-fermenting yeast Pichia stipitis. Nat Bio-technol 25:319–326

Kane PM, Yamashiro CT, Wolczyk DF, Neff N, Goebl M, Stevens TH (1990) Protein splicing converts the yeast TFP1 gene product to the 69-kD subunit of the vacu-olar H(+)-adenosine triphosphatase. Science 250:651–657

Klabunde T, Sharma S, Telenti A, Jacobs WRJ, Sacchettini JC (1998) Crystal structure of GyrA intein from Mycobacterium xenopi reveals structural basis of pro-tein splicing. Nat Struct Biol 5:31–36

Koonin EV (1995) A protein splice-junction motif in hedgehog family proteins. Trends Biochem Sci 20:141–142

Koufopanou V, Burt A (2004) Degeneration and domesti-cation of a selfish gene in yeast: molecular evolution versus site-directed mutagenesis. Mol Biol Evol 22:1535–1538

Koufopanou V, Goddard MR, Burt A (2002) Adaptation for horizontal transfer in a homing endonuclease. Mol Biol Evol 19:239–246

Kwon-Chung KJ, Boekhout T, Fell JW, Diaz M (2002) Proposal to conserve the name Cryptococcus gattii against C. hondurianus and C. bacillisporus (Basidio-mycota, Hymenomycetes, Tremellomycetidae). Taxon 51:804–806

Liu XQ (2000) Protein-splicing intein: Genetic mobility, origin, and evolution. Annu Rev Genet 34:61–76

Liu XQ, Yang J (2003) Split dnaE genes encoding multiple novel inteins in Trichodesmium erythraeum. J Biol Chem 278:26315–26318

Liu XQ, Yang J (2004) Prp8 intein in fungal pathogens: target for potential antifungal drugs. FEBS Lett 572:46–50

Liu XQ, Yang J, Meng Q (2003) Four inteins and three group II introns encoded in a bacterial ribonucleotide reductase gene. J Biol Chem 278:46826–46231

Luo HR, Moreau GA, Levin N, Moore MJ (1999) The human Prp8 protein is a component of both U2- and U12-dependent spliceosomes. RNA 5:893–908

Mills KV, Lew BM, Jiang S, Paulus H (1998) Protein splic-ing in trans by purified N- and C-terminal fragments of the Mycobacterium tuberculosis RecA intein. Proc Natl Acad Sci USA 95:3543–3548

Miyake T, Hiraishi H, Sammoto H, Ono B (2003) Involve-ment of the VDE homing endonuclease and rapamy-cin in regulation of the Saccharomyces cerevisiae GSH11 gene encoding the high affinity glutathione transporter. J Biol Chem 278:39632–39636

Mootz HD, Muir TW (2002) Protein splicing triggered by a small molecule. J Am Chem Soc 124:9044–9045

Mootz HD, Blum ES, Tyszkiewicz AB, Muir TW (2003) Conditional protein splicing: a new tool to control protein structure and function in vitro and in vivo. J Am Chem Soc 125:10561–10569

Moure CM, Gimble FS, Quiocho FA (2002) Crystal struc-ture of the intein homing endonuclease PI-SceI bound to its recognition sequence. Nat Struct Biol 9:764–770

Nagai Y, Nogami S, Kumagai-Sano F, Ohya Y (2003) Kar-yopherin-mediated nuclear import of the homing en-donuclease VMA1-derived endonuclease is required for self-propagation of the coding region. Mol Cell Biol 25:1726–1736

Nelson RT, Lodes JK (2006) Cryptococcus neoformans pathogenicity. In: Brown AJP (ed) The Mycota XIII. Springer, Heidelberg, pp 237–266

Noël AJ, Wende W, Pingoud A (2004) DNA recognition by the homing endonuclease PI-SceI involves a divalent metal ion cofactor-induced conformational change. J Biol Chem 279:6794–6804

Nolting N, Pöggeler S (2006) A MADS box protein interacts with a mating-type protein and is required for fruiting body development in the homothallic ascomycete Sor-daria macrospora. Eukaryot Cell 5:1043–1056

Noren CJ, Wang J, Perler FB (2000) Dissecting the chemis-try of protein splicing and its applications. Angew Chem Int Ed Engl 39:450–466

Ogata H, Raoult D, Claverie JM (2005) A new example of viral intein in mimivirus. Virol J 2:8

Okuda Y, Sasaki D, Nogami S, Kaneko Y, Ohya Y, Anraku Y (2003) Occurrence, horizontal transfer and degen-eration of VDE intein family in saccharomycete yeasts. Yeast 20:563–573

Ozawa T, Umezawa Y (2001) Detection of protein–protein interactions in vivo based on protein splicing. Curr Opin Chem Biol 5:578–583

Ozawa T, Nogami S, Sato M, Ohya Y, Umezawa Y (2000) A fluorescent indicator for detecting protein–protein interactions in vivo based on protein splicing. Anal Chem 72:5151–517

Ozawa T, Kaihara A, Sato M, Tachihara K, Umezawa Y (2001) Split luciferase as an optical probe for detect-ing protein–protein interactions in mammalian cells based on protein splicing. Anal Chem 73:2516–2521

Ozawa T, Sako Y, Sato M, Kitamura T, Umezawa Y (2003) A genetic approach to identifying mitochondrial pro-teins. Nat Biotechnol 21:287–293

Ozawa T, Nishitani K, Sako Y, Umezawa Y (2005) A high-throughput screening of genes that encode proteins transported into the endoplasmic reticulum in mammalian cells. Nucleic Acids Res 33:e34

Pasch JC, Nickelsen J, Schünemann D (2005) The yeast split-ubiquitin system to study chloroplast membrane protein interactions. Appl Microbiol Biotechnol 69:440–447

Paulus H (2000) Protein splicing and related forms of protein autoprocessing. Annu Rev Biochem 69:447–496

Pearl EJ, Tyndall JD, Poulter RT, Wilbanks SM (2007a) Sequence requirements for splicing by the *Cne* PRP8 intein. FEBS Lett 581:300–304

Pearl EJ, Bokor AA, Butler MI, Poulter RT, Wilbanks SM (2007b) Preceding hydrophobic and beta-branched amino acids attenuate splicing by the *Cne* PRP8 intein. Biochim Biophys Acta 1774:995–1001

Perler FB (1998) Protein splicing of inteins and hedgehog autoproteolysis: structure, function, and evolution. Cell 92:1–4

Perler FB (2002) InBase: the intein database. Nucleic Acids Res 30:383–384

Perler FB (2005) Inteins – a historical perspective. In: Belfort M, Derbyshire V, Stoddard BL, Wood DW (eds) Homing endonucleases and inteins. Springer, Heidelberg, pp 193–210

Perler FB (2007) Shining a light on protein expression in living organisms. Nat Chem Biol 3:17–18

Perler FB, Davis EO, Dean GE, Gimble FS, Jack WE, Neff N, Noren CJ, Thorner J, Belfort M (1994) Protein splicing elements: inteins and exteins – a definition of terms and recommended nomenclature. Nucleic Acids Res 22:1125–1127

Perler FB, Olsen GJ, Adam E (1997) Compilation and analysis of intein sequences. Nucleic Acids Res 25:1087–1093

Pietrokovski S (1994) Conserved sequence features of inteins (protein introns) and their use in identifying new inteins and related proteins. Protein Sci 3:2340–2350

Pietrokovski S (1998a) Identification of a virus intein and a possible variation in the protein-splicing reaction. Curr Biol 8:R634–R635

Pietrokovski S (1998b) Modular organization of inteins and C-terminal autocatalytic domains. Protein Sci 7:64–71

Posey KL, Koufopanou V, Burt A, Gimble FS (2004) Degeneration and domestication of a selfish gene in yeast: molecular evolution versus site-directed mutagenesis. Nucleic Acids Res 32:3947–3956

Poulter RT, Goodwin TJ, Butler MI (2007) The nuclear-encoded inteins of fungi. Fungal Genet Biol 44:153–179

Requena N, Mann P, Hampp R, Franken P (2002) Early developmentally regulated genes in the arbuscular mycorrhizal fungus *Glomus mosseae*: identification of GmGIN1, a novel gene with homology to the C-terminus of metazoan hedgehog proteins. Plant Soil 244:129–139

Rossi F, Charlton CA, Blau HM (1997) Monitoring protein-protein interactions in intact eukaryotic cells by beta-galactosidase complementation. Proc Natl Acad Sci USA 94:8405–8410

Saleh L, Perler FB (2006) Protein splicing in cis and in trans. Chem Rec 6:183–193

Schwartz EC, Muir TW, Tyszkiewicz AB (2003) "The splice is right": how protein splicing is opening new doors in protein science. Chem Commun (Camb) 2003:2087–2090

Schwartz EC, Saez L, Young MW, Muir TW (2007) Post-translational enzyme activation in an animal via optimized conditional protein splicing. Nat Chem Biol 3:50–54

Shingledecker K, Jiang SQ, Paulus H (1998) Molecular dissection of the *Mycobacterium tuberculosis* RecA intein: design of a minimal intein and of a trans-splicing system involving two intein fragments. Gene 207:187–195

Shingledecker K, Jiang S, Paulus H (2000) Reactivity of the cysteine residues in the protein splicing active center of the *Mycobacterium tuberculosis* RecA intein. Arch Biochem Biophys 375:138–144

Skretas G, Wood DW (2005) Regulation of protein activity with small-molecule-controlled inteins. Protein Sci 14:523–532

Southworth MW, Adam E, Panne D, Byer R, Kautz R, Perler FB (1998) Control of protein splicing by intein fragment reassembly. EMBO J 17:918–926

Southworth MW, Amaya K, Evans TC, Xu MQ, Perler FB (1999) Purification of proteins fused to either the amino or carboxy terminus of the *Mycobacterium xenopi* gyrase A intein. Biotechniques 27:110–114

Starokadomskyy PL (2007) Protein splicing. Mol Biol (Mosk) 41:314–330

Steuer S, Pingoud V, Pingoud A, Wende W (2004) Chimeras of the homing endonuclease PI-SceI and the homologous *Candida tropicalis* intein: a study to explore the possibility of exchanging DNA-binding modules to obtain highly specific endonucleases with altered specificity. Chembiochem 5:206–213

Stoddard BL (2005) Homing endonuclease structure and function. Q Rev Biophys 38:49–95

Sun P, Ye S, Ferrandon S, Evans TC, Xu MQ, Rao Z (2005) Crystal structures of an intein from the split *dnaE* gene of *Synechocystis* sp. PCC6803 reveal the catalytic model without the penultimate histidine and the mechanism of zinc Ion inhibition of protein splicing. J Mol Biol 353:1093–1105

Sun W, Yang J, Liu XQ (2004) Synthetic two-piece and three-piece split inteins for protein *trans*-splicing. J Biol Chem 279:35281–35286

Terpe K (2003) Overview of tag protein fusions: from molecular and biochemical fundamentals to commercial systems. Appl Microbiol Biotechnol 60:523–533

Theodoro RC, Bagagli E, Oliveira C (2008) Phylogenetic analysis of PRP8 intein in *Paracoccidioides brasiliensis* species complex. Fungal Genet Biol 45:1284–1291

Tyszkiewicz AB, Muir TW (2008) Activation of protein splicing with light in yeast. Nat Methods 5:303–305

Wende W, Grindl W, Christ F, Pingoud A, Pingoud V (1996) Binding, bending and cleavage of DNA substrates by the homing endonuclease Pl-*Sce*I. Nucleic Acids Res 24:4123–4132

Werner E, Wende W, Pingoud A, Heinemann U (2002) High resolution crystal structure of domain I of the *Saccharomyces cerevisiae* homing endonuclease PI-*Sce*I. Nucleic Acids Res 30:3962–3971

Wood DW, Wu W, Belfort G, Derbyshire V, Belfort M (1999) A genetic system yields self-cleaving inteins for bioseparations. Nat Biotechnol 17:889–892

Woods JP (2002) *Histoplasma capsulatum* molecular genetics, pathogenesis, and responsiveness to its environment. Fungal Genet Biol 35:81–97

Wu H, Xu MQ, Liu XQ (1998a) Protein trans-splicing and functional mini-inteins of a cyanobacterial dnaB intein. Biochim Biophys Acta 1387:422–432

Wu H, Hu Z, Liu XQ (1998b) Protein trans-splicing by a split intein encoded in a split DnaE gene of Synechocystis sp. PCC6803. Proc Natl Acad Sci USA 95:9226–9231

Xu MQ, Southworth MW, Mersha FB, Hornstra LJ, Perler FB (1993) In vitro protein splicing of purified precursor and the identification of a branched intermediate. Cell 75:1371–1377

4 Apoptosis in Fungal Development and Ageing

Diana Brust[1], Andrea Hamann[1], Heinz D. Osiewacz[1]

CONTENTS

I. General Description of Apoptosis

A. Apoptosis in Mammals

The term apoptosis, which originates from the ancient Greek word describing the fall of leaves from trees, was first suggested by Kerr et al. (1972) to describe the process of physiological cell death in liver cells. Later, it was more generally used as a functional definition of one type of programmed cell death (PCD). The process is characterised by specific morphological features including cell shrinkage, chromatin condensation and membrane blebbing, ultimately leading to phagocytosis of affected cells (reviewed by Hengartner 2000). The induction of apoptosis depends on specific signals from the extracellular (extrinsic) or the intracellular (intrinsic) micro-environment (Fig. 4.1). Stimulation of death receptors of the tumor necrosis factor (TNF) receptor superfamily in the plasma membrane results in activation of the extrinsic signalling pathway. After a death ligand is bound to a specific receptor, initiator caspase-8 becomes activated and transmits the apoptotic signal by cleavage of effector caspase-3 (reviewed by Walczak and Krammer 2000). Subsequently, activated caspase-3 leads to the cleavage of specific substrates, including the inhibitor of the caspase-activated DNase (ICAD) in the cytoplasm and the poly(ADP)ribose-polymerase (PARP) in the nucleus (reviewed by Degterev et al. 2003).

The intrinsic pathway of apoptosis is closely connected to mitochondria. Molecular damage of these organelles leads to mitochondrial outer membrane permeability (MOMP), which facilitates the release of apoptotic factors which are normally located in the intermembrane space. Two basic mechanisms inducing MOMP have been described. First, pore formation via the permeability transition pore complex (PTPC), a complex constituted by outer and inner membrane proteins as well as proteins in the matrix. Primarily the complex is formed by the voltage-dependent anion channel (VDAC), also termed porin, in the outer mitochondrial membrane, the adenine nucleotide transporter (ANT) in the inner membrane and cyclophilin D in the mitochondrial matrix, which allows water and solutes up to 1.5 kDa to cross through the outer and inner mitochondrial membrane. PTPC opening is highly sensitive to calcium ions and pro-oxidant agents. The opening results in mitochondrial swelling followed by rupture of the outer mitochondrial membrane. Second, MOMP can result from the formation of pores in the outer membrane by members of the BAX protein family which integrate into the mitochondrial outer membrane promoted by a plethora of apoptotic stimuli (reviewed by Kroemer et al. 2007).

The mammalian protein B-cell lymphoma 2 (BCL-2) is the prototype of a protein superfamily with over 20

[1]Institute for Molecular Biosciences, Department of Biosciences and Cluster of Excellence Macromolecular Complexes, Goethe-University, Max-von-Laue-Strasse 9, 60438 Frankfurt, Germany; e-mail: Osiewacz@bio.uni-frankfurt.de

Physiology and Genetics, 1st Edition
The Mycota XV
T. Anke and D. Weber (Eds.)
© Springer-Verlag Berlin Heidelberg 2009

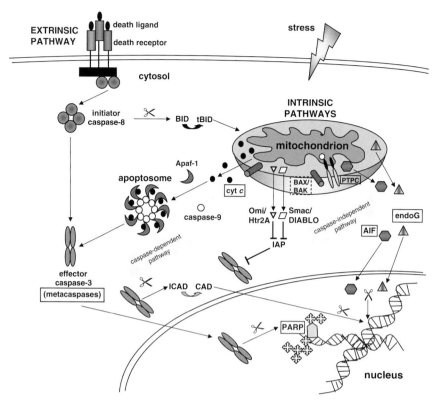

Fig. 4.1. Schematic representation of apoptotic signalling pathways in mammals. The extrinsic pathway is characterised by death receptors. After binding of death ligands, initiator caspase-8 becomes activated resulting in the cleavage of effector caspase-3 (reviewed by Walczak and Krammer 2000). Activated caspase-3 is able to cleave specific substrates like ICAD (inhibitor of the caspase-activated DNase, CAD) in the cytoplasm and the poly(ADP)ribose-polymerase (PARP) in the nucleus (reviewed by Degterev et al. 2003) triggering DNA damage. The intrinsic pathway depends on the release of apoptotic factors from mitochondria by formation of BAX/BAK pores or permeability transition pore complexes (PTPC) (reviewed by Kroemer et al. 2007). Liberated AIF and endoG translocate to the nucleus and execute the caspase-dependent pathway of apoptosis demonstrating common apoptotic markers like DNA fragmentation and chromatin condensation. The caspase-dependent pathway is characterised by released cytochrome c (cyt c) which is involved in the formation of the apoptosome consisting of Apaf-1, caspase-9 and cyt c. The apoptosome has the ability to activate effector caspase-3 initiating apoptosis as described before. Smac/DIABLO and Omi/Htr2A also released from mitochondria neutralise the negative effects of IAP (inhibitor of apoptosis proteins) on caspase-3 activity in the cytosol (reviewed by Saelens et al. 2004). The extrinsic and intrinsic pathways are linked by BID. Once BID is cleaved by initiator caspase-8, the cleavage product tBID supports the liberation of apoptosis proteins from mitochondria (reviewed by Cory and Adams 2002). Several factors of different apoptosis pathways also found in fungi are *boxed*. The *dashed box* of BAX/BAK implies that no structure homologues were found in fungi but heterologous expression of the mammalian proteins affects the cells in a pro-apoptotic way as well (e.g. Jürgensmeier et al. 1997)

relatives. Members of this family share at least one conserved BCL-2 homology (BH) domain. BCL-2 related proteins are divided into three subfamilies due to the number of their BH domains: (i) members of the BCL-2 family possess four BH domains, (ii) those of the BAX family three BH domains and (iii) the proteins of the BH3-only family contain only the shortest of the four BH motifs. While members of the BCL-2 family display an anti-apoptotic activity, proteins of the BAX and BH3-only family are pro-apoptotic. BH3-only proteins seem to act as direct antagonists of their anti-apoptotic homologues, whereas the integration of BAX proteins into the outer mitochondrial membrane leads to MOMP (reviewed by Cory and Adams 2002).

Once MOMP is triggered, pro-apoptotic molecules migrate from mitochondria into the cytosol, inducing two different signalling pathways: a caspase-dependent and a caspase-independent pathway. The main mediator of the caspase-dependent pathway is cytochrome c which directly initiates the activation of caspase-3 after leaving the mitochondrial intermembrane space via the formation of the apoptosome complex. Smac/DIABLO and Omi/Htr2A, two proteins which are also released from mitochondria, neutralise the negative effects of IAP (inhibitor of apoptosis proteins) on caspase-3 activity in the cytosol (reviewed by Saelens et al. 2004). The receptor and the mitochondrial pathway are connected via BID, a member of the BH3-only family (reviewed by Cory and Adams 2002). Activated caspase-8 of the extrinsic pathway is able to cleave BID, leading to the generation of tBID which subsequently migrates to mitochondria and causes cytochrome *c* release. The caspase-independent pathway is executed after the liberation of apoptosis-inducing factor (AIF) and endonuclease G (endoG) from mitochondria (reviewed by Saelens et al. 2004). After translocation of these proteins to the nucleus they cause DNA fragmentation and chromatin condensation, two typical apoptosis markers in mammals.

B. Apoptosis in Fungi

More than ten years ago the first evidence of a molecular apoptotic machinery emerged in *Saccharomyces cerevisiae* when Madeo et al. (1997) described typical apoptotic markers like chromatin condensation and DNA fragmentation. Subsequently, evidence for apoptosis was also reported in other fungi. For instance, in *Aspergillus fumigatus* the typical apoptotic DNA laddering was noticed after the application of stress to growing cultures (Mousavi and Robson 2004). The TUNEL assay was successfully applied to visualise DNA strand breaks in different fungal systems. For example, in yeast DNA breaks were demonstrated in a mutant defective in cell cycle (Madeo et al. 1997), in *A. fumigatus* protoplasts from mycelia entering the stationary phase tested TUNEL-positive (Mousavi and Robson 2003), and the TUNEL assay was also successfully applied in hyphae of *Colletotrichum trifolii* (Chen and Dickman 2005).

Furthermore, another apoptotic characteristic of mammalian cells, the relocalisation of phosphatidylserine (PS) from the inner to the outer layer of the plasma membrane was demonstrated by annexin V staining in *S. cerevisiae* (Madeo et al. 1997) and in several filamentous fungi (e.g. Hamann et al. 2007; Richie et al. 2007).

One of the most common apoptotic features is clearly the release of cytochrome *c* from mitochondria to trigger the activation of caspases, the main players of mammalian apoptosis. Until now, translocation of cytochrome *c* to the cytosol could only be demonstrated in yeast cells after treatment with acetic acid (Ludovico et al. 2002) but not in filamentous fungi. Silva et al. (2005) reported that deletion strains of cytochrome *c* are more resistant against hyperosmotic stress than the wild type, supporting the idea that cytochrome *c* plays a role in fungal apoptosis. Significantly, although cytochrome *c* release has not been demonstrated yet in filamentous fungi, several reports suggest caspase-like activity in total protein extracts from different fungi using available mammalian caspase substrates (Madeo et al. 2002; Thrane et al. 2004; Richie et al. 2007). Interestingly, contemporary degradation of PARP was observed in the presence of fungal protein extracts containing high caspase-like activity (Thrane et al. 2004). Specific proteases, termed metacaspases and sharing partial sequence homology with mammalian caspases, were suggested to be responsible for caspase-like activity determined in fungal cell extracts. However, today there is strong evidence for differences in the cleavage site in metacaspases and mammalian caspases (reviewed by Vachova and Palkova 2007). While caspases cleave aspartate-containing peptides, optimal cleavage activity of some metacaspases was detected using substrates with an arginine residue on P1 position of the peptide substrate (Gonzalez et al. 2007; Hamann et al. 2007).

In addition to a (meta)caspase-dependent pathway, a caspase-independent pathway of apoptosis appears to be active in fungi. Generally, in yeast AIF and endoG homologues were identified and characterised, indicating the same function of these proteins as in humans (Wissing et al. 2004; Büttner et al. 2007). Finally, it has been found that components of PTPC are well conserved in eukaryotes. Surprisingly, a yeast porin deletion strain was found to be more sensitive to acetic acid than the wild type. In contrast, deletion of the three ANT-encoding genes of *S. cerevisiae* resulted in a higher stress resistance, while deletion of the yeast cyclophilin D encoding gene displayed no effect on acetic acid-induced apoptosis (Pereira et al. 2007). However, the yeast cyclophilin D homologue seems to mediate the copper-induced apoptotic programme (Liang and Zhou 2007). Interestingly, in spite of the fact that fungi seem to lack sequence homologues of the BCL-2 family, heterologous expression in yeast of genes encoding the mammalian proteins let to pro- and anti-apoptotic effects, respectively (Jürgensmeier et al. 1997; Manon et al. 1997). It is thus possible that fungi possess functional homologues of the BCL-2 superfamily.

C. Differences Between Fungal and Mammalian Apoptosis

Although mammals and fungi share several apoptotic factors demonstrating programmed cell death, in lower eukaryotes the signalling pathways appear to be much simpler than in mammals (Fig. 4.1). Until now, the characteristic death receptors triggering the extrinsic pathway of mammalian apoptosis have not been identified in fungi. At most, mating pheromones might be considered as exterior signals recognized by specific G protein-coupled receptors which normally activate a well defined MAP kinase signalling cascade (Schrick et al. 1997). High pheromone concentrations lead to apoptotic cell death (Severin and Hyman 2002). In contrast, intrinsic pathways of apoptosis are clearly active in fungi, but in a simplified version when compared to mammals. Although a release of cytochrome c was at least demonstrated in yeast, the formation of the apoptosome could not be detected. In fungi metacaspases cleaving proteins at a site differing from those of mammalian caspases are part of the caspase-dependent intrinsic pathway (reviewed by Vachova and Palkova 2007). Interestingly, until now no more than two metacaspase genes could be identified in fungal genomes, while humans contain a set of caspases with different tasks involved in apoptotic and inflammatory processes (reviewed by Degterev et al. 2003). Similar to mammals, yeast contains an IAP homologue, called BIR1, which interacts with the Omi/Htr2A homologue NMA111 (Fahrenkrog et al. 2004; Walter et al. 2006). However, mammalian Omi/Htr2A is normally localised in mitochondria, while NMA111 is a nuclear protein, further stressing that mammalian and fungal apoptosis clearly differ in mechanistic details.

The caspase-independent pathways of apoptosis are poorly investigated in fungi. Only in yeast was evidence shown for AIF- and endoG-dependent cell death (reviewed by Liang et al. 2008). In the fungal kingdom no structural homologues of human BCL-2 proteins which play an important role in a pro- and anti- apoptotic manner could be detected, although this human protein class is active in yeast and other fungi

(reviewed by Eisenberg et al. 2007). Taken together, it is tempting to speculate about the rudimentary apoptosis signalling pathways of fungi as an evolutionary precursor of mammalian apoptosis. But why and in which context did an apoptotic programme evolve at this early stage of evolution?

II. Apoptosis in Fungal Development

A. Apoptosis in Host–Pathogen and Antagonistic Interactions

The observations described above demonstrate that apoptotic cell death is an evolutionarily conserved programme which is a common feature of both higher and lower eukaryotes. Hallmarks of apoptosis can even be observed in prokaryotes. This kind of cell death is therefore suggested to be a consequence of mitochondrial endosymbiosis and additional horizontal gene transfer events (Koonin and Aravind 2002). It is suggestive to assume that the ability to undergo apoptosis provides some selective advantage. Indeed, some experimental data point to vital roles of fungal apoptotic pathways. One example is a competition experiment in which a selective advantage of *S. cerevisiae* cells capable of metacaspase 1-dependent apoptosis was described (Herker et al. 2004). In a mixture of wild-type and *Yca1*-deletion cells, cells of the deletion strain were outgrown during long-term cultivation by wild-type cells, suggesting that wild-type cells dying via apoptosis stimulate the survival of other wild-type cells. Alternatively, apoptotic clearance of aged or harmed cells leads to a fitter, better adapted population in the long run.

Moreover, during the past ten years, programmed cell death and especially apoptosis has been identified to play an important role in processes induced via the attack of a host by a potential fungal pathogen. One of the first host responses is the so-called oxidative burst, the production of high levels of reactive oxygen species (ROS).

The oxidative burst (also known as respiratory burst in phagocytes of mammals) is defined as rapid production of high levels of ROS in response to external stimuli. It provides the first line of defence against pathogen attack. Upon recognition of a microbial pathogen, a plant or an animal host starts to release ROS. These ROS (mostly superoxide anion and hydrogen peroxide) are produced by different host enzyme systems. For instance, NADPH oxidases produce superoxide anions, which are converted to hydrogen peroxide by superoxide dismutases (SODs). Hydrogen peroxide can serve as precursor for hydroxyl radical generation via Fenton chemistry. The different types of ROS are able to oxidize proteins, lipids and DNA. Beside the role of directly attacking pathogens, it is clear that ROS are also important signalling molecules inducing several host pathways (Forman and Torres 2002; Gwinn and Vallyathan 2006).

Experimental data suggest that the oxidative burst might induce apoptosis in the pathogen and/or in the host. Animal pathogens often target and suppress the host's cell death pathways with different compounds to successfully establish infection (Table 4.1). The induction of programmed cell death is a critical step in the development of a successful infection and it is thus not surprising that the triggering of apoptosis is highly relevant for both host and pathogen.

One example for the role of apoptosis in the interaction of a pathogen with its host is the infection of humans by *A. fumigatus*. An early response to fungal attack is the induction of an oxidative burst directed against the germinating spores. Experimental data suggest that an apoptotic pathway is induced in this interaction. *A. fumigatus* cultures treated with 0.1 mM hydrogen peroxide display the classic hallmarks of apoptosis, including PS externalisation and DNA fragmentation (Mousavi and Robson 2004). This treatment might mimic the physiological scenario in the

pathogen. Comparable ROS levels were measured during oxidative bursts in *Arabidopsis thaliana*, revealing maximal levels of 0.025 mM hydrogen peroxide (Davies et al. 2006). The induction of apoptosis by hydrogen peroxide in *Aspergillus fumigatus* can completely be blocked by the addition of the protein synthesis inhibitor cycloheximide demonstrating dependency of this type of PCD on de novo protein synthesis.

It seems that not only does the host use apoptosis to get rid of the pathogen, but also the pathogen utilizes apoptosis to escape being killed by immune cells. Gliotoxin, a non-ribosomal peptide produced by *A. fumigatus* has been demonstrated to display apoptotic properties (Pahl et al. 1996; Suen et al. 2001; Kweon et al. 2003). The supernatant of *A. fumigatus* cultures, in which *gliZ* (encoding a transcriptional regulator of gliotoxin production) was deleted, was found to have a significantly lower ability to induce apoptotic cell death in polymorphonuclear leukocytes than wild-type supernatant (Bok et al. 2006). Obviously, successful establishment of infection in animal hosts depends strongly on the ability of the pathogen to escape killing by the reactive oxygen species released by immune cells during the oxidative burst. Moreover, to support the infection, the pathogen might profit via the ability to induce apoptosis specifically in those cells that are able to kill him, like e.g. neutrophils. The outcome of an infection event thus clearly depends on the

Table 4.1. Examples of apoptosis-inducing/inhibiting compounds implicated in interactions of fungi with other species

Compounds	Type of interaction	Putative biological function	References
Acetic acid	Antagonistic (acetic acid secreting bacteria vs yeast)	Selective advantage, induces apoptotic phenotype	Ludovico et al. (2002)
Farnesol	Antagonistic (*Candida albicans* vs *Aspergillus nidulans* and *Fusarium graminearum*)	Selective advantage, induces apoptotic phenotype	Semighini et al. (2006a, 2008)
Gliotoxin	Host (human)/pathogen (*A. fumigatus*)	Pathogen attack to eliminate immune response	Pahl et al. (1996), Suen et al. (2001), Kweon et al. (2003)
Killer toxins	Antagonistic (different yeast species vs other yeast species)	Selective advantage, induces apoptotic phenotype in susceptible yeasts	Klassen and Meinhardt (2005), Reiter et al. (2005)
PAF	Antagonistic (*Penicillium chrysogenum* vs *A. nidulans*)	Selective advantage, induces apoptotic phenotype	Leiter et al. (2005)
ROS	Host (human)/pathogen (*A. fumigatus*)	Respiratory burst, host defence mechanism	Mousavi and Robson (2004)
α-Tomatine	Host (tomato)/pathogen (*F. oxysporum*)	Host defence, induces apoptosis in the pathogen	Ito et al. (2007)
Transferrin	Host (*Protaetia brevitarsis*)/pathogen (*Beauveria bassiana*)	Prevents host from pathogen-induced apoptosis	Kim et al. (2008)

balance between apoptosis resistance and the ability to induce apoptosis.

Another example of a pathogenic fungus able to induce apoptotic cell death in the animal host is *Beauveria bassiana,* a common and serious insect pathogen. Recent studies with the white-spotted flower chafer (*Protaetia brevitarsis*) revealed an important role of the insect transferrin protein in apoptosis prevention. Insect larvae with reduced transferrin levels displayed a remarked increase in apoptosis induction upon exposure to different stressors, like heat stress and hydrogen peroxide stress, but also to fungal challenge (Kim et al. 2008). These data again support the idea that prevention of apoptosis is directly linked to host survival.

In plant/pathogen interactions the role of apoptosis is different. In plants attacked by a pathogen in principle two different scenarios can be observed. In one scenario, in plants which are able to recognise substances elicited during the pathogen attack, a defence reaction is induced, the so-called hypersensitive response, leading to the induction of programmed cell death in the infected parts. This response results in disease resistance. It starts with an oxidative burst and is thought not only to induce cell death in the host plant but also to act as signal and to attack the pathogen. In contrast, in plants that do not express this type of disease resistance, the pathogen is able to establish an infection (Greenberg and Yao 2004). Such a susceptible interaction also can depend on host cell death. Especially in interactions between plants and necrotrophic fungi the induction of programmed cell death in the host is clearly promoting pathogen growth (reviewed by Greenberg and Yao 2004).

ROS are not the only substances which can be released by plants and are directed against the pathogen (Table 4.1). One example is α-tomatine, a glycoside produced by tomatoes after infection with the pathogen *Fusarium oxysporum.* α-Tomatine efficiently induces cell death in the fungus. The process is accompanied by DNA fragmentation, depolarisation of the mitochondrial transmembrane potential and ROS accumulation. Treatment with the F_0F_1-ATPase inhibitor oligomycin, the protein synthesis inhibitor cycloheximide, or the caspase inhibitor D-VAD-fmk blocks apoptosis. This kind of cell death is thus not only dependent on de novo protein synthesis but also includes mitochondrial-dependent signalling and energy metabolism (Ito et al. 2007). Based on these data, it can be assumed that, in such a type of interaction, an increase in resistance of the fungus to undergo apoptosis improves the pathogen's capability to infect the host. In fact, this was clearly demonstrated by heterologous expression of the gene encoding the anti-apoptotic BCL-2 protein in the plant pathogen *Colletotrichum gloeosporioides.* Transgenic strains were hypervirulent and displayed an increased resistance against stresses induced by temperature, ultraviolet (UV) light and hydrogen peroxide (Barhoom and Sharon 2007). Vice versa, heterologous expression of the pro-apoptotic BAX protein induced apoptotic cell death in this fungus. Moreover, the few viable spores were only able to germinate in a non-pathogenic way with two germ tubes, even under conditions that should induce pathogenic germination (one germ tube) (Barhoom and Sharon 2007).

Apart from a role in host–pathogen interactions, PCD is also induced in antagonistic interactions, interactions in which fungi compete for nutrients with other species growing on the same substrate. Secretion of substances inducing apoptotic processes in a competitor provides a major selective advantage. A variety of substances of this type have been described and studied in some detail (Table 4.1). One example is the antifungal protein PAF produced by *Penicillium chrysogenum.* PAF efficiently induces an apoptotic phenotype in *A. nidulans* accompanied by hyperpolarisation of the plasma membrane, PS externalisation and the accumulation of DNA strand breaks (Leiter et al. 2005). Recently, it was postulated that selective ion permeability of the plasma membrane is relevant for the observed morphological changes in susceptible PAF-treated fungi (Marx et al. 2008). PAF and other similar secreted peptides produced by *Aspergillus* spp. (Gun et al. 1999; Theis et al. 2003) are not directed against a specific pathogen but efficiently inhibit growth of a large number of different fungi. Since neither PAF nor AFP, an antifungal peptide produced by *Aspergillus giganteus*, show detrimental effects on mammalian cells (Szappanos et al. 2005, 2006), these antifungal proteins are promising candidates for the development of therapies against human pathogenic fungi (reviewed by Marx et al. 2008). Moreover, PAF in concert with other antifungal compounds is able to inhibit growth of several members of the class zygomycetes, members of which are important post-harvest pathogens of agricultural products (Galgoczy et al. 2007).

Another well analysed apoptosis-inducing agent is farnesol, a molecule produced by *Candida albicans*. Farnesol accumulates at high cell density and mediates quorum sensing, thereby preventing *C. albicans* from switching from yeast to hyphal growth (Hornby et al. 2001). In a recent study, farnesol also provided a selective advantage in antagonistic interactions (Semighini et al. 2006a). Co-cultivation of *C. albicans* with *A. nidulans* displayed farnesol-dependent growth impairments and characteristic features of apoptosis, including nuclear condensation, accumulation of DNA strand breaks, PS externalisation and increased ROS production. Interestingly, the induction of apoptosis in *A. nidulans* by farnesol was dependent on poly(ADP-ribose)polymerase (PARP; Semighini et al. 2006b), an enzyme that in higher organisms is involved in DNA damage recognition, DNA repair and induction of apoptosis. Recently it has been demonstrated that the farnesol-dependent induction of apoptosis is not restricted to *A. nidulans* but is also observed in the phytopathogen *Fusarium graminearum*. Treatment of growing hyphae with 100–300 μM farnesol leads to nuclear condensation, DNA strand breaks and elevated ROS levels (Semighini et al. 2008).

One interesting type of antagonistic interaction of fungi, the so-called killer phenotype, is observed in several yeast species, e.g. in *Saccharomyces cerevisiae*, *Kluyveromyces lactis* (reviewed by Schmitt and Breinig 2006), in members of the genus *Pichia* and in *Wingea robertsiae* (Klassen and Meinhardt 2003; Klassen et al. 2004). The induction of cell death by these toxins appears to be linked to the execution of apoptotic processes. Depending on the genotype, some natural strains

of these yeasts produce a killer toxin that leads to the death of susceptible strains. The toxin-producing strains are resistant against their own toxin (reviewed by Schmitt and Breinig 2006). In a 2005 study, moderate doses of the *S. cerevisiae* toxins K1 and K28 and the *Zygosaccharomyces bailii* toxin zygocin trigger ROS production and apoptosis mediated via YCA1, the single metacaspase known in yeast, in susceptible *S. cerevisiae* strains (Reiter et al. 2005). The same holds true for the pPac1-2 toxin of *P. acaciae*. Treatment of *S. cerevisiae* with this toxin results in cell death accompanied by DNA strand breaks, PS externalisation and the formation of ROS (Klassen and Meinhardt 2005).

Not only fungi and other eukaryotes but also prokaryotes appear to be able to induce apoptosis in fungi. From investigations in which yeast cultures have been treated with acetic acid and found to undergo apoptosis, it has been speculated that acetic acid secreting bacteria have an advantage in nature when competing with yeast for available nutrients (Ludovico et al. 2002).

B. Apoptosis During Fungal Reproduction

Beside the well established role of apoptosis in fungal inter- and intraspecies interaction, only a few observations indicate a role of this suicide programme in fungal reproduction (Table 4.2). In yeast, the first step in sexual reproduction is the recognition of a pheromone secreted by a partner of one mating-type by a receptor on the surface of a partner of the opposite mating-type. Impairments of the subsequent step, fusion of two cells of the opposite mating-type, induce

Table 4.2. Influence of apoptosis factors on fungal development

Species	Apoptosis factor	Phenotype	References
A. nidulans	Over-expression of *Parp*	Impairment of conidiophore development	Semighini et al. (2006b)
C. gloeosporioides	Expression of human anti-apoptotic *Bcl-2*	Increased conidiospore production	Barhoom and Sharon (2007)
C. cinereus	Four different types of mutations	Impairment of spore formation, basidia-specific apoptosis	Lu (2000), Lu et al. (2003)
P. anserina	Deletion of both metacaspase genes	Impairment of fruiting body development	Hamann et al. (2007)
S. cerevisiae	pheromone treatment in the absence of mating partner	ROS production, DNA breakage, mitochondria-dependent apoptotic cell death	Severin and Hyman (2002)

mitochondria-dependent apoptotic cell death accompanied by ROS production and DNA breakage (Severin and Hyman 2002). Yeast cells of mating-type *a* were treated with the α pheromone of the opposite mating-type. After induction of the sexual reproduction pathway, the physical absence of the appropriate mating partner induced ROS production suggesting that failure of mating induces apoptosis. Death of cells has been suggested to represent an altruistic event eliminating in nature those cells that are unable to mate (Severin and Hyman 2002). Interestingly, both the formation of diploid yeast cells and the induction of apoptosis in cells induced by the α pheromone but not able to fuse with a mating partner are dependent on kinase STE20 function (Schrick et al. 1997; Severin and Hyman 2002). A mechanistic link of these two pathways, sexual reproduction and apoptosis, seems to be related to large-scale chromatin remodelling. Fragmentation of nuclear DNA during apoptosis is strictly related to phosphorylation of serine 10 of histone H2B, a process catalysed by STE20 (Ahn et al. 2005a). Interestingly, the same histone modification is observed during meiosis, the next step during sexual reproduction after the formation of diploids (Ahn et al. 2005b). H2B mutants mimicking a constitutive phosphorylation of Ser10 show a drastic reduction of growth. Surviving cells exhibit a cellular morphology similar to hydrogen peroxide-treated wild-type cells (Ahn et al. 2005a). Chromatin remodelling plasticity is speculated to be involved in cell survival (Ahn et al. 2005a). During apoptosis chromatin becomes condensed. This condensation ultimately leads to cell death. Significantly, cytological analyses revealed that chromatin condensation during meiosis is not a static event. Especially in prophase I of meiosis, chromatin globally cycles between expanded and contracted stages (reviewed by Kleckner et al. 2004). It thus appears that this type of cycling inhibits the induction of apoptosis. However, a more recent study questions the role of apoptosis in pheromone-induced cell death. In this study, the effect of a range of pheromone concentrations on yeast cells was investigated (Zhang et al. 2006). High pheromone concentrations induced rapid cell death, requiring the presence of the transmembrane protein FIG1, which promotes Ca^{2+}-independent cell death. Lower concentrations of pheromone induced a second wave of cell death and were dependent on cell wall degradation. Significantly,

several hallmarks of apoptosis were not observed in cells undergoing death (Zhang et al. 2006).

Additional data suggest an impact of apoptosis on development of fungi other than yeast. In the basidiomycete *Coprinopsis cinereus* four different types of mutants impaired in meiosis or spore formation were isolated and analysed. In group one, meiosis can be completed; however, the formation of spores is affected. In this group, apoptotic processes are induced at the tetrad stage of basidiospore formation. The three other groups of mutants have in common a defect in prophase of meiosis. In all groups typical markers of apoptosis were observed, including chromatin condensation, DNA fragmentation, cytoplasmic shrinking and DNA degradation (Lu et al. 2003). In another fungus, the filamentous ascomycete *Podospora anserina*, circumstantial evidence for a role of apoptosis factors in meiospore development was obtained in a mutant in which two metacaspase genes were deleted. This mutant is severely impaired in the formation of ascospores and fruiting bodies (Hamann et al. 2007).

Finally, some recent data also suggest a role of apoptotic processes in asexual reproduction. Using fungal extracts of conidiospore producing cultures of *A. nidulans*, in vitro degradation of PARP was demonstrated (Thrane et al. 2004). In these cultures, the amount of PARP is increased (Thrane et al. 2004) due to a strong transcriptional upregulation (Semighini et al. 2006b). A regulated expression of this gene appears to be essential for conidiospore development because overexpression of *Parp* leads to impairments of conidiophore development. As a consequence, a $\sim 10^4$-fold reduction in spore numbers was found in the corresponding mutants (Semighini et al. 2006b). Currently it is not clear whether the influence of PARP is due to apoptotic processes or is the result of an indirect role of this enzyme. However, a direct involvement of apoptosis in mitospore formation is supported by experiments with the plant pathogen *C. gloeosporioides*: overexpression of the gene encoding the human anti-apoptotic BCL-2 protein not only led to hypervirulence but also to an enhanced conidiospore production (Barhoom and Sharon 2007).

Beside the role in reproduction, apoptosis and autophagy also play an important role in fungal non-self-recognition between different isolates of the same species. After fusion of two genetically incompatible hyphae, a cell death reaction, termed 'heterokaryon incompatibility', is induced (reviewed by Glass and Dementhon 2006). In *Neurospora crassa* this reaction displays clear characteristics of apoptosis (Marek et al. 2003). In *P. anserina*, autophagy is strongly induced during the incompatibility reaction, antagonizing pro-death signals (Pinan-Lucarré et al. 2003; Pinan-Lucarré et al. 2005; reviewed by Pinan-Lucarré et al. 2007). Interestingly, *P. anserina* autophagy mutants are impaired in female fertility, linking this process to sexual development (reviewed by Pinan-Lucarré et al. 2007).

C. The Role of Apoptosis in Fungal Lifespan Control

While it is clear that apoptosis plays an important role in differentiation and development, much less is known about its specific role in ageing and lifespan control of biological systems. Only in the past decade have some clues for such a role emerged, some of which have been derived from investigations of fungi as experimentally tractable ageing models. In general, these studies revealed a connection between age-related symptoms and apoptosis induction (Table 4.3). Prevention of apoptosis has a clear lifespan-prolonging effect. One example for this connection is the increased viability of yeast cells in stationary phase that express the human

Bcl-2 gene (Longo et al. 1997). Another is that chronologically aged yeast cultures display characteristic markers of apoptosis, including DNA fragmentation, PS externalisation and chromatin condensation, markers which are not observed in exponentially growing cultures (Fabrizio et al. 2004; Herker et al. 2004). The involvement of apoptosis in chronological ageing is further stressed by the observation that overexpression of *Yap1*, one key regulator of tolerance against oxidative stress, leads to increased survival of the transgenic strains in stationary culture (Herker et al. 2004). The same holds true when the gene encoding yeast metacaspase YCA1 is deleted (Herker et al. 2004). Moreover, prolonged cultivation of *Yca1* overexpressing strains results in a pronounced reduction in viability after 44 h or 96 h of cultivation, accompanied by the appearance of different apoptotic markers. This effect is not observed when a mutant form of *Yca1* containing a serine residue instead of a cysteine in the catalytic centre of the protein is overexpressed (Madeo et al. 2002). During the past five years, additional findings have accumulated demonstrating that different apoptotic pathways are involved in the control of chronological ageing in yeasts (Wissing et al. 2004; Li et al. 2006; Barhoom and Sharon 2007; Büttner et al. 2007).

Interestingly, markers of apoptosis including PS externalisation and DNA strand breaks occur

Table 4.3. Factors implicated in apoptosis and fungal lifespan control

Species	Factor	Influence on lifespan	References
C. gloeosporioides	Expression of human anti-apoptotic Bcl-2	Increase in chronological lifespan	Barhoom and Sharon (2007)
P. anserina	Deletion of metacaspases	Increase in replicative lifespan	Hamann et al. (2007)
	Deletion of fission factor *PaDnm1*	Increase in replicative lifespan	Scheckhuber et al. (2007)
S. cerevisiae	Expression of human anti-apoptotic Bcl-2	Increase in chronological lifespan	Longo et al. (1997)
	Over-expression of *Yca1*	Increase in chronological lifespan	Madeo et al. (2002)
	Deletion of *Mmi1*	Increase in replicative lifespan	Rinnerthaler et al. (2006)
	Deletion of fission factors *Dnm1* and *Fis1*	Increase in replicative lifespan	Scheckhuber et al. (2007)
	Over-expression of *EndoG*	Decrease in chronological lifespan	Büttner et al. (2007)
	Deletion of *Aif1*	Increase in chronological lifespan	Wissing et al. (2004)
	Disruption of *Ndi1*	Increase in chronological lifespan	Li et al. (2006)
	Deletion of *Kex1*	Increase in chronological lifespan	Hauptmann and Lehle (2008)
	Deletion of *Isc1*	Decrease in chronological lifespan	Almeida et al. (2008)

also during replicative ageing of yeast cells (Laun et al. 2001). Moreover, an increase of yeast replicative lifespan by deletion of the yeast gene coding for the homologue of TCTP (a human protein interacting with BCL-XL) was reported, (Rinnerthaler et al. 2006). The details of the underlying mechanism are poorly understood.

Recently, data on the impact of apoptosis on ageing were reported in another fungal ageing model, *P. anserina*. Since this process occurs during active growth of cultures and depends on replication of the genetic material, it also represents a type of replicative ageing. In *P. anserina*, ageing can easily be followed macroscopically by analysing growth of cultures on solid medium. Depending on the genetic background and culture conditions, the mean lifespan of a wild-type culture is two to four weeks. Various molecular pathways are involved in lifespan control in this fungus. Among others these are: the type of respiration (Dufour et al. 2000; Borghouts et al. 2001; Lorin et al. 2001; Krause et al. 2004; Stumpferl et al. 2004; Sellem et al. 2005; Krause et al. 2006; Sellem et al. 2007), mitochondrial DNA stability (Stahl et al. 1978; Cummings et al. 1979; Belcour et al. 1981; Kück et al. 1981; Osiewacz and Esser 1984; Kück et al. 1985), dietary restriction (Tudzynski and Esser 1979; Maas et al. 2004; Maas et al. 2007), copper metabolism (Borghouts et al. 1997; Borghouts and Osiewacz 1998; Borghouts et al. 2000; Stumpferl et al. 2004), translation fidelity (Belcour et al. 1991; Silar and Picard 1994), mitochondrial protein import (Jamet-Vierny et al. 1997), mitochondrial dynamics (Jamet-Vierny et al. 1997; Scheckhuber et al. 2007). Interestingly, recent investigations identified apoptotic machinery in *P. anserina* which also was found to be involved in lifespan control. Deletion of the genes encoding the two metacaspases, *PaMca1* and *PaMca2*, resulted in an increase in mean lifespan, suggesting that apoptosis is also induced in the final stage of senescence in this fungus. Most pronounced is the effect of the *PaMca1* deletion, leading to a 2.5-fold lifespan increase. Importantly, protoplasts of 15-day-old cultures of the *PaMca1* deletion strain display a higher viability after treatment with hydrogen peroxide than protoplasts of the wild-type strain (Hamann et al. 2007). Moreover, *PaMca1* deletion leads to a pronounced resistance against the apoptosis-inducing agent etoposide (Scheckhuber et al. 2007), suggesting that metacaspases are involved in

apopototic cell death and fungal senescence. At least PaMCA1, one of the two metacaspases of *P. anserina*, becomes activated during senescence leading to apoptotic cell death. High levels of hydrogen peroxide can be detected in senescent mycelium and because this kind of ROS induces PS externalisation in *P. anserina*, PaMCA1 seems to be activated by ROS. *PaMca1* deletion strains are not characterised by reduced ROS levels, but rather by impairment to respond to ROS and to induce apoptosis (Hamann et al. 2007). This idea is supported not only by the lifespan extension of the deletion strains, but also by the demonstration of a metacaspase-dependent peptidase activity in senescent mycelium (Hamann et al. 2007).

One important issue that remains to be solved in order to better understand the impact of apoptosis on ageing and lifespan control is to unravel the intracellular signalling pathways involved in the induction of apoptosis. In yeast, the dynamics of the actin cytoskeleton plays an important role. First, a decrease in actin turnover, which can be experimentally induced by specific mutations in actin or treatment with actin-stabilising drugs, was observed in aged cells (Gourlay et al. 2004). Second, an increase in actin turnover via deletion of the *Scp1* gene encoding the actin bundling protein SCP1 leads to lifespan extension of dividing and non-dividing yeast cells (Gourlay et al. 2004). The role of actin dynamics is linked to the Ras-cAMP signalling, a pathway which is an important regulator of longevity. Constitutive Ras-cAMP activation impairs scavenging of mitochondrial deficiency and oxidative stress and reduces lifespan (Heeren et al. 2004). A link between Ras-cAMP signalling and actin is provided by the phenotype of yeast mutants lacking either of the two actin regulatory proteins SLA1 or END3. The characteristics of these strains resemble those of strains in which Ras-cAMP signalling is constitutive. The corresponding mutants are characterised by elevated ROS production, mitochondrial dysfunction and reduced longevity. Interestingly, ROS levels and lifespan of these strains can be reverted to wild-type characteristics via overexpression of *Pde1*, a negative regulator of the Ras-cAMP pathway (Gourlay and Ayscough 2005).

Recently, a direct link between oxidative stress and the actin cytoskeleton was obtained by analysis of the yeast old yellow enzymes OYE2 and OYE3. OYE2 was demonstrated to interact with actin (Haarer and Amberg 2004). An almost complete knockdown of *Oye2* resulted in defects in cytoskeletal organization (Haarer and Amberg 2004). Moreover, simultaneous deletion of *Oye2* and the gene encoding glutathione oxidoreductase GLR1 increases strongly the morphological defects (Odat et al. 2007), suggesting that OYE2 protects actin from oxidative damage. Based on these findings OYE proteins are considered to be important members of a signalling network connecting ROS generation, actin cytoskeleton dynamics and cell death.

In addition to actin cytoskeleton dynamics, microtubule dynamics seem to play a role in the induction of apoptosis. Deletion of *Mmi1*, the gene encoding the yeast homologue of the human BCL-XL binding protein TCTP, leads to increased resistance against hydrogen peroxide and a small, but significant increase in replicative lifespan (Rinnerthaler et al. 2006). This phenotype is accompanied by an increased sensitivity against the microtubule-destabilising agent benomyl suggesting that MMI1 interacts with microtubules.

Finally, another important contributor to a 'healthy' cellular status, dynamics of mitochondria, appears to be linked to apoptosis. This conclusion is derived from recent investigations in which the mitochondrial fusion and fission machinery was manipulated in both *S. cerevisiae* and *P. anserina* (Fannjiang et al. 2004; Scheckhuber et al. 2007). Deletion of genes encoding mitochondrial fission factors (DNM1 and FIS1 in yeast; DNM1 in *P. anserina*) retard ageing and increase 'health span', the healthy period of time within the life cycle of organisms. During the ageing of wild-type strains, mitochondria change their structure from networks to punctuate entities. However, this change occurs only in the very last period of the lifetime of this fungus. Although in yeast the deletion of fission factors has an effect on mean lifespan less pronounced than in *P. anserina*, yeast mutants maintain a generation time comparable to that of young cells, even after 30–40 generations, an indicator of a 'healthy' cell status. A clear link between mitochondrial dynamics and apoptotic processes is provided by an increased resistance of the *PaDnm1* deletion strain of *P. anserina* against etoposide, an elicitor of apoptosis independent of pro-apoptotic Ca^{2+} waves (Scheckhuber et al. 2007).

III. Concluding Remarks and Future Directions

Within the past decade, apoptosis in fungi emerged as a molecular mechanism that is linked to different biological processes. Of special applied interest is the impact of apoptotic processes of pathogenic fungi with their hosts because these kinds of interactions represent a major threat for human health as well as for food production. For example, in a recent study, the incidence of blood stream infections with species of the *Candida* genus was determined in a population-based active surveillance program of two areas in the United States (Hajjeh et al. 2004). Compared to earlier studies, the overall incidence of candidemia was higher. Another example is rice blast caused by the plant pathogen *Magnaporthe grisea*, which is responsible for 25% annual yield loss of rice production in Japan (Ribot et al. 2008). Research to understand the underlying basis of host-pathogen interactions (including the details involved in the induction of apoptotic processes) are of great relevance towards the development of efficient fungicides and effective antifungal therapies.

Apoptosis in fungi and mammals clearly share characteristic features; however, it is now also obvious that the underlying mechanisms differ in a number of details. As one example, fungi lack biochemical homologues of caspases, aspartate-specific cysteine proteases which are among the main players in mammalian apoptosis. Instead, they possess metacaspases, cysteine proteases which seem to originate from the same ancestral gene as caspases (Uren et al. 2000) and are probably functional homologues of caspases. Although classic caspase substrates are cleaved by fungal extracts (Madeo et al. 2002; Mousavi and Robson 2003; Thrane et al. 2004), it is unlikely that these substrates are metacaspase substrates. Plant, protozoan and fungal metacaspases display Arg-Lys protease specificity instead of the Asp specificity characteristic of caspases (Vercammen et al. 2004; Watanabe and Lam 2005; Gonzalez et al. 2007; Hamann et al. 2007). Obviously, additional work is required to elucidate the role of these proteases in fungal apoptosis in more detail. One important piece of information, the identity of the cellular substrate of metacaspases, could be obtained from a proteome analysis of wild-type and metacaspase-deletion strains. Because of the clear relationship between metacaspases and caspases, the study of fungal metacaspases could provide a key platform for understanding how apoptotic mechanisms evolved.

As described in this chapter, apoptotic pathways are linked to some other molecular and cellular pathways. However, the currently available data about these links are rather fragmented and descriptive. In particular, among other aspects, it will be important to elucidate the mechanistic details linking the role of the cytoskeleton, mitochondrial dynamics, ROS generating and/or scavenging pathways with apoptotic processes. In general, a more detailed view incorporating more of the individual components of the involved pathways will help to better understand apoptosis as a 'molecular module' of complex biological processes such as the interaction of fungi with other organisms, fungal development or ageing and lifespan control. Finally, investigations aimed at testing the general applicability of a newly elucidated pathway and its conservation seem to be relevant. In such approaches, microorganisms like fungi are experimentally tractable systems that can serve as model systems to decipher basic mechanisms, which – due to practical or ethical reasons – cannot be approached directly in higher systems.

Acknowledgements. The authors gratefully acknowledge the European Commission (LSHM-CT-2004-512020) and the Deutsche Forschungsgemeinschaft (Bonn, Germany) for supporting experimental work referenced from the authors' laboratory.

References

Ahn SH, Cheung WL, Hsu JY, Diaz RL, Smith MM, Allis CD (2005a) Sterile 20 kinase phosphorylates histone H2B at serine 10 during hydrogen peroxide-induced apoptosis in S. cerevisiae. Cell 120:25–36

Ahn SH, Henderson KA, Keeney S, Allis CD (2005b) H2B (Ser10) phosphorylation is induced during apoptosis and meiosis in S. cerevisiae. Cell Cycle 4:780–783

Almeida T, Marques M, Mojzita D, Amorim MA, Silva RD, Almeida B, Rodrigues P, Ludovico P, Hohmann S, Moradas-Ferreira P, Corte-Real M, Costa V (2008) Isc1p plays a key role in hydrogen peroxide resistance and chronological lifespan through modulation of iron levels and apoptosis. Mol Biol Cell 19:865–876

Barhoom S, Sharon A (2007) Bcl-2 proteins link programmed cell death with growth and morphogenetic adaptations in the fungal plant pathogen Colletotrichum gloeosporioides. Fungal Genet Biol 44:32–43

Belcour L, Begel O, Mosse MO, Vierny-Jamet C (1981) Mitochondrial DNA amplification in senescent cultures of Podospora anserina: Variability between the retained, amplified sequences. Curr Genet 3:13–21

Belcour L, Begel O, Picard-Bennoun M (1991) A site-specific deletion in mitochondrial DNA of Podospora is under the control of nuclear genes. Proc Natl Acad Sci USA 88:3579–3583

Bok JW, Chung D, Balajee SA, Marr KA, Andes D, Nielsen KF, Frisvad JC, Kirby KA, Keller NP (2006) GliZ, a transcriptional regulator of gliotoxin biosynthesis, contributes to Aspergillus fumigatus virulence. Infect Immun 74:6761–6768

Borghouts C, Kimpel E, Osiewacz HD (1997) Mitochondrial DNA rearrangements of Podospora anserina are under the control of the nuclear gene grisea. Proc Natl Acad Sci USA 94:10768–10773

Borghouts C, Osiewacz HD (1998) GRISEA, a copper-modulated transcription factor from Podospora anserina involved in senescence and morphogenesis, is an ortholog of MAC1 in Saccharomyces cerevisiae. Mol Gen Genet 260:492–502

Borghouts C, Kerschner S, Osiewacz HD (2000) Copper-dependence of mitochondrial DNA rearrangements in Podospora anserina. Curr Genet 37:268–275

Borghouts C, Werner A, Elthon T, Osiewacz HD (2001) Copper-modulated gene expression and senescence in the filamentous fungus Podospora anserina. Mol Cell Biol 21:390–399

Büttner S, Eisenberg T, Carmona-Gutierrez D, Ruli D, Knauer H, Ruckenstuhl C, Sigrist C, Wissing S, Kollroser M, Fröhlich KU, Sigrist S, Madeo F (2007) Endonuclease G regulates budding yeast life and death. Mol Cell 25:233–246

Chen C, Dickman MB (2005) Proline suppresses apoptosis in the fungal pathogen Colletotrichum trifolii. Proc Natl Acad Sci USA 102:3459–3464

Cory S, Adams JM (2002) The Bcl2 family: regulators of the cellular life-or-death switch. Nat Rev Cancer 2:647–656

Cummings DJ, Belcour L, Grandchamp C (1979) Mitochondrial DNA from Podospora anserina. II. Properties of mutant DNA and multimeric circular DNA from senescent cultures. Mol Gen Genet 171:239–250

Davies DR, Bindschedler LV, Strickland TS, Bolwell GP (2006) Production of reactive oxygen species in Arabidopsis thaliana cell suspension cultures in response to an elicitor from Fusarium oxysporum: implications for basal resistance. J Exp Bot 57:1817–1827

Degterev A, Boyce M, Yuan J (2003) A decade of caspases. Oncogene 22:8543–8567

Dufour E, Boulay J, Rincheval V, Sainsard-Chanet A (2000) A causal link between respiration and senescence in Podospora anserina. Proc Natl Acad Sci USA 97:4138–4143

Eisenberg T, Büttner S, Kroemer G, Madeo F (2007) The mitochondrial pathway in yeast apoptosis. Apoptosis 12:1011–1023

Fabrizio P, Battistella L, Vardavas R, Gattazzo C, Liou LL, Diaspro A, Dossen JW, Gralla EB, Longo VD (2004) Superoxide is a mediator of an altruistic aging program in Saccharomyces cerevisiae. J Cell Biol 166:1055–1067

Fahrenkrog B, Sauder U, Aebi U (2004) The *S. cerevisiae* HtrA-like protein Nma111p is a nuclear serine protease that mediates yeast apoptosis. J Cell Sci 117:115–126

Fannjiang Y, Cheng WC, Lee SJ, Qi B, Pevsner J, McCaffery JM, Hill RB, Basanez G, Hardwick JM (2004) Mitochondrial fission proteins regulate programmed cell death in yeast. Genes Dev 18:2785–2797

Forman HJ, Torres M (2002) Reactive oxygen species and cell signaling: respiratory burst in macrophage signaling. Am J Respir Crit Care Med 166:S4–S8

Galgoczy L, Papp T, Lukacs G, Leiter E, Pocsi I, Vagvolgyi C (2007) Interactions between statins and *Penicillium chrysogenum* antifungal protein (PAF) to inhibit the germination of sporangiospores of different sensitive Zygomycetes. FEMS Microbiol Lett 270:109–115

Glass NL, Dementhon K (2006) Non-self recognition and programmed cell death in filamentous fungi. Curr Opin Microbiol 9:553–558

Gonzalez IJ, Desponds C, Schaff C, Mottram JC, Fasel N (2007) *Leishmania major* metacaspase can replace yeast metacaspase in programmed cell death and has arginine-specific cysteine peptidase activity. Int J Parasitol 37:161–172

Gourlay CW, Ayscough KR (2005) Identification of an upstream regulatory pathway controlling actin-mediated apoptosis in yeast. J Cell Sci 118:2119–2132

Gourlay CW, Carpp LN, Timpson P, Winder SJ, Ayscough KR (2004) A role for the actin cytoskeleton in cell death and aging in yeast. J Cell Biol 164:803–809

Greenberg JT, Yao N (2004) The role and regulation of programmed cell death in plant–pathogen interactions. Cell Microbiol 6:201–211

Gun LD, Shin SY, Maeng CY, Jin ZZ, Kim KL, Hahm KS (1999) Isolation and characterization of a novel antifungal peptide from *Aspergillus niger*. Biochem Biophys Res Commun 263:646–651

Gwinn MR, Vallyathan V (2006) Respiratory burst: role in signal transduction in alveolar macrophages. J Toxicol Environ Health B Crit Rev 9:27–39

Haarer BK, Amberg DC (2004) Old yellow enzyme protects the actin cytoskeleton from oxidative stress. Mol Biol Cell 15:4522–4531

Hajjeh RA, Sofair AN, Harrison LH, Lyon GM, Arthington-Skaggs BA, Mirza SA, Phelan M, Morgan J, Lee-Yang W, Ciblak MA, Benjamin LE, Sanza LT, Huie S, Yeo SF, Brandt ME, Warnock DW (2004) Incidence of bloodstream infections due to *Candida* species and in vitro susceptibilities of isolates collected from 1998 to 2000 in a population-based active surveillance program. J Clin Microbiol 42:1519–1527

Hamann A, Brust D, Osiewacz HD (2007) Deletion of putative apoptosis factors leads to lifespan extension in the fungal ageing model *Podospora anserina*. Mol Microbiol 65:948–958

Hauptmann P, Lehle L (2008) Kex1 protease is involved in yeast cell death induced by defective N-glycosylation, acetic acid and during chronological aging. J Biol Chem 283:19151–19163

Heeren G, Jarolim S, Laun P, Rinnerthaler M, Stolze K, Perrone GG, Kohlwein SD, Nohl H, Dawes IW, Breitenbach M (2004) The role of respiration, reactive oxygen species and oxidative stress in mother cell-specific ageing of yeast strains defective in the RAS signalling pathway. FEMS Yeast Res 5:157–167

Hengartner MO (2000) The biochemistry of apoptosis. Nature 407:770–776

Herker E, Jungwirth H, Lehmann KA, Maldener C, Fröhlich KU, Wissing S, Büttner S, Fehr M, Sigrist S, Madeo F (2004) Chronological aging leads to apoptosis in yeast. J Cell Biol 164:501–507

Hornby JM, Jensen EC, Lisec AD, Tasto JJ, Jahnke B, Shoemaker R, Dussault P, Nickerson KW (2001) Quorum sensing in the dimorphic fungus *Candida albicans* is mediated by farnesol. Appl Environ Microbiol 67:2982–2992

Ito S, Ihara T, Tamura H, Tanaka S, Ikeda T, Kajihara H, Dissanayake C, bdel-Motaal FF, El-Sayed MA (2007) alpha-Tomatine, the major saponin in tomato, induces programmed cell death mediated by reactive oxygen species in the fungal pathogen *Fusarium oxysporum*. FEBS Lett 581:3217–3222

Jamet-Vierny C, Contamine V, Boulay J, Zickler D, Picard M (1997) Mutations in genes encoding the mitochondrial outer membrane proteins Tom70 and Mdm10 of *Podospora anserina* modify the spectrum of mitochondrial DNA rearrangements associated with cellular death. Mol Cell Biol 17:6359–6366

Jürgensmeier JM, Krajewski S, Armstrong RC, Wilson GM, Oltersdorf T, Fritz LC, Reed JC, Ottilie S (1997) Bax- and Bak-induced cell death in the fission yeast *Schizosaccharomyces pombe*. Mol Biol Cell 8:325–339

Kerr JF, Wyllie AH, Currie AR (1972) Apoptosis: a basic biological phenomenon with wide-ranging implications in tissue kinetics. Br J Cancer 26:239–257

Kim BY, Lee KS, Choo YM, Kim I, Je YH, Woo SD, Lee SM, Park HC, Sohn HD, Jin BR (2008) Insect transferrin functions as an antioxidant protein in a beetle larva. Comp Biochem Physiol B Biochem Mol Biol. 150:16–169

Klassen R, Meinhardt F (2003) Structural and functional analysis of the killer element pPin1-3 from *Pichia inositovora*. Mol Genet Genomics 270:190–199

Klassen R, Meinhardt F (2005) Induction of DNA damage and apoptosis in *Saccharomyces cerevisiae* by a yeast killer toxin. Cell Microbiol 7:393–401

Klassen R, Teichert S, Meinhardt F (2004) Novel yeast killer toxins provoke S-phase arrest and DNA damage checkpoint activation. Mol Microbiol 53:263–273

Kleckner N, Zickler D, Jones GH, Dekker J, Padmore R, Henle J, Hutchinson J (2004) A mechanical basis for chromosome function. Proc Natl Acad Sci USA 101:12592–12597

Koonin EV, Aravind L (2002) Origin and evolution of eukaryotic apoptosis: the bacterial connection. Cell Death Differ 9:394–404

Krause F, Scheckhuber CQ, Werner A, Rexroth S, Reifschneider NH, Dencher NA, Osiewacz HD (2004) Supramolecular organization of cytochrome c oxidase- and alternative oxidase-dependent respiratory chains in the filamentous fungus *Podospora anserina*. J Biol Chem 279:26453–26461

Krause F, Scheckhuber CQ, Werner A, Rexroth S, Reifschneider NH, Dencher NA, Osiewacz HD (2006)

OXPHOS Supercomplexes: respiration and life-span control in the aging model *Podospora anserina*. Ann NY Acad Sci 1067:106–115

Kroemer G, Galluzzi L, Brenner C (2007) Mitochondrial membrane permeabilization in cell death. Physiol Rev 87:99–163

Kück U, Stahl U, Esser K (1981) Plasmid-like DNA is part of mitochondrial DNA in *Podospora anserina*. Curr Genet 3:151–156

Kück U, Osiewacz HD, Schmidt U, Kappelhoff B, Schulte E, Stahl U, Esser K (1985) The onset of senescence is affected by DNA rearrangements of a discontinuous mitochondrial gene in *Podospora anserina*. Curr Genet 9:373–382

Kweon YO, Paik YH, Schnabl B, Qian T, Lemasters JJ, Brenner DA (2003) Gliotoxin-mediated apoptosis of activated human hepatic stellate cells. J Hepatol 39:38–46

Laun P, Pichova A, Madeo F, Fuchs J, Ellinger A, Kohlwein S, Dawes I, Fröhlich KU, Breitenbach M (2001) Aged mother cells of *Saccharomyces cerevisiae* show markers of oxidative stress and apoptosis. Mol Microbiol 39:1166–1173

Leiter E, Szappanos H, Oberparleiter C, Kaiserer L, Csernoch L, Pusztahelyi T, Emri T, Pocsi I, Salvenmoser W, Marx F (2005) Antifungal protein PAF severely affects the integrity of the plasma membrane of *Aspergillus nidulans* and induces an apoptosis-like phenotype. Antimicrob Agents Chemother 49:2445–2453

Li W, Sun L, Liang Q, Wang J, Mo W, Zhou B (2006) Yeast AMID homologue Ndi1p displays respiration-restricted apoptotic activity and is involved in chronological aging. Mol Biol Cell 17:1802–1811

Liang Q, Zhou B (2007) Copper and manganese induce yeast apoptosis via different pathways. Mol Biol Cell 18:4741–4749

Liang Q, Li W, Zhou B (2008) Caspase-independent apoptosis in yeast. Biochim Biophys Acta 1783:1311–1319

Longo VD, Ellerby LM, Bredesen DE, Valentine JS, Gralla EB (1997) Human Bcl-2 reverses survival defects in yeast lacking superoxide dismutase and delays death of wild-type yeast. J Cell Biol 137:1581–1588

Lorin S, Dufour E, Boulay J, Begel O, Marsy S, Sainsard-Chanet A (2001) Overexpression of the alternative oxidase restores senescence and fertility in a long-lived respiration-deficient mutant of *Podospora anserina*. Mol Microbiol 42:1259–1267

Lu BC (2000) The control of meiosis progression in the fungus *Coprinus cinereus* by light/dark cycles. Fungal Genet Biol 31:33–41

Lu BC, Gallo N, Kues U (2003) White-cap mutants and meiotic apoptosis in the basidiomycete *Coprinus cinereus*. Fungal Genet Biol 39:82–93

Ludovico P, Rodrigues F, Almeida A, Silva MT, Barrientos A, Corte-Real M (2002) Cytochrome c release and mitochondria involvement in programmed cell death induced by acetic acid in *Saccharomyces cerevisiae*. Mol Biol Cell 13:2598–2606

Maas MF, de Boer HJ, Debets AJ, Hoekstra RF (2004) The mitochondrial plasmid pAL2-1 reduces calorie restriction mediated life span extension in the filamentous fungus *Podospora anserina*. Fungal Genet Biol 41:865–871

Maas MF, Hoekstra RF, Debets AJ (2007) A mitochondrial mutator plasmid that causes senescence under dietary restricted conditions. BMC Genet 8:9

Madeo F, Fröhlich E, Fröhlich KU (1997) A yeast mutant showing diagnostic markers of early and late apoptosis. J Cell Biol 139:729–734

Madeo F, Herker E, Maldener C, Wissing S, Lächelt S, Herlan M, Fehr M, Lauber K, Sigrist SJ, Wesselborg S, Fröhlich KU (2002) A caspase-related protease regulates apoptosis in yeast. Mol Cell 9:911–917

Manon S, Chaudhuri B, Guerin M (1997) Release of cytochrome c and decrease of cytochrome c oxidase in Bax-expressing yeast cells, and prevention of these effects by coexpression of Bcl-xL. FEBS Lett 415:29–32

Marek SM, Wu J, Louise GN, Gilchrist DG, Bostock RM (2003) Nuclear DNA degradation during heterokaryon incompatibility in *Neurospora crassa*. Fungal Genet Biol 40:126–137

Marx F, Binder U, Leiter E, Pocsi I (2008) The *Penicillium chrysogenum* antifungal protein PAF, a promising tool for the development of new antifungal therapies and fungal cell biology studies. Cell Mol Life Sci 65:445–454

Mousavi SA, Robson GD (2003) Entry into the stationary phase is associated with a rapid loss of viability and an apoptotic-like phenotype in the opportunistic pathogen *Aspergillus fumigatus*. Fungal Genet Biol 39:221–229

Mousavi SA, Robson GD (2004) Oxidative and amphotericin B-mediated cell death in the opportunistic pathogen *Aspergillus fumigatus* is associated with an apoptotic-like phenotype. Microbiology 150:1937–1945

Odat O, Matta S, Khalil H, Kampranis SC, Pfau R, Tsichlis PN, Makris AM (2007) Old yellow enzymes, highly homologous FMN oxidoreductases with modulating roles in oxidative stress and programmed cell death in yeast. J Biol Chem 282:36010–36023

Osiewacz HD, Esser K (1984) The mitochondrial plasmid of *Podospora anserina*: a mobile intron of a mitochondrial gene. Curr Genet 8:299–305

Pahl HL, Krauss B, Schulze-Osthoff K, Decker T, Traenckner EB, Vogt M, Myers C, Parks T, Warring P, Mühlbacher A, Czernilofsky AP, Baeuerle PA (1996) The immunosuppressive fungal metabolite gliotoxin specifically inhibits transcription factor NF-κB. J Exp Med 183:1829–1840

Pereira C, Camougrand N, Manon S, Sousa MJ, Corte-Real M (2007) ADP/ATP carrier is required for mitochondrial outer membrane permeabilization and cytochrome c release in yeast apoptosis. Mol Microbiol 66:571–582

Pinan-Lucarré B, Paoletti M, Dementhon K, Coulary-Salin B, Clave C (2003) Autophagy is induced during cell death by incompatibility and is essential for differentiation in the filamentous fungus *Podospora anserina*. Mol Microbiol 47:321–333

Pinan-Lucarré B, Balguerie A, Clave C (2005) Accelerated cell death in *Podospora* autophagy mutants. Eukaryot Cell 4:1765–1774

Pinan-Lucarré B, Paoletti M, Clave C (2007) Cell death by incompatibility in the fungus *Podospora*. Semin Cancer Biol 17:101–111

Reiter J, Herker E, Madeo F, Schmitt MJ (2005) Viral killer toxins induce caspase-mediated apoptosis in yeast. J Cell Biol 168:353–358

Ribot C, Hirsch J, Balzergue S, Tharreau D, Notteghem JL, Lebrun MH, Morel JB (2008) Susceptibility of rice to the blast fungus, *Magnaporthe grisea*. J Plant Physiol 165:114–124

Richie DL, Miley MD, Bhabhra R, Robson GD, Rhodes JC, Askew DS (2007) The *Aspergillus fumigatus* metacaspases CasA and CasB facilitate growth under conditions of endoplasmic reticulum stress. Mol Microbiol 63:591–604

Rinnerthaler M, Jarolim S, Heeren G, Palle E, Perju S, Klinger H, Bogengruber E, Madeo F, Braun RJ, Breitenbach-Koller L, Breitenbach M, Laun P (2006) MMI1 (YKL056c, TMA19), the yeast orthologue of the translationally controlled tumor protein (TCTP) has apoptotic functions and interacts with both microtubules and mitochondria. Biochim Biophys Acta 1757:631–638

Saelens X, Festjens N, Vande WL, van GM, van LG, Vandenabeele P (2004) Toxic proteins released from mitochondria in cell death. Oncogene 23:2861–2874

Scheckhuber CQ, Erjavec N, Tinazli A, Hamann A, Nyström T, Osiewacz HD (2007) Reducing mitochondrial fission results in increased life span and fitness of two fungal ageing models. Nat Cell Biol 9:99–105

Schmitt MJ, Breinig F (2006) Yeast viral killer toxins: lethality and self-protection. Nat Rev Microbiol 4:212–221

Schrick K, Garvik B, Hartwell LH (1997) Mating in *Saccharomyces cerevisiae*: the role of the pheromone signal transduction pathway in the chemotropic response to pheromone. Genetics 147:19–32

Sellem CH, Lemaire C, Lorin S, Dujardin G, Sainsard-Chanet A (2005) Interaction between the oxa1 and rmp1 genes modulates respiratory complex assembly and life span in *Podospora anserina*. Genetics 169:1379–1389

Sellem CH, Marsy S, Boivin A, Lemaire C, Sainsard-Chanet A (2007) A mutation in the gene encoding cytochrome c1 leads to a decreased ROS content and to a long-lived phenotype in the filamentous fungus *Podospora anserina*. Fungal Genet Biol 44:648–658

Semighini CP, Hornby JM, Dumitru R, Nickerson KW, Harris SD (2006a) Farnesol-induced apoptosis in *Aspergillus nidulans* reveals a possible mechanism for antagonistic interactions between fungi. Mol Microbiol 59:753–764

Semighini CP, Savoldi M, Goldman GH, Harris SD (2006b) Functional characterization of the putative *Aspergillus nidulans* poly(ADP-ribose) polymerase homolog PrpA. Genetics 173:87–98

Semighini CP, Murray N, Harris SD (2008) Inhibition of *Fusarium graminearum* growth and development by farnesol. FEMS Microbiol Lett 279:259–264

Severin FF, Hyman AA (2002) Pheromone induces programmed cell death in S. cerevisiae. Curr Biol 12: R233–R235

Silar P, Picard M (1994) Increased longevity of EF-1 alpha high-fidelity mutants in *Podospora anserina*. J Mol Biol 235:231–236

Silva RD, Sotoca R, Johansson B, Ludovico P, Sansonetty F, Silva MT, Peinado JM, Corte-Real M (2005) Hyperosmotic stress induces metacaspase- and mitochondria-dependent apoptosis in *Saccharomyces cerevisiae*. Mol Microbiol 58:824–834

Stahl U, Lemke PA, Tudzynski P, Kück U, Esser K (1978) Evidence for plasmid like DNA in a filamentous fungus, the ascomycete *Podospora anserina*. Mol Gen Genet 162:341–343

Stumpferl SW, Stephan O, Osiewacz HD (2004) Impact of a disruption of a pathway delivering copper to mitochondria on *Podospora anserina* metabolism and life span. Eukaryotic Cell 3:200–211

Suen YK, Fung KP, Lee CY, Kong SK (2001) Gliotoxin induces apoptosis in cultured macrophages via production of reactive oxygen species and cytochrome *c* release without mitochondrial depolarization. Free Radic Res 35:1–10

Szappanos H, Szigeti GP, Pal B, Rusznak Z, Szucs G, Rajnavolgyi E, Balla J, Balla G, Nagy E, Leiter E, Pocsi I, Marx F, Csernoch L (2005) The *Penicillium chrysogenum*-derived antifungal peptide shows no toxic effects on mammalian cells in the intended therapeutic concentration. Naunyn Schmiedebergs Arch Pharmacol 371:122–132

Szappanos H, Szigeti GP, Pal B, Rusznak Z, Szucs G, Rajnavolgyi E, Balla J, Balla G, Nagy E, Leiter E, Pocsi I, Hagen S, Meyer V, Csernoch L (2006) The antifungal protein AFP secreted by *Aspergillus giganteus* does not cause detrimental effects on certain mammalian cells. Peptides 27:1717–1725

Theis T, Wedde M, Meyer V, Stahl U (2003) The antifungal protein from *Aspergillus giganteus* causes membrane permeabilization. Antimicrob Agents Chemother 47:588–593

Thrane C, Kaufmann U, Stummann BM, Olsson S (2004) Activation of caspase-like activity and poly (ADP-ribose) polymerase degradation during sporulation in *Aspergillus nidulans*. Fungal Genet Biol 41:361–368

Tudzynski P, Esser K (1979) Chromosomal and extrachromosomal control of senescence in the ascomycete *Podospora anserina*. Mol Gen Genet 173:71–84

Uren AG, O'Rourke K, Aravind LA, Pisabarro MT, Seshagiri S, Koonin EV, Dixit VM (2000) Identification of paracaspases and metacaspases: two ancient families of caspase-like proteins, one of which plays a key role in MALT lymphoma. Mol Cell 6:961–967

Vachova L, Palkova Z (2007) Caspases in yeast apoptosis-like death: facts and artefacts. FEMS Yeast Res 7:12–21

Vercammen D, van de CB, De JG, Eeckhout D, Casteels P, Vandepoele K, Vandenberghe I, Van BJ, Inze D, Van BF (2004) Type II metacaspases Atmc4 and Atmc9 of *Arabidopsis thaliana* cleave substrates after arginine and lysine. J Biol Chem 279:45329–45336

Walczak H, Krammer PH (2000) The CD95 (APO-1/Fas) and the TRAIL (APO-2L) apoptosis systems. Exp Cell Res 256:58–66

Walter D, Wissing S, Madeo F, Fahrenkrog B (2006) The inhibitor-of-apoptosis protein Bir1p protects against apoptosis in *S. cerevisiae* and is a substrate for the yeast homologue of Omi/HtrA2. J Cell Sci 119: 1843–1851

Watanabe N, Lam E (2005) Two *Arabidopsis* metacaspases AtMCP1b and AtMCP2b are arginine/lysine-specific cysteine proteases and activate apoptosis-like cell death in yeast. J Biol Chem 280:14691–14699

Wissing S, Ludovico P, Herker E, Büttner S, Engelhardt SM, Decker T, Link A, Proksch A, Rodrigues F, Corte-Real M, Fröhlich KU, Manns J, Cande C, Sigrist SJ, Kroemer G, Madeo F (2004) An AIF orthologue regulates apoptosis in yeast. J Cell Biol 166:969–974

Zhang NN, Dudgeon DD, Paliwal S, Levchenko A, Grote E, Cunningham KW (2006) Multiple signaling pathways regulate yeast cell death during the response to mating pheromones. Mol Biol Cell 17:3409–3422

5 Communication of Fungi on Individual, Species, Kingdom, and Above Kingdom Levels

Ursula Kües[1], Mónica Navarro-González[1]

CONTENTS

I. Introduction

In general terms, communication is understood as a process by which information is exchanged between living individuals through a common system of symbols, signs, tones or behaviour, i.e. messages are transmitted from a sender to a receiver by means of a suitable medium. Communication is an obvious instrument used to co-ordinate human and animal societies.

Communication is highest developed in the animal kingdom where it involves – to variable degrees – the five senses sight, hearing, touch, smell, and taste.

In a broader view, communication is not limited to the animal kingdom. Every interchange and exchange of information between living cells can be considered to be communication. Accordingly, there is communication in all types of living organisms, from the bacteria up to the highest developed eukaryotes. In nature, communication occurs manifold on different cellular levels: between cells of a same individual, between individuals of a species, and between organisms of different species within and above taxonomical clades up to interkingdom communication. Communication has many functions and serves e.g. in nutrient absorption and disposition, in reproduction and recombination, in distribution of species, and in the co-ordinated formation of ecological communities including also the targeted suppression of competitors.

Concerning bacteria, lower eukaryotes including the fungi, and plants, modes of intra- and interspecies communication appear to be mainly of chemical nature (Camilli and Bassler 2006; Dudareva et al. 2006; Bouwmeester et al. 2007; Thakeow et al. 2007; Lowery et al. 2008; Ryan and Dow 2008; this chapter). Electrophysical mechanisms exist also but are little studied. A new research area "plant neurobiology" focuses on intercellular long-distance communication in plants based on bioelectrochemical signals, which is seen as an analogous development to the neurosystems in the animal kingdom (Volkov 2000; Brenner et al. 2006; Barlow 2008). Electrical fields appear to be also used in communication between plants and oomycete

[1]Division of Molecular Wood Biotechnology and Technical Mycology, Büsgen-Institute, Georg-August-University Göttingen, Büsgenweg 2, 37077 Göttingen, Germany; e-mail: ukuees@gwdg.de, mnavarr@gwdg.de

Physiology and Genetics, 1st Edition
The Mycota XV
T. Anke and D. Weber (Eds.)
© Springer-Verlag Berlin Heidelberg 2009

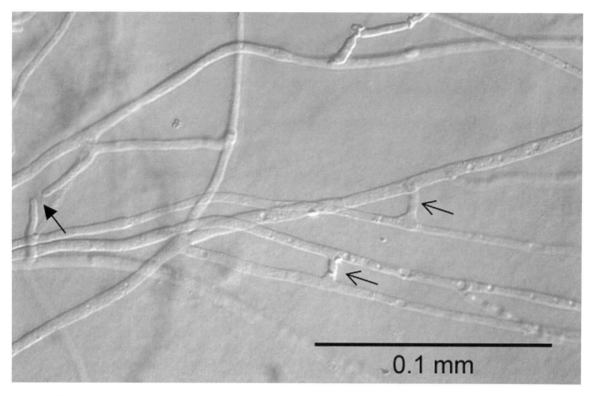

Fig. 5.1. Hyphal fusion (anastomosis) in a culture of *Coprinopsis stercorarius* strain CBS 477.70 occurred between short hyphal side branches (see *open arrows*) whilst other side branches grow with their tips towards each other (*filled arrow*)

plant pathogens, respectively between plants and arbuscular mycorrhizal (AM) fungi (Berbara et al. 1995; van West et al. 2002; Ramos et al. 2008). Bioelectric currents may also serve in intercellular communication in fungi, particularly in filamentous species (Harold et al. 1985, Gow 1989, Potapova 2004), but deeper biophysical research into this in fungi is necessary.

Research in fungi detected in recent years various types of extracellular signalling molecules ("infochemicals") that act or may act at a distance in biological signalling between fungal cells within a fungal individual, between fungal individuals of the same or other species and between fungi and other groups of organisms.

Biological aspects of signalling within and between fungal colonies were recently reviewed in a number of excellent chapters in the second edition of *The Mycota, Vol. 1* (Casselton and Challen 2006; Glass and Fleissner 2006; Feldbrügge et al. 2006; Pöggeler et al. 2006; Schimek and Wöstemeyer 2006; Ugalde 2006). For further reading on biological aspects on communication of fungi and other groups of organisms, the reader is referred to the topical reviews by Chiron and Michelot (2005), Bouwmeester et al. (2007), and Thakeow et al. (2007), respectively.

II. Communication Within and Between Fungal Colonies – Vegetative Interactions

A. Communication in Mediating Vegetative Fusions Within Fungal Colonies

Buller (1933) recognized already in the early last century that within mycelial fungal colonies there can be a functional division of hyphae – some of them act as leading hyphae in colony growth for occupying new substrates, others are formed by different order branching from the leading hyphae and their main branches lead to a network of hyphae in which communication and transport ways are dramatically shortened. To form a well interconnected network that eases nutritional translocation within the mycelium (Rayner 1991, Rayner et al. 1994, Olsson and Gray 1998), hyphae of higher branching order appear to grow within a colony towards each other in order to fuse by a process known as anastomosis (Fig. 5.1). The targeted growth towards each other – variously called positive autotropism, hyphal homing, or remote sensing - starts over certain distances, suggesting that the fungi are aware of other cells of the colony growing in their neighbourhood (Gregory 1984; Hickey et al. 2002; Glass et al. 2004; Glass and Fleissner 2006). Most likely, intercellular chemical signalling is taking place between different hyphae of a colony but the nature of such signalling molecules has not yet been discovered in any fungus (Ugalde 2006).

Bhuiyan and Arai (1993) reported as an indirect evidence for such communication that elimination of water soluble substances from the environment blocks hyphal attraction and fusion in the basidiomycete species *Rhizoctonia oryzae*.

B. Communication in Mediating Vegetative Fusions Between Different Individuals of Filamentous Ascomycetes

Substances acting in hyphal attraction within a colony may be the same than those functioning in homing between hyphae of genetically alike colonies or between hyphae and (germinated) spores of the same genotype (Glass et al. 2004). In the ascomycete *Neurospora crassa*, attraction between genetically identical spore germlings was studied by Read and colleagues, using micromanipulation by optical tweezers. With this technique, through application of laser beams, the position of a spore or a germling can be altered relative to another spore or spore germling (Fleissner et al. 2005; Roca et al. 2005a; Wright et al. 2007). In such way, homing of conidial anastomosis tubes (CATs) has unambiguously been documented in *N. crassa* (Roca et al. 2005a).

CATs are thin, short, and unbranched hyphae of determined growth that are only formed on ascomycete conidia when spores are present in high densities. CATs home towards each other and fuse, which is different from the larger germ tubes that avoid each other by growth (Roca et al. 2005b; Glass and Fleissner 2006).

In *N. crassa*, the tips of CATs were shown to be both the sites of chemo-attractant secretion and of perception. So far, analysis of a specific cAMP-defective *N. crassa* mutant (strain *cr-1*) excluded cAMP as the attractant (Roca et al. 2005a). Speculations are presented that the yet unknown chemo-attractants might be peptides, which is supported by the pertinent argument that peptides allow a higher specificity. CATs have been described in over 70 ascomycetous species (Roca et al. 2005b). CAT homing is however not restricted to one kind of spores but it can happen also between spores of different species, even up to CAT fusion (Köhler 1930; Ishitani and Sakaguchi 1956). At least in *N. crassa*, CAT homing as well as CAT fusion can occur between vegetative incompatible strains, albeit the latter occurs at a reduced level. Vegetative incompatibility reactions upon CAT fusions happen much later compared with those that follow fusion of vegetative hyphae in mature colonies (Roca et al. 2005a, b).

Vegetative incompatibility, also known as heterokaryon incompatibility, refers to a phenomenon where fusion between vegetative hyphal cells of different genetic background leads with the resulting heterokaryosis to cellular death (for further reading, see Glass et al. 2000; Esser 2006).

C. Communication in Mediating Vegetative Fusions Between Different Individuals of Basidiomycetes

Mating in the basidiomycetes is governed in the bipolar species by one mating type locus and in the tetrapolar species by two mating type loci (usually called *A* and *B*). For successful mating leading from sterile monokaryons to fertile dikaryons, two fusing strains need to be compatible, i.e. of different mating type (further information given by Casselton and Challen 2006; Casselton and Kües 2007; Sect. III.A). In species of higher basidiomycetes, hyphal fusions occur however between all kinds of monokaryotic mycelia, irrespectively of whether being of the same or of different mating types (Sicari and Ellingboe 1967; Ahmad and Miles 1970a, b; Leary and Ellingboe 1970; Fig. 5.2) and all attempts to isolate secreted substances from fungal cultures specifically acting in mating failed (Hiscock and Kües 1999). Accordingly, it was repeatedly assumed that there is no extracellular mating-signal reaction before cellular fusion (e.g. see Kothe 1996, 2008; Vaillancourt et al. 1997; Casselton and Olesnicky 1998). Frequencies of hyphal fusion in the tetrapolar *Schizophyllum commune* were reported to be higher between colonies with different *A* mating types than between strains sharing the same *A* mating type (Ahmad and Miles 1970a). In contrast, in *Coprinopsis cinerea* frequencies of hyphal fusion were said to be higher in confrontations of strains with different *B* mating types than between combinations of strains with the same *B* mating type (Smythe 1973). Other authors found with respect

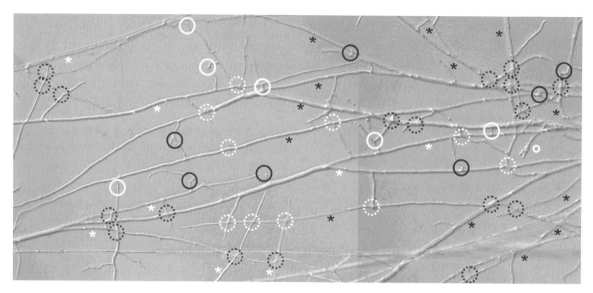

Fig. 5.2. Hyphal interactions in confronting mycelia of a compatible cross of *Coprinopsis cinerea* strains PS004-1 (mating type *A3*, *B42*; hyphae marked at the left or below with *white symbol* *) and PS004-2 (mating type *A42*, *B1*; hyphae marked at the left or below with *black symbol* *). Fusions can be seen between side branches of the same monokaryon (*black circles*) and between side branches of the compatible strains (*white circles*). Not all co-incidences between two hyphae of the same strain (*dotted black circles*) or of different monokaryons (*dotted white circles*) end in hyphal fusion – hyphae rather overgrow each other. For observing hyphal events in confrontations of mating compatible strains, the two monokaryons were inoculated at about 1 cm distance on a thin film of yeast extract/malt extract/glucose agar medium solidified on a microscope slide. A middle zone of agar about 0.7 mm in width was eliminated in between the two inocula prior to incubation at 37°C. After 24 h incubation both strains grew into the free window towards each other. The whole area was photographed in small sections (one after the other) which subsequently were composed into a larger collage that unambiguously allowed to follow up the origin of leading hyphae (the thicker hyphae in the photograph) and their thinner sub-branches and to correctly assign them to one of the two strains. The photograph shows a subsection of the collage where hyphae of the two strain intermingle and have undergone some self- or non-self-fusion events. Note at the right side the oidiophore (marked by an "o") that typically form in aerial mycelium monokaryons of *C. cinerea* (Polak et al. 2001)

to mating types no apparent differences in frequencies of hyphal fusions in *C. cinerea* (Sicari and Ellingboe 1967) and there are reports on mutual attraction of hyphae in *S. commune* irrespective of mating type (Ahmad and Miles 1970a; Raudaskoski 1998). Hyphal fusions in *S. commune* are stimulated by a secreted factor that acts equally well on combinations of mating-compatible strains and on combinations of mating-incompatible strains. The effective factor is however only induced in confrontations of mating-compatible strains whilst hyphal fusion is not required for its production (Ahmad and Miles 1970a). Furthermore, in *Lenzites betulinus* a secreted substance appears to cause formation of a "répulsion barrage" ("sexual barrier", a zone of poor growth between colonies caused by an aversion for one another) between colonies having the same *B* mating type (Hennebert et al. 1994).

Different categories of hyphal fusions are observed between confronting colonies: tip-to-tip (between two hyphal tips), tip-to-side (between a hyphal tip and a lateral hyphal wall), tip-to-peg (between a hyphal tip and a lateral swelling of a hypha), and peg-to-peg fusions (between lateral swellings of two neighbouring hyphae; Buller 1933; Smythe 1973). According to Raudaskoski (1998), tip-to-tip and tip-to-side fusions occur frequently in confrontations of *S. commune* strains independently of mating types and in these fusions one of the hyphae appears to be attracted by the partner hypha. At later stages, particularly in confrontation of strains with different *B* mating types, peg-to-peg fusions occur in higher frequencies. The two pegs at neighbouring hyphae appear to stimulate each other in growth. Raudaskoski (1998) suggested that such high frequency of peg-to-peg fusions may come from the action of mating type pheromones encoded in the *B* mating type locus. Thus, Raudaskoski (1998) distinguished potential vegetative hyphal interactions from potential sexual hyphal interactions.

As described in *S. commune* (Voorhees and Peterson 1986; Todd and Aylmore 1985), *Polyporus dryophilus* (Morton and French 1970), and *Laccaria laccata* (Fries 1983a), attraction is also possible between basidiospores and dikaryotic hyphae or monokaryotic and dikaryotic hyphae. In *C. cinerea* and other Coprini, uninucleate mitotic spores (oidia) are known to generally attract monokaryotic hyphae of the same and sometimes also of related species (Kemp 1975, 1977). Homing reactions within and between closely related species were described by Fries (1981, 1983b) between basidiospores and dikaryotic hyphae in the genus *Leccinum*. Interspecies homing by mitotic and meiotic spores results regularly in somatic incompatibility reactions, leading to the death of the foreign hyphae (Kemp 1977; Fries 1981), whereas in intraspecies reactions nuclei of one of the fusing partners might be degraded (Todd and Aylmore 1985). Particularly the observations of homing reactions argue for extracellular communication between different individuals of higher basidiomycetes. However, the controversial results and various types of observations do not allow us to deduce a clear-cut picture on the situation whether there is only vegetative communication prior to hyphal fusion or whether there is in addition a sexually determined communication (Raudaskoski 1998; Hiscock and Kües 1999). Vegetative and sexual reactions aiming at different targets (e.g. ease of translocation, supply of nutrients vs sexual reproduction) might just be overlapping and this would experimentally not be easy to separate (Hiscock and Kües 1999; Kües 2000). Figure 5.2 shows hyphae in the confrontation zone of two compatible monokaryons of *C. cinerea*. We found in close vicinity fusions between hyphae of the same and between hyphae of different mating type. The first case of fusions can be considered vegetative but whether the latter is vegetative or sexual cannot be distinguished. Hyphae may overgrow others without fusing to them regardless of whether they share the mating type or not (Fig. 5.2). The observations imply that not all hyphae are competent for fusion or that growing hyphae do not recognise all potential self or non-self partners due to a lack of communication between the hyphae.

Sexual development in higher basidiomycetes in part is regulated by the *B* mating type genes encoding small peptide pheromones and pheromone receptors of the seven-transmembrane G protein-coupled type (Casselton and Challen 2006; Casselton and Kües 2007). Mating-type specific peptide pheromones occur also in ciliates such as

Euplotes raikovi. The *E. raikovi* pheromones have both paracrine (non-self) functions by binding to mating-type-specific pheromone receptors of cells of different mating type as well as autocrine (self) functions by binding to pheromone receptors of the cell they are secreted from (Luporini et al. 2005, 2006). This finding provoked the idea that autocrine and paracrine reactions by pheromones might happen also in the higher basidiomycetes (Hiscock and Kües 1999; Kües 2000). A paradigm of autocrine and paracine reactions of mating-type pheromones has indeed subsequently been found in the basidiomycetous yeast *Cryptococcus neoformans* (Shen et al. 2002; Lin et al. 2005).

Binding of mating-type pheromones from higher basidiomycetes to pheromone receptors of another mating-type has been demonstrated by heterologous expression in baker's yeast (Fowler et al. 1999; Hegner et al. 1999; Olesnicky et al. 1999). It might be possible that in the native hosts, the mating-type specific pheromone-receptors bind the pheromones of the own mating type less stringent than competitive pheromones of different mating type, in a manner as described for autocrine and paracrine pheromone binding in the ciliate *E. raikovi* (Luporini et al. 2005, 2006). However, there are increasing reports on occurrence of genes in the basidiomycetes for potential pheromone receptors that are not mating-type-linked (James et al. 2004, 2006; Aimi et al. 2005; Niculita-Hirzel et al. 2008; Martinez et al. 2009). Moreover, in the genomes of *C. cinerea* and *Laccaria bicolor*, large families of genes for closely related, non-mating type, pheromone-like peptides were recently discovered (Niculita-Hirzel et al. 2008; Fig. 5.3). Both, the non-mating-type pheromone receptors and the pheromone-like peptides could be candidates for functioning in vegetative communication within a species and also between species. Experimental data will have to prove that.

Different from the highly dissimilar mating-type pheromones, the non-mating-type pheromone-like peptides of *C. cinerea* and *L. bicolor* are comparably well conserved in sequence (Niculita-Hirzel et al. 2008; Fig. 5.3). Sequence conservation could be of advantage in the case of interspecies communication, such as in non-self-competitive interaction for space and nutrients and hyphal interference, a phenomenon in which one species suppresses another by causing its hyphal death (Ikediugwu and Webster 1970). A foreign species might be attracted to the killing species (Kemp 1975) and hyphal interference might or might not include fusion between two species (Ikediugwu 1976; Traquair and McKeen 1977; Boddy 2000). There is certainly

Coprinopsis cinerea

```
MDSFTT..ITSFTNKPS.........VEVPSDLETGGGSGNGAYCVIA
MDSFTTTAITSFTNQPS.........VEVPSDLETGGGSGNGAYCVIA
MDSFTTIN...................IPTNFETGGGSGNGAYCVIA
MDSFT.LF..SNNVSAT.........TSVPAEEETGGGRGTGAYCTIA
MDFFTTIFASQVADLEP.........VSVPAEEETGGGRGTGAYCTIA
```

Laccaria bicolor

```
MDAFTFYITSVD.AIPIDEKTEV...ESLPTD.EESGGSGNNAYCVIA
MDSFTTYFSLAAKAEDISSQ........APVD.EESGGSGNNAYCVIA
MDSFATLSIFFA.AVSGPSD........IPTNADD.GGSGNNAYCVVA
MDSFSTISTSTD................IPTN.YEGGGSGNNAYCVIA
MDSFATLAIFFA.SVSGSSD........IPTNADD.GGSGNNAYCVVA
```

Cryptococcus neoformans

```
QSP1    M.SFTTLFTAALVLIAPALVAAAPAAEPAPSVKSQNFGAPGGAYPW
QSP2    M.SFTTLFTAALVLIAPALVAAAPAAEPAPSVKSQNFGAPGGASPI
QSP3    M.SFTTLFTAALVLIAPALVAAAPAAEPAPSVKSQNFGAPGGGKY
```

Fig. 5.3. Small peptides as possible messenger molecules in communication of basidiomycetes: Precursor sequences for a selection of small pheromone-like peptides of *Coprinopsis cinerea* and *Laccaria bicolor* are given from a multitude of similar small secreted peptides that appear to be encoded in the species' genomes (from Niculita-Hirzel et al. 2008) and precursor sequences from similar small secreted molecules from *Cryptococcus neoformans* of which QSP1 was documented to have quorum sensing function (Lee et al. 2007). In all cases, the (proposed) sequence of the mature peptides is shown in *bold* and putative processing and modification sides are *underlined*

room in these interesting antagonistic phenomenons for communication via pheromone-like molecules.

D. Communication in Population Growth Control and Germination

The families of pheromone-like peptides of *C. cinerea* and *L. bicolor* and their precursors have a sequence which somewhat resembles that of a family of small 10-mer and 11-mer peptides from *C. neoformans* (QSP1, QSP2, QSP3) that are generated from alternatively spliced transcripts of the gene *CQS1* (Fig. 5.3).

The predicted precursors of the families of small peptides of *C. cinerea* (33 different genes) and *L. bicolor* (45 different genes) have the typical structure of lipo-peptide pheromones encoded by respective mating type genes in the basidiomycetes: the precursors have an N-terminal signal sequence for protein secretion, there is an internal charged motif for putative processing by a endoprotease and at the C-terminal end, there is the typical CAAX (cysteine – aliphatic amino acid – aliphatic amino acid – any amino acid) motif for post-translational processing (Niculita-Hirzel et al. 2008). The precursors of the *C. neoformans* peptides have a similar N-terminal secretion signal and are internally processed at an analogous location but they do not have the typical CAAX-motif of fungal pheromone-precursors (Lee et al. 2007).

Most interestingly, without the 11-mer peptide QSP1 the basidiomycetous yeast is unable to grow at low cell densities. QSP1 apparently mediates a density-dependent growth phenotype which is reminiscent of bacterial quorum sensing (Lee et al. 2007).

The word quorum refers in politics to the minimum number of individuals to undergo a valid decision. In biology, quorum sensing (QS) is understood as cell-cell communication systems in species of prokaryotes that aid a concerted population response. QS as described in bacteria occurs in growth-phase- and cell-density-dependent manner. Populations of bacterial species are able to monitor cell density before expressing a defined phenotype. To this end, they make use of small specific diffusible signalling molecules (Whitehead et al. 2001; Daniels et al. 2004; Waters and Bassler 2005). More recent work suggests that bacterial QS has also capacity in interspecies and even in cross-kingdom communication (Lowery et al. 2008; see below).

QSP1 of *C. neoformans* is the only so far known fungal autoregulatory signalling molecule for density-dependent growth control on a peptide-basis. Autoregulatory signalling molecules of growth control might act positive as in case of QSP1 in *C. neoformans* but there can be others with negative effects. In the Saccharomycetales yeast *Candida albicans*, the phenolic alcohol tyrosol acts as a QS molecule for promoting induction of germ-tube formation on the yeast cells whilst another QS molecule, the acyclic sesquiterpene alcohol farnesol, inhibits this process in favour of typical yeast growth (Chen et al. 2004; see also Sect. II.E).

Depending on growth medium and yeast species, mathematical models deduced from experimental data suggest that reaching constant cell densities in cultures of unicellular yeasts can be the result of QS mediated by autoregulators (Walther et al. 2004). On agar medium, there is intracolony cell–cell communication as well as intercolony signalling in various ascomycetes yeast species, including the baker's yeast *Saccharomyces cerevisiae* mediated by the volatile alkaline compound ammonia. Ammonia production is oriented toward neighbouring colonies growing in close vicinity and results in growth inhibition (Palková et al. 1997; Gori et al. 2007). Via such communication, yeast colonies are able to synchronize with each other in physiology and development (Palková and Forstová 2000; Palková and Vachová 2003, 2006). In the human pathogenic yeast *Histoplasma capsulatum*, α-(1,3)-glucan incorporation in the cell wall occurs growth-dependent at high cell densities and this is under control of an unknown QS molecule (Kügler et al. 2000). Colonisation on agar medium of this fungus occurs only at a density $>10^6$ yeast cells/plate (Pine 1995). Iron siderophores and hydroxamic acid released by the yeast cells, acting as growth factors, regulate this inoculum size effect (Burt et al. 1981; Burt 1982).

Siderophores sequester required iron from the medium for the fungus and appear to counteract impurities of the agar medium. A certain concentration of the siderophores is required before the mechanism becomes effective – if the cell density of *H. capsulatum* is too low this concentration will not be reached but certain additions to the medium can overcome the inoculum size effect (Worsham and Goldman 1988; Woods et al. 1998). It is thus debatable whether such a cell-density-dependent mechanism as found in *H. capsulatum* is cellular communication in the strictest sense involving a sender and a receiver of a message. The case of *C. cryptococcus* appears to be different from this since it is medium-independent and unaffected by additions to the medium. The 11-mer peptide QSP1 might therefore be a true messenger molecule acting between cells (Lee et al. 2007).

Whilst most of the so far discussed cases of communication between fungal cells occur in growing vegetative cultures, growth can also be influenced by systems of QS at the stage of spore germination.

Many species produce self-inhibitors or autoinhibitors of germination when spores are present in high densities in the environment. Production of self-inhibitors of spore germination has been reported for more than 60 different species. Various types of secondary metabolites have been detected to exhibit self-inhibitor functions. Pathogens thereby appear to be more specific in the chemical compounds used than saprotrophs (reviewed in further detail by Ugalde 2006). For example, the basidiomycetous rusts *Uromyces phaseoli* and *Puccinia graminis* produce the aromatic substances methyl-cis-3,4-dimethoxycinnate and methyl-cis-ferulate, respectively (Allen 1972) and mitosporic *Colletotrichum* species have been shown to generate one or more self-inhibitors being such diverse organic compounds as mycosporine-alanine and the indol derivatives (*E*)- and (*Z*)-ethylidene-1,3-dihydroindol-2-one, colletofragarone B, (2*R*)-(3-indolyl)propionic acid, and (2*R*)-(3-indolyl) propionic acid (Leite and Nicholson 1992; Tsurushima et al. 1995; Inoue et al. 1996). Such very specific chemicals are likely produced during spore formation to be incorporated into the spore cell walls or the surrounding mucilage (Young and Patterson 1982) from which they later will successively be released (Ugalde 2006).

Other effective self-inhibitors of germination produced by conidia likely by fatty acid degradation are the volatile organic compounds (VOCs) types of C8-compounds, most often 1-octen-3-ol (Garrett and Robinson 1969; Chitarra et al. 2004). 1-Octen-3-ol in *Penicillium paneum* was shown to render membrane properties in suppressed spores and to change their internal pH (Chitarra et al. 2005). A similar outcome on cell membranes and internal pH of fungal spores was postulated for nonanoic acid (Breeuwer et al. 1997).

Studies in the higher basidiomycetes show that there can also be communication in spore populations for stimulation of germination. In the button mushroom *Agaricus bisporus* the volatile isovaleric acid was identified to be an autoinducer of basidiospore germination (Rast and Stäuble 1970), which however can be counteracted by self-inhibition through CO_2 produced by the spores (Rast et al. 1979). Amongst other unknown germination-inducing substances produced by the mycelium of the ectomycorrhizal species *Tricholoma robustum*, gluconic acid is inducing germination of its basidiospores (Iwase 1991). *n*-Butyric acid and related compounds are operative for *Tricholoma matsutake* (Ohta 1988). Volatile germination inducing factors acting in a species-dependent manner on basidiospores are also emitted from the mycelia of other basidiomycetes, e.g. from the ectomycorrhizal fungus *Leccinium aurantiacum* (Bjurman and Fries 1984).

As much as tested, volatiles emitted from mycelia of related species tend not to act in basidiospore germination in ectomycorrhizal species (Iwase 1992; Bjurman and Fries 1984). However, basidiospore germination can be triggered by exudates from non-ectomycorrhizal species such as the basidiomycetous yeast *Rhodotorula glutinis* (Fries 1978) or the filamentous ascomycete *Ceratocystis fagacearum* (Oort 1974).

E. Communication in Dimorphism and Asexual Reproduction

There is a growing body of evidence that QS occurs widespread in dimorphic fungi in order to positively or negatively regulate switching between morphological growth forms (Hogan 2006; Nickerson et al. 2006; Sprague and Winans 2006). For example, during growth at high cell densities ($>10^6$ cells/ml) of the dimorphic opportunistic human pathogen *C. albicans*, farnesol accumulates in the media and, in a QS manner, blocks the morphological transition from the yeast to the mycelial stage and with it biofilm formation (Hornby et al. 2001; Alem et al. 2006; see also Sect. II.D). Yeast–mycelium dimorphism in *Ceratocystis ulmi* (which causes Dutch elm disease) is regulated by an unknown QS factor that is not farnesol (Hornby et al. 2004) but very likely another oxygenated lipid molecule (Jensen et al. 1992). In some strains of the baker's yeast *S. cerevisiae*, colony density can negatively influence invasiveness into agar medium and pseudohyphal growth (Lucaccioni et al. 2007). Autosignalling aromatic alcohols (phenylethanol, tryptophol) were shown to suppress filamentous growth in strains having an active *FLO11*-flocculation gene (Chen and Fink 2006). Obviously, different chemical compounds are used by the various yeasts to regulate dimorphic morphological changes.

In the filamentous ascomycete *Aspergillus niger*, the oxygenated lipid farnesol is shown to suppress conidiation (Lorek et al. 2008). Farnesol appears not to be produced by *A. nidulans* but, when present in the environment, it induces apoptosis-related morphological changes (Semighini et al. 2006). In the related *A. flavus*, the formation of conidia and sclerotia is oppositely density-regulated. Sclerotia form at low cell density and conidia at high cell density and different autoregulatory factors seem to control these effects. Lipoxygenases appear to be involved in the production of the factors (Brown et al. 2008).

Lipoxygenases act in the synthesis of fatty acid metabolites by catalysing the hydroperoxidation of polyunsaturated fatty acids. Hydroperoxy fatty acids are then converted into a variety of degradation products, many of which can act as signalling compounds. Derivatives of fatty acids and their esters obtained by oxidation through lipoxygenase are known as oxylipins (Grechkin 1998; Liavonchanka and Feussner 2006). In fungi, oxylipins and other bioactive lipid metabolites are shown to affect development, QS, and effecter molecule production in interaction with other organisms (Erb-Downward and Huffnagle 2006).

A typical fungal oxylipin is the volatile 1-octen-3-ol, mediator of the characteristic mushroom odour (Combet et al. 2006). 1-Octen-3-ol is likely produced in the fungi from linoleic acid through oxidation by lipoxygenase into $8(E),12(Z),10(S)$-10-hydroxy-8,12-octadecadienoic acid that subsequently is cleaved to $3(R)$-1-octen-3-ol and $8(E)$-10-oxo-8-decenoic acid (Wurzenberger and Grosch 1984; Combet et al. 2006). 1-Octen-3-ol is a potent and often used signalling molecule in the fungi (see also Sects. III.B, VI). Application of the VOCs 1-octen-3-ol and its analogues 3-octanol and 3-octanone can overcome light-dependence in conidiation in *Trichoderma* spp. Sporulating colonies produce these compounds themselves and the VOCs mediate as signalling molecules, in minute concentrations, intercolony communication for the induction of sporulation on other colonies from the same or other *Trichoderma* species (Nemčovič et al. 2008).

Other oxylipins are the well studied endogenous precious sexual induction (psi) factors influencing conidiation in *A. nidulans* by governing the balance between asexual and sexual reproduction (reviewed by Ugalde 2006). psiAα [lactone of 5(S),8(R)-dihydroxy-9(Z),12(Z)-octadecadienoic acid] induces conidiation and, in turn, counteracts sexual sporulation. psiBα [8(R)-hydroxy-9(Z), 12(Z)-octadecadienoic acid] and psiCα [5(S),8(R)-dihydroxy-9(Z),12(Z)-octadecadienoic acid] are sporulation hormones that inbibit conidiation and stimulate ascospore formation (Champe et al. 1987; Calvo et al. 1999, 2001; Tsitsigiannis et al. 2004a, b, 2005; see also Sect. III.B). Moreover, a defect in production of the oxylipin 10(R)-hydroxy-8(E),12(Z)-hydroperoxyoctadecadienoic acid by the human pathogen *Aspergillus fumigatus* leads to a decrease in conidia production; the conidia have an altered size, shape and germination behaviour and, in addition, during fungal infestation in the human host there is an increased spore uptake and killing by primary alveolar macrophages (Dagenais et al. 2008).

Another important group of chemical compounds to be mentioned with respect to regulation of asexual development in filamentous ascomycetes are terpenes (Ugalde 2006). The best understood molecule is the diterpene conidiogenone from *Penicillium notatum*. Conidiogenone is produced throughout the whole growth phase of the fungus but when growing in submerged culture, the local conidiogenone concentration remains below a threshold required for sporulation. In contrast, when the fungus enters the aerial phase, conidiogenone is believed to accumulate on the hyphal cell walls at concentrations above threshold levels (>350 pM) so that sporulation can happen. In liquid culture, thresholds can be reduced fivefold by increasing Ca^{2+} concentrations with the consequence that sporulation is facilitated under submerged conditions (Roncal et al. 2002a, b; Ugalde 2006).

III. Communication Between Fungi in Sexual Interactions

A. Communication in Mating

A mating reaction is a prerequisite step in sexual reproduction of heterothallic fungi. Mating involves fusion of fungal cells of different mating type and, in many instances prior to that, a courtship period of communication between the cells of different mating type. Courtship for mating in ascomycetes and basidiomycetes involves mating-type specific pheromones, small peptides, and their also mating-type-specific seven-transmembrane G protein-coupled receptors.

Selective attraction between cells of opposite mating type has been observed for many yeasts and yeast-like species, both from the ascomycetes and from the basidiomycetes (see e.g. Bauch 1932; Bandoni 1965; Abe et al. 1975; Flegel 1981; Chenevert et al. 1994; Nielsen and Davey 1995; Snetselaar et al. 1996; Shen et al. 2002; Lockhart et al.

2003). Pheromones functioning in a mating-type-dependent manner have been isolated from various ascomycetous and basidiomycetous yeasts. Application of such pheromones to cells of opposite mating types induces an elongation of yeast cells directed towards the pheromone source and conjugation tube formation (Duntze et al. 1970; Imai and Yamamoto 1984; Wong and Wells 1985; Spellig et al. 1994; see works cited by Flegel 1981; Hiscock and Kües 1999).

In the filamentous ascomycetes, chemical attraction by mating-type pheromones is observed at fertilization when donor cells of nuclei (either spores or hyphae) are attracted and fuse with trichogynes, specialized hyphae differentiated in a mycelium of an opposite mating type from the ascogonia (the female reproductive structures) to act as receptors of the nuclei coming from the male-operating donor cells (Bistis 1983, 1998; Turina et al. 2003; Kim and Borkovich 2004, 2006; Coppin et al. 2005).

The situation differs from that in the filamentous higher basidiomycetes, where a reaction by secreted pheromones prior to hyphal fusion of mating-type competent monokaryons was never possible to verify (see Sect. II.C). Mating of two compatible monokaryons in the higher basidiomycetes leads to a dikaryon morphologically characterized by the clamp cells formed at the hyphal septa during the developmental course of nuclear and cellular division (Kües 2000; Casselton and Challen 2006). The arch-shaped clamp cells branch from the apical cell of hyphae, grow backwards, and appear to induce formation of a peg at the tip of the sub-apical cell to which the clamp cell eventually fuse (Buller 1931; Badalyan et al. 2004). Fusion of the clamp cells to the sub-apical cell is B-mating-type dependent. It is assumed but not proven that at least in this reaction the mating type pheromones mediate the communication between the clamp cell and the sub-apical cell and its peg, respectively (Brown and Casselton 2001; Badalyan et al. 2004).

Expression of pheromones and pheromone receptors in the bipolar ascomycetes is controlled by the mating type loci encoding types of transcription factors (for recent reviews, see e.g. Debuchy and Turgeon 2006; Tsong et al. 2007; Turgeon and Debuchy 2007). In the basidiomycetes in contrast, the pheromone and pheromone receptor genes are found included with genes for transcription factors within the single mating type locus of the bipolar species or, in the tetrapolar species, they are found in a second mating type locus separate from the mating type locus containing the transcription factor genes (see the recent reviews by Casselton and Challen 2006; Bakkeren and Kronstad 2007; Casselton and Kües 2007). In either way, cell-type specific communication systems are verified so that one mating type produces pheromones that by a distinctive amino acid sequence are only recognized by receptors of another mating type-specificity, which in turn are also characterized by unique amino-acid sequences. More information on this can be found in the above-cited reviews and in various further chapters in the book *Sex in Fungi* edited by Heitman et al. (2007).

In the Mucoromycotina (formerly classified within zygomycetes, now within the Fungi incertae sedis; Hibbett et al. 2007), there is a another mechanism of communication between cells of different mating type: mating type recognition and regulation of sexual reproduction (steps zygophore formation and zygotropism between such structures of different mating type) is governed by the β-carotene derivate trisporic acid as the major sex pheromone and its biosynthetic precursors. The biosynthesis of trisporic acid is shared between the two mating types (+) and (−) of a species. In short, 4-dihydrotrisporin obtained in both mating types from β-carotene in the (+) strains are converted to 4-dihydroxymethyl trisporate and in the (−) strains to trisporin. These metabolites are exchanged between the mating types upon which trisporin in the (+) mating type is converted to trisporol and finally trisporic acid and 4-dihydroxymethyl trisporate in the (−) mating type to methyl trisporate. Finally, both trisporol and methyl trisporate are transformed to trisporic acid (further reading in the recent reviews by Schimek and Wöstemeyer 2006; Wöstemeyer and Schimek 2007). It is interesting to note that the concerted mechanism of trisporic acid formation is not restricted to strains of one species but due to the widespread usage of the trisporic acid system in the Mucoromycotina, precursors can also be exchanged between different species (Schimek et al. 2003). Even more, strains of different species may fuse upon trisporic acid-regulated mating-type recognition such as the parasitic species *Parasitella parasitica* with *Absidia glauca* (Kellner et al. 1993; Schultze et al. 2005).

The first three enzymes for the co-operative biosynthesis of trisporic acid by Mucoromycotina have recently been described. As a first step confined to both mating types, β-carotene is cleaved by a β-carotene oxygenase. The final product of β-carotene degradation is likely 4-dihydrotrisporin as the last common precursor in both mating types (Burmester et al. 2007). An NADP-dependent 4-dihydrotrisporin-dehydrogenase as a (−) mating-type-specific enzyme transforms 4-dihydrotrisporin to trisporin which is released to reach the complementary mating partner. 4-Dihydromethyltrisporate dehydrogenase converts 4-dihydroxymethyl trisporate, a product of the (+) mating type, to methyl trisporate. Although both mating types carry the genes for the latter two enzymes and both transcribe the genes, active enzymes are only found in the sexually stimulated (−) mating type and they are thus strictly mating-type-specific (Schimek et al. 2005, Wetzel et al. 2009). Receptors for any of the intermediate compounds produced in trisporic acid synthesis pathway are not known (Schimek and Wöstemeyer 2006).

Reproduction in the Chytridiomycota such as *Allomyces macrogynus* involves fusion between motile uniflagellate male and female gametes that can be produced on the same mycelia but in different gametangia. The smaller male (−) gametes are sexually attracted to the larger female (+) gametes by sirenin, a bicyclic sesquiterpene with a cyclopropane ring, produced by the (+) gametes. In turn, the (−) gametes produce a compound parisin of unknown chemical structure that acts as attractant of the (+) gametes (Pommerville and Olson 1987; Pommerville et al. 1988; for detailed reviews see Gooday 1994; Schimek and Wöstemeyer 2006; Idnurm et al. 2007).

B. Communication in Fruiting Body Development

Fruiting bodies with various types of differentiated cells arranged in specific morphological patterns (Fig. 5.4A, B) are the most complex structures found within the fungal kingdom (for reviews see Kües et al. 2004; Pöggeler et al. 2006; Wösten and Wessels 2006). Fruiting body formation demands highly co-ordinated cellular programs of synchronized development and differentiation. This includes also a co-ordinated and specific fusion of cells such as occurring between the outer layer of enveloping hyphae in the formation of the peridia (outer rind tissue formed by melanized small and densely arranged cells) of ascomycetous fruiting bodies (Read 1994). In the early progress of "tissue" differentiation in the initials (nowadays known as secondary hyphal knots, i.e. small compact hyphal aggregates induced by a light signal; Kües et al. 2004) of fruiting bodies of the basidiomycete *C. cinerea*, anastomosis occurs frequently between cells in the core of the structure but not between the large hyphae of its outer layer (Van der Valk and Marchant 1978). Some "tissues" in the fully

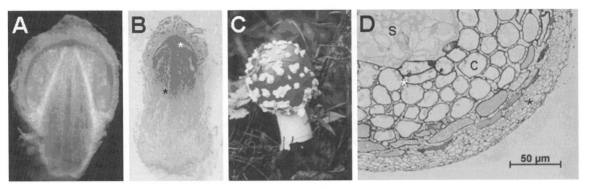

Fig. 5.4. A, B A hand-cut and a microtome cross-section through fully developed primordia of *Coprinopsis cinerea* directly prior to fruiting body maturation showing the complex tissue differentiation in the multicellular structure. A *white symbol* * marks the plectenchymatic mycelial regions in the pileus whilst a *black symbol* * indicates the position of the gills on which surface the prosenchymal hymenium is localized. **C** A young mushroom of the fly agaric *Amanita muscaria* (photograph kindly supplied by U. Helle) that can only develop in symbiosis with a woody host. **D** A cross-section through a pine root colonized by an unidentified ectomycorrhizal basidiomycete (preparation by E. Fritz, photograph by A. Olbrich). Marked are the central stele (*S*), the cortex (*C*), the Hartig net of fungal cells (*white symbol* *) in between plant cortex cells resulting in a large surface area for nutrient transfer and communication, and the ectomycorrhizal outer mantle (*black symbol* *) that covers the surface of the root

developed primordium are plectenchymal such as the inner parts of the gills and pileus with interwoven, multiple branched, and interconnected hyphae, whilst others such as the hymenium and subhymenium are prosenchymal with unbranched parallel arranged hyphal cells (Kües 2000; Fig. 5.4B). Obviously, a blueprint exists for the fruiting body which determines not only cell shapes and sizes and places but also which of the cells will interact by fusion and which not. To verify this blueprint, communication and coordination between neighbouring cells, respectively between different mycelial areas within the developing primordium should be required.

In the filamentous ascomycetes, diffusible factors including the mating-type pheromones can play a role in fruiting body development such as in frequency of fruiting body induction, in male-female communication and fertilization of the trichogyne, in ascocarp maturation and in ascospore formation (reviewed by Dyer et al. 1992; Pöggeler et al. 2006). In the heterothallic *N. crassa* as well as in the homothallic *Sordaria macrospora*, mating-type pheromone communication with their receptors is required for (efficient) fruiting body maturation (Kim et al. 2002; Mayrhofer et al. 2006), whilst in the also heterothallic *Podospora anserina* the pheromone function is restricted to fertilization (Coppin et al. 2005). Furthermore, in the homothallic *Gibberella zeae* the pheromone communication system is not essential for sexual development

but it can enhance its frequency (Kim et al. 2008; Lee et al. 2008).

In the homothallic *A. nidulans* (telemorph *Emericella nidulans*) producing its meiospore in cleistothecia (Pöggeler et al. 2006), the fruiting body production is determined by the availability of the already mentioned psi factors. Effects of the different psi factors on cleisthothecia production are opposite to those on conidiation (Tsitsigianni et al. 2004a, b, 2005; for more details see Sect. II.E). Exogenous application of certain types of lipids including linoleic acids can influence fruiting in the species (Champe et al. 1987; Calvo et al. 1999). Production of perithecia (Pöggeler et al. 2006) in *N. crassa*, *Nectria haematococca* as in plant pathogenic *Ceratocystis* species is also enhanced by addition of linoleic acids, even when added only in minute amounts (Nukina et al. 1981; Marshall et al. 1982; Dyer et al. 1993). The observations indicate that oxylipin control of ascocarp formation is very likely widely distributed within the ascomycetes.

VOCs as autoregulatory compounds emitted from the own mycelium can positively (Basith and Madelin 1968) or negatively (Moore-Landecker and Shropshire 1984) influence ascocarp formation. On the whole, however, information in ascomycetes on induction of fruiting, fruiting body differentiation and maturation by extracellular morphogens is sparse (Moore-Landecker 1992; Ugalde 2006; Pöggeler et al. 2006).

Little is yet known on cellular co-ordination of fruiting body development and maturation in the higher basidiomycetes (Kües 2000; Walser et al. 2003; Kües et al. 2004; Wösten and Wessels 2006), let alone on the participation of signalling molecules in this. In *C. cinerea*, work by Kamada and Tsuji (1979) showed that a diffusible darkness-induced factor from the fully developed primordia caps (Fig. 5.5) can bring about stipe elongation and fruiting body maturation. Also, there is a function of the *B* mating type pheromones and their receptors in inducing fruiting body maturation at the stage of karyogamy once the primordia were fully established in structure (Kües et al. 2002). In various instances, exogenous addition of surface-active compounds (lipids, detergents, amphipatic sugars and glycolipids, such as cerebrosides and other membrane-interacting compounds, specific cytolytic membrane-interacting proteins, such as ostreolysin from *Pleurotus ostreatus*) stimulated fruiting body initiation in species of higher basidiomycetes although not all such compounds are active (Kües and Liu 2000; Magae et al. 2005; Magae and Ohara 2006; Berne et al. 2007, 2008). From mutant analysis in *C. cinerea*, lipid-linked membrane-associated signalling processes are anticipated in the process of fruiting body initiation (secondary hyphal knot formation) occurring spatiotemporally defined within mycelial cultures

Fig. 5.5. **A** A slug grazing on tufts of fruiting bodies of *Coprinellus disseminatus* grown on a meadow. **B–F** An experiment for baiting animals to fresh fungal cultures with mushrooms: Petri dish cultures of *Coprinopsis cinerea* containing fully developed primordia ready to undergo overnight fruiting body maturation (see Fig. 5.4A) were moved either with covering lids or without lids at an evening in the late summer (in August) into a meadow underneath some bushes. Overnight, the fruiting bodies matured and snails and collembolans were apparently attracted to the cultures for consuming the fungus. Snails appeared to have consumed the complete culture with mycelium and fruiting bodies inclusive the meiotic spores as well as the agar. Faecal pellets (**B**) left in the Petri dishes contained hyphal fragments as well as masses of dark brown-stained basidiospores (**C**), suggesting that the fruiting bodies were fully developed before they were consumed. Whilst the lids on closed Petri dishes hindered larger animals reaching the fungal cultures, larvae and adult stages of collembolans were found in the cultures either at the underside of the opened mushroom cap (**D** see *tip of arrow*; *inset*: enlarged view on the transition between fruiting body gills and the centred stipe reveals many black-stained faecal pellets laid onto the mushroom tissues) or within the still present vegetative mycelium on the agar (see examples in **E**; the *arrow* in the figure points at a faecal pellet laid by the adult animal). Also the collembolans consumed the basidiospores and loads of intact spores are seen in squeezed faecal pellets (**F,G**)

(Liu et al. 2006). Following initiation at multiple places, a bigger part of young fruiting structures will be aborted in favour of one or a few, with translocation of nutrients towards those structures that eventually will mature as fully developed fruiting bodies (Madelin 1956; further discussed by Moore 1998). The concerted and locally directed action suggests that intra-colony communication happens.

C. cinerea mutants with defects in fruiting body initiation, in cap and stipe tissue formation, and in stipe elongation have been used to characterize responsible genes. In this way, four enzyme encoding genes were identified:

1. *cfs1* for a cyclopropane fatty acid synthase with a function in initiation of fruiting body development (Liu et al. 2006)
2. *eln2* for a new type of cytochrome P450 acting in stipe elongation and being distantly related to the oxidoreductases from *A. flavus* and *A. parasiticus* which convert *O*-methylsterigmatocystin to aflatoxin (Kües 2000; Muraguchi and Kamada 2000)
3. *ich1* for an *O*-methyltransferase acting in pileus differentiation and being distantly related to enzymes of *A. flavus* and *A. parasiticus* which are involved in the conversion of sterigmatocystin to *O*-methylsterigmatocystin and dihydrosterigmatocystin to dihydro-*O*-methylsterigmatocystin (Muraguchi and Kamada 1998; Kües 2000)
4. *eln3* for a potential glycosyltransferase probably involved in cell wall biogenesis (Arima et al. 2004).

For none of these enzymes are the substrates yet known, but the possibility of producing signalling molecules for development and communication has been discussed for the first three enzymes (Kües 2000). In this connection it is interesting to note that, in the *Aspergillus* species, aflatoxin production is tightly linked to conidiation although it is not essential for sporulation. Block of aflatoxin synthesis affects however the numbers of conidia that will be produced (Calvo et al. 2002).

In *A. bisporus*, the oxylipin (8*E*)-10-oxo-8-decenoic acid appears to stimulate stipe elongation as well as mycelial growth and fruiting initiation (Mau et al. 1992; Mau and Beelman 1996). In the young yet closed mushrooms (button stage 20 mm diam., with closed caps; medium stage 35 mm diam., with closed veil), 1-octen-3-ol accumulates at highest levels (Mau et al. 1993; Cruz et al. 1997), possibly to function in protection against moulds (Okull et al. 2003) and bacteria trying to infest the mushrooms (Beltran-Garcia et al. 1997). Generally, the concentration of these and other VOCs increases dramatically upon mushroom wounding, supporting a role in wound-activated chemical defence (Wadman et al. 2005; Spiteller 2008).

Many other VOCs as potential signalling molecules are produced by mushrooms – a literature account collected more than 250 different scent compounds that are either plain hydrocarbons, heterocycles, alcohols, phenols, acids, and derivatives, together with sulfur-containing molecules (Chiron and Michelot 2005).

IV. Communication Between Fungi and Bacteria

Cross-kingdom QS has been encountered between the gram-negative bacterium *Pseudomonas aeruginosa* and *C. albicans*. *P. aeruginosa* suppresses the filamentous pathogenic growth form of *C. albicans* by secreting its own cell-cell signalling molecule 3-oxo-C_{12} homoserine lactone (Hogan et al. 2004). In turn, farnesol produced by *C. albicans* inhibits the swarming motility of the bacterium (McAlester et al. 2008).

Taking signalling compounds developed for self QS and using them for inter-kingdom communication is not the only way how organisms within the same complex biotop may influence the intraspecies communication of others. Various fungi for example have evolved defence mechanisms against bacteria by blocking their QS systems. Within the genus *Penicillium* for example, 33 of 50 tested species were found to produce quorum sensing inhibitor (QSI) compounds against *P. aeruginosa*. The lacton mycotoxin patulin and the antibiotic penicillic acid are two potent QSI compounds negatively influencing QS-controlled gene expression in the bacterium (Rasmussen et al. 2005). As another mean of interrupting QS of competitive bacteria in the same environment, specific fungi in forest soils produce a lactonase in order to degrade N-acyl homoserine lactones used by bacteria in the rhizosphere as signalling compound (Uroz and Heinonsalo 2008).

Probably caused by competing for resources, bacteria can negatively influence growth, development and reproduction of fungi by releasing VOCs. For example, perithecial formation by *Gelasinospora cerealis* is inhibited by VOCs emitted by several kinds of bacteria (Moore-Landecker and Stotzky 1972), bacterial VOCs inhibit the ascogenous system of *Pyronema domesticum* (Moore-Landecker 1988), abnormal shortening of conidiophores of *Aspergillus giganteus* has been observed (Moore-Landecker and Stotzky

1973) and bacterial VOCs inhibit fungal spore germination (Barr 1976). Exudates from bacterial isolates from sporophores, mycorrhizae or soil in contrast stimulate germination of spores of ectomycorrhizal fungi (Ali and Jackson 1988). Most interesting are the mycorrhization helper bacteria acting positively in establishing the symbiosis between ectomycorrhizal basidiomycetes and woody plants. Amongst different effects exerted by the various species of helper bacteria, a *Streptomyces* strain was shown to produce an auxin-related compound termed auxofuran that stimulates growth of the fly agaric *Amanita muscaria* and suppresses growth of fungal root pathogens (Frey-Klett et al. 2007; Schrey and Tarkka 2008). A potent inducer of quinone-character (basidifferquinone) for the induction of fruiting in the saprotrophic mushroom *Flavolus arcularius* is known from another *Streptomyces* strain (Azuma et al. 1990).

V. Communication Between Fungi and Plants

Fungi and plants may undergo symbiotic reactions where both obtain a benefit from the interaction or they may exist in interactions where the selfish fungi live as a pathogen on the plants and cause damage to them. In both types of fungal life styles, communication between fungus and plant can play a crucial role.

The rhizosphere with its multiple root–root, root–microbe, and microbe–microbe interactions is a best ecological example for complex interspecies communications (Bais et al. 2004, 2006). Exudates from plant roots for example are shown to stimulate in a presymbiontic growth phase germinating spores of AM fungi in nutrient uptake (Bücking et al. 2008). Plant roots may synthesise certain types of apocarotenoids (e.g. strigolactones) as signalling compounds to the obligate symbiotic AM fungi in stimulating their growth by inducing hyphal branching and in the establishing the endomycorrhizae (Akiyama et al. 2005; Besserer et al. 2006; Sun et al. 2008). Flavenoids from root exudates also promote hyphal growth, differentiation and root colonization by AM fungi in a genus- and species-specific manner (Steinkellner et al. 2007). Plants in turn receive signals from the AM fungi – ungerminated and germinated

spores of *Gigaspora* and *Glomus* species from the Glomeromycota release diffusible molecules that in cells of a host plant such as soybean (*Glycine max*) are perceived through a Ca^{2+}-mediated signaling and thereby induce the expression of AM-related host genes (Navazio et al. 2007b). So far, the types of the chemical signals released from AM fungi have not been defined. Communication of the fungi with their hosts generally divides in two steps – prior to direct cellular contact and during the contact. Compounds used in the two stages of dialogue could indeed differ. Plants for example chance their flavenoid pattern from presymbiotic stages to root colonization (Genre and Bonfante 2007; Steinkellner et al. 2007).

Like AM fungi, the presence of non-mycorrhizal fungi in the environment also are sensed by the plant through intracellular Ca^{2+} changes upon exposure by secreted fungal molecules. Several pathogenic fungi elicit extremely specific variations of Ca^{2+} concentration (Lecourieux et al. 2006). Most interestingly, the intracellular Ca^{2+} responses on pathogens such as the nectrotrophic *Botrytis cinerea* and on simultaneous presence of biocontrol fungi such as *Trichoderma atroviride* are different. The plants can discriminate between the metabolites of the two different situations by different kinetics of the Ca^{2+} responses and by specific patterns of intracellular accumulation of reactive oxygen species (Navazio et al. 2007a). Different types of elicitors, messenger substances that cause defence reactions in plants, are known to be produced by pathogenic fungi attacking roots, leaves, and other parts of plants as well as by non-pathogens such as by the *Trichoderma* biocontrol fungi (for a review see Ebel and Scheel 1997; Harman et al. 2004; Berrocal-Lobo and Molina 2008). Upon host penetration, fungi secrete effector molecules into the apoplast or into host cells for suppressing plant defence responses. However, these effectors may then act also as signals to the plants to reinforce defence (see the recent review by Göhre and Robatzek 2008). Mostly, elicitors and effectors are of proteineous nature or oligo-carbohydrates but phytotoxic sesquiterpenoids are also described. For more details, the reader is referred to the respective cited reviews.

To illustrate the broad variety of fungal elicitors, peptaibol as non-ribosomally produced peptide (Viterbo et al. 2007), swollenin as a protein with a carbohydrate-binding domain (Brotman et al. 2008), small cysteine-rich proteins of the ceratoplatanin family of proteins (Djonovic et al. 2006; Vargas et al. 2008), and oligosaccharides released by *Trichoderma*-mediated enzymatic degradation of fungal and plant cell walls (Harman et al. 2004) are named at this place as characterized elicitors from *Trichoderma* species.

Compared to AM mycorrhiza, even less is known about communication between ectomycorrhizal fungal species and their plant hosts (Martin et al. 2001, 2008). For the truffle *Tuber borchii–Tilia americana* confrontation, 29 different VOCs specific to the premycorrhizal stage have been encountered as candidates mediating communication with its host roots over a distance of a few centimetres. Amongst these are e.g. the sesquiterpenes germacrene D, dehydroaromadendrene, β-cubebene, and longicyclene known to be produced by plants (Menotta et al. 2004). The particular VOCs emitted by the ascomycete *T. borchii* to mediate communications with its host plants are yet to be established. However, 1-octen-3-ol and trans-2-octenal from the fungal volatiles were shown to induce an oxidative burst of H_2O_2 in the non-host *Arabidopsis thaliana* and they inhibited development of the plant (Splivallo et al. 2007b). In nature, the burnt phenomenon referring to a zone of very meager vegetation around trees with roots colonized by the truffle may relate to this – some of the manifold VOCs emitted from the truffles (or from the bacteria commonly growing within the tubers) may exert phytotoxic effects on seed germination and reduce root growth of the vegetation (Splivallo 2008; Tarkka and Piechulla 2008).

Plants in biotrophic interactions may send out signals in turn to the fungi. The rust fungus *Uromyces fabae* stimulates volatile production by host bean plants. Of these VOCs, nonanal, decanal, and hexenyl acetate promote haustoria formation of the pathogen, whilst the terpenoid farnesyl acetate counteracts the effects by the VOCs (Mendgen et al. 2006).

In the case of ectomycorrhizal basidiomycetes, root exudates from the woody hosts stimulate germination of the poorly germinable basidiospores (Fries 1989; Ishida et al. 2008) and certain flavonoids from root exudates (hesperidin, morin, rutin, quercitrin, naringenin, genistein, chrysin) are nominated as candidates for this interaction (Kikuchi et al. 2007). The production of fruiting bodies of ectomycorrhizal species (Fig. 5.4C) is another example of fungus–plant communication where the fungus requires the plant in order to proceed in its development. Metabolites resulting from the photosynthesis of woody plants are transferred to the symbiontic fungi through interaction between root cells and the Hartig net (Fig. 5.4D) for general nutrition (Anderson and Cairney 2007) but this may not be all what is required from the host for fruiting body induction of the fungus. Otherwise, it should have been possible to produce mushrooms in the laboratory under good nutritional conditions without the plant host (Kües et al. 2007). As discussed in Sect. III.B, fruiting body development in saprotrophs is stimulated by surface-active compounds. There might be similar effects by specific plant metabolites on fruiting of ectomycorrhizal species.

Plant and fungi communicate with each other also above direct biotrophic interactions. Plant VOCs such as methyl salicylate for example triggers sporulation of entomopathogenic fungi (Hountondji et al. 2005, 2006) and plant oxilipins can replace endogenous oxylipins in control of growth and development in *Aspergillus* species (Mita et al. 2007; Brodhagen et al. 2008). Most fascinating are cases of hijacking plant signals addressed to pollinators by rust fungi for their spore distribution by insects (Kaiser 2006). Muscid and anthomyiid flies and halictid bees were shown to be attracted to phenylacetaldehyde, 2-phenylethanol, benzaldehyde, and methylbenzoate scents of pseudo-flowers induced by *Puccinia monoica* on crucifer hosts (Roy and Raguso 1997; Raguso and Roy 1998).

VI. Communication Between Fungi and Animals

Interactions between fungi and animals may serve different functions: the fungi may use animals as vectors for their distribution or for their nutrition, and the animals may enjoy the fungi as a resource of food for their living or use them as indicators for the availability of other food sources or also as indicators of places for breeding. Depending on the target, a fungus may send out signals for attraction or for repellence of animals. Usually, the infochemicals in fungus–animal communication are of a volatile and mostly organic nature (Thakeow et al. 2007). This is because fungi tend to be sessile whereas animals are commonly motile. Moreover, the large variety of different VOCs that can be formed by living cells offers multiple possibilities to develop and use very specific and discriminative chemical languages between interacting organisms according to need in a much more complex environment.

To give an example of how a sessile fungus takes advantage of mobile animals in an apparent symbiontic interaction, ascospores of the heterothallic grass endophyte *Epichloë typhina* are distributed by females of the fly *Brachypodium sylvaticum* that oviposit on the fungus (Bultman et al. 1998). Of the various volatiles produced simultaneously either by the fungus, its plant host, or by them collectively, the fungitoxic sesquiterpene alcohol chokol K of *Epichloe* species and the methylester methyl (*Z*)-3-methyldodec-2-enoate have been identified to be specific attractors of the female flies (Steinebrunner et al. 2008a, c). It is interesting to note that chokol K has a dual role for the fungus: one is attracting flies for spore distribution, another to repress growth of other fungi on the plant (Schiestl et al. 2006; Steinebrunner et al. 2008b).

Fungi may communicate with all types of animals up to the humans. Molds for example cause a damp-musty or an earthy-raunchy smell unpleasant for humans, resulting in avoidance reactions and refusal of infested food (Schnurer et al. 1999; Kuhn and Ghannoum 2003), whilst aromas from edible mushroom can be very attractive to humans (Chiron and Michelot 2006). Mycophagy within mammals is widely distributed and often they have a possible preference for hypogeal species such as truffles and false truffles, species that rely on mammals for spore distribution (Claridge and May 1994; Johnson 1996; Wheatley 2007). Strong and special aromas make them traceable for mammals below the covering soil (Gioacchini et al. 2005, 2008). Mycophagy is also found within mollusca (Silliman and Newell 2003; Fig. 5.5A), earthworms (Moody et al. 1996), nematodes (Mikola and Sulkava 2001), and various types of arthropods (Maraun et al. 2003; Fig. 5.5D, E; see below), combined with spore distribution by animal faeces and, for some fungal species, also with a positive effect on germination (see Fig. 5.5B, C, F, G and the cited papers).

Snails have been found to be attracted to fungi likely by their odours (Fig. 5.5A) but the nature of the VOCs causing their reaction needs to be elucidated. Certain mushrooms such as *Clitopilus prunulus* and *Clitocybe faccida* produce VOCs (e.g. 1-octene-3-ol, clitolactone) that protect them from being eaten by slugs (Wood et al. 2001, 2004). When being attracted to and consuming mushrooms, the snails also eat the spores which, together with undigested parts of hyphal filaments, are excreted with the faeces (Fig. 5.5B, C; Navarro-González, unpublished data). Currently, it remains to be shown whether spores and/or hyphal fragments from the mushrooms in faeces of snails will grow. However, faeces of the snail *Littoraria irrorata* grazing on fungal mycelium on infested cordgrass *Spartina alterniflora* are filled with hyphal fragments that appear able to grow. The snail distributes the faeces with fungi primarily of the ascomycetous genera *Phaeosphaeria* and *Mycosphaerella* onto new blades of cordgrass, where the fungi invade the cordgrass through wounds that the snails created purposely in the leafs (Silliman and Newell 2003).

Communication is best analysed between fungi and arthropods. Often it is 1-octen-3-ol as part of the typical mushroom odour that is used to specifically attract insects to mushrooms, such as the mycophagus beetle *Cis boleti* to the fruiting bodies of *Trametes gibbosa* (Thakeow 2008; Thakeow et al. 2008) and various wood-living beetles and the moth *Epinotia tedella* to fruiting bodies of wood-decaying Polypores (Fäldt et al. 1999). Strikingly, in case of the beetles it is preferentially again the females that are attracted by the fungal odour (Fäldt et al. 1999; Thakeow et al. 2008). This appears also be the case for the truffle beetle *Leiodes cinnamomea*: obviously only the females are attracted in a yet unknown way to *Tuber melanosporum* fruiting bodies, whilst the male insects are not attracted by the fungus but secondarily by the females infesting the truffles (Hochberg et al. 2003).

Fungal odours can also show the way for predators – the combination of 1-octen-3-ol and octan-3-none acts as a magnet for the predator on fungus-insects *Lordithon lunulatus* (Fäldt et al. 1999). In other instances, 1-octen-3-ol in combination with methyl cinnamate repel the mycophagus collembolan *Proisotoma minuta* and the production of these VOCs by *T. matsutake* protects its mushrooms against *P. minuta* (Sawahata et al. 2008). On its own as pure compound, 1-octen-3-ol acts as deterrent to both sexes of the phorid fly *Megaselia halteria*, a pest in the cultivation of the edible *A. bisporus* (Pfeil and Mamma 1993), whilst in combination with other C8 compounds (3-octanone, 1-octen-3-one) it may attract females to the mushroom culturing beds (Grove and Blight 1983).

Mites feared by mycological laboratories as pests in fungal cultures also use the aliphatic alcohol 1-octen-3-ol and related C8 compounds as signalling molecules to localize the emitting fungi in order to feed on them (van Haelen et al. 1978; Chaisaena 2008; Fig. 5.6); and a case where a nematode was attracted by *B. cinerea* through 1-octen-3-ol as well as 3-octanol has also been documented (Matsumori et al. 1989). In contrast, the banana slug *Ariolimax columbianus* reacts deterred when confronted with 1-octen-3-ol (Wood et al. 2001). Future studies need to reveal how much it is also a matter of concentration whether the different organisms sense 1-octen-3-ol as appealing or repellent.

Mushrooms of different species all have their own special aroma composed of different combinations of VOCs from chemical categories such as alcohols, aldehydes, ketones, and isoprenoids (e.g. see Mau et al. 1997; Venkateshwarlu et al. 1999; Boustie et al. 2005; Wu et al. 2005; Splivallo et al. 2007a; Thakeow 2008; see the review by Chiron and Michelot 2005). Therefore probably not surprisingly mushroom–insect interactions can be very specific, particularly for fresh mushrooms, whereas this is found not so much for decaying fruiting bodies (Guevara et al. 2000; Jonsell and Nordlander 2004; Orledge and Reynolds 2005). It is possibly the combination of different volatiles emitted by fresh specimens that leads to distinct attractions for insects. In contrast, decaying mushrooms have less striking VOC differences and become chemically more similar to each other (Jonsell and Nordlander 2004; Thakeow et al. 2007).

A paradigm for successive alterations in arthropode attraction is presented by the release of changing VOCs during the course of wood decay by fungi. Saproxylic insects are attracted by wood–rotting fungi to the resource wood – as the fungal wood decay progresses, an alteration in VOC patterns is detected which tells the insects the momentary levels of decay and whether it is at a suitable stage for a given insect to infest the wood for feeding and/or breeding (Holighaus and Schütz 2006; Thakeow et al. 2007).

VII. Conclusions

All the time, an ever-changing plethora of secondary metabolites augmented by proteinaceous compounds is found in the biosystems produced as a sum by the various bacteria, the lower eukaryotes such as fungi, the plants, and/or the animals living in a common habitat – not to forget that there are in addition also compounds coming from spontaneous chemical reactions occurring in the same environment. Specific compounds in such complex mixtures referred to as infochemicals have distinct communication functions within a species and between different species on dual

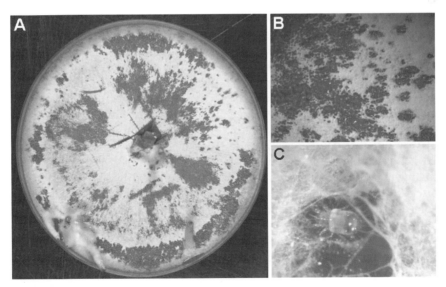

Fig. 5.6. Dust mites of the species *Tyrophagus putrescentiae* are easily attracted to fungal cultures in order to consume the obviously tasty mycelium of the fungus *Coprinopsis cinerea*. **A** Overview and **B** enlarged sector of a *C. cinerea* culture infested by mites: Perforations in the mycelial carpet indicate loss of mycelium by grazing mites. **C** A mite grazing at the mycelium

up to multiple organisms levels. The different functions might simultaneously come into action.

1-Octen-3-d and mushroom development may be taken as an example how at the same time one single compound may be used as a messenger for different targets: in the induction of mushroom development, in the attraction of animals for spore distribution, as a repellent of animals that otherwise might just graze on the fungus, and as a toxic compound reacting on mycotrophic moulds and bacteria that might infest and decompose mushrooms for their feed or that might suppress mushroom development. Using one and the same compound for all these different targets happening in parallel suggests a tight co-evolution between different organisms in a given biosystem. This reasonable avenue raises however the question: how does it come that ecology does not go wrong in such a complex environment if so many different organisms react in so different way on the same signal? Schiestl et al. (2006) recently suggested that defence mechanisms developed first in evolution and that defence chemicals were then appointed secondarily into a function of messenger molecules. To verify the different functions, one can expect that a fine-tuning of sensitivity via adaptation to specific compound concentrations will be very important and, connected to this, the steepness of the gradients of signalling molecules being sent out from a source to a receiver (Matsui 2006; Mita et al. 2007). Particularly the VOCs offer special qualities to function as messengers since in most instances they easily move in air and, by that, they are also very easily removed from a system once they are no longer required.

To fulfil their functions as infochemicals, the language of messenger molecules must be shared by a sender and a receiver. A multitude of chemical compounds of very different natures has already been identified acting as infochemicals. In most instances, signalling molecules are types of secondary metabolites. Like 1-octen-3ol, they are not always very specific to just one or a few combinations of communicating organisms. Specificity might be more enhanced with organic compounds of chemically higher complexity, such as for example sesquiterpenes. Of all, peptides are the most specific, due to the multitude of possible combinations of amino acids giving a distinct sequence. Not surprisingly therefore, many of the fungi appoint peptides in sexual attraction to each other, being one of the most specific biological tasks to fulfil in ecology and evolution.

Currently, we are only at the beginning of perceiving how complex chemical communication is in nature, both on the level of individuals and, generally, on the whole community level. We do not suggest that our literature overview is complete, neither in the examples where communication with a fungus plays a role nor in the type of messenger molecules used – there are too many individual observations reported which are often unlinked to each other and which often are not deeply enough evaluated for judging whether they present a true communication system. One reason for this is that it is difficult to consider the complete complexity within an ecosystem. Furthermore, too often evidence for the actions of signalling molecules is still only circumstantial. Better chemical definitions of signalling compounds and their production and better definitions of their concentration spectra of activity are a prerequisite for understanding the various chemical languages fungi may use in communication with their environments.

With the apparent exception of the mating systems using diffusible peptide pheromones (see above-cited reviews on fungal mating type systems), the way that signals are perceived by the receiver and the subsequent intracellular signalling pathways within the sensitized cells are in most instances fully unclear (Ugalde 2006). Establishing specific receptors for chemical communication and linking this to intracellular communication (signalling) cascades is part of evolving a chemical language in environments in which naturally a high noise of different chemical compounds exists. Elucidating these will be the exigent task of future work in molecular and cellular biology on the road to unravelling the various mechanisms of communication systems that exist in nature. A very exciting question remains for theoretical biologists and behavioural scientists (understood in a broader sense than in the usual limitation to animals) to go more deeply into the ecological and evolutionary functions and consequences of communication systems, including those in which fungi participate.

Acknowledgements The authors are very grateful to U. Helle, A. Olbrich, and E. Fritz for supplying the photos on ectomycorrhizal fungi. Research in our laboratory on fungal volatiles and on mites is done in collaboration with P. Thakeow, P. Plašil, and S. Schütz from the Division

Forest Zoology and Forest Conservation at the Büsgen-Institute Göttingen, and research on mating type genes in collaboration with international consortia on analysing genomes of higher basidiomycetes supplied by the DOE Joint Genome Institute in Walnut Creek, California. Research on *L. bicolor* is done within the framework of the Evoltree programme of the European Community.

References

Abe K, Kusaka I, Fukui S (1975) Morphological change in the early stages of the mating process of *Rhodosporidium toruloides*. J Bacteriol 122:710–718

Ahmad SS, Miles PG (1970a) Hyphal fusions in the wood-rotting fungus *Schizophyllum commune*. I. The effect of incompatibility factors. Genet Res Camb 15:19–28

Ahmad SS, Miles PG (1970b) Hyphal fusions in *Schizophyllum commune*. 2. Effects of environmental and chemical factors. Mycologia 62:1008–1017

Aimi T, Yoshida R, Ishikawa M, Bao DP, Kitamoto Y (2005) Identification and linkage mapping of the genes for the putative homeodomain protein (*hox1*) and the putative pheromone receptor protein homologue (*rbc1*) in a bipolar basidiomycete, *Pholiota nameko*. Curr Genet 48:184–194

Akiyama K, Matsuzaki K, Hayashi H (2005) Plant sesquiterpenes induce hyphal branching in arbuscular mycorrhizal fungi. Nature 435:824–827

Alem MAS, Oteef MDY, Flowers TH, Douglas LJ (2006) Production of tyrosol by *Candida albicans* biofilms and its role in quorum sensing and biofilm development. Eukaryot Cell 5:1770–1779

Ali NA, Jackson RM (1988) Effects of plant roots and their exudates on germination of spores of ectomycorrhizal fungi. Trans Br Mycol Soc 91:253–260

Allen PJ (1972) Specificity of the cis-isomers of inhibition of uredospores germination in the rust fungi. Proc Natl Acad Sci USA 69:3497–3500

Anderson IC, Cairney JWG (2007) Ectomycorrhizal fungi: exploring the mycelial frontier. FEMS Microbiol Rev 31:388–406

Arima T, Yamamoto M, Hirata A, Kawano S, Kamada T (2004) The *eln3* gene involved in fruiting body morphogenesis of *Coprinus cinereus* encodes a putative membrane protein with a general glycosyltransferase domain. Fungal Genet Biol 41:805–812

Azuma M, Hori K, Ohashi Y, Yoshida M, Horinouchi S, Beppu T (1990) Basidifferquinone, a new inducer for fruiting-body formation of a basidiomycete *Favolus arcularius* from a *Streptomyces* strain. II. Structure of basidifferquinone. Agric Biol Chem 54:1447–1452

Badalyan SM, Polak E, Hermann R, Aebi M, Kües U (2004) Role of peg formation in clamp cell fusion of homobasidiomycete fungi. J Basic Microbiol 44:167–177

Bais HP, Park SW, Weir TL, Callaway RM, Vivanco JM (2004) How plants communicate using the underground information superhighway. Trends Plant Sci 9:26–32

Bais HP, Weir TL, Perry LG, Gilroy S, Vivanco JM (2006) The role of root exudates in rhizosphere interactions with plants and other organisms. Ann Rev Plant Biol 57:233–266

Bakkeren G, Kronstad JW (2007) Bipolar and tetrapolar mating systems in the Ustilaginales. In: Heitman J, Kronstad JW, Taylor JW, Casselton LA (eds) Sex in fungi. Molecular determination and evolutionary implications. ASM, Washington, D.C., pp 389–404

Bandoni RJ (1965) Secondary control of conjugation in *Tremella mesenterica*. Can J Bot 41:467–474

Barlow PW (2008) Reflections on 'plant neurobiology'. Biosystems 92:132–147

Barr JG (1976) Effects of volatile bacterial metabolites on growth, sporulation and mycotoxin production of fungi. J Sci Food Agricult 27:324–330

Basith M, Madelin MF (1968) Studies on production of perithecial stromata by *Cordyceps militaris* in artificial culture. Can J Bot 46:473–480

Bauch R (1932) Untersuchungen über die Entwicklungsgeschichte und Sexualphysiologie der *Ustiloga bromivera* und *Ustilago grandis*. Z Bot 17:129–177

Beltran-Garcia MJ, Estarron-Espinosa M, Ogura T (1997) Volatile compounds secreted by the oyster mushroom (*Pleurotus ostreatus*) and their antibacterial activities. J Agric Food Chem 45:4049–4052

Berbara RLL, Morris BM, Fonseca HMAC, Reid B, Gow NAR, Daft MJ (1995) Electrical currents associated with arbuscular mycorrhizal interactions. New Phytol 129:433–438

Berne S, Pohleven J, Vidic I, Rebolj K, Pohleven F, Turk T, Maček P, Sonnenberg A, Sepčić K (2007) Ostreolysin enhances fruiting initiation in the oyster mushroom (*Pleurotus ostreatus*). Mycol Res 111:1431–1436

Berne S, Pohleven F, Turk T, Sepčić K (2008) Induction of fruiting in oyster mushroom (*Pleurotus ostreatus*) by polymeric 3-alkylpyridinium salts. Mycol Res 112:1085–1087

Berrocal-Lobo M, Molina A (2008) *Arabidopsis* defense response against *Fusarium oxysporum*. Trends Plant Sci 13:145–150

Besserer A, Puech-Oagès V, Kiefer P, Gomez-Roldan V, Jauneau A, Roy S, Portais JC, Roux C, Bécard G, Séjalon-Delmas N (2006) Strigolactones stimulate arbuscular mycorrhizal fungi by activating mitochondria. PLOS Biol 4:1239–1247

Bhuiyan MKA, Arai K (1993) Physiological factor affecting hyphal growth and fusion of *Rhizoctonia oryzae*. Trans Mycol Soc Jpn 32:389–397

Bistis GN (1983) Evidence for diffusible, mating-type-specific trichogyne attractants in *Neurospora crassa*. Exp Mycol 7:292–295

Bistis GN (1998) Physiological heterothallism and sexuality in euascomycetes: a partial history. Fungal Genet Biol 23:213–222

Bjurman J, Fries N (1984) Purification and properties of the germination-inducing factor in the ectomycorrhizal fungus *Leccinium aurantiacum* (Boletaceae). Physiol Plant 62:465–471

Boddy L (2000) Interspecific combative interactions between wood-decaying basidiomycetes. FEMS Microbiol Ecol 31:185–194

Boustie J, Rapior S, Fortin H, Tomasi S, Bessiere JM (2005) Chemotaxonomic interest of volatile components in

Lepista inversa and *Lepista flaccida* distinction. Cryptogam Mycol 26:27–35

Bouwmeester HJ, Roux C, Lopez-Raez JA, Bécard G (2007) Rhizosphere communication of plants, parasitic plants and AM fungi. Trends Plant Sci 12:224–230

Brenner ED, Stahlberg R, Mancuso S, Vivanco J, Baluska F, Van Volkenburgh E (2006) Plant neurobiology: an integrated view of plant signalling. Trans Plants Sci 11:413–419

Breeuwer P, de Reu JC, Drocourt JL, Rombouts FM, Abee T (1997) Nonanoic acid, a fungal self-inhibitor, prevents germination of *Rhizopus oligosporus* sporangiospores by dissipation of the pH gradient. Appl Environ Microbiol 63:178–185

Brodhagen M, Tsitsigiannis DI, Hornung E, Goebel C, Feussner I, Keller NP (2008) Reciprocal oxylipin-mediated cross-talk in the *Aspergillus*-seed pathosystem. Mol Microbiol 67:378–391

Brotman Y, Briff E, Viterbo A, Chet I (2008) Role of swollenin, an expansin-like protein from *Trichoderma*, in plant root colonization. Plant Physiol 147:779–789

Brown AJ, Casselton LA (2001) Mating in mushrooms: increasing the chances but prolonging the affair. Trends Genet 17:393–400

Brown SH, Zarnowski R, Sharpee WC, Keller NP (2008) Morphological transitions governed by density dependence and lipoxygenase activity in *Aspergillus flavus*. Appl Env Microbiol 74:5674–5686

Bücking H, Abubaker J, Govindarajulu M, Tala M, Pfeffer PE, Nagahashi G, Lammers P, Sachar-Hill Y (2008) Root exudates stimulate uptake and metabolism of organic carbon in perminating spores of *Glomus intraradices*. New Phytol 180:684–695

Buller AHR (1931) Researches on fungi. IV. Further observations on the Coprini together with some investigations on social organisation and sex in the Hymenomycetes. Longmans, London

Buller AHR (1933) Researches on fungi. V. Hyphal fusions and protoplasmic streaming in the higher fungi, together with an account of the production and liberation of spores in *Sporobolomyces*, *Tilletia*, and *Sphaerobolus*. Longmans, London

Bultman TL, White JF, Bowdish TI, Welch AM (1998) A new kind of mutualism between fungi and insects. Mycol Res 102:235–238

Burmester A, Richter M, Schultze K, Voelz K, Schachtschabel D, Boland W, Wöstemeyer J, Schimek C (2007) Cleavage of β-carotene as the first step in sexual hormone synthesis in zygomycetes is mediated by a trisporic acid regulated β-carotene oxygenase. Fungal Genet Biol 44:1096–1108

Burt WR (1982) Identification of coprogen B and its breakdown products from *Histoplasma capsulatum*. Infect Immun 35:990–996

Burt WR, Underwood AL, Appleton GL (1981) Hydroxamic acid from *Histoplasma capsulatum* that displays growth factor activity. Appl Env Microbiol 42:560–563

Calvo AM, Hinze L, Gardner HW, Keller NP (1999) Sporogenic effect of polyunsaturated fatty acid on development of *Aspergillus* spp. Appl Env Microbiol 65:3668–3673

Calvo AM, Gardner HW, Keller NP (2001) Genetic connection between fatty acid metabolism and sporulation in *Aspergillus nidulans*. J Biol Chem 276:25766–25774

Calvo AM, Wilson RA, Bok JW, Keller NP (2002) Relationship betwen secondary metabolism and fungal development. Microbiol Mol Biol Rev 66:447–459

Camilli A, Bassler BL (2006) Bacterial small-molecule signalling pathways. Science 311:1113–1116

Casselton LA, Challen MP (2006) The mating type genes of the basidiomycetes. In: Kües U, Fischer R (eds) The mycota, vol I, 2nd edn. Growth, differentiation and sexuality. Springer, Heidelberg, pp 357–374

Casselton LA, Kües U (2007) The origin of multiple mating types in the model mushrooms *Coprinopsis cinerea* and *Schizophyllum commune*. In: Heitman J, Kronstad JW, Taylor JW, Casselton LA (eds) Sex in fungi. Molecular determination and evolutionary implications. ASM, Washington, D.C., pp 283–300

Casselton LA, Olesnicky NS (1998) Molecular genetics of mating recognition in basidiomycete fungi. Microbiol Mol Biol Rev 62:55–70

Chaisaena W (2008) Light effects on fruiting body development of wildtype in comparison to light-insensitive mutant strains of the basidiomycete *Coprinopsis cinerea*, grazing of mites (*Tyrophagus putrescentiae*) on the strains and production of volatile organic compounds during fruiting body development. Dissertation, Georg-August-University of Göttingen

Champe SP, Rao P, Chang A (1987) An endogenous inducer of sexual development in *Aspergillus nidulans*. J Gen Microbiol 133:1383–1387

Chen H, Fink GR (2006) Feedback control of morphogenesis in fungi by aromatic alcohols. Genes Dev 20:1150–1161

Chen H, Fujita M, Feng QH, Clardy J, Fink GR (2004) Tyrosol is a quorum-sensing molecule in *Candida albicans*. Proc Natl Acad Sci USA 1001:5048–5052

Chenevert J, Valtz N, Herskowitz I (1994) Identification of genes required for normal pheromone-induced cell polarization in *Saccharomyces cerevisiae*. Genetics 136:1287–1296

Chiron N, Michelot D (2005) Mushroom odors, chemistry and role in the biotic interactions – a review. Cryptogam Mycol 26:299–364

Chitarra GS, Abee T, Rombouts FM, Posthumus MA, Dijksterhuis J (2004) Germination of *Penicillium paneum* conidia is regulated by 1-octen-3-ol, a volatile self-inhibitor. Appl Environ Microbiol 70:2823–2829

Chitarra GS, Abee T, Rombouts FM, Dijksterhuis J (2005) 1-Octen-3-ol inhibits conidia germination of *Penicillium paneum* despite of mild effects on membrane permeability, respiration, intracellular pH, and changes the protein composition. FEMS Microbiol Ecol 54:67–75

Claridge AW, May TW (1994) Mycophagy among Australian mammals. Australian J Ecol 19:251–275

Combet E, Henderson J, Eastwood DC, Burton KS (2006) Eight-carbon volatiles in fungi: properties, analysis, and biosynthesis. Mycoscience 47:317–326

Coppin E, de Renty C, Debuchy R (2005) The function of the coding sequences for the putative pheromone precursors in *Podospora anserina* is restricted to fertilization. Eukaryot Cells 4:407–420

Cruz C, Noël-Suberville C, Montury M (1997) Fatty acid content and some flavour compound release in two strains of *Agaricus bisporus*, according to three stages of development. J Agric Food Chem 45:64–67

Dagenais TRT, Chung D, Giles SS, Hull CM, Andes D, Keller NP (2008) Defects in conidiophore development and conidium–macrophage interactions in a dioxygenase mutant of *Aspergillus fumigatus*. Infect Immun 76:3214–3220

Daniels R, Vanderleyden J, Michiels J (2004) Quorum sensing and swarming migration in bacteria. FEMS Microbiol Rev 28:261–289

Debuchy R, Turgeon BG (2006) Mating-type structure, evolution, and function in Euascomycetes. In: Kües U, Fischer R (eds) The mycota, vol I, 2nd edn. Growth, differentiation and sexuality. Springer, Heidelberg, pp 293–324

Djonovic S, Pozo MJ, Dangott LJ, Howell CR, Kenerly CM (2006) Sm1, a proteinaceous elicitor secreted by the biocontrol fungus *Trichoderma virens* induces plant defense responses and systemic resistance. Mol Plant Microb Interact 19:838–853

Dudareva N, Negre F, Nagegowda DA, Orlova I (2006) Plant volatiles: recent advances and future perspectives. Crit Rev Plant Sci 25:417–440

Duntze W, Mackay V, Manney TR (1970) *Saccharomyces cerevisiae* – a diffusible sex factor. Science 168:1472–1473

Dyer PS, Ingram DS, Johnston K (1992) The control of sexual morphogenesis in the ascomycetes. Biol Rev 67:421–458

Dyer PS, Ingram DS, Johnstone K (1993) Evidence for the involvement of linoleic-acid and other endogenous lipid factors in perithecial development of *Nectria haematococca* mating population 6. Mycol Res 97:485–496

Ebel J, Scheel D (1997) Signals in host-parasite interactions. In: Carroll GC, Tudzynski P (eds) The mycota, vol V. Plant relationships, part A. Springer, Heidelberg, pp 85–196

Erb-Downward JR, Huffnagle GB (2006) Role of oxylipins and other lipid mediators in fungal pathogenesis. Future Microbiol 1:219–227

Esser K (2006) Heterogenic incompatibility in fungi. In: Kües U, Fischer R (eds) The Mycota, vol I, 2nd edn. Growth, differentiation and sexuality. Springer, Heidelberg, pp 141–166

Fäldt J, Jinsell M, Nordlander G, Borg-Karlson AK (1999) Volatiles of bracket fungi *Fomitopsis pinicola* and *Fomes fomentarius* and their functions as insect attractants. J Chem Soc 25:567–590

Feldbrügge M, Bölker M, Steinberg G, Kämper J, Kahmann R (2006) Regulatory and structural networks orchestrating mating, dimorphism, cell shape, and pathogenesis in *Ustilago maydis*. In: Kües U, Fischer R (eds) The mycota, vol I, 2nd edn. Growth, differentiation and sexuality. Springer, Heidelberg, pp 375–392

Flegel TW (1981) The pheromonal control of mating in yeast and its phylogenetic implication, a review. Can J Microbiol 27:373–389

Fleissner A, Sarkar S, Jacobson DJ, Roca MG, Read ND, Glass NL (2005) The *so* locus is required for vegetative cell fusion and postfertilization events in *Neurospora crassa*. Eukaryot Cell 4:920–930

Fowler TJ, DeSimone SM, Mitton MF, Kurjan J, Raper CA (1999) Multiple sex pheromones and receptors of a mushroom-producing fungus elicit mating in yeast. Mol Biol Cell 10:2559–2572

Frey-Klett P, Garbaye J, Tarkka M (2007) Tansley review. The mycorrhiza helper bacteria revisited. New Phytol 176:22–36

Fries N (1978) Basidiospore germination in some mycorrhiza-forming hymenomycetes. Trans Br Mycol Soc 70:319–324

Fries N (1981) Recognition reactions between basidiospores and hyphae in *Leccinum*. Trans Br Mycol Soc 77:91–94

Fries N (1983a) Spore germination, homing reaction, and intersterility groups in *Laccaria laccata* (Agaricales). Mycologia 75:221–227

Fries N (1983b) Intra- and interspecific basidiospore homing reactions on *Leccinum*. Trans Brit Mycol Soc 81:559–561

Fries N (1989) The influence of tree roots on spore germination of ectomycorrhizal fungi. Agric Ecosyst Environ 28:139–144

Garrett MK, Robinson PM (1969) A stable inhibitor of germination produced by fungi. Arch Microbiol 67:370–377

Genre A, Bonfante P (2007) Check-in procedures for plant cell entry by biotrophic microbes. Mol Plant Microb Interact 20:1023–1020

Gioacchini AM, Menotta M, Bertini L, Rossi I, Zeppa S, Zambonelli A, Piccoli G, Stocchi V (2005) Solid-phase microextraction gas chromatography/mass spectrometry: a new method for species identification of truffles. Rapid Commun Mass Spec 19:2365–2370

Gioacchini AM, Menotta M, Guescini M, Saltarelli R, Ceccaroli P, Amicucci A, Barbieri E, Giomaro G, Stocchi V (2008) Geographical traceability of Italian white truffle (*Tuber magnatum* Pico) by the analysis of volatile organic compounds. Rapid Commun Mass Spec 22:3147–3153

Glass NJ, Fleissner A (2006) Re-wiring the network: understanding the mechanism and function of anastomosis in filamentous ascomycete fungi. In: Kües U, Fischer R (eds) The mycota, vol I, 2nd edn. Growth, differentiation and sexuality. Springer, Heidelberg, pp 123–140

Glass NL, Jacobson DJ, Shiu PKT (2000) The genetics of hyphal fusion and vegetative incompatibility in filamentous ascomycete fungi. Annu Rev Genet 34:165–186

Glass NL, Rasmussen C, Roca MG, Read ND (2004) Hyphal homing, fusion and mycelial interconnectedness. Trends Microbiol 12:135–141

Göhre V, Robatzek S (2008) Breaking the barriers: Microbial effector molecules subvert plant immunity. Annu Rev Phytopathol 46:189–215

Gooday DW (1994) Hormones in mycelial fungi. In: Wessels JGH, Meinhardt F (eds) The Mycota, vol I, 1st edn. Growth, differentiation and sexuality. Springer, Heidelberg, pp 401–411

Gori K, Mortensen HD, Arneborg N, Jespersen L (2007) Ammonia production and its possible role as a mediator of communication for *Debaryomyces hansenii* and other cheese-relevant yeast species. J Dairy Sci 90:5032–5041

Gow NAR (1989) Circulating ionic currents in microorganisms. Adv Microb Physiol 30:89–123

Grechkin A (1998) Recent developments in biochemistry of the plant lipoxygenase pathway. Prog Lipid Res 37:317–352

Gregory P (1984) The fungal mycelium: a historical perspective. Trans Br Mycol Soc 82:1–11

Grove JF, Blight MM (1983) The oviposition attractant for the mushroom phorid *Megasiella halterata*: the identification of volatiles present in mushroom house air. J Sci Food Agric 34:181–185

Guevara R, Hutcheson KA, Mee AC, Rayner ADM, Reynolds SE (2000) Resource partitioning of the host fungus *Coriolus versicolor* by two ciid beetles: the role of odour compounds and host aging. Oikos 91:184–194

Harman GE, Howell CR, Viterbo A, Chet I, Lorito M (2004) *Trichoderma* species – opportunistic, avirulent plant symbionts. Nature Rev 2:43–56

Harold FM, Schreurs WJ, Harold RL, Caldwell JH (1985) Electrobiology of the fungal hyphae. Microbiol Sci 2:363–366

Hegner J, Siebert-Bartholmei C, Kothe E (1999) Ligand recognition in multiallelic pheromone receptors from the basidiomycete *Schizophyllum commune* studied in yeast. Fungal Genet Biol 26:190–197

Heitman J, Kronstad JW, Taylor JW, Casselton LA (eds) (2007) Sex in fungi. Molecular determination and evolutionary implications. ASM, Washington, D.C.

Hennebert GL, Pascal S, Cosyny M (1994) Interactions d'incompatibilité entre homocaryons du basidiomycéte bifactoriel *Lenzites betulinus*. Une pheromone de repulsion sexuelle?. Cryptogam Mycol 15:83–116

Hibbett DS, Binder M, Bischoff JF, Blackwell M, Cannon PF, Eriksson OE, Hundorf S, James T, Kirk PM, Lücking R, Lumbsch HAT, Lutzoni F, Matheny PB, McLaughlin DJ, Powell MJ, Redhead S, Schoch CL, Spatafora JW, Stalpers JA, Vilgalys R, Aime MC, Aptroot A, Bauer R, Begerow D, Benny GL, Castlebury LA, Crous PW, Dai YC, Gams W, Geiser DM, Griffith GW, Gueidan C, Hawskworth DL, Hestmark G, Hosaka K, Humber RA, Hyde KD, Ironside JE, Koljalg U, Kurtzman CP, Larsson KH, Lichtwardt R, Longcore J, Miadlikowska J, Miller A, Moncalvo JM, Mozley-Standrige S, Oberwinkler F, Parasto E, Reeb V, Rogers JD, Roux C, Ryvarden L, Sampaio JP, Schüssler A, Sugiyama J, Thorn RG, Tibell L, Untereiner WA, Walker C, Wang Z, Weir A, Weiss M, White MM, Winka K, Yoa YJ, Zhang N (2007) A higher-level phylogenetic classification of the Fungi. Mycol Res 111:509–547

Hickey PC, Jacobson DJ, Read ND, Glass NL (2002) Live-cell imaging of vegetative hyphal fusion in *Neurospora crassa*. Fungal Genet Biol 37:109–119

Hiscock SJ, Kües U (1999) Cellular and molecular mechanisms of sexual incompatibility in plants and fungi. Int Rev Cytol 193:165–295

Hochberg ME, Bertault G, Poitrineau K, Janssen A (2003) Olfactory orientation of the truffle beetle, *Leiodes cinnamomea*. Entomol Exp Appl 109:147–153

Hogan DA (2006) Talking to themselves: autoregulatory and quorum sensing in fungi. Eukaryot Cell 5:613–619

Hogan DA, Vik A, Kolter R (2004) A *Pseudomonas aeruginosa* quorum-sensing molecule influences *Candida albicans* morphology. Mol Microbiol 54:1212–1223

Holighaus G, Schütz S (2006) Odours of wood decay as semiochemicals for *Xyloterus demesticus*, L. (Col., Scolytidae). Mittlg Deutsch Ges allg angew Entomol 15:161–165

Hornby JM, Jensen EC, Lisec AD, Tasto JJ, Jahnke B, Shoemaker R, Dussault P, Nickerson KW (2001) Quorum sensing in the dimorphic fungus *Candida albicans* is mediated by farnesol. Appl Env Microbiol 67:2982–2992

Hornby JM, Jacobitz-Kizzier SM, McNeel DJ, Jensen EC, Treves DS, Nickerson KW (2004) Inoculum size effect in dimorphic fungi: extracellular control of yeast-mycelium dimorphism in *Ceratocystis ulmi*. Appl Env Microbiol 70:1356–1359

Hountondji FCC, Sabelis MW, Hanna R, Janssen A (2005) Herbovore-induced plant volatiles trigger sporulation in entomopathogenic fungi: The case of *Neozygites tanajoae* infecting the cassava green mite. J Chem Ecol 31:1003–1021

Hountondji FCC, Hanna R, Sabelis MW (2006) Does methyl salicylate, a component of herbivore-induced plant odour, promote sporulation of the mite-pathogenic fungus *Neozygites tanajoae*? Exp Appl Agrarol 39:63–74

Idnurm A, James TY, Vilgalys R (2007) Sex in the rest: mysterious mating in the Chytridiomycota and Zygomycota. In: Heitman J, Kronstad JW, Taylor JW, Casselton LA (2007) Sex in fungi. Molecular determination and evolutionary implications. ASM, Washington, D.C., pp 407–418

Ikediugwu FEO (1976) Interface in hyphal interference by *Peniophora gigantea* against *Heterobasidion annosum*. Trans Br Mycol Soc 66:291–296

Ikediugwu FEO, Webster K (1970) Hyphal interference in a range of coprophilus fungi. Trans Br Mycol Soc 54:205–210

Imai Y, Yamamoto M (1994) The fission yeast mating pheromone P-factor, its molecular structure, gene structure, and ability to induce gene expression and $G(1)$ arrest in the mating partner. Genes Dev 8:328–338

Inoue M, Mori N, Yamanaka H, Tsurushima T, Miyagawa H, Ueno T (1996) Self-germination inhibitors from *Colletotrichum fragariae*. J Chem Ecol 22:2111–2122

Ishida TA, Nara K, Tanaka M, Kinoshita A, Hogetsu T (2008) Germination and infectivity of ectomycorrhizal fungal spores in relation to their ecological traits during primary succession. New Phytol 180:491–500

Ishitani C, Sakaguchi K-I (1956) Hereditary variation and recombination in koji-molds (*Aspergillus oryzae* and

Asp. sojae). V. Heterocaryosis. J Gen Appl Microbiol 2:345–400

Iwase K (1992) Induction of basidiospore germination by gluconic acid in the ectomycorrhizal fungus *Tricholoma robustum*. Can J Bot 70:1234–1238

James TY, Liou SR, Vilgalys R (2004) The genetic structure and diversity of the *A* and *B* mating-type genes from the tropical oyster mushroom, *Pleurotus djamor*. Fungal Genet Biol 41:813–825

James TY, Srivilai P, Kües U, Vilgalys R (2006) Evolution of the bipolar mating system of the mushroom *Coprinellus disseminatus* from its tetrapolar ancestors involves loss of mating-type-specific pheromone receptor function. Genetics 172:1877–1891

Jensen EC, Ogg C, Nickerson KW (1992) Lipoxygenae inhibitors shift the yeast/mycelium dimorphism in *Ceratocystis ulmi*. Appl Environ Microbiol 58:2505–2508

Johnson CN (1996) Interactions between mammals and ectomycorrhizal fungi. Trends Ecol Evol 11:503–507

Jonsell M, Nordlander G (2004) Host selection patterns in insects breeding in bracket fungi. Ecol Entomol 29:697–705

Kaiser R (2006) Flowers and fungi use scents to mimic each other. Science 311:806–807

Kamada T, Tsuji M (1979) Darkness-induced factor affecting basidiocarp maturation in *Coprinus macrorhizus*. Plant Cell Physiol 20:1445–1448

Kellner M, Burmester A, Wöstemeyer A, Wöstemeyer J (1993) Transfer of genetic information from the mycoparasite *Parasitella parasitica* to its host *Absidia glauca*. Curr Genet 23:334–337

Kemp ROF (1975) Breeding biology of *Coprinus* species in the section Lanatuli. Trans Br Mycol Soc 65:375–388

Kemp ROF (1977) Oidial homing and the taxonomy and speciation of basidiomycetes with special reference to the genus *Coprinus*. In: Clémençon H (ed) The species concept in Hymenomycetes. Cramer, Vaduz, pp 259–276

Kikuchi K, Matsushita N, Suzuki K, Hogetsu T (2007) Flavonoids induce germination of basidiospores of the ectomycorrhizal fungus *Suillus bovinus*. Mycorrhiza 17:563–570

Kim H, Borkovich KA (2004) A pheromone receptor gene, *pre-1*, is essential for mating type-specific directional growth and fusion of trichogynes and female fertility in *Neurospora crassa*. Mol Microbiol 52:1781–1798

Kim H, Borkovich KA (2006) Pheromones are essential for male fertility and sufficient to direct chemotropic polarized growth of trichogynes during mating in *Neurospora crassa*. Eukaryot Cell 5:544–554

Kim H, Metzenberg RL, Nelson MA (2002) Multiple functions of *mfa-1*, a putative pheromone precursor gene of *Neurospora crassa*. Eukaryot Cell 1:987–999

Kim H-J, Lee T, Yun SW-H (2008) A putative pheromone signaling pathway is dispensible for self-fertility in the homothallic ascomycete *Gibberella zeae*. Fungal Genet Biol 45:1188–1196

Köhler E (1930) Zur Kenntnis der vegetativen Anastomosen der Pilze. II. Mitteilung. Planta 10:495–522

Kothe E (1996) Tetrapolar fungal mating types: Sexes by the thousands. FEMS Microbiol Rev 18:65–87

Kothe E (2008) Sexual attraction: On the role of fungal pheromone/receptor systems (a review). Acta Microbiol Immunol Hung 55:125–143

Kügler S, Schurtz Sebghati T, Groppe Eissenberg L, Goldman WE (2000) Phenotypic variation and intracellular parasitism by *Histoplasma capsulatum*. Proc Natl Acad Sci USA 97:8794–8799

Kuhn DM, Ghannoum MA (2003) Indoor mold, toxigenic fungi, and *Stachybotrys chartarum*: Infectious disease perspective. Clin Microbiol Rev 16:144–172

Kües U (2000) Life history and developemental processes in the basidiomycete *Coprinus cinereus*. Microbiol Mol Biol Rev 64:316–353

Kües U, Liu Y (2000) Fruiting body production in basidiomycetes. Appl Microbiol Biotechnol 54:141–152

Kües U, Walser PJ, Klaus MJ, Aebi M (2002) Influence of activated *A* and *B* mating-type pathways on developmental processes in the basidiomycete *Coprinus cinereus*. Mol Gen Genom 268:262–271

Kües U, Künzler M, Bottoli APF, Walser PJ, Granado PJ, Liu Y, Bertossa RC, Ciardo D, Clergeot PH, Loos S, Ruprich-Robert G, Aebi M (2004) Mushroom development in higher basidiomycetes: implication for human and animal health. In: Kushwaha RKS (ed) Fungi in human and animal health. Scientific Publishers, Jodhpur, pp 431–470

Kües U, Navarro-González M, Srivilai P, Chaisaena W, Velagapudi R (2007) Mushroom biology and genetics. In: Kües U (ed) Wood production, wood technology, and biotechnological impacts. Universitätsverlag Göttingen, Göttingen, pp 587–608, http://webdoc.sub.gwdg.de/univerlag/2007/wood_production.pdf

Leary JV, Ellingboe AH (1970) The kinetics of initial nuclear exchange in compatible and noncompatible matings of *Schizophyllum commune*. Am J Bot 57:19–23

Lecourieux D, Ranjewa R, Pugin A (2006) Calcium in plant defence-signalling pathways. New Phytol 171:249–269

Lee H, Chang YC, Nardone G, Kwon-Chung KJ (2007) *TUP1* disruption in *Cryptococcus neoformans* uncovers a peptide-mediated density-dependent growth phenomenon that mimics quorum sensing. Mol Microbiol 64:591–601

Lee JK, Leslie JF, Bowden RL (2008) Expression and function of sex pheromones and receptors in the homothallic ascomycete *Gibberella zeae*. Eukaryot Cell 7:1211–1221

Leite B, Nicholson RJ (1992) Mycosporine alanine: A self-inhibitor of germination from the conidial mucilage of *Colletotrichum graminicola*. Exp Mycol 16:76–86

Liavonchanka A, Feussner I (2006) Lipoxygenases: occurrence, functions and catalysis. J Plant Physiol 163:348–357

Lin XR, Hull CM, Heitman J (2005) Sexual reproduction between partners of the same mating type in *Cryptococcus neoformans*. Nature 434:1017–1021

Liu Y, Srivilai P, Loos S, Aebi M, Kües U (2006) An essential gene for fruiting body initiation in the basidiomycete *Coprinopsis cinerea* is homologous to bacterial cyclopropane fatty acid synthase genes. Genetics 172:873–884

Lockhart SR, Zhao R, Daniels KJ, Soll DR (2003) α-Pheromone-induced "shmooing" and gene regulation require white opaque switching during *Candida albicans* mating. Eukaryot Cell 2:847–855

Lorek J, Pöggeler S, Weide MR, Breves R, Bockmühl DP (2008) Influence of farnesol on the morphogenesis of *Aspergillus niger*. J Basic Microbiol 48:99–103

Lowery CA, Dickerson TJ, Janda KD (2008) Interspecies and interkingdom communication mediated by bacterial quorum sensing. Chem Soc Rev 37:1337–1346

Lucaccioni A, Morpurgo G, Achilli A, Barberio C, Casalone E, Baduri N (2007) Colony density influences invasive and filamentous growth in *Saccharomyces cerevisiae*. Folia Microbiol 52:35–38

Luporini P, Alimenti C, Ortenzi C, Vallesi A (2005) Ciliate mating types and their specific protein pheromones. Acta Protozool 44:89–101

Luporini P, Vallesi A, Alimenti C, Ortenzi C (2006) The cell type-specific signal proteins (pheromones) of protozoan ciliates. Curr Pharmacol Res 12:3015–3024

Madelin MF (1956) Studies on the nutrition of *Coprinus lagopus* Fr., especially as affecting fruiting. Ann Bot 20:307–330

Magae Y, Ohara S (2006) Structure–activity relationships of triterpenoid saponins on fruiting body induction in *Pleurotus ostreatus*. Biosci Biotech Biochem 70:1979–1982

Magae Y, Nishimura T, Ohara S (2005) 3-*O*-alkyl-D-glucose derivatives induce fruit bodies of *Pleurotus ostreatus*. Mycol Res 109:374–376

Maraun M, Martens H, Migge S, Theenhaus A, Scheu S (2003) Adding to 'the enigma of soil animal diversity': fungal feeders and saprophagous soil invertebrates prefer similar food substrates. Eur J Soil Biol 39:85–95

Marshall MR, Hindal DF, Macdonald WL (1982) Production of perithecia in culture by *Ceratocystis ulmi*. Mycologia 74:376–381

Martin F, Duplessis S, Ditengou F, Larange H, Voiblet C, Lapeyrie F (2001) Developmental cross talking in the ectomycorrhizal symbiosis: signals and communication genes. New Phytol 151:145–154

Martin F, Aerts A, Ahrén D, Brun A, Danchin EGJ, Durchaussoy F, Gibon J, Kohler A, Lindquist E, Pereda V, Salamov A, Shapiro HJ, Wuyts J, Blaudez D, Buée M, Brokstein P, Canbäck B, Cohen D, Courty PE, Coutinho PM, Delaruelle C, Detter JC, Deveau A, DiFazio S, Duplessis S, Fraissinet-Tachet L, Lucic E, Frey-Klett P, Fourrey C, Feussner I, Gay G, Grimwood J, Hoegger PJ, Jain P, Kilaru S, Labbé J, Lin YC, Legué V, Le Tacon F, Marmeisse R, Melayah D, Montanini B, Muratet M, Nehl U, Niculita-Hirzel H, Oudet-Le Secq MP, Peter M, Quesneville H, Rajashekar B, Reich M, Rouhier N, Schmutz J, Yin T, Chalot M, Henrissat B, Kües U, Lucas S, de Perr YV, Podila GK, Polle A, Pukkila PJ, Richardson PM, Rouzé, Sander IR, Stajich JE, Tunlid A, Tuskan G, Grigoriev IV (2008) The genome of *Laccaria bicolor* provides insights into mycorrhizal symbiosis. Nature 452:88–92

Martinez D, Challacombe J, Morgenstern I, Hibbett D, Schmoll M, Kubicek CP, Ferreira P, Ruiz-Duenas FJ, Martinez AT, Kersten P, Hammel KE, Vanden Wymelenberg A, Gaskell J, Lindquist E, Sabat G, Splinter BonDurant S, Larrondo LF, Canessa P, Vicuna R, Yadav J, Doddapaneni H, Subramanian V, Pisabarro AG, Lavín JL, Oguiza JA, Master E, Henrissat B,

Coutinho PM, Harris P, Magnuson JK, Baker SE, Bruno K, Kenealy W, Hoegger PJ, Kües U, Ramaiya P, Lucas S, Salamov A, Shapiro H, Tu H, Chee CL, Misra M, Xie G, Teter S, Yaver D, James T, Mokrejs M, Pospisek M, Grigoriev IV, Brettin T, Rokhsar D, Berka R, Cullen D (2009) Genome, transcriptome, and secretome analysis of wood decay fungus *Postia placenta* supports unique mechanisms of lignocellulose conversion. Proc Natl Acad Sci USA 106:1954–1959

Matsui K (2006) Green leaf volatiles: hydroperoxide lyase pathway of oxylipin metabolism. Curr Opin Plant Biol 9:274–280

Matsumori K, Izumi S, Watanabe H (1989) Hormone-like action of 3-octanol and 1-octen-3-ol from *Botrytis cinerea* on the pine wood nematode, *Bursaphelenchus xylophilus*. Agric Biol Chem 53:1777–1781

Mau JL, Beelman ML (1996) Role of 10-oxo-trans-8-decenoic acid in the cultivated mushroom, *Agaricus bisporus*. In: Royse DJ (ed) Mushroom biology and mushroom products. Pennsylvania State University, University Park, pp 553–562

Mau JL, Beelman RB, Ziegler GR (1992) Effect of 10-oxo-trans-8-decenoic acid on growth of *Agaricus bisporus*. Phytochem 31:4059–4064

Mau JL, Beelman RB, Ziegler GR (1993) Factors affecting 1-octen-3-ol in mushrooms at harvest and during postharvest storage. J Food Sci 58:331–334

Mau JJ, Chyau CC, Li JY, Tseng YH (1997) Flavor compounds in straw mushroom *Volvariella volvaceae* harvested at different stages of maturity. J Agric Food Chem 45:4726–4729

Mayrhofer S, Weber JM, Pöggeler S (2006) Pheromone and pheromone receptors are required for proper sexual development in the homothallic ascomycete *Sordaria macrospora*. Genetics 172:1521–1533

McAlester G, O'Gara F, Morrissey JP (2008) Signal-mediated interactions between *Pseudomonas aeruginosa* and *Candida albicans*. J Med Microbiol 57:563–569

Mendgen K, Wirsel SGR, Jux A, Hoffmann J, Boland W (2006) Volatiles modulate the development of plant pathogenic rust fungi. Planta 224:1353–1361

Menotta M, Gioacchini AM, Amicucci A, Buffalini M, Sisti D, Stocchi V (2004) Headspace solid-phase microextraction with gas chromatography and mass spectrometry in the investigation of volatile organis compounds in an ectomycorrhizae synthesis system. Rapid Commun Mass Spec 18:206–210

Mikola J, Sulkava P (2001) Responses of microbial-feeding nematodes to organic matter distribution and predation in experimental soil habitat. Soil Biol Biochem 33:811–817

Mita G, Fasano P, De Domenico S, Perrone G, Epifani F, Iannacone R, Casey R, Santino A (2007) 9-Lipoxygenase metabolism is involved in the almond/*Aspergillus carbonarius* interaction. J Exp Bot 58:1803–1811

Moody SAQ, Pierce TG, Dighton J (1996) Fate of some fungal spores associated with wheat straw decomposition on passage through the guts of *Lumbricus terrestris* and *Aporrectodea longa*. Soil Biol Biochem 28:533–537

Moore D (1998) Fungal morphogenesis. Cambridge University Press, Cambridge

Moore-Landecker E (1988) Response of *Pyronema domesticum* to volatiles from microbes, seeds, and natural substrata. Can J Bot 66:194–198

Moore-Landecker E (1992) Physiology and biochemistry of ascocarp induction and development. Mycol Res 96:705–716

Moore-Landecker E, Shropshire W Jr (1984) Effects of ultraviolet A radiation and inhibitory volatile substances on the discomycete, *Pyronema domesticum*. Mycologia 76:820–829

Moore-Landecker E, Stotzky G (1972) Inhibition of fungal growth and sporulation by volatile metabolites from bacteria. Can J Microbiol 18:957–962

Moore-Landecker E, Stotzky G (1973) Morphological abnormalities of fungi induced by volatile microbial metabolites. Mycologia 65:519–536

Morton HL, French DW (1970) Attraction toward and penetration of *Polyporus dryophilus* var. *vulpinus* basidiospores by hyphae of the same species. Mycologia 62:714–720

Muraguchi H, Kamada T (1998) The *ich1* gene of the mushroom *Coprinus cinereus* is essential for pileus formation in fruiting. Development 125:3133–3141

Muraguchi H, Kamada T (2000) A mutation in the *eln2* gene encoding a cytochrome P450 of *Coprinus cinereus* affects mushroom morphogenesis. Fungal Genet Biol 29:49–59

Navazio L, Baldan B, Moscatiello R, Zuppini A, Woo SL, Mariani P, Lorito M (2007a) Calcium-mediated perception and defense responses activated in plant cells by metabolite mixtures secreted by the biocontrol fungus *Trichoderma atroviride*. BMC Plant Biol 7: article 41

Navazio L, Moscatiello R, Genre A, Novero M, Balden B, Bonfante P, Mariani P (2007b) A diffusible signal from arvuscular mycorrhizal fungi elicits a transient cytosolic calcium elevation in host plant cells. Plant Physiol 144:673–681

Nemčovič M, Jakubáková L, Viden I, Farkaš V (2008) Induction of conidiation by endogenous volatile compounds in *Trichoderma* spp. FEMS Microbiol Lett 284:231–236

Niculita-Hirzel H, Labbé J, Kohler A, le Tacon F, Martin F, Sanders IR, Kües U (2008) Gene organization of the mating type regions in the ectomycorrhizal fungus *Laccaria bicolor* reveals distinct evolution between two mating type loci. New Phytol 180:329–324

Nickerson KW, Atkin AL, Hornby JM (2006) Quorum sensing in dimorphic fungi: Farnesol and beyond. Appl Env Microbiol 72:3805–3813

Nielsen O, Davey J (1995) Pheromone communication in the fission yeast *Schizosaccharomyces pombe*. Semin Cell Biol 6:95–104

Nukina M, Sassa T, Ikeda M, Takahashi K, Toyota S (1981) Linoleic-acid enhances perithecial production in *Neurospora crassa*. Agric Biol Chem 45:2371–2373

Ohta A (1988) Effects of butyric acid and related compounds on basidiospore germination of some ectomycorrhizal fungi. Trans Mycol Soc Jpn 29:375–381

Okull DO, Beelman RB, Gourama H (2003) Antifungal activity of 10-oxo-trans-8-decenoic acid and 1-octen-3-ol against *Penicillum expansum* in potato dextrose agar medium. J Food Prot 66:1503–1505

Olesnicky NS, Brown AJ, Dowell SJ, Casselton LA (1999) A consitutively active G-protein-coupled receptor causes mating self-compatbility in the mushroom *Coprinus*. Genetics 18:2756–2763

Olsson S, Gray SN (1998) Patterns and dynamics of ^{32}P-phosphate and labelled 2-aminoisobutyric acid (^{14}C-AIB) translocation in intact basidiomycete mycelia. FEMS Microbiol Ecol 26:109–120

Oort AJP (1974) Activation of spore germination in *Lactarius* species by volatile compounds of *Ceratocystis fagacearum*. Proc K Ned Akad Wet Ser C 77:301–307

Orledge GM, Reynolds SE (2005) Fungivore host-use groups from cluster analysis: patterns of utilisation of fungal fruiting bodies by ciid beetles. Ecol Entomol 30:620–641

Palková Z, Forstová J (2000) Yeast colonies synchronise their growth and development. J Cell Sci 113:1923–1928

Palková Z, Vachová L (2003) Ammonia signalling in yeast colony formation. Int Rev Cytol 225:229–272

Palková Z, Vachová L (2006) Life within a community benefit to yeast long-term survival. FEMS Microbiol Lett 30:806–824

Palková Z, Janderová B, Gabriel J, Zikánová B, Pospíšek M, Forstová J (1997) Ammonia mediates communication between yeast colonies. Nature 390:532–536

Pfeil RM, Mamma RO (1993) Bioassay for evaluating attraction of the phorid fly, *Megaselia halterata* to compost colonized by the commercial mushroom, *Agaricus bisporus* and to 1-octen-3-ol and 3-octanone. Entomol Exp Appl 69:137–144

Pine L (1955) Studies on the growth of *Histoplasma capsulatum*. II. Growth of the yeast phase on agar media. J Bascteriol 70:375–381

Pöggeler S, Nowrousian M, Kück U (2006) Fruiting-body development in ascomycetes. In: Kües U, Fischer R (eds) The Mycota Vol I, 2nd edn. Growth, differentiation and sexuality. Springer, Heidelberg, pp 325–356

Polak E, Aebi M, Kües U (2001) Morphological variations in oidium formation in the basidiomycete *Coprinus cinereus*. Mycol Res 105:603–610

Pommerville J, Olson LW (1987) Evidence for a male-produced pheromone in *Allomyces macrogynus*. Exp Mycol 11:145–248

Pommerville JC, Strickland B, Romo D, Harding KE (1988) Effects of analogs of the fungal sexual pheromone sirenin on male gamete motility in *Allomyces macrogynus*. Plant Physiol 88:139–142

Potapova TV (2004) Intercellular interactions in the *Neurospora crassa* hyphae – Twenty years later. Biol Memb 21:163–191

Raguso RA, Roy B (1998) 'Floral' scent production by *Puccinia* rust fungi that mimic flowers. Mol Ecol 7:1127–1136

Ramos AC, Facanha AR, Feijo JA (2008) Proton (H^+) flux signature for the presymbiotic development of the arbuscular mycorrhizal fungi. New Phytol 178:177–188

Rasmussen TB, Skindersoe ME, Bjarnsholt T, Phipps RK, Christensen KB, Jensen PO, Andersen JB, Koch B,

Larsen TO, Hentzer M, Eberl L, Hoiby N, Givskov M (2005) Identity and effects of quorum-sensing inhibitors produced by *Penicillum* species. Microbiol 151:1325–1340

Rast D, Stäuble EJ (1970) On the mode of action of isovaleric acid in stimulating the germination of *Agaricus bisporus* spores. New Phytol 69:557–566

Rast D, Stüssi H, Zobrist P (1979) Self-inhibition of the *Agaricus bisporus* spore by CO_2 and/or γ-glutaminyl-4-hydroxybenzene and γ-glutaminyl-3,4-benzoquinone: a biochemical analysis. Physiol Plant 46:227–234

Raudaskoski M (1998) The relationship between B-mating-type genes and nuclear migration in *Schizophyllum commune*. Fungal Genet Biol 24:207–227

Rayner ADM (1991) The challenge of the individualistic mycelium. Mycologia 83:48–71

Rayner ADM, Griffith GS, Ainsworth AM (1994) Mycelial interconnectedness. In: Gow NAR, Gadd GM (eds) The growing fungus. Chapman & Hall, London, pp 21–40

Read ND (1994) Cellular nature and multicellular morphogenesis of higher fungi. In: Ingram D, Hudson A (eds) Shape and form in plants and fungi. Academic, London, pp 254–271

Roca MG, Arlt J, Jeffree CE, Read ND (2005a) Cell biology of conidial anastomosis tubes in *Neurospora crassa*. Eukaryot Cell 4:911–919

Roca MG, Read ND, Wheals AE (2005b) Conidial anastomosis tubes in filamentous fungi. FEMS Microbiol Lett 249:191–198

Roncal T, Cordobes S, Steiner O, Ugalde U (2002a) Conidiation in *Penicillium cyclopium* is induced by conidiogenone, an endogenous diterpene. Eukaryot Cell 1:823–829

Roncal T, Cordobes S, Ugalde U, He YH, Steiner O (2002b) Novel diterpenes with potent conidiation inducing activity. Tetrahedron Lett 43:6799–6802

Roy BA, Raguso RA (1997) Olfactory versus visual cue in a floral mimicry system. Oecologia 109:414–426

Ryan RP, Dow JM (2008) Diffusible signals and interspecies communication in bacteria. Microbiology 154:1845–1858

Sawahata T, Shimano S, Suzuki M (2008) *Tricholoma matsutake* 1-octen-3-ol and methyl cinnamate repel mycophagus *Proisotoma minuta* (Collembola: Insecta). Mycorrhiza 18:111–114

Schiestl FP, Steinebrunner F, Schulz C, von Reuss S, Francke W, Weymuth C, Leuchtmann A (2006) Evolution of 'pollinator'-attracting signals in fungi. Biol Lett 2:401–404

Schimek C, Wöstemeyer J (2006) Pheromone action in the fungal groups Chytridiomycota, and Zygomycota, and in the Oomycota. In: Kües U, Fischer R (eds) The Mycota, vol I, 2nd edn. Growth, differentiation and sexuality. Springer, Heidelberg, pp 215–232

Schimek C, Kleppe K, Saleem AR, Voigt K, Burmester A, Wöstemeyer J (2003) Sexual reactions in Mortierellales are mediated by the trisporic acid system. Mycol Res 107:736–747

Schimek C, Petzold A, Schultze K, Wetzel J, Wolschendorf F, Burmester A, Wöstemeyer J (2005) 4-Dihydromethyltrisporate dehydrogenase, an enzyme of the sex hormone pathway in *Mucor mucedo*, is constitutively transcribed but its activity is differentially regulated in (+) and (−) mating types. Fungal Genet Biol 42:804–812

Schnurer J, Olsson J, Borjesson T (1999) Fungal volatiles as indicators of food and feeds spoilage. Fungal Genet Biol 27:209–217

Schrey SD, Tarkka M (2008) Friends and foes: Streptomycetes as modulators of plant disease and symbiosis. Antonie van Leeuwenhoek 94:11–19

Schultze K, Schimek C, Wöstemeyer J, Burmester A (2005) Sexuality and parasitism share common regulatory pathways in the fungus *Parasitella parasitica*. Gene 348:33–44

Semighini CP, Hornby JM, Dumitru R, Nicherson KW, Harris SD (2006) Farnesol-induced apoptosis in *Aspergillus nidulans* reveals a possible mechanism for antagonistic interactions between fungi. Mol Microbiol 59:753–764

Shen WC, Davidson RC, Cox GM, Heitman J (2002) Pheromones stimulate mating and differentiation via paracrine and autocrine signalling in *Cryptococcus neoformans*. Eukaryot Cell 1:366–377

Sicari LM, Ellingboe AH (1967) Microscopical observation of initial interactions in various matings of *Schizophyllum commune* and of *Coprinus cinereus*. Am J Bot 54:437–439

Silliman BR, Newell SY (2003) Fungal farming in a snail. Proc Natl Acad Sci USA 100:15643–15648

Smythe R (1973) Hyphal fusions in the basidiomycete *Coprinus lagopus* sensu Buller. 1. Some effects of incompatibility factors. Heredity 31:107–111

Snetselaar KM, Bölker M, Kahmann R (1996) *Ustilago maydis* mating hyphae orient their growth toward pheromone sources. Fungal Genet Biol 20:299–312

Spellig T, Bölker M, Lottspeich FR, Frank RW, Kahmann R (1994) Pheromones trigger filamentous growth in *Ustilago maydis*. EMBO J 13:1620–1627

Spiteller P (2008) Chemical defense strategies of higher fungi. Chem Eur J 14:9100–9110

Splivallo R (2008) Biological significance of truffle secondary metabolites. In: Karlovsky P (ed) Secondary metabolites in soil biology. Soil biology 14. Springer, Heidelberg, pp 141–165

Splivallo R, Bosso S, Maffei M, Bonfante P (2007a) Discrimination of truffle fruiting bodies versus mycelial aromas by stir bar sorptive extraction. Phytochemistry 68:2548–2598

Splivallo R, Novero M, Bertea CM, Bossi S, Bonfante P (2007b) Truffle volatiles inhibit growth and induce an oxidative burst in *Arabidopsis thaliana*. New Phytol 175:417–424

Sprague GF, Winans SC (2006) Eukaryotes learn how to count: quorum sensing by yeast. Genes Dev 20:1045–1049

Steinebrunner F, Twele R, Francke W, Leuchtmann A, Schiestl FP (2008a) Role of odour compounds in the attraction of gamete vectors in endophytic *Epichloë* fungi. New Phytol 178:401–411

Steinebrunner F, Schiestl FP, Leuchtmann A (2008b) Ecological role of volatiles produced by *Epichloë*: differences in antifungal toxicity. FEMS Microbiol Ecol 64:307–316

Steinebrunner F, Schiestl FP, Leuchtmann A (2008c) Variation of insect attractants odor in endophytic *Epichloë* fungi: Phylogenetic constraints versus host influence. J Chem Ecol 34:772–782

Steinkellner S, Lendzemo V, Langer I, Schweiger P, Khaosaad T, Toussaint J-P, Vierheilig (2007) Flavenoids and strigolactones in root exudates as signals in symbiotic and pathogenic plant–fungus interactions. Molecules 12:1290–1306

Sun Z, Hans J, Walter MH, Matusova R, Beekwilder J, Verstappen FWA, Ming Z, van Echtell E, Strack D, Bisseling T, Bouwmeester HJ (2008) Cloning and characterization of a maize carotenoid cleavage dioxygenase (ZmCCD1) and its involvement in the biosynthesis of apocarotenoids with various roles in mutatualistic and parasitic interactions. Planta 228:789–801

Tarkka M, Piechulla B (2008) Aromatic weapons: truffles attack plants by the production of volatiles. New Phytol 175:381–383

Thakeow P (2008) Development of a basic biosensor system for wood degradation using volatile organic compounds. Dissertation, Georg-August-University of Göttingen

Thakeow P, Holighaus G, Schütz S (2007) Volatile compounds for wood assessment. In: Kües U (ed) Wood production, wood technology, and biotechnological impacts. Universitätsverlag Göttingen, Göttingen, pp 197–228, http://webdoc.sub.gwdg.de/univerlag/2007/wood_production.pdf

Thakeow P, Angeli S, Weissbecker B, Schütz S (2008) Antennal and behavioral responses of *Cis boleti* to fungal odor of *Trametes gibbosa*. Chem Senses 33:379–387

Todd NK, Aylmore RC (1985) Cytology of hyphal interactions and reactions in *Schizophyllum commune*. In: Moore D, Casselton LA, Wood DA, Frankland JC (eds) Development biology of higher fungi. Cambridge University Press, Cambridge, pp 231–230

Traquair JA, McKeen WE (1977) Hyphal interference by *Trametes hispida*. Can J Microbiol 23:1675–1662

Tsitsigiannis DI, Kowieski TM, Zarnowski R, Keller NP (2004a) Endogenous lipogenic regulators of spore balance in *Aspergillus nidulans*. Eukaryot Cell 3:1398–1411

Tsitsigiannis DI, Zarnowski R, Keller NP (2004b) The lipid body protein, PpoA, coordinates sexual and asexual sporulation in *Aspergillus nidulans*. J Biol Chem 279:11344–11353

Tsitsigiannis DI, Kowieski TM, Zarnowski R, Keller NP (2005) Three putative oxylipin biosynthetic genes integrate sexual and asexual development in *Aspergillus nidulans*. Microbiology 151:1809–1821

Tsong AE, Tuch BB, Johnson AD (2007) Rewiring transcriptional circuity: Mating-type regulation in *Saccharomyces cerevisiae* and *Candida albicans* as a model for evolution. In: Heitman J, Kronstad JW, Taylor JW, Casselton LA (2007) Sex in fungi. Molecular determination and evolutionary implications. ASM, Washington, D.C., pp 75–89

Tsurushima T, Ueno T, Fukami H, Irie H, Inoue M (1995) Germination self-inhibitors from *Colletotrichum gloeosporioides* f. sp. *jussiaea*. Mol Plant Microb Interact 8:652–657

Turgeon BG, Debuchy R (2007) *Cochliobolus* and *Podospora*: mechanisms of sex determination and the evolution of reproductive lifestyle. In: Heitman J, Kronstad JW, Taylor JW, Casselton LA (2007) Sex in fungi. Molecular determination and evolutionary implications. ASM, Washington, D.C., pp 93–121

Turina M, Prodi A, Alfen NK (2003) Role of the *Mf1-1* pheromone precursor gene in the filamentous ascomycete *Cryphonectria parasitica*. Fungal Genet Biol 40:242–251

Ugalde U (2006) Autoregulatory signals in mycelial fungi. In: Kües U, Fischer R (eds) The Mycota. vol I, 2nd edn. Growth, differentiation and sexuality. Springer, Heidelberg, pp 203–214

Uroz S, Heinonsalo J (2008) Degradation of N-acyl homoserine lactone quorum sensing signal molecules by forest root-associated fungi. FEMS Microbiol Ecol 65:271–278

Vaillancourt LJ, Raudaskoski M, Specht CA, Raper CA (1997) Multiple genes encoding pheromones and a pheromone receptor define the $B\beta 1$ mating-type specificity in *Schizophyllum commune*. Genetics 146:541–551

Van der Valk P, Marchant R (1978) Hyphal ultrastructure in fruitbody primordia of the basiduiomycete *Schizophyllum commune* and *Coprinus cinereus*. Protoplasma 95:57–72

van Haelen M, van Haelen-Fastre R, Geeraerts J, Wirthlin T (1978) Cis-octa-1,5-dien-3-ol and trans-octa-1,5-dien-3-ol, new attractants to the cheese mite *Tyrophagus putrescentiae* (Schrank) (Acarina, Acaridae) identified in *Trichothecium roseum* (Fungi Imperfecti). Microbios 23:199–212

van West P, Morris BM, Reid B, Appiah AA, Osborne MC, Campbell TA, Shepherd SJ, Gow NAR (2002) Oomycete plant pathogens use electric fields to target roots. Mol Plant Microb Interact 15:790–798

Vargas WA, Djonović S, Sukno SA, Kenerley CM (2008) Dimerization controls the activity of fungal elicitors that trigger systematic resistance in plants. J Biol Chem 283:19804–19815

Venkateshwarlu G, Chandravadana MV, Tewari RP (1999) Volatile flavour components of some edible mushrooms (Basidiomycetes). Flavour Fragrance J 14:191–194

Viterbo A, Wiest A, Brotman Y, Chet I, Kenerley C (2007) The 18mer peptaibols from *Trichoderma virens* plant defense responses. Mol Plant Pathol 8:737–746

Volkov AG (2000) Green plants: electrochemical interfaces. J Electroanal Chem 483:150–156

Voorhees DA, Peterson JL (1986) Hyphae–spore attraction in *Schizophyllum commune*. Mycologia 78:762–765

Wadman MW, van Zadelhoff G, Hamberg M, Visser T, Veldink GA, Vliegenthart JFG (2005) Conversion of linoleic acid into novel oxylipins by the mushroom *Agaricus bisporus*. Lipids 40:1163–1170

Walser PJ, Velagapudi R, Aebi M, Kües U (2003) Extracellular matrix proteins in mushroom development. Rec Res Dev Microbiol 7:381–415

Walther T, Reibsch H, Grosse A, Ostermann K, Deutsch A, Bley T (2004) Mathematical modelling of regulatory mechanisms in yeast colony development. J Theoret Biol 229:327–338

Waters CM, Bassler BL (2005) Quorum sensing: cell-to-cell communication in bacteria. Annu Rev Cell Dev Biol 21:319–346

Wetzel J, Scheibner O, Burmester A, Schimek C, Wöstemeyer J (2009) 4-Dihydrotrisporin-dehydrogenase, an enzyme of the sex hormone pathway of *Mucor mucedo*: purification, cloning of the corresponding gene, and developmental expression. Eukaryot Cell 8:88–95

Wheatley M (2007) Fungi in summer diets of northern flying squirrels (*Glaucomys sabrinus*) within managed forests of western Alberta, Canada. Northwest Sci 81:265–273

Whitehead NA, Barnard AML, Slater H, Simpson NJL, Salmond GPC (2001) Quorum-sensing in gram-negative bacteria. FEMS Microbiol Rev 25:365–404

Wong GJ, Wells K (1985) Modified bifactorial incompatibility in *Tremella mesenterica*. Trans Br Mycol Soc 84:95–109

Wood WF, Archer CL, Largent DL (2001) 1-Octen-3-ol, a banana slug antifeedant from mushrooms. Biochem Syst Ecol 29:531–533

Wood WF, Clark TJ, Bradshaw DE, Foy BD, Largent DL, Thompson BL (2004) Clitolactone, a banana slug antifeedant from *Clitocybe flaccida*. Mycologia 96:23–25

Woods JP, Heinecke EL, Goldman WE (1998) Electrotransformation and expression of bacterial genes encoding hygromycin phosphotransfgerase and β-galactosidase

in the pathogenic fungus *Histoplasma capsulatum*. Infect Immun 66:1697–1707

Worsham PL, Goldman WE (1988) Quantitative plating of *Histoplasma capsulatum* without addition of conditioned medium or siderophores. J Med Vet Mycol 26:137–143

Wöstemeyer J, Schimek C (2007) Trisporic acid and mating in zygomycetes. In: Heitman J, Kronstad JW, Taylor JW, Casselton LA (2007) Sex in fungi. Molecular determination and evolutionary implications. ASM, Washington, D.C., pp 431–443

Wösten HAB, Wessels JGH (2006) The emergence of fruiting bodies in basidiomycetes. In: Kües U, Fischer R (eds) The Mycota, vol I, 2nd edn. Growth, differentiation and sexuality. Springer, Heidelberg, pp 393–414

Wright GD, Arlt J, Poon WCK, Read ND (2007) Optical tweezer micromanipulation of filamentous fungi. Fungal Genet Biol 44:1–13

Wu SM, Zorn H, Krings U, Berger RG (2005) Characteristic volatiles from young and aged fruiting of wild *Polyporus sulfureus* (Bull.: Fr.) Fr. J Agric Food Chem 53:4524–4528

Wurzenberger M, Grosch W (1984) Sterochemistry of the cleavage of the 10-hydroperoxide isomer of linoleic acid to 1-octen-3-ol by a hydroperoxide lyase from mushrooms (*Psalliota bispora*). Biochim Biophys Acta 795:163–165

Young H, Patterson VJ (1982) A UV protective compound from *Glomerella cingulata* – a mycosporine. Phytochemistry 21:1075–1077

6 Yeast Killer Toxins: Fundamentals and Applications

FRIEDHELM MEINHARDT[1], ROLAND KLASSEN[1]

CONTENTS

I. Introduction

Microbes competing for limited resources have established a broad arsenal of lethal compounds aiming at the inhibition or destruction of competitors. Toxic matters released by microbial killers include classic antibiotics and low/medium molecular weight substances belonging to various chemical classes, but – rather frequently – there are antibiotic peptides and even large protein complexes termed bacteriocines (bacteriocidic proteins), mycocines, zymocines and killer toxins (fungizidic proteins). The discovery of antibiotics dates back to the beginning of the twentieth century and revolutionized medical anti-infective therapies; however, the secretion of fungicidal proteins did not become obvious until the 1960s

with the discovery of a so-called killer strain of the brewer's yeast *Saccharomyces cerevisiae* (Bevan and Makower 1963). At present, we are aware of a panoply of killer toxins originating from both ascomycetous and basidiomycetous yeasts. Toxin sizes range from small chromosomally encoded polypeptides to larger di- or trimeric protein complexes, the latter being encoded by extranuclear DNA or RNA elements of viral origin which parasitize the host cell's cytoplasm. Continually growing understanding of the mode of action of various killer toxins along with the corresponding immunity mechanisms not only facilitates the recognition of new toxic principles but also significantly contributes to our understanding of precarious mechanisms during cell proliferation. Reviews addressing yeast killer toxins (generally or more specifically dsRNA- or dsDNA-encoded ones) have been published over the years (Stark et al. 1990; Bussey 1991; Magliani et al. 1997; Schmitt and Breinig 2002, 2006; Schaffrath and Meinhardt 2004; Golubev 2006; Klassen and Meinhardt 2007). However, significant progress in the identification of toxic principles and strategies, immunity mechanism and promising approaches for the application of such proteins has been made only rather recently. In this contribution, we summarize current knowledge on killer toxins and toxic principles, starting from chromosomally encoded ones, followed by dsRNA- and dsDNA-encoded toxins and finally we discuss the state of the art of toxin applications.

II. Chromosomally Encoded Killer Toxins

A. *Williopsis*

A large variety of yeast species secrete chromosomally encoded killer toxins. However, these appear to be exceptionally frequent among *Williopsis* strains (Table 6.1). Among the most extensively studied toxins are HM-1 produced by

[1]Institut für Molekulare Mikrobiologie und Biotechnologie, Westfälische Wilhelms-Universität Münster, Corrensstrasse 3, 48149 Münster, Germany; e-mail: meinhar@uni-muenster.de, roland.klassen@uni-muenster.de

Physiology and Genetics, 1st Edition
The Mycota XV
T. Anke and D. Weber (Eds.)
© Springer-Verlag Berlin Heidelberg 2009

Table 6.1. Chromosomally encoded yeast killer toxins

Species	Strain	Toxin	Size	Receptor	Toxic activity	References
Ascomycetes						
Pichia anomala	NCYC434	K5/panomycocin	49.0	β-1-3 Glucan	Glucanase	Izgü and Altinbay (2004)
	WC65		83.3	β-1-6-Glucan		Sawant et al. (1989)
	ATCC 96603, K36, UP25F	PaKT/PKT	85.0	β-Glucan (?)		Guyard et al. (1999), Polonelli and Morace (1986)
	YF07b	Pikt	47.0		Glucanase	Wang et al. (2007a)
	DBVPG 3003		>3.0			Comitini et al. (2004a)
P. farinosa	KK1	SMKT	α (6.6), β (7.9)		Membrane permeabilisation	Suzuki and Nikkuni (1994)
P. kluyveri	1002	PMKT	19.0	β-1-6-Glucan	Membrane permeabilisation	Middlebeek et al. (1979)
P. membranifaciens	CYC1106		18.0		Membrane permeabilisation	Santos et al. (2000)
Williopsis californica	DSM 12865	Wicaltin	34.0			Theisen et al. (2000)
W. saturnus	IFO 0117	HYI	9.5		Inhibition of β-1,3-glucan synthase	Komiyama et al. (1995, 1998)
W. saturnus var. mrakii	DBVPG 4561	KT4561	62.0	β-Glucan	Glucanase/membrane permeabilization	Buzzini et al. (2004)
	MUCL 41968	WmKT	85.0			Guyard et al. (2002a, 2002b)
	IFO 0895	HM-1	10.7	β-Glucan	Inhibition of β-1,3-glucan synthase	Ashida et al. (1983), Kasahara et al. (1994)
W. mrakii	NCYC500	K500	1.8–5.0	β-Glucan	Membrane permeabilisation	Hodgson et al. (1995)
Debaryomyces hansenii	CYC1021		23.0	β-1-6-Glucan		Santos et al. (2002)
Kluyveromyces waltii	IFO1666T		>10.0			Kono and Himeno (1997)
K. wickerhamii	DBVPG 6077	Kwkt	>10.0		Glucanase	Comitini et al. (2004a)
K. phaffii	DBVPG 6076	KpKt	33.0	β-1,6-/β-1,3-Glucan		Comitini et al. (2004b)
K. marxianus	NCYC587	K6	42.0	Mannan	Membrane permeabilisation	Izgü et al. (1999)
Schwanniomyces occidentalis	ATCC 44252		α (7.4), β (4.9)			Chen et al. (2000)
Saccharomyces cerevisiae	111	KHR	20.0			Goto et al. (1990)
	115	KHS	75.0			Goto et al. (1991)
Candida noadenis		CnKT				da Silva et al. (2008)
Basidiomycetes						
Filobasidium capsuligenum	IFM 40078	FC-1		β-1,6-Glucan	Membrane permeabilisation	Keszthelyi et al. (2006)
Cryptococcus humicola	VKM Y-1439	cellobiose lipid	<1.0		Membrane permeabilisation	Golubev and Shabalin (1994), Puchkov et al. (2001, 2002)

W. *saturnus* var. *mrakii* IFO0895 (previously known as *Hansenula mrakii*) and WmKT from strain MUCL41968 of the same species (Ashida et al. 1983; Kasahara et al. 1994; Guyard et al. 2002a).

HM-1 (known also as HMK) is a 10.7-kDa toxin consisting of 88 amino acids (aa). Five internal disulfide bridges probably contribute to the remarkable stability of the protein (Yamamoto et al. 1986a, b). HM-1, the geneproduct of chromosomal *HMK*, is produced as a precursor consisting of 125 aa, that is post-translationally processed upon secretion by the Kex2 endopeptidase (Fig 6.1; Kimura et al. 1993). HSK, an almost identical toxin, is produced by W. *saturnus* IFO0117 (Kimura et al. 1993). Such toxins exert their effects on *Saccharomyces cerevisiae* by inhibiting the β-glucan synthase (Yamamoto et al. 1986; Komiyama et al. 1996), a plasma membrane-bound enzyme catalysing the formation of β-1,3-glucan, a major polysaccharide of yeast cell walls (Lesage and Bussey 2006), eventually resulting in pore formation in actively growing cells in regions in which cell wall synthesis is normally active. As a consequence, intracellular material is released and cells lyse concomitantly at budding regions (Takasuka et al. 1995; Komiyama et al. 1996). Accordingly, osmotic stabilization in isotonic medium reduced HM-1 toxicity (Komiyama et al. 1996). In fact, cell wall glucan is a prerequisite

for HM-1 action since *S. cerevisiae* mutants displaying a reduced glucan content (*kre6*) and spheroplasts are toxin-protected (Kasahara et al. 1994; Komiyama et al. 2002).

Since HM-1 toxicity could be antagonized by exogenously applied β-1-6- and β-1-3-glucan, such polysaccharides were assumed to represent the cell wall receptors (Kasahara et al. 1994). HM-1 has a basic isoelectric point (pI = 9.1) and is rather unique in its broad pH and temperature stability, as it retains its activity even after treatment at 100°C for 10 min and between pH 2 and pH 11 (Ashida et al. 1983; Lowes et al. 2000).

The three-dimensional structure of HM-1 was elucidated by determining the nuclear magnetic resonance spectrum (Antuch et al. 1996). The toxin turned out to be a close structural homologue of γB-crystallin, a monomeric protein originally isolated from bovine eye lenses.

Interestingly, HM-1 and γB-crystallin are not detectably homologous at the aminoacid level (Antuch et al. 1996). Other than γB crystallin and further eye lens crystallins, which consist of two domains of the crystallin fold, HM-1 is a single domain representative (Antuch et al. 1996). Presumably, eye lens crystallins and HM-1 have evolved from an ancestral single domain precursor that underwent gene duplication in the case of the crystallins. Resembling the extreme thermal and pH characteristics of HM-1, crystallins too exhibit outstanding stability features; they occur in high concentration in eye lenses in which almost no protein turnover takes place (Wistow and Piatigorsky 1988).

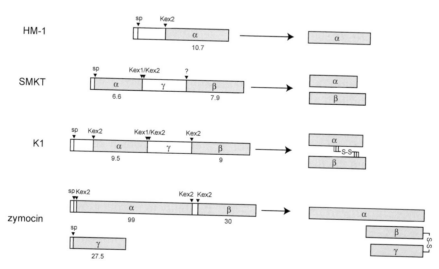

Fig. 6.1. Processing of toxin precursors. Toxin precursors are schematically depicted on the *left*, mature proteins on the *right*. Parts constituting mature toxins are given in *grey*, parts being removed during processing are in *white*. Subunit sizes are given in kDa; processing sites by Kex1/Kex2 peptidases as well as signal peptidase (*sp*) are indicated as such. –S–S– Disulfide bridge in mature toxins; K1 contains multiple disulfide bridges

The WmKT toxin of *W. saturnus* var. *mrakii* strain MUCL41968 differs significantly from HM-1 in structure and function; it is a 85-kDa protein attacking a number of yeast species but, in contrast to HM-1, it has a relatively narrow pH optimum and a low temperature optimum (i.e. around pH 4.6 and 26–28°C; Guyard et al. 2002a), attributes common to many other yeast killer toxins (see below). Toxicity suppression by a glucosidase inhibitor and in vitro glucosidase activity suggests that WmKT kills target cells by degradation of cell wall β-glucans (Guyard et al. 2002a). Consistently, both spheroplasts and *S. cerevisiae kre1* and *kre4* mutants (the latter being defective in β-1-3- or β-1-6-glucan synthesis) are protected from toxic WmKT effects (Guyard et al. 2002a, b).

Other killer strains of the genus include *W. saturnus* DBVPG 4561 and IFO 0117, *W. californica* DSM 12865 and *W. mrakii* NCYC500 (Komiyama et al. 1995; Hodgson et al. 1995; Theisen et al. 2000; Buzzini et al. 2004; Table 6.1). While the IFO 0117 toxin termed HYI is similar to HM-1 in size, sequence and its inhibitory effect on β-1-3-glucan synthase (Komiyama et al. 1995, 1998), the DBVPG4561 toxin is significantly larger, with an estimated size of 62 kDa. It displays heat and pH stability, as it is equally active at pH values between 4.5 and 8.0 and tolerates up to 50 g/l NaCl. Elevated temperatures (30–45°C) decreased its activity only slightly; even treatments of the toxin at 100°C – though administered rather briefly – did not negatively influence its lethal action (Buzzini et al. 2004). The exceptionally small toxin of *W. mrakii* NCYC500 (1.8–5.0 kDa) was found to be unstable at pH values above 4.0 and incubation temperatures exceeding 30°C (Hodgson et al. 1995). Wicaltin, a toxin from *W. californica* is a 34-kDa protein that displays a broad spectrum of activity against pathogenic and non-pathogenic yeast species (Theisen et al. 2000). As toxicity is partially suppressed in sorbitol-stabilized yeast cells and the toxin dramatically affects protoplast regeneration (Theisen et al. 2000), wicaltin may inhibit β-1,3-glucan synthesis, as for HM-1 of *W. saturnus* var. *mrakii*.

B. *Pichia*

Several strains of *Pichia anomala* secrete chromosomally encoded toxins of varying molecular masses (Table 6.1). Strain NCYC434 produces K5,

a 49-kDa toxin alternativley termed panomomycin; it has exo-β-1,3-glucanase activity prefering pH values between 3.0 and 5.5 and temperatures up to 37°C (Izgü and Altinbay 2004; Izgü et al. 2005). Such enzymatic activity and the acidic pH optimum reminds one of WmKT of *W. saturnus* var. *mrakii* (Guyard et al. 2002a). Interestingly, a monoclonal antibody raised against another *P. anomala* killer toxin (strain ATCC 96603) displayed cross-reactivity with WmKT (Guyard et al. 2001), additionally supporting the assumption that toxins of both taxa (*Pichia*, *Williopsis*) may be structurally and functionally related, though it still remains uncertain whether the toxin of *P. anomala* ATCC 96603 is similar to the well characterized K5 toxin (Guyard et al. 2001, 2002a, b; Izgü and Altinbay 2005; Izgü et al. 2005). Since at least the experimentally deduced molecular masses of the toxins differ significantly (Table 6.1), both may only be distantly related.

Additional *P. anomala* killers include strains WC65, K36, YF07b and DBVPG3003 (Table 6.1; Sawant et al. 1989; Guyard et al. 1999; Comitini et al. 2004; Wang et al. 2007a). As for K5, the toxin of strain YF07b (isolated from marine environment for biocontrol of crab-pathogenic yeast species in the food industry) displays β-1,3-glucanase activity and has a similar molecular mass (47 kDa), raising the possibility that K5 and YF07b toxin are cognate proteins (Wang et al. 2007a, b). Immunofluorescence analysis of the 83.3-kDa toxin of strain WC65 revealed binding to both the cell wall and the membranes of sensitive *Candida albicans* cells (Sawant et al. 1989; Sawant and Ahearn 1990). Since a *S. cerevisae kre1-1* mutant with reduced β-1-6 glucan content displayed a significant degree of resistance, β-1-6 glucan was suggested to be the primary cell wall receptor; possibly there is a second, unidentified receptor in the plasma membrane, which could explain the toxin-binding capability and sensitivity of spheroplasts (Sawant and Ahearn 1990).

P. anomala strain DBVPG3003 was selected for its ability to restrict growth of wine spoilage yeasts (Comitini et al. 2004a). The toxin termed PiKt has an acidic pH optimum (around pH 4.4) and is inactivated above 40°C. However, it retained its activity during prolonged incubation (10 days) in a wine environment. The preliminary characterization by ultrafiltration revealed a small molecular mass toxin of 3–10 kDa (Comitini et al. 2004a).

P. kluyveri strain 1002 (Middelbeek et al. 1979) produces a killer toxin which is stable between pH 2.5 and pH 4.7 and at temperatures below 40°C (Middelbeek et al. 1979, 1980a). The toxin was shown to form ion-permeable channels in phospholipid bilayer membranes in vitro (Kagan 1983). In vivo, cell death was accompanied by cell shrinkage, leakage of potassium ions and adenosine 5'-triphosphate, decrease of intracellular pH and inhibition of the active uptake of amino acids (Middelbeek et al. 1980a, b). Consistent with membrane permeabilization as the lethal principle, toxicity could be suppressed when toxin treated cells were plated on media containing physiological concentrations of K^+ and H^+, whereas higher ion concentrations had a clear negative effect on toxin tolerance (Middlebeek et al. 1980b).

The two closely related halotolerant yeasts *P. membranifaciens* and *P. farinosa* secrete toxins termed *P. membranifaciens* killer toxin (PMKT) and salt-mediated killer toxin (SMKT), respectively, which are stimulated in the presence of salt (Suzuki and Nikkuni 1989; Marquina et al. 1992; Lorente et al. 1997). SMKT, encoded by the chromosomal *SMK1* gene, is a preprotoxin of 222 aa (Suzuki and Nikkuni 1994). The precursor is processed during secretion by the action of the signal peptidase, as well as Kex1/2-like endo/carboxypeptidases, resulting in mature α- and β-subunits of 6.6 kDa and 7.9 kDa, respectively; simultaneously the interstitial γ-polypeptide is liberated (Fig. 6.1; Suzuki and Nikkuni 1994). At pH values above 5.0, α- and β-subunits of mature SMKT dissociate, resulting in the complete loss of its activity (Suzuki et al 1997).

Determination of the crystal structure of SMKT revealed that α and β subunits are folded together in a single ellipsoidal domain (Kashiwagi et al. 1997). The primary receptor of the toxin remains unknown, however, sensitivity of *S. cerevisiae kre1* and *kre5* mutants, being defective in β-1-6 glucan synthesis suggest that molecules other than β-1-6 glucan are required for target cell binding (Suzuki and Nikkuni 1994).

Analysis of the interaction of SMKT with artificial liposomes or intact yeast cells revealed that the toxin disrupts membrane integrity (Suzuki et al. 2001).

The *P. membranifaciens* toxin PMKT is a monomeric 18 kDa protein which kills sensitive yeast cells such as *Candida boidinii* at temperatures below 20°C and pH values below 4.8 (Santos et al. 2000). Following binding to the primary receptor β-1-6 glucan, the toxin is assumed to form channels in the membrane, eventually causing leakage of ions and low molecular weight metabolites such as glycerol (Santos et al. 2000; Santos and Marquina 2004a). Thus, it resembles the mode of action of *P. kluyveri* and *S. cerevisiae* K1 toxins (see above and below). The characterization of the genome wide transcriptional response of *S. cerevisiae* target cells to PMKT exposure revealed the induction of genes of the high glycerol (HOG) pathway (Santos et al. 2005), which is implemented in the general environmental stress response and which results in increased levels of compatible solutes (such as glycerol) to adapt the cellular osmotic pressure. Hence, PMKT induces a co-ordinated transcriptional response in target cells resembling the response to osmotic stress (Santos et al. 2005; Rep et al. 2000). Indeed, consistent with HOG induction, intracellular and extracellular glycerol accumulation were shown to be induced by PMKT, as enhanced compatible solutes simultaneously leak out through PMKT-mediated membrane channels (Santos et al. 2005). Strains deficient in Hog1, the mitogen-activated protein kinase (MAPK) integral to the HOG signalling pathway as well as mutants defective in osmoadaptive glycerol synthesis display PMKT hypersensitivity, clearly revealing compatible solute accumulation to counteract PMKT toxicity (Santos et al. 2005). Before reaching the plasma membrane, PMKT was shown to interact with Cwp2, a cell wall mannoprotein, the precursor of which is attached to the plasma membrane via a glycosylphosphatidylinositol (GPI) anchor, while the mature Cwp2 localizes to the cell wall, being covalently attached to β-1-6 glucan (Santos et al. 2007). It was, thus, proposed that PMKT first binds to the primary cell wall receptor β-1-6 glucan, followed by interaction with the mature form of Cwp2, which is attached to β-1-6 glucan. Subsequently, a third interaction with the GPI-anchored form of Cwp2 facilitates close contact with the plasma membrane, allowing lethal ion channel formation (Fig. 6.2; Santos et al. 2007).

As for *P. membranifaciens*, a killer strain of *Debaryomyces hansenii* (Table 6.1) isolated from high salt substrate (olive brine) produces a toxin that is stimulated by high salt concentrations and which also binds to β-1-6 glucan (Lorente et al. 1997; Santos et al. 2002); however, the mechanism of cell killing remains unknown.

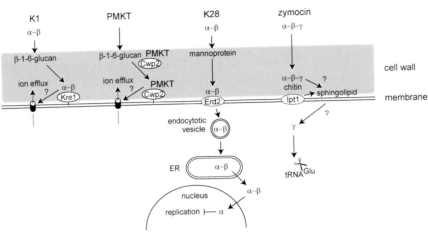

Fig. 6.2. Schematic representation of killer toxin mode of action. β-1-6 glucan is the primary cell wall receptor for chromosomally encoded PMKT and dsRNA encoded K1, manno-protein serves as the receptor for dsRNA encoded K28 and chitin, located close to the membrane is the receptor for the dsDNA encoded zymocin. PMKT and K1 utilize GPI-anchored Kre1 and Cwp2 as secondary membrane receptors and induce membrane channel formation. Erd2 was proposed as the secondary membrane receptor for K28, which gains access to the target cell's cytoplasm by retrograde transport and finally inhibits DNA synthesis in the nucleus. The secondary membrane receptor for zymocin is unknown, however, Ipt1-synthesized sphingolipid is required for cellular uptake of the toxin's γ-subunit which cleaves cellular tRNA^{Glu}. See text for references and further details

C. *Kluyveromyces*

The genus *Kluyveromyces* harbours killer strains secreting both, extranuclearly and chromosomally encoded toxins (Tables 6.1, 6.2). While *K. lactis* represents the most thougoughly investigated plasmid encoded toxin (see below), chromosomally encoded toxins are described for *K. marxianus*, *K. wickerhamii*, *K. waltii* and *K. phaffii* (Young and Yagiu 1978; Kono and Himeno 1997; Izgü et al. 1999; Ciani and Fatichenti 2001; Comitini et al. 2004a, b). The best characterized of the latter, i.e. chromosomally encoded *Kluyveromyces* toxins, is the 33-kDa KpKt of *K. phaffii* (Comitini et al. 2004a). Based on competitive inhibition of toxin activity, β-1-6/β-1,3 glucan was again identified as the cell wall receptor. The purified toxin displayed glucanase activity, resembling the properties of *W. saturnus* var. *mrakii* WmKT and *P. anomala* K5/panomomycin (Guyard et al. 2002a; Comitini et al. 2004a; Izgü and Altinbay 2004; Izgü et al. 2005). However, KpKt is much smaller than WmKT or K5 and other than WmKT, it did not induce rapid cell permeabilization as judged from cell staining with propidium iodide, suggesting differences in cell killing mediated by the two glucanase toxins (Comitini et al. 2004a).

Cellular receptors or killing mechanisms for the remaining chromosomally encoded *Kluyveromyces* toxins are currently unknown. Nonetheless, toxins of *K. phaffii*, *K. waltii* and *K. wickerhamii* have been investigated for their potential application for biocontrol during wine-making (see below).

For several other yeast species killer toxins are known, a comprehensive compilation is given by Golubev (2006); instances in which detailed information about structural and mechanistic details are available are presented in Table 6.1 Other toxins have been isolated solely based on their application potential, and basic information about cell targeting and activity remains to be established. In some cases, mycocinogenic activity may resemble those of killer (protein-)toxins, but in fact it is due to the release of toxic glycolipids (Purchov et al. 2002, reviewed by Golubev 2006).

III. Extrachromosomally Encoded Toxins

A. dsRNA Virus Toxins

Shortly after the discovery of the first yeast killer in *S. cerevisiae*, it was realized that dsRNA viruses constitute the molecular basis of toxin secretion (Bevan et al. 1973). The *S. cerevisiae* killer viruses belong to the *totiviridae* family. They persist as a pair of two separately encapsulated virus like

Table 6.2. Extrachromosomally encoded yeast killer toxins

Species	Strain	Toxin	Size	Receptor	Toxic activity	References
Ascomycetes dsRNA-encoded						
S. cerevisiae	KL88	K1	α (9.5), β (9.0)	β-1-6-Glucan	Membrane permeabilisation	Young and Yagiu (1978)
	CBS8112	K28	α (10.0), β (11.0)	Mannoprotein	Blocking DNA synthesis	Schmitt and Tipper (1990)
Hanseniaspora uvarum	470		18.0	β-1-6-Glucan		Schmitt and Neuhausen (1994)
Zygosaccharomyces bailii	412	zygocin	10.0	Mannoprotein	Membrane permeabilization	Schmitt and Neuhausen (1994)
dsDNA-encoded						
Kluyveromyces lactis	IFO1267	zymocin	α (99.0), β (30.0), γ (28.0)	Chitin	tRNA^{Glu}-specific tRNase	Gunge et al. (1981)
Pichia inositovora	NRRL Y-18709		>100.0	Chitin		Hayman and Bolen (1991)
P. acaciae	NRRL Y-18665	PaT	α (110.0), β (39.0), γ (38.0)	Chitin	tRNA^{Gln}-specific tRNase	Worsham and Bolen (1990)
Debaryomyces robertsiae	CBS6693		>100.0	Chitin		Klassen and Meinhardt (2002)
Basidiomycetes dsRNA-encoded						
Ustilago maydis	P1	KP1	13.4			Park et al. (1996a)
	P4	KP4	13.6		Blocking calcium uptake	Park et al. (1994)
	P6	KP6	α (8.6), β (9.1)		K$^+$ depletion	Tao et al. (1990)
Cryptococcus aquaticus	VKM Y-2428					Pfeiffer et al. (2004)
Cystofilobasidium infirmominiatum	VKM Y-2897		>15.0			Golubev et al. (2003)
Trichosporon pullulans	VKM Y-2303		>15.0			Golubev et al. (2002)
T. insectorum	CBS10422					Fuentefria et al. (2008)

particles in the host's cytoplasm (for reviews, see Wickner 1996, 1992; Schmitt and Breinig 2002, 2006). One of these is the L-A helper virus, which encodes the major capsid protein (Gag) and a RNA-dependent RNA polymerase.

The 5′ end of the ORF encoding the RNA polymerase overlaps in the −1 frame with the Gag-encoding ORF and is expressed as a Gag-Pol fusion protein by a −1 ribosomal frameshift event (Icho and Wickner 1989; Dinman et al. 1991). The L-A virus may be associated with different satellite (non-autonomous) viruses, termed M-1, M-2 or M-28, which encode a single preprotoxin that is processed during secretion to yield the mature toxin. The capsid encoded by L-A consists of 60 asymmetric Gag-dimers and two Gag-Pol fusion proteins. The capsid is perforated allowing the

exit of (+)ssRNA transcribed from the viral genome within the capsid as well as facilitating influx of host metabolites but excluding entry of degradative enzymes or leakage of dsRNA genomic molecules (Caston et al. 1997).

Replication of L-A and M viruses occurs in a conservative manner (Fig. 6.3), with a (+)ssRNA copy being generated in the capsid by the Gag-Pol protein. Following export to the cytoplasm, such ssRNA serves as the messenger that is translated into preprotoxin (M-virus), or Gag and Gag-Pol fusion proteins (L-A virus). The (+)ssRNA is subsequently incorporated into newly formed capsids, in which synthesis of the complementary (−)RNA strand is carried out by Gag-Pol, closing the replicative cycle.

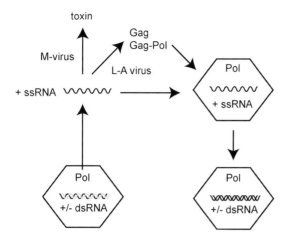

Fig. 6.3. Conservative replication cycle of the dsRNA toti-viridae. Encapsulated virus like particles (VLP) are schematically represented by a hexagon, they consist of Gag, the major capsid protein and two Gag-RNA-polymerase fusion proteins (Gag-Pol). *+/− dsRNA* double-stranded RNA virus genome; *+ ssRNA* single-stranded RNA transcribed from the virus genome within the capsid by Gag-Pol, which is encapsidated in new VLPs, followed by synthesis of the complementary strand

L-A and M-viruses are considered virus like particles, since they lack, in contrast to classic viruses, an extracellular route of transmission. Propagation occurs exclusively by cell to cell passage (anastomosis) during vegetative or sexual reproduction. The most thoroughly studied viral toxins from *S. cerevisiae* are K1 (encoded by M1 virus) and K28 (encoded by M28 virus, see also Schmitt and Breinig 2002, 2006).

1. K1

K1 is a dimer with subunit sizes of 9.5 kDa (α) and 9.0 kDa (β), which are covalently linked by three disulfide bridges (Bostian et al. 1984). Both of the subunits are formed by processing of a single precursor protein termed preprotoxin, first by the signal peptidase (giving rise to protoxin) and subsequently by Kex1 (carboxypeptidase) and Kex2 (endopeptidase) during secretion (Bostian et al. 1984; Zhu et al. 1992). Kex2-processing removes the γ-peptide of the protoxin, a process strikingly similar to the one occuring with the chromosomally encoded SMKT (see above and Fig. 6.1) and K28 (see below), despite the fact that the toxins themselves do not display similarities at the sequence level and also differ significantly in their mode of action (de la Pena et al. 1981; Suzuki and Nikkuni 1994; Schmitt and Tipper 1995). K1 toxin, like many chromosomally encoded toxins, first binds to β-1,6 glucan of the

cell wall (Hutchins and Bussey 1983). Subsequently, K1 interacts with Kre1, a GPI-anchored cell wall protein, thereby inducing formation of (cation selective) ion channels within the plasmamembrane (Fig. 6.2 de la Pena et al. 1978; Breinig et al. 2002, 2004). Mutational analysis revealed that both, α and β are involved in glucan binding, while solely α is required for membrane interaction, which agrees well with the presence of two remarkably hydrophobic regions (Bussey 1991; Zhu and Bussey 1991).

Whether ion channel formation is brought about by K1 alone (Martinac et al. 1990) or requires interaction with primary membrane associated effector protein remains controversial (Breinig et al. 2002; Breinig and Schmitt 2006). *kre1* mutant spheroplasts were shown to be sensitive to excess K1, raising the possibility of receptor-independent membrane channel formation (Martinac et al. 1990) or, alternatively, Kre1-independent access to a hypothetic primary membrane effector (Breinig et al. 2002). Although evidence has been presented that the increase in membrane permeability by K1 is due to the activation of Tok1 potassium channels (Ahmed et al. 1999), full K1 sensitivity of *tok1* cells indicates Tok1 activation to be (at best) a late secondary effect of the toxin (Breinig et al. 2002; Pagé et al. 2003).

Besides ionophoric action, K1 mode of action strikingly resembles the chromosomally encoded PMKT from *Pichia membranifaciens*, which has also been shown to utilize a GPI-anchored cell wall protein (Cwp2) as the secondary membrane receptor (Fig. 6.2 see above). As for Cwp2, Kre1 is localized to both the cell membrane and cell wall, as is typical for cell wall GPI-proteins which become attached to glucan by transglycosylation (Breinig et al. 2002). Thus, it appears possible that both, K1 and PMKT use an identical strategy for reaching the plasma membrane, involving sequential interaction with cell wall glucan, cell wall-localized secondary receptors and, finally, membrane-associated secondary receptor proteins.

As for PMKT, *S. cerevisiae* cells with a defect in HOG signalling due to the loss of Hog1 display extreme K1 sensitivity (Pagé et al. 2003), which suggests a transcriptional response similar to other osmotic stresses being effective in suppressing K1 toxicity. However, while strains deficient in intracellular glycerol accumulation, such as *gpd1* (glycerol-3-phosphate dehydrogenase1) mutants react hypersensitively to PMKT (Santos et al. 2005), this was not the case for K1 and no downstream effectors of Hog1 relevant for K1 resistance could be identified (Pagé et al. 2003). Thus, while the PMKT resistance promoting effect of HOG induction involves osmo-adaptive compatible solute production, its relevance for K1 resistance remains to be elucidated (Pagé et al. 2003).

A characteristic trait of most extrachromosomally encoded killer toxins is the appurtenant-specific toxin resistance (immunity), which is genetically linked to toxin production (Schmitt and Breinig 2006). In the case of K1, viral M1 RNA displays a single ORF and the encoded preprotoxin is not only processed to the mature secreted toxin but also confers K1 immunity. Expression of a cDNA copy of M1 in mutants which are disabled in Kex2 function and therefore unable either to secrete an active toxin or to carry out normal protoxin processing and γ-release confers K1 immunity. Further, expression of K1-α along with 31 N-terminal residues of γ is sufficient for K1 immunity (Zhu et al. 1993). However, the exact immunity strategy remains to be elucidated (Schmitt and Breinig 2006) as the proposed mechanism based on internal inhibition of Tok1 potassium channels by K1 precursors is questionable (see above).

Another *S. cerevisiae* dsRNA virus termed M2 encodes the K2 toxin, which is assumed to exhibit a rather similar toxic activity and apparently also uses Kre1 as the plasma membrane receptor; however, the primary K2 sequence is unrelated to K1, and K2 displays a slightly more acidic pH optimum (Young and Yagui 1978; Pfeiffer and Radler 1984; Dignard et al. 1991; Schmitt and Breinig 2002; Novotna et al. 2004). As K2 killers remain susceptible to K1 (and vice versa) and differential resistance effects of a *kre2* mutation to both toxins were observed, functional difference with respect to target cell interactions and immunity mechanisms are obvious, the details of which, however, remain to be uncovered (Dignard et al. 1991; Meskauskas and Citavicius 1992; Novotna et al. 2004).

2. K28

The other well characterized RNA viral toxin in *Saccharomyces* is K28, encoded by the M28 virus. Very similar to K1, K28 is encoded by a single ORF as a preprotoxin, which is processed during secretion by the signal peptidase as well as Kex1 and Kex2 peptidases (Schmitt and Tipper 1990, 1995; Riffer et al. 2002; Schmitt and Breinig 2006). As for K1, the α-β intervening γ-peptide of K28 protoxin is deleted by the Kex2 endopeptidase. Mature K28 consists of two subunits (α, β) of 10

kDa and 11 kDa, which are covalently linked by a single disulfide bridge (Schmitt and Tipper 1995). Despite this striking similarity, subsequent steps in target cell attack are completely different for K1 and K28. In fact, the latter is so far the only instance of a yeast killer toxin gaining access to the target cell by endocytosis and subsequent retrograde passage of the secretory pathway, followed by exit from the endoplasmic reticulum (ER) and entry into the nucleus, where it blocks DNA replication (Fig. 6.2 Schmitt et al. 1996; Eisfeld et al. 2000; Heiligenstein et al. 2006).

K28 first binds to mannoprotein in the cell wall, most likely followed by interaction with Erd2 which functions as membrane receptor (Schmitt and Radler 1987; Schmitt and Breinig 2006). Erd2 is an integral membrane protein that binds to the HDEL motif in proteins destined for retention in the endoplasmic reticulum (Semenza et al. 1990) and it is assumed that a low number of Erd2 also reach the cytoplasmic membrane where they may contact the K28 β-subunit (Schmitt and Breinig 2006). Kex1-processing of K28 β at the C-terminal HDELR motif uncovers the ER retention signal, which is essential for re-entry into the secretory pathway of the target cell. Consistently, K28 derivatives lacking the HDELR motif are inactive and cells defective in Erd2 no longer internalize the toxin (Eisfeld et al. 2000). Endocytotic uptake probably involves (mono-)ubiquitination of the membrane-receptor by Uba1, Udc4 and Rsc5 and, subsequently, retrograde passage of K28 from endosomes to the ER occurs, from where it exits to the cytoplasm via the Sec61 complex (Eisfeld et al. 2000; Heiligenstein et al. 2006). The Sec61 translocon is a bidirectional membrane channel which mediates translocation of secretory proteins into the ER but also removes misfolded proteins from the secretory pathway for subsequent degradation by the ER-associated protein degradation (ERAD) pathway (reviewed by Nakatsukasa and Brodsky 2008). K28 apparently masquerades as an ERAD substrate and the ER chaperones Kar2, Pdi1, Scj1, Jem1 and Pmr1 assist the exit of partially unfolded α-β subunits from the ER. Once the cytosol is reached, β is ubiquitinated and degraded by the proteasome, while α enters the nucleus and inhibits DNA synthesis (Schmitt et al. 1996; Heiligenstein et al. 2006).

As for K1, K28 preprotoxin not only confers the killer phenotype to M28-infected cells but also renders toxin producers immune against their own toxin. Moreover, K28 immunity is also established in a *kex2* mutant, which is unable to release active toxin, as described above for K1 (Schmitt and Tipper 1992; Zhu et al. 1992). While the mechanism of K1 immunity is still unknown, details for K28 were recently uncovered (Breinig et al. 2006). Immune K28 killer cells reinternalize their own toxin and subsequent interaction of the latter with unprocessed preprotoxin in the cyto-plasm is the key step for immunity, which involves ubiquitination and selective proteasomal degradation of mature K28 (Breinig et al. 2006). Interestingly, partial immunity could also be established when only the α-subunit of the toxin is expressed in the cytoplasm and full immunity requires only a non-specific sequence extension to the α-subunit, which strikingly resembles the sit-uation in K1 immunity (Breinig et al. 2006).

3. Other dsRNA Virus Toxins

Shortly after the discovery of the killer phenotype in *S. cerevisiae* an extranuclear inherited killer phe-nomenon was recognized in the dimorphic fungus *Ustilago maydis* (Puhalla 1968), a basidiomycete which can grow like a budding yeast or switch to filamentous growth (reviewed by Steinberg and Perez-Martin 2007). Three different immunity spe-cificities (P1, P4, P6) are reported to exist in natural killer isolates, all of which are encoded by dsRNA viruses (Koltin and Day 1976; Tipper and Bostian 1984). While KP1 and KP4 are monomeric proteins of 13.4 kDa and 13.6 kDa, respectively (Park et al. 1994, 1996a), KP6 is a dimer consisting of subunits of 8.6 kDa and 9.1 kDa (Tao et al. 1990). Structure determination of KP4 revealed that the toxin belongs to the α/β-sandwich fold, displaying weak similarities to scorpion toxins, which exert their neurological effects by acting on Na^+ channels (Gu et al. 1995). Rescuing KP4 treated sensitive *U. maydis* cells by supplying exogenous Ca^{2+} sug-gested toxin action by inhibition of Ca^{2+} channels (Gu et al. 1995). Such a hypothesis was further substantiated by demonstrating KP4-mediated in-hibition of voltage-gated Ca^{2+} channels in several types of mammalian neuronal cells (Gu et al. 1995). Crystallization of SMKT from *Pichia farinosa* (see above) surprisingly revealed close structural simi-larities of KP4 and SMKT (Kashiwagi et al. 1997), despite the fact that the latter is a chromosomally encoded dimeric protein, assumed to interact and disrupt cytoplasmic membranes of target cells (Suzuki et al. 2001).

In contrast, KP6 is composed of two subunits as for the *S. cerevisiae* dsRNA encoded toxins; however, the two subunits are not covalently linked by disulfide bonds (Peery et al. 1987; Tao et al. 1990). The toxic effect, which probably involves K^+ depletion, requires both of the subunits but,

interestingly, they do not directly interact in solu-tion (Peery et al. 1987). Crystallization of the smal-ler KP6 (α)-subunit revealed a third killer toxin member of the α/β sandwich fold; however, other than SMKT and KP4, KP6α subunits form a hex-americ assembly with a central pore which may provide a structure of the appropriate dimension to generate a membrane K^+ channel (Li et al. 1999).

Basidiomyceteous yeasts belonging to the genera *Trichosporon*, *Cryptococcus* and *Cystofili-basidium* harbour virus-like particles with dsRNA genomes, which are associated with killer (alter-natively termed mycocinogenic) phenotypes (Golubev et al. 2002, 2003; Pfeiffer et al. 2004; Fuentefria et al. 2008). Though basic characteris-tics of the toxins, including spectra of target yeasts and pH or temperature optima were investigated, it remains to be elucidated whether their struc-tures or mode of action resemble known dsRNA or chromosomally encoded toxins.

The ascomycetous yeasts *Hanseniaspora uvarum* and *Zygosaccharomyces bailii* also contain virus-like particles with toxin-encoding dsRNA genomes (Zorg et al. 1988; Radler et al. 1990). Similar to *S. cerevisiae*, both of the former contain autonomous L- and accompanying toxin-encoding M-viruses (Schmitt and Neuhausen 1994; Schmitt et al. 1997; Weiler et al. 2002). The *H. uvarum* toxin is a 18-kDa monomer, which uses β-1-6 glucan as the cell wall receptor; it displays a broader killing spectrum than *S. cerevisiae* killers (Radler et al. 1990; Schmitt and Neuhausen 1994; Schmitt et al. 1997). The *Z. bailii* toxin (termed zygocin) likewise exhibits a broad action spectrum, including human and phytopathogenic fungi (Weiler and Schmitt 2003). The 10-kDa zygocin rapidly induces per-meabilization, probably due to membrane channel formation (Weiler and Schmitt 2003; Schmitt and Breinig 2006). In contrast to the other dsRNA-encoded toxins, zygocin immunity is not mediated by the preprotoxin precursor. Since intact cells and protoplasts of *Z. bailii* are naturally resistant to zygocin, specific immunity is obviously dispens-able, a situation that is also realized for chro-mosomally encoded toxins (Weiler et al. 2003).

B. Linear Plasmid-Encoded Toxins

Linear dsDNA plasmids which persist in the cytoplasm are related to viruses with respect to gene content and mode of replication; they were originally detected in a killer strain of *Kluyvero-myces lactis* (Gunge et al. 1981). Today, such cyto-plasmic DNA elements are known to occur in several ascomycetous yeasts belonging to different genera (such as *Pichia, Candida, Debaryomyces, Saccharomycopsis, Schwanniomyces, Botryoascus*) but also in the basidiomycete *Trichosporon pull-ulans* (Kitada and Hishinuma 1987; Ligon et al. 1989; Worsham and Bolen 1990; Hayman and Bolen 1991; Bolen et al. 1992; Cong et al. 1994; Fukuhara 1995; Chen et al. 2000).

In addition to *K. lactis*, three other species (*P. acaciae, P. inositovora, D. robertsiae*; Table 6.2) contain a set of two or three linear plasmids confer-ring a killer phenotype (Fig. 6.4; Worsham and Bolen 1990; Hayman and Bolen 1991; Klassen and Meinhardt 2002). Though such dsDNA molecules are probably not encapsulated, a general functional partition among the plasmids of each species resem-bles the L- and M-viruses of *Saccharomyces cerevi-siae*. Invariably, there is a highly conserved autonomous element, which provides factors essential for cytoplasmic gene expression and repli-cation (reviewed by Jeske et al. 2006a; Klassen and Meinhardt 2007). Such element can exist solely or together with a non-autonomous element encod-ing a killer toxin. As expression of toxin genes and replication of the element requires functions provided by the autonomous element, it reminds one of the L-dependency of M-viruses (see above).

Replication of both autonomous and non-autonomous elements is initiated by protein priming, involving a viral B-type DNA polymerase fused to the terminal protein that remains covalently bound to the 5′ ends of the plasmids

(Hishinuma et al. 1984; Stark et al. 1984; Sor and Fukuhara 1985; Tommasino et al. 1988; Hishinuma and Hirai 1991; Klassen et al. 2001; Klassen and Meinhardt 2003; Jeske and Meinhardt 2006).

Cytoplasmic transcription of linear plasmid-encoded genes involves a rather unique plasmid-encoded RNA polymerase and a mRNA-capping enzyme which resem-bles the capping enzyme of the cytoplasmic vaccinia virus (Wilson and Meacock 1988; Larsen et al. 1998; Tiggemann et al. 2001; Klassen and Meinhardt 2007). The plasmid-encoded RNA polymerase recognizes unique cytoplasmic promoters with a short consensus sequence (6 nt) which is incompatible with the nuclear transcription machinery (Romanos and Boyd 1988; Kämper et al. 1989a, b; Kämper et al. 1991; Schickel et al. 1996).

The four known toxin-encoding killer plasmids consist of two subgroups judged from overall similarity, gene content and genetic prerequisites for toxin function (Fig. 6.4 see below). Group I comprises the *K. lactis* and *P. inositovora* killer plasmids while group II harbours the *P. acaciae* and *D. robertsiae* killers (Schaffrath and Mein-hardt 2004; Jeske et al. 2006b; Klassen and Mein-hardt 2007).

1. Group I

The *K. lactis* killer system consists of the autono-mous element pGKL2 and the toxin (zymocin)-encoding non-autonomous pGKL1 (for a review, see Stark et al. 1990). Zymocin is a heterotrimeric (α, β, γ) glycoprotein, the subunits of which dis-play molecular masses of 99, 30 and 28 kDa, re-spectively (Stark and Boyd 1986). While the smallest subunit (γ) is encoded separately, the

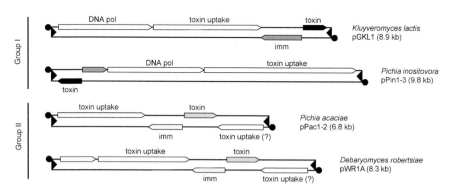

Fig. 6.4. Schematic representation of virus like dsDNA plasmids from yeasts encoding killer toxins. Based on sequence similarity and gene content, two subgroups are defined. *Arrows* indicate ORFs and their transcriptional direction; terminal proteins are depicted as *filled circles*; terminal inverted repeats correspond to *filled triangles*.

Known or proposed functions of encoded proteins are indicated. *toxin* Intracellular toxic subunit, *imm* immunity proteins. Similarity at the sequence level for either toxic subunits or immunity proteins is indicated by *grey shading*. The predicted toxin uptake protein is homolo-gous in all systems

two larger ones are formed from a single geneproduct (Orf2) by endopeptidase processing during secretion (involving the *S. cerevisiae* Kex2 homologue of *K. lactis*, Kex1; see Fig. 6.1 Hishinuma et al. 1984; Stark et al. 1984; Sor and Fukuhara 1985; Stark and Boyd 1986; Tokunaga et al. 1987; Stark et al. 1990). In the mature zymocin, β and γ are covalently linked via a disulfide bond and the α-subunit contains at least one internal disulfide bridge (Stark and Boyd 1986; Stark et al. 1990).

Zymocin differs from many other yeast killer toxins in having a neutral pH optimum and additionally exhibits relatively high thermal stability (no loss of activity following incubation at 50°C; Butler et al. 1991a). Cell killing by zymocin involves cell wall binding of the complex by virtue of its largest (α)-subunit, which has a cysteine-rich chitin binding and a LysM domain, the latter also found in peptidoglycan binding proteins (Stark et al. 1990; Butler et al. 1991b; Jablonowski et al. 2001; Jeske et al. 2006a). The α-subunit further exhibits clear similarities to family 18 chitinases and displays exochitinase activity which is apparently essential for toxicity (Butler et al. 1991b; Jablonowski et al. 2001). Cell-killing activity, however, resides in the smallest subunit of zymocin, as conditional intracellular expression of solely this subunit is toxic and also results in a G1 cell cycle arrest induced by exo-zymocin (Tokunaga et al. 1989; White et al. 1989; Stark et al. 1990; Butler et al. 1991c; Lu et al. 2005). Cellular uptake of the γ-subunit occurs in an unknown manner, probably with participation of the α-subunit's chitin binding and chitinase activity and the remarkably hydrophobic β-subunit (Fig. 6.2). Also involved is the plasmamembrane sphingolipid mannosyl-diinositolphospho-ceramide M(IP)$_2$C (synthesized by the Kti6/Ipt1 protein) and the plasma membrane ATPase Pma1; both are conjointly required for toxin action from outside (Mehlgarten and Schaffrath 2004; Zink et al. 2005). In contrast to chitin-deficient (*chs3*) cells, which lack zymocin binding, an *ipt1* mutant is proficient for binding but denies γ-import (Zink et al. 2005), revealing the M(IP)$_2$C-involving step to occur subsequent to initial toxin recruitment to the cell wall. Whether M(IP)$_2$C has a function similar to the secondary membrane receptor for other toxins (see above), remains unknown at present.

Following membrane passage, γ apparently exists in a transiently dormant form which requires intracellular activation by the plasma membrane H$^+$ATPase Pma1 or the membrane potential generated by this enzyme (Mehlgarten and Schaffrath 2004; Zink et al. 2005). In the absence of functional Pma1, γ-activation could be induced experimentally by acidification of the cell exterior, though it remains unclear which process is affected by H$^+$ flux (Mehlgarten and Schaffrath 2004).

Once inside the cell, γ specifically cleaves tRNA^{Glu} within the anticodon loop, at the 3' side of the wobble uridine (U$_{34}$; Lu et al. 2005). A posttranscriptional modification, which is present at the uracil-C5 position of U$_{34}$ in tRNA$^{Glu}_{UUG}$, a 5-methoxy-carbonyl-methyl (mcm^5) residue, is crucial for cleavage (Lu et al. 2005; Jablonowski et al. 2006).

mcm^5 synthesis in tRNA^{Glu} and several other tRNAs with an uridine residue at the wobble position requires the six-subunit Elongator complex (Elp1–6) as well as the tRNA methyltransferase Trm9 which is assumed to function downstream of Elongator in the final step of mcm^5 synthesis (Kalhor and Clarke 2003; Huang et al. 2005; Lu et al. 2005). Consistent with the importance of mcm^5 for cell killing by the tRNase, loss of *ELP1-6* or *TRM9* prevents zymocin toxicity (Frohloff et al. 2001; Lu et al. 2005; Jablonowski et al. 2006) and complementation of spontaneous killer toxin insensitive (*kti*) *S. cerevisiae* mutants identified further loci with a previously unknown role in tRNA mcm^5 modification (Fichtner and Schaffrath 2002; Fichtner et al. 2002; Huang et al. 2005; Lu et al. 2005; Jablonowski and Schaffrath 2007; Zabel et al. 2008).

In vitro analysis revealed additional cleavage of two other mcm^5-modified tRNAs, i.e. tRNA^{Gln} and tRNA^{Lys}, by the heterologously expressed toxic subunit of zymocin, although the efficiency was much lower when compared to tRNA^{Gln} (Lu et al. 2005). Contrary to the latter, no depletion of tRNA^{Gln} or ^{Lys} was seen in cells treated with zymocin. However, as combined overexpression of tRNA$^{Glu, Gln}$ and ^{Lys} intensified zymocin resistance when compared to a strain overexpressing tRNA^{Glu} solely, the glutamine and lysine tRNAs were assumed to represent zymocin targets in vivo, in addition to tRNA^{Glu} (Lu et al. 2005).

A killer strain of *Pichia inositovora* contains a set of three plasmids (pPin1-1, pPin1-2, pPin1-3) of which pPin1-2 is dispensable for killer activity (Hayman and Bolen 1991); As for *K. lactis*, the larger pPin1-1 probably represents the autonomous element whereas the toxin is encoded by the smaller pPin1-3 (Klassen and Meinhardt 2003). The *P. inositovora* toxin displays a molecular mass larger than 100 kDa and probably consists of at least two subunits (Klassen and Meinhardt 2003). The largest subunit (∼100 kDa) is similar to zymocin α- and β and – as for zymocin – was proven to bind chitin in vitro. Disruption of the chitin synthase III-encoding gene in a sensitive *S. cerevisiae* strain resulted in insensitivity to the *P. insitovora* toxin as well. As the pPin1-3 killer plasmid also encodes a homologue of the zymocin's tRNase subunit and toxicity requires the Elongator complex involved in tRNA-mcm^5 modification, a similar mode of action of the *P. inositovora* toxin and zymocin is likely to occur (Klassen and Meinhardt 2003).

Despite such striking similarities, zymocin and the *P. inositovora* toxin clearly differ with respect to target cell spectra and pH optima; in contrast to the broad spectrum of activity with a neutral pH optimum of zymocin, *P. inositovora* toxin has an acidic pH optimum and a clearly narrower spectrum of target yeasts (Hayman and Bolen 1991). Sensitive reaction to *P. inositovora* toxin is apparently not species-specific, as two related *S. cerevisiae* strains are killed by the toxin, while several other laboratory strains sensitive to zymocin display resistance to *P. inositovora* toxin (Hayman and Bolen 1991; Klassen and Meinhardt, unpublished data).

The zymocin-encoding plasmid (pGKL1) harbours (in addition to the toxin-encoding ORF) a gene responsible for the specific related immunity, the mechanism of which, however, remains to be elucidated (Tokunaga et al. 1987). Despite the fact that a gene homologous to the zymocin immunity factor is located on the *P. inositovora* killer plasmid pPin1-3 (Klassen and Meinhardt 2003), the latter is not required for immunity in this species (Hayman and Bolen 1991). Possibly, intrinsic immunity has evolved subsequent to the acquisition of the killer plasmid system, which concomitantly made a plasmid-encoded immunity function redundant. In *K. lactis*, the plasmid-encoded immunity function establishes an auto-selection system for the entire plasmid set. Spontaneous loss of the non-autonomous element alone or together with the autonomous element would give rise to zymocin (−) and sensitive cells which are eliminated due to the stable zymocin toxin in the environment. A very similar autoselection system is also present in most dsRNA- and other dsDNA-encoded killer systems (see above and below). It cannot be decided whether toxin/immunity functions are recruited by viral or virus-like genomes persisting in the yeast cell's cytoplasm for the purpose of creating a selective advantage for the host (elimination of competitors) or stable propagation of the virus or both. However, in cases where intrinsic immunity is established, a toxin-dependent autoselection effect on the respective element is no longer present.

2. Group II

The two killer plasmids of homology group II, pPac1-2 from *P. acaciae* and pWR1A from *D. robertsiae* are similar with respect to genome organization (Klassen et al. 2004). pWR1A harbours one additional gene, which is not present in pPac1-2 (Fig. 6.4). The pPac1-2 encoded toxin (PaT) was purified and found to consist of three subunits with molecular masses of 110, 39 and 38 kDa, respectively (McCracken et al. 1994). The pWR1A-encoded toxin too is a larger protein of above 100 kDa and both, PaT and *D. robertsiae* toxin display neutral pH optima (McCracken et al. 1994; Klassen and Meinhardt 2002). The similarity of the toxin-encoding plasmids to homology group I elements (pGKL1, pPin1-3) is restricted to the ORF encoding the chitin-binding polypeptide (α-subunit), whereas separately encoded and potentially secreted proteins (pPac1-2 Orf2p, pWR1A Orf3p) are not related to zymocin γ and/or the *P. inositovora* equivalent (Klassen et al. 2004). However, as intracellular expression of these genes mimics the effects of extracellularly applied toxins on target cells, the toxic activity resides in such separately encoded toxin subunits, which act intracellularly on target cells, as for zymocin γ (Klassen et al. 2004). Other than zymocin, neither of the toxins encoded by group II plasmids requires the Elongator complex for function; however, PaT was recently shown to target cellular tRNA, as for zymocin (Klassen et al. 2008). The tRNase activity of PaT is specific for tRNA^{Gln} (in contrast to zymocin, which attacks tRNA^{Glu}) and loss of mcm^5 modification does not prevent substrate cleavage in vivo or in vitro. Interestingly, however, an mcm^5 biosynthesis intermediate (carboxyl-methyl-residue; cm^5), which presumably exists in the tRNA^{Gln} of cells deficient in the tRNA methyltransferase Trm9, strongly inhibits cleavage and, thus, largely prevents PaT toxicity. In vitro analysis of cleavage products revealed two alternative cleavage sites in tRNA^{Gln}, one of which is the mcm^5 modified wobble nucleoside U$_{34}$ and the other is most probably U$_{32}$. Loss of mcm^5 prevented cleavage at position U$_{34}$, but not at U$_{32}$, explaining the sensitive phenotype of Elongator mutants to PaT. The negative charge of cm^5 might be the reason for its inhibitory effect on cleavage at both positions (Klassen et al. 2008).

As for zymocin, the heterologously expressed toxic subunit of PaT displayed an extended substrate spectrum, not only accepting tRNA^{Gln} but with reduced efficiency also mcm^5-modified tRNA^{Glu} and ^{Lys} (Klassen at al. 2008). Cleavage of all these substrates was (at least partially) inhibited by cm^5, but not by the entire loss of the mcm^5 modification of U$_{34}$, which is in marked contrast to zymocin's tRNase activity and may be explained by the existence of an alternative cleavage site (see above).

Interestingly, cell killing by both zymocin and PaT involves the occurrence of DNA damage, presumably as a downstream consequence of tRNA depletion (Klassen et al. 2007, 2008). Though not systematically analysed yet, it is speculated that tRNase toxin exposure may lead to deficiency in translation of messengers relevant for genome surveillance during replication. Consistently, genetic analysis revealed strong evidence for the occurrence of replication-derived DNA double-strand breaks in both PaT- and zymocin-exposed cells and DSB repair by homologous recombination was found to confer resistance against both toxins. An alternative DSB repair pathway (non-homologous end-joining), which differs from homologous recombination in its inability to operate on replication-derived DSBs, is even disadvantageous for PaT/zymocin survival and this effect is due to inhibition of homologous recombination (Klassen et al. 2006, 2007, 2008). Clearly, however, further work is required to understand the cellular consequences of tRNA attack by specific endonucleases in yeast, as such a toxic principle was only quite recently recognized in eukaryotic toxins (Lu et al. 2005; Klassen et al. 2008). Interestingly, PaT and zymocin both do not deplete their target tRNAs completely and, though both significantly impair translation, there is no complete shut-off of protein biosynthesis (Butler et al. 1991a; Klassen and Meinhardt, unpublished data). Moreover, PaT was shown to induce apoptotic cell death under conditions where only a fraction of target cells is rapidly killed; the remaining initially surviving fraction later executed an active cell death program involving protein biosynthesis and the classic apoptotic hallmarks, including nuclear fragmentation, ROS production and phosphatidyl-serine externalization (Klassen and Meinhardt 2005). As other toxins, such as K1, K28 and zygocin also induce apoptotic cell death when applied in moderate doses (Reiter et al. 2005); such cellular response is apparently not functionally connected to the primary lethal principle of the individual toxins, but rather a common reaction of individual cells to life-threatening conditions.

With the exception of pPin1-3, all killer plasmids carry an additional gene which is responsible for a specific immunity phenotype (Tokunaga et al. 1987; Paluszynski et al. 2007; Paluszynski and Meinhardt, unpublished data). Though the mode of action of the encoded immunity proteins remains to be elucidated, they disable the effects of the respective toxic subunit at the intracellular level, as for the only other toxin known to act intracellularly on target cells, the M28-encoded K28 toxin (Tokunaga et al. 1989; Breinig et al. 2006; Paluszynski et al. 2007). Thus, in all known killer/immunity systems where toxicity requires cellular uptake of the toxin (zymocin, PaT, K28), toxin uptake apparently also occurs in the toxin producer and the reinternalized toxin is specifically degraded by the host cell's proteasome (K28) or disposed by yet unknown mechanisms (zymocin, PaT).

IV. Applications

A. Antifungals for Human Therapy

Several yeast killer toxins, such as *W. mrakii* var. *mrakii* HM-1, *W. saturnus* toxin, *P. anomala* PaKT and *Zygosaccharoymces bailii* zygocin, exhibit a broad spectrum of activity including human pathogenic yeasts, such as *Candida albicans* and *Cryptococcus neoformans*. Consequently, they are suggested to be potentially useful for therapy of human infections caused by fungal pathogens (Yamamoto et al. 1988; Walker et al. 1995; Magliani et al. 1997a; Theisen et al. 2000; Weiler and Schmitt 2002; Buzzini et al. 2004; Izgü et al. 2007a). However, direct application of killer toxins is of limited practical importance because many of these proteins are unstable or inactive at temperatures around 37°C or neutral pH. Furthermore, antigenicity and toxicity may prohibit application in the human bloodstream (Magliani et al. 2004). Yet, some killer toxins (e.g. the *Williopsis* toxins) display a remarkable temperature stability, which might facilitate their use as topical applications on superficial skin lesions (Buzzini et al. 2004). *P. anomala* K5 toxin exhibiting acceptable temperature stability was studied against dermatophytes and several pathogenic *Candida* species. All clinical isolates as well as type strains belonging to the genera *Trichosporon*, *Microsporum* and *Candida* were found to be susceptible to K5, suggesting this toxin to be applicable as a topical antifungal agent (Izgü et al. 2007a, b).

A rather promising approach to overcome the above-mentioned problems associated with the direct application of killer toxins was initiated by Polonelli and Morace (1988). A monoclonal antibody (mAbKT4), which neutralized the in vitro activity of *P. anomala* UCSC 25F (= ATCC 96603; Table 6.1) PaKT (Polonelli and Morace 1987) was used to raise anti-idiotypic antibodies displaying an internal image of the toxin's active side. Strikingly, such natural polyclonal, as well as subsequently developed monoclonal antibodies or single-chain variable fragments (scFv) derived from a phage display library were able to interact with the cell wall and kill yeast cells susceptible to the original *P. anomala* toxin (Polonelli and Morace 1988; Polonelli et al. 1990, 1997; Magliani et al. 1997a; Magliani et al. 2004). Vaccination with mAbKT4 in the mouse model resulted in the production of killer toxin-like antibodies, which conferred significant protection against experimental candidiasis (Polonelli et al. 1993, 1994).

A completely novel approach for the treatment of vaginal candidiasis was realized with the construction of genetically engineered *Streptococcus gordonii* strains, being able to colonize vaginal mucosal sites and either displaying killer toxin-like antibodies (ScFv) at the cell surface or secreting them, resulting in anti-candida activity of potential therapeutic value (Magliani et al. 1997b). Decapeptides derived from scFv and displaying *P. anomala* toxin-like activity (so-called mimotopes) were subsequently identified and further optimized by alanine scanning, resulting in the extremely stable fungizidal killer peptide KP (Polonelli et al. 2003), which is active against a broad variety of pathogenic micro-organisms, including *Cryptococcus neoformans* and *Paracoccidioides brasiliensis* (Cenci et al. 2004; Travassos et al. 2004). Killer toxin-like antibodies or killer peptides derived from PaKT were shown to additionally kill a variety of pathogenic prokaryotic micro-organisms, such as *Mycobacteria*, *Staphylococcus*, *Streptococcus* species and plant-pathogenic *Pseudomonas* strains (reviewed by Magliani et al. 2004). It was proposed that β-glucan serves as the yeast cell wall receptor for *P. anomala* toxin, as for the related WmKT from *W. saturnus* var. *mrakii* (see also above) and

thus, glucan or glucan-like molecules in the cell wall of susceptible pro- and eukaryotic micro-ogranisms were proposed to provide the basis for the observed extremely broad spectrum of toxin activity, idiotypic antibodies and derived decapeptides (Magliani et al. 2004).

In addition to *P. anomala* toxin, HM-1 from *W. saturnus* var. *mrakii* was used to produce toxin neutralizing antibodies that were subsequently employed in idiotypic vaccination and the production of killer toxin-like antibodies, which display an internal image of HM-1's active site and inhibit the target cell's glucan synthase activity (Selvakumar et al. 2006a, b, c). As for the *P. anomala* toxin, such killer activity-bearing antibodies may have application potential in the treatment of human fungal infections and may possibly also be developed to recombinant antimycotic-producing bacteria or mimotopes displaying the desired activity. Small peptide mimotopes can potentially be produced much more economically when compared to immunological killer toxin derivatives. Consequently, the *P. anomala* toxin-derived KP was patented and has entered clinical trials, possibly ending up as the first member of a new generation of antifungal agents derived from yeast killer toxins (Magliani et al. 2004).

B. Antifungals in Agriculture, Food and Feed Industry

There is a clear application potential for yeast killer toxins in the wine industry. The wine environment is characterized by low pH ($\sim$3.5), at which some of the chromosomally or dsRNA-encoded toxins are active. In large-scale wine production, fermentation is usually carried out using defined *S. cerevisiae* starter cultures optimized for fermentation performance and able to dominate native yeasts in the grape must (Pretorius 2000). To achieve the latter, wine and also sake fermentation starter yeasts were engineered by cytoduction to possess the L-A and M viruses and the corresponding killer phenotype (Ouchi and Akiyama 1976; Hara et al. 1980; Seki et al. 1985; Boone et al. 1990; Sulo and Michalcakova 1992; Sulo et al. 1992; Michalcakova et al. 1994). Among the *S. cerevisiae* dsRNA toxins, K2 was considered to be most suitable for biocontrol in the wine environment, since it is similar to K1 (see above) but displays higher activity at wine pH (~3.5) compared to K1 (Pfeiffer and Radler 1984). The obtained industrial *S. cerevisiae* killer strains retain desired flavour and fermentation characteristics but are additionally able to suppress indigenous *S. cerevisiae* strains due to toxin production. Since K2 killer strains are frequent among the natural population on grape surfaces, the use of defined K2 killer-positive fermentation starters which also display K2 immunity additionally prevents overgrowth by the indigenous killer (Jacobs and Van Vuuren 1991).

As an alternative to engineered killer strains, fermentation starters which naturally express the K2-type killer phenotype as well as desired fermentation characteristics can be selected from the population of indigenous yeasts (Lopes et al. 2007). However, non-*Saccharomyces* yeasts present at grape surfaces are routinely insensitive to the *S. cerevisiae* killer toxins (Young and Yagui 1978) and thus are largely restricting the biocontrol potential to *Saccharomyces* contaminants. Additionally, indigenous non-*Saccharomyces* yeasts often produce toxins, potentially enabling the producer to overgrow the inoculated starter culture and spoil the fermentation.

The killer toxin KpKt from *K. phaffii* displays extensive activity against apiculate yeast species, such as *Hanseniaspora uvarum*, which dominate on grapes and grape juice and therefore KpKt was analysed for its potential as a biopreservative agent in wine (Ciani and Fatichenti 2001). During experimental vinifications, KpKt was found to display anti-*H. uvarum* activity comparable to the routinely applied SO₂, which is used to inhibit apiculate yeasts at the prefermentation stage. It was suggested that KpKt could substitute for SO₂, thereby eliminating undesired or harmful residual traces of SO₂ in the final product (Ciani and Fatichenti 2001; Comitini et al. 2004b). The killer toxins PiKt and KwKt from *P. anomala* and *K. wickerhamii* were shown to be active and stable in wine environment and are both capable of inhibiting spoilage yeasts belonging to the genus *Dekkera/Brettanomyces*, which can grow during wine aging and represent a major problem due to unpleasant odour and taste development (Comitini et al. 2004a).

As several killer toxins, including those of *P. anomala*, *P. membranifaciens* and *D. hansenii* are not only active against yeast cells but also may inhibit filamentous fungi, they are proposed as

potential agents for biocontrol of pre- and post-harvest diseases caused by plant pathogenic fungi (Walker et al. 1995; Santos and Marquina 2004b; Santos et al. 2004). In particular, the toxin from *P. membranifaciens* is active against *Botrytis cinerea*, the causal agent of grey mould disease in grapes and treatment of *Vitis vinifera* plants with either purified toxin or the *P. membranifaciens* killer strain protected against *B. cinerea* (Santos and Marquina 2004b). The same toxin was also active in suppressing *B. cinerea* growth on apples following harvest, suggesting that *P. membranifaciens* might be developed into a novel biocontrol agent for grey mould disease (Santos et al. 2004). A killer strain of *P. anomala* (YF07b) was selected from a marine environment based on its ability to kill *Metschnikowia bicuspidate*, a yeast species pathogenic to *Portunus trituberculatus*, the most widely fished crab species (Wang et al. 2007a, b). It was suggested that the toxin of strain YF07b could be used for controlling growth and infection of the crabs by the pathogenic yeast (Wang et al. 2006a, b).

Silage, a fermentation product of field crops and lactic acid bacteria used for feeding of ruminant livestock is prone to aerobic spoilage by lactic acid-utilizing yeasts. Two killer toxins were analysed for their applicability in controlling silage spoilage (Kitamoto et al. 1993; 1999; Lowes et al. 2000). The *K. lactis* zymocin was found to be active against a variety of silage spoilage yeasts and the producer strain was further optimized for biocontrol in silage by generating a zymocin-producing phosphoenolpyruvate carboxykinase-negative mutant, which can no longer grow on lactic acid and thus cannot contribute to spoilage (Kitamoto et al. 1993, 1998). Co-inoculation of model silage fermentations with the optimized *K. lactis* strain and *Lactobacillus plantarum* indeed reduced aerobic spoilage and, thus, *K. lactis* was suggested to be useful in prolonging the aerobic stability of silage (Kitamoto et al. 1999). The other toxin selected for control of silage and also dairy product spoilage is HM-1 from *Williopsis saturnus* var. *mrakii*, which exhibits a broad spectrum of activity and exceptional stability as well as pH and temperature stability (see above). The HM-1 (HMK)-encoding *HMK* gene was heterologously expressed in *Aspergillus niger* and partially purified HM-1 was found to suppress aerobic spoilage of model silage fermentations as well as growth of *Candia krusei* and *S. cerevisiae* in yoghurt. Thus,

HM-1 is generally suited for the control of unde-sired yeasts in both fermented feed and food pro-ducts. Heterologous expression of HM-1 in lactic acid bacteria might further extend the practicabil-ity of such approaches (Lowes et al. 2000).

In order to protect plants against pathogenic fungi and bacteria, the killer toxins KP4 and KP6 from *Ustilago maydis* and the synthetic killer pep-tide KP (derived from *P. anomala* toxins; see above) were heterologously expressed in tobacco (Kinal et al. 1995; Park et al. 1996b; Donini et al. 2005). Authentic and active KP6 and KP4 could be recovered from tobacco; however, the expression level of KP6 was diminished compared to the original toxin-producing *U. maydis* strain (Kinal et al. 1995; Park et al. 1996b). KP4, in contrast, was efficiently expressed; the expression was so effi-cient that a small piece of leaf secreted enough toxin to kill susceptible *U. maydis* (Park et al. 1996b). Therefore, heterologous expression of KP4 was suggested to be potentially valuable for the control of KP4 susceptible fungi in economi-cally important crop plants (Park et al. 1996b). By using a potato virus X-derived vector, the syn-thetic killer peptide KP was heterologously expressed in *Nicotiana benthaminana* as a fusion to the viral coat protein (Donini et al. 2005). Such purified chimeric virus particles displayed broad-spectrum killer activity against human and plant pathogens, as for KP (see above), and their expression in planta conferred resistance against plant-pathogenic *Pseudomonas syringae*, thereby identifying KP as a promising molecule not only for the treatment of fungal infections in man (see above) but also to confer broad-spectrum resistance to phytopathogens in plants (Donini et al. 2005).

C. Killer Toxins in Biotyping

In 1983, it was realized that killer toxin sensitivity patterns (sensitivity or insensitivity to a panel of selected killer strains) can be used to distinguish different *Candida albicans* isolates (Polonelli et al. 1983). Subsequently, the principle was extended by the inclusion of more killer strains, the use of partially purified toxins and computer-aided data analysis, resulting in an improvement in repro-ducibility (Polonelli et al. 1985; Buzzini and Mar-tini 2000; Buzzini et al. 2003, 2007). Killer toxin

sensitivity patterns (KSPs) were shown to dis-criminate not only different *C. albicans* strains but also different species of other pathogenic yeasts (Morace et al. 1984; Polonelli et al. 1985). KSPs also proved useful for ecological studies on the population diversity of indigenous wine yeasts (*S. cerevisiae*; Sangorrin et al. 2002; Lopes et al. 2005). Since some yeast killer toxins are also ac-tive against filamentous fungi or even bacteria (see above), killer toxin-based biotyping has suc-cessfully been extended to micro-organsims other than yeast, including *Penicillium camembertii*, *Staphylococcus aureus*, *Neisseria meningitidis* and *Mycobacterium tuberculosis* (Morace et al. 1987, 1989; Polonelli et al. 1987).

However, as outlined by Buzzini et al. (2007), the determination of KSP is currently not in routine use for the differentiation of industrial or clinical yeasts or other micro-organisms, mainly due to problems associated with the limited reproducibili-ty and automatization potential of phenotypic tests involving living killer yeasts. Promising approaches to achieve a higher reproducibility and discrimina-tory power of KSP-based biotyping include the use of a broader spectrum of killer toxins with known function and their use as a standardized protein preparation rather than as toxin-producing strains, which would also enable automatization of the pro-cedure (Buzzini et al. 2007). Further, the combina-tion of KSP-based and molecular biotyping methods, such as mtDNA-RFLP increases the dis-ciminatory power of both and, thus, may represent a promising perspective for the future application of yeast killer toxins in biotyping (Lopes et al. 2005).

V. Concluding Remarks

The group of proteins subsumed as killer toxins display considerable heterogeneity concerning structure and evolutionary origin. Many are chro-mosomally encoded and provide a selective ad-vantage over sensitive non-killer yeasts in the environment. In other cases toxin production is conferred by viruses or virus-like elements, in most of which the ability of toxin production is genetically linked with specific immunity. In such cases it can hardly be decided whether toxin pro-duction evolved for the purpose of elimination of competitors or for autoselection of the virus or virus-like element. The fact that, at least for

S. cerevisiae, such toxins display a narrow range of activity, being essentially specific for non-virus-carrying *S. cerevisiae* cells, may support an auto-selective purpose of the toxins. Mechanistically, a few toxic principles are repeatedly realized in chromosomally and virally encoded killer toxins, such as degradation or biosynthesis inhibition of cell wall polysaccharides and direct membrane permeabilization. In addition to lethal principles, similar structures and biogenesis involving processing of toxin precursors to mature toxins are repeatedly encountered in toxins originating from the chromosome or parasitizing viruses, despite the fact that similarity in primary sequences is lacking. Thus, convergent evolution may have repeatedly invented functionally similar toxins. However, other lethal mechanisms, such as DNA synthesis inhibition by the RNA virus-encoded K28 toxin or tRNA-cleaving toxins encoded by virus-like cytoplasmic dsDNA elements appear unique to these elements yet. Nevertheless, only a small fraction of recognized killer toxins has been characterized in detail so far, raising the possibility of repeated realization of DNA synthesis-inhibiting or tRNA-cleaving activities or even further yet unknown toxic mechanisms in other killer yeasts. As for the antibiotics which are the constituents of a similar antagonistic interaction, killer toxins may potentially be developed into a new generation of antifungal agents or find other useful applications in the food and feed industry.

Acknowledgement. Financial support by the Deutsche Forschungs Gemeinschaft (DFG) grant no. ME 1142/5-1 is gratefully acknowledged.

References

Ahmed A, Sesti F, Ilan N, Shih TM, Sturley SL, Goldstein SA (1999) A molecular target for viral killer toxin: TOK1 potassium channels. Cell 99:283–291

Antuch W, Güntert P, Wüthrich K (1996) Ancestral beta gamma-crystallin precursor structure in a yeast killer toxin. Nat Struct Biol 3:662–665

Ashida S, Shimazaki T, Kitano K, Hara S (1983) New killer toxin of *Hansenula mrakii*. Agric Biol Chem 47:2953–2955

Bevan EA, Makower M (1963) The physiological basis of the killer character in yeast. Proc XI Int Congr Genet 1:202–203

Bevan EA, Herring AJ, Mitchell DJ (1973) Preliminary characterization of two species of dsRNA in yeast and their relationship to the "killer" character. Nature 245:81–86

Bolen PL, Kurtzman CP, Ligon JM, Mannarelli BM, Bothast RJ (1992) Physical and genetic characterization of linear DNA plasmids from the heterothallic yeast *Saccharomycopsis crataegensis*. Antonie Van Leuwenhoek 61:195–295

Boone C, Sdicu AM, Wagner J, Degré R, Sanchez C, Bussey H (1990) Integration of the yeast K1 killer toxin gene into the genome of marked wine yeasts and its effect on vinifcation. Am J Enol Vitic 41:37–42

Bostian KA, Elliott Q, Bussey H, Burn V, Smith A, Tipper DJ (1984) Sequence of the preprotoxin dsRNA gene of type I killer yeast: multiple processing events produce a two-component toxin. Cell 36:741–751

Breinig F, Tipper DJ, Schmitt MJ (2002) Kre1p, the plasma membrane receptor for the yeast K1 viral toxin. Cell 108:395–405

Breinig F, Schleinkofer K, Schmitt MJ (2004) Yeast Kre1p is GPI-anchored and involved in both cell wall assembly and architecture. Microbiology 150:3209–3218

Breinig F, Sendzik T, Eisfeld K, Schmitt MJ (2006) Dissecting toxin immunity in virus-infected killer yeast uncovers an intrinsic strategy of self-protection. Proc Natl Acad Sci USA 103:3810–3815

Bussey H (1991) K1 killer toxin, a pore-forming protein from yeast. Mol Microbiol 5:2339–2343

Butler AR, White JH, Stark MJR (1991a) Analysis of the response of *Saccharomyces cerevisiae* cells to *Kluyveromyces lactis* toxin. J Gen Microbiol 137:1749–1757

Butler AR, O'Donnell RW, Martin VJ, Gooday GW, Stark MJ (1991b) *Kluyveromyces lactis* toxin has an essential chitinase activity. Eur J Biochem 199:483–488

Butler AR, Porter M, Stark MJR (1991c) Intracellular expression of *Kluyveromyces lactis* toxin γ subunit mimics treatment with exogenous toxin and distinguishes two classes of toxin-resistant mutant. Yeast 7:617–625

Buzzini P, Martini A (2000) Differential growth inhibition as a tool to increase the discriminating power of killer toxin sensitivity in fingerprinting of yeasts. FEMS Microbiol Lett. 193:31–36

Buzzini P, Berardinelli S, Turchetti B, Cardinali G, Martini A (2003) Fingerprinting of yeasts at the strain level by differential sensitivity responses to a panel of selected killer toxins. Syst Appl Microbiol 26:466–470

Buzzini P, Corazzi L, Turchetti B, Buratta M, Martini A (2004) Characterization of the in vitro antimycotic activity of a novel killer protein from *Williopsis saturnus* DBVPG 4561 against emerging pathogenic yeasts. FEMS Microbiol Lett 238:359–365

Buzzini P, Turchetti B, Vaughan-Martini AE (2007) The use of killer sensitivity patterns for biotyping yeast strains: the state of the art, potentialities and limitations. FEMS Yeast Res 7:749–760

Castón JR, Trus BL, Booy FP, Wickner RB, Wall JS, Steven AC (1997) Structure of L-A virus: a specialized compartment for the transcription and replication of double-stranded RNA. J Cell Biol 138:975–985

Cenci E, Bistoni F, Mencacci A, Perito S, Magliani W, Conti S, Polonelli L, Vecchiarelli A (2004) A synthetic peptide as a novel anticryptococcal agent. Cell Microbiol 6:953–961

Chen WB, Han JF, Jong SC, Chang SC (2000) Isolation, purification, and characterization of a killer protein from *Schwanniomyces occidentalis*. Appl Environ Microbiol 66:5348–5352

Ciani M, Fatichenti F (2001) Killer toxin of *Kluyveromyces phaffii* DBVPG 6076 as a biopreservative agent to control apiculate wine Yeasts. Appl Environ Microbiol 67:3058–3063

Comitini F, De Ingeniis J, Pepe L, Mannazzu I, Ciani M (2004a) *Pichia anomala* and *Kluyveromyces wickerhamii* killer toxins as new tools against *Dekkera/Brettanomyces* spoilage yeasts. FEMS Microbiol Lett 238:235–240

Comitini F, Di Pietro N, Zacchi L, Mannazzu I, Ciani M (2004b) *Kluyveromyces phaffii* killer toxin active against wine spoilage yeasts: purification and characterization. Microbiology 150:2535–2541

Cong YS, Yarrow D, Li YY, Fukuhara H (1994) Linear DNA plasmids from *Pichia etchellsii*, *Debaryomyces hansenii* and *Wingea robertsiae*. Microbiology 140:1327–1335

da Silva S, Calado S, Lucas C, Aguiar C (2008) Unusual properties of the halotolerant yeast Candida nodaensis Killer toxin, CnKT. Microbiol Res 163:243–251

de la Peña P, Barros F, Gascón S, Lazo PS, Ramos S (1981) Effect of yeast killer toxin on sensitive cells of *Saccharomyces cerevisiae*. J Biol Chem 256:10420–10425

Dignard D, Whiteway M, Germain D, Tessier D, Thomas DY (1991) Expression in yeast of a cDNA copy of the K2 killer toxin gene. Mol Gen Genet 227:127–136

Dinman JD, Icho T, Wickner RB (1991) A -1 ribosomal frameshift in a double-stranded RNA virus of yeast forms a gag-pol fusion protein. Proc Natl Acad Sci USA 88:174–178

Donini M, Lico C, Baschieri S, Conti S, Magliani W, Polonelli L, Benvenuto E (2005) Production of an engineered killer peptide in *Nicotiana benthamiana* by using a potato virus X expression system. Appl Environ Microbiol 71:6360–6367

Eisfeld K, Riffer F, Mentges J, Schmitt MJ (2000) Endocytotic uptake and retrograde transport of a virally encoded killer toxin in yeast. Mol Microbiol 37:926–940

Fuentefria AM, Suh SO, Landell MF, Faganello J, Schrank A, Vainstein MH, Blackwell M, Valente P. Trichosporon insectorum sp. nov., a new anamorphic basidiomycetous killer yeast. Mycol Res 112:93–99

Fichtner L, Schaffrath R (2002) *KTI11* and *KTI13*, *Saccharomyces cerevisiae* genes controlling sensitivity to G1 arrest induced by *Kluyveromyces lactis* zymocin. Mol Microbiol 44:865–875

Fichtner L, Frohloff F, Burkner K, Larsen M, Breunig KD, Schaffrath R (2002) Molecular analysis of *KTI12/TOT4*, a *Saccharomyces cerevisiae* gene required for *Kluyveromyces lactis* zymocin action. Mol Microbiol 43:783–791

Fukuhara H (1995) Linear DNA plasmids of yeasts. FEMS Microbiol Lett 131:1–9

Frohloff F, Fichtner L, Jablonowski D, Breuning KD, Schaffrath R (2001) *Saccharomyces cerevisiae* elongator mutations confer resistance to the *Kluyveromyces lactis* zymocin. EMBO J 20:1993–2003

Golubev WI (2006) Antagonistic interactions among yeasts. In: Rosa CA, Péter G (eds) The yeast handbook. Biodiversity and ecophysiology of yeasts. Springer, Heidelberg, pp 197–219

Golubev W, Shabalin Y (1994) Microcin production by the yeast *Cryptococcus humicola*. FEMS Microbiol Lett 119:105–110

Golubev WI, Pfeifer I, Golubeva E (2002) Mycocin production in *Trichosporon pullulans* populations colonizing tree exudates in the spring. FEMS Microbiol Ecol 40:151–157

Golubev WI, Pfeiffer I, Churkina LG, Golubeva EW (2003) Double-stranded RNA viruses in a mycocinogenic strain of *Cystofilobasidium infirmominiatum*. FEMS Yeast Res 3:63–68

Goto K, Iwatuki Y, Kitano K, Obata T, Hara S (1990) Cloning and nucleotide sequence of the KHR killer gene of *Saccharomyces cerevisiae*. Agric Biol Chem 54:979–984

Goto K, Fukuda H, Kichise K, Kitano K, Hara S (1991) Cloning and nucleotide sequence of the KHS killer gene of *Saccharomyces cerevisiae*. Agric Biol Chem 55:1953–1958

Gu F, Khimani A, Rane SG, Flurkey WH, Bozarth RF, Smith TJ (1995) Structure and function of a virally encoded fungal toxin from *Ustilago maydis*: a fungal and mammalian Ca^{2+} channel inhibitor. Structure 3:805–814

Gunge N, Tamaru A, Ozawa F, Sakaguchi K (1981) Isolation and characterization of linear deoxyribonucleic acid plasmids from *Kluyveromyces lactis* and the plasmid-associated killer character. J Bacteriol 145:382–390

Guyard C, Séguy N, Lange M, Ricard I, Polonelli L, Cailliez JC (1999) First steps in the purification and characterization of a *Pichia anomala* killer toxin. J Eukaryot Microbiol 46:144S

Guyard C, Evrard P, Corbisier-Colson AM, Louvart H, Dei-Cas E, Menozzi FD, Polonelli L, Cailliez J (2001) Immuno-cross reactivity of an anti-*Pichia anomala* killer toxin monoclonal antibody with a *Williopsis saturnus* var. *mrakii* killer toxin. Med Mycol 39:395–400

Guyard C, Séguy N, Cailliez JC, Drobecq H, Polonelli L, Dei-Cas E, Mercenier A, Menozzi FD (2002a) Characterization of a *Williopsis saturnus* var. *mrakii* high molecular weight secreted killer toxin with broad-spectrum antimicrobial activity. J Antimicrob Chemother. 49:961–971

Guyard C, Dehecq E, Tissier JP, Polonelli L, Dei-Cas E, Cailliez JC, Menozzi FD (2002b) Involvement of β-glucans in the wide-spectrum antimicrobial activity of *Williopsis saturnus* var. *mrakii* MUCL 41968 killer toxin. Mol Med 8:686–694

Hara S, Iimura Y, Otsuka K (1980) Breeding of useful killer wine yeasts. Am J Enol Vitic 31:28–33

Hayman GT, Bolen BL (1991) Linear DNA plasmids of *Pichia inositovora* are associated with a novel killer toxin activity. Curr Genet 19:389–393

Heiligenstein S, Eisfeld K, Sendzik T, Jimenéz-Becker N, Breinig F, Schmitt MJ (2006) Retrotranslocation of a viral A/B toxin from the yeast endoplasmic reticulum is independent of ubiquitination and ERAD. EMBO J 25:4717–4727

Hishinuma F, Hirai K (1991) Genome organization of the linear plasmid, pSKL, isolated from *Saccharomyces kluyveri*. Mol Gen Genet 226:97–106

Hishinuma F, Nakamura K, Hirai K, Nishizawa R, Gunge N, Maeda T (1984) Cloning and nucleotide sequence of the DNA killer plasmids from yeast. Nucleic Acids Res 12:l7581–7597

Hodgson VJ, Button D, Walker GM (1995) Anti-Candida activity of a novel killer toxin from the yeast *Williopsis mrakii*. Microbiology 141:2003–2012

Huang B, Johansson MJ, Bystrom AS (2005) An early step in wobble uridine tRNA modification requires the Elongator complex. RNA 11:424–436

Hutchins K, Bussey H (1983) Cell wall receptor for yeast killer toxin: involvement of (1-6)-β-D-glucan. J Bacteriol 154:161–169

Icho T, Wickner RB (1989) The double-stranded RNA genome of yeast virus L-A encodes its own putative RNA polymerase by fusing two open reading frames. J Biol Chem 264:6716–6723

Izgü F, Altinbay D, Sağiroğlu AK (1999) Isolation and characterization of the K6 type yeast killer protein. Microbios 99:161–172

Izgü F, Altinbay D (2004) Isolation and characterization of the K5-type yeast killer protein and its homology with an exo-β-1,3-glucanase. Biosci Biotechnol Biochem 68:685–693

Izgü F, Altinbay D, Sertkaya A (2005) Enzymic activity of the K5-type yeast killer toxin and its characterization. Biosci Biotechnol Biochem 69:2200–2206

Izgü F, Altinbay D, Türeli AE (2007a) In vitro susceptibilities of *Candida* spp. to Panomycocin, a novel exo-β-1,3-glucanase isolated from *Pichia anomala* NCYC 434. Microbiol Immunol 51:797–803

Izgü F, Altinbay D, Türeli AE (2007b) In vitro activity of panomycocin, a novel exo-β-1,3-glucanase isolated from *Pichia anomala* NCYC 434, against dermatophytes. Mycoses 50:31–34

Jacobs CJ, and Van Vuuren HJJ (1991) Effects of different killer yeasts on wine fermentations Am J Enol Vitic 42:4:295–300

Jablonowski D, Schaffrath R (2007) Zymocin, a composite chitinase and tRNase killer toxin from yeast. Biochem Soc Trans 35:1533–1537

Jablonowski D, Fichtner L, Martin VJ, Klassen R, Meinhardt F, Stark MJR, Schaffrath R (2001) *Saccharomyces cerevisiae* cell wall chitin, the potential *Kluyveromyces lactis* zymocin receptor. Yeast 18:1285–1299

Jablonowski, D, Zink, S, Mehlgarten, C, Daum, G, and Schaffrath, R (2006) tRNA^Glu wobble uridine methylation by Trm9 identifies Elongator's key role for zymocin-induced cell death in yeast. Mol Microbiol 59:677–688

Jeske S, Meinhardt F (2006) Autonomous cytoplasmic linear plasmid pPac1-1 of *Pichia acaciae*: molecular structure and expression studies. Yeast 23:479–486

Jeske S, Meinhardt F, Klassen R (2006a) Extranuclear inheritance: virus-like DNA-elements in yeast. In: Esser K, Lüttge U, Kadereit J, Beyschlag W (eds) Progress in botany, vol 68. Springer, Heidelberg, pp 98–129

Jeske S, Tiggemann M, Meinhardt F (2006b) Yeast autonomous linear plasmid pGKL2: ORF9 is an actively transcribed essential gene with multiple transcription start points. FEMS Microbiol Lett 255:321–327

Kalhor HR, Clarke S (2003) Novel methyltransferase for modified uridine residues at the wobble position of tRNA. Mol Cell Biol 23:9283–9292

Kämper J, Meinhardt F, Gunge N, Esser K (1989a) New recombinant linear DNA-elements derived from *Kluyveromyces lactis* killer plasmids. Nucleic Acids Res 17:1781

Kämper J, Meinhardt F, Gunge N, Esser K (1989b) In vivo construction of linear vectors based on killer plasmids from *Kluyveromyces lactis*: selection of a nuclear gene results in attachment of telomeres. Mol Cell Biol 9:3931–3937

Kämper J, Esser K, Gunge N, Meinhardt F (1991) Heterologous gene expression on the linear DNA killer plasmid from *Kluyveromyces lactis*. Curr Genet 19:109–118

Kasahara S, Ben Inoue S, Mio T, Yamada T, Nakajima T, Ichishima E, Furuichi Y, Yamada H (1994) Involvement of cell wall β-glucan in the action of HM-1 killer toxin. FEBS Lett 348:27–32

Kashiwagi T, Kunishima N, Suzuki C, Tsuchiya F, Nikkuni S, Arata Y, Morikawa K (1997) The novel acidophilic structure of the killer toxin from halotolerant yeast demonstrates remarkable folding similarity with a fungal killer toxin. Structure 5:81–94

Keszthelyi A, Ohkusu M, Takeo K, Pfeiffer I, Litter J, Kucsera J (2006) Characterisation of the anticryptococcal effect of the FC-1 toxin produced by *Filobasidium capsuligenum*. Mycoses 49:176–183

Kimura T, Kitamoto N, Matsuoka K, Nakamura K, Iimura Y, Kito Y (1993) Isolation and nucleotide sequences of the genes encoding killer toxins from *Hansenula mrakii* and *H. saturnus*. Gene 137:265–270

Kinal H, Park CM, Berry JO, Koltin Y, Bruenn JA (1995) Processing and secretion of a virally encoded antifungal toxin in transgenic tobacco plants: evidence for a Kex2p pathway in plants. Plant Cell 7:677–688

Kitada K, Hishinuma H (1987) A new linear plasmid isolated from the yeast *Saccharomyces kluyveri*. Mol Gen Genet 206:377–381

Kitamoto HK, Ohmomo S, Nakahara T (1993) Selection of killer yeasts (*Kluyveromyces lactis*) to prevent aerobic deterioration in silage making. J Dairy Sci 76:803–811

Kitamoto HK, Ohmomo S, Iimura Y (1998) Isolation and nucleotide sequence of the gene encoding phosphoenolpyruvate carboxykinase from *Kluyveromyces lactis*. Yeast 14:963–967

Kitamoto HK, Hasebe A, Ohmomo S, Suto EG, Muraki M, Iimura Y (1999) Prevention of aerobic spoilage of maize silage by a genetically modified killer yeast, *Kluyveromyces lactis*, defective in the ability to grow on lactic acid. Appl Environ Microbiol 65:4697–4700

Klassen R, Meinhardt F (2002) Linear plasmids pWR1A and pWR1B of the yeast *Wingea robertsiae* are associated with a killer phenotype. Plasmid 48:142–148

Klassen R, Meinhardt F (2003) Structural and functional analysis of the killer element pPin1-3 from *Pichia inositovora*. Mol Genet Genomics 270:190–199

Klassen R, Meinhardt F (2005) Induction of DNA damage and apoptosis in *Saccharomyces cerevisiae* by a yeast killer toxin. Cell Microbiol 7:393–401

Klassen R, Meinhardt F (2007) Linear protein-primed replicating plasmids in eukaryotic microbes. In: Meinhardt F, Klassen R (eds) Microbial linear plasmids. Microbiology monographs, vol 7. Springer, Heidelberg, pp 187–226

Klassen R, Tontsidou L, Larsen M, Meinhardt F (2001) Genome organization of the linear cytoplasmic element pPE1B from *Pichia etchellsii*. Yeast 18:953–961

Klassen R, Teichert S, Meinhardt F (2004) Novel yeast killer toxins provoke S-phase arrest and DNA damage checkpoint activation. Mol Microbiol 53:263–273

Klassen R, Jablonowski D, Stark MJR, Schaffrath R, Meinhardt F (2006) Mating type locus control of killer toxins from *Kluyveromyces lactis* and *Pichia acaciae*. FEMS Yeast Res 6:404–413

Klassen R, Krampe S, Meinhardt F (2007) Homologous recombination and the yKu70/80 complex exert opposite roles in resistance against the killer toxin from *Pichia acaciae*. DNA Repair (Amst) 6:1864–1875

Klassen R, Paluszynski JP, Wemhoff Sa, Pfeiffer A, Fricke J, Meinhardt F (2008) The primary target of the killer toxin from *Pichia acaciae* is tRNA^{Gln}. Mol Microbiol 69:681–697

Koltin Y, Day PR (1976) Inheritance of killer phenotypes and double-stranded RNA in *Ustilago maydis*. Proc Natl Acad Sci USA 73:594–598

Komiyama T, Ohta T, Furuichi Y, Ohta Y, Tsukada Y (1995) Structure and activity of HYI killer toxin from *Hansenula saturnus*. Biol Pharm Bull 18:1057–1059

Komiyama T, Ohta T, Urakami H, Shiratori Y, Takasuka T, Satoh M, Watanabe T, Furuichi Y (1996) Pore formation on proliferating yeast *Saccharomyces cerevisiae* cell buds by HM-1 killer toxin. J Biochem 119:731–736

Komiyama T, Shirai T, Ohta T, Urakami H, Furuichi Y, Ohta Y, Tsukada Y (1998) Action properties of HYI killer toxin from *Williopsis saturnus* var. *saturnus*, and antibiotics, aculeacin A and papulacandin B. Biol Pharm Bull 21:1013–1019

Komiyama T, Kimura T, Furuichi Y (2002) Round shape enlargement of the yeast spheroplast of *Saccharomyces cerevisiae* by HM-1 toxin. Biol Pharm Bull 25:959–965

Kono I, Himeno K (1997) A novel killer yeast effective on *Schizosaccharomyces pombe*. Biosci Biotechnol Biochem. 61:563–564

Larsen M, Gunge N, Meinhardt F (1998) *Kluyveromyces lactis* killer plasmid pGKL2: Evidence for a viral-like capping enzyme encoded by OFR3. Plasmid 40:243–246

Lesage G, Bussey H (2006) Cell wall assembly in *Saccharomyces cerevisiae*. Microbiol Mol Biol Rev 70:317–343

Li N, Erman M, Pangborn W, Duax WL, Park CM, Bruenn J, Ghosh D (1999) Structure of Ustilago maydis killer toxin KP6 alpha-subunit. A multimeric assembly with a central pore. J Biol Chem. 4:20425–20431

Ligon JM, Bolen PL, Hill DS, Bothast RJ, Kurtzman CP (1989) Physical and biological characterization of

linear DNA plasmids of the yeast *Pichia inositovora*. Plasmid 2:185–194

Lopes CA, Lavalle TL, Querol A, Caballero AC (2006) Combined use of killer biotype and mtDNA-RFLP patterns in a Patagonian wine *Saccharomyces cerevisiae* diversity study. Antonie Van Leeuwenhoek 89:147–156

Lopes CA, Rodríguez ME, Sangorrín M, Querol A, Caballero AC (2007) Patagonian wines: the selection of an indigenous yeast starter. J Ind Microbiol Biotechnol 34:539–546

Llorente P, Marquina D, Santos A, Peinado JM, Spencer-Martins I (1997) Effect of salt on the killer phenotype of yeasts from olive brines. Appl Environ Microbiol 63:1165–1167

Lowes KF, Shearman CA, Payne J, MacKenzie D, Archer DB, Merry RJ, Gasson MJ (2000) Prevention of yeast spoilage in feed and food by the yeast mycocin HMK. Appl Environ Microbiol 66:1066–1076

Lu J, Huang B, Esberg A, Johansson MJ, Bystrom AS (2005) The *Kluyveromyces lactis* γ-toxin targets tRNA anticodons. RNA 11:1648–1654

Magliani W, Conti S, Gerloni M, Bertolotti D, Polonelli L (1997a) Yeast killer systems. Clin Microbiol Rev 10:369–400

Magliani W, Conti S, de Bernardis F, Gerloni M, Bertolotti D, Mozzoni P, Cassone A, Polonelli L (1997b). Therapeutic potential of antiidiotypic single chain antibodies with yeast killer toxin activity. Nat Biotechnol 15:155–158

Magliani W, Conti S, Salati A, Vaccari S, Ravanetti L, Maffei DL, Polonelli L (2004). Therapeutic potential of yeast killer toxin-like antibodies and mimotopes. FEMS Yeast Res 5:11–18

Marquina D, Peres C, Caldas FV, Marques JF, Peinado JM, Spencer-Martins I (1992) Characterization of the yeast populations in olive brines. Lett Appl Microbiol 14:279–283

Martinac B, Zhu H, Kubalski A, Zhou XL, Culbertson M, Bussey H, Kung C (1990) Yeast K1 killer toxin forms ion channels in sensitive yeast spheroplasts and in artificial liposomes. Proc Natl Acad Sci USA 87:6228–6232

McCracken DA, Martin VJ, Stark MJ, Bolen PL (1994) The linear-plasmid-encoded toxin produced by the yeast *Pichia acaciae*: characterization and comparison with the toxin of *Kluyveromyces lactis*. Microbiology 140:425–431

Mehlgarten C, Schaffrath R (2004) After chitin docking, toxicity of *Kluyveromyces lactis* zymocin requires *Saccharomyces cerevisiae* plasma membrane H+-ATPase. Cell Microbiol 6:569–580

Meskauskas A, Citavicius D (1992) The K2-type killer toxin- and immunity-encoding region from *Saccharomyces cerevisiae*: structure and expression in yeast. Gene 111:135–139

Michalcáková S, Sturdík E, Sulo P (1994) Construction and properties of K2 and K3 type killer *Saccharomyces wine* yeasts. Wein-Wissenschaft 49:130–132

Middelbeek EJ, Hermans JM, Stumm C (1979) Production, purification and properties of a *Pichia kluyveri* killer toxin. Antonie Van Leeuwenhoek 45:437–450

Middelbeek EJ, van de Laar HH, Hermans JM, Stumm C, Vogels GD (1980a) Physiological conditions affecting the sensitivity of *Saccharomyces cerevisiae* to a *Pichia kluyveri* killer toxin and energy requirement for toxin action. Antonie Van Leeuwenhoek 46:483–497

Middelbeek EJ, Crützen QH, Vogels GD (1980b) Effects of potassium and sodium ions on the killing action of a *Pichia kluyveri* toxin in cells of *Saccharomyces cerevisiae*. Antimicrob Agents Chemother 18:519–524

Morace G, Archibusacci C, Sestito M, Polonelli L (1984) Strain differentiation of pathogenic yeasts by the killer system. Mycopathologia 84:81–85

Morace G, Dettori G, Sanguinetti M, Manzara S, Polonelli L (1988) Biotyping of aerobic actinomycetes by modified killer system. Eur J Epidemiol 4:99–103

Morace G, Manzara S, Dettori G, Fanti F, Conti S, Campani L, Polonelli L, Chezzi C (1989) Biotyping of bacterial isolates using the yeast killer system. Eur J Epidemiol 5:303–310

Nakatsukasa K, Brodsky JL (2008) The recognition and retrotranslocation of misfolded proteins from the endoplasmic reticulum. Traffic 9:861–870

Novotná D, Flegelová H, Janderová B (2004) Different action of killer toxins K1 and K2 on the plasma membrane and the cell wall of *Saccharomyces cerevisiae*. FEMS Yeast Res 4:803–813

Ouchi K, Akiyama H (1976) Breeding of useful killer sake yeasts by repeated back-crossing. J Ferment Technol 54:615

Pagé N, Gérard-Vincent M, Ménard P, Beaulieu M, Azuma M, Dijkgraaf GJ, Li H, Marcoux J, Nguyen T, Dowse T, Sdicu AM, Bussey H (2003) A *Saccharomyces cerevisiae* genome-wide mutant screen for altered sensitivity to K1 killer toxin. Genetics 163:875–894

Paluszynski JP, Klassen R, Meinhardt F (2007) *Pichia acaciae* killer system: genetic analysis of toxin immunity. Appl Environ Microbiol 73:4373–4378

Park CM, Bruenn JA, Ganesa C, Flurkey WF, Bozarth RF, Koltin Y (1994) Structure and heterologous expression of the *Ustilago maydis* viral toxin KP4. Mol Microbiol 11:155–164

Park CM, Banerjee N, Koltin Y, Bruenn JA (1996a) The *Ustilago maydis* virally encoded KP1 killer toxin. Mol Microbiol 20:957–963

Park CM, Berry JO, Bruenn JA (1996b) High-level secretion of a virally encoded anti-fungal toxin in transgenic tobacco plants. Plant Mol Biol 30:359–366

Peery T, Shabat-Brand T, Steinlauf R, Koltin Y, Bruenn J (1987) Virus-encoded toxin of *Ustilago maydis*: two polypeptides are essential for activity. Mol Cell Biol 7:470–477

Pfeiffer P, Radler F (1984) Comparison of the killer toxin of several yeasts and the purification of a toxin of type K2. Arch Microbiol 137:357–361

Pfeiffer I, Golubev WI, Farkas Z, Kucsera J, Golubev N (2004) Mycocin production in *Cryptococcus aquaticus*. Antonie Van Leeuwenhoek 86:369–375

Polonelli L, Morace G (1986) Reevaluation of the yeast killer phenomenon. J Clin Microbiol 24:866–869

Polonelli L, Morace G (1987) Production and characterization of yeast killer toxin monoclonal antibodies. J Clin Microbiol 25:460–462

Polonelli L, Morace G (1988) Yeast killer toxin-like anti-idiotypic antibodies. J Clin Microbiol 26:602–604

Polonelli L, Archibusacci C, Sestito M, Morace G (1983) Killer system: a simple method for differentiating *Candida albicans* strains. J Clin Microbiol 17:774–780

Polonelli L, Castagnola M, Rossetti DV, Morace G (1985) Use of killer toxins for computer-aided differentiation of *Candida albicans* strains. Mycopathologia. 91:175–179

Polonelli L, Dettori G, Cattel C, Morace G. Biotyping of micelial fungus cultures by the killer system. Eur J Epidemiol 3:237–242

Polonelli L, Fanti F, Conti S, Campani L, Gerloni M, Castagnola M, Morace G, Chezzi C (1990) Detection by immunofluorescent anti-idiotypic antibodies of yeast killer toxin cell wall receptors of *Candida albicans*. J Immunol Methods 132:205–209

Polonelli L, Lorenzini R, De Bernardis F, Gerloni M, Conti S, Morace G, Magliani W, Chezzi C (1993) Idiotypic vaccination: immunoprotection mediated by anti-idiotypic antibodies with antibiotic activity. Scand J Immunol 37:105–110

Polonelli L, De Bernardis F, Conti S, Boccanera M, Gerloni M, Morace G, Magliani W, Chezzi C, Cassone A (1994) Idiotypic intravaginal vaccination to protect against candidal vaginitis by secretory, yeast killer toxin-like anti-idiotypic antibodies. J Immunol 152:3175–3182

Polonelli L, Séguy N, Conti S, Gerloni M, Bertolotti D, Cantelli C, Magliani W, Cailliez JC (1997) Monoclonal yeast killer toxin-like candidacidal anti-idiotypic antibodies. Clin Diagn Lab Immunol 4:142–146

Polonelli L, Magliani W, Conti S, Bracci L, Lozzi L, Neri P, Adriani D, De Bernardis F, Cassone A (2003) Therapeutic activity of an engineered synthetic killer antiidiotypic antibody fragment against experimental mucosal and systemic candidiasis. Infect Immun 71:6205–6212

Pretorius IS (2000) Tailoring wine yeast for the new millennium: novel approaches to the ancient art of winemaking. Yeast 16:675–729

Puhalla JE (1968) Compatibility reactions on solid medium and interstrain inhibition in *Ustilago maydis*. Genetics 60:461–474

Puchkov EO, Wiese A, Seydel U, Kulakovskaya TV (2001) Cytoplasmic membrane of a sensitive yeast is a primary target for *Cryptococcus humicola* mycocidal compound (microcin). Biochim Biophys Acta 1512:239–250

Puchkov EO, Zähringer U, Lindner B, Kulakovskaya TV, Seydel U, Wiese A (2002) The mycocidal, membrane-active complex of *Cryptococcus humicola* is a new type of cellobiose lipid with detergent features. Biochim Biophys Acta 1558:161–170

Radler F, Schmitt MJ, Meyer B (1990) Killer toxin of *Hanseniaspora uvarum*. Arch Microbiol 154:175–178

Reiter J, Herker E, Madeo F, Schmitt MJ (2005) Viral killer toxins induce caspase-mediated apoptosis in yeast. J Cell Biol 168:353–358

Rep M, Krantz M, Thevelein JM, Hohmann S (2000) The transcriptional response of *Saccharomyces cerevisiae* to osmotic shock. Hot1p and Msn2p/Msn4p are required for the induction of subsets of high osmolarity glycerol pathway-dependent genes. J Biol Chem 275:8290–8300

Riffer F, Eisfeld K, Breinig F, Schmitt MJ (2002) Mutational analysis of K28 preprotoxin processing in the yeast *Saccharomyces cerevisiae*. Microbiology 148:1317–1328

Romanos M, Boyd A (1988) A transcriptional barrier to expression of cloned toxin genes of the linear plasmid k1 of *Kluyveromyces lactis*: evidence that native k1 has novel promoters. Nucleic Acids Res 16:7333–7350

Sangorrín M, Zajonskovsky I, van Broock M, Caballero A (2002) The use of killer biotyping in an ecological survey of yeast in an old patagonian winery World J Microbiol Biotechnol 18:115–120

Santos A, Marquina D (2004a). Ion channel activity by *Pichia membranifaciens* killer toxin. Yeast 21:151–162

Santos A, Marquina D (2004b) Killer toxin of *Pichia membranifaciens* and its possible use as a biocontrol agent against grey mould disease of grapevine. Microbiology 150:2527–2534

Santos A, Marquina D, Leal JA, Peinado JM (2000) (1-6)-β-D-glucan as cell wall receptor for *Pichia membranifaciens* killer toxin. Appl Environ Microbiol 66:1809–1813

Santos A, Marquina D, Barroso J, Peinado JM (2002) (1-6)-β-D-glucan as the cell wall binding site for *Debaryomyces hansenii* killer toxin. Lett Appl Microbiol 34:95–99

Santos A, Sánchez A, Marquina D (2004) Yeasts as biological agents to control *Botrytis cinerea*. Microbiol Res 159:331–338

Santos A, Del Mar Alvarez M, Mauro MS, Abrusci C, Marquina D (2005) The transcriptional response of *Saccharomyces cerevisiae* to *Pichia membranifaciens* killer toxin. J Biol Chem 280:41881–41892

Santos A, San Mauro M, Abrusci C, Marquina D (2007) Cwp2p, the plasma membrane receptor for *Pichia membranifaciens* killer toxin. Mol Microbiol 64:831–843

Sawant AD, Abdelal AT, Ahearn DG (1989) Purification and characterization of the anti-*Candida* toxin of *Pichia anomala* WC 65. Antimicrob Agents Chemother 33:48–52

Schaffrath R, Meinhardt F (2004) *Kluyveromyces lactis* zymocin and other plasmid-encoded yeast killer toxins. In: Topics in Current Genetics Vol. 11, M. Schmitt, R. Schaffrath (eds.): Microbial Protein Toxins 133–155

Schickel J, Helmig C, Meinhardt F (1996) *Kluyveromyces lactis* killer system. Analysis of cytoplasmic promoters of linear plasmids. Nucleic Acids Res 24:1879–1886

Schmitt M, Radler F (1987) Mannoprotein of the yeast cell wall as primary receptor for the killer toxin of *Saccharomyces cerevisiae* strain 28. J Gen Microbiol 133:3347–3354

Schmitt MJ, Tipper DJ (1987) Genetic analysis of maintenance and expression of L and M double-stranded RNAs from yeast killer virus K28. Yeast 8:373–384

Schmitt MJ, Tipper DJ (1990) K28, a unique double-stranded RNA killer virus of *Saccharomyces cerevisiae*. Mol Cell Biol 10:4807–4815

Schmitt MJ, Neuhausen F (1994) Killer toxin-secreting double-stranded RNA mycoviruses in the yeasts *Hanseniaspora uvarum* and *Zygosaccharomyces bailii*. J Virol 68:1765–1772

Schmitt MJ, Tipper DJ (1995) Sequence of the M28 dsRNA: preprotoxin is processed to an α/β heterodimeric protein toxin. Virology 213:341–351

Schmitt MJ, Breinig F (2002). The viral killer system in yeast: from molecular biology to application. FEMS Microbiol Rev 26:257–276

Schmitt MJ, Breinig F (2006) Yeast viral killer toxins: lethality and self-protection. Nat Rev Microbiol 4:212–221

Schmitt MJ, Klavehn P, Wang J, Schönig I, Tipper DJ (1996) Cell cycle studies on the mode of action of yeast K28 killer toxin. Microbiology 142:2655–2662

Schmitt MJ, Poravou O, Trenz K, Rehfeldt K (1997) Unique double-stranded RNAs responsible for the anti-Candida activity of the yeast Hanseniaspora uvarum. J Virol 71:8852–8855

Seki T, Choi EH, Ryu D (1985) Construction of killer wine yeast strain. Appl Environ Microbiol 49:1211–1215

Semenza JC, Hardwick KG, Dean N, Pelham HR (1990) *ERD2*, a yeast gene required for the receptor-mediated retrieval of luminal ER proteins from the secretory pathway. Cell 61:1349–1357

Selvakumar D, Zhang QZ, Miyamoto M, Furuichi Y, Komiyama T (2006a) Identification and characterization of a neutralizing monoclonal antibody for the epitope on HM-1 killer toxin. J Biochem 139:399–406

Selvakumar D, Miyamoto M, Furuichi Y, Komiyama T (2006b) Inhibition of fungal β-1,3-glucan synthase and cell growth by HM-1 killer toxin single-chain anti-idiotypic antibodies. Antimicrob Agents Chemother 50:3090–3097

Selvakumar D, Miyamoto M, Furuichi Y, Komiyama T (2006c) Inhibition of β-1,3-glucan synthase and cell growth of *Cryptococcus* species by recombinant single-chain anti-idiotypic antibodies. J Antibiot (Tokyo) 59:73–79

Sulo P, Michalcáková S (1992) The K3 type killer strains of genus *Saccharomyces* for wine production. Folia Microbiol (Praha) 37:289–294

Sulo P, Michalcakova S, Reiser V (1992) Construction and properties of K1 type killer wine yeasts. Biotechnology Letters 14, 55–60

Sor F, Fukuhara H (1985) Structure of a linear plasmid of the yeast *Kluyveromyces lactis*: compact organization of the killer genome. Curr Genet 9:147–155

Stark MJR, Boyd A (1986) The killer toxin of *Kluyveromyces lactis*: characterization of the toxin subunits and identification of the genes which encode them. EMBO J 5:1995–2002

Stark MJR, Boyd A, Mileham AJ, Romanos MA (1990) The plasmid encoded killer system of *Kluyveromyces lactis*: a review. Yeast 6:1–29

Stark MJ, Mileham AJ, Romanos MA, Boyd A (1984) Nucleotide sequence and transcription analysis of a linear DNA plasmid associated with the killer character of the yeast *Kluyveromyces lactis*. Nucleic Acids Res 12:6011–6030

Steinberg G, Perez-Martin J (2008) *Ustilago maydis*, a new fungal model system for cell biology. Trends Cell Biol 18:61–67

Suzuki C, Nikkuni S (1989) Purification and properties of the killer toxin produced by a halotolerant yeast, *Pichia farinosa*. Agricultural Biol Chem 53:2599–2604

Suzuki C, Nikkuni S (1994) The primary and subunit structure of a novel type killer toxin produced by a halotolerant yeast, *Pichia farinosa*. J Biol Chem 269:3041–3046

Suzuki C, Kashiwagi T, Tsuchiya F, Kunishima N, Morikawa K, Nikkuni S, Arata Y (1997) Circular dichroism analysis of the interaction between the α and β subunits in a killer toxin produced by a halotolerant yeast, *Pichia farinosa*. Protein Eng 10:99–101

Suzuki C, Ando Y, Machida S (2001) Interaction of SMKT, a killer toxin produced by *Pichia farinosa*, with the yeast cell membranes. Yeast 18:1471–1478

Takasuka T, Komiyama T, Furuichi Y, Watanabe T (1995) Cell wall synthesis specific cytocidal effect of *Hansenula mrakii* toxin-1 on *Saccharomyces cerevisiae*. Cell Mol Biol Res 41:575–581

Tao J, Ginsberg I, Banerjee N, Held W, Koltin Y, Bruenn JA (1990) *Ustilago maydis* KP6 killer toxin: structure, expression in *Saccharomyces cerevisiae*, and relationship to other cellular toxins. Mol Cell Biol 10:1373–1381

Theisen S, Molkenau E, Schmitt MJ (2000) Wicaltin, a new protein toxin secreted by the yeast *Williopsis californica* and its broad-spectrum antimycotic potential. J Microbiol Biotechnol 10:547–550

Tiggemann M, Jeske S, Larsen M, Meinhardt F (2001) *Kluyveromyces lactis* cytoplasmic plasmid pGKL2: Heterologous expression of Orf3p and prove of guanylyltransferase and mRNA-triphosphatase activities. Yeast 18:815–825

Tipper DJ, Bostian KA (1984) Double-stranded ribonucleic acid killer systems in yeasts. Microbiol Rev 48:125–156

Tokunaga M, Wada N, Hishinuma F (1987) Expression and identification of immunity determinants on linear DNA killer plasmids pGKL1 and pGKL2 in *Kluyveromyces lactis*. Nucleic Acids Res 15:1031–1046

Tokunaga M, Kawamura A, Hishinuma F (1989) Expression of pGKL killer 28K subunit in *Saccharomyces cerevisiae*: identification of 28K subunit as a killer protein. Nucleic Acids Res 17:3435–3446

Tommasino M, Ricci S, Galeotti C (1988) Genome organization of the killer plasmid pGKL2 from *Kluyveromyces lactis*. Nucleic Acids Res 16:5863–5978

Travassos LR, Silva LS, Rodrigues EG, Conti S, Salati A, Magliani W, Polonelli L (2004) Therapeutic activity of a killer peptide against experimental paracoccidioidomycosis. J Antimicrob Chemother 54:956–958

Walker GM, McLeod AH, Hodgson VJ (1995) Interactions between killer yeasts and pathogenic fungi. FEMS Microbiol Lett 127:213–222

Wang X, Chi Z, Yue L, Li J, Li M, Wu L (2007a) A marine killer yeast against the pathogenic yeast strain in crab (*Portunus trituberculatus*) and an optimization of the toxin production. Microbiol Res 162:77–85

Wang X, Chi Z, Yue L, Li J (2007b) Purification and characterization of killer toxin from a marine yeast *Pichia anomala* YF07b against the pathogenic yeast in crab. Curr Microbiol 55:396–401

Weiler F, Schmitt MJ (2003) Zygocin, a secreted antifungal toxin of the yeast *Zygosaccharomyces bailii*, and its effect on sensitive fungal cells. FEMS Yeast Res 3:69–76

Weiler F, Rehfeldt K, Bautz F, Schmitt MJ (2002) The *Zygosaccharomyces bailii* antifungal virus toxin zygocin: cloning and expression in a heterologous fungal host. Mol Microbiol 46:1095–1105

White JH, Butler AR, Stark MJR (1989) *Kluyveromyces lactis* toxin does not inhibit yeast adenylyl cylcase. Nature 341:666–668

Wickner RB (1992) Double-stranded and single-stranded RNA viruses of *Saccharomyces cerevisiae*. Annu Rev Microbiol 46:347–375

Wickner RB (1996) Double-stranded RNA viruses of *Saccharomyces cerevisiae*. Microbiol Rev 60:250–265

Wilson DW, Meacock PA (1988) Extranuclear gene expression in yeast: evidence for a plasmid encoded RNA-polymerase of unique structure. Nucleic Acids Res 16:8097–8112

Wistow GJ, Piatigorsky J (1988) Lens crystallins: the evolution and expression of proteins for a highly specialized tissue. Annu Rev Biochem 57:479–504

Worsham PL, Bolen PL (1990) Killer toxin production in *Pichia acaciae* is associated with linear DNA plasmids. Curr Genet 18:77–80

Yamamoto T, Iratani T, Hirata H, Imai M, Yamaguchi H (1986a) Killer toxin from *Hansenula mrakii* selectively inhibits cell wall synthesis in a sensitive yeast. FEBS Lett 197:50–54

Yamamoto T, Imai M, Tachibana K, Mayumi M (1986b) Application of monoclonal antibodies to the isolation and characterization of a killer toxin secreted by *Hansenula mrakii*. FEBS Lett 195:253–257

Yamamoto T, Uchida K, Hiratani T, Miyazaki T, Yagiu J, Yamaguchi H (1988) In vitro activity of the killer toxin from yeast *Hansenula mrakii* against yeasts and molds. Antibiot (Tokyo) 41:398–403

Young TW, Yagiu M (1978) A comparison of the killer character in different yeasts and its classification. Antonie Van Leeuwenhoek 44:59–77

Zabel R, Bär C, Mehlgarten C, Schaffrath R (2008) Yeast alpha-tubulin suppressor Ats1/Kti13 relates to the Elongator complex and interacts with Elongator partner protein Kti11. Mol Microbiol 69:175–187

Zhu H, Bussey H (1991) Mutational analysis of the functional domains of yeast K1 killer toxin. Mol Cell Biol 11:175–181

Zhu YS, Zhang XY, Cartwright CP, Tipper DJ (1992) Kex2-dependent processing of yeast K1 killer preprotoxin includes cleavage at ProArg-44. Mol Microbiol 6:511–520

Zhu YS, Kane J, Zhang XY, Zhang M, Tipper DJ (1993) Role of the gamma component of preprotoxin in expression of the yeast K1 killer phenotype. Yeast 9:251–266

Zink S, Mehlgarten C, Kitamoto HK, Nagase J, Jablonowski D, Dickson RC, Stark MJR, Schaffrath R (2005) Mannosyl-diinositolphospho-ceramide, the major yeast plasma membrane sphingolipid, governs toxicity of *Kluyveromyces lactis* zymocin. Eukaryot Cell 4:879–889

Zorg J, Kilian S, Radler F (1988) Killer toxin producing strains of the yeasts *Hanseniaspora uvarum* and *Pichia kluyveri*. Arch Microbiol 149:261–267

7 Evolutionary and Ecological Interactions of Mould and Insects

Marko Rohlfs[1], Monika Trienens[1], Ulrike Fohgrub[2], Frank Kempken[2]

CONTENTS

I. Introduction

Besides plants, fungi are among the most prodigious producers of secondary metabolites, of which some belong to the most toxic compounds in the living world. In contrast to plants, however,

[1]Zoological Institute, Department of Animal Ecology, Christian-Albrechts-University of Kiel, Olshausenstrasse 40, 24098 Kiel, Germany

[2]Botanical Institute, Department of Plant Genetics and Molecular Biology, Christian-Albrechts-University of Kiel, Olshausenstrasse 40, 24098 Kiel, Germany; e-mail: fkempken@bot.uni-kiel.de

where the function of many secondary metabolites is regarded as a direct defence against pathogens and herbivores that reduce their survival and reproduction (Steppuhn et al. 2004), the benefits of producing low molecular weight compounds to fungi are often obscure. In contrast to this critical gap in knowledge about fungal biology, an enormous research effort world-wide has been devoted to identifying biochemical pathways and the underlying genetic mechanisms leading to the biosynthesis of fungal secondary metabolites, especially those with detrimental (e.g. toxins) or beneficial (e.g. antibiotics) impact on humans and animals (Keller et al. 2005; Hoffmeister and Keller 2007). Although molecular biologists have just scratched the surface of secondary metabolism, filamentous fungi, such as the model genus *Aspergillus*, appear to have evolved sophisticated ways of producing and regulating a plethora of secondary metabolites (Bok and Keller 2004; Yu and Keller 2005; Herrmann et al. 2006; Perrin et al. 2007; Shwab et al. 2007). The ability to regulate secondary metabolism has often been suggested to reflect an evolutionary adaptation to ensure efficient exploitation of environmental resources and to synthesise the energetically costly compounds only when the ecological conditions demand the employment of these 'chemical weapons' against their natural enemies and competitors (Vining 1990; Demain and Fang 2000). Therefore, chemical diversity and its regulation in fungi may be considered as a key life-history trait that is favoured by natural selection because it ensures survival and reproduction in a variable environment. In this review, we focus on the role of fungal secondary metabolites in interactions with antagonistic insects. In particular, we address the question when, how and why should fungi use secondary metabolites as adaptive responses to interactions within a fungus natural environment, and we hope to stimulate cross-disciplinary communication among ecologists, mycologists and molecular biologists.

Physiology and Genetics, 1st Edition
The Mycota XV
T. Anke and D. Weber (Eds.)
© Springer-Verlag Berlin Heidelberg 2009

A number of chapters in this volume also cover mycotoxins or deal with related subjects in more detail: Cyclic peptides and depsipeptides from fungi by Heidrun Anke, non-ribosomal peptide synthetases of fungi by Kathrin Eisfeld, physiological and molecular aspects of ochratoxin A biosynthesis by Rolf Geisen and Markus Schmidt-Heydt, and polyketide biosynthesis in fungi by Julia Schümann and Christian Hertweck.

II. Genetics of Secondary Metabolites in Mycelial Fungi

A. Different Types of Fungal Secondary Metabolites

Many if not all fungi produce specific secondary metabolites which are not required for primary metabolism. These include toxins like aflatoxins or fumonisins and pharmaceutically useful drugs such as penicillin, cephalosporin or cyclosporine (Cole and Schweikert 2003). Without doubt, many fungal secondary metabolites have not yet been detected and further efforts to discover new fungal secondary metabolites may provide important drugs in the future. Secondary metabolites are species-specific and often accumulate in certain stages of development or during specific morphogenic differentiation (Bennett and Bentley 1989).

Usually fungi can survive without secondary metabolites, at least in the laboratory. However, secondary metabolites may be of vital importance to the fungus in its particular ecological niche. A recently performed survey included some 1 500 secondary metabolites and provided evidence, that more than 50% of the metabolites showed antibacterial or antifungal activities. Some even showed antitumor activity (Pelaez 2005). The latter may be due to the fact that antifungal molecules may also harm other eukaryotic cells, including tumour cells.

Fungal secondary metabolites may be separated in four groups, i.e. alkaloids, non-ribosomal peptides, polyketides and terpenes. Examples of some well characterized metabolites are given in Table 7.1.

1. Alkaloids

Indole alkaloids are most often synthesized from dimethylallyl pyrophosphate and tryptophan. Sometimes other amino acids may be involved. The ergotamine pathway of *Claviceps purpurea* is particularly well characterised. The first enzymatic step is prenylation of tryptophan by the dimethylallyl trypthophan synthetase, followed by methylation of dimethylallyl tryptophan and several oxidation steps leading to lysergic acid. By the action of two different non-ribosomal peptide synthetases finally ergotamine is synthesised (Tudzynski et al. 1999, 2001b).

2. Non-Ribosomal Peptides

A database of non-ribosomal peptides of fungal and bacterial origin was established and published recently (Caboche et al. 2008). These molecules are synthesized by a class of multidomain and multimodule fungal enzymes called non-ribosomal peptide synthetases, which catalyse a variety of non-ribosomal peptides using proteinogenic and non-proteinogenic amino acids. Each module in a

Table 7.1. Some important secondary metabolites

Main group	Secondary metabolite	Organism	References
Alkaloids	Ergotamine	*Claviceps purpurea*	(Haarmann et al. 2005)
Peptides	Gliotoxin	*Aspergillus fumigatus*	(Keller et al. 2005; Bok et al. 2006a; Fox and Howlett 2008)
	Cyclosporin	*Tolypocladium inflatum*	(Hoppert et al. 2001; Keller et al. 2005)
Polyketides	Aflatoxin B1	*Aspergillus flavus*	(Katz and Donadio 1993; Keller et al. 2005)
	Lovastatin	*Aspergillus terreus*	(Keller et al. 2005; Collemare et al. 2008)
Terpenes	Gibberellin GA3	*Gibberella fujikuroi*	(Keller et al. 2005; Appleyard et al. 1995; Kawaide 2006)

non-ribosomal peptide synthetase has several domains for recognition, activation and covalent binding of a module-specific amino acid. The diversity of non-ribosomal peptides is based on varying length of peptides, linear or cyclic peptides and variations in the function of enzyme domains (Keller et al. 2005a). The first non-ribosomal peptide synthetase which has been characterised in detail catalyses the formation of β-lactam antibiotics (Smith et al. 1990). The immuno-suppressive drug cyclosporine, acyclic undecapeptide, is also synthesized by non-ribosomal peptide synthetase in the filamentous fungus *Tolypocladium inflatum* (Zocher et al. 1986; Weber et al. 1994). Non-ribosomal peptide synthetases are also discussed with respect to fungal pathogenicity (Collemare et al. 2008).

3. Polyketides

Polyketides resemble the most abundant fungal secondary metabolites. Well known examples are aflatoxins (Sweeney and Dobson 1999) and the anti-cholesterol substance lovastatin (Kennedy et al. 1999). Fungal polyketides are synthesized by type I polyketide synthases (PKS) which are multidomain enzymes similar to eukaryotic fatty acid synthases. Three domains are essential in all fungal PKS, i.e. acyl carrier, acyltransferase, and ketoacyl CoA synthase, while other domains such as enoyl reductase, dehydratase, or ketoreductase may or may not be present. Like fatty acid synthases, fungal PKS condense short-chain carboxylic acids (acetylCoA, malonylCoA) to form carbon chains of different length. However, while fatty acid synthases fully reduce β-carbon, this is optional with fungal PKS. The diversity of fungal polyketides is due to varying iteration reactions, varying number of reduction reactions, type of extender units used, and a variety of post-polyketide synthesis steps (Keller et al. 2005a).

4. Terpenes

Terpenes are typically associated with plants. Consequently a vast amount of terpenes are known from plants. However, there are also fungi which produce terpens. Several fungal terpene cyclases have been identified and characterised, i.e. from *Gibberella fujikuroi* (Tudzynski et al. 2001a) or a *Fusarium* species (Rynkiewicz et al. 2001). All terpenes are made of several

isoprene units, can be linear or cyclic, saturated or unsaturated, and may be modified in several ways.

B. Fungal Secondary Metabolite Clusters

In bacteria, genes encoding enzymes of a specific pathway are usually combined in operons. This ensures a rather simple regulation which allows for expression or repression of all pathway specific genes at the same time. Although thought to be restricted to bacteria, similar structures have been observed in fungi. Secondary metabolite genes typically are organised in gene clusters (Hull et al. 1989; Keller and Hohn 1997). In this review we give an overview of two important secondary metabolite clusters and their regulation. In addition a compilation of secondary metabolite gene classes in selected *Aspergillus* species are shown in Table 7.2.

1. Aflatoxin and Sterigmatocystin Clusters

The aflatoxin/sterigmatocystin gene clusters are particularly well characterised. While in *Aspergillus nidulans* only sterigmatocystin is produced, in *A. flavus* and *A. parasiticus* sterigmatocystin is the penulatimate precursor to aflatoxin synthesis. The biosynthesis of sterigmatocytin and aflatoxin is shown in Fig. 7.1 (Hamasaki et al. 1973; Barnes et al. 1994; Hicks et al. 2002; Yu et al. 2004). The aflatoxin gene cluster has a size of about 70 kb, and 21 genes of this cluster have been functionally characterized in more detail, while the functions of six other genes have yet to be elucidated (Yu et al. 2004). Most of the functionally assigned genes encode different biosynthetic enzymes (Yu and Keller 2005). One gene, *aflR*,

Table 7.2. Secondary metabolite gene classes in selected *Aspergillus* species (modified from Keller et al. 2005b)

Gene	A. fumigatus	A. nidulans	A. oryzae
DMATS (dimethylallyl tryptophan synthetase)	7	2	2
FAS (fatty acid synthase)	1	6	5
NRPS (non-ribosomal peptide synthetase)	14	14	18
PKS (polyketide synthase)	14	27	30
SesCyc (sesquiterpene cyclase)	0	1	1

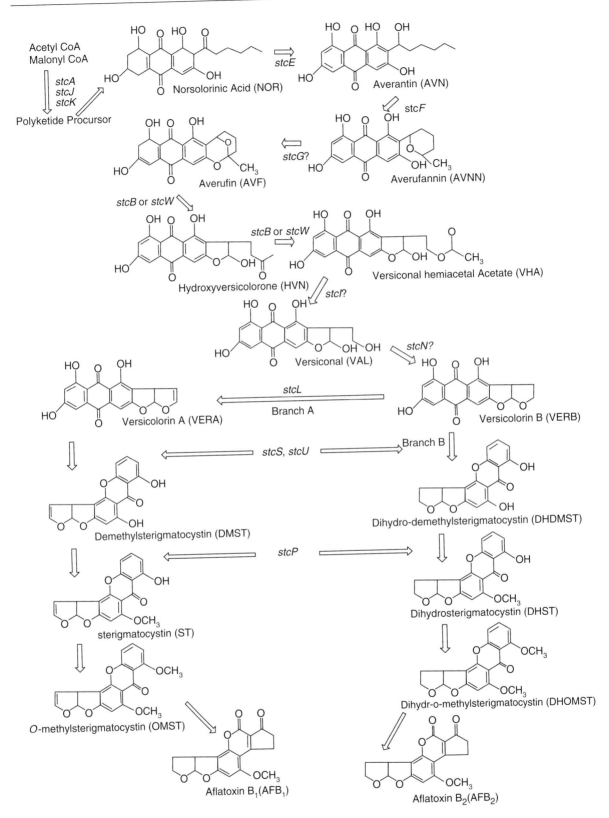

Fig. 7.1. Aflatoxin/sterigmatocystin biosynthetic pathway. Figure modified from Hicks et al. (2002) with permission

encodes a binuclear zinc cluster transcription factor (Chang et al. 1993; Woloshuk et al. 1994; Yu et al. 1996a; Keller and Hohn 1997; Shimizu et al. 2003), which is specific for fungi (Keller and Hohn 1997; Fernandes et al. 1998; Yu and Keller 2005). The regulatory functions of these genes are discussed below. The sterigmatocystin cluster of *A. nidulans* is somewhat smaller (60 kb), but still contains 25 genes (Hicks et al. 2002). A comparison of the two gene clusters is given in Fig. 7.2. Homologous genes in both clusters are shown in the same colours. In addition Table 7.3 gives a compilation of all genes in the two clusters, their accession numbers and encoded enzymes.

2. Epipolythiodioxopiperazine clusters

Epipolythiodioxopiperazines are toxic secondary metabolites, almost exclusively produced by ascomycetes. This class of secondary metabolites is characterised by the presence of a sulfur-bridged dioxopiperazine ring synthesised from two amino acids (see Fig. 7.3). Toxicity results from cross-linking proteins via cysteine residues and generating reactive oxygen species through redox cycling (Fox and Howlett 2008). So far 14 different epipolythiodioxopiperazines are known, with gliotoxin produced by *Aspergillus fumigatus*, *A. flavus*, or *A. niger* being the best characterized example (Fox and Howlett 2008). Another important epipolythiodioxopiperazines is sirodesmin PL from *Leptosphaeria maculans*, which is involved in plant pathology and causes necrosis on plants (Rouxel et al. 1988). The biosynthesis routes of sirodesmin PL and gliotoxin were deduced decades ago by the use of labelling experiments and analysis of intermediates (see literature cited by Fox and Howlett 2008). The pathway for gliotoxin biosynthesis is shown in Fig. 7.4. More recently the gene clusters for both metabolites were characterized (Fig. 7.2). The sirodesmin cluster was characterised first (Gardiner et al. 2004). Then, based on sequence comparison of the *L. maculans* sirodesmin genes the gliotoxin gene cluster was predicted for *A. fumigatus* (Gardiner and Howlett 2005). This prediction was confirmed by knock-out experiments (Cramer et al. 2006; Kupfahl et al. 2006).

C. Regulation of Secondary Metabolite Synthesis

1. Pathway-Specific Regulation

As mentioned above, *aflR* encodes a transcription factor. Both, sterigmatocystin and aflatoxin clusters contain an *aflR* gene. This gene is essential for expression of all other cluster genes and is post-transcriptionally regulated (Yu et al. 1996a). Furthermore, the AflR protein is necessary for the positive co-regulation of AF/ST biosynthetic gene expression (Feng and Leonard 1998; Meyers et al. 1998; Hicks et al. 2002; Shimizu et al. 2003; Price et al. 2006). Deletion or mutation of *aflR* led to strong down regulation of sterigmatocystin and aflatoxin gene expression. A second gene *aflJ* also is involved in gene regulation, although it does not encode a typical transcription factor. Deletion of *aflJ* results in a failure to produce aflatoxin, and certain pathway intermediates cannot be used by the fungus when applied exogenously. However some transcripts of the pathway genes still accumulate and therefore *aflJ* does not appear to influence aflatoxin regulation on a transcriptional level (Meyers et al. 1998; Yu and Keller 2005; Du et al. 2007). Likewise the *gliZ* gene, a $ZN(II)_2Cys_6$ (see Fig. 7.2) regulates gene expression of the gliotoxin gene cluster, as an *A. fumigatus* mutation in this gene led to the unability to produce gliotoxin (Bok et al. 2006a). This mutant is also characterised by a reduced virulence.

2. Global Regulation

Secondary metabolite biosynthesis is influenced by a number of environmental factors such as nitrogen or carbon sources, temperature or light. There is evidence that such environmental signals are mediated through general global transcription factors, e.g. for carbon CreA and for nitrogen AreA (Ehrlich et al. 2003; Mihlan et al. 2003). Several secondary metabolite gene clusters are positively or negatively regulated by these transcription factors.

In *Aspergillus* species a novel nuclear protein, LaeA, was discovered as a global regulator of secondary metabolism. Deletion mutants have lost expression of metabolic gene clusters, including the sterigmatocystin, lovastatin and penicillin clusters. Over-expression of *laeA* increases transcription and subsequent product

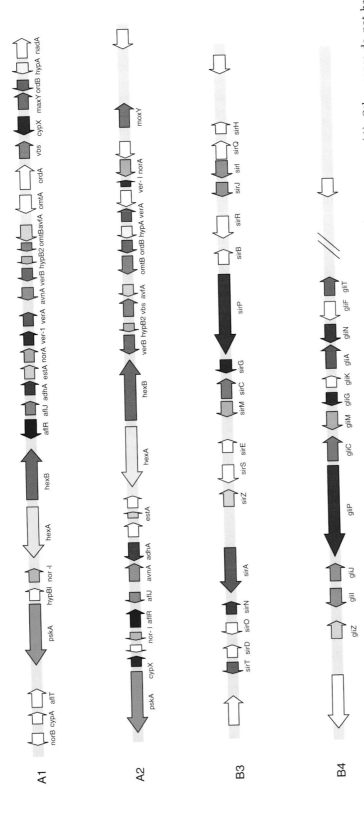

Fig. 7.2. Secondary metabolite gene clusters: **A1** aflatoxin gene cluster from *A. flavus*; **A2** sterigmatocystin gene cluster from *A. nidulans*; homologous genes in **A1** and **A2** are given in the same colour. Gene names and encoded peptides are given in Table 7.3. Figure modified from Ehrlich et al. (2008) with permission. **B3** *Leptosphaeria maculans* sirodesmin PL; **B4** *Aspergillus fumigatus* gliotoxin biosynthetic gene clusters. Common ETP genes include non-ribosomal peptide synthetase (*P*), thioredoxin reductase (*T*), methyl transferases (*M, N*), glutathione S-transferase (*G*) and cytochrome P450 mono-oxygenase (*J*), dipeptidase (*J*), as well as a cyclopropane carboxylate synthase (ACCS; *I*), dipeptidase (*J*), as well as a transcriptional regulator (*Z*) and a transporter (*A*). Other genes do not have homologues in the other cluster and encodes changes in the side chains of the ETP molecule structure, including cytochrome P450 mono-oxygenases (*F, B, E*), a prenyl transferase (*D*), an acetyl transferase (*H*), epimerases (*Q, S, R*), an oxidoreductase (*O*) and a hypothetical protein (*K*; Gardiner and Howlett 2005). Genes shaded in *grey* encode proteins with best matches to proteins with no potential roles in ETP biosynthesis. The *forward slash* marks represent a 17-kb region of repetitive DNA. Homologous genes in **B3** and **B4** are given in the same colour. Figure modified from Fox and Howlett (2008) with permission

Table 7.3. Comparison of aflatoxin and sterigmatocystin pathway gene

Gene	Original name and other names used (accession no.)[a]	ST gene homologue[b]	Enzyme or product
aflA	fas-2 (hexA; AF391094)	stcJ	FAS alpha subunit
aflB	fas-1 (hexB; AF391094), uvm8, fas1, fas-1A (L48183)	stcK	FAS beta subunit
aflC	pksA (Z47198), pksL1 (L42765, L42766)	stcA	PKS
aflD	nor-1 (L27801)	stcE	Reductase
aflE[c]	norA, aad (U24698), adh-2 in A. flavus (U32377)	stcV	NOR reductase/dehydrogenase
aflF[c]	norB		Dehydrogenase
aflG	avnA (U62774), ord-1 (L40839)	stcF	P450 monooxygenase
aflH	adhA (U76621)	stcG	Alcohol dehydrogenase
aflI	avfA (AF154050), ord-2 (L40840), (AF159789 in A. flavus)	stcO	Oxidase
aflJ	estA (AF417002)	stcI	Esterase
aflK	vbs (AF169016, U51327)	stcN	VERB synthase
aflL	verB (AF106958), (AF106959 and AF106960 in A. flavus)	stcL	Desaturase
aflM	ver-1 (M91369)	stcU	Dehydrogenase/ketoreductase
aflN[c]	verA	stcS (verA)	Monooxygenase
aflO	dmtA (mt-I) (AB022905, AB022906), omtB (AF154050), (AF159789 in A. flavus)	stcP	O-methyltransferase I or O-methyltransferase B
aflP	omtA (L25834), omt-1 cDNA (L22091), (L25836 in A. flavus)		O-methyltransferase A or O-methyltransferase II
aflQ	ordA (AF017151, AF169016), A. flavus ord-1 (U81806, U81807)		Oxidoreductase/P450 monooxygenase
aflR	aflR (L26222), apa-2 (L22177), afl-2, (AF427616, AF441429)	aflR	Transcription activator
aflS	aflJ (AF002660), (AF077975 in A. flavus)		Transcription enhancer
aflT	aflT (AF268071)	Unnamed	Transmembrane protein
aflU	cypA		P450 monooxygenase
aflV	cypX (AF169016)	stcB	P450 monooxygenase
aflW	moxY (AF169016)	stcW	Monooxygenase
aflX	ordB	stcQ	Monooxygenase/oxidase
aflY	hypA		Hypothetical protein
aflR2[d]	aflR2 (AF452809)	Second copy	Transcription activator
aflS2	aflJ2 (AF452809, AF295204)	Second copy	Transcription enhancer
aflH2	adhA2 (AF452809)	Second copy	Alcohol dehydrogenase
aflJ2	estA2 (AF452809)	Second copy	Esterase
aflE2	norA2 (AF452809)	Second copy	Dehydrogenase (early terminated)
aflM2	ver-1B (AF452809)	Second copy	Dehydrogenase (missing N-terminal)
aflO2	omtB2 (AF452809)	Second copy	Methyltransferase B (missing N-terminal)

[a]The accession number of the complete 82 081-bp aflatoxin gene cluster, including a sugar utilization gene cluster, in A. parasiticus is AY391490 and updates the sequences of the underlined accession numbers. The genes and their accession numbers are from A. parasiticus unless otherwise noted.

[b]The accession number of the ST gene cluster in A. nidulans is U34740, and the corresponding contig number is 1.132 (from 183 018 to 242 843) in the Whitehead database.

[c]The placements of aflE (norA), aflF (norB), and aflN (verA) in the pathway were based on their homologies to aflatoxin or ST genes and their functions have not been experimentally confirmed.

[d]The aflR2, aflS2, aflH2, aflJ2, aflE2, aflM2, and aflO2 genes are partially duplicated cluster genes (second copy) in A. parasiticus, and their functions and chromosomal locations in the genome have not yet been clarified.

[e]Arrows signify conversion.

formation (Bok and Keller 2004). The *laeA* gene is negatively regulated by AflR and also regulated by two signal transduction factors, protein kinase A and RasA. Interestingly, spore formation in delta *laeA* is rather similar to the wild type, indicating a primary role of LaeA in regulation of metabolic gene clusters (Bok and Keller 2004). LaeA is also known to regulate virulence in *A. fumigatus*, as deletion mutants show reduced virulence (Bok et al. 2005) and reduced gliotoxin levels (Sugui et al. 2007). Genes in 13 of 22 secondary metabolite clusters appear to be

Fig. 7.3. Structure of epipoly-thiodioxopiperazines. **A** Generic structure of an epidithiodioxop-iperazine (ETP). R = any atom or group. **B** Gliotoxin

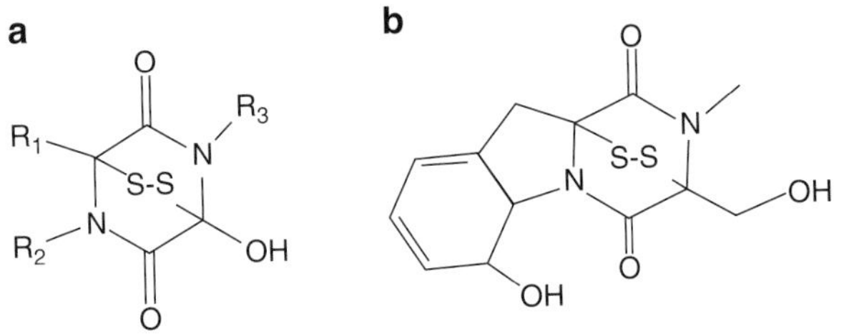

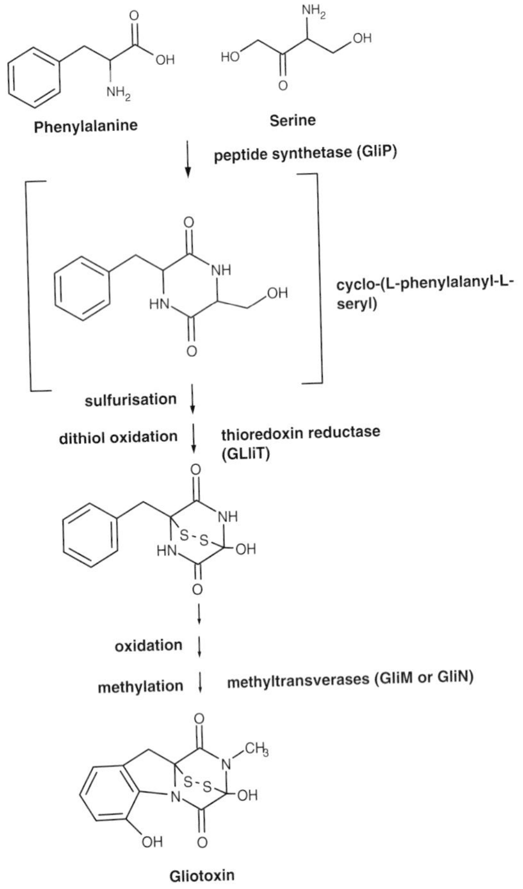

a

b

Fig. 7.4. Biosynthesis of gliotoxin. Figure modified from Fox and Howlett (2008) with permission

Phenylalanine

Serine

peptide synthetase (GliP)

cyclo-(L-phenylalanyl-L-seryl)

sulfurisation

dithiol oxidation

thioredoxin reductase (GLliT)

oxidation

methylation | methyltransferases (GliM or GliN)

Gliotoxin

proteins, VeA and VelB, in which VeA bridges VelB to LaeA. Deletion in either VeA or VelB results in defects in both, sexual development and secondary metabolites. VeA is light-regulated. The VeA-VelB-LaeA complex is formed in the dark only and significantly increases secondary metabolism (Bayram et al. 2008). In addition to LaeA an histone deacetylase (HDAC) appears to be involved in epigenetic regulation of secondary metabolite expression (Shwab et al. 2007). While positive regulation of *Aspergillus* secondary metabolite clusters by LaeA has been shown to be less spatially limited, suppression of secondary metabolite biosynthesis by expression of *hdaA* is confined to telomere-proximal clusters that encode sterigmatocystin and penicillin (Shwab et al. 2007). Such different regulatory mechanisms able to independently activate or disable secondary chemical production may provide fungi with an efficient means of adaptively responding to multiple natural enemies and competitors.

3. Regulation by Signal Transduction

Secondary metabolite synthesis and regulation of secondary metabolite gene clusters are also controlled by upstream signalling mechanisms (Yu and Keller 2005). As shown in Fig. 7.5, a complex pathway controls vegetative growth, conidiation, and production of secondary metabolites in *A. nidulans*, in which heterotrimeric G protein and its regulation are of crucial importance. External signals at a seven transmembrane receptor lead to the dissociation of Gα from the heterotrimeric G protein. As in other fungi, three Gα genes are present in *Aspergillus* species. One Gα, encoded by *FadA*, activates the penicillin biosynthesis and via cAMP and protein kinase A induces hyphal growth. Gβγ subunits inhibit asexual development. The G protein regulator FlbB plays a key role by converting Gα-GTP to Gα-GDP.

regulated by LaeA (Perrin et al. 2007). Most interestingly, LaeA regulation of the sterigmatocystin cluster appears to be location-specific. Genes next to the cluster are not affected by LaeA, but genes artificially introduced into the cluster are subject to LaeA regulation. LaeA appears to be a methyltransferase which may directly interact with chromatin structure (Bok et al. 2006a). Recently it was shown that LaeA forms a heterotrimeric complex with two

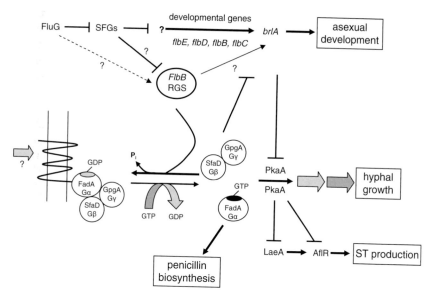

Fig. 7.5. Antagonistic regulation by signal transduction of vegetative growth, conidiation and production of secondary metabolites. *AflR* Sterigmatocystin transcription factor. *brl A, fluG, flbE, flbD, flB, flbC* Genes required for conidiation (Lee and Adams 1996). *LaeA* Global secondary metabolite regulator, *PkaA* protein kinase A, *RGS* regulator of G protein signaling. Figure modified from Yu and Keller (2005) with permission

Activation of asexual development requires partial inhibition of the G protein-mediated signalling. This requires activity of *fluG* and *flbA* (Yu et al. 1996b; Hicks et al. 1997; Rosen et al. 1999; Tag et al. 2000). Protein kinase A induces hyphal growth and inhibits both asexual development and sterigmatocystin biosynthesis (Shimizu et al. 2003). The latter is due to inhibition of LaeA and AflR. For a detailed review see Yu and Keller (2005). Signal transduction will also be necessary to coordinate and regulate other secondary metabolite genes in response to stresses and confrontation with other micro-organisms or antagonistic insects. Yet literally nothing is known about this type of regulation.

4. Activation of Silent Secondary Metabolite Clusters

Genomic analysis has led to the identification of a large number of secondary metabolite genes and gene clusters (some included in Table 7.2). However, many of these clusters remain silent under laboratory conditions. This may be due to a lack of appropriate signals which will lead to cluster expression under certain natural conditions. Recently, a silent gene cluster was activated by artificially expression of its regulatory gene (Bergmann et al. 2007). While this approach will be useful for discovery of new drugs (Brakhage et al. 2008), it will also be useful to identify the natural signals which led to activation of silent clusters. This may be caused by interaction with specific chemicals, or by contact with other micro-organisms or natural enemies, such as insects (see below).

III. Three Types of Fungus–Insect Interactions and the Role of Secondary Metabolites

There are three basic effects of species interaction: (1) one species may cause an increase in the survival, growth and fecundity of another species, (2) it may cause a decrease, or (3) it may have no effect at all. The most obvious example of the second type of interaction (in symbolic form: '+ −') is a predator–prey relationship, that is one species is at least partly eaten by the other. Besides host-parasite interactions and herbivory, fungivory also falls in this category. In addition to commensalism (+ 0), amensalism (− 0), and mutualism (+ +), which are probably only minor categories, interspecific competition (– –) is the most prevalent form of species interaction. The '– –' symbolism stresses its essential aspect, namely that two species cause a demonstrable reduction in each other's fitness parameters (i.e. survival, growth, and/or reproduction). Nevertheless, it should be noted that whenever two species compete there are some conditions under which one species is significantly more affected than the other. For instance, the dominance relationship between two competitors is altered when there is a temporal variation between the settlement of the two species in a habitat. In general, species that become established earlier have a greater chance of being competitively superior over later arrivals. These historical effects are called priority effects and could result from size- or stage-dependent competition. These detrimental effects of inter-specific competition can be expected to act through some combination of reproductive success and survival. It can be expected that there is a competition for a resource that is limited (e.g. food supply), that the effects will be density-dependent, and that this type of interaction is reciprocal (Begon et al. 1996).

In this review, we concentrate on antagonistic interactions between fungi and insects and highlight two types of the '+ -' interrelationship, namely: (1) where fungi are pathogens of insects and (2) where fungi are a nutritional source for arthropods. Special attention is paid to the competitive interplay (- -) between saprophytic fungi and saprophagous insects since competition between distantly related organisms has repeatedly been suggested to be widespread but has only rarely been addressed in ecological research and is virtually absent from functional fungal and insect molecular biology.

A. Host–Pathogen

Pathogens are considered as biological agents that cause suffering of animal hosts upon infection. This definition is medical rather than biological, i.e. an evolutionary definition, because it states nothing about the reproductive success of the pathogenic organism invading an animal's body. For example, filamentous fungi, such as *Aspergillus fumigatus*, may be serious vertebrate pathogens, though this type of host is assumed to be a 'dead end' since fungi are unable to reproduce and disperse to new hosts (Tekaia and Latgé 2005). Therefore, the evolution of virulence factors responsible for causing, e.g. aspergillosis, are

unlikely to have taken place in vertebrates and it is still unknown what kind of selective forces have turned opportunistic *Aspergillus* species into dreaded human pathogens. In contrast, many mould-like fungi have been selected for successfully exploiting insects as hosts (Vega and Blackwell 2005). After germination of spores on the insect's cuticle, fungal hyphae start penetrating the cuticle, followed by invasion of epidermis and hypodermis. Insect death is the consequence of tissue invasion by hyphae or hyphal bodies proliferating in the insect's hemolymph (Clarkson and Charnley 1996). However, fungi, after having successfully conquered the exoskeleton as a first line of insect defence, are additionally faced with the challenge to overcome insect cellular and humoral immune responses before being able to grow in their hosts (Gillespie et al. 2000; Gottar et al. 2006). Finally, the dead insect body allows saprotrophic growth and fungi burst the insect's exoskeleton, which eventually leads to the formation of reproductive organs and spores which are dispersed by wind or phoresy. As recently summarised by Furlong and Pell (2005), entomopathogenic fungi seriously impact trophic interactions within insect communities and hence biological control programmes.

Interestingly, there is a huge variation among and within different fungus species in the ability to infect hosts, which may be attributed to species- or strain-specific differences in the ability to express virulence factors (Clarkson and Charnley 1996), but additionally depends on the host's immune system and hence the ability to defend oneself against the fungal invader (Tinsley et al. 2006). Given inherited variation in virulence, insect hosts may represent a selective environment that favours different degrees of expression of virulence factors, such as immunosuppressive agents like some mycotoxins (see below), which might also be of relevance when fungi enter vertebrate 'hosts' (Reeves et al. 2004). It should be noted, however, that well adapted entomopathogenic fungi (e.g. *Beauveria bassiana*) are not among the most prominent vertebrate pathogens, though invasive infection of humans is possible (Henke et al. 2002). This may indicate that pathogen evolution in arthropods does not correlate with pathogenicity in vertebrates, i.e. virulence appears to be based on specific components required for infection of either insect or vertebrate hosts, rather than on non-specific components.

Identification of the evolutionary origin of virulence factors may become even more difficult when we consider the fact that opportunistic pathogens, such as *Aspergillus flavus* or *A. fumigatus*, but also facultative entomopathogenic fungi, can perform vital growth and reproduction as saprophytes. While fungi have to overcome passive and active defence mechanisms in insect hosts, decaying organic matter may be occupied by a large number and high density of competing micro-organisms (yeast, bacteria, other filamentous fungi) and fungi are likely to encounter antagonistic insects that can seriously hamper fungal growth (see Sect. III.C); thus, in comparison to living hosts, decaying organic matter poses a different ecological challenge to the fungi. Interestingly, the first data point in possible trade-offs between a saprophytic and a pathogenic life style is *A. flavus* (Scully and Bidochka 2005), i.e. there may be constraints on being equally successful in both environments. The latter comprises the ideas of the life history paradigm, namely that organism cannot simultaneously maximise all components of fitness (Stearns 1992). As suggested by the results of the study by Scully and Bidochka (2005), improvement of the pathogenic lifestyle was paid for with a decrease in the ability to exploit dead organic material. Nonetheless, comprehensive studies addressing the extent to which selection pressures in these different environments are affecting virulent and saprobic capacities in fungi remain to be carried out.

Toxic secondary metabolites have often been suggested to constitute important virulence factors enabling successful colonisation of insect hosts by fungi (Clarkson and Charnley 1996), such as destruxins or beauvericin from the potent entomopathogens *Metarhizium anisophilae* and *Beauveria bassiana*, respectively. Recently, production of gliotoxin, immunosuppressive agent in vertebrates (Bok et al. 2005), has been correlated with virulence of *Aspergillus fumigatus* in the greater wax moth *Galleria mellonella* (Reeves et al. 2004); and the injection of aflatoxin B1 into larvae of the Egyptian cotton leaf worm (*Spodoptera littoralis*) had a huge impact on insect development (Abdou et al. 1984). However, given that the arthropod exoskeleton provides an efficient physical barrier against pathogens, several studies imply that selection for virulence of entomopathogenic fungi is related to the proteinase and chitinase secretion required to degrade the insect cuticle (St Leger et al. 1997; Bagga et al. 2004; Wang et al. 2005; Gottar et al. 2006; Fan et al. 2007). Yet the production of toxic metabolites may become important when the fungus has successfully entered the insect's body cavity (Wang et al. 2005), e.g. to impair the host's immune response. Thus, for opportunistic as well as for facultative entomopathogenic fungi, the causal relationship between toxic secondary metabolite expression and virulent phenotypes still remains elusive. However, studies using an experimental evolutionary approach, similar to the work by Scully and Bidochka (2005), will provide valuable sources of phenotypically diverged fungal populations, which can be thoroughly screened for correlated responses in fitness traits under different environmental conditions; and coupled with genetic manipulation and biochemical analyses one can expect to progress in shedding light on the function of toxic secondary metabolites as virulence factors.

B. Predator–Prey

Fungi, as an important part of decomposer communities, contain high amounts of nitrogen and phosphorus, one reason why all fungal structures, e.g. hyphae and spores, are susceptible to predation by animals, especially the soil fauna (Ruess and Lussenhop 2005). Common soil arthropods, such as mites, collembola, diplura, protura, and nematode worms, possess specialised mouthparts, making them excellent fungal feeders (Ruess and Lussenhop 2005) and some studies demonstrate changes in mycelial growth and respiration of various soil fungi in response to collembolan grazing (Hedlund et al. 1991; Bengtsson et al. 1993; Kampichler et al. 2004; Harold et al. 2005), thus providing evidence of fitness loss in fungi due to fungivory (Guevara et al. 2000; Tordoff et al. 2006). In contrast, fungivorous arthropods have been selected for abilities that allow them to discriminate between low- and high-quality fungal diets, which positively correlates with important fitness parameters (e.g. Scheu and Simmerling 2004; Jørgensen et al. 2005) and fungivores may base their food choice decisions on volatiles emitted by the fungi (Bengtsson et al. 1991; Hedlund et al. 1995). Such behavioural plasticity in fungivores has repeatedly been linked to the toxicity of the fungal diet (Shaw 1988; Ruess et al. 2000) and frequently observed dietary mixing strategies in soil arthropods may be adaptive in terms of diluting the amount of fungal toxic compounds (Freeland and Janzen 1974; Singer et al. 2002). From the fungal perspective, optimal foraging strategies in fungivores (i.e. avoidance of toxic fungi) may provide relief from fungivory and hence a fitness benefit to the fungi. For this reason, chemical defence based on toxic secondary metabolite production is a widespread and often-cited hypothesis to explain the existence of insecticidal compounds in fungi. Although, as for the proposed role of secondary metabolites in pathogenic fungi, a causal relationship between toxin production and reduced fitness in the fungus' predator coupled with a benefit to the fungus remains to be demonstrated. A recent investigation used a transgenic strain of *A. nidulans* that lacked the expression of *laeA* and hence the ability to synthesise several known and unknown secondary metabolites (Bok and Keller 2004). As described above, the Δ*laeA* mutant has been demonstrated to be unable to produce various secondary metabolites including penicillin, hyphal pigments, and the mycotoxin sterigmatocystin (Bok and Keller 2004). In an ecological experiment, fungivorous soil-dwelling springtails (*Folsomia candida*) displayed a distinct preference for this chemical-deficient mutant of *A. nidulans* and the transgenic fungus, compared with a wild-type strain capable of producing the entire arsenal of secondary chemicals, suffered from a significant biomass loss (Rohlfs et al. 2007; Fig. 7.6). Moreover, as can be seen from Fig. 7.6, groups of springtails displayed a significantly stronger aggregation (group formation on one of two fungal colonies offered) on the wild type than on the chemical-deficient *A. nidulans* strain. In analogy to plant–herbivore systems (Fordyce 2003), this behaviour may reflect an adaptive strategy for the arthropods to manipulate fungal suitability as a food source. Because the tendency to form a large group of foraging individuals on one fungal diet was less strong when springtails were offered two colonies of the Δ*laeA* mutant, chemical diversity as regulated by *laeA* appears to play an important role. The behaviour of the springtails in this set-up clearly demonstrate that

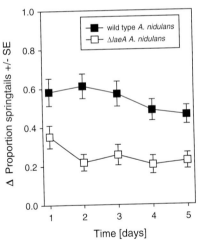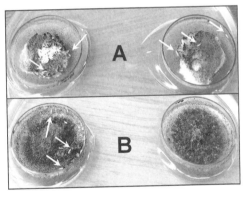

Fig. 7.6. Effect of chemical deficient mutant Δ*laeA Aspergillus nidulans* on the feeding behaviour of the springtail *Folsomia candida*. The *left panel* shows the tendency of 20 springtails to form groups on one of two colonies of the wild type or the Δ*laeA* strain of *A. nidulans* over several days. The measurement 'Δ proportion springtails' indicates the degree of group-formation (values approaching '0' indicate regular distribution between the colonies whereas '1' indicates that all animals gathered on one colony; 'Δ proportion springtails' was calculated by dividing the difference between the number of animals on colony one and the number on colony two by the total number of animals on both fungal colonies). There was a significant effect of the fungal diet on the degree of group formation, but no influence of time (repeated measurement ANOVA; fungal strain: $F = 44.97$, d.f. = 1, $P < 0.0001$, time: $F = 2.13$, d.f. = 4, $P = 0.0791$, fungal strain × time: $F = 0.86$, d.f. = 4, $P = 0.4877$). The *right panel* illustrates the effect of springtail feeding on fungal development (*A* Δ*laeA*, *B* wild-type *A. nidulans*) and the influence of the fungal strain on the distribution of the arthropods between the two colonies (*A* springtails feed on both colonies; *B* springtails feed only on the left colony; *arrows* point at some individual springtails). See Rohlfs et al. (2007) for methodological details

they base their food choice decisions on cues related to the ability to synthesise secondary metabolites and, as suggested by Rohlfs et al. (2007), the secondary metabolites themselves. These observations provide genetic support for the hypothesis that fungi may contain secondary metabolites as an anti-predator adaptation and illustrate how the lack of fungal chemicals is driving the food choice behaviour of fungivorous arthropods; yet, since *laeA* globally regulates secondary metabolism in *Aspergillus*, the key fungal compounds involved in deterring fungivores and improving chemical protection in this ecological interaction are still unknown.

C. Interspecific Competition

Although micro-organisms, including filamentous fungi, are considered to function as important decomposers and recyclers in almost all ecosystems, they can behave like classic consumers of critical energy and nutrient resources. They may be engaged in competition with other microbes but may also compete against representatives from higher trophic levels, i.e. animals (Janzen 1977; Cipollini and Stiles 1993; Crist and Friese 1993; Burkepile et al. 2006). Indeed, competition has been documented to occur between various mould species and saprophagous insect larvae, such as drosophilid flies (Atkinson 1981; Hodge 1996; Hodge et al. 1996; Hodge and Arthur 1997; Rohlfs 2005a, b, 2008; Rohlfs and Hoffmeister 2005; Rohlfs et al. 2005). In line with the general assumption about interspecific competition (see above), the outcome of *Drosophila*–fungus competition strongly depends on priority effects (i.e. the age of fungal colonies at the time insect larvae enter a resource patch) and insect density (Rohlfs 2005b; Rohlfs et al. 2005; Fig. 7.7). Moreover, the negative effect on insect development varies among fungal species (Fig. 7.8), even with some fungi causing no mortality in the insects, but themselves suffering from the presence of the insect larvae (e.g. *Penicillium* sp.; Rohlfs et al. 2005). Impairment of rapid expansion of a fungal colony (or entire suppression of fungal growth) on a resource patch appears to be due to physical (and/or chemical?) destruction of tissue in the mould's active growth zone by the insect

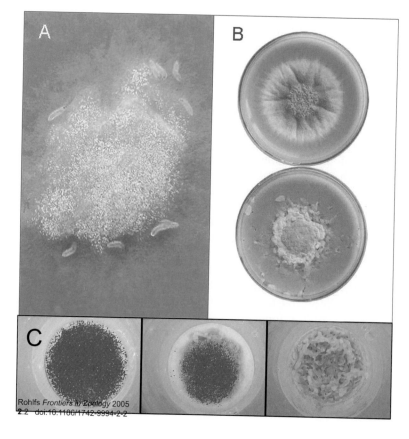

Fig. 7.7. Aggregation of *Drosophila melanogaster* larvae on fungal colonies and its effect on fungal development. **A** Several *Drosophila* larvae 'attacking' a two-day-old colony of *Aspergillus fumigatus*. **B** Effect of ten *Drosophila* larvae (*lower* Petri dish) on the growth of *Aspergillus nidulans* compared with an undisturbed colony (*upper* Petri dish). Colonies were four days old, growing on *Drosophila* corn meal/yeast hydolysate medium at 25°C. **C** Impact of one (*left*), five (*middle*) and ten (*right*) *Drosophila* larvae on the growth of *Aspergillus niger* ten days after inoculation on *Drosophila* medium

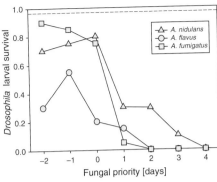

Fig. 7.8. Fungus-specific effects on *Drosophila* survival and experimental setup. *Left panel* Impact of three *Aspergillus* species on the survival of *Drosophila* larvae to the adult stage as a function of fungal priority (e.g. '0' denotes simultaneous transfer of larvae and fungal conidia, '1' and '−1' fungal or insect 'head start' of one day, respectively. *Dashed line* indicates insect survival under mould-free condition. For each treatment N = 20. For clarity, error bars are omitted. *Right panel Aspergillus–Drosophila* confrontation takes place in 2 ml microtubes filled with 1 ml standard *Drosophila* corn meal/yeast hydolysate medium and sealed with cotton plug (dental roll). Storing these miniaturised experimental units in racks allows preparation of high numbers of simultaneous replicates. Following a standard protocol, experimental units are inoculated with ∼1000 conidiospores in 1 µl Ringer solution. Subsequently, the experimental setup is incubated at 25°C and a 16-h light cycle

larvae (Fig. 7.7). Despite the eventually negative effect of, e.g. *Aspergillus* fungi, on an insect feeding substrate, *Drosophila* larvae display conspicuous aggregation behaviour on fungal colonies (Rohlfs 2005a; Fig. 7.7). This behavioural response has been correlated with retarded mould growth, a relationship that becomes especially pronounced when the effect of increasing insect density is taken into account (Fig. 7.7). Importantly, fungal growth as determined by *Drosophila* larval density has been shown to be negatively correlated to insect survival (Hodge 1996; Rohlfs et al. 2005). Although at the moment we cannot exclude any nutritional benefit from feeding on young fungal tissue to the insects, one interpretation of this behaviour and its consequences for both the insect and the fungus is that it is part of an animal defensive response to the harmful mould (Rohlfs 2005a). The clearly active interaction between the fungi and the animal antagonists indicates a strong interference rather than exploitation competition. The latter signifies competition without direct antagonistic contact between competitors. However, organisms compete by capturing resources faster than their competitors.

Interestingly some fungi, such as *A. flavus*, are still able to successfully invade resources and kill a significant proportion of larvae in which the immature insects have already started feeding (Fig. 7.8). Since *A. flavus* and *A. fumigatus* have been described as opportunistic pathogens (see

Sect. III.A), *Drosophila* larval mortality might be attributable to a pathogenic effect; however, there is as yet no evidence that these fungi do infect the insect larvae. Continuous exposure to fungal spores during larval development did not induce higher mortality rates than development on a sterile substrate (Fig. 7.9). The interpretation of these results is based on the current knowledge on how insect get infected by pathogenic fungi (Clarkson and Charnley 1996; see also Sect. III.A). We suppose that infection of the *Drosophila* larval stage is, if at all, only possible via attachment of spores to the gut epithelium, since the physical conditions due to the foraging and digging behaviour of the larvae in the feeding substrates appears to be unlikely to lead to a durable spore–host contact that is long enough to allow the fungal infection process to occur via the larval exoskeleton. As indicated by the data shown in Fig. 7.9, however, also infection via the larval gut does not seem to occur (for non-pathogenic effects of injection of *A. fumigatus* spores into adult flies, see Chamilos et al. 2008). It seems to be a general phenomenon that *Drosophila* is rather insusceptible to fungal infections during the larval stage, even if fungi are classified as entomopathogens, such as *B. bassiana* (Kraaijeveld et al. 2008).

Such non-pathogenic but competitive interactions between fungi and insects may have far-reaching consequences for trophic interactions within the animal communities. For instance, infestation of larval insect feeding substrates with

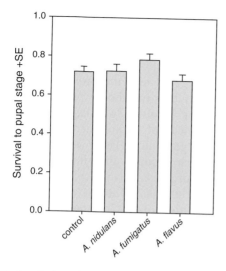

Fig. 7.9. Survival to the pupal stage of *Drosophila melanogaster* larvae as a function of continuous exposure to spore of various *Aspergillus* species. For a period of one week, larvae were transferred each day to fresh medium containing new conidiospores (∼ one million). These data illustrate that *Drosophila* larvae are quite resistant to *Aspergillus* via the common way of infection, as described in Sect. III.A, supporting the proposed competitive interaction between mould and saprophagous insects

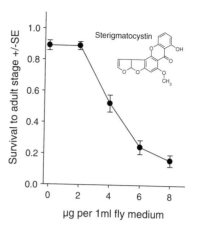

Fig. 7.10. Effect of sterigmatocystin at varying concentrations on the survival of *Drosophila melanogaster* larvae to the adult stage

A. niger had a strong negative effect on *Drosophila* larvae carrying the eggs/larvae of a parasitic wasp (Rohlfs 2008). When parasitised *Drosophila* larvae were forced to develop on fungus-infested substrates, they suffered (depending on larval density; see above) from higher mortality than healthy larvae. This could negatively influence the population growth of parasitic wasps. Moreover, foraging wasps abandoned searching for host larvae significantly earlier on mould-infested patches than they did on mould-free substrate (Rohlfs 2008), implying that the fungus allows the *Drosophila* larvae to feed in a refuge which protects the larvae from parasite attacks. Interestingly, because *A. niger* had significantly enhanced conidiospore production when the fungus was confronted not with unparasitised but with parasitised *Drosophila* larvae (Rohlfs 2008), antagonistic processes within the insect communities have the potential to control reproduction of the fungal competitor and hence microbial community structure.

Besides these immediate ecological effects of saprophytic mould on saprophagous insect communities, fungi can be expected to influence the evolution of *Drosophila* behaviour and life history strategies. In the view of the severe direct effects on insect development, the insects are likely to be selected to resist the fungal competitor. Yet by using the isofemale line technique Rohlfs (2006) demonstrated genetic variation in the ability of *Drosophila* to successfully develop in the presence of *A. niger*. Heritable variation in the survival of insect larvae ranged from less than 20% to almost 100%, which reveals the potential of fungi to select for improved insect development when confronted with the fungal competitor. From an adaptationist's perspective, the fact that insect populations display variation in the ability to resist fungal competitors suggests that the animals have to 'pay' some cost for the increased resistance. Indeed replicated fly populations selected for resistance to *A. nidulans* during the larval stage show a reduced ability to withstand abiotic stress and to fend off parasites (Rohlfs, Trienens, Wölfle, unpublished data).

Given the diverse fatal but non-pathogenic effects of mould on saprophagous insects, fungal secondary metabolites are prime candidates for explaining the ecological and evolutionary impact of fungi on *Drosophila* and vice versa. First evidence of a critical role of toxic secondary metabolites comes from various pharmacological tests. Prominent mycotoxins, such as aflatoxin, patulin, kojic acid (Reiss 1975; Dobias et al. 1977; Chinnici et al. 1979; Rohlfs 2008), and sterigmatocystin (Fig. 7.10) are toxic to *Drosophila* during larval development. Second, insect mortality drastically increases with increasing age of fungal colonies prior to larval settlement in a resource patch

(Fig. 7.8). These strong priority effects (see above) might be attributable to the positive relationship between fungal development and toxic secondary metabolite production (Calvo et al. 2002). However, as for the pharmacological studies this interpretation is based on observations on correlative rather than causal relationships. For instance, pharmacological studies are confounded by the altered ecological context and chemical milieu in which both insect and fungal resistance traits may evolve: fungi critically modify the resource (e.g. pH), secrete numerous extracellular enzymes (hydrolases), and occupy space on a limited resource patch that cannot be exploited by the insects. Moreover, fungi such as *Aspergillus* sp. are well known for producing many different secondary metabolites, which may lead to unforeseeable synergistic effects (Creppy et al. 2004) that may drive both the efficiency of the chemical defence and the evolution of resistance in competing insects. And the most exciting question in this context is whether fungi are able to adjust their chemical armoury to the corresponding ecological challenge since biosynthesis of e.g. mycotoxins can be considered costly for the organism (Shwab et al. 2007), or if temporal patterns in secondary metabolite production represent an inflexible constitutive defence system that is maintained even in the absence of animal antagonists. Even though fungi have been shown to possess the mechanistic prerequisites for regulating their chemical diversity, it has not yet been demonstrated whether: (1) this machinery is capable of responding to e.g. competing fungi or predatory arthropods and (2) such proposed responses are adaptive in terms of reducing the opponent's impact on fungal development. From a fungal molecular perspective, the major goal is to understand whether genes and gene clusters involved in secondary metabolite biosynthesis function as 'environmental response genes' that exhibit condition and temporal specific expression (c.f. Berenbaum 2002).

IV. Melding Ecology and Molecular Biology

Given the widespread ability to produce some of the most potent toxins in the living world, secondary metabolite biosynthesis can be viewed as a successful evolutionary outcome. Yet the ecological and evolutionary mechanisms that influence or underlie this success still remain elusive. To resolve the interplay between fungal secondary metabolite genes and the ecological process in which secondary metabolite production is proposed to be involved, we suggest to apply the emerging field of evolutionary and ecological functional genomics (Feder and Mitchell-Olds 2003), which requires the simultaneous use of molecular, organismal, and ecological approaches (Jackson et al. 2002; Thomas and Klaper 2004). The application of this perspective provides an integrative understanding of the interrelationship of genes, the associated pathways, and the corresponding phenotypes involved in the ecological process. We believe we have identified the *Aspergillus–Drosophila* system as being suitable to reveal the ecological and evolutionary importance of fungal secondary metabolites, since both organisms fulfil the basic prerequisite of completely sequenced genomes for being models in genomics (Adams et al. 2000; Galagan et al. 2005). Also the rapid advances in developing high-throughput microarrays enable global gene expression analyses in both fungi and insects (Gupta and Oliver 2003; Andersen et al. 2008) and, from an ecological point of view, this offers the unique opportunity to study simultaneous changes in gene expression in both competitors. A preliminary genome-wide screening for changes in transcriptional activity in *D. melanogaster* larvae confronted with *A. nidulans* indicates the up-regulation of several genes, e.g. those involved in antimicrobial immune defence and insecticide resistance (Trienens and Rohlfs, unpublished data). Likewise quantitative RT-PCR of RNA from *A. nidulans* confronted with *D. melanogaster* larvae indicates the up-regulation of secondary metabolite regulator genes (Fohgrub and Kempken, unpublished data). To our current knowledge, the possibilities of using expression profiling techniques for *Aspergillus* have not yet been applied in the context of species interaction. But for differentially expressed genes in *A. fumigatus* in response to variation in vertebrate immune response, see Sugui et al. (2008); and for a recent review on microarray studies in filamentous fungi, see Breakspear and Momany (2007).

The second reason why organisms can become models in genomics and hence ecological genomics is the possibility of genetic manipulation.

Being able to knock-out, down-regulate, or over-express candidate genes against a genetic background that is the same as that of the wild type will be necessary to validate the causal relationship between e.g. expression of secondary metabolites genes in fungi and *Drosophila* larval development (van Straalen and Roelofs 2006; Kammenga et al. 2007). Transposon-based mutagenesis may be useful as an additional tool (Kempken 1999; Pöggeler and Kempken 2004; Braumann et al. 2007). Based on a priori selection of the candidate gene *laeA*, Rohlfs et al. (2007) have already provided evidence of the usefulness of this approach in a fungus-fungivore system; however, this study did not distinguish between the causal agent(s) driving fungivore behaviour and unimportant secondary metabolite pathways, since *laeA* is a regulator of many secondary metabolites (see above). Critical advances may be achieved by manipulating gene expression based on differentially expressed fungal genes as stimulated by insect attack, and testing whether transgenic fungi confirm the hypothesised role of genes involved in secondary metabolite biosynthesis in the corresponding ecological setting. A comprehensive molecular toolbox is available to achieve this goal in *Aspergillus* (Goldman and Osmani 2008). The same experimental strategy can be applied to the insects. By using the binary GAL4/UAS system to generate even tissue-specific post-transcriptional silencing of candidate genes in *Drosophila* (Dietzl et al. 2007) will provide unparalleled insights into the mechanistic basis of the role fungal defences against competing insects and the animals' counter-adaptations. Moreover, coupling this molecular approach with experimental evolution as described above will enable us to thoroughly study the possible role of co-evolutionary processes in this ecological interaction. However, such studies are currently still under development. As pointed out by Jackson et al. (2002), integrating molecular and ecological research calls for interdisciplinary approaches and hence intensive collaboration between molecular biologists and ecologists.

typical mould species and various insect groups. Independently of whether fungi and insects are engaged in parasitism, predation, or competition, the production of secondary metabolites has been proposed to incur a key function within fungi in attacking insects or defending against them. Many studies provide correlative or pharmacological evidence of such a role of mycotoxins. Future molecular genomic approaches in combination with experimental evolutionary ecology will forge a mechanistic basis for fungus–insect interactions that will be able to answer fundamental questions regarding the biological significance of fungal secondary metabolism. We are still at the stage of having only vague arguments for fungal toxins being part of a chemical defence system against insects; but the use of carefully designed transgenic fungi coupled with biochemical analyses will shed light on such a function of possibly still unknown secondary metabolites. An evolutionary ecological genomic approach will also help us proving that secondary metabolite production is indeed costly to fungi and that the biosynthesis of e.g. mycotoxins is related to the ecological context, such as insect attacks. It remains an open question whether the signalling of changes in environmental condition in fungi, as controlled by oxylipins (Tsitsigiannis and Keller 2007), have its ecological analogy in plant–insect systems (Kessler et al. 2004). To our knowledge the molecular regulatory machinery of secondary metabolite synthesis and its proposed function in controlling the allocation of limited resources to costly insecticidal molecules in relation to the threat of being attacked by natural enemies have yet to be implemented. Having a good ecological model system at hand (e.g., *Aspergillus–Drosophila*) will finally allow us to draw an evolutionary line between the organisation of fungal defence against natural enemies and virulence in invertebrate and vertebrate hosts.

Acknowledgement. The laboratory work of F.K. and M.R. is funded by the German Research Council (DFG).

References

V. Conclusions

Fungus–insect relationships range across symbiosis, mutualism, and antagonism. Here we concentrate on various antagonistic interactions between

Abdou RF, Megalla SE, Azab SG (1984) Mutagenic effects of aflatoxin B-1 and G-1 on the Egyptian cotton leafworm, *Spodoptera littoralis* (Boisd.). Mycopathologia 88:23–26

Adams MD et al. (2000) The genome sequence of *Drosophila melanogaster*. Science 287:2185–2195

Andersen MR, Vongsangnak W, Panagiotou G, Salazar MP, Lehmann L, Nielsen J (2008) A trispecies *Aspergillus* microarray: comparative transcriptomics of three *Aspergillus* species. Proc Natl Acad Sci USA 105:4387–4392

Appleyard VC, Unkles SE, Legg M, Kinghorn JR (1995) Secondary metabolite production in filamentous fungi displayed. Mol Gen Genet 247:338–342

Atkinson WD (1981) An ecological interaction between citrus fruit, *Penicillium* moulds and *Drosophila immigrans* Sturtevant (Diptera: Drosophilidae). Ecol Entomol 6:339–344

Bagga S, Hu G, Screen SE, St Leger RJ (2004) Reconstructing the diversification of subtilisins in the pathogenic fungus *Metarhizium anisopliae*. Gene 324:159–169

Barnes SE, Dola TP, Bennett JW, Bhatnagar D (1994) Synthesis of sterigmatocystin on a chemically defined medium by species of *Aspergillus* and *Chaetomium*. Mycopathologia 125:173–178

Bayram O, et al (2008) VelB/VeA/LaeA complex coordinates light signal with fungal development and secondary metabolism. Science 320:1504–1506

Begon M, Mortimer M, Thompson DJ (1996) Population ecology - a unified study of animals and plants. Blackwell Science, Oxford

Bengtsson G, Hedlund K, Rundgren S (1991) Selective odor perception in the soil collembola *Onychirus armatus*. Journal of Chemical Ecology 17:2113–2125

Bengtsson G, Hedlund K, Rundgren S (1993) Patchiness and compensatory growth in a fungus–Collembola system. Oecologia 93:296–302

Bennett JW, Bentley R (1989) What's in a name? Microbial secondary metabolism. Adv Appl Microbiol 34:1–28

Berenbaum MR (2002) Postgenomic chemical ecology: from genetic code to ecological interactions. J Chem Ecol 28:873–896

Bergmann S, Schumann J, Scherlach K, Lange C, Brakhage AA, Hertweck C (2007) Genomics-driven discovery of PKS-NRPS hybrid metabolites from *Aspergillus nidulans*. Nat Chem Biol 3:213–217

Bok JW, Keller NP (2004) LaeA, a regulator of secondary metabolism in *Aspergillus* spp. Eukaryot Cell 3:527–535

Bok JW, et al (2005) LaeA, a regulator of morphogenetic fungal virulence factors. Eukaryot Cell 4:1574–1582

Bok JW, et al (2006a) GliZ, a transcriptional regulator of gliotoxin biosynthesis, contributes to *Aspergillus fumigatus* virulence. Infect Immun 74:6761–6768

Bok JW, Noordermeer D, Kale SP, Keller NP (2006b) Secondary metabolic gene cluster silencing in *Aspergillus nidulans*. Mol Microbiol 61:1636–1645

Brakhage AA, Schuemann J, Bergmann S, Scherlach K, Schroeckh V, Hertweck C (2008) Activation of fungal silent gene clusters: a new avenue to drug discovery. Prog Drug Res 66:1, 3–12

Braumann I, van den Berg M, Kempken F (2007) Transposons in biotechnologically relevant strains of *Aspergillus niger* and *Penicillium chrysogenum*. Fungal Genet Biol 44:1399–1414

Breakspear A, Momany M (2007) The first fifty microarray studies in filamentous fungi. Microbiology 153:7–15

Burkepile DE, et al (2006) Chemically mediated competition between microbes and animals: microbes as consumers in food webs. Ecology 87:2821–2831

Caboche S, Pupin M, Leclere V, Fontaine A, Jacques P, Kucherov G (2008) NORINE: a database of nonribosomal peptides. Nucleic Acids Res 36:D326–D331

Calvo AM, Wilson RA, Bok JWW, Keller NP (2002) Relationship between secondary metabolism and fungal development. Microbiol Mol Biol Rev 66:447–459

Chamilos G, et al (2008) *Drosophila melanogaster* as a model host to dissect the immunopathogenesis of zygomycosis. Proc Natl Acad Sci USA 105:9367–9372

Chang PK, et al (1993) Cloning of the *Aspergillus parasiticus apa-2* gene associated with the regulation of aflatoxin biosynthesis. Appl Environ Microbiol 59:3273–3329

Chinnici JP, Erlanger L, Charnock M, Jones M, Stein J (1979) Sensitivity differences displayed by *Drosophila melanogaster* larvae of different ages to the toxic effects of growth on media containing aflatoxin B_1. Chem Biol Interact 24:373–380

Cipollini ML, Stiles EW (1993) Fruit rot, antifungal defense, and palatability of fleshy fruits for fungivorous birds. Ecology 74:751–762

Clarkson JM, Charnley AK (1996) New insights into the mechanisms of fungal pathogenesis in insects. Trends Microbiol 4:197–203

Cole R, Schweikert M (2003) Handbook of secondary fungal metabolites (vol 1–3). Elsevier, Amsterdam

Collemare J, Billard A, Bohnert HU, Lebrun MH (2008) Biosynthesis of secondary metabolites in the rice blast fungus *Magnaporthe grisea*: the role of hybrid PKS-NRPS in pathogenicity. Mycol Res 112:207–215

Cramer RA Jr, et al (2006) Disruption of a nonribosomal peptide synthetase in *Aspergillus fumigatus* eliminates gliotoxin production. Eukaryot Cell 5:972–980

Creppy EE, Chiarappa P, Baudrimont I, Borracci P, Moukha S, Carratù MR (2004) Synergistic effects of fumonisin B1 and ochratoxin A: are in vitro cytotoxicity data predictive of in vivo acute toxicity? Toxicology 201:115–123

Crist TO, Friese CF (1993) The impact of fungi on soil seeds: implication for plants and granivores in a semiarid shrub-steppe. Ecology 74:2231–2239

Demain AL, Fang A (2000) The natural function of secondary metabolites. In: Sheper T (ed) Advances in biochemical engineering/biotechnology. Springer, Heidelberg, pp 1–39

Dietzl G, et al (2007) A genome-wide transgenic RNAi library for conditional gene inactivation in *Drosophila*. Nature 448:151–156

Dobias J, Nemec P, Brtko J (1977) The inhibitory effect of kojic acid and its two derivatives on the development of *Drosophila melanogaster*. Biologia (Brat) 32:417–421

Du W, Obrian GR, Payne GA (2007) Function and regulation of *aflJ* in the accumulation of aflatoxin early pathway intermediate in *Aspergillus flavus*. Food Addit Contam 24:1043–1050

Ehrlich KC, Montalbano BG, Cotty PJ (2003) Sequence comparison of *aflR* from different *Aspergillus* species provides evidence for variability in regulation of aflatoxin production. Fungal Genet Biol 38:63–74

Ehrlich KC (2008) Genetic diversity in Aspergillus flavus and its implications for agriculture. In: Varga J, Samson RA (eds) Aspergillus in the genomic era, Wageningen, pp 233–247

Fan Y, et al (2007) Increased insect virulence in *Beauveria bassiana* strains overexpressing an engineered chitinase. Appl Environ Microbiol 73:295–302

Feder ME, Mitchell-Olds T (2003) Evolutionary and ecological functional genomics. Nat Genet 4:649–655

Feng GH, Leonard TJ (1998) Culture conditions control expression of the genes for aflatoxin and sterigmatocystin biosynthesis in *Aspergillus parasiticus* and *A. nidulans*. Appl Environ Microbiol 64:2275–2277

Fernandes M, Keller NP, Adams TH (1998) Sequence-specific binding by *Aspergillus nidulans* AflR, a C6 zinc cluster protein regulating mycotoxin biosynthesis. Mol Microbiol 28:1355–1365

Fordyce JA (2003) Aggregative feeding of pipevine swallowtail larvae enhances hostplant suitability. Oecologia 135:250–257

Fox EM, Howlett BJ (2008) Biosynthetic gene clusters for epipolythiodioxopiperazines in filamentous fungi. Mycol Res 112:162–169

Freeland WJ, Janzen DH (1974) Strategies in herbivory by mammals: the role of plant secondary compounds. American Naturalist 108:269–289

Furlong MJ, Pell JK (2005) Interactions between entomopathogenic fungi and arthropod natural enemies. In: Vega FE, Blackwell M (eds) Insect–fungal associations – ecology and evolution. Oxford University Press, Oxford, pp 51–73

Galagan JE, et al (2005) Sequencing of Aspergillus nidulans and comparative analysis with *A. fumigatus* and *A. oryzae*. Nature 438:1105–1115

Gardiner DM, Howlett BJ (2005) Bioinformatic and expression analysis of the putative gliotoxin biosynthetic gene cluster of *Aspergillus fumigatus*. FEMS Microbiol Lett 248:241–248

Gardiner DM, Cozijnsen AJ, Wilson LM, Pedras MS, Howlett BJ (2004) The sirodesmin biosynthetic gene cluster of the plant pathogenic fungus *Leptosphaeria maculans*. Mol Microbiol 53:1307–1318

Gillespie JP, Bailey AM, Cobb B, Vilcinskas A (2000) Fungi as elicitors of insect immune responses. Arch Insect Biochem Physiol 44:49–68

Goldman GH, Osmani SA (eds) (2008) The Aspergilli – genomics, medical aspects, biotechnology and research methods. CRC, Boca Raton

Gottar M, et al (2006) Dual detection of fungal infections in *Drosophila* via recognition of glucans and sensing of virulence factors. Cell 127:1425–1437

Guevara R, Rayner ADM, Reynolds SE (2000) Effects of fungivory by two specialist ciid beetles (*Octotemnus glabriculus* and *Cis boleti*) on the reproductive fitness of their host fungus, *Coriolus versicolor*. New Phytol 145:137–144

Gupta V, Oliver B (2003) *Drosophila* microarray platforms. Brief Funct Genomics Proteomics 2:97–105

Haarmann T et al. (2005) The ergot alkaloid gene cluster in *Claviceps purpurea*: extension of the cluster sequence and intra species evolution. Phytochemistry 66:1312–1320

Hamasaki T, Matsui K, Isono K, Hatsuda Y (1973) A new metabolite from *Aspergillus nidulans* cellular developmental. Agric Biol Chem 37:1769–1770

Harold S, Tordoff GM, Jones TH, Boddy L (2005) Mycelial responses of *Hypholoma fasciculare* to collembola grazing: effect of inoculum age, nutrient status and resource quality. Mycol Res 109:927–935

Hedlund K, Boddy L, Preston CM (1991) Mycelial responses of the soil fungus, *Mortierella isabellina*, to grazing by *Onychiurus armatus* (collembola). Soil Biol Biochem 23:361–366

Hedlund K, Bengtsson G, Rundgren S (1995) Fungal odour discrimination in two sympatric species of fungivorous collembolans. Funct Ecol 9:869–875

Henke MO, de Hoog GS, Gross U, Zimmermann G, Kraemer D, Weig M (2002) Human deep tissue infection with an entomopathogenic *Beauveria* species. J Clin Microbiol 40:2698–2702

Herrmann M, Sprote P, Brakhage AA (2006) Protein kinase C (PkcA) of *Aspergillus nidulans* is involved in penicillin production. Appl Environ Microbiol 72:2957–2970

Hicks JK, Yu JH, Keller NP, Adams TH (1997) *Aspergillus* sporulation and mycotoxin production both require inactivation of the FadA G alpha protein-dependent signaling pathway. EMBO J 16:4916–4923

Hicks JK, Shimizu K, Keller NP (2002) Genetic and biosynthesis of aflatoxin and sterigmatocystin. In: Esser K, Bennett JW (eds) Agricultural applications. The Mycota XI. Springer, Heidelberg, pp 55–69

Hodge S (1996) The relationship between *Drosophila* occurrence and mould abundance on rotting fruit. Brit J Entomol Nat Hist 9:87–91

Hodge S, Arthur W (1997) Direct and indirect effects of *Drosophila* larvae on the growth of moulds. Entomologist 116:198–204

Hodge S, Wallace A, Mitchell P (1996) Effect of temporal priority on interspecific interactions and community development. Oikos 76:350–358

Hoffmeister D, Keller NP (2007) Natural products of filamentous fungi: enzymes, genes, and their regulation. Nat Prod Rep 24:393–416

Hoppert M, Gentzsch C, Schorgendorfer K (2001) Structure and localization of cyclosporin synthetase, the key enzyme of cyclosporin biosynthesis in Tolypocladium inflatum. Arch Microbiol 176:285–293

Hull EP, Green PM, Arst HN, Jr, Scazzocchio C (1989) Cloning and physical characterization of the L-proline catabolism gene cluster of *Aspergillus nidulans*. Mol Microbiol 3:553–559

Jackson RB, Linder CR, Lynch M, Puruggannan M, Somerville S, Thayer SS (2002) Linking molecular insight and ecological research. Trends Ecol Evol 17:409–414

Janzen DH (1977) Why fruits rot, seeds mold and meat spoils. Am Nat 111:691–713

Jørgensen HB, Johansson T, Canbäck B, Hedlund K, Tunlid A (2005) Selective foraging of fungi by collembolans in soil. Biol Lett 1:243–246

Kammenga JE, Herman MA, Ouborg NJ, Johnson L, Breitling R (2007) Microarray challenges in ecology. Trends Ecol Evol 22:273–279

Kampichler C, Rolschewski J, Donnelly DP, Boddy L (2004) Collembolan grazing affects the growth strategy of the cord-forming fungus *Hypholoma fasciculare*. Soil Biol Biochem 36:591–599

Katz L, Donadio S (1993) Polyketide synthesis: prospects for hybrid antibiotics. Annu Rev Microbiol 47:875–912

Kawaide H (2006) Biochemical and molecular analyses of gibberellin biosynthesis in fungi. Biosci Biotechnol Biochem 70:583–590

Keller NP, Hohn TM (1997) Metabolic pathway gene clusters in filamentous fungi. Fungal Genet Biol 21:17–29

Keller NP, Turner G, Bennett JW (2005) Fungal secondary metabolism – from biochemistry to genomics. Nat Rev Microbiol 3:937–947

Kempken A (1999) Fungal transposons: from mobile elements towards molecular tools. Appl Microbiol Biotechnol 52:756–760

Kennedy J, Auclair K, Kendrew SG, Park C, Vederas JC, Hutchinson CR (1999) Modulation of polyketide synthase activity by accessory proteins during lovastatin biosynthesis. Science 284:1368–1372

Kessler A, Halitschke R, Baldwin IT (2004) Silencing the jasmonate cascade: induced plant defenses and insect populations. Science 30:665–668

Kraaijeveld AR, Barker CL, Godfray HCJ (2008) Stage-specific sex differences in *Drosophila* immunity to parasites and pathogens. Evol Ecol 22:217–228

Kupfahl C, et al (2006) Deletion of the *gliP* gene of *Aspergillus fumigatus* results in loss of gliotoxin production but has no effect on virulence of the fungus in a low-dose mouse infection model. Mol Microbiol 62:292–302

Lee BN, Adams TH (1996) *FluG* and *flbA* function interdependently to initiate conidiophore development in *Aspergillus nidulans* through *brlAβ* activation. EMBO J 15:299–309

Meyers DM, Obrian G, Du WL, Bhatnagar D, Payne GA (1998) Characterization of *aflJ*, a gene required for conversion of pathway intermediates to aflatoxin. Appl Environ Microbiol 64:3713–3717

Mihlan M, Homann V, Liu TW, Tudzynski B (2003) AREA directly mediates nitrogen regulation of gibberellin biosynthesis in *Gibberella fujikuroi*, but its activity is not affected by NMR. Mol Microbiol 47:975–991

Pelaez F (2005) Biological activities of fungal metabolites. In: An Z (ed) Handbook of industrial mycology. Dekker, New York, pp 49–92

Perrin RM, et al (2007) Transcriptional regulation of chemical diversity in *Aspergillus fumigatus* by LaeA. PLoS Pathogens 3:e50

Pöggeler S, Kempken F (2004) Mobile genetic elements in mycelial fungi. In: Kück U (ed) The Mycota II, genetics and biotechnology, 2nd edn. Springer, Heidelberg, pp 165–198

Price MS, et al (2006) The aflatoxin pathway regulator AflR induces gene transcription inside and outside of the aflatoxin biosynthetic cluster. FEMS Microbiol Lett 255:275–279

Reeves EP, Messina CGM, Doyle S, Kavanagh K (2004) Correlation between gliotoxin production and virulence of *Aspergillus fumigatus* in *Galleria mellonella*. Mycopathology 158:73–79

Reiss J (1975) Insecticidal and larvicidal activities of the mycotoxins aflatoxin B1, rubratoxin B, patulin and diacetoxyscirpenol towards *Drosophila melanogaster*. Chem Biol Interact 10:339–342

Rohlfs M (2005a) Clash of kingdoms or why *Drosophila* larvae positively respond to fungal competitors. Front Zool 2:2

Rohlfs M (2005b) Density-dependent insect–mold interactions: effects on fungal growth and spore production. Mycologia 97:996–1001

Rohlfs M (2006) Genetic variation and the role of insect life history traits in the ability of *Drosophila* larvae to develop in the presence of a filamentous fungus. Evol Ecol 20:271–289

Rohlfs M (2008) Host–parasitoid interaction as affected by interkingdom competition. Oecologia 155:161–168

Rohlfs M, Hoffmeister TS (2005) Maternal effects increase survival probability in *Drosophila subobscura* larvae. Entomol Exp Appl 117:51–58

Rohlfs M, Obmann B, Petersen R (2005) Competition with filamentous fungi and its implications for a gregarious life-style in insects living on ephemeral resources. Ecol Entomol 30:556–563

Rohlfs M, Albert M, Keller NP, Kempken F (2007) Secondary chemicals protect mould from fungivory. Biol Lett 3:523–525

Rosen S, Yu JH, Adams TH (1999) The *Aspergillus nidulans sfaD* gene encodes a G protein beta subunit that is required for normal growth and repression of sporulation. EMBO J 18:5592–5600

Rouxel T, Chupeau Y, Fritz R, Kollmann A, Bousquet J-F (1988) Biological effects of sirodesmin PL, a phytotoxin produced by *Leptosphaeria maculans*. Plant Sci 57:45–53

Ruess L, Lussenhop J (2005) Trophic interactions of fungi and animals. In: Dighton J, White JK, Oudemans P (eds) The fungal community – its organization and role in the ecosystem. CRC, Boca Raton, pp 581–598

Ruess L, Garcia Zapata EJ, Dighton J (2000) Food preferences of a fungal-feeding *Aphelenchoides* species. Nematology 2:223–230

Rynkiewicz MJ, Cane DE, Christianson DW (2001) Structure of trichodiene synthase from *Fusarium sporotrichioides* provides mechanistic inferences on the terpene cyclization cascade. Proc Natl Acad Sci USA 98:13543–13548

Scheu S, Simmerling F (2004) Growth and reproduction of fungal feeding Collembola as affected by fungal species, melanin and mixed diets. Oecologia 139:347–353

Scully LR, Bidochka MJ (2005) Serial passage of the opportunistic pathogen *Aspergillus flavus* through an insect host yields decreased saprobic capacity. Can J Microbiol 51:185–189

Shaw PJA (1988) A consistent hierarchy in the fungal feeding preference of the collembola *Onychiurus armatus*. Pedobiologia 39:179–187

Shimizu K, Hicks JK, Huang TP, Keller NP (2003) Pka, Ras and RGS protein interactions regulate activity of AflR, a Zn(II)2Cys6 transcription factor in *Aspergillus nidulans*. Genetics 165:1095–1104

Shwab EK, Bok JW, Tribus M, Galehr J, Graessle S, Keller NP (2007) Histone deacetylase activity regulates chemical diversity in *Aspergillus*. Eukaryot Cell 6:1656–1664

Singer MS, Bernays EA, Carrière Y (2002) The interplay between nutrient balancing and toxin dilution in foraging by a generalist insect herbivore. Anim Behav 64:629–643

Smith DJ, Earl AJ, Turner G (1990) The multifunctional peptide synthetase performing the first step of penicillin biosynthesis in *Penicillium chrysogenum* is a 421 073 dalton protein similar to *Bacillus brevis* peptide antibiotic synthetases. EMBO J 9:2743–2750

St Leger R, Joshi L, Roberts D (1997) Adaptation of proteases and carbohydrates of saprophytic, phytopathogenic and entomopathogenic fungi to the requirements of their ecological niches. Microbiology 143:1983–1992

Stearns SC (1992) The evolution of life histories. Oxford University Press, Oxford

Steppuhn A, Gase K, Krock B, Halitschke R, Baldwin IT (2004) Nicotine's defensive function in nature. PLoS Biology 2:e217

Sugui JA, et al (2007) Role of *laeA* in the regulation of *alb1*, *gliP*, conidial morphology, and virulence in *Aspergillus fumigatus*. Eukaryot Cell 6:1552–1561

Sugui JA, et al (2008) Genes differentially expressed in conidia and hyphae of *Aspergillus fumigatus* upon exposure to human neutrophils. PloS ONE 3:e2655

Sweeney MJ, Dobson AD (1999) Molecular biology of mycotoxin biosynthesis. FEMS Microbiol Lett 175: 149–163

Tag A, et al (2000) G-protein signalling mediates differential production of toxic secondary metabolites. Mol Microbiol 38:658–665

Tekaia F, Latgé J-P (2005) *Aspergillus fumigatus*: saprophyte or pathogen? Curr Opin Microbiol 8:385–392

Thomas MA, Klaper R (2004) Genomics for the ecological toolbox. Trends Ecol Evol 19:439–445

Tinsley MC, Blanford S, Jiggins FM (2006) Genetic variation in *Drosophila melanogaster* pathogen susceptibility. Parasitology 132:767–773

Tordoff GM, Boddy L, Jones TH (2006) Grazing by *Folsomia candida* (Collembola) differentially affects mycelial morphology of the cord-forming basidiomycetes *Hypholoma fasciculare*, *Phanerochaete velutina* and *Resinicium bicolor*. Mycol Res 110:335–345

Tsitsigiannis DI, Keller NP (2007) Oxylipins as developmental and host–fungal communication signals. Trends Microbiol 15:109–118

Tudzynski P, Holter K, Correia T, Arntz C, Grammel N, Keller U (1999) Evidence for an ergot alkaloid gene cluster in *Claviceps purpurea*. Mol Gen Genet 261:133–141

Tudzynski B, Hedden P, Carrera E, Gaskin P (2001a) The P450-4 gene of *Gibberella fujikuroi* encodes ent-kaurene oxidase in the gibberellin biosynthesis pathway. Appl Environ Microbiol 67:3514–3522

Tudzynski P, Correia T, Keller U (2001b) Biotechnology and genetics of ergot alkaloids. Appl Microbiol Biotechnol 57:593–605

van Straalen NM, Roelofs D (2006) An introduction to ecological genomics. Oxford University Press, Oxford

Vega FE, Blackwell M (eds) (2005) Insect–fungal associations – ecology and evolution. Oxford University Press, New York

Vining LC (1990) Function of secondary metabolites. Ann Rev Microbiol 44:395–427

Wang C, Hu G, St. Leger RJ (2005) Differential gene expression by *Metarhizium anisopliae* growing in root exudate and host (*Manduca sexta*) cuticle or hemolymph reveals mechanisms of physiological adaptation. Fungal Genet Biol 42:704–718

Weber G, Schorgendorfer K, Schneider-Scherzer E, Leitner E (1994) The peptide synthetase catalyzing cyclosporine production in *Tolypocladium niveum* is encoded by a giant 45.8-kilobase open reading frame. Curr Genet 26:120–125

Woloshuk CP, Foutz KR, Brewer JF, Bhatnagar D, Cleveland TE, Payne GA (1994) Molecular characterization of *aflR*, a regulatory locus for aflatoxin biosynthesis. Appl Environ Microbiol 60: 2408–2414

Yu JH, Keller N (2005) Regulation of secondary metabolism in filamentous fungi. Annu Rev Phytopathol 43:437–458

Yu JH, Butchko RA, Fernandes M, Keller NP, Leonard TJ, Adams TH (1996a) Conservation of structure and function of the aflatoxin regulatory gene *aflR* from *Aspergillus nidulans* and *A. flavus*. Curr Genet 29:549–555

Yu JH, Wieser J, Adams TH (1996b) The *Aspergillus* FlbA RGS domain protein antagonizes G protein signaling to block proliferation and allow development. EMBO J 15:5184–5190

Yu JH, et al (2004) Clustered pathway genes in aflatoxin biosynthesis. Appl Environ Microbiol 70:1253–1262

Zocher R, et al (1986) Biosynthesis of cyclosporin A: partial purification and properties of a multifunctional enzyme from *Tolypocladium inflatum*. Biochemistry 25:550–553

8 Endophytic Fungi, Occurrence and Metabolites

Daniela Weber[1]

CONTENTS

[1]Institute of Biotechnology and Drug Research, Erwin-Schröd-inger-Strasse 56, 67663, Kaiserslautern, Germany; *present address:* MIP International Pharma Research GmbH, Mühlstrasse 50, 66386 St. Ingbert, Germany; e-mail: weber-daniela@gmx.de

Physiology and Genetics, 1st Edition
The Mycota XV
T. Anke and D. Weber (Eds.)
© Springer-Verlag Berlin Heidelberg 2009

I. Introduction

Symbiosis is defined in the original sense as the living together of different organisms (de Bary 1879). Perhaps two-thirds of the known species of fungi form intimate relationships with other living organisms in parasitic, commensalistic, or mutualistic symbiosis (Pirozynski and Hawksworth 1988). There exist different fungus–plant interactions, which range from mycorrhiza to plant pathogens. Arbuscular mycorrhizal fungi are ancient Zygomycetes and are associated with the roots of about 80% of plant species (Bonfante and Perotto 1995). This fact supports the hypothesis of Pirozynski und Malloch (1975) that the colonization of land and the evolution of plants were possible only through the establishment of symbiotic association of a semi-aquatic ancestral alga and an aquatic fungus.

Of the estimated 1.5×10^6 fungal species on earth (Hawksworth 2001) about 1.30×10^6 species are estimated to live as endophytes in plants (Dreyfuss and Chapela 1994). So most of the fungi are endophytes. They are ubiquitous, and probably all vascular and non-vascular terrestrial plants harbor them (Weber and Anke 2006). The term "endophyte" includes all organisms that, at some time of their life cycle, live symptomlessly within plant tissue (Petrini 1996). This definition also includes pathogens with extended latency periods. Endophyte association may range from intimate contact where the fungus inhabits the intercellular spaces

and xylem vessels within the plant, to more or less superficial colonization of peripheral, often dying or dead tissues such as bark layers on plants with secondary growth (Petrini 1996). Ascomycetes, Basidiomycetes, Deuteromycetes, and Oomycetes were isolated as endophytes (Sinclair and Cerkauskas 1996); of these the Ascomycetes and imperfect fungi constitute the largest group.

In some cases this fungus–plant interaction has mutualistic features. Enhanced rate of growth (Bradshaw 1959; Kackley et al. 1990; Hill et al. 1991), resistance and toxicity against herbivores (Wallner et al. 1983; Read and Camp 1986), deterrence of insects (Clay 1988b), resistance against nematodes (Kimmons et al. 1990; West et al. 1990) and pathogens (White and Cole 1985; Yshihara et al. 1985), and enhanced drought tolerance (Read and Camp 1986; Arechavaleta et al. 1989; West et al. 1993) are observed as advantages to the host plants. The benefits for the fungal symbionts can be greater access to nutrients, protection against desiccation, insects, and parasitic fungi, and competition from other microbes (White et al. 2000). During the coevolution, endophytic microbes improved the resistance of the host plants to adverse conditions by secreting bioactive secondary metabolites (Ge et al. 2008).

Although the first discovery of an endophytic fungus was in 1902 by Freeman, little attention was paid to this group of fungi until the recent realization of their ecological relevance and their potential as source for new bioactive metabolites (Gunatilaka 2006).

II. The Ecological Relevance of Endophytic Fungi

Two ecological groups of endophytes can be distinguished: grass endophytes and endophytes of woody plants. Endophytic fungi were found in all woody plants examined for endophytes (Saikkonen et al. 1998). Despite the great diversity and abundancy of endophytes in woody plants, these endophytes and their interaction with their host plants receive less attention than those of grass endophytes (Saikkonen et al. 1998). The production of alkaloids in infected grasses (in planta) can cause intoxication of grazing livestock. Because of this economic importance, endophytes of poaceous grasses are the best-examined (Weber and Anke 2006).

In the family Clavicipitaceae there is a development from parasitic grass endophytes like *Claviceps* to mutualistic species like *Acremonium* (anamorphic form of *Epichloë*; Clay 1988a). Species of the genus *Claviceps* cause local infections of the inflorescences of grasses. Sclerotia develop at the site of infection instead of seeds of the host plant. The local infection evolved to a systemic infection, which can be found in *Acremonium* species (Clay 1988a). Reduction of sexual reproduction can be observed in this group of obligate endophytic fungi (Saikkonen et al. 1998). The asexual or anamorphic form is transmitted vertically by the growth of fungal hyphae into the host plant's seeds (Arnold et al. 2003). Vogl already discovered the mycelium of an endophytic fungus in seeds of *Lolium temulentum* in 1898. Infection of the inflorescenses by parasitic species results in sexual sterilization of the host plant. This is in contrast to the loss of sexual reproduction of endophytes in mutualistic symbiosis. In these cases propagation of the fungus is ensured by viable seeds of the host plant.

The endophytic fungi of woody plants show a high diversity and belong to different taxonomic groups. These fungal endophytes are horizontally transmitted via spores and are not known to grow into seeds and systemically infecting the plant following seed germination (Wilson 2000). The infections are often highly localized within leaves, petioles, bark, or stems (Saikkonen et al. 1998).

Many endophytes of woody plants are closely related to pathogenic species. Freeman and Rodriguez (1993) were able to show that the mutation at a single genetic locus can change an isolate of *Colletotrichum magna* from a pathogen to a non-pathogenic endophytic mutualist. The wild-type fungus and the mutant of the species (path 1) were capable to infect the host plant and systemic growth was observed. The host plants infected by path 1 produced no visible effect on any plant and were protected from the wild-type fungus. Freeman and Rodriguez (1993) assumed that a portion of the host defense system is activated by path 1 so that, when exposed to the wild-type fungus, there was no delay in the defense response.

In contrast to the fungi of the Clavicipitaceae the endophytes of woody plants are not intensively investigated. One exception is the family of the Xylariaceae, which attracts attention because of its fruiting bodies and its host specificity. Most xylariaceous species are collected from living or dead angiospermous plants (Rogers 1979); most members of the family are wood inhabitants but representatives are also found in litter, soil, dung, and associated with insects (Whalley 1996). The greatest diversity of xylariaceous fungi can be found in the tropics (Petrini et al. 1995). Whalley (1996) distinguishes three groups: saprophytes, phytopathogens, and endophytes. Endophytic species, e.g. *Hypoxylon* and related genera, can colonize the xylem of host plants (Stone et al. 2000). Other xylem-colonizing endophytes have been identified as species of the Diaporthales (e. g. *Phragmoporthe*, *Amphiporthe*, *Phomopsis*), Hypocreales (*Nectria* spp.), and a few Basidiomycetes (e.g. *Coniophora*; Stone et al. 2000).

Water deficiency and drought in the host plant can cause a change in the lifecycle of endophytic species. Many plant-pathogenic species of the Xylariaceae have typical characteristics of endophytes; they live as "latent invaders". Water stress triggers the switch to the pathogenic phase of the fungus (Whalley 1996). A switch from the latent state to the saprophytic stage is observed in *Hypoxylon fragiforme* when the colonized host tissue dies. The death of the tissue is accompanied by loss of water, which triggers the switch to the saprophytic stage of fungus (Weber and Anke 2006).

Some of the xylariaceous fungi have developed host specificity, e.g. *Hypoxylon fragiforme* is consistently isolated from healthy beech trees in Europe and *H. mammatum* is isolated from aspen trees in the northeast of the United States (Chapela and Boddy 1988; Chapela 1989). The wide distribution of the Xylariaceae, as endophytes and as saprophytes and their ability to produce bioactive compounds (cytochalasins, melleins, xylarin) suggest that these fungi have an important ecological role (Petrini et al. 1995).

Compounds from the bark of beech trees can trigger the eclosion of *Hypoxylon fragiforme* ascospores, which is a precondition for germination of the spores (Chapela et al. 1993). This can partly explain the host specificity of *H. fragiforme*. The eclosion of the spores can be observed 10 min after the addition of the beech extract (Webster and Weber 2004). Two substances have been identified which trigger the release of the spores at micromolar concentrations: the two monolignol glucosides Z-syringin and Z-isoconiferin (Fig. 8.1; Chapela et al. 1991).

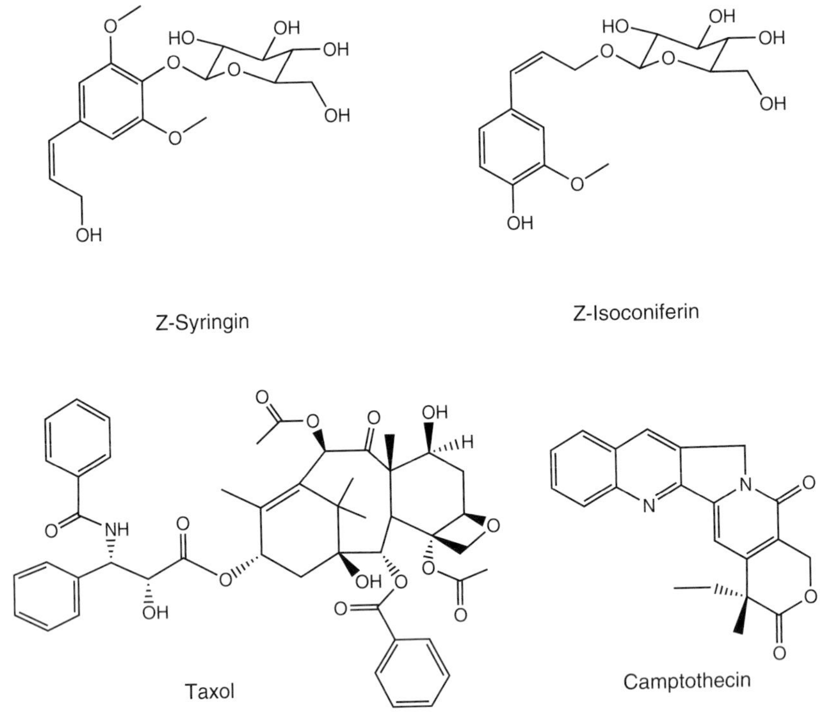

Fig. 8.1. Structures of Z-syringyn, Z-isoconiferin, taxol, and camptothecin

III. Metabolites Isolated from Host Plants and Their Endophytic Fungi

A. Taxol

Endophytes are a rich source for bioactive metabolites and therefore have opened an interesting field of natural product research (Strobel et al. 2004; Ding et al. 2006). A well-known example is the antitumor agent taxol (paclitaxel) which binds to β-tubulin and promotes the stabilization of microtubules by inhibiting depolymerization of microtubules to soluble tubulin (Jennewein and Croteau 2001). Especially, the fast proliferation of tumor cells is inhibited by this mode of action. The original targeted diseases for taxol were ovarian and breast cancer, but now it is also used for the treatment of other human tumors (Strobel et al. 2004). The diterpenoid taxol (Fig. 8.1) was first isolated from the bark of the pacific yew (*Taxus brevifolia*; Wani et al. 1971). All *Taxus* species produce taxol, or related taxane diterpenoids (taxanoids) and over 350 different taxoid structures are already known (Jennewein and Croteau 2001; Strobel 2003). In 1993 Stierle and cowrkers reported that the endophyte *Taxomyces*

andreanae, which was isolated from the inner bark of *T. brevifolia*, produced the antitumor agent taxol. Since then, a variety of endophytic fungi belonging to different categories isolated from *Taxus* species have been reported to be capable of producing taxol and/or taxane derivates in culture (Tan and Zou 2001).

Since pure cultures of endophytic fungi as well as cells of *Taxus* species accumulate taxol, both must have the genes for taxol synthesis in their genome. It seems unlikely that the biosynthesis of an elaborate molecule like taxol developed repeatedly during evolution (Kucht et al. 2004). So it is assumed that a horizontal (lateral) gene transfer occurred (Kucht et al. 2004). Walker and Croteau (2001) isolated the genes specific to taxol biosynthesis and noticed that the cDNA sequence of the taxadiene synthase includes a N-terminal targeting sequence for localization and processing in the plastids. This fact supports the hypothesis that this gene is plant-derived rather than a fungal product (Cassady et al. 2004). Ge et al. (2008) assume that, during the long co-evolution of endophytes and their host plants, endophytes have adapted themselves to their microenvironments by genetic variation, including uptake of some plant DNA into their genomes. This could lead to in the isolation of microbial metabolites originally known as phytochemicals.

The ecologic relevance of taxol and its synthesis can be deduced from its antifungal properties. Oomycetes like *Phytophthora*, *Pythium*, and *Aphanomyces*, which can cause diseases in plants, are extremely sensitive to taxol (Strobel 2003). So it is almost impossible to find *Taxus* species that show any infections caused by any of these Oomycetes (Strobel 2003). Additionally, the production of taxol in a *T. chinensis* cell suspension culture was stimulated by addition of a fungal homogenate (Wang et al. 2001). *Aspergillus niger*, an endophytic fungus isolated from the inner bark of a *T. chinensis* tree, was used as an elicitor to stimulate the taxol production. Taxol yield was increased to about seven times the amount obtained in the non-elicited culture (Wang et al. 2001). This supports the assumption that taxol is part of the plant defense against fungi.

B. Camptothecin

Another important antitumor agent is the pentacyclic quinoline alkaloid camptothecin (Fig. 8.1). Camptothecin, first isolated from *Camptotheca acuminata* (Wall et al. 1966), has antineoplastic properties by inhibition of topoisomerases I. These enzymes relax DNA torsional strain generated during replication, transcription, recombination, repair, and chromosome condensation (Meng et al. 2003). Analogues of camptothecin are used to treat ovarian, colorectal, and small-cell lung cancers, and several second-generation analogues are in clinical trials (Oberlies and Kroll 2004).

The quinoline alkaloid was also isolated from the Indian tree *Nothapodytes foetida* (formely *Mappia foetida*; Govindachari et al. 1974). The compound is available only in low concentrations in the root of *N. foetida*, which requires uprooting of rare and 50- to 75-year-old trees (Puri et al. 2005). In search of alternatives the endophytic fungi of the tree were investigated and a fungus producing camptothecin under culture conditions was identified (Puri et al. 2005). Molecular analysis of the fungus revealed similarity to *Entrophosphora infrequens* and other related taxa, e.g. *Rhizopus oryzae*. This was the first report of a camptothecin-producing microorganism.

C. Ergot Alkaloids

Poisoning by the alkaloids of *Claviceps* species has been an important problem in human history (Schardl et al. 2004). These pathogenic fungi infect grain, and especially rye is very susceptible because of its open-pollinated grain (Schardl et al. 2004). After the infection of the host floret the fungus forms its sclerotium, commonly called an ergot (Schardl et al. 2006). Sclerotia are a rich source of secondary metabolites, e.g. ergot alkaloids like ergotamine (Parker and Scott 2004). During the Middle Ages ergot contamination of rye flour caused mass ergot poisoning (ergotism). The human intoxication acquired by eating cereals infected with ergot sclerotia, usually in the form of bread made from contaminated flour, is called ergotism or St. Anthony's fire (Bennett and Klich 2003). Ergotamine (Fig. 8.2) and its derivates share structural similarities with the adrenergic, dopaminergic, and serotonergic neurotransmitters and therefore have wide-ranging effects on the physiological processes that they mediate (Silberstein and McCrory 2003). Ergot was used in folk medicine as an abortifacient and a drug to accelerate uterine contractions (Bennett and Klich 2003). Today ergotamine and dihydroergotamine are used for the treatment of migraine (Aktories et al. 2005). Other ergot derivatives are used as prolactin inhibitors, in the treatment of Parkinsonism and in cases of cerebrovascular insufficiency (Bennett and Klich 2003).

D. Mycotoxins in *Baccharis* Species

In some cases a direct contribution of the endophytic fungi to the metabolites of the host plant was observed. The shrubs *Baccharis coridifolia* and *B. artemisioides*, which are widespread in Argentinia, cause toxic effects in livestock, particularly during the flowering season (Rizzo et al. 1997). Macrocyclic thrichothecenes, roridins and verrucarins, are the reason for these toxic effects of healthy *B. coridifolia* plants (Busam and Habermehl 1982). These metabolites are mycotoxines (e.g. miotoxin; Fig. 8.2) with cytotoxic properties, formerly known from different fungi, particularly of the genus *Myrothecium*. The isolation of endophytic fungi from *B. coridifolia* resulted in the discovery of a new species, *Ceratopycnidium baccharidicola*. One-third of *C. baccharidicola* from Argentinian plants produced roridin and

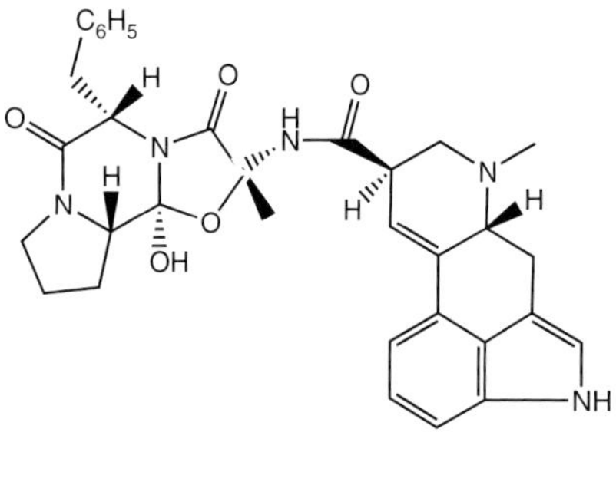

158   Daniela Weber

Fig. 8.2. Structures of ergotamine, miotoxin A, hypericin, and podophyllotoxin

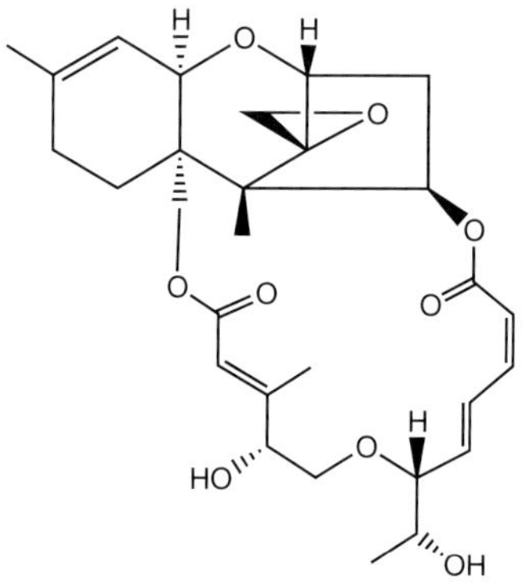

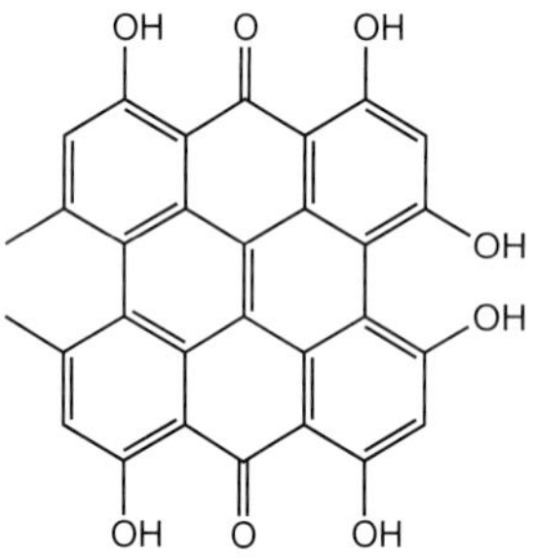

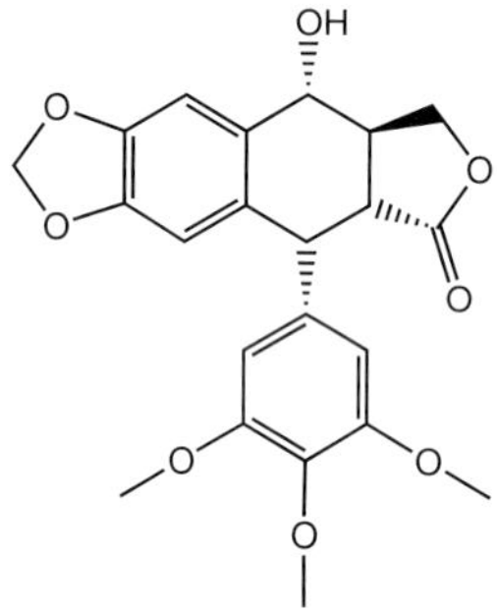

verrucarin toxins in pure culture (Rizzo et al. 1997). It can be assumed that the toxins are produced by the fungi in their host plants, accumulate in the plant tissue, and may protect both the plant and the fungus against herbivores.

E. Hypericin

Hypericum perforatum (St. John's Wort) has been used for centuries in the treatment of burns, bruises, swelling, inflammation, and anxiety, as well as bacterial and viral infections (Hashida et al. 2008). Traditional indications and uses include enhancement of wound healing, anti-inflammatory, and analgesic activities (Miller 1998a). The medicinal plant *H. perforatum* is commonly used for the treatment of mood disorders, since its effectiveness in the therapy for mild to moderate depression was claimed to have smaller side effects than the traditional antidepressant medications (Hashida et al. 2008). Numerous studies have proven the clinical efficacy of

H. perforatum in both human and animal behavioral models of depression (Hashida et al. 2008).

Hypericin (Fig. 8.2) is one of the main constituents of *Hypericum* species and was first isolated from *H. perforatum* (Kusari et al. 2008). The compound is responsible for photosensitivity of patients during therapy with *H. perforatum* preparations (Roth et al. 1994) and has only been available in the plants of *Hypericum* species (Kusari et al. 2008). Recently, an endophytic fungus was isolated from the stems of *H. perforatum*, which produced the naphthodianthrone derivate hypericin and the putative precursor emodin (Kusari et al. 2008). An unequivocally identification of the fungal endophyte was not possible, but molecular analysis revealed that it is related to *Chaetomium globosum* (Kusari et al. 2008).

F. Podophyllotoxin

The lignan podophyllotoxin (Fig. 8.2) is highly valued as a precursor to clinically useful anticancer drugs (Eyberger et al. 2006). The consequence of the destruction of the primary source *Podophyllum emodi* was an investigation of the endophytic fungi of *Podophyllum* species (Eyberger et al. 2006; Puri et al. 2006). Recently, Puri et al. (2006) and Eyberger et al. (2006) described the production of podophyllotoxin by *Trametes hirsuta* isolated from *P. emodi* and *Phialocephala fortinii* from rhizomes of the plant *Podophyllum peltatum*, respectively.

IV. Metabolites from Endophytic Fungi

The review of Tan and Zou (2001) "Endophytes: a rich source of functional metabolites" gives an overview of endophytic metabolites characterized before 2000. In the review "Natural products from plant-associated microorganisms: distribution, structural diversity, bioactivity, and implications of their occurrence" of Gunatilaka (2006) 230 metabolites from over 70 plant-associated microbial strains are discussed. Since these two articles give a broad and comprehensive overview of the metabolites from endophytic fungi, the compounds from endophytic fungi in this article concentrate on publications of the past two years.

The following presents metabolites produced by endophytic fungi from medicinal plants and plants not used in medicine and describes the medicinal plants and their traditional use in ethnopharmacology. Table 8.1 gives an overview of the discussed endophytes, their host plants, and their metabolites.

A. Metabolites from Endophytic Xylariaceous Fungi

1. 7-Amino-4-Methylcoumarin from *Xylaria* sp. of *Ginkgo biloba*

Traditional Chinese medicine uses the *Ginkgo biloba* leaves for brain disorders, circulatory disorders, asthma, and as an antiparasitic agent (Dugoua et al. 2006). *G. biloba* is commonly used for the treatment of dementia, but also for memory improvement, cerebrovascular and aterial insufficiency, tinnitus, vertigo, asthma, and allergies (Bent et al. 2005). The mode of action of this herb is not completely clear, it is believed to improve cerebral and peripheral blood flow through nitric oxid-induced vasodilatation (Bent et al. 2005). In addition, *G. biloba* possess

antioxidant, anti-inflammatory, and antitumor activities (Bent et al. 2005; Chan et al. 2007).

Recently, Liu and coworkers (2008) reported the isolation of an endophytic *Xylaria* sp. from healthy twigs of *G. biloba*. Extracts of the fungal culture fluid exhibited broad antimicrobial activity and bioactivity-guided purification yielded 7-amino-4-methylcoumarin (Fig. 8.3; Liu et al. 2008). Strong antibacterial and antifungal activities of this coumarin derivate were observed against several bacteria and fungi, respectively.

2. Metabolites from *Xylaria* sp. from *Sandoricum koetjape*

Sandoricum koetjape Merr. is a medicinal plant widely distributed in tropical areas of Asia (Ismail et al. 2003). The bark and the whole plant are used in folk medicine as a tonic after childbirth and for the treatment of colic and leucorrhea in Malaysia and Indonesia (Ismail et al. 2003). Phytochemical studies of this plant reported the isolation of triterpenoids and limonoids from the heartwood, bark, and seeds and the limonoids sandrapins A, B, and C from the leaves (Ismail et al. 2004). Anti-inflammatory, ichthyotoxic, and anticarcinogenic effects were described for extracts of *S. koetjape* and some of its constituents, respectively (Ismail et al. 2004; Rasadah et al. 2004).

Tansuwan and coworkers (2007) isolated the endophytic fungus PB-30 from healthy leaves of *S. koetjape* collected from Prachinburi Province in Thailand. The ITS sequence suggests that this endophytic fungus belongs to the genus *Xylaria* (Tansuwan et al. 2007). Two novel benzoquinone metabolites, 2-chloro-5-methoxy-3-methylcyclohexa-2,5-diene-1,4-dione and 7-hydroxy-8-methoxy-3,6-dimethyldibenzofuran-1,4-dione, named xylariaquinone A, together with the known natural products 2-hydroxy-5-methoxy-3-methylcyclohexa-2,5-diene-1,4-dione and 4-hydroxymellein were obtained from extracts of fungal cultures (Fig. 8.3; Tansuwan et al. 2007). The antimalarial activity of the four compounds was tested against *Plasmodium falciparum*. 2-Chloro-5-methoxy-3-methylcyclohexa-2,5-diene-1,4-dione and xylariaquinone A exhibited IC_{50} values of 1.84 µM and 6.68 µM, whereas the two known compounds were inactive (Tansuwan et al. 2007). The cytotoxic activity of the cyclohexa-diene-dione and xylariaquinone A against African green monkey kidney fibroblasts was investigated by using a colorimetric method and the IC_{50} values were respectively determined as 1.35 µM and >184 µM (Tansuwan et al. 2007).

Table 8.1. Metabolites from endophytic fungi

Microbial strain	Host plant; plant part or tissue	Metabolites
Alternaria sp.	*Polygonum senegalense*; fresh healthy leaves	Alternariol 5-*O*-sulfate
		Alternariol 5-*O*-methyl ether-4′-*O*-sulfate
		Alternariol
		Alternariol 5-*O*-methylether
		Altenusin
		2,5-Dimetyl-7-hydroxy-chromone
		Tenuazonic acid
		Altertoxin I
		3′-Hydroxyalternariol 5-*O*-methylether
		Desmethylaltenusin
		Alterlactone
		Alternaric acid
		Alternariol
		Alternariol 5-*O*-methyl ether
		Altenusin
		Talaroflavone
		4′-Epialtenuene
Alternaria sp. (P 0506)	*Vinca minor*; leaf	Altersetin
		Alternariol
		Alternariol monomethylether
		Tenuazonic acid
Alternaria sp. (P 0535)	*Euonymus europaeus*; fruit	Altersetin
		Alternariol
		Alternariol monomethylether
		Tenuazonic acid
Ascochyta sp. (6651)	*Meliotus dentatus*	(4*S*)-(+)-Ascochin
		(*S*,*S*)-(+)-Ascodiketone
		(3*R*,4*R*)-(−)-4-Hydroxymellein
		Ent-α-cyperone
		(3*S*,4*R*)-(−)-Dihydroxy-(6*S*)-undecyl-α-pyranone
Botryosphaeria mamane (PSU-M76)	*Garcinia mangostana*; leaves	Botryomaman
		2,4-Dimethoxy-6-pentylphenol
		(*R*)-(−)-Mellein
		Primin
		Cis-4-hydroxymellein
		Trans-4-hydroxymellein
		4,5-Dihydroxy-2-hexenoic acid
Botryosphaeria parva	*Schinus molle*; small woody twigs	2,7-Dimethoxy-6-(1-acetoxyethyl)-juglon
Chaetomium sp.	*Nerium oleander* L.	Phenolics (e. g. phenolic acids and their derivates)
		Volatile and aliphatic compounds
Chaetomium globosum (IFB-E019)	*Imperata cylindrica*; stem	Chaetoglobosin U
		Chaetoglobosin C
		Chaetoglobosin F
		Chaetoglobosin E
		Penochalasin A
Colletotrichum dematium	*Pteromischum* sp.	Colutellin
Colletotrichum gloeosporioides	*Artemisia mongolica*	Colletotric acid
Colletotrichum gloeosporioides (E01141)	*Schinus molle*; leaves	DW E01141-1
Coniothyrium sp. (7303)	*Artemisia maritima*	Massarilactone E
		Massarilactone F
		Massarilactone G
		Massarilactone acetonide
Diaporthe sp.		

(*continued*)

Table 8.1. (continued)

Microbial strain	Host plant; plant part or tissue	Metabolites
	Camellia sinensis L.; young stems	(+)-Epicytoskyrin (+)-1,1′-Bislunatin
Dothiodeomycete sp. (LRUB20)	*Leea rubra*; stem	2-Hydroxymethyl-3-methylcyclopent-2-enone *Cis*-2-hydroxymethyl-3-methyl-cyclo-pentanone Asterric acid
Emericella variecolor	*Croton oblongifolius*; mature petioles	Shamixanthone 14-Methoxytajixanthone-25-acetate Taijixanthone methanoate Tajixanthone hydrate
Endophytic ascomycete (related to *Phaeosphaeria avenaria*)	Plant samples	Phaeosphaeride A Phaeosphaeride B
Epichloe typhina	*Phleum pretense* L.	Epichlicin
Fusarium sp. (CR377)	*Selaginella pallescens*; stem	CR377
Fusarium sp. (YG-45)	*Maackia chinensis*; twig segment	Fusaristatin A Fusaristatin B
Hormonema dematioides	Pinus sp.; needle base	Hormonemate
Hypoxylon truncatum (IFB-18)	*Artemisia annua*; stem tissue	Daldinone C Daldinone D Altechromone A (4*S*)-5,8-Dihydroxy-4-methoxy-α-tetralone
Leptosphaeria sp. (IV403)	*Artemisia annua*	Leptosphaeric acid
Mitosporic, hyaline fungus E-3	*Prymnopitys andina*; wood pieces	4-(2-Hydroxyethyl)phenol *p*-Hydroxybenzaldehyde Mellein
Nodulisporium sp.	*Junipercus cedre*; twigs	3-Hydroxy-1-(2,6-dihydroxyphenyl)-butan-1-one 1-(2-Hydroxy-6-methoxyphenyl)butan-1-one 2,3-Dihydro-5-methoxy-2-methylchomen-4-one Nodulisporin A Nodulisporin B Nodulisporin C (4*E*,6*E*)-2,4,6-Trimethyloacta-4,6-dien-3-one 1-(2,6-Dihydroxyphenyl)butan-1-one 5-Hydroxy-2-methyl-4*H*-chromen-4-one 2,3-Dihydro-5-hydroxy-2-methylchomen-4-one 8-Methoxynaphthalen-1-ol 1,8-Dimethoxynaphthalene Daldinol Helicascolide A (1*S*,4*S*,4a*S*,6*R*,6a*R*)-Decahydro-1,4-dimethyl-6-(prop-1-en-2-yl)naphthalene-1-ol Ergosterol 5α,8α-Epidioxyergosterol
Penicillium sp. (IFB-E22)	*Quercus variabilis*; stem	Penicidone A Penicidone B Penicidone C Lumichrome Physicion Emodin-1,6-dimethylether
Penicillium sp. (GQ-7)	*Aegiceras corniculatum* L.; inner bark	Penicillenol A_1 Penicillenol A_2 Penicillenol B_1 Penicillenol B_2 Penicillenol C_1 Penicillenol C_2 Citrinin Phenol A acid Phenol A Dihydrocitrinin

(*continued*)

Table 8.1. (continued)

Microbial strain	Host plant; plant part or tissue	Metabolites
Penicillium chrysogenum (No. 005)	*Cistanche deserticola*; root	Chrysogenamide A Circumdatin G 2-[(2-Hydroxypropionyl) amino] benzamide 2′,3′-Dihydrosorbicillin (9Z,12Z)-2,3-Dihydroxypropyl octadeca-9, 12-dienoate
Penicillium janczewskii K.M.Zalesky	*Prymnopitys andina*; wood pieces	Peniprequinolone Gliovictin
Penicillium janczewskii K.M.Zalesky	*Prymnopitys andina*	Pseurotin A Cycloaspeptide A
Penicillium paxilli (PSU-A71)	*Garcinia atroviridis*; leaves	Penicillone Paxilline Pyrenocine A Pyrenocine B
Phoma medicaginis	*Medicago sativa*, *M. lupulina*	Brefeldin A
Phomopsis sp.	*Camptotheca acuminata*	8-*O*-Acetylmultiplolide A 8-*O*-Acetyl-5,6-dihydro-5,6-epoxymultiplolide A 5,6-Dihydro-5,6-epoxymultiplolide A 3,4-Deoxy-3,4-didehydromultiplolide A Multiplolide A (4E)-6,7,9-Trihydroxydec-4-enoic acid (4E)-6,7,9-Trihydroxydec-4-enoate
Phomopsis spp.	*Erythrina crista-galli*; twigs, leaves	Clavatol Hydroxymellein Mellein Mevalonolactone Mevinic acid Nectriapyrone Scytalone Tyrosol Phomol DW E02018-5 DW E02018-9 Phomopyronol 3-Phenylpropane-1,2-diol 4-(2,3-Dihydroxypropoxy)benzoic acid 2-(Hydroxymethyl)-3-propylphenol 2-(Hydroxymethyl)-3-(1-hydroxy-propyl)-phenol Compound 6 8-(Hydroxymethyl)-2,2-dimethyl-7-propyl-chroman-3-ol (4E,10E)-Trideca-4,10,12-triene-2,8-diol
Phomopsis sp.	*Eupatorium arnotianum*; leaves, stems	Phomopsidone Mellein Nectriapyrone
Phomopsis sp.	*Phytolacca dioica*; twig	Cytosporone D
Phomopsis sp. (BCC 1323)	*Tectonia grandis* L.; leaf	Phomaxanthone A Phomaxanthone B
Phomopsis sp. (YM3111483)	*Azadirachate indica*; fresh stems	Multiplolide A 8α-Acetoxy-5α-hydroxy-7-oxodecan-9-olide 7α,8α-Dihydroxy-3,5-decadien-10-olide 7α-Acetoxymultiplolide A 8α-Acetoxymultiplolide A
Phomopsis cassiae	*Cassia spectabilis*; leaves	Two diastereoisomeric 3,9,12-trihydroxycalamenenes 3,12-Dihydroxycalamenene 3,12-Dihydroxycadalene 3,11,12-Trihydroxycadalene

(*continued*)

Table 8.1. (continued)

Microbial strain	Host plant; plant part or tissue	Metabolites
Phomopsis longicolla	Dicerandra frutescens; stem segment	Dicerandrol A
		Dicerandrol B
		Dicerandrol C
Phyllosticta spinarum	Platycladus orientalis; healthy foliage	(+)-(5S,10S)-4′-Hydroxymethylcyclozonarone
		3-Ketotauranin
		3α-Hydroxytauranin
		12-Hydroxytauranin
		Phyllospinarone
		Tauranin
Pestalotiopsis sp.	Pinus taeda	Pestalotiopsolide A
		Taedolidol
		6-Epitaedolidol
Pestalotiopsis foedan	Unidentified tree	Pestafolide A
		Pestaphthalide A
		Pestaphthalide B
Pestalotiopsis microspora	Terminalia morobensis; small tree stem	Isopectacin
Pestalotiopsis theae	Unidentified tree	
		Pestalotheol A
		Pestalotheol B
		Pestalotheol C
		Pestalotheol D
Xylaria sp.	Ginkgo biloba; healthy twigs	7-Amino-4-methyl-coumarin
Xylaria sp.	Sandoricum koetjape; healthy leaves	
		2-Chloro-5-methoxy-3-methylcyclohexa-2, 5-diene-1,4-dione
		Xylariaquinone A
		2-Hydroxy-5-methoxy-3-methylcyclohexa-2, 5-diene-1,4-dione
		4-Hydroxymellein
Xylaria sp. (no. 2508)	Seed of angiosperm tree	Xyloketal J
		Xyloester A
		Xyloallenolide B
		Dihydrobenzofuran derivate
Xylariaceous endophyte (YUA-026)	Twigs and petiols of plant samples	Eremoxylarin A
		Eremoxylarin B

3. Sesquiterpenoids from a Xylariaceous Fungus

Shiono and Murayama (2005) isolated the xylariaceous endophyte YUA-026 from twigs and petiols of plant samples collected from Mount Takadate in Japan. Two new eremophilane-type sesquiterpenoids, eremoxylarin A and B, were obtained from rice culture of the strain YUA-026 (Fig. 8.3; Shiono and Muranyama 2005). The purification of the metabolites was guided by their antimicrobial activity against *Pseudomonas aeruginosa* and analytical methods (Shiono and Muranyama 2005). Using agar dilution methods, the minimum inhibition concentration (MIC) of eremoxylarins A and B were evaluated against the gram-positive bacterium *Staphylococcus aureus*

(12.5, 25.0 µg/ml), the gram-negative bacterium *Pseudomonas aeruginosa* (6.25, 12.5 µg/ml), and the fungi *Candida albicans* and *Aspergillus clavatus* (Shiono and Muranyama 2005). The MICs of the two eremoxylarins against the two tested fungi were over 100 µg/ml (Shiono and Muranyama 2005).

4. Metabolites from *Xylaria* sp.

The ascomycete *Xylaria* sp. (2508) was isolated from seed of an angiosperm tree in Mai Po mangrove, Hong Kong (Xu et al. 2008). Fermentations of fungus yielded three new metabolites, named xyloketal J, xyloester A, and xyloallenolide B,

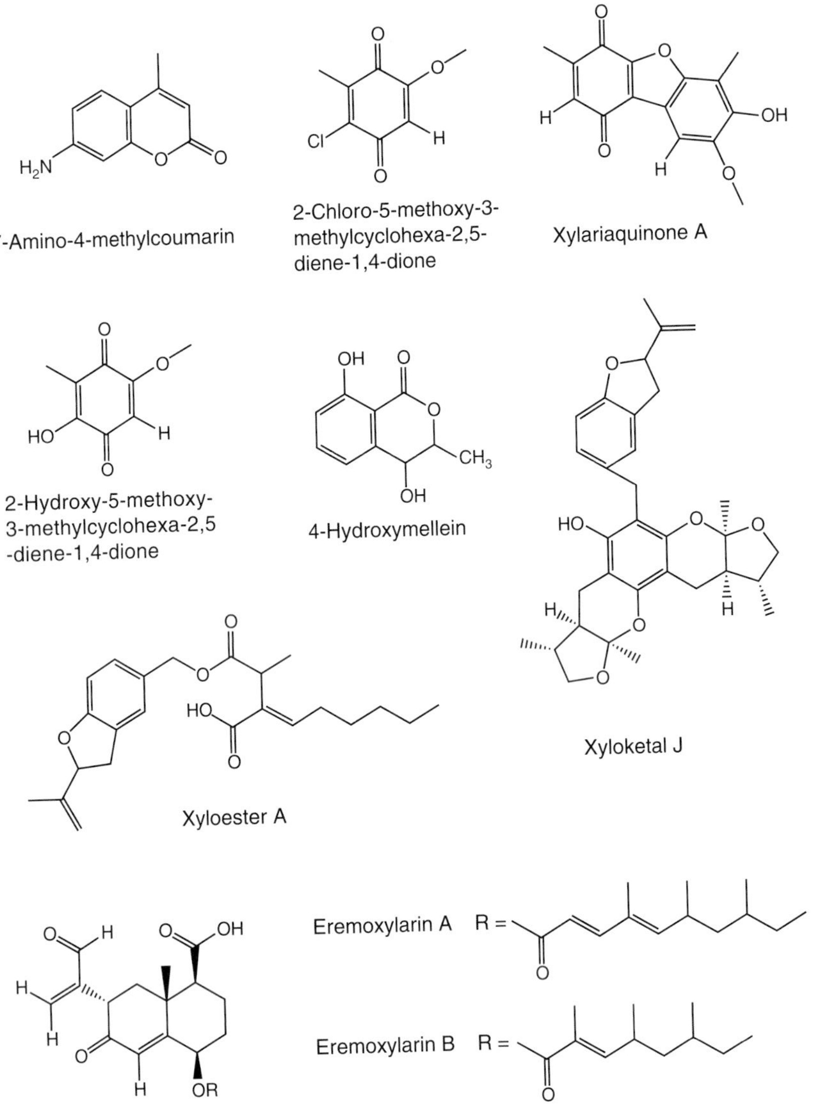

7-Amino-4-methylcoumarin

2-Chloro-5-methoxy-3-
methylcyclohexa-2,5-
diene-1,4-dione

Xylariaquinone A

2-Hydroxy-5-methoxy-
3-methylcyclohexa-2,5
-diene-1,4-dione

4-Hydroxymellein

Xyloketal J

Xyloester A

Eremoxylarin A   R =

Eremoxylarin B   R =

Fig. 8.3. Metabolites from endophytic xylariaceous fungi

along with the known substituted dihydrobenzofuran (Figs. 8.3, 8.4; Xu et al. 2008). The structures
of all the compounds possess a substituted dihydrobenzofuran unit (Xu et al. 2008).

B. Metabolites from Endophytic *Phomopsis* or *Diaporthe* Species

1. Lactones from *Phomopsis* sp. from *Azadirachata indica*

Neem (*Azadirachata indica*) is used as a traditional medicinal plant in India (Biswas et al. 2002). Each part of the
tree is used as traditional medicine, e.g. the leaves, twigs,

flowers, and fruits are used for the elimination of intestinal
worms, and the bark is used as an analgesic and curative of
fever (Biwas et al. 2002). Neem preparations are used economically as insecticides, pesticides, and agrochemicals
(Brahmachari 2004). Neem seed extracts provide a potential control of *Anopheles* larvae that could be complementary to other eradication methods (Gianotti et al. 2008).

Wu et al. (2008) isolated the endophytic fungus
Phomopsis sp. YM311483 from fresh stems of an
apparently healthy *Azadirachata indica* collected
in Yunnana Province, China. Four new 10-
membered lactones and the known lactone,
multiplolide A (Fig. 8.4), were isolated from the
culture broth of this *Phomopsis* strain. The new

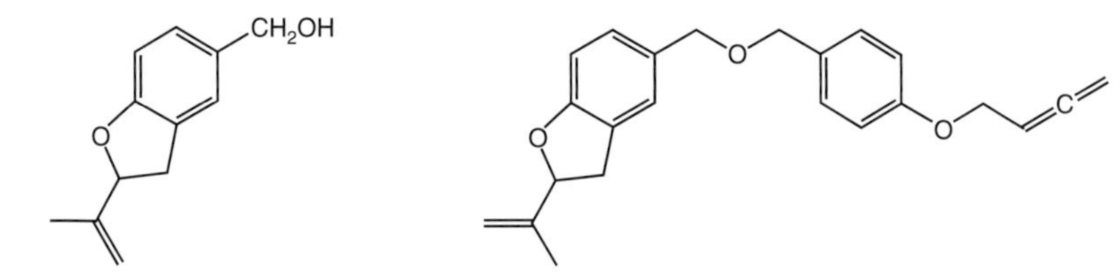

Dihydrobenzofuran derivate

Xyloallenolide B

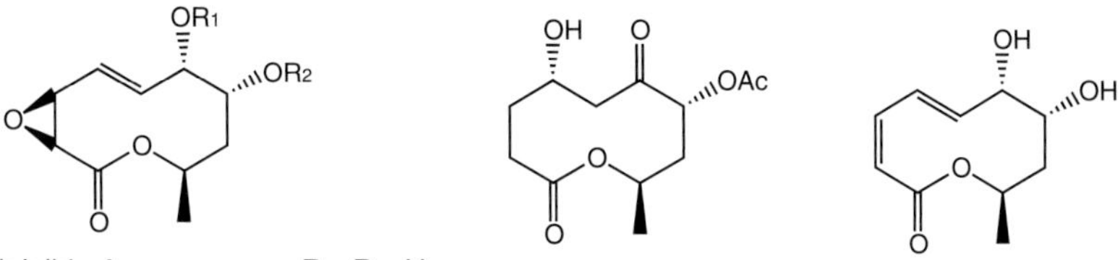

Multiplolide A $R_1=R_2=H$
8α-Acetoxymultiplolide A $R_1=H, R_2=Ac$
7α-Acetoxymultiplolide A $R_1=Ac, R_2=H$

8α-Acetoxy-5α-hydroxy-
7-oxodecan-9-olide

7α,8α-dihydroxy-3,5-
decadien-10-olide

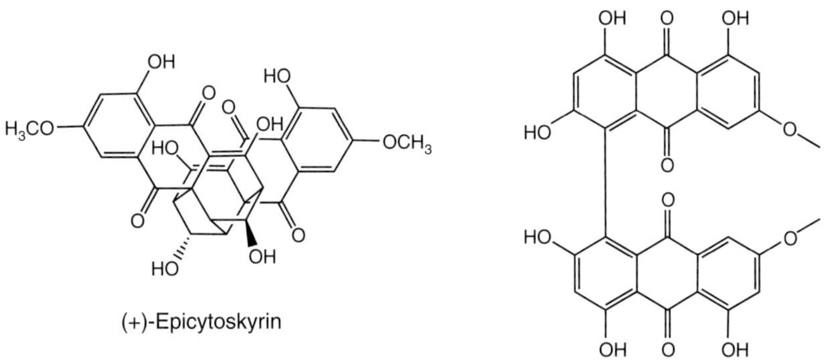

(+)-Epicytoskyrin

(+)-1,1′-Bislunatin

Fig. 8.4. Metabolites from endophytic xylariaceous fungi and from endophytic *Phomopsis* and *Diaporthe* species

compounds were determined as 8α-acetoxy-5α-hydroxy-7-oxodecan-9-olide, 7α,8α-dihydroxy-3,5-decadien-10-olide, 7α-acetoxymultiplolide A, and 8α-acetoxymultiplolide A (Wu et al. 2008). Antifungal activity against seven plant pathogens was examined using the paper disk diffusion method. All five metabolites showed weak antifungal activity, whereas 8α-acetoxymultiplolide A was the most potent, with MIC values of 31.25–500.0 µg/ml (Wu et al. 2008).

2. Metabolites from *Diaporthe* sp. from *Camellia sinensis* L.

Green, black, and oolong teas are prepared from leaves of *Camellia sinensis* L., a member of the family Theaceae

cultivated widely in China, India, Japan, and Indonesia (Brown 1999, Luper 1999). It is the production process which makes the difference of the types of tea (Brown 1999). Consumption of tea (*C. sinensis*) is suggested to prevent cancer, heart disease, and other diseases (Lambert and Yang 2003). Yang et al. (2001) describe tea as one of the few chemoprotective agents known to have protective effects at different stages of the carcinogenic process. It is assumed that tea constituents (like polyphenolic components) may inhibit this process by modulating signal transduction pathways leading to the inhibition of cell proliferation and transformation and enhancement of apoptosis (Yang et al. 2001). In addition, hepatoprotective activity against a variety of toxic compounds (e.g. alcohol) and detoxification activity by enhancement of glucuronidation in the liver are described for tea (Luper 1999). The antimicrobial activity against a variety of bacterial species (*Staphylococcus*, *Salmonella*, *Shigella*, *Vibrio*, and others; Hamilton-Miller 1995) may explain the observed reduction of incidence and severity of caries by regular tea drinking in a limited number of clinical trails in man (Hamilton-Miller 2001).

Agusta and coworkers (2005) isolated 35 filamentous endophytic fungi from young stems of *C. sinensis* collected in West Java, Indonesia. These fungi were classified into six species by morphological considerations and RAPD analysis (Agusta et al. 2005). One isolate, a *Diaporthe* sp., was able to transform catechins with a 2*R*-phenyl substitution into the corresponding 3,4-*cis*-dihydroxyflavan derivatives (Agusta et al. 2006). A second *Diaporthe* sp., closely related to *Diaporthe phaseolorum* srain sw-93-13, produced two bisanthraquinones named (+)-epicytoskyrin and (+)-1,1'-bislunatin (Fig. 8.4, Agusta et al. 2006). (+)-Epicytoskyrin is an epimer of cytoskyrin A, which was isolated from an endophytic *Cytospora* sp. isolated from *Conocarpus erecta* (Brady et al. 2000b). (+)-1,1'-Bislunatin is a homodimeric compound of lunatin joined together at C-1 and C-1' and the relative structure of bislunatin has been reported from *Verticillium lecanii* (Agusta et al. 2006). (+)-Epicytoskyrin and (+)-1,1'-bislunatin showed a moderate cytotoxic activity against KB cells, with an IC$_{50}$ value of 0.5 µg/ml and 3.5 µg/ml, respectively (Agusta et al. 2006).

3. Metabolites from *Phomopsis* sp. from *Camptotheca acuminata*

The antitumor agent camptothecin was originally isolated from the medicinal plant *Camptotheca acuminata*. Lin et al. (2007) isolated 174 endophytic fungi from this medicinal plant. From fermentations of one strain, which was identified by morphologic and genetic properties as *Phomopsis* sp., several new compounds were purified, including the 10-membered macrolides 8-*O*-acetylmultiplolide A, 8-*O*-acetyl-5,6-dihydro-5,6-epoxymultiplolide A, 5,6-dihydro-5,6-epoxymultiplolide A, and 3,4-deoxy-3,4-didehydromultiplolide A, and their known parent compound multiplolide A (Figs. 8.4, 8.5), as well as the ansaturated fatty acid (4*E*)-6,7,9-trihydroxydec-4-enoic acid and its methyl ester (Fig. 8.5; Tan et al. 2007). The 10-membered macrolides and multiplolide A exhihibited no evident antifungal activities against *Candida albicans*.

4. Sesquiterpenoids from *Phomopsis cassiae* from *Cassia spectabilis*

Several species of *Cassia* are widely used in traditional medicine for their antimicrobial, laxative, antiulcerogenic, analgesic, and anti-inflammatory properties (Viegas et al. 2008). The endophytic fungus *Phomopsis cassiae* was isolated from leaves of the Leguminosae *Cassia spectabilis* collected in Brazil (Silva et al. 2006). Bioassay-guided fractionation of crude extracts of the fungal strain resulted in the isolation of the two diastereoisomeric 3,9,12-trihydroxycalamenenes, 3,12-dihydroxycalamenene, 3,12-dihydroxycadalene, and 3,11,12-trihydroxycadalene (Fig. 8.5; Silva et al. 2006). Of these new cadinane sesquiterpenoids, 3,11,12-trihydroxycadalene, was the most active against the phytopathogenic fungi *Cladosporium sphaerospermum* and *C. cladosporioides* (Silva et al. 2006). 3,12-Dihydroxycalamenene, 3,12-dihydroxycadalene, and 3,11,12-trihydroxycadalene exhibited cytotoxic activity against the human cervical tumor cell line HeLa with IC$_{50}$ values of 100, 20, and 110 µmol/l, respectively (Silva et al. 2006).

5. Metabolites from *Phomopsis* spp. from *Erythrina crista-galli*

Erythrina crista-galli (Fabaceae) is used in Argentinean ethnopharmacology as anti-inflammatory medication, narcotic, desinfectant, and for the treatment of wounds. Anti-inflammatory (Miño et al. 2002) and antibacterial (Mitscher et al. 1988) activities have been described for *E. crista-galli*. This Argentinean medicinal plant can be found in the tropical and subtropical regions of America and is commonly used as an ornamental plant. In Argentina the wood is used in infusions or decoctions as astringent, narcotic, and sedative (Toursarkissian 1980).

The majority of the fungi isolated so far from different collections of *E. crista-galli* belong to the genus *Phomopsis* (Weber et al. 2005). This genus comprises a multitude of species widely distributed as pathogens, endophytes or even symbionts of plants (Uecker 1988). The metabolites of 12 different *Phomopsis* isolates were investigated and from four strains metabolites were isolated. Besides the known compounds clavatol, hydroxymellein, mellein, mevalonolactone, mevinic acid, nectriapyrone, scytalone, and

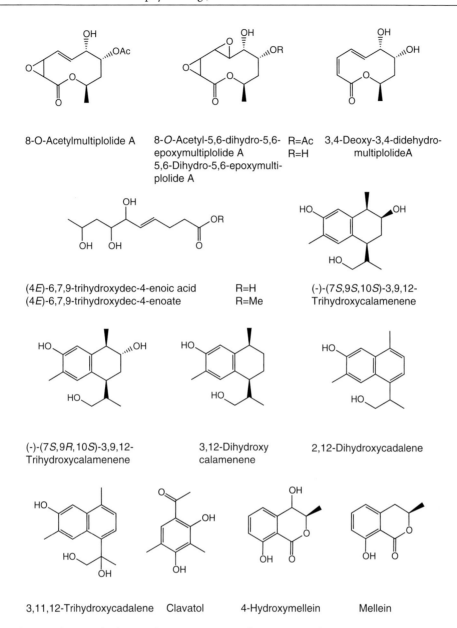

Fig. 8.5. Metabolites from endophytic *Phomopsis* species from *Camptotheca acuminata*, *Cassia spectabilis*, and *Erythrina crista-galli*

tyrosol (Figs. 8.5, 8.6), the new compounds pho-mol, two derivates of phomol, DW E02018-5 and DW E02018-9, phomopyronol, 3-phenylpropane-1,2-diol, 4-(2,3-dihydroxypropoxy)benzoic acid, 2-(hydroxymethyl)-3-propylphenol, 2-(hydroxy-methyl)-3-(1-hydroxypropyl)phenol, compound 6, 8-(hydroxymethyl)-2,2-dimethyl-7-propylchro-man-3-ol, and (4E,10E)-trideca-4,10,12-triene-2,8-diol were identified (Figs. 8.6, 8.7; Weber et al. 2004, 2005; Weber 2006).

Mellein was described from fermentations of *Aspergillus melleus* (Nishikawa 1933), *Fusarium larvarum* (Grove and Pople 1979), *Cercospora tai-wanensis* (Camarda et al. 1976), and many other fungi. The compound has phytotoxic, antibacter-ial, and antifungal activities (Takeuchi et al. 1992; Wenke 1993). Mellein and nectriapyrone were isolated by Claydon et al. (1985) from *Phomopsis oblonga*, commonly found on the bark of trees of the genus *Ulmus*. Trees infected by *P. oblonga* are

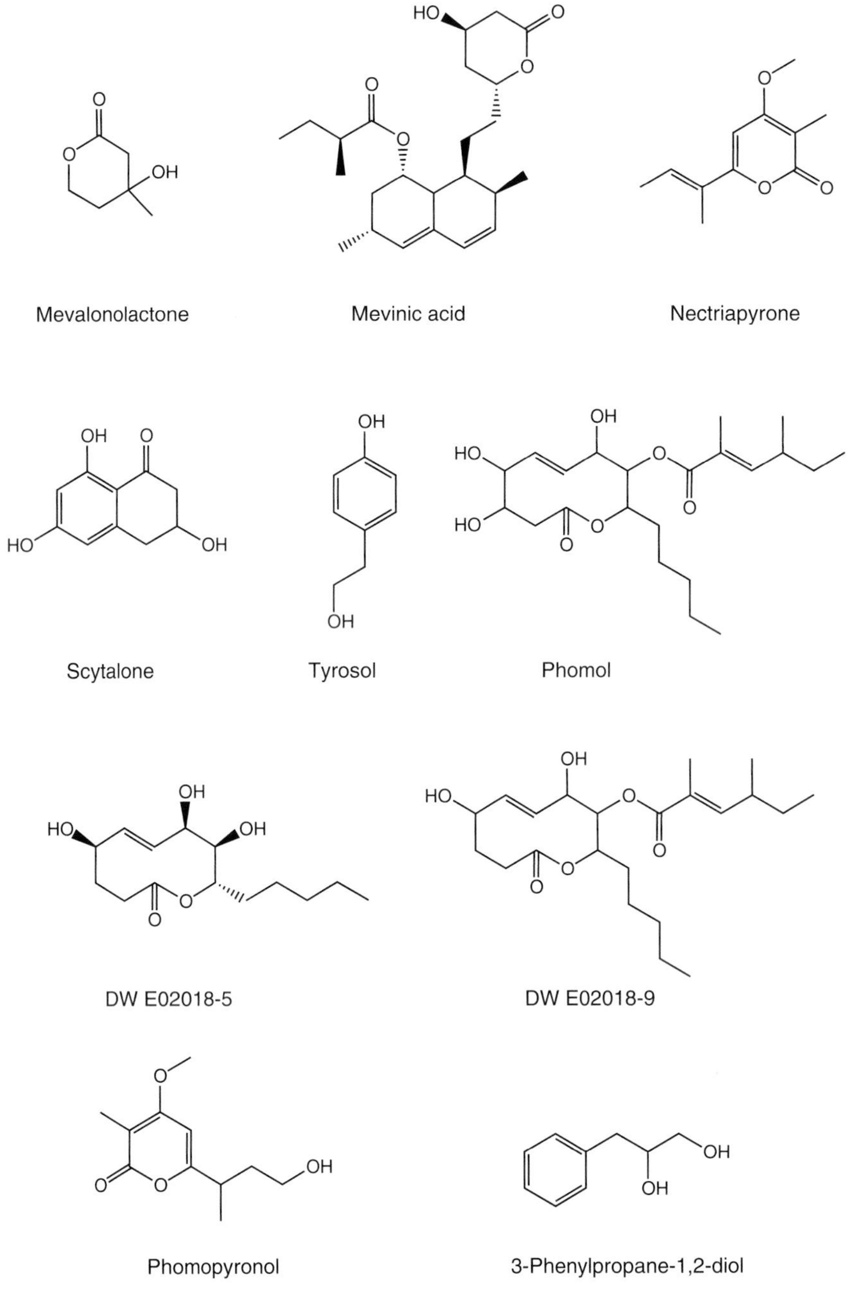

Fig. 8.6. Metabolites from endophytic *Phomopsis* species from *Erythrina crista-galli*

protected from the attack of insects of the genus *Scolytus* (bark beetle; Webber 1981).

Moderate or no antimicrobial, nematicidal, or cytotoxic activities were observed for the new compounds isolated from *Phomopsis* spp. (Weber et al. 2005; Weber 2006). In the mouse ear assay however, two compounds, mevinic acid and phomol, showed significant activities (Weber et al. 2004, 2005). Mevalonic acid is an intermediate in cholesterol biosynthesis. Mevinic acid is an inhibitor of HMG-CoA reductase, the key enzyme of cholesterol biosynthesis (Endo and Hasumi 1997). Derivatives are used as cholesterol-lowering drugs. Phomol was the only fungal compound, which was detected in small amounts in plant extracts of *E. crista-galli* by HPLC-MS analysis and thus can contribute to the anti-inflammatory activity of the plant (Weber 2006).

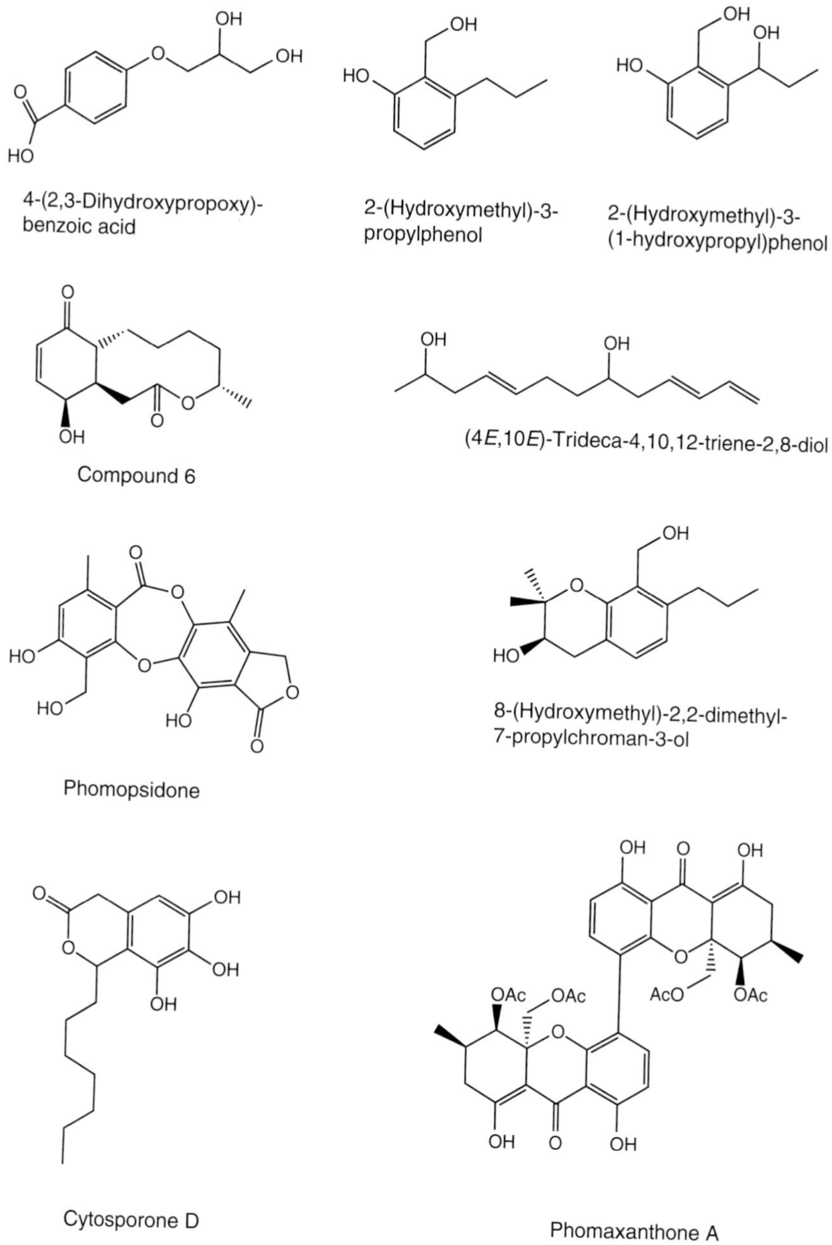

4-(2,3-Dihydroxypropoxy)-
benzoic acid

2-(Hydroxymethyl)-3-
propylphenol

2-(Hydroxymethyl)-3-
(1-hydroxypropyl)phenol

Compound 6

(4*E*,10*E*)-Trideca-4,10,12-triene-2,8-diol

Phomopsidone

8-(Hydroxymethyl)-2,2-dimethyl-
7-propylchroman-3-ol

Cytosporone D

Phomaxanthone A

Fig. 8.7. Metabolites from endophytic *Phomopsis* species from *Erythrina crista-galli, Eupatorium arnottianum, Phyto-lacca dioica,* and *Tectona grandis* L.

6. Metabolites from *Phomopsis* sp. from *Eupatorium arnottianum*

Eupatorium arnottianum occurs in the northeast and center of Argentina and in the south of Bolivia and is used for the treatment of gastric pains (Iharlegui and Hurell 1992), asthma, bronchitis, and colds (Girault 1987; Clavin et. al. 1999). Infusions of *E. arnottianum* showed analgesic (Clavin et. al. 2000a), antiviral (against Herpes simplex type 1;

Clavin et. al. 2000b), and antimicrobial activities (Penna et. al. 1997).

An endophytic *Phomopsis* was isolated from plant material collected in Argentina (Meister et al. 2007). Besides mellein and nectriapyrone (Figs. 8.5, 8.6), cultures yielded phomopsidone, a novel depsidone derivate (Fig. 8.7; Meister et al., 2007). The new metabolite showed no antimicrobial, cytotoxic, nematicidal, or phytotoxic activities (Meister et al. 2007).

7. Cytosporone D from *Phomopsis* sp. of *Phytolacca dioica*

Phytolacca dioica is used in ethnopharmacology as a vulnerary, vermifuge, and antipyretic medicine (Quiroga et al. 2001). Aqueous and alcoholic extracts from *P. dioica* showed antimicrobial activity against gram-positive and gram-negative microorganisms (Quiroga et al. 2001). *Phytolacca* species are used as antirheumatics and have molluscicidal activity (Soliman et al. 2001).

The known metabolite cytosporone D was isolated from an endophytic *Phomopsis* sp., isolated from a twig of *P. dioica* (Fig. 8.7; Weber 2006). The antibacterial octaketide was already described from *Cytospora* sp. and *Diaporthe* sp., which were isolated as endophytes from *Conocarpus erecta* and *Forsteronia spicata* (Brady et al. 2000a).

8. Xanthone Dimers from *Phomopsis* sp. from *Tectona grandis* L.

The endophyte *Phomopsis* sp. BCC 1323 was isolated from a teak leaf of *Tectona grandis* L. collected in Northern Thailand. (Isaka et al. 2001). The search for bioactive and diverse metabolites resulted in the isolation and identification of two new xanthone dimers, phomaxanthones A and B (Figs. 8.7, 8.8; Isaka et al. 2001). The two compounds exhibited activity against *Plasmodium falciparum* (K1, multidrug-resistant strain) and against *Mycobacterium tuberculosis* (H37Ra strain) with IC_{50} values of 0.11–0.33 µg/ml and 0.50–6.25 µg/ml, respectively (Isaka et al. 2001). The deacetylated derivate of phomaxynthone A was inactive in all assays.

9. Dicerandrols from *Phomopsis longicolla* from the Mint *Dicerandra frutescens*

Dicerandra frutescens is a rare mint plant that showed to be free of injury from insects (Wagenaar and Clardy 2001). A fungus identified as *Phomopsis longicolla* by ITS sequencing was isolated from a stem segment of *D. frutescens* (Wagenaar and Clardy 2001). Bioassay-guided fractionation yielded three related antibiotics, named dicerandrols A, B, and C (Fig. 8.8). These new natural products are structurally related to the

ergochromes and secalonic acids and exhibited antibacterial activity against *Staphylococcus aureus* and *Bacillus subtilis* as well as moderate activity against human cancer cell lines (A549 and HCT-116; Wagenaar and Clardy 2001).

C. Metabolites from Endophytic *Penicillium* Species

1. Metabolites from *Penicillium* sp. from *Aegiceras corniculatum* L.

Aegiceras corniculatum L. is used as a traditional medicine for treatment of inflammation and liver injury (Roome et al. 2008). Furthermore Dahdouh-Guebas and coworkers (2006) reported that the bark of *A. corniculatum* is converted into a paste and used as a fish poison.

Lin et al. (2008a) isolated the endophytic *Penicillium* sp. GQ-7 from the inner bark of *A. corniculatum* collected in China. They were able to purify six new tetramic acids derivatives, along with the mycotoxin citrinin, phenol A acid, phenol A, and dihydrocitrinin from ethyl acetate extracts of *Penicillium* sp. (Fig. 8.9). The six new compounds were named penicillenols A_1, A_2, B_1, B_2, C_1, and C_2 (Fig. 8.8; Lin et al. 2008a). The cytotoxic activity of the penicillenols was tested against A-549, BEL-7402, P388, and HL-60 cells by the MTT method. Penicillenols C_1 and C_2 showed no cytotoxicity, whereas the other four compounds exhibited cytotoxic activity with IC_{50} values ranging from 0.76 µM to 16.26 µM against HL-60 cells (Lin et al. 2008a).

2. Metabolites from *Penicillium chrysogenum* from *Cistanche deserticola*

Species of the genus *Cistanche* are used in traditional medicine and are also known as "Ginseng of the desert". The *Cistanche* herbs are used as a superior tonic for the treatment of kidney deficiency, impotence, female infertility, morbid leucorrhea, profuse metrorrhagia, and senile constipation (Jiang and Tu 2009). *Cistanche deserticola* is a holoparasitic plant and a valuable traditional Chinese medicine (Dong et al. 2007).

Penicillium chrysogenum No. 005 was isolated from the root of *Cistanche deserticola* collected from Inner Mangolia in northwest China (Lin et al. 2008b). A new member of the marcfortine

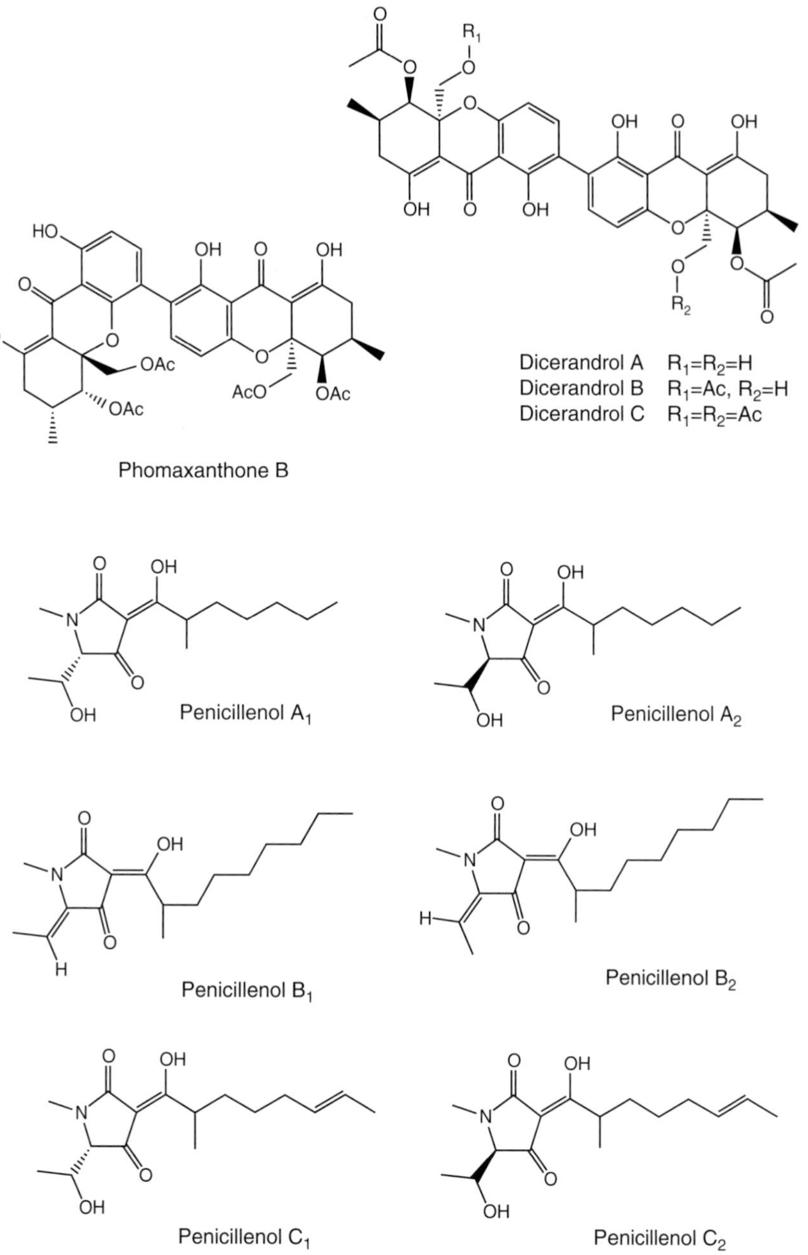

Dicerandrol A R₁=R₂=H
Dicerandrol B R₁=Ac, R₂=H
Dicerandrol C R₁=R₂=Ac

Phomaxanthone B

Penicillenol A₁

Penicillenol A₂

Penicillenol B₁

Penicillenol B₂

Penicillenol C₁

Penicillenol C₂

Fig. 8.8. Metabolites from endophytic *Phomopsis* species from *Tectona grandis* L. and *Dicerandra frutescens* and from *Penicillium* sp. from *Aegiceras corniculatum* L.

alkaloids, named chrysogenamide A (Fig. 8.9), was isolated from the culture broth of *P. chrysogennum* 005 (Lin et al. 2008b). In addition, four known compounds were isolated and identified as cir-cumdatin G, 2-[(2-hydroxypropionyl)amino]ben-zamide, 2′,3′-dihydrosorbicillin, and (9Z,12Z)-2,3-dihydroxypropyl octadeca-9,12-dienoate by com-parision of their spectroscopic data with literature data (Fig. 8.9; Lin et al. 2008b). Chrysogenamide A

showed a neurocyte-protecting effect against oxi-dative stress-induced cell death in SH-SY5Y cells (Lin et al. 2008b).

3. Metabolites from *Penicillium* spp. from *Prumnopitys andina*

Schmeda-Hirschmann and coworkers (2005) isolated two endophytic fungi from wood pieces

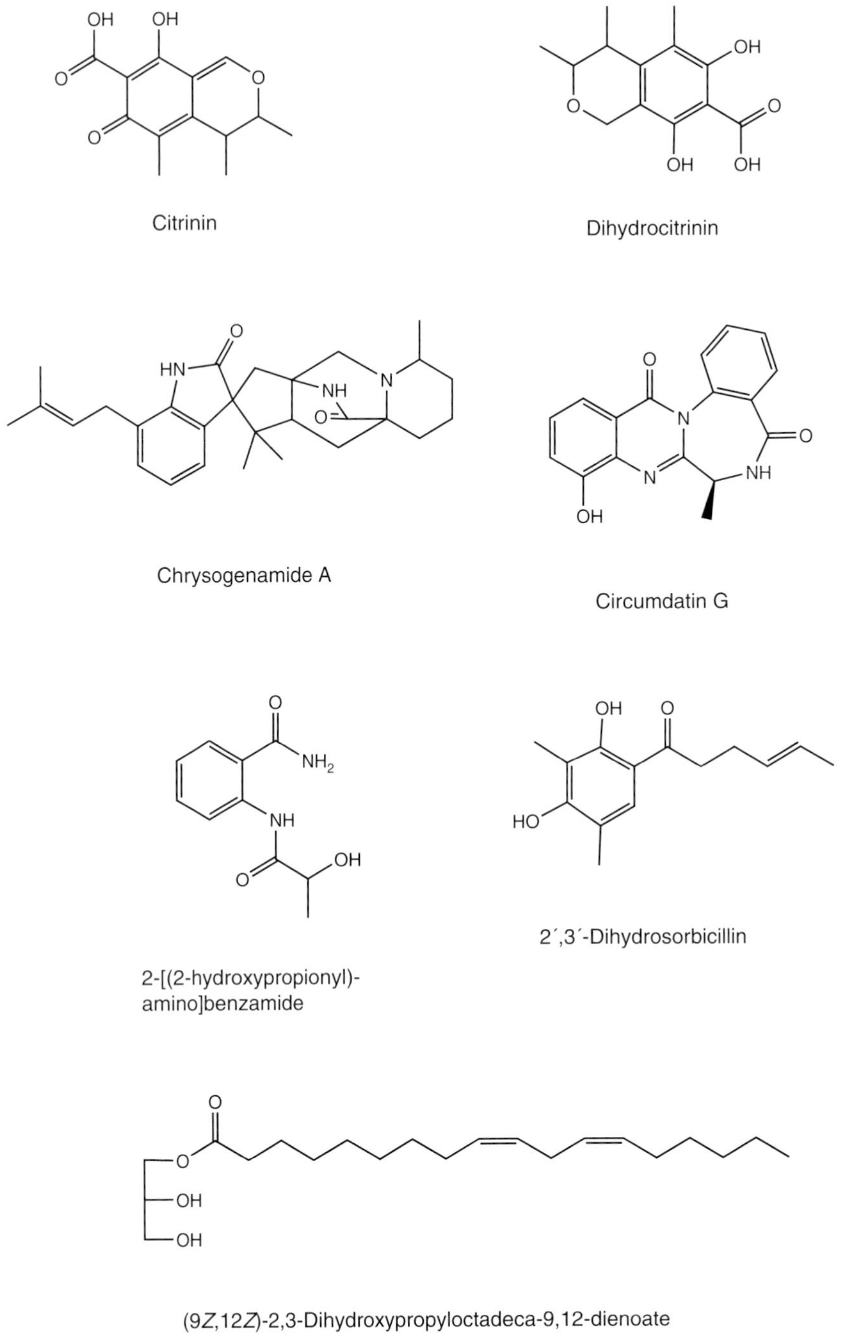

Citrinin

Dihydrocitrinin

Chrysogenamide A

Circumdatin G

2-[(2-hydroxypropionyl)-
amino]benzamide

2´,3´-Dihydrosorbicillin

(9Z,12Z)-2,3-Dihydroxypropyloctadeca-9,12-dienoate

Fig. 8.9. Metabolites from endophytic *Penicillium* species from *Aegiceras corniculatum* L. and *Cistanche deserticola*

of *Prumnopitys andina* (Lleuque). The mitosporic and hyaline fungus E-3, which could not be identified, yielded 4-(2-hydroxyethyl)phenol, *p*-hydroxybenzaldehyde, and the isochromanone mellein (Fig. 8.10; Schmeda-Hirschmann et al. 2005). Peniprequinolone and gliovictin were obtained from the second fungus (Fig. 8.10), which was identified

as *Penicillium janczewskii* K.M. Zalessky. The metabolites from the fungus E-3 exhibited weak activity against several gram-negative and -positive bacteria (Schmeda-Hirschmann et al. 2005). The known compounds, peniprequinolone and gliovictin, showed cytotoxic activity against AGS cells (IC_{50} = 89 µM, 475 µM) and fibroblasts

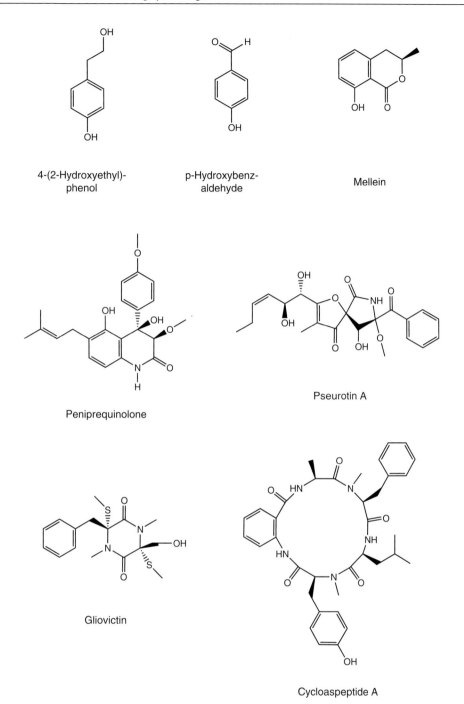

4-(2-Hydroxyethyl)-
phenol

p-Hydroxybenz-
aldehyde

Mellein

Peniprequinolone

Pseurotin A

Gliovictin

Cycloaspeptide A

Fig. 8.10. Metabolites from endophytic *Penicillium* species from *Prumnopitys andina*

(IC$_{50}$ = 116 µM, 681 µM) and antibacterial activity against *Bacillus brevis* and *B. subtilis* in the agar diffusion assay (Schmeda-Hirschmann et al. 2005). Recently, Schmeda-Hirschmann and coworkers (2008) reported the purification of pseurotin A and cycloaspeptide A (Fig. 8.10) from cultures of *P. janczewskii* K.M. Zalessky isolated from

P. andina. The two known compounds have been described for several other *Penicillium* species, like the psychrotolerant species *P. jamesonlandense* or *P. soppii* (Frisvad et al. 2006). For pseurotin A, which is known as a mono amine oxidase inhibitor, effective nematicidal activity against *Bursaphelenchus xylophilus* was described, but it did not show any nematicidal activity against *Pratylenchus penetrans* and *Caenorhabditis elegans* (Hayashi et al. 2007). Cycloaspeptide A exhibited moderate activity (IC_{50} = 3.5 µg/ml) against *Plasmodium falciparum* (Dalsgaard et al. 2005). Pseurotin A exhibited moderate antimicrobial activity and moderate cytotoxic activity against human lung fibroblasts with an IC_{50} value of 1000 µM, whereas the IC_{50} value of cycloaspeptide A was >1000 µM (Schmeda-Hirschmann et al. 2008).

4. Penicidones from *Penicillium* sp. from *Quercus variabilis*

The three new and cytotoxic γ-pyridone alkaloids, penicidones A-C (Fig. 8.11), were isolated from extracts of the endophytic fungus *Penicillium* sp. IFB-E22 from the stem of *Quercus variabilis* (Ge et al. 2008). The three new compounds, along with the known metabolites lumichrome, physicion, and emodin-1,6-dimethylether, were purified from the extracts (Fig. 8.11). One of

the compounds was identified as (*R,E*)-5-(5,7-dimethoxy-3-oxo-1,3-dihydroisobenzofuran-1-yl)-2-(prop-1-enyl)pyridin-4(1H)-one, shortly named penicidone A, and the two others penicidones B and C (Ge et al. 2008). The penicidones exhibited moderate cytotoxicity against SW1116, K562, KB, and HeLa cells with IC_{50} values between 21.11 µM and 80.8 µM. The biosynthesis of the penicidones is thought via a polyketide pathway together with several biotransformations like cyclization and transamination (Ge et al. 2008).

D. Metabolites from Endophytic *Alternaria* Species

1. Metabolites from *Alternaria* sp. from *Polygonum senegalense*

Polygonum species are known in traditional medicine for their diuretic, cholagic, antihemorrhagic, and antiseptic properties (Aly et al. 2008). *P. senegalense* is a traditional

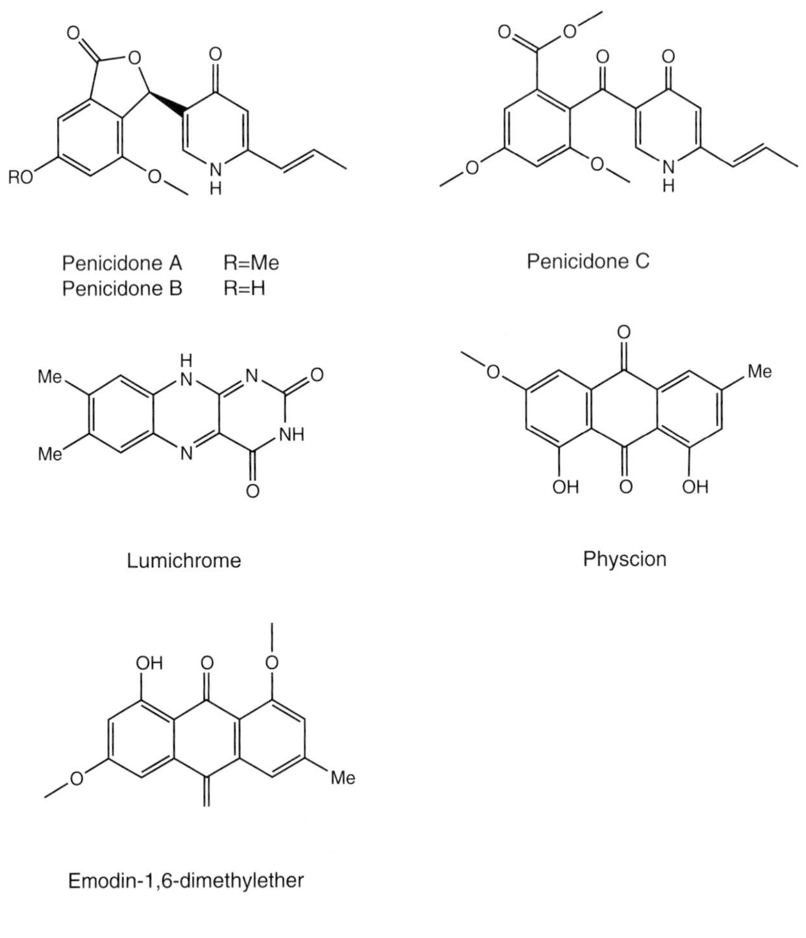

Penicidone A   R=Me
Penicidone B   R=H

Penicidone C

Lumichrome

Physcion

Emodin-1,6-dimethylether

Fig. 8.11. Metabolites from endophytic *Penicillium* sp. from *Quercus variabilis*

medicinal plant growing in Egypt which is used for the treatment of skin diseases.

The endophyte *Alternaria* sp. was isolated from fresh healthy leaves of *P. senegalense* collected near Alexandria, Egypt (Aly et al. 2008). In liquid Wickerham medium this fungus produced new sulfated derivatives of alternariol and its monomethyl ethers (alternariol 5-*O*-sulfate and alternariol 5-*O*-methyl ether-4'-*O*-sulfate) besides the known compounds alternariol, alternariol 5-*O*-methylether, altenusin, 2,5-dimetyl-7-hydroxy-chromone, tenuazonic acid, and altertoxin I (Fig. 8.12; Aly et al. 2008). From solid rice cultures the four new compounds 3'-hydroxyalternariol 5-*O*-methylether, desmethylaltenusin, alterlactone, and alternaric acid together with alternariol, alternariol 5-*O*-methyl ether, altenusin, talaroflavone, and a new stereoisomer of altuene, named 4'-epialtenuene, were obtained (Aly et al. 2008). The cytotoxic activity of the isolated compounds was tested against L5178Y mouse lymphoma cells. Alternariol was the most active compound and exhibited an EC_{50} value of 1.7 µg/ml, whereas alternariol 5-*O*-sulfate and alternariol 5-*O*-methyl ether showed EC_{50} values of 4.5 µg/ml and 7.8 µg/ml, respectively (Aly et al. 2008). Alternariol 5-*O*-methyl ether, altenusin, and alternariol were detected in crude MeOH extracts of the host plant *P. senegalense* by LC-MS (Aly et al. 2008).

2. Metabolites from *Alternaria* spp. from *Vinca minor* and *Euonymus europaeus*

The endophytic *Alternaria* species P 0506 and P 0535 were isolated from a leaf of *Vinca minor* and from a fruit of *Euonymus europaeus*, respectively (Hellwig et al. 2002).

The species of the genus *Vinca* are well known because of the isolation of the vinca alkaloids, e.g. vincristine or vinblastine. These alkaloids bind specifically to tubulin, and therefore the polymerization of the microtubuli and the formation of the spindle apparatus are blocked, resulting in an inhibition of mitosis (Aktories et al. 2005). The vinca alkaloids are used for the treatment of various tumors like malign lymphomas or acute leukemia.

Several species of the genus *Euonymus* are used as traditional medicinal plants, e.g. *E. hederaceus* or *E. alatus* (Hu et al. 2005; Park et al. 2005). *E. alatus* has been used for the treatment of diabetes and tumors in China and Korea (Kim et al. 2006; Fang et al. 2008). *E. europaeus* is the sole representative of this genus in central and western Europe (Descoins et al. 2002). The fruits of the plant are toxic for insects and ingestion of the fruits induces gastrointestinal discomfort in humans (Roth et al. 1994).

The two *Alternaria* species from *V. minor* and *E. europaeus* produced similar secondary metabolites in shake flasks and stirred fermentors (Hellwig et al. 2002). A new metabolite, named altersetin was produced together with the known metabolites alternariol, alternariol monomethylether, and tenuazonic acid by both fungal strains (Fig. 8.12; Hellwig et al. 2002). Altersetin exhibited antibacterial activity against various human pathogenic bacteria in the serial agar dilution assay (Hellwig et al. 2002).

E. Xanthones from *Emericella variecolor* from *Croton oblongifolius*

Emericella variecolor was isolated from mature petioles of the Thai medicinal plant *Croton oblongifolius* (Pornpakapul et al. 2006). Four xanthones were isolated from the mycelium of the endophytic fungus. The metabolites were identified by spectroscopic analysis as shamixanthone, 14-methoxytajixanthone-25-acetate, taijixanthone methanoate, and tajixanthone hydrate (Fig. 8.13). The cytotoxic activitiy of the four compounds was tested against several human tumor cell lines and only tajixanthone hydrate exhibited modest activity against all tested cell lines including gastric carcinoma (KATO3; IC50 = 10.9 mM), colon carcinoma (SW620; IC50 = 13.6 mM), breast carcinoma (BT474; IC50 = 12.3 mM), human hepatocarcinoma (HEP-G2; IC50 = 16.4 mM), and lung carcinoma (CHAGO; IC50 = 11.6 mM; Pornpakapul et al. 2006).

F. Cyclopentenons from *Dothideomycete* sp. from *Leea rubra*

The endophyte *Dothideomycete* sp. LRUB20 was isolated from the stem of the Thai medicinal plant *Leea rubra* (Chomcheon et al. 2006). Two new natural products, 2-hydroxymethyl-3-methylcyclopent-2-enone (synthetically known) and cis-2-hydroxymethyl-3-methylcyclopentanone, and the known compound asterric acid were isolated from extracts of the culture broth of the endophyte (Fig. 8.13; Chomcheon et al. 2006). In the microplate Alamar Blue assay 2-hydroxymethyl-3-methylcyclopent-2-enone and asterric acid exhibited weak antimicrobial activity against *Mycobacterium tuberculosis* with a MIC value of 200 µg/ml (Chomeon et al. 2006).

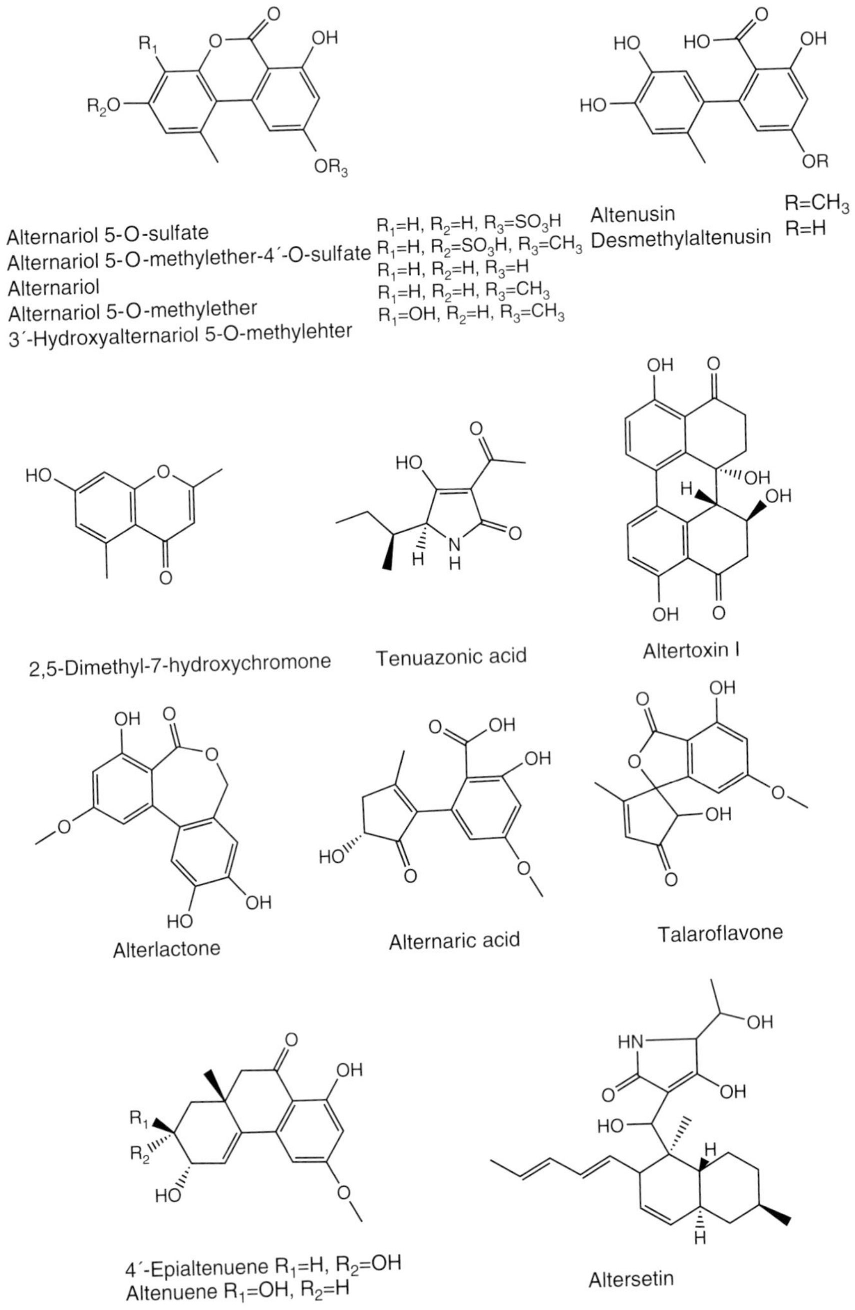

Alternariol 5-O-sulfate R_1=H, R_2=H, R_3=SO$_3$H   Altenusin   R=CH$_3$
Alternariol 5-O-methylether-4′-O-sulfate R_1=H, R_2=SO$_3$H, R_3=CH$_3$ Desmethylaltenusin R=H
Alternariol R_1=H, R_2=H, R_3=H
Alternariol 5-O-methylether R_1=H, R_2=H, R_3=CH$_3$
3′-Hydroxyalternariol 5-O-methylehter R_1=OH, R_2=H, R_3=CH$_3$

2,5-Dimethyl-7-hydroxychromone   Tenuazonic acid   Altertoxin I

Alterlactone   Alternaric acid   Talaroflavone

4′-Epialtenuene R_1=H, R_2=OH   Altersetin
Altenuene R_1=OH, R_2=H

Fig. 8.12. Metabolites from endophytic *Alternaria* species

G. Metabolites from Endophytic *Chaetomium* Species

1. Metabolites from *Chaetomium* sp. from *Nerium oleander* L.

Oleanders (Apocynaceae), which contain the inotropic cardenolides (e.g. oleanderin), have been exploited therapeutically and as an instrument of suicide since the antiquity (Langford and Boor 1996). The pink oleander, *Nerium oleander*, is one of the principle oleander representatives (Langford and Boor 1996). Certain parts of the plant are used in Chinese folk medicine (Huang et al. 2007). *N. oleander* is supposed to have cardiotonic, antibacterial, antileprotic, anti-inflammatory, anticancer, and antiplatelet aggregation activities, insecticidal activity, mammalian cytotoxicity, and act as a depressant of the central nerve system (Huang et al. 2007).

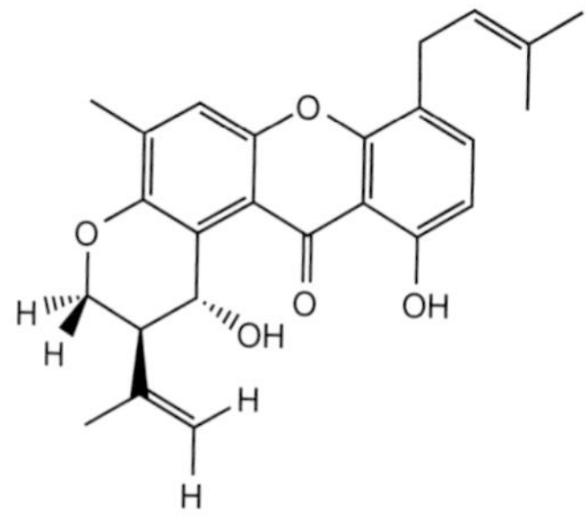

Shamixanthone

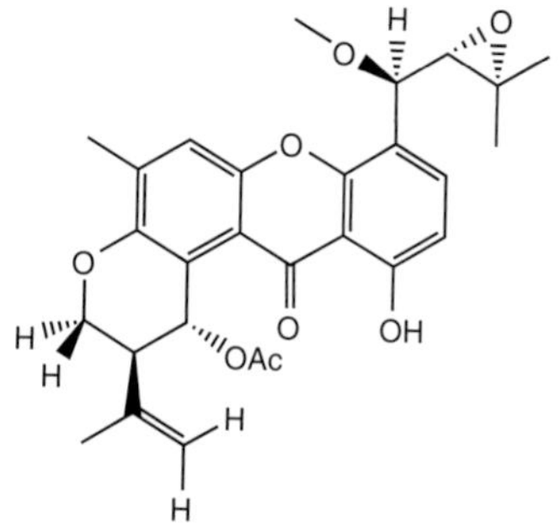

14-Methoxytajixanthone-25-acetate

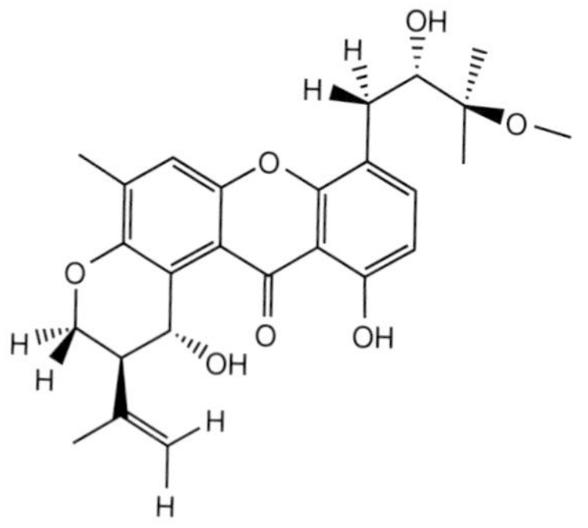

Tajixanthone methanoate

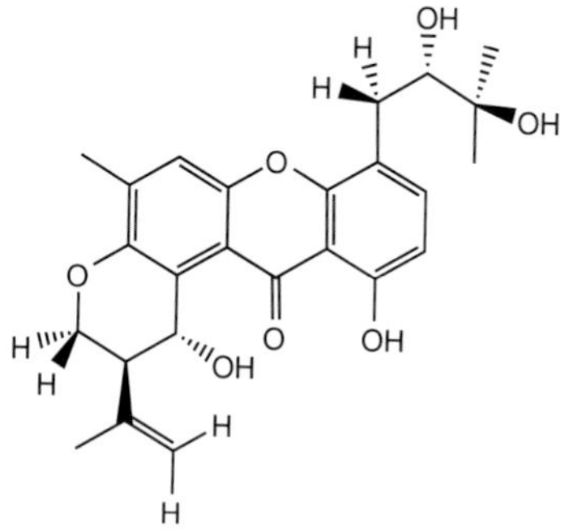

Tajixanthone hydrate

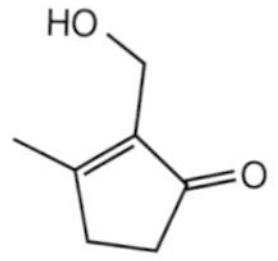

2-Hydroxymethyl-3-methyl-cyclopent-2-enone

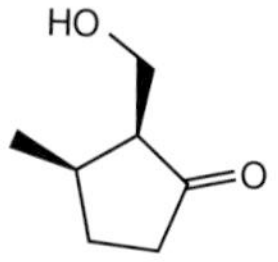

cis-2-Hydroxymethyl-3-methylcyclopentenone

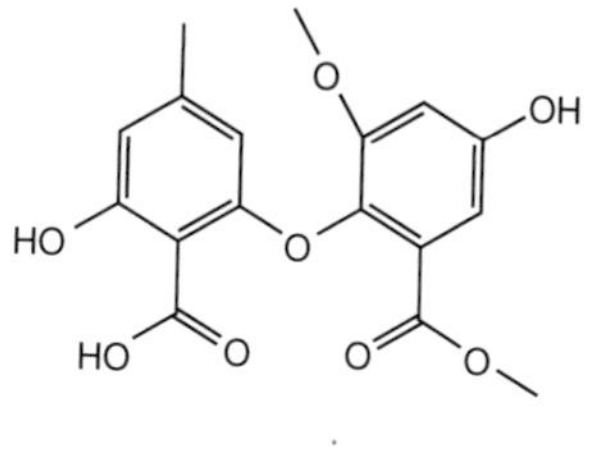

Asterric acid

Fig. 8.13. Metabolites from endophytic *Emericella variecolor* and *Dothideomycete* sp.

Huang and coworkers (2007) isolated, among 42 endophytic fungi, a *Chaetomium* sp. with the strong antioxidant capacity from *N. oleander*. The main bioactive metabolites from the fungal cultures were phenolics (e.g. phenolic acids and their derivates) and volatile and aliphatic compounds which were detected by LC-ESI-MS and GC-MS (Huang et al. 2007). Huang et al. (2007) were not able to find the same compounds in the host plant and its endophytic fungi.

2. Cytochalasan Alkaloids from *Chaetomium globosum* from *Imperata cylindrica*

The endophyte *Chaetomium globosum* IFB-E019 was isolated from the stem of *Imperata cylindrica*

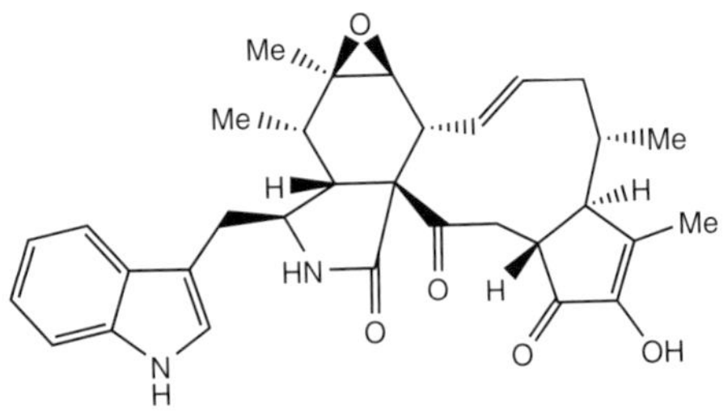

Chaetoglobosin U

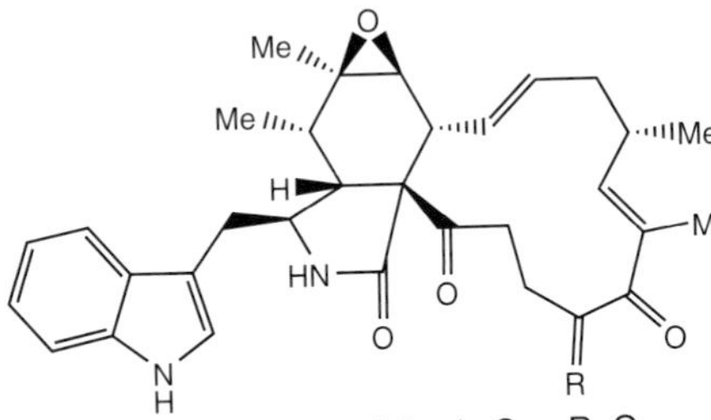

Chaetoglobosin C   R=O
Chaetoglobosin F   R=OH

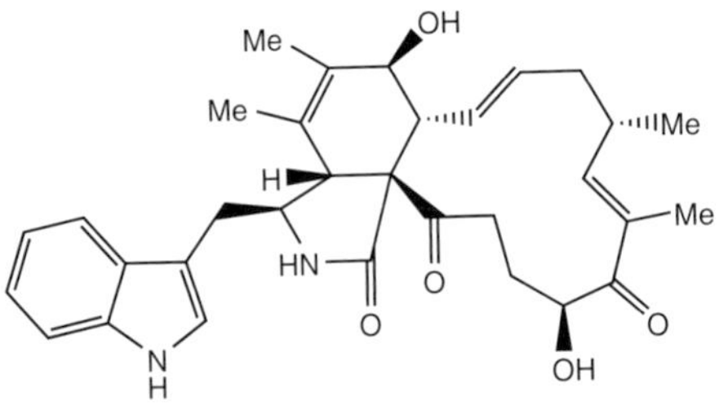

Chaetoglobosin E

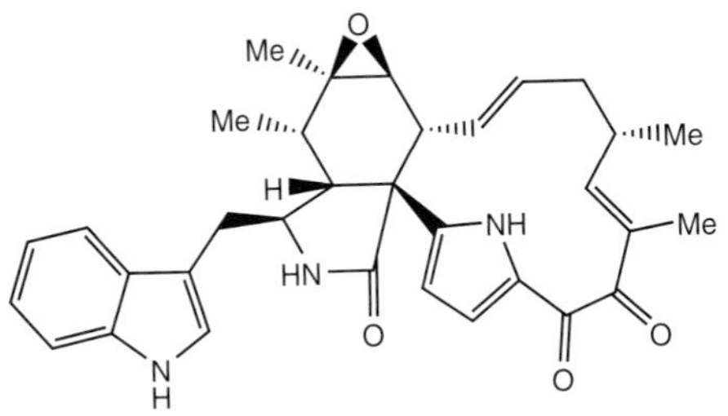

Phenochalasin A

Fig. 8.14. Cytochalasan alkaloids from *Chaetomium globosum* from *Imperata cylindrica*

(Ding et al. 2006). The new cytotoxic cytochalasan-based alkaloid chaetoglobosin U and four known analogues chaetoglobosins C, F, E, and penochalasin A were isolated from exracts of the fungal biomass (Fig. 8.14; Ding et al. 2006). The cytotoxic activity of all five compounds was tested against the

human nasopharyngeal epidermoid tumor KB cell line. Chaetoglobosin U was the most cytotoxic, with an IC$_{50}$ value of 16 µM, whereas the four known compounds exhibited moderate activity with IC$_{50}$ values of 34–52 µM (Ding et al. 2006).

H. Metabolites from Endophytic *Pestalotiopsis* Species

1. Isopestacin from *Pestalotiopsis microspora* from *Terminalia morobensis*

Species of the genus *Terminalia* are used as medicinal plants, e.g. the bark of *Terminalia arjuna* has a long history of use as a cardiac tonic and is indicated for the treatment of coronary artery disease, heart failure, hypercholesterolemia, and for relief of anginal pain (Miller 1998b).

Strobel and coworkers (2002) isolated an endophytic *Pestalotiospis microspora* from a small tree stem of *Terminalia morobensis* collected on the north coast of Papua New Guinea. Isopectacin (Fig. 8.15), an isobenzofuranone, was obtained from the culture broth of the fungal isolate (Strobel et al. 2002). The compound possessed antioxidant activity against a variety of free radicals and a moderate antimycotic activity against the plant pathogen *Pythium ultimum* (Strobel et al. 2002).

2. Sesquiterpenes from *Pestalotiopsis* sp. from *Pinus taeda*

Pestalotiopsis sp. was the predominant endophyte in a collection of microorganisms from *Pinus taeda* from São Paulo State in Brazil (Magnani et al. 2003). Three new highly oxidized caryophyllene sequiterpenes, named pestalotiopsolide A, taedolidol, and 6-epitaedolidol, were purified from liquid culture of this endophytic fungus (Fig. 8.15; Magnani et al. 2003). The carbon skeleton of pestalotiopsolide A is similar to that of pestalotiopsin C (Magnani et al. 2003). The pestalotiopsins A, B, and C were originally described from endophytic *Pestalotiopsis* species isolated from *Taxus brevifolia* (Pulici et al. 1996, 1997). Pestalotiopsin A exhibited immunosuppressive and cytotoxic activity in preliminary assays (Johnston et al. 2003). The pestalotiopsins were not isolated by Magnani and coworkers, but they assume that these compounds could be precursors of pestalotiopsolide A, taedolidol, and 6-epitaedolidol.

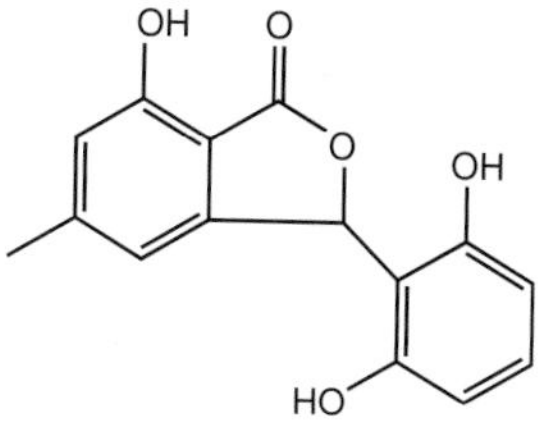

Isopectacin

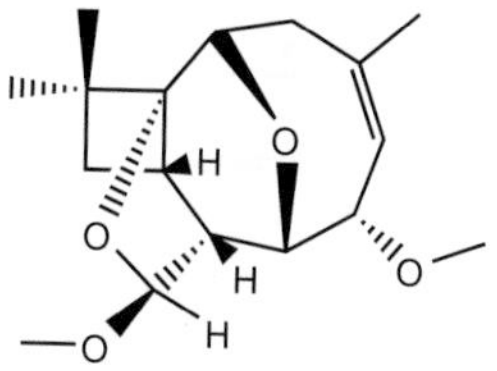

Pestalotiopsolide A

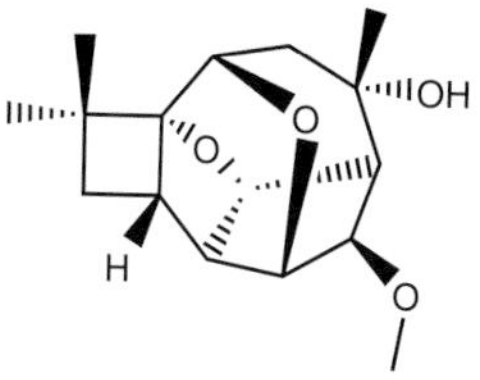

Taedolidol

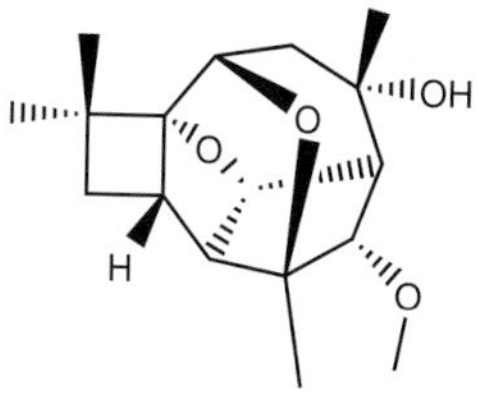

6-Epitaedolidol

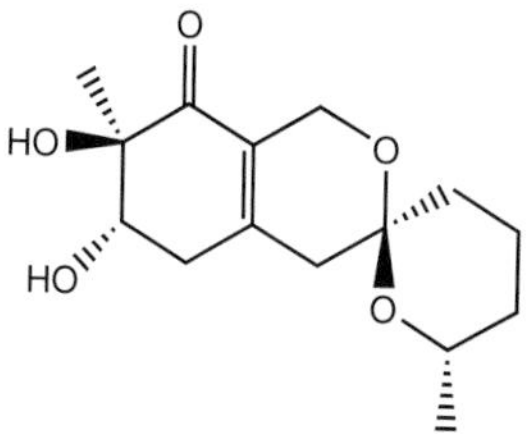

Pestafolide A

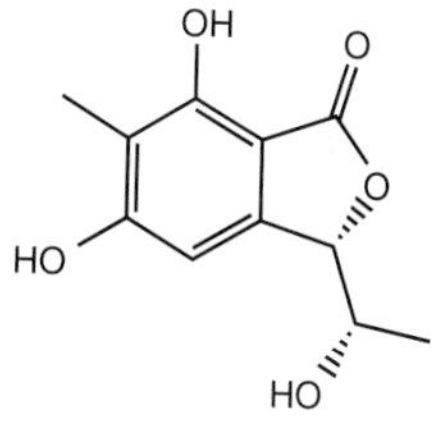

Pestaphthalide A

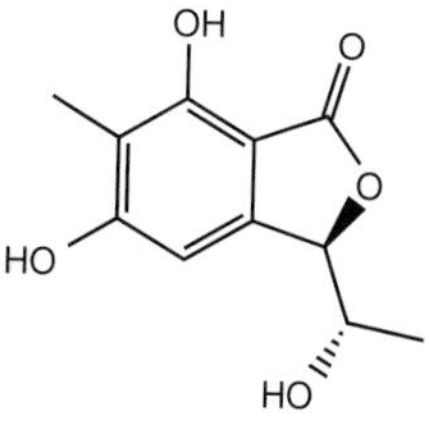

Pestaphthalide B

Fig. 8.15. Metabolites from endophytic *Pestalotiopsis* species

3. Metabolites from *Pestalotiopsis foedan*

The endophytic fungus *Pestalotiopsis foedan* was isolated from an unidentified tree (Ding et al. 2008). Bioassay-guided fractionation led to the isolation of three new metabolites, a reduced spiro azaphilone derivative named pestafolide A, and the two isobenzofuranones pestaphthalides A and B (Fig. 8.15; Ding et al. 2008). The antifungal activity of the three compounds was investigated against *Candida albicans*, *Geotrichum candidum*, and *Aspergillus fumigatus* in agar diffusion assays. Pestafolide A exhibited antifungal activity against *A. fumigatus*,

pestaphthalide A was active against *C. albicans*, and pestaphthalide B showed antifungal activity against *G. candidum* (Ding et al. 2008).

4. Metabolites from *Pestalotiopsis theae*

The endophytic *Pestalotiopsis theae* was isolated from an unidentified tree by Li et al. (2008). Four new compounds, pestalotheols A–D (Fig. 8.16), were purified from fungal rice cultures by bioassay-guided fractionation (Li et al. 2008). It is assumed that the chromenones pestalotheols A, B, and C are derived from two units of isoprenoids and a polyketide, whereas pestalotheol C could be

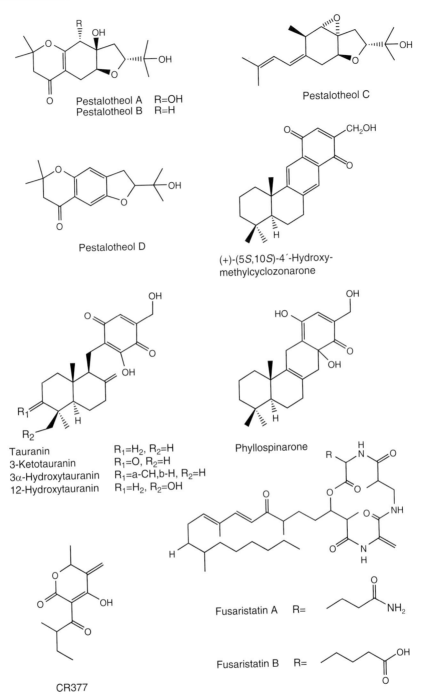

Fig. 8.16. Metabolites from endophytic *Pestalotiopsis theae*, *Phyllosticta spinarum* from *Platycladus orientalis*, and *Fusarium* sp.

their biosynthetic precursor. Of these four compounds pestalotheol C showed an inhibitory effect on HIV-1 replication in C8166 cells, but none of the compounds exhibited noticeable antimicrobial activity (Li et al. 2008).

J. Metabolites from *Phyllosticta spinarum* from *Platycladus orientalis*

Platycladus orientalis (Cupressaceae) is a traditional Chinese herb and food additive used for the treatment of gout, rheumatism, diarrhea, and chronic tracheitis (Lu et al. 2006). The plant is also cultivated as an ornamental in southeastern Arizona (Wijeratne et al. 2008).

The fungal endophyte *Phyllosticta spinarum* was isolated from healthy foliage of *Platycladus orientalis* collected in Arizona (Wijeratne et al. 2008). Five new sesquiterpene quinones and related derivates, (+)-(5S,10S)-4'-hydroxymethyl-cyclozonarone, 3-ketotauranin, 3α-hydroxytauranin, 12-hydroxytauranin, and phyllospinarone, together with tauranin were isolated from cultures of *P. spinarum* (Fig. 8.16; Wijeratne et al. 2008). The in vitro antiproliferative activity of the six compounds was tested against five cell lines: NCI-H460, MCF-7, SF-268, PC-3M, and MIA Pa Ca-2. Only tauranin showed antiproliferative activity against the cancer cell lines tested; all other compounds were found to be inactive at concentrations up to 5 µM (Wijeratne et al. 2008).

K. Metabolites from Endophytic *Fusarium* Species

1. CR377 from *Fusarium* sp. from *Selaginella pallescens*

Species of the genus *Selaginella* are used in traditional Chinese medicines, e.g. *S. tamariscina* and *S. uncinata* (Ma et al. 2003; Cheng et al. 2008). *S. unicata* is used to treat infectious diseases and tumors and ethanol extracts of the whole plant show significant inhibitory activities against respiratory syncytial virus (Ma et al. 2003). *S. tamariscina* is a medicinal plant used for the treatment of advanced cancer in the Orient (Ahn et al. 2006) and for the therapy of chronic tracheitis (Yang et al. 2007). The plant exhibits antitumor and vasorelaxant activity (Kang et al. 2004; Yang et al. 2007).

CR377, a *Fusarium* species, was collected in the Guanacaste Conversation Area in Costa Rica and isolated from a piece of a stem from another *Selaginella* species, *S. pallescens* (Brady and Clardy 2000). A new pentaketide, CR377 (Fig. 8.16), was isolated from the culture broth of this *Fusarium* strain CR377. The compound showed potent antifungal activity against strains of *Candida albicans* in the agar diffusion assay (Brady and Clardy 2000).

2. Lipopeptides from *Fusarium* sp. from *Maackia chinensis*

Fungal strain *Fusarium* sp. YG-45 was isolated from a twig segment of *Maackia chinensis* (Shiono

et al. 2007). The plant material was collected in the botanical garden in Göttingen, Germany. From fungal rice cultures two new cyclic lipopeptides, fusaristatins A and B, were obtained (Fig. 8.16; Shiono et al. 2007). These two compounds did not exhibit antimicrobial activity at a concentration of 100 µg/ml against *Staphylococcus aureus* NBRC 13276, *Pseudomonas aeruginosa* ATCC 15442, *Candida albicans* ATCC 2019, and *Aspergillus clavatus* F 318a (Shiono et al. 2007). Since the fusaristatins are structurally related to the DNA topoisomerase inhibitor, topostatin, the inhibitory effect on topoisomerases was investigated: fusaristatin B showed a moderate inhibition of topoisomerase I and II with an IC_{50} value of 73 µM and 98 µM, respectively (Shiono et al 2007). Fusaristatin A and B exhibited cytototoxic activity against the human lung cancer cells LU 65 with an IC_{50} value of 23 µM and 7 µM, respectively, whereas no activity could be detected against the colon cancer cells COLO 201 up to a concentration of 100 µM (Shiono et al. 2007).

L. Epichlicin from *Epichloe typhina* from *Phleum pretense* L.

Infection by the choke disease endophytic fungus, *Epichloe typhina* (Pers. Ex. Fr.) Tul., whose immature form is *Acremonium typhinum*, in *Phleum pretense* induces resistance of leaf spot disease caused by the pathogen, *Cladosporium phlei* (Gregory) de Vries (Seto et al. 2007). In order to explain the beneficial mutual relatioship between fungus and host plant, Seto et al. (2007) investigated the metabolites of *E. typhina*, since culture broth of the endophytic fungus was able to inhibit the spore germination in *C. phlei*. So they were able to isolate the germination inhibitor epichlicin (Fig. 8.17). This novel cyclic peptide with the amino acid sequence 3-amino tetradecanoic acid (β-amino acid), L-Asn, D-Tyr, L-Asn, L-Gln, L-Ser, L-Asn, and D-Pro is a diastereomer of mixirin A (Seto et al. 2007). Spore germination of *C. phlei* was inhibited by epichlicin with an IC_{50} value of 22 nM (Seto et al. 2007).

M. Metabolite from *Colletotrichum dematium* from *Pteromischum* sp.

Colletotrichum dematium was isolated from a *Pteromischum* sp. growing in a tropical forest in

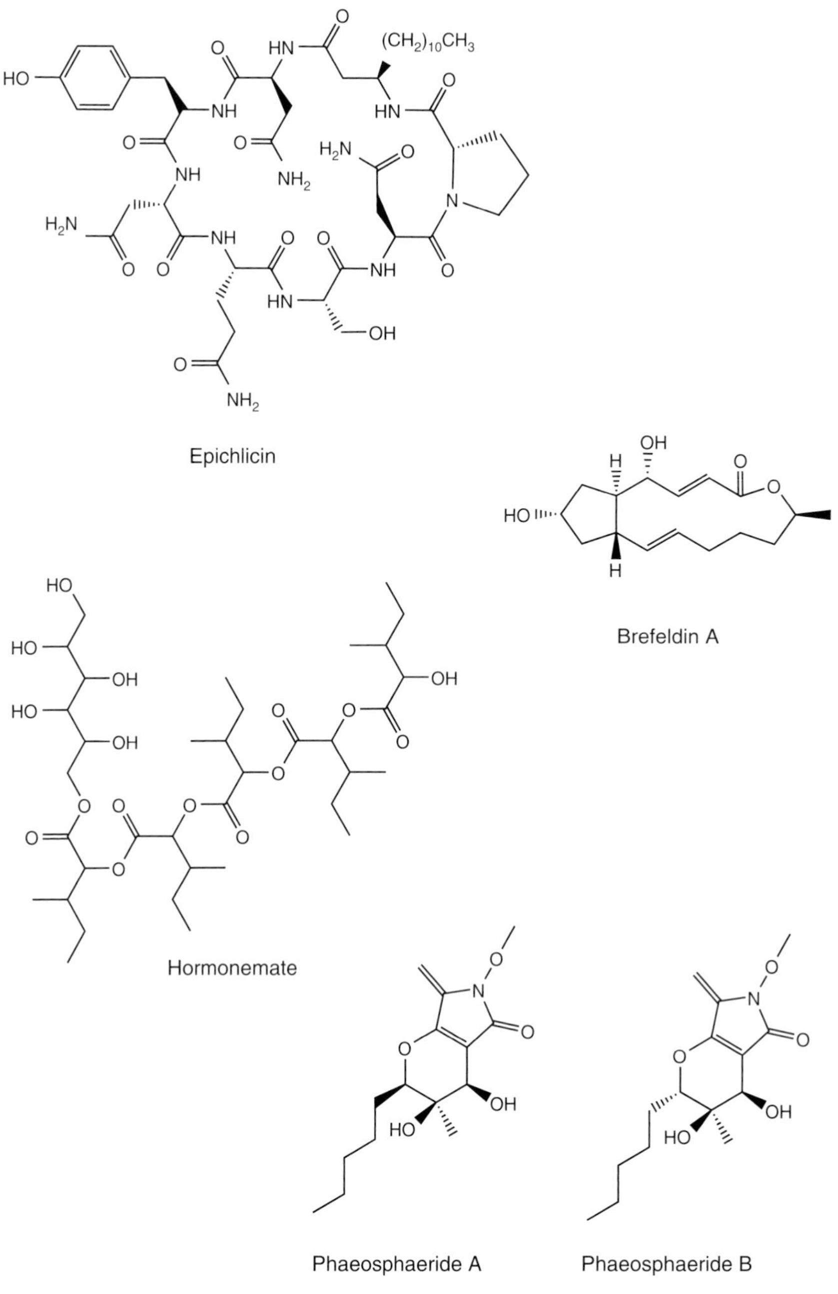

Epichlicin

Brefeldin A

Hormonemate

Phaeosphaeride A   Phaeosphaeride B

Fig. 8.17. Structures of epichlicin, brefeldin A, hormonemate, and phaeospaerides A and B

Costa Rica (Ren et al. 2008). The fungus produced a novel antimycotic peptide, colutellin A, which contains residues of Ile, Var, Ser, N-methyl-Val, and beta-aminoisobutryic acid in nominal molar ratios of 3:2:1:1:1 (Ren et al. 2008). Colutellin A is assumed to have potential immunosuppressive activity, since it inhibited CD4(+) T-cell activation of interleukin 2 production with an IC_{50} value of 167.3 nM, wherease

cyclosporine A in the same test yielded a value of 61.8 nM (Ren et al. 2008).

N. Hormonemate from *Hormonema dematioides* from *Pinus* sp.

The fungal strain E99156 was isolated as a symptomless endophyte from the needle base of a *Pinus* species collected in Portugal (Filip et al. 2003). The endophyte was identified as *Hormonema dematioides* by microscopic and genetic characteristics. The new metabolite hormonemate was isolated from submerged cultures of the endophytic fungus (Fig. 8.17; Filip et al. 2003). The compound exhibited cytotoxic acitivity against the tested cancer cell lines COLO 320, DLD-1, HT-29, Jurkat, and HL-60 cells, with IC_{50} values of 3.5 µg/ml and 3.75 µg/ml, and induced apoptosis in COLO-320 cells (Filip et al. 2003).

O. Brefeldin A from *Phoma medicaginis* from *Medicago lupulina*

Phoma medicaginis was the dominating endophyte in *Medicago sativa* and *M. lupulina* (Weber et al. 2004). The macrocyclic lactone brefeldin A (Fig. 8.17), an inhibitor of protein trafficking in the endomembrane system of mammalian cells (Nebenführ et al. 2002), was isolated from pure fungal cultures and detected in infected and artificially inoculated plant material of *M. sativa* (Weber et al. 2004). Brefeldin A was not found in freshly harvested plant material, so the authors assumed that this secondary metabolite is produced to defend the dead plant against other saprophytic organisms after the switch from the endophytic to the saprotrophic phase, following the death of infected host tissue (Weber et al. 2004).

P. Phaeosphaerides from an Endophytic Fungus

Maloney and coworkers (2006) isolated endophytic fungi from plant samples collected in the Archbold Biological Station, a 5000-acre (20.2 km²) preserve in Lake Placid in Florida. The endophyte FA 39, having 97% identity to the ascomycete *Phaeosphaeria avenaria*, exhibited interesting activity against STAT3 (Maloney et al. 2006). Two compounds, phaeosphaeride A and B, were purified from fungal material and have a new carbon skeleton related to the fungal curvupallides and the lipid-lowering spirostaphylotrichins/triticones (Fig. 8.17; Maloney et al. 2006). These three families share the functional group *O*-methyl hydroxamic acid (Maloney et al. 2006). In an ELISA-based screening phaeosphaeride A

inhibited STAT3, with an IC_{50} value of 0.61 mM, whereas the diastereomer, phaeosphaerid B, exhibited no activity against STAT3 (Manoley et al. 2006). Phaeosphaeride A is an inhibitor of STAT3 signaling and selective for STAT3 over other STAT proteins (Manoley et al. 2006).

Q. Metabolites from *Nodulisporium* sp. from *Junipercus cedre*

The fungus *Nodulisporium* sp. 7080, belonging to the Xylariaceae, was isolated from twigs of the plant *Juniperus cedre* (Dai et al. 2006). Searching for biologically active compounds from endophytic fungi, Dai and coworkers (2006) found that the crude extracts of *Nodulisporium* sp. 7080 exhibited antimicrobial and algicidal activities. Seven new natural products, 3-hydroxy-1-(2,6-dihydroxyphenyl)-butan-1-one, 1-(2-hydroxy-6-methoxyphenyl) butan-1-one, 2,3-dihydro-5-methoxy-2-methyl-chomen-4-one, the dimeric naphthalenes nodulisporin A and B, the first naturally occurring dimeric indanone nodulisporin C, and (4E,6E)-2,4,6-trimethylocta-4,6-dien-3-one, and ten known compounds were isolated by bioassay guided fractionation (Fig. 8.18; Dai et al. 2006). The known compounds were identified by NMR spectroscopy as 1-(2,6-dihydroxyphenyl)butan-1-one, 5-hydroxy-2-methyl-4*H*-chromen-4-one, 2,3-dihydro-5-hydroxy-2-methylchomen-4-one, 8-methoxynaphthalen-1-ol, 1,8-dimethoxynaphthalene, daldinol, helicascolide A, the eudesmane derivative (1*S*,4*S*,4a*S*,6*R*,6a*R*)-decahydro-1,4-dimethyl-6-(prop-1-en-2-yl)naphthalene-1-ol, ergosterol, and 5α,8α-epidioxyergosterol (Fig. 8.18, 8.19; Dai et al. 2006). 1-(2-Hydroxy-6-methoxyphenyl) butan-1-one and 1-(2,6-dihydroxyphenyl)butan-1-one exhibited biological activity at a concentration of 0.25 mg/filter in the agar diffusion assay against the fungi *Microbotryum violaceum* and *Septoria tritici* and the green alga *Chlorella fusca* (Dai et al. 2006).

R. Metabolites from *Ascochyta* sp. from *Meliotus dentatus*

The endophytic fungus *Ascochyta* sp. 6651 was isolated from the plant *Meliotus dentatus* (Krohn et al. 2007b). The plant material was collected at the

Fig. 8.18. Metabolites from *Nodulisporium* sp. from *Junipercus cedre* – part 1

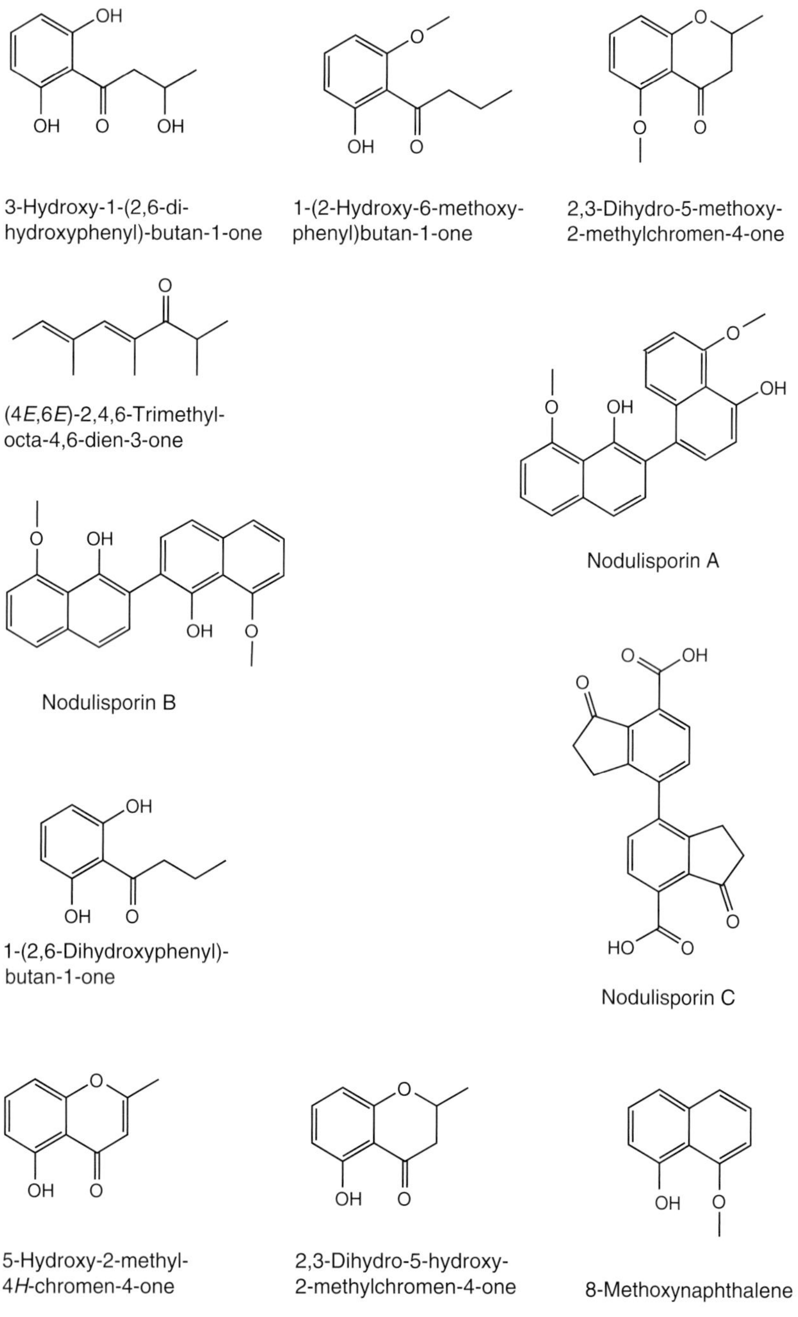

3-Hydroxy-1-(2,6-di-
hydroxyphenyl)-butan-1-one

1-(2-Hydroxy-6-methoxy-
phenyl)butan-1-one

2,3-Dihydro-5-methoxy-
2-methylchromen-4-one

(4*E*,6*E*)-2,4,6-Trimethyl-
octa-4,6-dien-3-one

Nodulisporin A

Nodulisporin B

1-(2,6-Dihydroxyphenyl)-
butan-1-one

Nodulisporin C

5-Hydroxy-2-methyl-
4*H*-chromen-4-one

2,3-Dihydro-5-hydroxy-
2-methylchromen-4-one

8-Methoxynaphthalene

shores of the Baltic Sea. Two new compounds, (4*S*)-(+)-ascochin and (*S*,*S*)-(+)-ascodiketone, and three known compounds, (3*R*,4*R*)-(−)-4-hydroxymellein, *ent*-α-cyperone, and (3*S*,4*R*)-(−)-dihydroxy-(6*S*)-undecyl-α-pyranone, were purified from fungal cultures of *Ascochyta* sp. 6651 (Fig. 8.20; Krohn et al. 2007b). Ascochin, 4-hydroxymellein, and dihydroxy-undecyl-pyranone showed activity against the gram-positive bacterium *Bacillus megasterium*, the fungus *Microbotryum violaceum*, and

the alga *Chlorella fusca* in the agar diffusion test at a concentration of 0.5 mg/filter (Krohn et al. 2007b).

S. Metabolites from Endophytic Fungi from *Garcinia* Species

Species of the genus *Garcinia* are used in ethnopharmacology, e.g. *Garcinia mangostana* L. (mangosteen) is a tree widespread in Southeast Asian countries with medicinal properties (Suksamrarn et al. 2003). The fruit hulls are

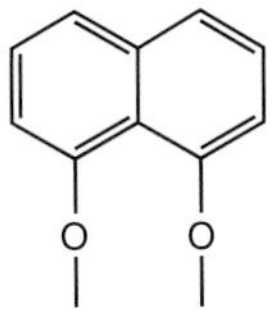

1,8-Dimethoxynaphthalene

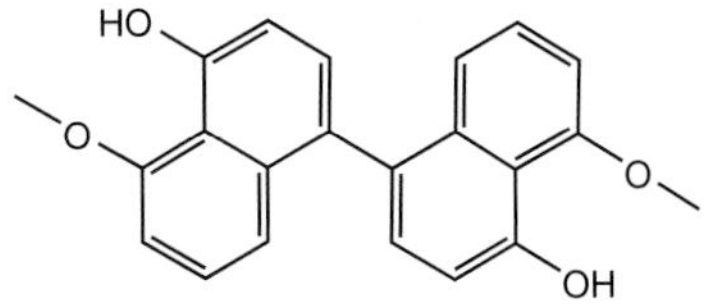

Daldinol

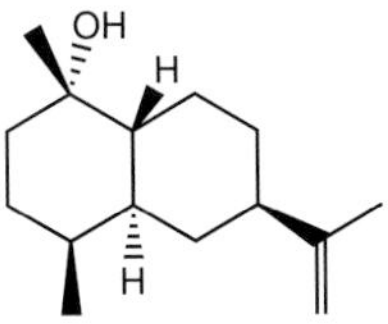

(1S,4S,4aS,6R,8aR)-Deca-
hydro-1,4-dimethyl-6-(prop-
1-en-2-yl)-naphthalen-1-ol

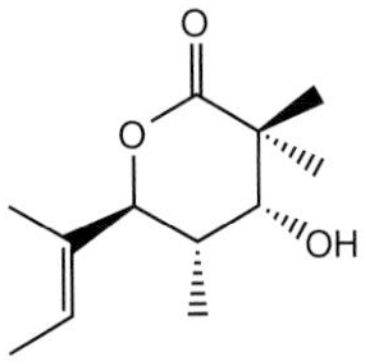

Helicascolide A

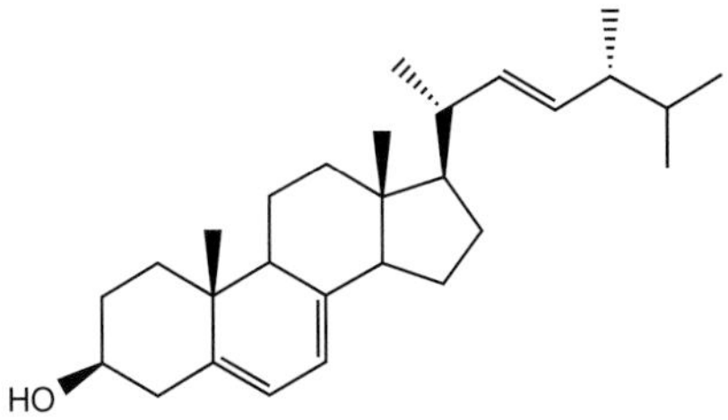

Ergosterol

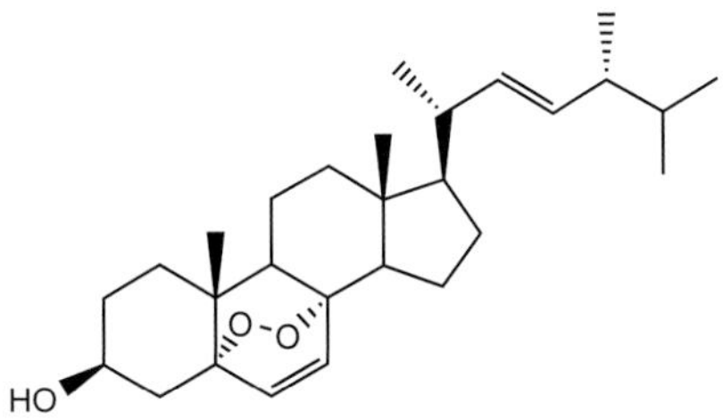

5α,8α-Epidioxyergosterol

Fig. 8.19. Metabolites from *Nodulisporium* sp. from *Junipercus cedre* – part 2

used in Thai folk medicine for the treatment of skin infections, wounds, and diarrhea (Suksamrarn et al. 2003) and
as an anti-inflammatory drug (Chen et al. 2008). Pedraza-
Chaverri et al. (2008) reported that the pericarp (peel, rind,
hull, or ripe) of *G. mangostana* is used in traditional
medicine for the treatment of abdominal pain, diarrhea,
dysentery, infected wound, suppuration, and chronic
ulcer. Experimental studies demonstrated that extracts of
the plant have antioxidant, antitumoral, antiallergic, anti-
inflammatory, antibacterial, and antiviral activities (Pedraza-Chaverri et al. 2008).

The endophytic fungus *Botryosphaeria mamane*
PSU-M76 was isolated from the leaves of

G. mangostana collected in Thailand (Pongcharoen et al. 2007). A new dihydrobenzofuran,
botryomaman, was isolated along with six
known compounds, 2,4-dimethoxy-6-pentylphe-
nol, (R)-(−)-mellein, primin, *cis*-4-hydroxymel-
lein, *trans*-4-hydroxymellein, and 4,5-dihydroxy-
2-hexenoic acid (Figs. 8.20, 8.21 Pongcharoen
et al. 2007). Pongcharoen and coworkers (2007)
assumed that botryomaman is derived from the
condensation of the demethylated 2,4-dimethoxy-
6-pentylphenol and 4,5-dihydroxy-2-hexenoic
acid. Primin exhibited the best activity against

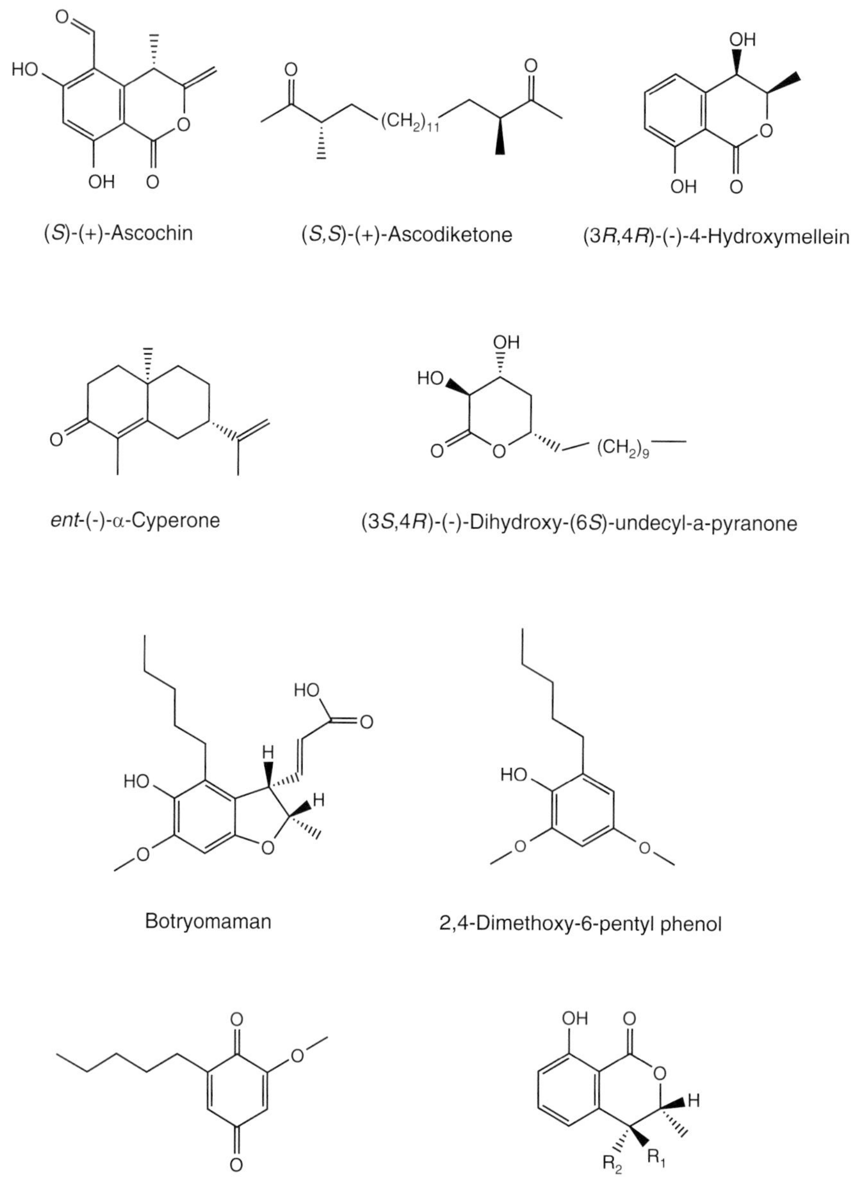

(S)-(+)-Ascochin   (S,S)-(+)-Ascodiketone   (3R,4R)-(-)-4-Hydroxymellein

ent-(-)-α-Cyperone   (3S,4R)-(-)-Dihydroxy-(6S)-undecyl-a-pyranone

Botryomaman   2,4-Dimethoxy-6-pentyl phenol

Primin

(R)-(-)-Mellein   R₁=R₂=H
cis-4-Hydroxymellein   R₁=H, R₂=OH
trans-4-Hydroxymellein   R₁=OH, R₂=H

Fig. 8.20. Metabolites from *Ascochyta* sp. from *Meliotus dentatus* and from endophytic fungi from *Garcinia* species

Staphylococcus aureus and methicillin-resitant *S. aureus*, with MIC values of 8 µg/ml, while all the other compounds were inactive, with MIC values >128 µg/ml (Pongcharoen et al. 2007).

The endophyte *Penicillium paxilli* PSU-A71 was isolated from leaves of *G. atroviridis* collected in Thailand (Rukachaisirikul et al. 2007). A new pyrone derivative, penicillone, together with paxilline and

pyrenocines A and B were obtained from extracts of fungal fermentations (Fig. 8.21; Rukachaisirikul et al. 2007). Pyrenocine B showed mild antifungal activity against the human pathogenic fungus *Microsporum gypseum* with a MIC value of 32 µg/ml, whereas penicillone and pyrenocine A were much less active with MIC values of 64 µg/ml and 128 µg/ml, respectively (Rukachaisirikul et al. 2007).

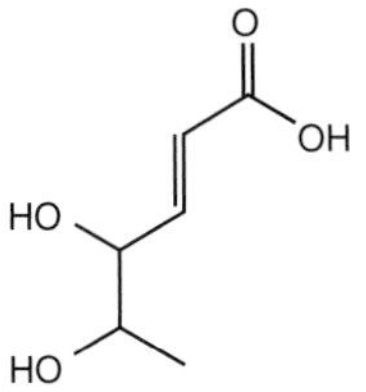

4,5-Dihydroxy-2-hexenoic acid

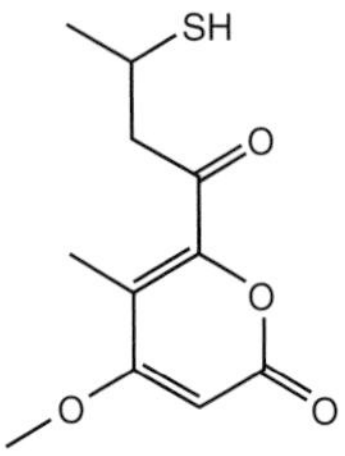

Penicillone

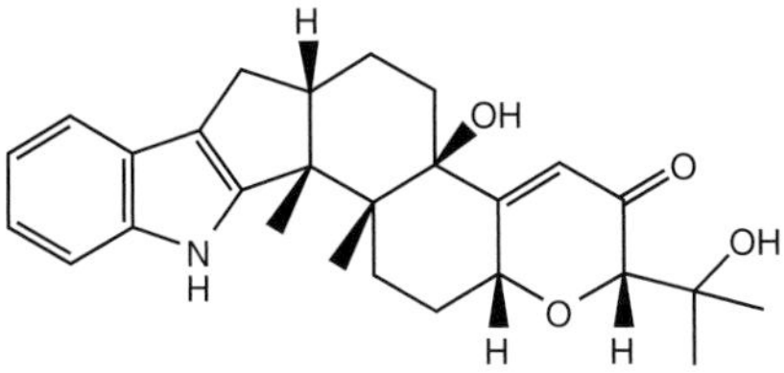

Paxilline

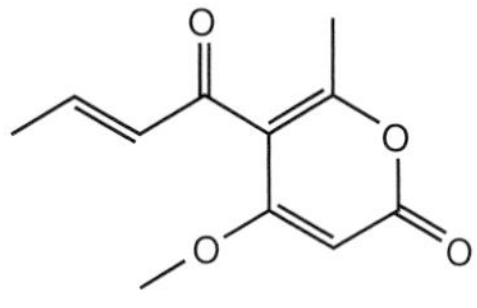

Pyrenocine A

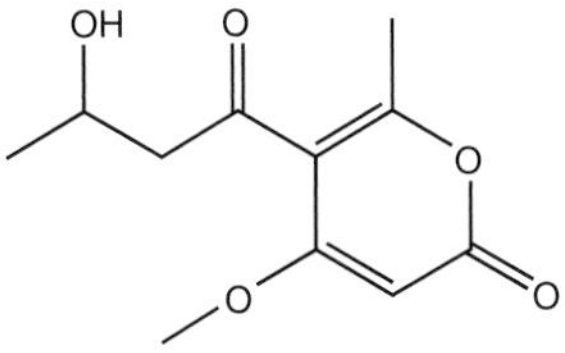

Pyrenocine B

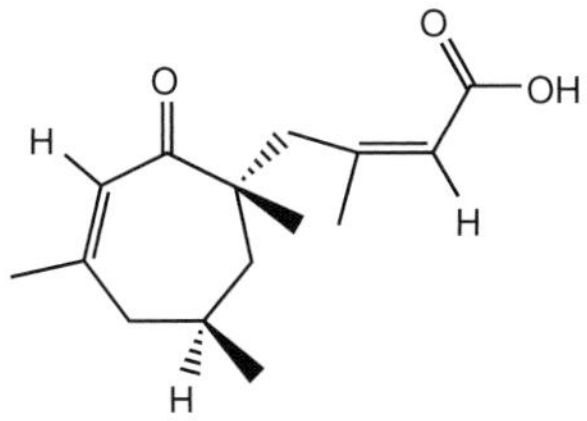

Leptosphaeric acid

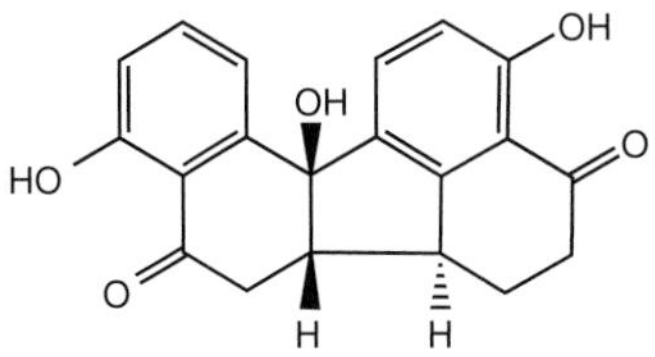

Daldinone C

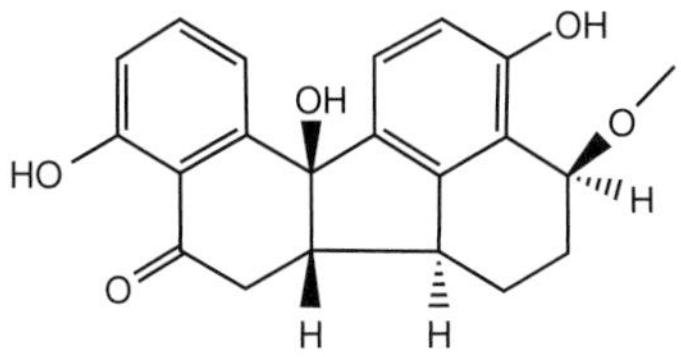

Daldinone D

Fig. 8.21. Metabolites from endophytic fungi from *Garcinia* and *Artemisia* species

T. Metabolites from Endophytic Fungi from *Artemisia species*

Artemisia species are widespread in nature and are frequently used for the treatment of malaria, hepatitis, cancer, inflammation, and infections by fungi bacteria and viruses (Tan et al. 1998). The endoperoxides sequiterpene lactone artemisinin was isolated from *A. annua*. The highest artemisinin concentrations can be found in the leaves and inflorescences (Woerdenbag et al. 1990). Artemisinin and its derivatives are a potent new class of antimalarials (Balint 2001), since artemisinin is effective against both drug-resistant and cerebral malaria caused by strains of *Plasmodium falciparum* (Abdin et al. 2003).

The endophyte *Leptosphaeria* sp. IV 403 was isolated from the plant *Artemisia annua*. Liu and coworkers (2003) purified a metabolite, leptosphaeric acid

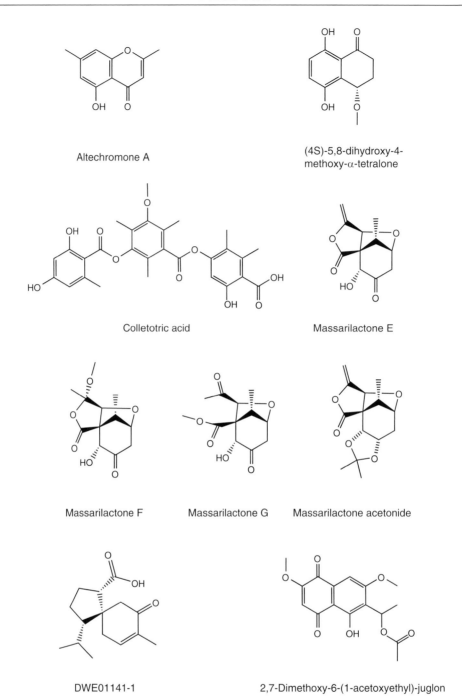

Altechromone A

(4S)-5,8-dihydroxy-4-
methoxy-α-tetralone

Colletotric acid

Massarilactone E

Massarilactone F   Massarilactone G   Massarilactone acetonide

DWE01141-1   2,7-Dimethoxy-6-(1-acetoxyethyl)-juglon

Fig. 8.22. Metabolites from endophytic fungi from *Artemisia* species and *Schinus molle*

(Fig. 8.21), with a novel carbon skeleton and unrelated to artemisin from extracts of the culture of the endophyte.

The endophytic *Hypoxylon truncatum* IFB-18 was isolated from symtomless stem tissue of *A. annua* by Gu et al. (2007). Two new benzo[*j*]fluoranthene-based natural products, daldinone C and daldinone D, and the known metabolites altechromone A and (4S)-5,8-dihydroxy-4-methoxy-α-tetralone were isolated from extracts of a solide culture of the fungal strain (Figs. 8.21, 8.22; Gu et al. 2007). Daldinones C and D exhibited cytotoxic ativity against the SW1116 cell line with IC$_{50}$ values of 49.5 μM and 41.0 μM (Gu et al. 2007).

Biosynthetically these compounds could be derived from a polyketide precusor.

Artemisia mongolica is known to be strongly resistant to insects and pathogens (Zou et al. 2000). From apparently healthy plants the fungus *Colletotrichum gloeosporioides* was isolated (Zou et al. 2000). From the culture liquid of this fungus a new metabolite, named colletotric acid (Fig. 8.22), was isolated. *C. gloeosporioides* is known as a host-specific pathogen causing anthracnose of host plants and some strains are of value for biological control of weeds (Zou et al. 2000). Antimicrobial bioassays revealed that colletotric acid was antibacterial to *Bacillus subtilis*, *Staphylococcus aureus*, and *Sarcina lutea*, with MICs of 25, 50, and 50 µg/ml and antifungal to *Helmithosporium sativum* with a MIC of 50 µg/ml (Zou et al. 2000).

The fungal isolate 7303, a *Coniothyrium* species, was isolated from the halotolerant plant *Artemisia maritima* (Krohn et al. 2007a). The plant material was collected from Mandø in Denmark. Three new metabolites, massarilactones E-G and massarilactone acetonide (Fig. 8.22), named according to related metabolites from the freshwater fungus *Massarina tunicata*, were isolated from extracts of fungal cultures (Krohn et al. 2007a).

U. Metabolites from Endophytes from *Schinus molle*

The genus *Schinus* comprises 30 species, which are originally widespread from Mexico to Argentina (Roth et al. 1994). *S. molle* is used in ethnopharmacology as a vulnerary and anti-inflammatory agent (Quiroga et al. 2001) and the fruits are used as spice. Extracts of *S. molle* showed antifungal activities against several fungi (Quiroga et al. 2001). The oil of *S. molle* exhibited fungitoxic activity against some common storage and animal pathogenic fungi (Dikshit et al. 1986).

Artemisia new compound, DW E01141-1, was isolated from an endophytic *Colletotrichum gloeosporioides*, isolated from leaves of *S. molle* (Fig. 8.22; Weber 2006). 2,7-Dimethoxy-6-(1-acetoxyethyl)-juglon was purified from *Botryosphaeria parva*, which was isolated from small woody twigs of *S. molle* (Weber 2006). DW E01141-1 exhibited no antimicrobial, cytotoxic, phytotoxic, or nematicidal activities (Weber 2006). 2,7-Dimethoxy-6-(1-acetoxyethyl)-juglon

is an orange pigment described from *Hendersonula toruloidea* (Fig. 8.22; van Eijik and Roeymans 1978) and *Kirschsteiniothelia* sp. (Poch et al. 1992).

V. Conclusion

Endophytic fungi are still an interesting source for new natural products. New metabolites together with already known compounds are isolated from cultures of endophytic fungi. This confirms the hypothesis of Seymour et al. (2004) that endophytic isolates which are genotypically similar show considerable metabolite variability. The assessment of the potential metabolic diversity of a fungal species cannot be adequately performed by investigating a single "representative" isolate (Seymour et al. 2004). Fungal species possess an interesting potential for secondary metabolic variation.

There is an increasing number of reports of endophytic fungi that produce metabolites originally known from their host plants. This was first described for taxol produced by *Pestalotiopsis andreanae* and *Taxus brevifolia*. It can be assumed that there occurred a horizontal gene transfer, but the details of this gene transfer are still unknown. These new publications support the hypothesis that endophytic fungi adapt to the microenvironment in the host plants and are able to uptake plant DNA into their genome.

The isolation of endophytes from plants is described in various publications. In the past years many authors reported in detail the isolation of endophytic fungal communities from host plants (Kumar and Hyde et al. 2004; Nalini et al. 2005; Raviraja 2005; Tejesvi et al. 2005; Gond et al. 2007; El-Zayat et al. 2008) or the screening of crude extracts of endophytic fungi for bioactivity (Wiyakrutta et al. 2004; Li et al. 2005). Since it is often mentioned that the isolation is performed in order to investigate the secondary metabolites of the endophytic fungi, it can be expected that the literature on natural compounds from endophytes will increase constantly in the next years.

References

Abdin MZ, Israr M, Rehman RU, Jain SK (2003) Artemisinin, a novel antimalarial drug: biochemical and molecular approaches for enhanced production. Planta Med 69:289–299

Agusta A, Maehara S, Ohashi K, Simanjuntak P, Shibuya H (2005) Stereoselective oxidation at C-4 of flavans by the endophytic fungus *Diaporthe* sp. isolated from a tea plant. Chem Pharm Bull 53:1565–1569

Agusta A, Ohashi K, Shibuya H (2006) Bisanthraquinone metabolites produced by the endophytic fungus *Diaporthe* sp. Chem Pharm Bull 54:579–582

Ahn SH, Mun YJ, Lee SW, Kwak S, Choi MK, Baik SK, Kim YM, Woo WH (2006) *Selaginella tamariscina* induces apoptosis via a caspase-3-mediated mechanism in human promyelocytic leukema cells. J Med Food 9:138–144

Aktories K, Förstermann U, Hofmann F, Starke K (2005) Allgemeine und spezielle Pharmakologie und Toxikologie. Elsevier, Urban und Fischer, München Jena

Aly AH, Edrada-Ebel R, Indriani ID, Wray V, Müller WEG, Totzke F, Zirrgiebel U, Schächtele C, Kubbutat MHG, Lin WH, Proksch P, Ebel R (2008) Cytotoxic metabolites from the fungal endophyte *Alternaria* sp. and their subsequent detectionin its host plant . J Nat Prod 71: 972–980

Arechavaleta M, Bacon CW, Hoveland CS, Radcliff DE (1989) Effect of the tall fescue endophyte on plant response to environmental stress. Agron J 81:83–90

Arnold AE, Mejia LC, Kyllo D, Rojas EI, Maynard Z, Robbins N, Herren EA (2003) Fungal endophytes limit pathogen damage in a tropical tree. Ecology 100:15649–15654

Balint GA (2001) Artemisinin and its derivatives; an important new class of antimalarial agents. Pharmacol Ther 90:261–265

Bennett JW, Klich M (2003) Mycotoxins. Clin Microbiol Rev 16:497–516

Bent S, Goldberg H, Padula A, Avins AL (2005) Spontaneous Bleeding associated with *Ginkgo biloba*: A case report and systematic review of the literatur. J Gen Intern Med 20:657–661

Biswas K, Chattopadhyay I, Banerjee RK, Bandyopadhyay U (2002) Biological activities and medicinal properties of neem (*Azadirachata indica*). Curr Sci 82: 1336–1345

Bonfante P, Perotto S (1995) Strategies of arbuscular mycorrhizal fungi when infecting host plants. New Phytol 130:3–21

Bradshaw AD (1959) Population differentiation in *Agrostis tenuis* Sibth. II The incidence and significance of infection by *Epichloe typhina*. New Phytol 58:310–315

Brady SF, Clardy J (2000) CR377, a new pentaketide antifungal agent isolated from an endophytic fungus. J Nat Prod 63:1447–1448

Brady SF, Wagenaar MM, Singh MP, Janso JE, Clardy J (2000a) The cytosporones, new octaketide antibiotics isolated from an endophytic fungus. Org Lett 2:4043–4046

Brady SF, Singh MP, Janso JE, Clardy J (2000b) Cytoskyrins A and B, new BIA active bisanthraquinones isolated from an endophytic fungus. Org Lett 2:4047–4049

Brahmachari G (2004) Neem – an omnipotent plant: a retrospection. Chembiochem 5:408–421

Brown MD (1999) Green tea (*Camellia sinensis*) extract and its possible role in the prevention of cancer. Altern Med Rev 4:360–370

Busam L, Habermehl GG (1982) Accumulation of mycotoxins by *Baccharis coridifolia*: A reason of livestock poisoning. Naturwissenschaften 69:392–393

Camarda L, Merlini L, Nasini G (1976) Metabolites of *Cercospora taiwapyrone*, an α-pyrone of unusual structure from *Cercospora taiwanensis*. Phytochemistry 15:537–539

Cassady JM, Chan KK, Floss HG, Leistner E (2004) Recent developments in the maytansinoid antitumor agents. Chem Pharm Bull 52:1–26

Chan PC, Xia Q, Fu PP (2007) *Ginkgo biloba* leave extract: biological, medicinal, and toxicological effects. J Environ Sci Health C Environ Carcinog Ecotoxicol Rev 25:211–244

Chapela IH (1989) Fungi in healthy stems and branches of American beech and aspen: a comparative study. New Phytol 113:65–75

Chapela IH, Boddy L (1988) Fungal colonization of attached beech branches I. Early stages of development of fungal communities. New Phytol 110:39–45

Chapela IH, Petrini O, Hagmann L (1991) Monolignol glucosides as specific recognition messengers in fungal/plant symbiosis. Physiol Mol Plant Pathol 39:289–298

Chapela IH, Petrini O, Biesle G (1993) The physiology of ascospore eclosion in *Hypoxylon fragiforme*; mechanisms in the early recognition and establishment of an endophytic symbiosis. Mycol Res 97:157–162

Chen LG, Yang LL, Wang CC (2008) Anti-inflammatory activity of mangostins from *Garcinia mangostana*. Food Chem Toxicol 46:688–693

Cheng XL, Ma SC, Yu JD, Yang SY, Xiao XY, Hu JY, Lu Y, Shaw PC, But PPH, Lin RC (2008) Selaginellin A and B, two novel natural pigments isolated from *Selaginella tamariscina*. Chem Pharm Bull 56:982–984

Chomcheon P, Sriubolmas N, Wiyakrutta S, Ngamrojanavanich N, Chaicit N, Mahidol C, Ruchirawat S, Kittakoop P (2006) Cyclopentenones, scaffolds for organic syntheses produced by the endophytic fungus mitosporic *Dothideomycete* sp. LRUB20. J Nat Prod 69:1351–1353

Clavin M, Lorenzen K, Mayer A, Martino V, Anke T (1999) Biological activities in medicinal species of *Eupatorium*. Acta Hort 501:277–281

Clavin M, Gorzalczany S, Mino J, Kadarian C, Martino V, Ferraro G, Acevedo C (2000a) Antinociceptive effect of some Argentine medicinal species of *Eupatorium*. Phytother Res 14:275–277

Clavin M, Ferraro G, Coussio J, Martino V, Garcia G, Campos R (2000b) Actividad antiherpetic "in vitro" en extractos de 5 especies de Eupatorium. Anales de Saipa N° 16:131–134

Clay K (1988a) Clavicipitaceous fungal endophytes of grasses: Coevolution and the change from parasitism to mutualism. In: Pirozynski KA, Hawksworth DL (eds) Coevolution of fungi with plants and animals, Academic Press, London. pp 79–105

Clay K (1988b) Fungal endophytes of grasses: a defensive mutualism between plants and fungi. Ecology 69:10–16

Claydon N, Grove JF, Pople M (1985) Elm bark beetle boring and feeding deterrents from *Phomopsis oblonga*. Phytochemistry 24:937–943

Dahdouh-Guebas F, Collin S, Seen DL, Rönnbäck P, Depommier D, Ravishankar T, Koedam N (2006) Analysing ethnobotanical and fishery-related importance of mangroves of the East-Godavari Delta (Andhhra Pradesh, India) for conservation and management purposes. J Ethnobiol Ethnomed 2:24

Dai J, Krohn K, Flörke U, Draeger S, Schulz B, Kiss-Szikszai A, Antus S, Kurtán T, van Ree T (2006) Metabolites from the endophytic fungus *Nodulisporium* sp. from *Juniperus cedre*. Eur J Org Chem 15:3498–3506

Dalsgaard PW, Larsen TO, Christophersen C (2005) Bioactive cyclic peptides from the psychrotolerant fungus *Penicillium algidum*. J Antibiot 58:141–144

de Bary A (1879) Die Erscheinungen der Symbiose. Verlag von Karl J. Trübner, Strassbourg

Descoins C Jr, Bazzocchi IL, Ravelo AG (2002) New sesquiterpenes from *Euonymus europaeus* (Celastraceae). Chem Pharm Bull 50:199–202

Dikshit A, Naqvi AA, Husain A (1986) *Schinus molle*, a new source of natural fungitoxicant. Appl Environ Microbiol 51:1085–1088

Ding G, Song YC, Chen JR, Xu C, Ge HM, Wang XT, Tan RX (2006) Chaetoglobosin U, a cytochalasan alkaloid from endophytic *Chaetomium globosum* IFB-E019. J Nat Prod 69:302–304

Ding G, Liu S, Guo L, Zou Y, Che Y (2008) Antifungal metabolites from the plant endophytic fungus *Pestalotiopsis foedan*. J Nat Prod 71:615–618

Dong Q, Jao J, Fang JN, Ding K (2007) Structural characterization and immunological activity of two cold-water extractable polysaccharides from *Cistanche deserticola* Y. C. Ma. Carbohydr Res 342:1343–1349

Dreyfuss MM, Chapela IH (1994) Potential of fungi in the discovery of novel, low-molecular weight pharmaceuticals. In: Gullo VP (ed) The discovery of natural products with therapeuthic potential. Butterworth-Heinemann, London, pp 49–80

Dugoua JJ, Mills E, Perri D, Koren G (2006) Safety and efficacy of Ginkgo (*Ginkgo biloba*) during pregnancy and lactation. Can J Clin Pharmacol 13:277–284

El-Zayat SA, Nassar MSM, El-Hissy FT, Abdel-Motaal FF, Ito SI (2008) Mycoflora associated with *Hyoscyamus muticus* growing under an extremely arid desert environment (Aswan region, Egypt). J Basic Microbiol 48:82–92

Endo A, Hasumi K (1997) Mevinic acids. In: Anke T (ed) Fungal biotechnology. Chapman & Hall, London, pp 162–172

Eyberger AL, Dondapati R, Porter JR (2006) Endophyte fungal isolates from *Podophyllum peltatum* produce podophyllotoxin. J Nat Prod 69:1121–1124

Fang XK, Gao Y, Yang HY, Lang SM, Wang QJ, Yu BY, Zhu DN (2008) Alleviating effects of active fraction of *Euonymus alatus* abundant in flavonoids on diabetic mice. Am J Chin Med 36:125–140

Filip P, Weber RWS, Sterner O, Anke T (2003) Hormonemate, a new cytotoxic and apoptosis-inducing compound from the endophytic fungus *Hormonema dematioides*. I. Identification of the producing strain, and isolation and biological properties of hormonemate. Z Naturforsch 58c:547–552

Freeman EM (1902) The seed fungus of *Lolium temulentum*, L., the Darnel. Philos Trans R Soc London Bio1:1–27

Freeman S, Rodriguez RJ (1993) Genetic conversion of a fungal plant pathogen to a nonpathogenic, endophytic mutualist. Science 260:75–78

Frisvad JC, Larsen TO, Dalsgaard PW, Seifert KA, Louis-Seize G, Lyhne EK, Jarvis BB, Fettinger JC, Overy DP (2006) Four psychrotolerant species with high chemical diversity consistently producing cycloaspeptide A, *Penicillium jamesonlandense* sp. nov., *Penicillim ribium* sp. nov., *Penicillium soppii* and *Penicillium lanosum*. Int J Syst Evol Microbiol 56:1427–1437

Ge HM, Shen Y, Zhu HC, Tan SH, Ding H, Song YS, Tan RX (2008) Penicidones A-C, three cytotoxic alkaloidal metabolites of an endophytic *Penicillium* sp. Phytochemistry 69:571–576

Gianotti RL, Bomblies A, Dafalla M, Issa-Arzika I, Duchemin JB, Eltahir EAB (2008) Efficacy of local neem exracts for sustainable malaria vector control in an African village. Malar J 7:138–148

Girault L (1987) Kallawaya. Curanderos itinerantes de los Andes. Unicef/OPS/OMS, La Paz, p. 462

Gond SK, Verma VC, Kumar A, Kumar V, Kharwar RN (2007) Study of endophytic fungal community from different parts of *Aegle marmelos* Correae (Rutaceae) from Varanasi (India). World J Microbiol Biotechnol 23:1371–1375

Govindachari TR, Ravindranath KR, Viswanathan N (1974) Mappicine, a minor alkaloid from *Mappia foetida* Miers. J Chem Soc Perkin Trans I 11:2215–1217

Grove JF, Pople M (1979) Metabolic products of *Fusarium lavarum* Fuckel. The fusarentins and the absolute configuration of moncerin. J. C. S. Perkin 1:2048–2051

Gu W, Ge M, Song YM, Ding H, Zhu HL, Zhao XA, Tan RX (2007) Cytotoxic benzo[j]fluoranthene metabolites from *Hypoxylon truncatum* IFB-18, an endophyte of *Artemisia annua*. J Nat Prod 70:114–117

Gunatilaka AAL (2006) Natural products from plant-associated microorganisms: Distribution, structural diversity, bioactivity, and implications of their occurence. J Nat Prod 69:509–526

Hamilton-Miller JMT (1995) Antimicrobial properties of tea (*Camellia sinensis* L.). Antimicrob Agents Chemother 39:2375–2377

Hamilton-Miller JMT (2001) Anti-carciogenic properties of tea (*Camellia sinensis*). J Med Microbiol 50:299–302

Hashida C, Tanaka N, Kashiwada Y, Ogawa M, Takaishi Y (2008) Prenylated chloroglucinol derivates from *Hypericum perforatum* var. *angustifolium*. Chem Pharm Bull 56:1164–1167

Hawksworth DL (2001) The magnitude of fungal diversity: the 1.5 million species estimated revisited. Mycol Res 105:1422–1432

Hayashi A, Fujioka S, Nukina M, Kawano T, Shimada A, Kimura Y (2007) Fumiquinones A and B, nematicidal quinones produced by *Aspergillus fumigatus*. Biosci Biotechnol Biochem 71:1697–1702

Hellwig V, Grothe T, Mayer-Bartschmid A, Endermann R, Geschke FU, Henkel T, Stadler M (2002) Altersetin, a new antibiotic from cultures of endophytic *Alternaria* spp. Taxonomy, fermentation, isolation, structure elucidation and biologic activities. J Antibiot 55:881–892

Hill NS, Belesky DP, Stringer WC (1991) Competitiveness of tall fescue as influenced by *Acremonium coenophialum*. Crop Sci 31:185–190

Hu HJ, Wang KW, Wu B, Sun CR, Pan YJ (2005) Chemical shift assignments of two oleanane triterpenes from *Euronymus hederaceus*. J Zhejiang Univ Sci B 6:719–721

Huang WY, Cai YZ, Hyde KD, Corke H, Sun MEI (2007) Endophytic fungi from *Nerium oleander* L (Apocynaceae): main constituents and antioxidant activity. World J Microbiol Biotechnol 23:1253–1263

Iharlegui L, Hurrel J (1992) Asteraceae de interes ethnobotanico de los departamentos de Santa Victoria e Iruya (Salta, Argentinien). Ecognicion 3:3–18

Isaka M, Jaturapat A, Rukseree K, Danwisetkanjana K, Tanticharoen M, Thebtaranonth Y (2001) Phomaxanthones A and B, novel xanthone dimers from the endophytic fungus *Phomopsis* species. J Nat Prod 64:1015–1018

Ismail IS, Ito H, Mukainaka T, Higashihara H, Enja F, Tokuda H, Nishina H, Yoshida T (2003) Ichthyotoxic and anticarcinogenic effects of triterpenoids from *Sandoricum koetjape* Bark. Biol Pharm Bull 26:1351–1353

Ismail IS, Ito H, Hatano T, Taniguchi S, Yoshida T (2004) Two new analogues of trijugin-type limonoids from the leaves of *Sandoricum koetjape*. Chem Pharm Bull 52:1145–1147

Jennewein S, Croteau R (2001) Taxol: biosynthesis, molecular genetics, and biotechnological applications. Appl Microbiol Biotechnol 57:13–19

Jiang Y, Tu PF (2009) Analysis of chemical constituents in *Cistanche* species. J Chromatogr A 1216:1970–1979

Johnston D, Couché E, Edmonds DJ, Muir KW, Procter DJ (2003) The first synthetic studies on pestalotiopsis A. A stereocontrolled approach to the functionalised bicyclic core. Org Biol Chem 1:328–337

Kackley KE, Grybauskas AP, Dernoeden PH, Hill RL (1990) Role of drought stress in the development of summer patch in field-inoculated Kentucky bluegrass. Phytopathology 80:665–658

Kang DG, Yin MH, Oh H, Lee DH, Lee HS (2004) Vasorelaxation by amentoflavone isolated from *Selaginella tamariscina*. Planta Med 70:718–722

Kim CH, Kim DI, Kwon CN, Kang SK, Jin UH, Suh SJ, Lee TK, Lee IS (2006) *Euonymus alatus* (Thunb.) Sieb induces apoptosis via mitochondrial pathway as prooxidant in human uterine leiomyomal smooth muscle cells. Int J Gynecol Cancer 16:843–848

Kimmons CA, Gwinn KD, Bernard EC (1990) Nematode reproduction on endophyte-infected and endophyte-free tall fescue. Plant Dis 74:757–761

Krohn K, Ullah Z, Hussain H, Flörke U, Schulz B, Draeger S, Pescitelli G, Salvadori P, Antus S, Kurtán T (2007a)

Massarilactones E-G, new metabolites from the endophytic fungus *Coniothyrium* sp., associated with the plant *Artemisia maritima*. Chirality 19:464–470

Krohn K, Kock I, Elsässer B, Flörke U, Schulz B, Draeger S, Pescitelli G, Antus S, Kurtán T (2007b) Bioactive natural products from the endophyic fungus *Ascochyta* sp. from Meliotus dentatus – configurational assignment by solid-state CD and TDDFT calculations. Eur J Org Chem 7:1123–1129

Kucht S, Groß J, Hussein Y, Grothe T, Keller U, Basar S, König WA, Steiner U, Leistner E (2004) Elimination of ergoline alkaloids following treatment of *Ipomoea asarifolia* (Convolvulaceae) with fungicides. Planta 219:619–625

Kumar DS, Hyde KD (2004) Biodiversity and tissue-recurrence of endophytic fungi in *Tripterygium wilfordii*. Fungal Divers 17:69–90

Kusari S, Lamshöft M, Zühlke S, Spiteller M (2008) An endophytic fungus from *Hypericum perforatum* that produces hypericin. J Nat Prod 71:159–162

Lambert JD, Yang CS (2003) Mechanisms of cancer prevenion by tea constituents. J Nutr 133:3262–3267

Langford SD, Boor PJ (1996) Oleander toxicity: an examination of human and animal toxic exposures. Toxicology 3:1–13

Li E, Tian R, Liu S, Chen X, Guo L, Che Y (2008) Pestalotheols A-D, bioactive metabolites from the plant endophytic fungus *Pestalotiopsis theae*. J Nat Prod 71:664–668

Li H, Qing C, Zhang Y, Zhao Z (2005) Screening for endophytic fungi with antitumour and antifungal activities from Chinese medicinal plants. World J Microbiol Biotechnol 21:1515–1519

Lin X, Lu C, Huang Y, Zheng Z, Su W, Shen Y (2007) Endophytic fungi from a pharmaceutical plant, *Camptotheca acuminata*: isolation, identification and bioactivity. Worl J Microbiol Biotechnol 23:1037–1040

Lin ZJ, Lu ZY, Zhu TJ, Fang YC, Gu QQ, Zhu WM (2008a) Penicillenols from *Penicillium* sp. GQ-7, an endophytic fungus associated with *Aegiceras corniculatum*. Chem Pharm Bull 56:217–221

Lin ZJ, Wen JG, Zhu TJ, Fang YC, Gu QQ, Zhu WM (2008b) Chrysogenamide A from an endophytic fungus associated with *Cistanche deserticola* and its neuroprotective effect on SH-SY5Y cells. J Antibiot 61:81–85

Liu JY, Liu CH, Zou WX, Tan RX (2003) Leptosphaeric acid, a metabolite with a novel carbon skeleton from *Leptosphaeria* sp. IV403, an endophytic fungus in *Artemisia annua*. Helv Chim Acta 86:657–660

Liu X, Dong M, Chen X, Jiang M, Lv X, Zou J (2008) Antimicrobial activity of an endophytic *Xylaria* sp. YX-28 and identification of its antimicrobial compound 7-amino-4-methylcoumarin. Appl Microbiol Biotechnol 78:241–247

Lu YH, Liu ZY, Wang ZT, Wei DZ (2006) Quality evaluation of *Platycladus orientalis* (L.) Franco through simultaneous determination of four bioactive flavonoids by high-performance liquid chromatography. J Pharm Biomed Anal 41:1186–1190

Luper S (1999) A review of plants used in the treatment of liver disease: part two. Altern Med Rev 4:178–188

Ma LY, Ma SC, Wei F, Lin RC, But PPH, Lee SHS, Lee SF (2003) Uncinoside A and B, two new antiviral chormone glycosides from *Selaginalla uncinata*. Chem Pharm Bull 51:1264–1267

Magnani RF, Rodigues-Fo E, Daolio C, Ferreira AG, de Souza AQL (2003) Three highly oxygenated caryophyllene sequiterpenes from *Pestalotiopsis* sp., a fungus isolated from bark of *Pinus taeda*. Z Naturforsch 58c:319–324

Maloney N, Hao W, Xu J, Gibbons J, Hucul J, Roll D, Brady SF, Schroeder FC, Clardy J (2006) Phaeosphaeride A, an inhibitor o STAT3-dependent signalling isolated from an endophytic fungus. Org Lett 8:4067–4070

Meister J, Weber D, Martino V, Sterner O, Anke T (2007) Phomopsidone, a novel depsidone from an endophyte of the medicinal plant *Eupatorium arnottianum*. Z Naturforsch C 62:11–15

Meng LH, Liao ZY, Pommier Y (2003) Non-camptothecin DNA topoisomerase I inhibitors in cancer therapy. Curr Top Med Chem 3:305–320

Miller AL (1998a) St. John's wort (*Hypericum perforatum*): clinical effects on depression and other conditions. Altern Med Rev 3:18–26

Miller AL (1998b) Botanical influences on cardiovascular disease. Altern Med Rev 3:422–431

Miño J, Gorzalczany S, Moscatelli V, Ferraro G, Acevedo C, Hnatyszyn O (2002) Actividad antinociceptiva y antinflamatoria de *Erythrina crista-galli* (Ceibo). Acta Farm Bonaerense 21:93–98

Mitscher LA, Gollapudi RSR, Gerlach DC, Drake SD, Veliz EA, Ward JA (1988) Erycristin, a new antimicrobial petrocarpan from *Erythrina crista-galli*. Phytochemistry 27:381–385

Nalini MS, Madesh B, Tejesvi MV, Prakash HS, Subbaiah V, Kini KR, Shetty HS (2005) Fungal endophytes from the three-leaved caper, *Crataeva magna* (Lour.) DC. (Capparidaceae). Mycopathologia 159:245–249

Nebenführ A, Ritzenthaler C, Robinson DG (2002) Brefeldin A: deciphering an enigmatic inhibitor of secretion. Plant Physiology 130:1102–1108

Nishikawa W (1933) A metabolic product of *Aspergillus melleus* Yukawa. J Agric Chem Soc Japan Chem 9:772

Oberlies NH, Kroll DJ (2004) Camptothecin and taxol: historic achievements in natural product research. J Nat Prod 67:129–135

Park SH, Ko SK, Chung SH (2005) *Euronymus alatus* prevents the hyperglycemia and hyperlipidemia induced by high-fat diet in ICR mice. J Ethnopharmacol 102:326–335

Parker EJ, Scott DB (2004) Indole-diterpene biosynthesis in ascomycetous fungi. In: An Z (ed) Handbook of industrial mycology. Dekker, New York, pp 405–426

Pedraza-Chaverri J, Cárdenas-Rodríguez N, Orozo-Ibarra M, Pérez-Rojas JM (2008) Medicinal properties of mangosteen (*Garcinia mangostana*). Food Chem Toxicol 46:3227–3239

Penna C, Marino S, Gutkind G, Clavin M, Ferraro G, Martino V (1997) Antimicrobial activity of *Eupatorium* species growing in Argentina. J Herbs Spices Med Plants 5:21–28

Petrini O (1996) Ecological and physiological aspects of host-specificity in endophytic fungi. In: Redlin SC, Carris LM (eds) Endophytic funig in grasses and woody plants. Systematics, ecology, and evolution. APS, St Paul, pp 87–100

Petrini O, Petrini LE, Rodrigues KF (1995) Xylariaceous endophytes: an exercise in biodiversity. Fitopatol Bras 20:531–539

Pirozynski KA, Hawksworth DL (1988) Coevolution of fungi with plants and animals: Introduction and overview. In: Pirozynski KA, Hawksworth DL (eds.) Coevolution of Fungi with Plants and Animals. Academic Press, London, pp 1–29

Pirozynski KA, Malloch DW (1975) The origin of plants: a matter of mycotrophism. Biosystems 6:153–164

Poch GK, Gloer JB, Shearer CA (1992) New bioactive metabolites from a freshwater isolate of the fungus *Kirschsteiniothelia* sp. J Nat Prod 55:1093–1099

Pongcharoen W, Rukachaisirikul V, Phongpaichit S, Sakayaroj J (2007) A new didydrobenzofuran derivative from the endophytic fungus *Botryosphaeria mamane* PSU-M76. Chem Pharm Bull 55:1404–1405

Pornpakakul S, Liangsakul J, Ngamrojanavanich N, Roengsumran S, Sihanonth P, Piapukiew J, Sangvichien E, Puthong S, Petsom A (2006) Cytotoxic activity of four xanthones from *Emericella variecolor*, an endophytic fungus isolated from *Croton oblongifolius*. Arch Pharm Res 29:140–144

Pulici M, Sugawara F, Koshino H, Uzawa J, Yoshida S (1996) Pestalotiopsins A and B: new caryophyllenes from an endophytic fungus of *Taxus brevifolia*. J Org Chem 61:2122–2124

Pulici M, Sugawara F, Koshino H, Okada G, Esumi Y, Uzawa J, Yoshida S (1997) Metabolites of *Pestalotiopsis* spp., endophytic fungi of *Taxis brevifolia*. Phytochemistry 46:313–319

Puri SC, Verma V, Amna T, Qazi GN, Spiteller M (2005) An endophytic fungus from *Nothapodytes foedida* that produces camptothecin. J Nat Prod 68:1717–1719

Puri SC, Nazir A, Chawia R, Arora R, Riyaz-ul-Hasan S, Amna T, Ahmed B, Verma V, Singh S, Sagar R, Sharma A, Kumar R, Sharma RK, Oazi GN (2006) The endophytic fungus *Trametes hirsuta* as a novel alternative source of podophyllotoxin and related aryl tetralin lignans. J Biotechnol 122:494–510

Quiroga EN, Sampietro AR, Vattuone MA (2001) Screening antifungal activities of selected medicinal plants. J Ethnopharmacol 74:89–96

Rasadah MA, Khozirah S, Aznie AA, Nik MM (2004) Antiinflammatory agents from *Sandoricum koetjape* Merr. Phytomedicine 11:262–263

Raviraja NS (2005) Fungal endophytes in five medicinal plant species from Kudremukh Range, Western Ghats of India. J Basic Microbiol 45:230–235

Read JC, Camp BJ (1986) The effect of fungal endophyte *Acremonium coenophialum* in tall fescue on animal performance, toxicity, and stand maintenance. Agron J 78:848–850

Ren Y, Strobel GA, Graff JC, Jutila M, Park SG, Gosh S, Teplow D, Condrom M, Pang E, Hess WM, Moore E (2008) Colutellin A, an immunosuppressive peptide from *Colletotrichum dematium*. Microbiology 154:1973–1979

Rizzo I, Varsavky E, Haidukowski M, Frade H (1997) Macrocyclic trichothecenes in *Baccharis coridifolia* plants and endophytes and *Baccharis artemisioides* plants. Toxicon 35:753–757

Rogers JD (1979) The Xylariaceae: Systematic, biological and evolutionary aspects. Mycologia 71:1–42

Roome T, Dar A, Ali S, Naqvi S, Choudhary MI (2008) A study on antioxidant, free radical scavenging, anti-inflammatory and hepatoprotective actions of *Aegiceras corniculatum* (stem) extracts. J Ethnopharmacol 13:514–521

Roth L, Daunderer M, Kormann K (1994) Giftpflanzen Pflanzengifte. Nikol, Hamburg

Rukachaisirikul V, Kaeobamrung J, Panwiriyarat W, Saitai P, Sukpondma Y, Phongpaichit S, Sakayaroj J (2007) A new pyrone derivative from the endophytic fungus *Penicillium paxilli* PSU-A71. Chem Pharm Bull 55:1383–1384

Saikkonen K, Faeth SH, Helander M, Sullivan TJ (1998) Fungal endophytes: a continuum of interactions with host plants. Annu Rev Ecol Syst 29:319–343

Schardl CL, Blankenship JD, Spiering MJ, Machado C (2004) Loline and ergot alkaloids in grass endophytes. In: An Z (ed) Handbook of industrial mycology. Dekker, New York, pp 427–448

Schardl CL, Panaccione DG, Tudzynski P (2006) Ergot alkaloids – biology and molecular biology. Alkaloids 63:47–88

Schmeda-Hirschmann G, Hormazabal E, Astudillo L, Rodriguez J, Theoduloz C (2005) Secondary metabolites from endophytic fungi isolated from the Chilean gymnosperm *Prumnopitys andina* (Lleuque). World J Microbiol Biotechnol 21:27–32

Schmeda-Hirschmann G, Hormazabal E, Rodriguez JA, Theoduloz C (2008) Cycloaspeptide A and pseurotin A from the endophytic fungus *Penicillium janczewskii*. Z Naturforsch 63c:383–388

Seto Y, Takahashi K, Matsuura H, Kogami Y, Yada H, Yoshihara T, Nabeta K (2007) Novel cyclic peptide, epichlicin, from the endophytic fungus, *Epichloe typhina*. Biosci Biotechnol Biochem 71:1470–1475

Seymour FA, Cresswell JE, Fisher PJ, Lappin-Scott HM, Haag H, Talbot NJ (2004) The influence of genotypic variation of two endophytic fungal species. Fungal Genet Biol 41:721–734

Shiono Y, Muranyama T (2005) New eremophilane-type sequiterpenoids, eremoxylarins A and B from xylariaceous endophytic fugus YUA-026. Z Naturforsch 60b:885–890

Shiono Y, Tsuchinari M, Shimanuki K, Miyajima T, Murayama T, Koseki T, Laatsch H, Funakoshi T, Takanami K, Suzuki K (2007) Fusaristatins A and B, two new cyclic lipopeptides from an endophytic *Fusarium* sp. J Antibiot 60:309–316

Silberstein SD, McCrory DC (2003) Ergotamine and dihydroergotamine: history, pharmacology, and efficacy. Headache 43:411–166

Silva GH, Teles HL, Zanardi LM, Marx Young MC, Eberlin MN, Hadad R, Pfenning LH, Costa-Neto CM, Castro-Gamboa I, da Silva Bolzani V, Araújo AR (2006) Cadinane sesquiterpenoids of *Phomopsis cassiae* an endophytic fungus associated with *Cassia spectabilis* (Leguminosae). Phytochemistry 67:1964–1969

Sinclair JB, Cerkauskas RF (1996) Latent infection vs endophytic colonization by fungi. In: Redlin SC, Carris LM (eds) Endophytic fungi in grasses and woody plants. systematics, ecology, and evolution. APS, St. Paul, pp 3–29

Soliman HSM, Simon A, Toth G, Duddeck H (2001) Identification and structure determination of four triterpene saponins from some Middle-East plants. Magn Reson Chem 39:567–576

Stierle A, Strobel G, Stierle D (1993) Taxol and taxane production by *Taxomyces andreanae*, an endophytic fungus of pazific yew. Science 260:214–216

Stone JK, Bacon CW, White JF Jr (2000) An overview of endophytic microbes: endophytism defined. In: Bacon CW, White JF (eds) Microbial endophytes. Dekker, New York, pp 389–420

Strobel G (2003) Endophytes as sources of bioactive products. Microbes Infect 5:535–544

Strobel G, Ford E, Worapong J, Harper JK, Arif AM, Grant DM, Fung PCW, Chau RMW (2002) Isopestacin, an isobenzofuranone from *Pestalotiopsis microspora*, possessing antifungal and antioxidant activities. Phytochemistry 60:179–183

Strobel G, Daisy B, Castillo U, Harper J (2004) Natural products from endophytic microorganisms. J Nat Prod 67:257–268

Suksamrarn S, Suwannapoch N, Phakhodee W, Thanuhiranlert J, Ratananukul P, Chimnoi N, Suksamrarn A (2003) Antimycobacterial activity of prenylated xanthones from the fruits of *Garcinia mangostana*. Chem Pharm Bull 51:857–859

Takeuchi N, Goto K, Sasaki Y, Fujeta T, Okazaki K, Kamata K, Tobinaga S (1992) Synthesis of (+)- and (−)-mellein utilizing an annelation reaction of isoxazoles with dimethyl-3-oxoglutarate. Heterocycles 33:357–374

Tan Q, Yan X, Lin X, Huang Y, Zheng Z, Song S, Lu C, Shen Y (2007) Chemical constituents of the endophytic fungal strain *Phomopsis* sp. NXZ-05 of *Camptotheca acuminata*. Helv Chim Acta 90:1811–1817

Tan RX, Zou WX (2001) Endophytes: a rich source of functional metabolites. Nat Prod Rep 18:448–459

Tan RX, Zheng WF, Tang HQ (1998) Biologically active substances from the genus *Artemisia*. Planta Med 64:295–302

Tansuwan S, Pornpakakul S, Roengsumran S, Petsom A, Muangsin N, Sihanonta P, Chaichit N (2007) Antimalarial benzoquinones from an endophytic fungus *Xylaria* sp. J Nat Prod 70:1620–1623

Tejesvi MV, Madesh B, Nalini MS, Prakash HS, Kini KR, Subbiah V, Shetty HS (2005) Endophytic fungal assemblages from inner bark and twig of *Terminalia arjuna* W. & A. (Combretaceae). World J Microbiol Biotechnol 21:1535–1540

Toursarkissian M (1980) Plantas Medicinales de la Argentina. Hemisferio Sur, Buenos Aires, p. 70

Uecker FA (1988) A world list of *Phomopsis* names with notes on nomenclature, morphology and biology. Cramer, Berlin van Eijik GW, Royemans HJ (1978) Naphthoquinone derivatives from the fungus *Hendersonula toruloidea*. Experientia 34:1257–1258

Viegas C Jr, Alexandre-Moreira MS, Fraga CAM, Barreiro EJ, da Silva Bolzani V, de Miranda ALP (2008) Antinociceptive profile of 2,3,6-trisubstituted piperidine alkaloids: 3-*O*-acetyl-spectaline and semi-synthetic derivatives of (−)-spectaline. Chem Pharm Bull 56:407–412

Vogl AE (1898) Mehl und die anderen Mehlprodukte der Cerealien und Leguminosen. Nahrungsm Unters Hyg Warenk 12:25–29

Wagennaar MM, Clardy J (2001) Dicerandrols, new antibiotic and cytotoxic dimers produced by the fungus *Phomopsis longicolla* isolated from an endangered mint. J Nat Prod 64:1006–1009

Walker K, Croteau R (2001) Molecules of interest. Taxol biosynthetic genes. Phytochemistry 58:1–7

Wall ME, Wani MC, Cook CE, Palmer KH, McPhail AT, Sim GA (1966) Plant antitumor agents. I The isolation and structure of camptothecin, a novel alkaloid leukemia and tumor inhibitor from *Camptotheca acuminata*. J Am Chem Soc 88:3888–3890

Wallner BM, Booth NH, Robbins JD, Bacon CW, Porter JK, Kiser TE, Wilson RW, Johnson B (1983) Effect of an endophytic fungus isolated from toxic pasture grass on serum prolactin concentrations in the lactation cow. Am J Vet Res 44:1317–1322

Wang C, Wu J, Mei X (2001) Enhancement of taxol production and excretion in *Taxus chinensis* cell culture by fungal elicitation and medium renewal. Appl Microbiol Biotechnol 55:404–410

Wani MC, Taylor HL, Wall ME, Coggon P, McPhail AT (1971) Plant antitumor agents. VI. The isolation and structure of taxol, a novel antileukemic and antitumor agent from *Taxus brevifolia*. J Am Chem Soc 93:2325–2327

Webber J (1981) A natural biological control of dutch elm disease. Nature 292:449–451

Weber D (2006) Biologisch aktive Metabolite endophytischer Pilze argentinischer Medizinalpflanzen und anderer Pilze. Dissertation, University of Kaiserslautern

Weber D, Sterner O, Anke T, Gorzalczancy S, Martino V, Acevedo C (2004) Phomol, a new antiinflammatory metabolite from an endophyte of the medicinal plant *Erythrina crista-galli*. J Antibiot 57:559–563

Weber D, Gorzalczany S, Martino V, Acevedo C, Sterner O, Anke T (2005) Metabolites from endophytes of the medicinal plant *Erythrina crista-galli*. Z Naturforsch C 60:467–477

Weber RWS, Anke H (2006) Effects of endophytes on colonisation by leaf surface microbiota. In: Bailey MS, Lilley AK, Timms-Wilson TM, Spencer-Phillips PT (eds) Microbial ecology of plant surfaces. CABI, Wallingford

Weber RWS, Stenger E, Meffert A, Hahn M (2004) Brefeldin A productiuon by *Phoma medicaginis* in dead pre-colonized plant tissue: a strategy for habitat conquest? Mycol Res 108:662–671

Webster J, Weber RWS (2004) Teaching techniques for mycology: 23. Eclosion of *Hypoxylon fragiforme* ascospores as a prelude to germination. Mycologist 18:170–173

Wenke J (1993) Isolierung und Charakterisierung neuer Inhibitoren der Chitinsynthase aus höheren Pilzen. Dissertation, University of Kaiserslautern

West CP, Izekor E, Turner KE, Elmi AA (1993) Endophytes effects on growth and persistence of tall fescue along a water-supply gradient. Agron J 85:264–270

Whalley AJS (1996) The xylariaceous way of life. Mycol Res 100:897–922

White JF Jr, Cole GT (1985) Endophyte–host associations in forage grasses. III. In vitro inhibition of fungi by *Acremonium coenophialum*. Mycologia 77:487–489

White JF Jr, Reddy PV, Bacon CW (2000) Biotrophic endophytes of grasses: A systematic appraisal. In: Bacon CW, White JF (eds) Microbial endophytes. Dekker, New York, pp 49–62

Wijeratne EMK, Paranagama PA, Marron MT, Gunatilaka MK, Arnold AE, Gunatilaka AAL (2008) Sesquiterpene quinines and related metabolites from *Phyllosticta spinarum*, a fungal strain endophytic in *Platycladus orientalis* of the Sonoran Dessert. J Nat Prod 71:218–222

Wilson D (2000) Ecology of woody plant endopyhtes. In: Bacon CW, White JF (eds) Microbial endophytes. Dekker, New York, pp 389–420

Wiyakrutta S, Sriubolmas N, Panphut W, Thongon N, Danwiset-kanjana K, Ruangrungsi N, Meevootisom V (2004) Endophytic fungi with anti-microbial, anti-cancer and anti-malarial activities isolated from Thai medicinal plants. World J Microbiol Biotechnol 20:265–272

Woerdenbag HJ, Lugt CB, Pras N (1990) *Artemisia annua* L.: a source of novel antimalarial drugs. Pharm Weekbl Sci 19:169–181

Wu SH, Chen YW, Shao SC, Wang LD, Li ZY, Yang LY, Li SL, Huang R (2008) Ten-membered lactones from *Phomopsis* sp, an endophytic fungus of *Azadirachta indica*. J Nat Prod 71:731–734

Xu F, Zhang Y, Wang J, Pang J, Huang C, Wu X, She Z, Vrijmoed LLP, Jones EBG, Lin Y (2008) Benzofuran derivatives from the mangrove endophytic fungus *Xylaria* sp. (#2508). J Nat Prod 71:1251–1253

Yang SF, Prabhu S, Landau J (2001) Prevention of carcinogenesis by tea phenols. Drug Metab Rev 33:237–253

Yang SF, Chu SC, Liu SJ, Chen YC, Chang YZ, Hsieh YS (2007) Antimetastatic activities of *Salaginella tamariscina* (Beauv.) on lung cancer cells in vitro and in vivo. J Ethnopharmacol 110:483–489

Yshihara T, Togiya S, Koshino H, Sakamura S, Shimanuki T, Sato T, Tajimi A (1985) Three fungitoxic sesquiterpenes from stromata of *Epichloe typhina*. Tetrahedron Lett 26:5551–5554

Zou WX, Meng JC, Lu H, Chen GX, Shi GX, Zhang TY, Tan RX (2000) Metabolites of *Colletotrichum gloeosporioides*, an endophytic fungus in *Artemisia mongolica*. J Nat Prod 63:1529–1530

9 Fungal Origin of Ergoline Alkaloids Present in Dicotyledonous Plants (Convolvulaceae)

ECKHARD LEISTNER[1], ULRIKE STEINER[2]

CONTENTS

Dedicated to Prof. Dr. Detlef Gröger on the occasion of his 80th birthday

I. The Ecological Role of Natural Products

Microorganisms and plants have one thing in common: both are equipped with a frequently elaborate biosynthetic machinery responsible for the formation of an almost unlimited variety of natural products. Typically, natural products –

which are also called secondary metabolites – are characteristic of a limited number of microbial or plant taxa, e.g. an order, a family, a species or even a subspecies only. Many of the natural products exhibit physiological activities which is the basis for their use in medical applications (Clardy and Walsh 2004).

The high physiological activities of many natural products triggered a now historical dispute about the role of natural products in the producing organism. It was proposed that "the multiplicity of natural products is caused by random processes of mutations, i.e. it reflects the gambling of nature rather than a sophisticated strategy" (Mothes 1981; Mothes et al. 1985).

This hypothesis, however, neglects the possibility that mutations may turn out to be detrimental or advantageous to the mutated organism. In the former case a mutated organism may be eliminated, or in the latter case may benefit from an increased fitness and a better chance to survive in a certain ecological setting (Zenk 1967). Today the ecological role of natural products is well accepted in the scientific community (Eisner 2003; White Jr. et al. 2003; Harborne 2004).

Natural product research entered a new era when it was discovered that plants and fungi elaborated during evolution another way to acquire natural products: Not only may they be formed in biosynthetic processes by one particular organism itself, but a host organism may instead harbor a natural product-producing microorganism: A plant may be associated with a bacterium (Piel 2004; Strobel 2004; Gunatilaka 2006) or a fungus (Strobel 2004; Puri et al. 2005; Gunatilaka 2006), while a fungus may harbor a bacterium (Partida-Martinez and Hertweck 2005).

It is remarkable that three important and frequently used cytostatic compounds employed in today's tumor therapy apparently are synthesized by plant-associated microorganisms: Vincristine

[1]Institut für Pharmazeutische Biologie, Rheinische Friedrich Wilhelm-Universität Bonn, Nussallee 6, 53115 Bonn, Germany; e-mail: eleistner@uni-bonn.de
[2]Institut für Nutzpflanzenwissenschaften und Ressourcenschutz (INRES), Rheinische Friedrich Wilhelm-Universität Bonn, Nussallee 9, 53115 Bonn, Germany; e-mail: u-steiner@uni-bonn.de

Physiology and Genetics, 1st Edition
The Mycota XV
T. Anke and D. Weber (Eds.)
© Springer-Verlag Berlin Heidelberg 2009

is produced by *Fusarium oxysporum* associated with *Catharanthus roseus* (Zhang et al. 2000; cited by Gunatilaka 2006), camptothecin, a lead compound in cancer research and therapy is produced by a fungal endophyte present in *Nothapodytes foetida* (Puri et al. 2005) and paclitaxel is believed to be formed by different fungi such as *Taxomyces andreanae* and different *Taxus* species, including *T. brevifolia* (Strobel et al. 2004; Leistner 2005). The latter case, however, may not yet be setteled because the genes in *Taxomyces andreanae* have not yet been reported whereas they are well known from the *Taxus brevifolia* plant (Croteau et al. 2006).

At present it is unclear whether these natural products and their biosynthetic machinery occur exclusively in the associated microorganism or in both the microorganism and the host plant. In such associations both organisms may be part of a symbiotum in which the associated microorganism benefits by receiving nutrients, protection, reproduction and dissemination, whereas the host takes advantage of physiologically active compounds which may promote plant growth, herbivore deterrence and/or increased fitness (Arnold et al. 2003; White Jr. et al. 2003; Saikkonen et al. 2004).

A point in case is the beneficial activity of ergoline alkaloids which are products of clavicipitaceous fungi (Clay and Schardl 2002; White Jr. et al. 2003; Schardl et al. 2006) colonizing grasses like Poaceae, Juncaceae and Cyperaceae.

In the current literature (including our own publications) two different adjectives for the fungi described herein are used: These are "clavicipitaceous" and "clavicipitalean". The former term refers to the family Clavicipitaceae but the latter to the old order of Clavicipitales. Since the family Clavicipitaceae is now generally accepted to belong to the order Hypocreales (Sung et al. 2007) only the term "clavicipitaceous" should be used. We are grateful to Dr. Chris Schardl (Lexington, Kentucky, USA) for bringing this to our attention.

Ergoline alkaloids, however, are also present in higher dicotyledonous plants of the family Convolvulaceae (Hofmann 1961, 2006). This disjointed occurrence of a group of natural products in evolutionarily unrelated taxa (fungi, Convolvulaceae plants) seemed to contradict the generally accepted principle of chemotaxonomy that similar or even identical natural products are present in related taxa. It was therefore assumed that during evolution a horizontal transfer of genes responsible for ergoline alkaloid biosynthesis might have occurred from fungi to higher plants (Groeger and Floss 1998; Tudzynski et al. 2001; Clay and Schardl 2002). Alternatively it was discussed that ergoline alkaloid biosynthesis was repeatedly invented during evolution (Mothes et al. 1985). In a recent review in this series Keller and Tudzinsky (2002) dealt with the pharmacological aspects, biochemistry, genetics and biotechnology of ergoline alkaloids in fungi associated with Poaceae. We show in the present review that neither the horizontal transfer of genes encoding the ergoline alkaloid biosynthesis, nor the repeated invention of a rather complicated biosynthetic pathway took place during evolution, but rather that clavicipitaceous fungi not only live on different grasses but also colonize plant species of the dicotyledonous family of Convolvulaceae (Kucht et al. 2004; Steiner et al. 2006; Ahimsa-Mueller et al. 2007; Markert et al. 2008; Steiner et al. 2008). This indicates that ergoline alkaloids are components in a fungus/plant symbiotum characterized by mutual defense.

II. The Symbiosis Between Poaceae and Clavicipitaceous Fungi

A rather well investigated experimental system consists of clavicipitaceous fungi which colonize Juncaceae, Cyperaceae and Poaceae plants. In these symbiota ergoline alkaloids play an important role (Keller and Tudzynski 2002). The symbiotic fungi belong either to the tribe Clavicipeae or Balanseae within the family Clavicipitaceae (Bacon and Lyons 2005). The morphological associations of the fungi with grasses is either epicuticular, epibiotic or endophytic (Bacon and Lyons 2005). In epiphytic growth the fungal mycelium is concentrated on the surface of young leaves, buds, meristematic regions and reproductive structures (Clay and Schardl 2002). The association between fungi and their plant hosts is likely to be an example of host-symbiont codivergence (Schardl et al. 2008).

The fungus may be asexual belonging to the group of fungi imperfecti, show a sexual lifestyle or switch between sexual and asexual propagation. In the sexual lifestyle fungi parasitize a wide range of grasses where they form infections of single grass florets and replace the seed with individual sclerotia (Clay and Schardl 2002).

The asexual fungi are vertically transmitted through seeds. They have never been known to produce infectious spores and rely entirely on seed transmission. Especially the asexual fungi exhibit high host specificity. Most interestingly, sexual and asexual fungi may interact in parasexual processes contributing to a high diversity of fungal asexual endophytes (Tsai et al. 1994).

In general, grasses are poor producers of natural products that assist other plants in their long-term strategy to gain an ecological advantage. Grasses, however, have the ability to compensate for this deficiency by acquiring fungi notorious for their poisonous natural products. In some cases fungi can be considered the lifestock of grasses.

Fungi associated with plants may produce different classes of alkaloids among which toxic ergoline alkaloids are an important group (Schardl et al. 2004, 2007). The main ecological roles of ergoline alkaloids in nature are probably to protect the fungi from consumption by vertebrate and invertebrate animals (Schardl et al. 2006). Ergoline alkaloids benefit the fungus by protecting the health and productivity of the host (Schardl et al. 2006). Other benefits include growth of the plant, competitive abilities, resistence to drought (Malinowski and Belesky 2000), pests and fungal pathogens (Brem and Leuchtmann 2002; White Jr. et al. 2003). In some cases clavicipitaceous fungi are culturable in vitro (Keller and Tudzynski 2002). This helped to identify the fungus as the producer of ergoline alkaloids and revealed that the host plant is not the site of ergoline alkaloid biosynthesis.

It was therefore somewhat unexpected when Hofmann (1961, 2006) found that dicotyledonous plants belonging to the family Convolvulaceae contained ergoline alkaloids and that these alkaloids were responsible for the hallucinogenic properties enjoyed by Meso- and South American indians in religious ceremonies.

The idea that a fungus could be responsible for the alkaloid occurrence was discussed but no evidence for the presence of such a fungus was found (Hofmann 2006). This seemed to be in agreement with the notion that plant tissue cultures which are believed to be germ-free, i.e. devoid of any microbes, were reported to produce ergoline alkaloids (Dobberstein and Staba 1969; see below).

III. Epibiotic Clavicipitaceous Fungi Associated with Convolvulaceae

A. Identification of Clavicipitaceous Fungi

1. Microscopic and Electron Microscopic Characterization

The infestation of the clavicipitaceous fungi on members of the family Convolvulaceae is systemic. Evidence of systemic infection came from demonstrations that the fungi are seed transmitted, that surface-sterilized seeds grown in vitro and under germ-free conditions result in plantlets which are colonized exclusively by the respective clavicipitaceous fungi and that they are transmitted through vegetative propagation (Steiner et al. 2008). It is an unusual type of systemic infection in that there are no signs of penetration into the host tissue and growth on the host plant is superficial. Attempts made to visualize the fungus within the stem and leaf tissue, using methodologies commonly employed to detect endophytes in grasses (Bacon and White Jr. 1994), were not successful. Up to now the fungi have proved to be non-detectable using these procedures. Among the Clavicipitaceae, *Atkinsonella hypoxylon*, *Balansia cyperi*, *B. pilulaeformis* and *Myriogenospora atramentosa* are examples of epibiotic species that grow on the meristematic tissues of host plants (Luttrell and Bacon 1977; Rykard et al. 1985; Leuchtmann and Clay 1988, 1989; Clay and Frentz 1993). The clavicipitaceous fungi colonizing members of the Convolvulaceae inhabit an epibiotic niche and thus seem most comparable to the epibiotic members of the grass borne Clavicipitaceae. The mutualistic endophyte *Neothyphodium typhinum* also forms a stable external mycelial net on the leaves of the host plant (Moy et al. 2000). This suggested a possible alternative pathway of fungal dispersal and transmission to hosts, i.e. through epiphyllously produced conidia.

The clavicipitaceous fungi form colonies on the upper surfaces of young unfolded leaves which are visible to the naked eye, as shown for *Turbina corymbosa* (Fig. 9.1A) and *Ipomoea asarifolia*. They are also detectable by molecular biological techniques in seeds of *I. violacea* (Ahimsa-Müller et al. 2007). On *T. corymbosa* colony distribution mainly follows the veins of the leaves (Fig. 9.1A, B), in contrast to the distribution on *I. asarifolia* which is more

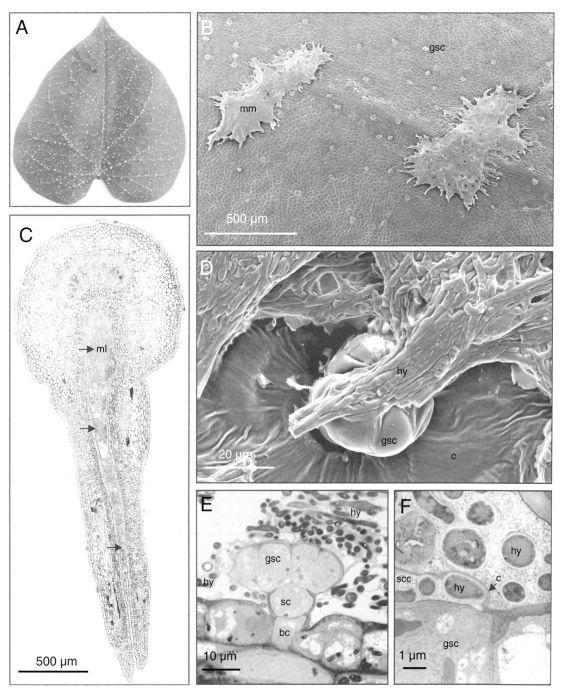

Fig. 9.1. Colonization of *Turbina corymbosa* with the clavicipitaceous fungus (provisionally named TcorF01). **A** Colonies formed by white mycelium on the adaxial surface of a young unfolded leaf. Preferential development on the veins is visible with the naked eye. **B** Aggregated hyphae differentiating typical mycelium mats (*mm*) consisting of several layers which cover leaf areas with peltate glandular trichomes and are adhered to the cuticle. **C** Cross-section of a closed leaf bud showing that the fungus is well established on the adaxial leaf surfaces at this early stage of plant development. The mycelium is formed by tightly packed hyphae as a mycelium layer (*ml, arrows*) in the cavity between the halves of the leaf. **D** Close association of secretory cells (*gsc*) on the adaxial leaf surface with hyphae (*hy*) which often encircle the peltate glandular trichomes of the plants. **E** Cross-section of a peltate glandular trichome composed of basal cell (*bc*), stalk cell (*sc*) and secretory cells (*gsc*) showing the epiphytic development of mycelium embedded in a mucilage matrix concentrated on the cuticle over a subcuticular oil storage cavity. **F** Electron microscopic view of secretory cells with hyphae outside and inside of the subcuticular oil storage cavity (*scc*) bordered by the cuticle (*c, arrow*). No evidence for direct penetration of the plant cells is visible

random. These colonies differ in size and mycelium density, and depending on the developmental stage, the fungi produce synnemata-like structures. No stromata with perithecia and ascospores were detected in the mycelium mats. Maybe the environmental conditions are not suitable for the development of the sexual stage of the fungi, or they lost the ability to reproduce sexually, or the mating type is lacking. No traces of mycelium were detected on the lower side of the leaves. Visual inspection of leaf buds which were opened by manipulation showed that the fungus was well established as dense white mycelial layers on the adaxial leaf surfaces of both *I. asarifolia* and *T. corymbosa* plants (Fig. 9.1C) at this early stage of plant development. The mycelium is formed by tightly packed hyphae in the cavity between the leaf halves. Sections through colonized tissue revealed that the fungal mycelium is entirely superficial. The hyphae measured approximately 1.0–1.5 µm across, were hyaline, thin-walled and septated. Chlamydospore-like structures are produced. As indicated by the intense mycelium development, the space between the upper surfaces of folded leaves probably offer a refuge of protection to the fungus. As leaves expand and mature the hyphae are evident as isolated clumps only visible microscopically, often near or around peltate glandular trichomes, and the ends of the hyphae often appeared broken (Fig. 9.1D).

The epibiotic fungi of *I. asarifolia* and *T. corymbosa* (Fig. 9.1D, E) are closely associated with secretory glands on the adaxial leaf surface, an anatomic feature which may be essential for ergoline alkaloid biosynthesis in the epibiotic fungus/plant association (Steiner et al. 2008). In cell cultures which harbor the fungus no ergoline alkaloids are synthesized and no secretory glands are developed.

Members of the Convolvulaceae like *I. asarifolia* and *T. corymbosa* (Fig. 9.1E) form peltate glandular trichomes, which consist of one basal cell, one stalk cell, up to eight glandular secretory cells and a subcuticular oil storage cavity that is derived from the cuticle of the secretory cells. Metabolites are released after rupture of the cuticle. As indicated by staining with the lipophilic dye Nile red these specialized structures contain essential oils (Kucht et al. 2004). The secretory glands and their specific metabolites may be the basis of a metabolic dialog between fungus and plant (Steiner et al. 2008). The fungi may feed on the volatile oil and derive precursors like terpenes for ergoline alkaloid biosynthesis from the oil. The fungi inhabit the epibiotic niche of glandular cells on the upper surface of leaves. This observation is supported by showing hyphae of the clavicipitaceous fungus on *T. corymbosa* both outside of the subcuticular oil storage cavity and inside of this compartment embedded in an electron-dense matrix (Fig. 9.1F). The localization of mycelium with the glandular cells ensures the close association between fungus and host tissues. A continuous maintenance of the symbiotic relationship requires that the fungus derives energy from the host plant. In clavicipitaceous epibiotic fungi, substrate utilization depends on the availability of organic material from the waxy cuticle covering the plant surface and exuded compounds, lipids, amino acids and vitamins. The main energy-yielding compounds are simple sugars that in the case of endophytic mycelia are derived from the apoplasm through intercellular fungal hyphae (White and Morgan-Jones 1996). In clavicipitaceous fungi present on *I. asarifolia* and *T. corymbosa*, superficial fungal hyphae have been observed with tip enlargements tightly adherent to the glandular cells as well as to the cuticle. Probably, mechanisms for a selective and efficient transfer of carbohydrates to the fungus could be present.

Physiological changes paralleled by morphological adaptations of the host have been described for some endophytic associations (Bacon and White Jr. 2000). In *M. atramentosa*, plant host changes in the epidermal cell size and shape suggest the activity of growth regulatory substances which are either produced by the fungus or secreted into the host or are produced by the host in response to the fungal symbiont (Bacon and White Jr. 2000).

The epiphytic proliferation of hyphae on the cuticle may be additionally enabled through degradation of the cuticular layers of the leaf surfaces. Previous ultrastructural studies of the host–fungus interfaces of the clavicipitaceous fungi on *I. asarifolia* and *T. corymbosa* revealed progressive cuticular disintegration. Substrate utilization studies have shown that epiphytic members of the Balansiaceae such as *Atkinsonella hypoxylon*, possess the capacity to colonize and degrade paraffin wax droplets (White Jr. et al. 1991). *A. hypoxylon* grows superficially on young leaves of grasses as an epiphyte, perhaps degrading wax in the cuticle to obtain nutrients for epiphytic growth (White Jr.

et al. 1991). Leaves and inflorescence primordia within the stroma never develop a cuticular layer that would impede flow of nutrients and moisture to the fungus. Through these modifications of the host tissues, the endophyte removes the barriers which obstruct nutrient flow into the mycelium. Very similar to this situation, the cuticle covering the glandular cells of the Convolvulaceae appears thinner and therefore more permeable than the cuticle on epidermal cells.

Clavicipitaceous fungi have evolved to survive both as saprophytes, degrading organic material, and as biotrophs of plants, fungi, nematodes and insects. They are described to have become particularly successful as endophytes and epibionts of grasses. The associations between clavicipitaceous fungi and their hosts constitute unique biotrophic symbioses where the stages of physiological adaptation to the plant host may yield an understanding of how evolution among these fungi and their hosts has progressed (Schardl et al. 2008). The detection on Convolvulaceae of clavicipitaceous fungi able to synthesize ergoline alkaloids known to play a role in enhanced resistance to diseases, pests and tolerance to drought has shown that such associations have evolutionary value not only in grass hosts but also in dicots. The colonization of a unique plant niche, the clavicipitaceous fungi on Convolvulaceae, represents a novel finding among beneficial plant–fungus symbioses in non-graminaceous plants.

2. Phylogenetic Trees

The unusual colony-forming fungus (provisionally named IasaF13) on the leaf surface of *I. asarifolia* (white blooming) was found to belong to the family Clavicipitaceae. Proof, however, was not possible by conventional techniques because all attempts to cultivate the fungus on synthetic media usually supporting fungal growth were negative. This indicates that the leaf material contains factors or structures essential for growth of the fungus IasaF13. All experiments to characterize these fungi in terms of taxonomy are therefore based on molecular biological techniques (Steiner et al. 2006).

Construction of phylogenetic trees has been repeatedly and successfully employed in the systematic classification of grass-borne clavicipitaceous fungi (Spatafora and Blackwell 1993; Glenn et al. 1996; Kuldau et al. 1997; Reddy et al. 1998;

Lewis et al. 2002; Bischoff and White Jr. 2005; Sung et al. 2007). Removal of fungal mycelium from the leaf surface of convolvulaceous plants was possible by ultrasonic treatment. After DNA from the fungus was extracted and sequenced phylogenetic trees were constructed from 18SrDNA and internal transcribed spacer. This showed that the fungus clustered with clavicipitaceous fungi. Confirmation was obtained by partial sequencing with phylogenetic analysis of the gene (partially) responsible for the committed step in ergoline alkaloid biosynthesis encoding the 4-[γ,γ-dimethylallyl]tryptophan synthase.

Essentially the same results were observed when the fungi associated with *I. asarifolia* (red variety), *T. corymbosa*, and *I. violacea* were investigated (Ahimsa-Mueller et al. 2007). This showed that our observations are not restricted to one single plant taxon (*I. asarifolia*, white blooming) and its associated fungus IasaF13 but are of a broader significance at least within the family Clavicipitaceae. The clavicipitaceous fungi present on representatives of the four convolvulaceous taxa turned out to be not identical, although closely related. In-depth investigation of the biosynthetic pathway leading to ergoline alkaloids in the fungus present on *I. asarifolia* (white blooming) provided further evidence for the clavicipitaceous nature of the fungus IasaF13 (Markert et al. 2008).

B. Fungicidal Treatment

Discovery of the fungus on the adaxial leaf surface was a surprise. The hypothesis that an associated fungus was responsible for the presence of alkaloids in Convolvulaceae had been tested but the fungus was not found for unexplained reasons (Hofmann 2006).

The presence of the clavicipitaceous fungus alone, however, was not evidence enough to postulate the fungal origin of ergoline alkaloids in the symbiotic association. Twelve endophytic fungi had been isolated from the *I. asarifolia* Roem. et Schult plant and an epibiotic fungus – provisionally named IasaF13 – was discovered. Not only was it possible that one of the fungi colonizing the plant (Steiner et al. 2006) would have been responsible for the occurrence of alkaloids in Convolvuaceae plants, it was also possible that the plant itself was the synthesizing organism. To clarify this situation *I. asarifolia* and *T.*

corymbosa (L.) Raf (syn. *Rivea corymbosa* (L) Hall. f.) plants were treated with four different fungicides in a regime with two-week intervals. After 18 weeks, microscopic inspection and alkaloid analysis of plants belonging to both species revealed that systemic azole fungicides were most effective in the removal of both epibiotic fungus and ergoline alkaloids. Simultaneously, the volatile oil present in *I. asarifolia* was isolated by steam destillation, and as opposed to ergoline alkaloids, was found not to be removed during the fungicide treatment (Kucht et al. 2004). Thus, the removal of alkaloids is a specific process and the fungicide does not interfer with the supply of hemiterpenoid biosynthetic building stones which are precursors of both ergoline alkaloids (Groeger and Floss 1998; Keller and Tudzynski 2002) and terpenoid compounds. This result is a frequent observation: ergoline alkaloids and the epibiotic fungus always cooccur because they are both part of a functional entity (Ahimsa-Mueller et al. 2007).

C. Plant Growth Under Germ-Free Conditions

The notion that fungicides eliminate ergoline alkaloids from the plant is a clear indication that ergoline alkaloids in Convolvulaceae plants are of fungal origin. This observation is somewhat unusual because it had been reported that ergoline alkaloids are produced by plant cell cultures established from different Convolvulaceae plants (Dobberstein and Staba 1969). Plant cell cultures are usually germ-free; they should not contain any microbes and can therefore be used as a test system to probe the biosynthetic capacities of plant cells.

Numerous attempts, however, to reproduce this result (Dobberstein and Staba 1969) and to find a plant cell culture raised from *Ipomoea asarifolia*, *Turbina corymbosa* and *I. violacea* (L) (Convolvulaceae) showing ergoline alkaloid production were unsuccessful in our hands (Hussein 2004; Kucht et al. 2004). Indeed, thin-layer chromatography combined with vanUrk's spray reagent were used by Dobberstein and Staba (1969) to detect ergoline alkaloids, techniques which are of limited reliability in the identification of natural products (Jenett-Siems et al. 1994; Kucht et al. 2004).

Again, it was a surprise when we found that the epibiotic fungus lived together with the plant cells in the callus and cell suspension culture.

Microscopic examination, single-strand conformation polymorphism (SSCP) and sequencing of the internal transcribed spacer revealed the presence in the cell culture of the epibiotic fungus (IasaF13) previously detected on the leaf surface of *I. asarifolia*. Fungi contaminating plant tissue cultures usually cause plant cells to react in a hypersensitive response.

This, however, was not observed in our cultures where the fungus coexisted asymptomatically and undetected by the naked eye in association with the plant cells. Other endophytic fungi which had been isolated from intact *I. asarifolia* plants were not detectable by SSCP within the callus and cell suspension culture (Steiner et al. 2006).

When a callus culture was subjected to a new hormone regime (the amount of benzylaminopurine was lowered from 2.0 mg/l to 0.01 mg/l) a plantlet regenerated from the callus. This plantlet was colonized by the fungus and contained ergoline alkaloids (Steiner et al. 2006, 2008).

These observations show also that an intact *I. asarifolia* plant colonized by the fungus IasaF13 is required for the successful synthesis of ergoline alkaloids and gives an idea about the extreme specificity in the interaction between the epibiotic fungus and the *I. asarifolia* plant (Steiner et al. 2008). It is in line with these conclusions that we were hitherto unable to grow the fungus IasaF13 in vitro (Steiner et al. 2006). Apparently the plant contains some kind of component essential for fungal growth. The specificity between the plant and its associated fungus is also evident from the fact that different plant taxa within the Convolvulaceae (*I. asarifolia*, *I. violacea*, *T. corymbosa*) are colonized by related but different clavicipitaceous fungi (Ahimsa-Mueller et al. 2007).

This raises the question as to how the specific interaction between the fungus and the host plant is brought about (Steiner et al. 2008). Interestingly, the fungus apparently has a very high affinity to the secretory glands on the adaxial leaf surface (Kucht et al. 2004).This seems to be unusual because essential oils may have an antifungal activity (Chang et al. 2008). It is conceivable that during evolution clavicipitaceous fungi were able to overcome this barrier and to take advantage of volatile oil components, using these compounds as mediators of specificity and even as substrates to feed upon.

The volatile oil of *I. asarifolia* consists of many minor but five major components, the latter of which are sesquiterpenes (Kucht et al. 2004). Sesquiterpenes play an important role in ecological interactions between plants and insects (Schnee et al. 2006; Gershenzon and Dudareva 2007). Our observations raise the question whether this class of terpenoids is also essential for the interaction between different Convolvulaceae species and their associated clavicipitaceous fungi.

D. Biosynthesis and Accumulation of Ergoline Alkaloids in the Fungus/Plant Symbiotum

Ergoline alkaloids are natural products of high physiological activity. They are likely to confer drought resistance, herbivore deterrence and fitness to the host plant (Malinowski and Belesky 2000; White Jr. et al. 2003; Bacon and Lyons 2005; Gershenzon and Dudareva 2007). This raises the question as to how this may be brought about when plant-associated clavicipitaceous fungi are the site of ergoline alkaloid biosynthesis. Indeed, Convolvulaceae plants do not seem to have the biosynthetic capacity to produce ergoline alkaloids: neither the genes nor the enzymic machinery were detectable in the shoots. The genetic material responsible for ergoline alkaloid biosynthesis was clearly found in the associated fungi present on *I. asarifolia* and *T. corymbosa* (Markert et al. 2008). The determinant step in ergoline alkaloid biosynthesis is the prenylation in the 4 position of tryptophan catalyzed by 4-[γ,γ-dimethylallyl] tryptophan synthase (DmaW; Groeger and Floss 1998; Keller and Tudzinski 2002). The encoding gene – which has different synonyms, i.e. dmaW or cpd1 (Schardl et al. 2006) or fgaPT2 (Unsöld and Li 2005) – is clearly present in the fungus and is part of a cluster in which the ergoline alkaloid genes are oriented. This is found in *Claviceps* but is different from *Aspergillus* and *Neotyphodium* species (Markert et al. 2008). A reverse genetics experiment showed that the fungus is also the site of transcription of the dmaW gene (Markert et al. 2008).

Initial attempts to detect ergoline alkloids in the fungal mycelium present on *I. asarifolia* and *T. corymbosa* failed although two different analytical approaches were used (Markert et al. 2008). When a sample of the mycelium found on *T. corymbosa* was directly placed into the injection port of a GC/MS system a trace of agroclavine was detectable and clearly identified by comparison with an au-

thentic sample. No alkaloid was detectable when a mycelial sample from *I. asarifolia* was checked in the same way (W. Boland, personal communication). When the leaf material was analyzed for ergoline alkaloids after removal of the mycelium by ultrasonic treatment, alkaloids were qualitatively and quantitatively detected in the plant material, showing that the plant leaf material contained almost all alkaloids whereas the producing fungus provisionally named TcorF01 contained only a trace of agroclavine (Markert et al. 2008).

Thus, biosynthesis of alkaloids takes place in the mycelium; however, ergoline alkaloids accumulate in the host plant. We have to postulate a transport system that translocates ergoline alkaloids from the mycelium into the plant tissue. In an experimental system similar to the one discussed here transport was postulated to occur through the apparently intact cuticle (Smith et al. 1985).

E. Seed Transmittance of Epibiotic Fungi Colonizing Convolvulaceae

The genus *Ipomoea* comprises 600–700 pantropical species. Over half of them are concentrated in the Americas. The American species are mostly native but a few have been introduced (Austin and Huáman 1996). The classification of *Ipomoea* species is still under discussion (Amor-Prats and Harborne 1993; Austin and Huáman 1996). There may be multiple reasons for this: Some species are only endemic and the description of species is incomplete, especially those native to Brazil. Many have not been validly described or are even undiscovered. The genus *Ipomoea* consists of three subgenera, i.e. subgenus *Eriospermum*, subgenus *Ipomoea* and subgenus *Quamoclit*. Amor-Prats and Harborne (1993) attempted to find chemotaxonomic support for an infrageneric classification of the genus *Ipomoea* by analyzing seeds from 43 species for their ergoline alkaloid content. The alkaloid-bearing species fall, however, into each of the taxonomically defined subgenera.

Hence, there is no clear relationship between the distribution of alkaloids and the infrageneric classification of the genus *Ipomoea*. The whole problem is clouded by the inability to reproduce published analytical data in many cases. Consequently, Eich

(2008) lists ergoline positive versus ergoline negative reports and Amor-Prats and Harborne (1993) believe that "the methods of analysis varied leading to some uncertainty".

The reason for the inconsistent picture very likely depends on the (until recently unknown) presence of clavicipitaceous fungi on convolvulacous plants and within seeds (Kucht et al. 2004; Steiner et al. 2006; Ahimsa-Mueller et al. 2007) and the capacity of these fungi to synthesize ergoline alkaloids (see below).

A freshly harvested and surface-sterilized seed grown under germ-free conditions gives a plant colonized by the epibiotic clavicipitaceous fungus. This plant contains ergoline alkaloids. The epibiotic fungus is the only fungus that is detectable by SSCP on this particular plant. Such a fungus is detectable in seeds of *I. asarifolia* and *I. violacea* (Steiner et al. 2006; Ahimsa-Mueller et al. 2007). This shows that the fungus is seed-transmitted and points to the host specificity typical of asexual clavicipitaceous fungi (see above).

The viability of the seed-transmitted fungus very likely is limited and depends on seed age and storage (Schardl 1994) as well as moisture and storage temperature (Welty et al. 1987). An *I. violacea* plant devoid of ergoline alkaloids and derived from an alkaloid- and clavicipitaceous fungus-containing seed was recently described (Ahimsa-Mueller et al. 2007). In this particular case the viability of the seed exceeded the viability of the inhabiting fungus.

It follows that the presence in a convolvulaceous plant of ergoline alkaloids may be an unsuitable character for taxonomic classifications.

IV. Additional Fungus/Plant Symbiota in Dicotyledonous Plants

An interesting association consisting of *Ipomoea batatas* (L.) Lam. (i.e. sweet potato; Convolvulaceae) and *Fusarium lateritium* Nees: Fr has also been reported. As described for our clavicipitaceous fungi (Sect. III) *F. lateritium* is primarily located between the halves of young unfolded leaves of the *I. batatas* plant (Hyun and Clark 1998). Yet there is another feature of this fungus/plant association which we also observed in our system (Sect. III): The fungus is associated on the phylloplane with pearl glands and is located

around bases of trichomes (Clark 1992). The fungus apparently produces trichothecenes and protects the host plant against infection by pathogenic *F. oxysporum* f. sp. *batatas* (Wollenw.) W.C. Snyder & H.N. Hans. However, the associated *F. lateritium* may also be the cause of chlorotic leaf distortion (CLD) disease mediated by trichothecenes (Clark 1994). After light activation of trichotecenes during prolonged exposure of the plant to sunlight CLD occurs. Plants usually recover when cloudy weather prevails. Thus, the associated fungus may exert a beneficial and a detrimental effect on the host plant and in both cases trichothecenes are likely to be the causative agent.

Two new clavicipitaceous fungi belonging to a newly established genus (*Hyperdermium*) were isolated from an unidentified Asteraceae plant (genus *Bernonia*).The fungi were named *H. bertonii* (Speg.) J. White, R. Sullivan, G. Bills et N. Hywel-Jones and *H. pulvinatum* J. White, R. Sullivan, G. Bills et N. Hywel-Jones. As with the clavicipitaceous fungi described in Sect. III, the fungi are epibiotic. They belong to the subfamily Cordycipitoideae (Sullivan et al. 2000).

An entirely superficial mycelium was observed on a South American Asteraceae plant, *Baccharis coridifolia* D.C. The endophyte belongs to the Hypocreales, an order which accommodates also the family Clavicipitaceae. The fungus occurs not only epibiotically but also in meristematic tissue of leaf primordia. No reproductive structures were detectable. The plant is toxic and it was assumed that the epibiont is a trichothecene producer. Since this fungus and graminaceous Clavicipitaceae (Sect. II) are not closely related, colonizations (that must have occurred during evolution) were assumed to be distinct events (Bertoni et al. 1997).

The same conclusion was drawn for a *Mentha piperita* L. plant colonized by a pyrenomycete which is also associated with glandular trichomes (Mucciarelli et al. 2002), a striking observation which led to speculations about the possible function of the secretory glands and trichomes in the establishment of a symbiotic association: it may be that the glandular trichomes are entry gates for the fungus in its attempt to establish a molecular dialog with the host plant (Steiner et al. 2008).

Another interesting fungus/plant association has been described for locoweed plants belonging to the family Fabaceae. *Astragalus molissimus*,

Oxytropis lambertii and *O. sericea* are collectively called locoweed and are colonized by endophytes which seem to be closely related to the genus *Embellisia*. Locoism, as observed in cattle intoxicated by locoweed plants, is a neurological disease resulting in a staggering walk and lack of muscular coordination. The causative agent seems to be the indolizidine alkaloid swainsonine. This alkaloid is also known to be a product of in vitro grown *Rhizoctonia leguminicola* cultures. Again there are observations that parallel those described in Sect. III:

1. Some collections of host plants were devoid of natural products.
2. The fungus and the alkaloid do not occur in the root system.
3. The alkaloid exerts an ecological function in protecting the fungus against insects and animal feeding (Braun et al. 2003).

V. Conclusions

The data described in Sect. III solve a historical mystery and explain why ergoline alkaloids occur in disjointed taxa, clavicipitaceous fungi and convolvulaceous plants. They dispute the possibility that during evolution a horizontal transfer of genes responsible for the synthesis of ergoline alkaloids occurred from fungi to plants. They also show that there is no necessity to invoke a repeated invention of the ergoline alkaloid biosynthetic pathway during evolution. In fact genes present in IasaF13, TcorF01, *Claviceps purpurea*, *C. fusiformis*, *Balansia obtecta*, *Neotyphodium coenophialum* as well as *Asperillus fumigatus* involved in the biosynthesis of ergoline alkaloids share a high similarity (Markert et al. 2008). It is now evident that clavicipitaceous fungi do not only colonize Poaceae and related monocots but also Convolvulaceae. It is also clear that the association between fungus and convolvulaceous plant is asymptomatic and that a molecular dialog occurs between associated fungi and convolvulaceous plants, indicating that both are members of a symbiotum in which biosynthesis and accumulation of ergoline alkaloids is spatially separated and sequestered in different but associated organisms. One of the unsolved questions is whether there are also sexual forms of these vertically transmitted asexual clvicipitaceous fungi. It is also unknown how the plant-associated fungi spread within the plant. Despite repeated attempts to localize hyphea, spores or propagules within the host plants, endophytic structures of the fungi have remained undetected until now.

Some of these observations parallel those made with fungus/plant associations occurring in the plant families Asteraceae and Fabaceae and in the plant species *I. batatas*, as outlined in Sect. IV. In each case the associated fungus is the producer of poisonous natural products (trichothecenes, swainsonine) which may benefit and protect the host plant.

References

Ahimsa-Mueller MA, Markert A, Hellwig S, Knoop V, Steiner U, Drewke C, Leistner E (2007) Clavicipitaceous fungi associated with ergoline alkaloid-containing Convolvulaceae. J Nat Prod 70:1955–1960

Amor-Prats D, Harborne JB (1993) New sources of ergoline alkaloids within the genus *Ipomoea*. Biochem Syst Ecol 21:455–461

Arnold AE, Mejia LC, Kyllo D, Rojas EI, Maynard Z, Robbins N, Herre EA (2003) Fungal endophytes limit pathogen damage in a tropical tree. Proc Natl Acad Sci USA 100:15649–15654

Austin DF, Huáman Z (1996) A synopsis of *Ipomoea* (Convolvulaceae) in the Americas. Taxon 45:3–38

Bacon CW, Lyons P (2005) Ecological fitness factors for fungi within the Balansiae and Clavicipiteae. In: Dighton J, White JF Jr, Oudemans P (eds) The fungal community, its organisation and role in the ecosystem, 3rd edn. CRC Taylor and Francis, Boca Raton, pp 519–532

Bacon CW, White JF Jr (1994) Stains, media, and procedures for analyzing endophytes. In: Bacon CW, White JF Jr (eds) Biotechnology of endophytic fungi of grasses. CRC, Boca Raton, pp 47–58

Bacon CW, White JF Jr (2000) Microbial endophytes. Dekker, New York, pp 341–388

Bertoni M, Romero N, Reddy PV, White JF Jr (1997) A hypocralean epibiont on meristems of *Baccharis cordifolia*. Mycologia 89:375–382

Bischoff JF, White JF Jr (2005) Evolutionary development of the Clavicipitaceae. In: Dighton J, White JF Jr, Oudemans P (eds) The fungal community – its organisation and role in the ecosystems, 3rd edn. CRC Taylor and Francis, Boca Raton, pp 505–518

Braun U, Romero J, Liddell C, Creamer R (2003) Production of swainsonine by fungal endophytes of locoweed. Mycol Res 107:980–988

Brem D, Leuchtmann A (2002) Intraspecific competition of endophyte infected vs uninfected plants in two woodland grass species. Oikos 96:281–290

Chang H-T, Cheng Y-H, Wu C-L, Chang S-T, Chang T-T, Su Y-C (2008) Antifungal activity of essential oil and its constituents from *Calocedrus macrolepis* var. *formosana* Florin leaf against plant pathogenic fungi. Bioresour Technol 99:6266–6270

Clardy J, Walsh C (2004) Lessons from natural molecules. Nature 432:829–837

Clark CA (1992) Histological evidence that *Fusarium lateritium* is an exopathogen on sweetpotato with chlorotic leaf distortion. Phytopathology 82:656–663

Clark CA (1994) The chlorotic leaf distortion pathogen, *Fusarium lateritium*, cross protects sweetpotato against *Fusarium* wilt caused by *Fusarium oxysporum* f.sp.batatas. Biol Control 4:59–66

Clay K, Frentz IC (1993) *Balansia pilulaeformis*, an epiphytic species. Mycologia 85:527–534

Clay K, Schardl C (2002) Evolutionary origins and ecological consequences of endophyte symbiosis with grasses. Am Nat 160:S99–S127

Croteau R, Ketchum REB, Long RM, Kaspera R, Wildung MR (2006) Taxol biosynthesis and molecular genetics. Phytochem Rev 5:75–97

Dobberstein RH, Staba EJ (1969) *Ipomoea*, *Rivea* and *Argyreia* tissue cultures: influence of various chemical factors on indole alkaloid production and growth. Lloydia 32:141–177

Eich E (2008) Solanaceae and Convolvulaceae: secondary metabolites – biosynthesis, chemotaxonomy, biological and economic significance (a handbook). Springer, Heidelberg

Eisner T (2003) For love of insects. Harvard University Press, Cambridge, Mass.

Gershenzon J, Dudareva N (2007) The function of terpene natural products in the natural world. Nat Chem Biol 3:408–414

Glenn AE, Bacon CW, Price R, Hanlin RT (1996) Molecular phylogeny of *Acremonium* and its taxonomic implications. Mycologia 88:369–383

Groeger D, Floss HG (1998) Biochemistry of ergot alkaloids – achievements and challenges. In: Cordell GA (ed) The alkaloids: chemistry and biology, vol 50. Academic, New York, pp 171–218

Gunatilaka AAL (2006) Natural products from plant-associated microorganisms: distribution, structural diversity, bioactivity, and implications of their occurrence. J Nat Prod 69:509–526

Harborne JB (2004) Introduction to ecological biochemistry, 4th edn, Academic, London

Hofmann A (1961) Die Wirkstoffe der mexikanischen Zauberdroge "Ololuiqui". Planta Med 9:354–367

Hofmann A (2006) LSD – mein Sorgenkind, die Entdeckung einer Wunderdroge. Deutscher Taschenbuchverlag, Munich

Hussein YHA (2004) Biochemical analysis of Convolvulaceae plant tissue cultures for the presence of ergoline alkaloids. Dissertation, Zagazig University

Hyun J-W, Clark CA (1998) Analysis of *Fusarium lateritium* using RAPD and rDNA RFLP techniques. Mycol Res 102:1259–1264

Jenett-Siems K, Kaloga M, Eich E (1994) Ergobalansine/ergobalansinine, a proline-free peptide type alkaloid of the fungal genus *Balansia* is a constituent of *Ipomoea piurensis*. J Nat Prod 57:1304–1306

Keller U, Tudzynski P (2002) Ergot alkaloids. In: Osiewacz HD (ed) The Mycota, industrial applications, vol X. Springer, Heidelberg, pp 157–181

Kucht S, Gross J, Hussein Y, Grothe T, Keller U, Basar S, Koenig WA, Steiner U, Leistner E (2004) Elimination of ergoline alkaloids following treatment of *Ipomoea asarifolia* (Convolvulaceae) with fungicides. Planta 219:619–625

Kuldau GA, Liu JS, White JF Jr, Siegel MR, Schardl CL (1997) Molecular systematics of Clavicipitaceae supporting monophyly of genus *Epichloë* and form genus *Ephelis*. Mycologia 89:431–441

Leistner E (2005) Die Biologie der Taxane. Pharm Unserer Zeit 34:98–103

Leuchtmann A, Clay K (1988) *Atkinsonella hypoxylon* and *Balansia cyperi*, epiphytic members of the Balansiae. Mycologia 80:192–199

Leuchtmann A, Clay K (1989) Morphological, cultural and mating studies on *Atkinsonella*, including *A. texensis*. Mycologia 81:692–701

Lewis EA, Bills GF, Heredia G, Reyes M, Arias RM, White JF Jr (2002) A new species of endophytic balansia from Veracruz, Mexico. Mycologia 94:1066–1070

Luttrell ES, Bacon CW (1977) Classification of *Myriogenospora* in the Clacicipitaceae. Can J Bot 55:2090–2097

Malinowski DP, Belesky DP (2000) Adaptations of endophyte-infected cool-season grasses to environmental stresses. Crop Sci 40:923–940

Markert A, Steffan N, Ploss K, Hellwig S, Steiner U, Drewke C, Li S-M, Boland W, Leistner E (2008) Biosynthesis and accumulation of ergoline alkaloids in a mutualistic association between *Ipomoea asarifolia* (Convolvulaceae) and a clavicipitalean fungus. Plant Physiol 147:296–305

Mothes K (1981) The problem of chemical convergence in secondary metabolism. Sci Scientists 1981:323–326

Mothes K, Schütte HR, Luckner M (1985) Biochemistry of alkaloids. VEB, Berlin

Moy M, Belanger F, Duncan R, Freehoff A, Leary C, Meyer W, Sullivan R, White JF Jr (2000) Identification of epiphyllous mycelial nets on leaves of grasses infected by clavicipitaceous endophytes. Symbiosis 28:291–302

Mucciarelli M, Scannerini S, Bertea CM, Maffei M (2002) An ascomycetous endophyte isolated from *Mentha piperita* L.: biological features and molecular studies. Mycologia 94:28–39

Partida-Martinez LP, Hertweck C (2005) Pathogenic fungus harbours endosymbiotic bacteria for toxin production. Nature 437:884–888

Piel J (2004) Metabolites from symbiotic bacteria. Nat Prod Rep 21:519–538

Puri SC, Verma V, Amna T, Qazi GN, Spiteller M (2005) An endophytic fungus from *Nothapodytes foetida* that produces camptothecin. J Nat Prod 68:1717–1719

Reddy PV, Bergen MS, Patel R, White JF Jr (1998) An examination of molecular phylogeny and morphology of the grass endophyte *Balansia claviceps* and similar species. Mycologia 90:108–117

Rykard DM, Bacon CW, Luttrell ES (1985) Host relations of *Myriogenospora atramentosa* and *Balansia epichloë* (Clavicipitaceae). Phytopathology 75:950–956

Saikkonen K, Wäli P, Helander M, Feath SH (2004) Evolution of endophyte-plant symbiosis. Trends Plant Sci 9:275–280

Schardl CL (1994) Molecular and genetic methodologies and transformation of grass endophytes. In: Bacon CW, White JF Jr (eds) Biotechnology of endophytic fungi of grasses. CRC Taylor and Francis, Boca Raton, pp 151–166

Schardl CL, Craven KD, Speakman S, Stromberg A, Lindstrom A, Yoshida R (2008) A novel test for host-symbiotum codivergence indicates ancient origin of fungal endophytes in grasses. Syst Biol 57:483–498

Schardl CL, Grossman RB, Nagabhyru P, Faulkner JR, Mallik UP (2007) Loline alkaloids: currencies of mutualism. Phytochemistry 68:980–996

Schardl CL, Leuchtmann A, Spiering MJ (2004) Symbiosis of grasses with seedborne fungal endophytes. Annu Rev Plant Biol 55:315–340

Schardl CL, Panaccione DG, Tudzynski P (2006) Ergot alkaloids - biology and molecular biology. In: Cordell GA (ed) The alkaloids: chemistry and biology, vol 63. Academic, New York, pp 45–86

Schnee C, Köllner TG, Held M, Turlings TCJ, Gershenzon J, Degenhardt J (2006) The products of a single maize sesquiterpene synthase form a volatile defense signal that attracts natural enemies of maize herbivores. Proc Natl Acad Sci USA 103:1129–1134

Smith KT, Bacon CW, Luttrell ES (1985) Reciprocal translocation of carbohydrates between host and fungus in Bahiagrass infected with *Myriogenospora atramentosa*. Phytopathology 75:407–411

Spatafora JW, Blackwell M (1993) Molecular systematics of unitunicate perithecial ascomycetes: the Clavicipitales–Hypocreales connection. Mycologia 85:912–922

Steiner U, Ahimsa-Mueller MA, Markert A, Kucht S, Gross J, Kauf N, Kuzma M, Zych M, Lamshoeft M, Furmanowa M, Knoop V, Drewke C, Leistner E (2006) Molecular characterisation of a seed transmitted clavicipitaceous fungus occurring on dicotyledonous plants (Convolvulaceae). Planta 224:533–544

Steiner U, Hellwig S, Leistner E (2008) Specificity in the interaction between an epibiotic clavicipitalean fungus and its convolvulaceous host in a fungus/plant symbiotum. Plant Signal Behav 3:704–706

Strobel G, Daisy B, Castillo U, Harper J (2004) Natural products from endophytic microorganisms. J Nat Prod 67:257–268

Sullivan RF, Bills GF, Hywel-Jones NL, White JF Jr (2000) Hyperdermium: a new clavicipitalean genus for some tropical epibionts of dicotyledonous plants. Mycologia 92:908–918

Sung GH, Sung JM, Hywel-Jones NL, Spatafora JW (2007) A multi-gene phylogeny of Clavicipitaceae (Ascomycota, Fungi): Identification of localized incongruence using a combinational bootstrap approach. Mol Phyl Evol 44:1204–1223

Tsai H-F, Liu J-S, Staben C, Christensen MJ, Latch CM, Siegel MR, Schardl CL (1994) Evolutionary diversification of fungal endophytes of tall fescue grass by hybridisation with *Epichloë* species. Proc Natl Acad Sci USA 91:2542–2546

Tudzynski P, Correia T, Keller U (2001) Biotechnology and genetics of ergot alkaloids. Appl Microbiol Biotechnol 57:593–605

Unsöld IA, Li S-M (2005) Overproduction, purification and characterization of FgaPT2, a dimethylallyltryptophan synthase from *Aspergillus fumigatus*. Microbiology 151:1499–1505

Welty RE, Azevedo MD, Cooper TM (1987) Influence of moisture content, temperature, and length of storage on seed germination and survival of endophytic fungi in seeds of tall fescue and perennial ryegrass. Phytopathology 77:893–900

White JF Jr, Morgan-Jones G (1996) Morphological and physiological adaptations of Balansieae and trends in the evolution of grass endophytes. In: Redlin SC, Carris LM (eds) Endophytic fungi in grasses and woody plants, APS, St. Paul, pp 133–154

White JF Jr, Bacon CW, Hinton DM (1991) Substrate utilization in selected *Acremonium*, *Atkinosella*, and *Balansia* species. Mycologia 83:601–610

White JF Jr, Bacon CW, Hywel-Jones NL, Spatafora JW (2003) Clavicipitalean fungi, evolutionary biology, chemistry, biocontrol, and cultural impacts. Mycology series, vol 19, Dekker, New York

Zenk MH (1967) Biochemie und Physiologie sekundärer Pflanzenstoffe. Ber Dtsch Bot Ges 80:573–591

Zhang L, Guo B, Li H, Zeng S, Shao H, Gu S, Wei R (2000) Zhongcaoyao 31:805–807

10   Secondary Metabolites of Basidiomycetes

Anja Schüffler[1], Timm Anke[1]

CONTENTS

[1]Institut für Biotechnologie und Wirkstoff-Forschung e.V./Institute for Biotechnology and Drug Research, Erwin-Schroedinger-Strasse 56, 67663 Kaiserslautern, Germany; e-mail: anke@rhrk.uni-kl.de

Physiology and Genetics, 1st Edition
The Mycota XV
T. Anke and D. Weber (Eds.)
© Springer-Verlag Berlin Heidelberg 2009

I. Introduction

Basidiomycetes, a major class of higher fungi adapted to many different climates, habitats, and substrates have developed a rich and very diverse secondary metabolism. Its products differ in biogenetic origin and structure remarkably from the metabolites of ascomycetes or other prolific producers of secondary metabolites like actinomycetes or myxobacteria. There are, however, some similarities to the products of plants, especially with regard to some polyketides, acetylenes, and sesquiterpenoids. The first systematic investigations of basidiomycete metabolites originated after the discovery and introduction into clinical practise of penicillin, the first antibiotic not derived by chemical synthesis. Its great success initiated a real "gold rush" in the search for new antibiotics and prospective producers. From 1940 until the early 1950s mycelial cultures or fruiting bodies of more than 2000 basidiomycetes were screened for the production of antibiotics by the pioneering groups of M. Anchel, A. Hervey, F. Kavanagh, W. J. Robbins, and W. H. Wilkins (for reviews, see Florey et al. 1949; Wilkins and Harris 1944). Their investigations resulted in the discovery of pleuromutilin, the lead compound for the semisynthetic tiamulin used in veterinary practise and recently also in humans (Kavanagh et al. 1951; Högenauer 1979; Daum et al. 2007). This systematic search came to an end after Waksman's dicovery of the streptomycetes as the most prolific producers of antibiotic metabolites. These soil bacteria are easily obtained and can be grown in simple media in large fermenters which greatly facilitates the discovery and production process. Meanwhile the basidiomycetes and mainly their fruiting bodies attracted the interest of natural products chemists as a source of toxins (Bresinsky and Besl 1985), hallucinogens (Schultes and Hofmann 1980), and pigments (Gill and Steglich 1987; Gill 1999). A systematic screening of mycelial cultures only regained interest when it became increasingly difficult to discover new chemical entities among the metabolites of actinomycetes and other bacterial sources. This was furthered by the recent progress in cultivation techniques and the development of fast methods for the isolation and structural determination of natural compounds. The following chapter mainly deals with compounds described from 1998 on and focuses on biologically active metabolites. Some classes of fungal metabolites which presently are used as medical, veterinary, or agricultural antibiotics were dealt with in the chapter "Non-β-Lactam Antibiotics" of *The Mycota, Vol. X* (2002). For earlier reviews on bioactive natural products of basidiomycetes and bioactive sesquiterpenes of fungi the reader is referred to Lorenzen and Anke (1998), Abraham (2001), and Liu (2005).

II. Secondary Metabolites and Their Biological Activities

Basidiomycetes like other saprophytic or soil-inhabiting microorganisms are prolific producers of secondary metabolites. Their mycelia and fruiting bodies are exposed to a number of predators or competitors, which in some cases can explain the production of antibiotics, insecticides, or feeding deterrents. Among the antibiotics the α-methylene lactones and ketones have a very broad and unspecific action against prokaryots and eukaryots while others are highly selective with regard to their biochemical targets and organisms. As demonstrated for the producers of antifungal strobilurins, antibiotic production can be stimulated severalfold in the presence of other, possibly competing, fungi (Kettering et al. 2004). In many cases, however, a possible benefit for the producing fungus is not obvious. This is especially true for secondary metabolites for which up to now no antibiotic or other biological activities have been detected. According to Zähner et al. (1983) these could be part of a still ongoing "evolutionary playground" providing new chemical solutions for an improved fitness of the producing organisms. This contribution covers metabolites derived from basidiomycetes and their biological activities published from 1998 until 2008. The compounds are arranged by their presumed biogenetic origin and the producing fungi. Synthetic approaches are included when they have contributed to elucidate the (absolute) configuration or structure–activity studies.

A. Terpenoids

Terpenoids are among the prominent metabolites of basidiomycetes. Most of these have structures not encountered elsewhere in nature, with the notable exception that some sesquiterpenes have ring structures also found in higher plants, e.g. caryophyllanes and acoranes (Lorenzen and Anke 1998; Abraham 2001). Geosmin (Fig. 10.1), with its characteristic musty-earthy odor and first described as a typical streptomycete metabolite, has now been identified as responsible for the characteristic odor of *Cortinarius herculeus* (Cortinariaceae), *Cystoderma amianthinum* (Agaricaceae), and *Cy. carcharias* (Breheret et al. 1999).

1. Sesquiterpenoids

Basidiomycetes are prolific producers of sesquiterpenoids. Many of these are readily detected and isolated in different screenings because of their unusually high antibiotic and cytotoxic activities. These are often due to a high chemical reactivity which, however, makes these compounds less desirable as possible lead structures.

a) *Resupinatus leightonii* (Tricholomataceae)
1(10),4-Germacradiene-2,6,12-triol, a new germacrane sesquiterpene, and 1,6-farnesadiene-3,10, 11-triol, a known nerolidol derivative, were isolated from submerged cultures of *Resupinatus leightonii*. Both compounds (Fig. 10.1) inhibited the cAMP-induced appressorium formation in *Magnaporthe grisea* and showed cytotoxic activity. The formation of melanized appressoria is a prerequisite for the invasion of host plants by *M. grisea*. Inhibitors are considered to be valuable tools for investigating the mechanisms and pathways leading to infection and could contribute to the lead finding process for novel fungicides. 1(10),4-Germacradiene-2,6,12-

Geosmin   1(10),4-Germacradiene-2,6,12-triol   1,6-Farnesadiene-3,10,11-triol

Conocenol A   Conocenol B   Conocenol C

Conocenolide A   Conocenolide B   5-Demethylovalicin

Riparol A   Riparol B   Riparol C

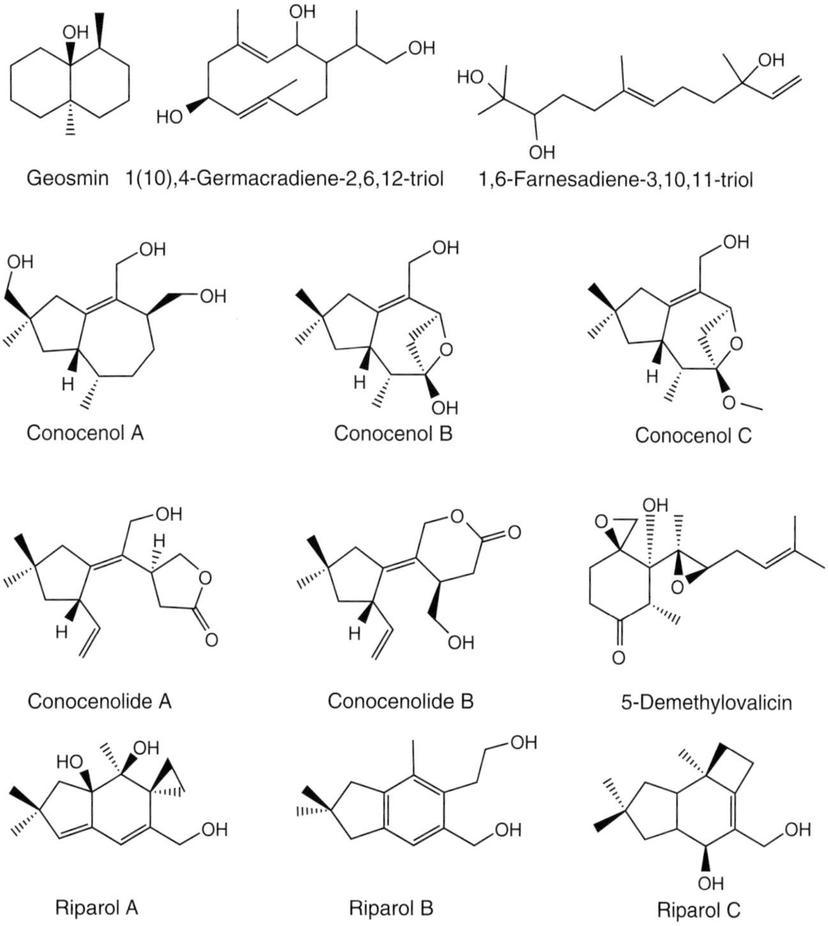

Fig. 10.1. Sesquiterpenoids from *Resupinatus leightonii*, *Conocybe siliginea*, and riparols A–C from *Ripartites* spp.

triol and 1,6-farnesadiene-3,10,11-triol are the first natural compounds interfering with the cAMP-mediated signal transduction leading to appressorium formation in *M. grisea*. Antifungal or antibacterial activities were not detected, but moderate cytotoxic activities were described (Eilbert et al. 2000).

b) *Conocybe siliginea* (Bolbitiaceae)
Six new tremulane sesquiterpenes, conocenols A–C (Fig. 10.1) and D, conocenolide A, and conocenolide B (Fig. 10.1), have been isolated from cultures of a Chinese collection of *Conocybe siliginea* (Liu D-Z et al. 2007). So far no biological activities have been reported for these compounds. The first tremulanes have been reported from the wood-rotting *Phellinus tremulae* by Ayer and Cruz (1993).

c) *Ripartites tricholoma* and *R. metrodii* (Paxillaceae)
In a screening for new metabolites antibiotic and cytotoxic activities were detected in extracts of mycelial cultures of *Ripartites tricholoma* and *R.*

metrodii. As so far no secondary compounds have been described from the genus *Ripartites* a closer look at the metabolites seemed to be warranted. Three new compounds, the illudane riparol A, the illudalane riparol B and the protoilludane riparol C (Fig. 10.1), together with 13-oxo-9(Z),11(E)-octadecadienoic acid (a compound of sunflowers), psathyrellon A, 5-desoxyilludosin (Fig. 10.2), the illudane compound 8 (Fig. 10.2), and 5-demethylovalicin (Fig. 10.1) were isolated by Weber et al. (2006). 5-Demethylovalicin was first described by Ito et al. (1999). Its production by *Chrysosporium luchnowense* was reported by Son et al. (2002). It is a sesquiterpene and inhibits the human methionine aminopeptidase-2 and the growth of human endothelial cells. 5-Demethylovalicin was the first ovalicin derivate reported from a basidiomyete. Psathyrellon A (illudin C) is a constituent of *Psathyrella pseudogracilis* (Bastian 1985) and *Clitocybe illudens* (*Omphalotus olearius*; Arnone et al. 1991). 5-Desoxyilludosin and compound 8 were also described from a *Bovista* sp. (Rasser et al. 2002). Riparol C exhibits weak antibacterial

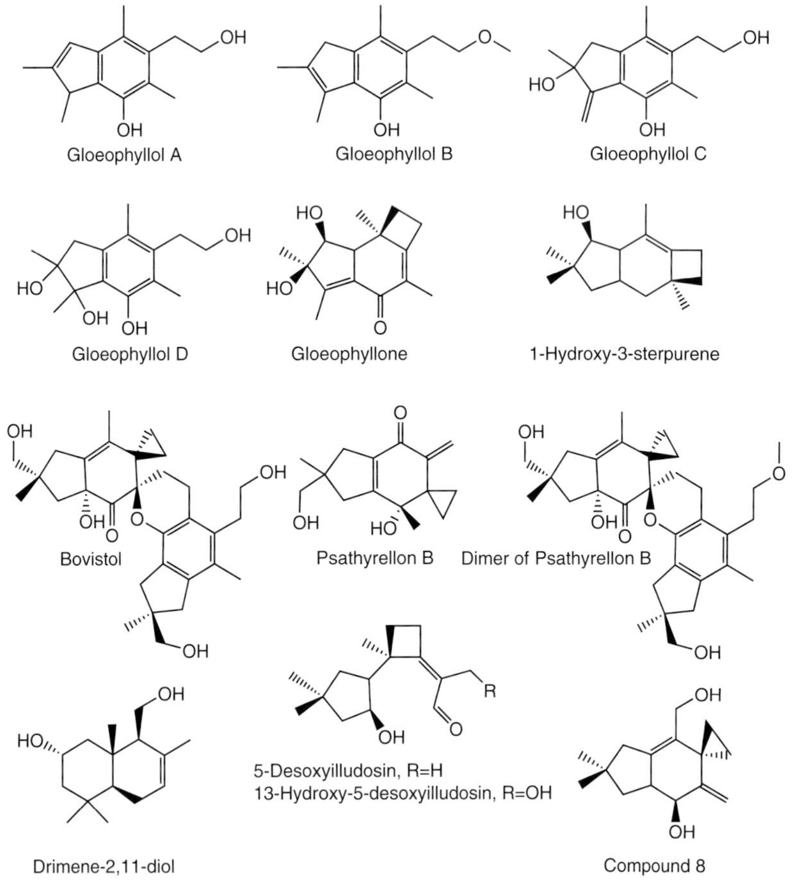

Fig. 10.2. Sesquiterpenoids from *Omphalotus olearius*, *O. nidiformis*, *Mycena leaiana*, *Radulomyces confluens*, and the synthetic (−)-irofulven

and antifungal activities while riparol A inhibits the growth of MDA-MB-231 (human breast adenocarcinoma) and MCF-7 (human breast adenocarcinoma) cells at concentrations of 1 µg/ml.

d) *Omphalotus olearius, O. nidiformis* (Omphalotaceae)

Omphalotus olearius (syn. *Clitocybe illudens, O. illudens*) is a rich source of sesquiterpenoids with carbon skeletons named after the original producer (e.g. illudane, protoilludane, illudalane). To these a new sesquiterpenoid, omphadiol (Fig. 10.3), with a carbon skeleton and also found among the metabolites of the soft coral *Lemnalia africana* and the ascomycete *Leptographium lundbergii* was added recently (McMorris et al. 2000). It is highly remarkable that a synthetic analog of illudin S, (−)-irofulven (Fig. 10.3; Mc Morris et al. 2004), has entered

phase II clinical trials and demonstrated activity in ovarian, gastrointestinal, and non-small cell lung cancer. As compared to the natural product, irofulven has a much better therapeutic index and pharmacological profile (Paci et al. 2006). Leaianafulven (Fig. 10.3), a natural product of *Mycena leaiana* (Harttig et al. 1990), shares high cytotoxic activities and a preferential inhibition of nucleic acid synthesis with irofulven. *O. nidiformis* is native to Australia and usually found in eucalypt forests. From submerged cultures of this species three new iludins (illudin F, G, H; Fig. 10.3) so far not found in the North American species have been isolated (Burgess 1999).

e) *Radulomyces confluens* (Corticiaceae s. lat.)

Radulone A, radulone B, and radudiol, the illudalane radulactone and the illudane radulol

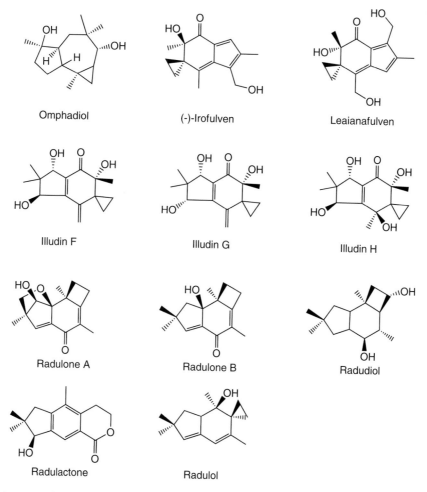

Fig. 10.3. Sesquiterpenes from *Gloeophyllum* sp., *Bovista* sp., and 5-desoxyilludosin and compound 8 from *Ripartites* spp.

(Fig. 10.3) were isolated from fermentations of *Radulomyces confluens* (Fabian et al. 1998). Radulone A is a potent inhibitor of human and bovine platelet aggregation stimulated by different agonists, inhibiting preferentially the aggregation of human platelets induced by ADP with an IC_{50} value of 2 µM. In addition, radulone A exhibits cytotoxic and antimicrobial activities. The other four compounds show weak antimicrobial and cytotoxic activities.

f) *Gloeophyllum* sp. (Gloephyllaceae)

Six new sesquiterpenoids, four rearranged illudalanes, one rearranged protoilludane, and one sterpurane (Fig. 10.2), were isolated from fermentations of a *Gloeophyllum* species (Rasser et al. 2000). The strain was isolated from fruiting bodies collected from a mangrove in Florida, United States. The sporophores showed the characteristics of the genus (Jülich 1984). The morphological characteristics of the sporophores closely resembled *G. sepiarium* for which, however, an occurrence on mangrove wood was considered unlikely. Gloeophyllol A, gloeophyllol D, and gloeophyllone did not exhibit antibacterial or antifungal activity. Gloeophyllols B and C showed weak antifungal activity, while 1-hydroxy-3-sterpurene showed weak antifungal, antibacterial, and cytotoxic activities.

g) *Bovista* sp. (Lycoperdaceae)

Mycelial cultures of *Bovista* sp. 96042 were derived from tissue plugs of a young fruiting body. The saprophytic soil inhabiting species showed all characteristics of the genus; the species however could not be identified. The strain yielded several cytotoxic sesquiterpenoids (Rasser et al. 2002). Besides the known illudane psathyrellon B (illudin C3) and protoilludane armillol, 5-desoxyilludosin and 13-hydroxy-5-desoxyilludosin, two new secoprotoilludanes, the illudane compound 8 and drimene-2,11-diol, the novel hexacyclic metabolite bovistol were obtained (Fig. 10.2). This can formally be regarded as a triterpene, but appears to be formed from psathyrellon B by a heteroatom Diels–Alder dimerization. Indeed, the spontaneous dimerization of psathyrellon B in methanol resulted in the formation of the expected dimer (Fig. 10.2).

h) *Marasmius* sp. (Tricholomataceae)

6,9-Dihydroxy-3(15)-caryophyllen-4,8-dione (Fig. 10.4), a new caryophyllane sesquiterpene, was isolated from fermentations of a tropical *Marasmius* species. Caryophyllane sesquiterpenoids have been described as widespread constituents of several plants, e.g. *Jasminum*, *Lavendula*, and *Juniperus*, and basidiomycetes. 6,9-Dihydroxy-3 (15)-caryophyllen-4,8-dione showed strong cytotoxic activities on L1210 (mouse lymphocytic leukemia) and HL-60 (human promyelocytic leukemia) cells with IC_{50} values of 1.9 µM and 3.8 µM but no antimicrobial activity. The aggregation of human thrombocytes stimulated with ADP or collagen was inhibited with IC_{50} values of 47 µM while no inhibition of thrombin induced aggregation was observed (Fabian et al. 1999).

i) *Dichomitus squalens* (Polyporaceae)

From mycelial cultures of a Chinese collection of *Dichomitus squalens* three new sesquiterpenes dichomitol (dichomitin A), 2β,13-dihydroxyledol (dichomitin B), and dichomitone (dichomitin C) were described by Huang et al. (2004). The structure of dichomitol is interesting because C-15 is attached to C-6 instead of C-4 as in other hirsutane sesquiterpenes. Dichomitol is the first example of 1,10-seco-2,3-seco-aromandendrane sesquiterpenes. The nomenclature of the compounds is confusing because in the same paper the three compounds were given two different names each (above in brackets; Fig. 10.4). Of the three compounds 2β,13-dihydroxyledol exhibited nematicidal activity against *Bursaphelenchus xylophilus*, a *Pinus* pathogen with a LC_{50} of 35.6 µg/ml.

j) *Macrocystidia cucumis* (Tricholomataceae)

Eight new triquinane-type sesquiterpenoids were isolated from fermentations of *Macrocystidia cucumis* (Hellwig et al. 1998). Cucumins A–D are highly unsaturated hirsutanes while cucumins E–G possess a novel carbon skeleton for which the name cucumane is proposed. Cucumin H is a new ceratopicane (Fig. 10.4). Three further metabolites were identified as *cyclo*(phenylalanylprolyl), *cyclo*(leucylprolyl), and arthrosporone. Cucumin A exhibits high antibiotic activity against *Bacillus subtilis*, *Nematospora coryli*, and *Mucor miehei* at minimal inhibitory concentrations (MIC) of 1–10 µg/ml, while cucumin C preferentially inhibits the growth of *N. coryli* and *Saccharomyces cerevisiae*. The antibacterial and antifungal activities of cucumin B are very weak. However, all three compounds are highly cytotoxic with IC_{100} = 0.5–1.0 µg/ml. As is the case with other sesquiterpenes the biological

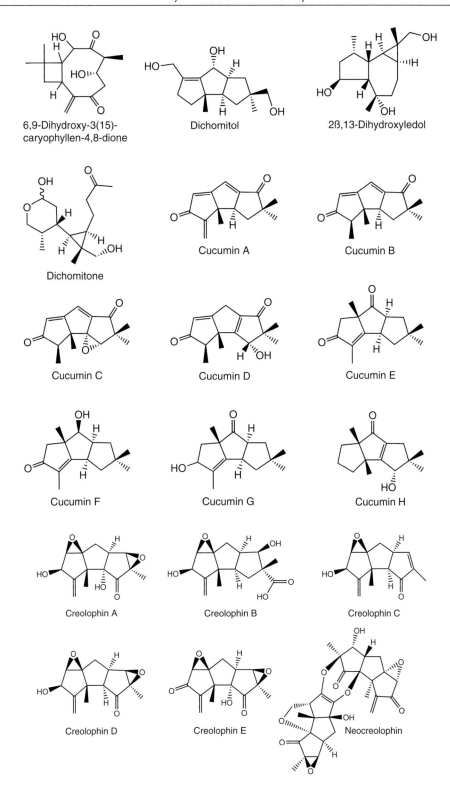

6,9-Dihydroxy-3(15)-caryophyllen-4,8-dione

Dichomitol

2ß,13-Dihydroxyledol

Dichomitone

Cucumin A

Cucumin B

Cucumin C

Cucumin D

Cucumin E

Cucumin F

Cucumin G

Cucumin H

Creolophin A

Creolophin B

Creolophin C

Creolophin D

Creolophin E

Neocreolophin

Fig. 10.4. Sesquiterpenes from *Marasmius* sp., *Dichomitus squalens, Macrocystidia cucumis* and *Creolophus cirrhatus*

activities are closely correlated to their reactivity towards nucleophiles like cysteine.

k) *Creolophus cirrhatus* (Hericiaceae)

Five new norhirsutanes with 14 carbon atoms, instead of 15 as in the hirsutanes, were isolated from fermentations of *Creolophus cirrhatus* isolated from a fruiting body collected in France (Opatz et al. 2007; Birnbacher et al. 2008). Creolophins A–E (Fig. 10.4) were isolated together with the known hirsutane complicatic acid. Only creolophin E and its dimer neocreolophin (Fig. 10.4B) which forms as an artefact during purification showed antibiotic and cytotoxic activity. In contrast to creolophins A–D, which exhibit no antibiotic and cytotoxic activities, creolophin E and its dimer possess a highly reactive exomethylene ketone moiety which readily reacts with nucleophiles like cysteine, resulting in a complete loss of activity (Birnbacher et al. 2008).

l) *Coprinus* sp. (Coprinaceae)

Coprinol, a new cuparane (Fig. 10.5), was isolated from fermentations of a wood inhabiting *Coprinus* sp. (Johansson et al. 2001). The new antibiotic exhibited activity against Gram-positive multiresistant strains, including penicillin-resistant pneumococci (PRSP), methicillin- and quinolone-resistant staphylococci (MRSA, QRSA), vancomycin-resistant enterococci (VREF), and vancomycin intermediate-resistant staphylococci (VISA). Coprinol exhibited no activity against Gram-negative bacteria and fungi.

m) *Dacrymyces* sp. (Dacrymycetaceae)

Two new antibiotic metabolites, the eremophilane dacrymenone and VM 3298-2 (Fig. 10.5) were isolated from fermentations of a *Dacrymyces* sp. collected in La Réunion, France (Mierau et al. 2003). Dacrymenone inhibits the growth of bacteria with MIC values of 25-100 µg/ml (86–343 µM).

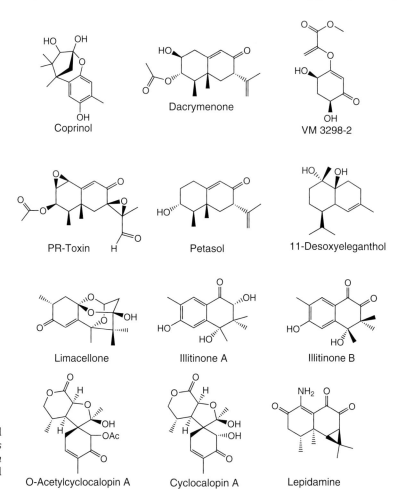

Fig. 10.5. PR toxin, petasol, and sesquiterpenes from *Coprinus* sp., *Dacrymyces* sp., *Limacella illinata*, *Boletus calopus*, and *Russula lepida*

VM 3298-2 exhibits weak antifungal activity against yeasts and filamentous fungi, with the highest activity on *Candida parapsilosis* at 25 μg/ml (110 μM) and *S. cerevisiae* at 10 μg/ml (44 μM). PR toxin (Fig. 10.5) from *Penicillium roquefortii* is described to have strong cytotoxic, fungicidal, and bactericidal properties (Moulé et al. 1977). The greatly reduced antibiotic activities and the lack of cytotoxicity of dacrymenone are likely due to the lack of the aldehyde group and the oxirane rings present in PR toxin. VM 3298-2 is cytotoxic towards Colo-320 (human colon adenocarcinoma) with IC_{50} values of 10 μg/ml (44 μM) and 5 μg/ml (22 μM) for HL-60 and L1210 cells. Related eremophilans, e.g. petasol (Fig. 10.5), produced by fungi and plants have phytotoxic effects (Sugawara et al. 1993). When tested in an assay using leaf cuts of *Hordeum sativum*, concentrations as low as 1 μM dacrymenone produced green islands (20 mm^2) and necrotic lesions on leafs, whereas in a similar assay petasol was reported to produce green islands on monocot leafs at much higher concentrations (10–20 μM; Sugawara et al. 1993).

n) *Limacella illinita* (Amanitaceae)
In a search for new bioactive compounds from basidiomycetes, four new compounds (Fig. 10.5) were isolated from fermentations of *Limacella illinita* (Gruhn et al. 2007). Limacellone has a new C15 carbon skeleton. Limacellone and illinitones A and B share large parts of the carbon skeleton and are therefore considered to be biogenetically related. 11-Desoxyeleganthol belongs to a group of sesquiterpenes which have been described from higher plants, but are quite rarely isolated from fungi. It is similar to eleganthol from *Clitocybe elegans* (Arnone et al. 1993). Illinitone A exhibits weak phytotoxic and moderate nematicidal activities against *Caenorhabditis elegans*, illinitone B is moderately cytotoxic, while limacellone exhibits weak cytotoxic and phytotoxic activities. 11-Desoxyeleganthol was found to be inactive in all assays. No biological activities have been described for eleganthol so far, which is in accordance with our findings.

o) *Boletus calopus* (Boletaceae)
Bitter tasting boletes are a nuisance to all mushroom hunters and gourmets who inadvertently add a fruiting body of, e.g. *Boletus calopus*, to a meal of *B. erythropus* or *B. edulis*. In fruiting bodies of *B. calopus* and closely related mushrooms (*B. radicans*, *B. coniferarum*, *B. rubripes*,

B. peckii) O-acetylcyclocalopin A (Fig. 10.5) was identified as the main bitter principal by Hellwig et al. (2002). It was accompanied by the slightly bitter cyclocalopin A and a number of other metabolites. Interestingly, no calopins could be detected in extracts of the bitter bolete *Tylopilus felleus*, a look-alike of *B. edulis*.

p) *Russula lepida* (Russulaceae)
A number of new terpenoids have been isolated from fruiting bodies of *Russula lepida*. Among them are lepidamine (Fig. 10.5), the first natural aristolane-type sesquiterpene alkaloid (Tan et al. 2003), and among other triterpenoids lepidolide (Tan et al. 2002) a cucurbitane (Fig. 10.6). No data on biological activity are available.

2. Diterpenoids

Diterpenoids are less frequently encountered among the basidiomycete metabolites. However some of them have interesting ring systems so far not found elsewhere. A semisynthetic pleuromutilin has even become a product used in veterinary medicine.

a) *Sarcodon scabrosus* (Thelephoraceae)
The cyathins and striatins are prominent products of *Cyathus* spp. Some of them have very high cytotoxic activities. Interesting recent additions to this class of compounds are the sarcodonins (Shibata et al. 1998) and the scabronines (Kita et al. 1998; Ohta et al. 1998; Ma et al. 2004) from *Sarcodon scabrosus*. The sarcodonins (e.g. sarcodonin A; Fig. 10.6) have a bitter taste, antiinflammatory and antibacterial activities (Shibata et al. 1998; Hirota et al. 2002; Kamo et al. 2004). Interesting pharmacological activities have been reported for scabronines. Scabronines A and G (Fig. 10.6) have been shown to promote the secretion of neurotrophic factors, including nerve growth factor from 1321N1 (human astrocytoma) cells, causing the enhancement of differentiation (neurite outgrowth) of PC-12 (pheochromocytoma of rat adrenal medulla) cells (Obara et al. 1999). Obara et al. (2001) suggested that scabronine G and its methylester would enhance the secretion of neurotrophic factors from 1321N1 cells by activation of protein kinase C-ζ.

b) *Mycena tintinnabulum* (Tricholomataceae)
Fruiting bodies and mycelia of *Mycena tintinnabulum* grown in complex media or on oak wood

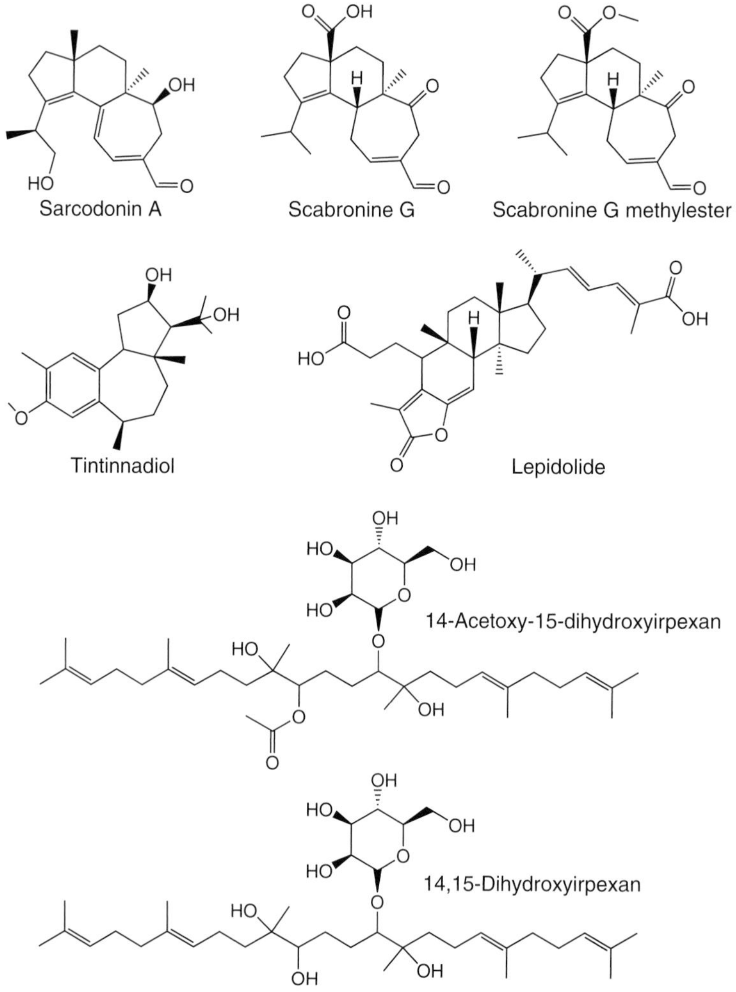

Fig. 10.6. Diterpenoids and triterpenoids from *Sarcodon scabrosus*, *Mycena tintinnabulum*,
Russula lepida, and *Irpex* sp.

Sarcodonin A

Scabronine G

Scabronine G methylester

Tintinnadiol

Lepidolide

14-Acetoxy-15-dihydroxyirpexan

14,15-Dihydroxyirpexan

contained antifungal strobilurins, whereas tintinnadiol (Fig. 10.6), a new sphaeroane diterpene, was
isolated only from the fruiting bodies. Tintinnadiol
exhibited cytotoxic activities towards HL-60 cells
($IC_{50} = 10$ µg/ml) and L1210 cells ($IC_{50} = 40$ µg/ml).
In the agar diffusion assay, the compound showed
neither antifungal nor antibacterial activity at concentrations up to 50 µg/disc (Engler et al. 1998).

3. Triterpenoids

Basidiomycetes are prolific producers of triterpenoids, among them many with interesting biological
activities.

a) *Irpex* sp. (Steccherinaceae)
The irpexans were isolated in the course of a screening for new inhibitors of AP-1 and NF-κB mediated

signal transduction pathways (Silberborth et al.
2000). Both pathways play major roles in inflammation and cancer. The irpexans consist of a triterpenoid chain with an attached mannose moiety.
14-acetoxy-15-dihydroxyirpexan (Fig. 10.6) inhibited the phorbol ester induced expression of an
AP-1 or NF-κB dependent reporter gene with
IC_{50} values of 6–7 µg/ml followed by 14,15-irpexan-
oxide, 14,15-dihydroxyirpexan (Fig. 10.6) and
14-acetoxy-22,23-dihydro-15,23-dihydroxyirpexan,
whereas irpexan exhibited no activity. All isolated
compounds showed no antibacterial or antifungal
activity. Cytotoxic effects were observed starting at
20 µg/ml.

b) *Favolaschia* spp. (Favolaschiaceae)
The genus *Favolaschia* is a rich source of biologically active compounds. Among them are

strongly antifungal strobilurins and oudeman-sins (Zapf et al. 1995; Wood et al. 1996; Nicholas et al. 1997), (+)-10α-hydroxy-4-muurolen-3-one, an inhibitor of leukotriene biosynthesis (Zapf et al. 1996), favolon A (Anke et al. 1995), and laschiatrion (Anke et al. 2004). Laschiatrion (Fig. 10.7), a new antifungal antibiotic, was isolated from fermentations of *Favolaschia* sp. 87129. It possesses a new steroid skeleton. Laschia-trion exhibits broad in vitro antifungal activity against *Candida albicans*, *Cryptococcus neofor-mans*, *Aspergillus flavus*, *Fusarium verticillioides*, *Trichophyton mentagrophytes*, and *Microsporum gypseum* at concentrations of 10–50 µg/ml. No antibacterial and cytotoxic activities could be detected. By comparison, favolon A, an antifungal triterpenoid with a different ring system and substitution pattern exhibits higher activities at concentrations starting from 1 µg/ml. Recently, favolon B (Fig. 10.7) was isolated from a Chilean *Mycena* species (Aqueveque et al. 2005). It differs from favolon A in the replacement of favolon's epoxide group in the B ring by a double bond. It shares the high antifungal activities of the parent compound.

c) *Fomitella fraxinea* (Polyporaceae)

3β-Hydroxylanosta-8,24-dien-21-oic acid and fomi-tellic acids A–D (Fig. 10.7 7), four new lanostanes, were isolated from the mycelia of *Fomitella fraxinea* in a search for new inhibitors of DNA polymerases (Tanaka et al. 1998). All five metabolites inhibited calf DNA polymerase α and rat DNA polymerase β at concentrations of 35–75 µM and 90–130 µM respectively. Data on other biological activities are not given.

d) *Grifola frondosa* (Polyporaceae s. lat.)

Many different biological activities have been ascribed to the common fungal metabolite ergos-terol, its oxidation products as well as some deri-vatives. New to the list are inhibitory activities of ergosterol and ergostra-4,6,8(14),22-tetraen-3-one (Fig. 10.8) on cyclooxygenase-1 and -2 at rather high concentrations (250 µg/ml). Both com-pounds were isolated from cultured mycelia of *Grifola frondosa* by Zhang et al. (2002).

e) *Leucopaxillus gentianeus* (Tricholomataceae)

The cucurbitanes leucopaxillones A and B (Fig. 10.8) have been isolated from fruiting bodies of

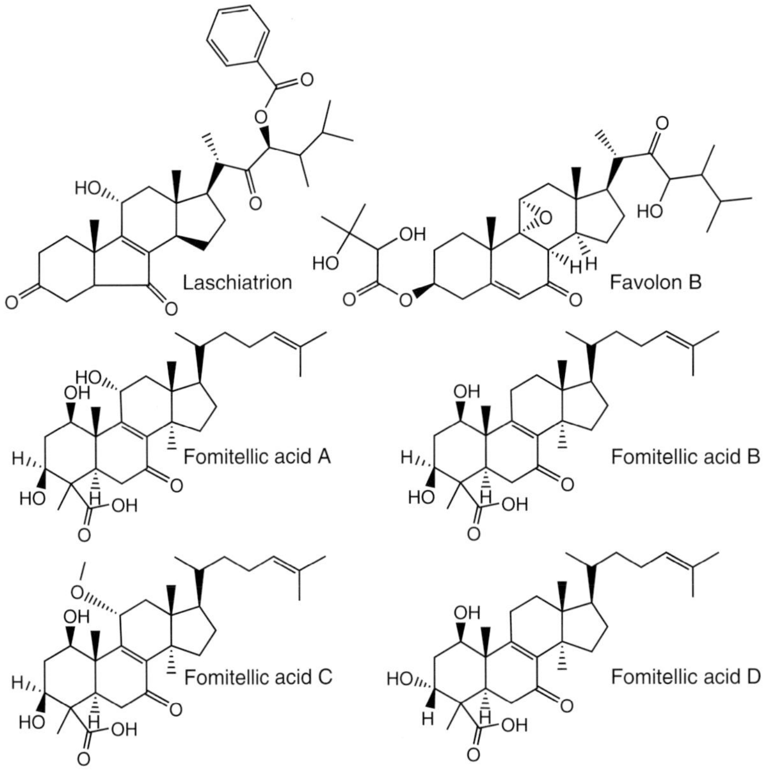

Fig. 10.7. Triterpenoids from *Favolaschia* sp. and *Fomitella fraxinea*

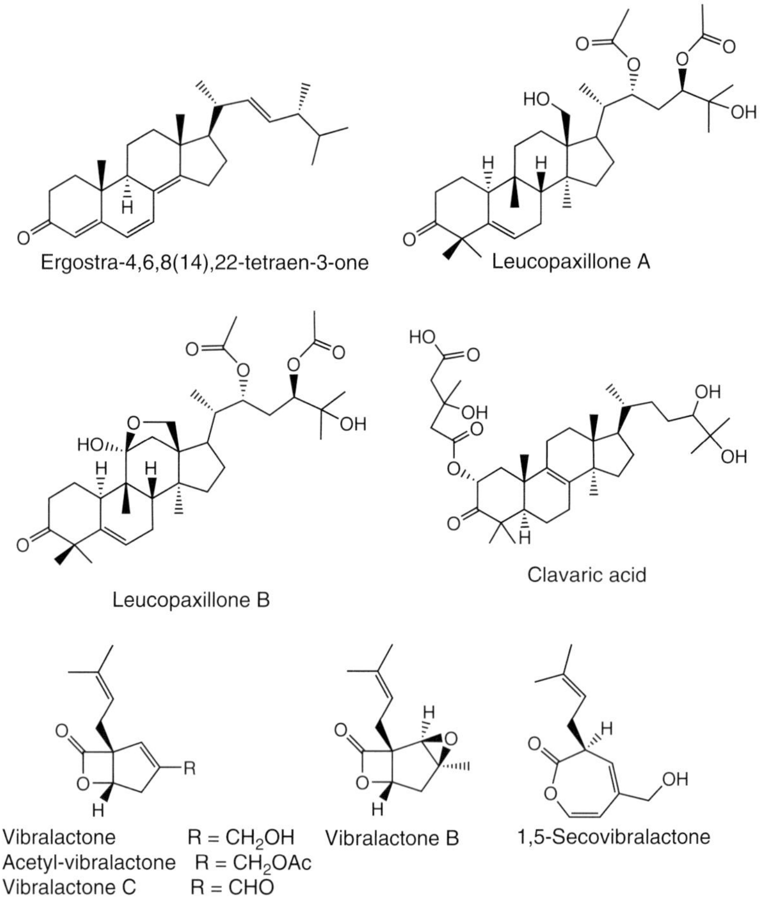

Fig. 10.8. Triterpenoids from *Grifola frondosa, Leucopaxillus gentianeus, Clavariadelphus truncatus,* and vibralactones from *Boreostereum vibrans*

Ergostra-4,6,8(14),22-tetraen-3-one

Leucopaxillone A

Leucopaxillone B

Clavaric acid

Vibralactone   R = CH$_2$OH   Vibralactone B   1,5-Secovibralactone
Acetyl-vibralactone R = CH$_2$OAc
Vibralactone C   R = CHO

Leucopaxillus gentianeus, together with cucurbitacin B, a known plant metabolite (Clericuzio et al. 2004). When tested for cytotoxic activities towards the cell lines A549 (human lung carcinoma), CAKI 1 (human kidney carcinoma), Hep-G2 (human liver carcinoma), and MCF-7 both leucopaxillones compared unfavorably with the known cytotoxic activities of cucurbitacin B.

f) *Clavariadelphus truncatus* (Clavariaceae)
Lingham et al. (1998) identified clavaric acid (24,25-dihydroxy-2-(3-hydroxy-3-methylglutaryl) anostan-3-one; Fig. 10.8) as a novel metabolite of *Clavariadelphus truncatus*. The compound exhibits very interesting activities. It inhibits selectively human farnesyl-protein transferase (FPTase) with an IC$_{50}$ value of 1.3 μM without interfering with geranylgeranyl-protein transferase-I (GGPTase-I) or squalene synthase activity. It is competitive with respect to Ras and is a reversible inhibitor of FPTase.

B. Polyketides, Fatty Acid Derivatives

a) *Boreostereum vibrans* (Stereaceae)
The vibralactones (Fig. 10.8), unusual fused β-lactone-type metabolites were isolated from cultures of *Boreostereum vibrans* (syn. *Stereum vibrans*; Liu et al. 2006; Jiang et al. 2008). Vibralactone is a potent inhibitor of pancreatic lipase with an IC$_{50}$ value of 0.4 μg/ml. A structurally closely related metabolite, percyquinnin, had previously been isolated from fermentations of *Stereum complicatum* as a potent lipase-inhibitor (Hopmann et al. 2001). It later turned out to be identical with vibralactone. Four new related metabolites, 1,5-secovibralactone, vibralactone B, vibralactone C, and acetylated vibralactone have now been described from the same fungus. Inhibitors of pancreatic lipase like orlistat, a derivative of the streptomycete metabolite lipstatin (Weibel et al. 1987) are now medications for obese patients (Bray and Greenway 2007). Its mode of action was

elucidated by Hadvary et al. (1991). Tetrahydro-lipstatin binds covalently to the putative active site serine by means of the β-lactone moiety. The inhibitory activity of vibralactone may be due to the same mechanism.

b) *Junghuhnia nitida* (Steccherinaceae)

Many polyacetylenes have been described from basidiomycetes. Nitidon (Fig. 10.9), a highly oxi-dized pyranone derivative produced by *Junghuhnia nitida*, exhibits antibiotic and cytotoxic activities and induces morphological and physio-logical differentiation of tumor cells at nanomolar concentrations (Gehrt et al. 1998). At a concentra-tion of 100 ng/ml (0.46 µM) nitidon induces the differentation of 25–30% of the HL-60 cells into granulocyte-monocyte-like cells and a differenta-tion of 20% of the U-937 cells (human histiocytic leukemia) into monocyte-like cells. The biological activity of nitidon is at least in part due to its high chemical reactivity. Addition of cysteine yielded adducts which were almost devoid of differentia-tion inducing activity.

c) *Ceriporia subvermispora* (Polyporaceae s. lat.)

Ceriporia subvermispora is a white rot fungus which degrades lignin without substantially dam-aging the remaining cellulose. This bears implica-tions, e.g. for the cellulose and paper industry. The reason for this selective degadation of lignin was investigated by Watanabe's group (Ohashi et al. 2007). They found that alkylitaconic acids, e.g. ceriporic acid B (Fig. 10.9), suppress the produc-tion of hydroxyl radicals by the Fenton reaction (Rahmawati et al. 2005) even in the presence of reductants for Fe^{3+}. The alkyl side-chain and the two carboxyl groups are essential for redox

silencing and high stability against oxidative deg-radation by OH radicals.

d) *Suillus luteus* (Boletaceae)

Suillumide, (2R)-N-[(3S*,4S*,5R*)-5-dodecyl-4-hydroxytetrahydrofuran-3-yl]-2-hydroxyheptade-canamide, (Fig. 10.9), was isolated from fruiting bodies of *Suillus luteus* collected in Colombia (Léon et al. 2008). So far, ceramides with a tetrahy-drofuranyl ring have been found only in fungi. Suil-lumide inhibits the growth of SK-MEL-1 (human melanoma) cells with an IC_{50} value of 10 µM.

e) *Hygrophorus* spp. (Hygrophoraceae)

The hygrophorones (e.g. hygrophorones D12, F12, A12; Fig. 10.10), new acylpentenones, were isolated from fruiting bodies of *Hygrophorus latitabundus, H. olivaceoalbus, H. persoonii*, and *H. pustulatus* (Lübken 2006; Lübken et al. 2004, 2006). All hygro-phorones exhibited modest antifungal activity against *Cladosporium cucumerinum*.

Chrysotriones A and B (Fig. 10.10) were isolated from the fruiting bodies of the *H. chryso-don* (Gilardoni et al. 2007). Both compounds are new 2-acylcyclopentene-1,3-dione derivatives. The chrysotriones exhibited modest antifungal activity against the phytopathogen *Fusarium verticillioides*.

f) *Trametes menziesii* (Polyporaceae s. lat.)

The trametenamides A and B (Fig. 10.11), two new ceramides were isolated from fruiting bodies of *Trametes menziesii* collected in Columbia. The absolute stereochemistry of trametamide B was determined by total synthesis. Acetyl-trameta-mide A was cytotoxic towards SK-MEL-1 cells with an IC_{50} value of 8 µM by induction of apoptosis measured by DNA fragmentation, poly-(ADP-ribose) polymerase cleavage, and

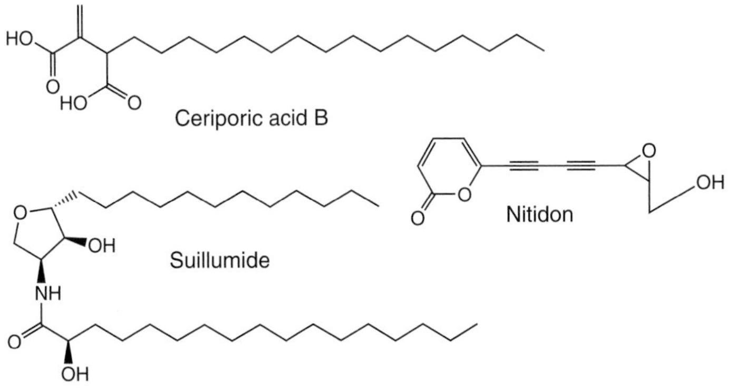

Ceriporic acid B

Nitidon

Suillumide

Fig. 10.9. Nitidon, ceriporic acid B, and suillumide from *Junghuhnia nitida, Ceriporia subvermispora*, and *Suillus luteus*

Fig. 10.10. Hygrophorones and chrysotriones from *Hygrophorus* spp.

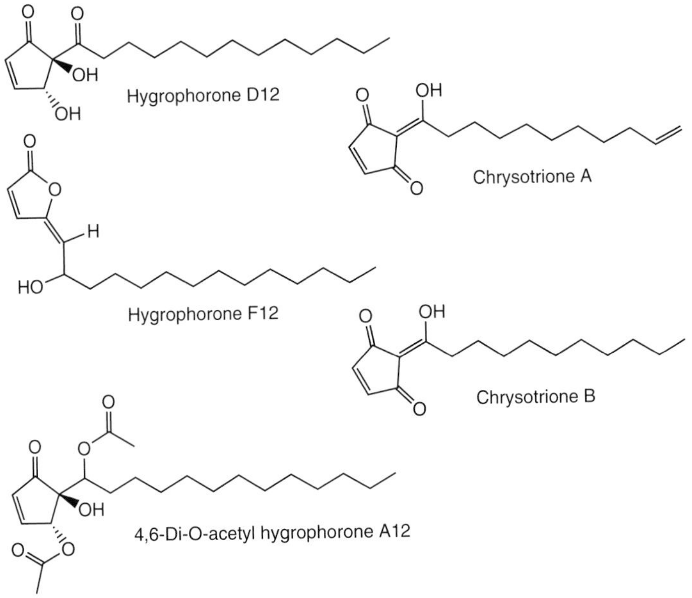

Hygrophorone D12

Chrysotrione A

Hygrophorone F12

Chrysotrione B

4,6-Di-O-acetyl hygrophorone A12

procaspase-9 and -8 processing (Léon et al. 2006). Ceramides are thought to play a role in the induction of apoptosis by permeabilization of the mitochondrial membrane, allowing the exit of proapoptotic factors like cytochrome *c*.

g) *Gerronema* sp. (Tricholomataceae)
The gerronemins (Fig. 10.11) were isolated from fermentations of a *Gerronema* species collected in the United States (Silberborth et al. 2002). They are composed of a C12–C16 alkane or alkene substituted at both ends by 2,3-dihydroxyphenyl groups. The gerronemins blocked the inducible expression of a human COX-2 and iNOS promoter driven reporter gene with IC_{50} values of 1–5 µg/ml. In addition, cytotoxic activities were observed which were due to the inhibition of cellular macromolecular syntheses.

h) Imperfect Basidiomycete Strain 96244
5-(2′-Oxoheptadecyl)-resorcinol and 5-(2′-oxononadecyl)-resorcinol (Fig. 10.12) were isolated from fermentations of an imperfect basidiomycete (Filip et al. 2002). This was obtained from pieces of a pinkish mycelium covering a piece of rotting wood in Provence, France. The mycelia bore clamp connections but neither basidia nor conidia could be found. Both resorcinols exhibit cytotoxic effects against the human colon tumor cell lines Colo-320, DLD-1 and HT-29 and the human promyeloid leukemia cell line HL-60, the human leukemia T cell JURKAT, the human hepatocellular carcinoma cell line Hep-G2 as well as the J774 mouse macrophage cell line. The compounds induce morphological and physiological differentiation of HL-60 cells into granulocytes, which subsequently die by apoptosis. Both compounds show no antibacterial or antifungal activity. Interestingly 5-(2′-oxoheptadecyl)-resorcinol was reported from rye grains (Kozubek and Tyman 1995) and both compounds are suggested to be produced by rice (Suzuki et al. 1996).

i) *Pterula* sp. (Pterulaceae), *Mycena* spp. (Tricholomataceae)
A new antifungal (E)-β-methoxyacrylate, noroudemansin A (Fig. 10.12), was isolated from cultures of *Pterula* sp. 82168. As compared to oudemansin A, its antifungal activities are much lower (Engler-Lohr et al. 1999). Strobilurin N (Fig. 10.12) was isolated from fruiting bodies of *Mycena crocata* (Buchanan et al. 1999). Interestingly, strobilurin N is the first strobilurin without antifungal activity. The new antifungal strobilurins I from *Agaricus* sp. 89139 and K from *M. tintinnabulum* (Fig. 10.12) are 3,4-dihydro-2H-benzo[b](1,4)dioxepin derivatives (Hellwig et al. 1999). Both compounds strongly inhibit yeasts and filamentous fungi.

j) Terphenyls from *Thelephora aurantiotincta, T. terrestris, Hydnellum caeruleum, Sarcodon leucopus, S. scabrosus* (all Thelephoraceae), and *Paxillus curtisii* (Paxillaceae)
Terphenyls are derived from the shikimate–chorismate pathway and are among the frequently

encountered basidiomycete pigments (Gill and Steglich 1987; Gill 1999; Liu 2006). More recent additions are the thelephantins (e.g. A and H; Fig. 10.13) from *Thelephora aurantiotincta* and *Hydnellum caeruleum* (Quang et al. 2003a, b, 2004), terrestins from *T. terrestris* (Radulović et al. 2005), and curtisians (e.g. A and C; Fig. 10.13) from *Paxillus curtisii* which has inhibitory activity against lipid peroxidation with IC_{50} values of 0.15, 0.17, 0.24, and 0.14 µg/ml (Yun et al. 2000). A number of biological activities have been ascribed to members of this group. Thelephoric acid (Fig. 10.13), the characteristic pigment of many Thelephoraceae, is claimed to be an antitumor (Kawai et al. 2005) and antiallergic agent

(Tateishi et al. 2005). Sarcodonin (Fig. 10.13), a very interesting nitrogen-containing terphenyl derivative with a modified diketopiperazine attached to it was isolated from fruiting bodies of *Sarcodon leucopus* (Geraci et al. 2000). *S. scabrosus* collected in China yielded sarcodonin δ (Fig. 10.13; Ma and Liu 2005).

C. Compounds of Unclear Biogenetic Origin

a) *Tremella aurantialba* (Tremellaceae)

Tremellin (Fig. 10.14), a simple, highly symmetric structure was isolated from fruiting bodies of *Tremella aurantialba*, an edible mushrooms (Ding

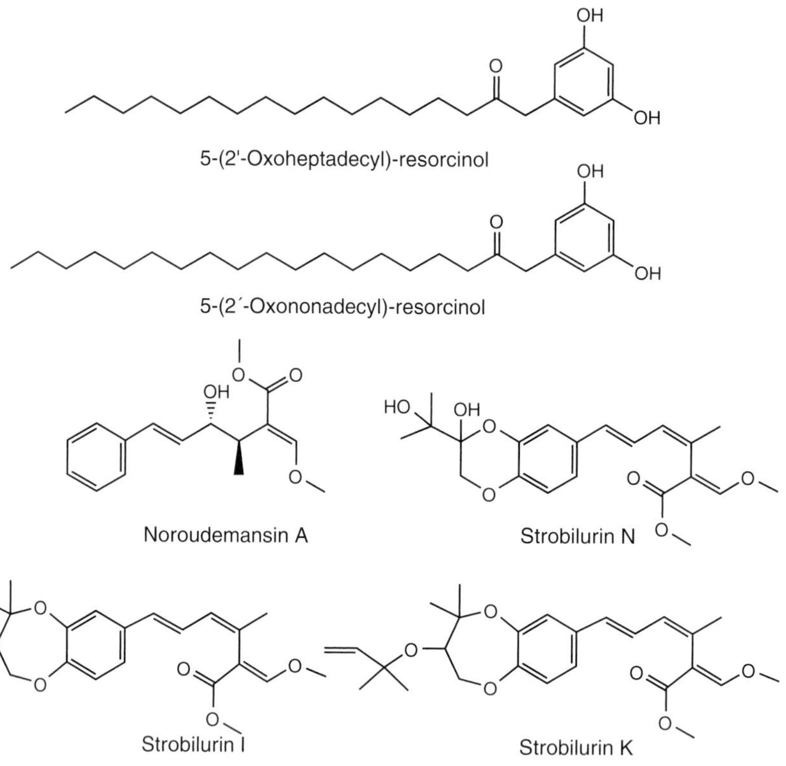

Fig. 10.12. Resorcinols, noroudemansin A and strobilurins from an imperfect basidiomycete, *Pterula* sp., and *Mycena* spp.

et al. 2002). No biological activities have been reported so far.

b) *Stereum hirsutum* (Stereaceae)
Scavengers of free radicals are considered as possible protectants of cells and tissues against oxidative damage and resulting diseases. In a search for new antioxidants Yun et al. (2002) isolated sterins A and B (Fig. 10.14) from fermentations of a strain of *Stereum hirsutum* collected in Korea. Sterin A was found to inhibit lipid peroxidation in an assay using rat liver microsomes at concentrations of 8 µg/ml. In the same assay vitamin E was ten times more effective.

c) *Antrodia serialis* (Polyporaceae s. lat.)
Serialynic acid (Fig. 10.14) was isolated from agar cultures of *Antrodia serialis* in a screening for new antifungal compounds (Kokubun et al. 2007). Phenols with an isopentenyne side-chain like siccayne (Kupka et al. 1981) are sometimes produced by both ascomycetes and basidiomycetes. Serialynic acid has been found to be weakly active against some phytopathogenic fungi.

d) *Aporpium caryae* (Tremellaceae)
The aporpinones (Fig. 10.14), four furanones containing an acetylene moiety, were isolated from cultures of *Aporpium caryae* (Levy et al. 2003). Aporpinone B and its acetyl derivative exhibited modest antibacterial activity against *Bacillus subtilis*, *Staphylococcus aureus*, and *Escherichia coli*.

e) *Bondarzewia montana* (Bondarzewiaceae)
When exposed to aqueous KOH the root of *Bondarzewia montana* turns bright yellow. The main chromogen was isolated and turned out to be a new compound which was named montadial (Fig. 10.14). When tested for biological activity montadial exhibited weak phytotoxic and stronger cytotoxic activities, the latter starting from 5 µg/ml (Sontag et al. 1999).

f) *Cortinarius* sp. (Cortinariaceae)
Cortamidine oxide (Fig. 10.14), an interesting asymmetrical disulfide metabolite, was isolated from the fruiting bodies of a *Cortinarius* sp. collected in New Zealand (Nicholas et al. 2001).

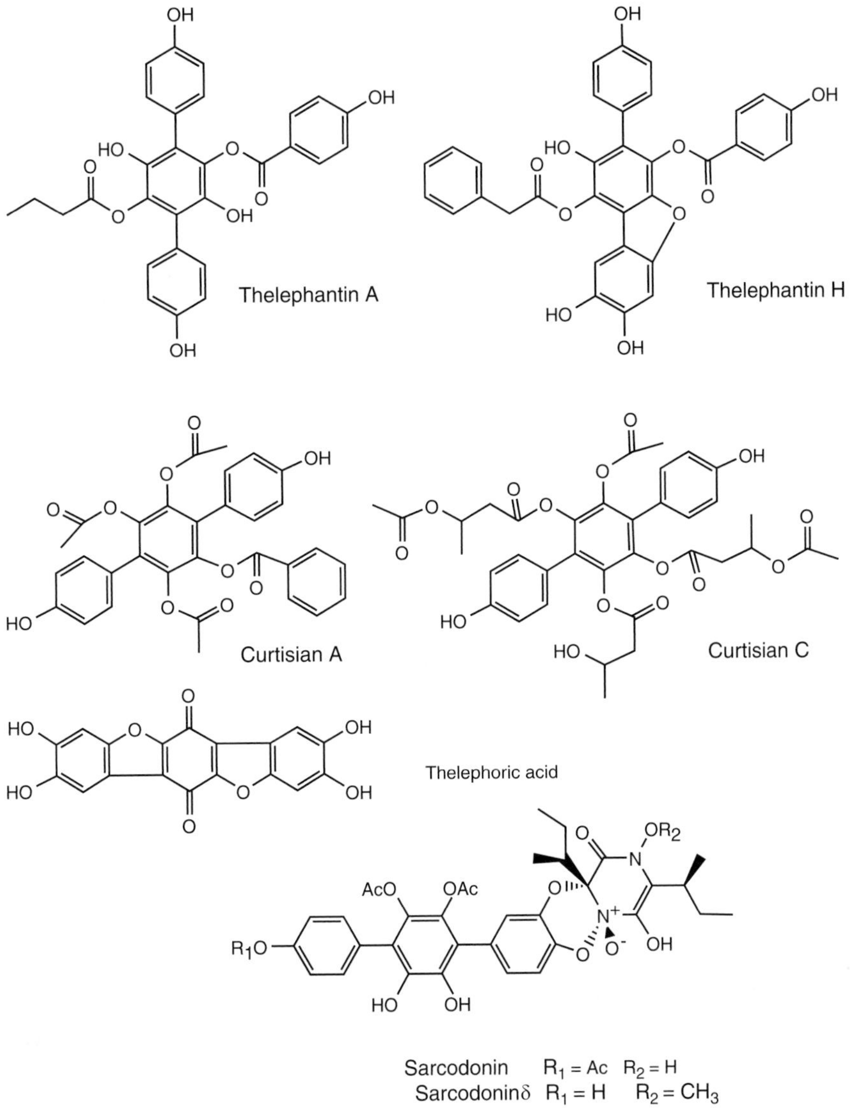

Fig. 10.13. Terphenyls from *Thelephora aurantiotincta*, *T. terrestris*, *Hydnellum caeruleum*, *Sarcodon leucopus*, *S. scabrosus*, and *Paxillus curtisii*

Cortamidine oxide has both significant antimicrobial and cytotoxic activity. Other compounds containing a pyridine *N*-oxide functionality have been reported from *Cortinarius* species before, namely, the toxin orellanine.

g) *Chamonixia pachydermis* (Boletaceae)
An unusual oxalylated tetramic acid, pachydermin (Fig. 10.15), was isolated from the gastromycete *Chamonixia pachydermis* collected in New Zealand (Lang et al. 2006). When tested for antibacterial activity pachydermin did not exhibit antibiotic activity.

h) *Serpula himantoides* (Coniophoraceae)
The himanimides (Fig. 10.15) were isolated from fermentations of a *Serpula himantoides* strain collected in Chile (Aqueveque et al. 2002). All four compounds are succinimide and maleimide derivatives, of which two are N-hydroxylated. Only himanimide C exhibits broad antibacterial, antifungal, and cytotoxic activity, suggesting a link to the N-hydroxylated maleinimide moiety.

i) *Pholiota spumosa* (Strophariaceae)
(*R*)-2-hydroxyputrescine dicinnamamide and (2*R*)-2-[(*S*)-3-hydroxy-3-methylglutaryloxy]putrescine-dicinnamamide (Fig. 10.15), two new cinnamic

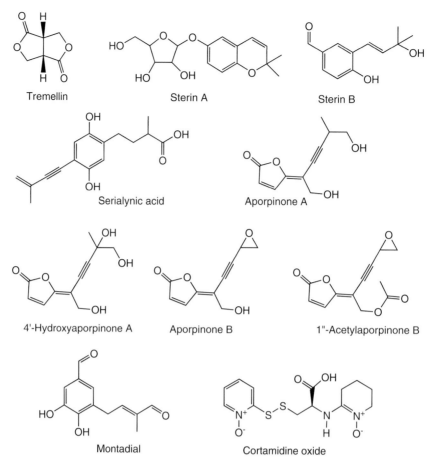

Fig. 10.14. Tremellin from *Tremella aurantialba*, sterins from *Stereum hirsutum*, serialynic acid from *Antrodia serialis*, aporpinones from *Aporpium caryae*, montadial from *Bondarzewia montana*, and cortamidine oxide from *Cortinarius* sp.

acid were isolated from fruiting bodies of *Pholiota spumosa* in addition to the known maytenine (*N*1,*N*8-dicinnamoyl spermidine) by Clericuzio et al. (2007). Both compounds exhibit modest cytotoxic activities. Unfortunately, the second compound was named pholiotic acid, a name earlier given to another (illudalane) metabolite of *P. destruens* (Becker et al. 1994).

D. Amino Acid Derivatives

Basidiomycetes produce a number of cyclopeptides with very interesting biological activities. These are being dealt with in Chap. 13.

a) *Aporpium caryae* (Tremellaceae)

A culture of *Aporpium caryae* yielded isolates of 1H-indole-3-carboxylic acid, 1-(1,1-dimethyl-2-propenyl) methyl ester, and 1H-indole-3-carboxylic acid, 1-(2,3-dihydroxy-1,1-dimethylpropyl)

methyl ester (Fig. 10.16), two metabolites of mixed biogenetic origin (Levy et al. 2000). Both metabolites exhibited modest antifungal activities against *Cladosporium cucumerinum*.

III. Conclusions

As can be deduced from the numerous new structures, interest in basidiomycete secondary metabolism has gained momentum. The biological activities are interesting and may help to define new lead compounds offering structures not easily detected by the random screening of libraries derived from combinatorial chemical synthesis. The availability of basidiomycete metabolites is facilitated by important progress in fermentation technologies and genetics, opening access to templates for chemical syntheses and providing new chemical approaches to yet unexplored biological targets.

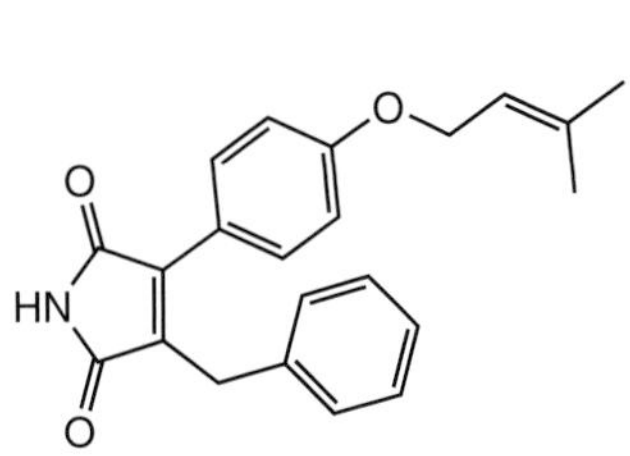

Fig. 10.15. Himanimides, dicinnamamides, and pachydermin from *Serpula himantoides*, *Pholiota spumosa*, and *Chamonixia pachydermis*

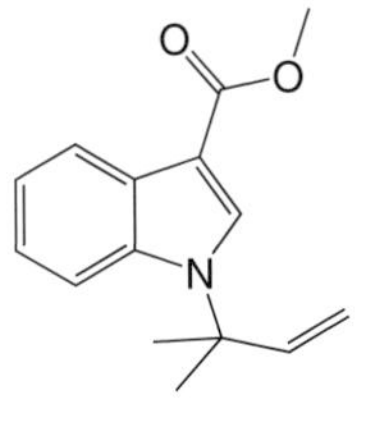

Himanimide A

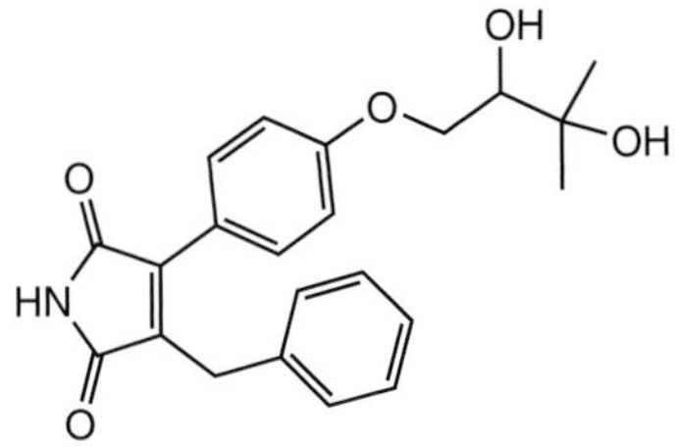

Himanimide B

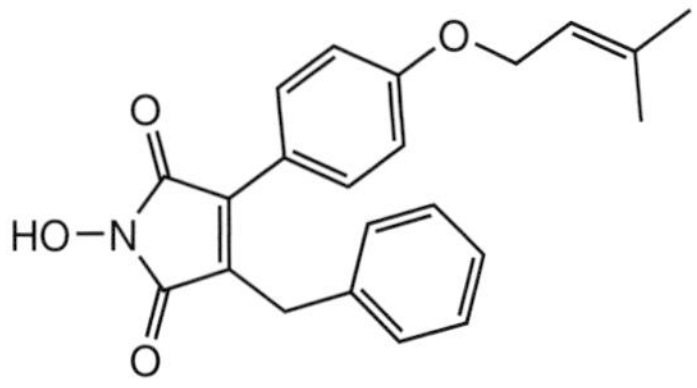

Himanimide C

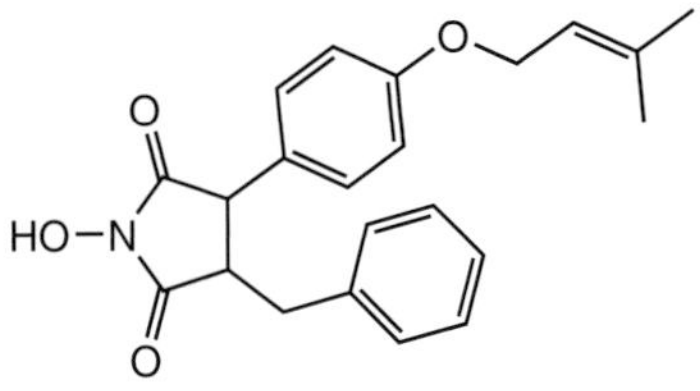

Himanimide D

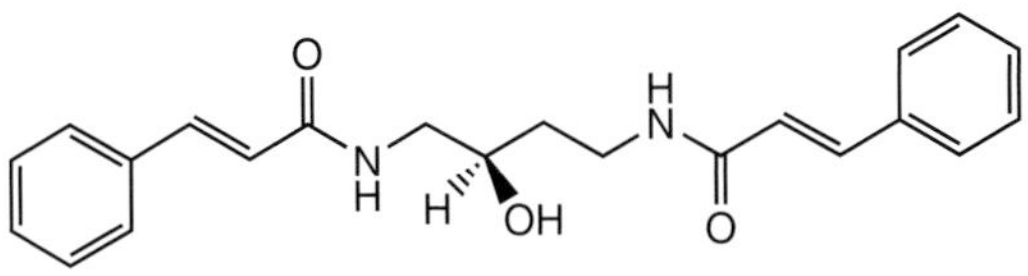

(R)-2-Hydroxyputrescine dicinnamamide

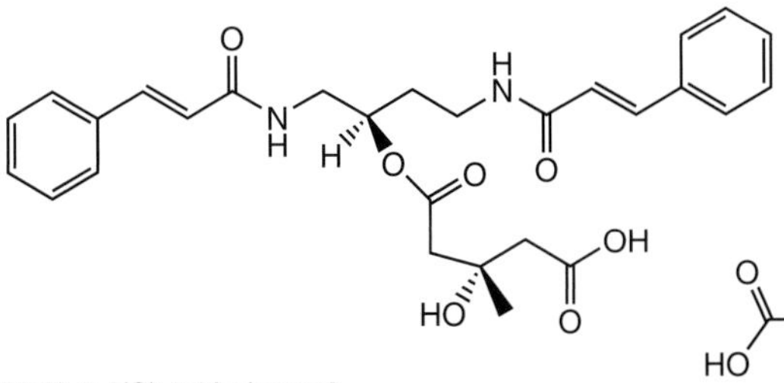

{(2R)-2-[(S)-3-Hydroxy-3-methylglutaryloxy]putrescine dicinnamamide}
(pholiotic acid)

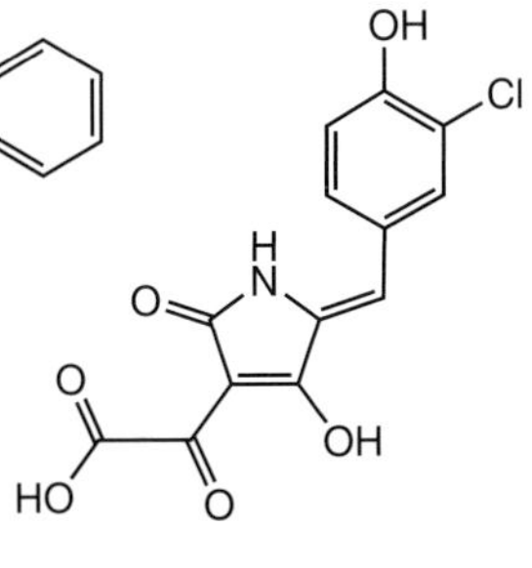

Pachydermin

1H-indole-3-carboxylic acid, 1-(1,1-dimethyl-2-propenyl) methyl ester

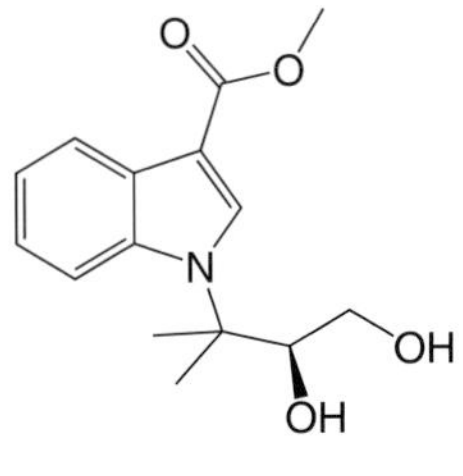

1H-indole-3-carboxylic acid, 1-(2,3-dihydroxy-1,1-dimethylpropyl) methyl ester

Fig. 10.16. Amino acid derivatives from *Aporpium caryae*

References

Abraham WR (2001) Bioactive sesquiterpenes produced by fungi: are they useful for humans as well? Curr Med Chem 8:583–606

Anke T, Werle A, Kappe R, Sterner O (2004) Laschiatrion, a new antifungal agent from a *Favolaschia* species (Basidiomycetes) active against human pathogens. J Antibiot 57:496–501

Anke T, Werle A, Zapf S, Velten R, Steglich W (1995) Favolon, a new antifungal triterpenoid from a *Favolaschia* species. J Antibiot 48:725–726

Aqueveque P, Anke T, Anke H, Sterner O, Becerra J, Silva M (2005) Favolon B, a new triterpenoid isolated from the chilean *Mycena* sp. strain 96180. J Antibiot 58:61–64

Aqueveque P, Anke T, Sterner O (2002) The himanimides, new bioactive compounds from *Serpula himantoides* (Fr)Karst. Z Naturforsch 57c:257–262

Arnone A, Cardillo R, di Modugno V, Nasini G (1991) Secondary mould metabolites. XXXIII. Structure elucidation of illudins C–E, novel illudane sesquiterpenoids isolated from *Clitocybe illudens*, using one- and two-dimensional NMR spectroscopy. Gazz Chim Ital 121:345–348

Arnone A, Colombo A, Nasini G, Meille SV (1993) Eleganthol, a sesquiterpene from *Clitocybe elegans*. Phytochemistry 32:1493–1497

Ayer WA, Cruz ER (1993) The tremulanes, a new group of sequiterpenes from the aspen rotting fungus *Phellinus tremulae*. J Org Chem 58:7529–7534

Bastian W (1985) Vergleichende Untersuchung zum Sekundärstoffwechsel von coprophilen und erd- oder holzbewohnenden Basidiomyceten. Dissertation, University of Kaiserslautern

Becker U, Anke T, Sterner O (1994) A novel illudalane isolated from the fungus *Pholiota destruens* (Brond.) Quel. Nat Prod Lett 5:171–174

Birnbacher J, Schüffler A, Deininger F, Opatz T, Anke T (2008) Isolation and biological activity of new norhirsutanes from *Creolophus cirrhatus*. Z Naturforsch 63c:203–206

Bray GA, Greenway FL (2007) Pharmacological treatment of the overweight patient. Pharmacol Rev 59:151–184

Breheret S, Talou T, Rapior S, Bessiere J-M (1999) Geosmin a sesquiterpenoid compound responsible for the musty-earthy odor of *Cortinarius herculeus*, *Cystoderma amianthinum* and *Cy. carcharias*. Mycologia 91:117–120

Bresinsky A, Besl H (1985) Giftpilze. Wissenschaftliche Verlagsgesellschaft, Stuttgart

Buchanan MS, Steglich W, Anke T (1999) Strobilurin N and two metabolites of chorismic acid from the fruit-bodies of *Mycena crocata* (Agaricales). Z Naturforsch 54c:463–468

Burgess ML, Zhang YL, Barrow KD (1999) Characterization of new illudanes illudins F, G, and H from the basidiomycete *Omphalotus nidiformis*. J Nat Prod 62:1542–1544

Clericuzio M, Mella M, Vita-Finzi P, Zema M, Vidari G (2004) Cucurbitane triterpenoids from *Leucopaxillus gentianeus*. J Nat Prod 67: 1823–1828

Clericuzio M, Tabasso S, Garbarino JA, Piovano M, Cardile V, Russo A, Vidari G (2007) Non-phenolic dicinnamamides from *Pholiota spumosa*: isolation synthesis and antitumour activity. Eur J Org Chem 33:5551–5559

Daum RS, Kar S, Kirkpatrick P (2007) Fresh from the pipeline: retapamulin. Nat Rev Drug Discov 6:865–866

Ding Z-H, Li J-P, Liu J-K, Yang L, Wang C, Zheng Q-T (2002) Tremellin a novel symmetrical compound from the basidiomycete *Tremella aurantialba*. Helv Chim Acta 85:882–884

Eilbert F, Engler-Lohr M, Anke H, Sterner O (2000) Bioactive sesquiterpenes from the basidiomycete *Resupinatus leightonii*. J Nat Prod 63:1286–1287

Engler M, Anke T, Sterner O (1998) Tintinnadiol, a new sphaeroane diterpene from fruiting bodies of *Mycena tintinnabulum*. Phytochemistry 49:2591–2593

Engler-Lohr M, Anke T, Hellwig V, Steglich W (1999) Noroudemansin A, a new antifungal antibiotic from *Pterula* species 82168 and three semisynthetic derivatives. Z Naturforsch 54c:163–168

Fabian K, Anke T, Sterner O (1999) 69-Dihydroxy-3(15)-caryophyllen-48-dione – a new antibiotic from a *Marasmius* species. Z Naturforsch 54c:469–473

Fabian K, Lorenzen K, Anke T, Johansson M, Sterner O (1998) Five new bioactive sesquiterpenes from *Radulomyces confluens* (Fr) Christ. Z Naturforsch 53c:939–945

Filip P, Anke T, Sterner O (2002) 5-(2′-Oxoheptadecyl)-resorcinol and 5-(2′-oxononadecyl)-resorcinol, cytotoxic metabolites from a wood-inhabiting basidiomycete. Z Naturforsch 57c:1004–1008

Florey HW, Chain E, Heatley NG, Jennings MA, Sanders AG, Abraham EP, Florey ME (1949) In: Florey HW (ed) Antibiotics, vol 1. Oxford University Press, Oxford, pp 248–272

Gehrt A, Erkel G, Anke T, Sterner O (1998) Cycloepoxydon, 1-hydroxy-2-hydroxymethyl-3-pent-1-enylbenzene, and 1-hydroxy-2-hydroxymethyl-3-pent-1,3-dienylbenzene, new inhibitors of eukaryotic signal transduction. J Antibiot 51:455–463

Gehrt A, Erkel G, Anke T, Sterner O (1998) Nitidon a new bioactive metabolite from the basidiomycete *Junghuhnia nitida*(Pers:Fr)Ryv. Z Naturforsch 53c:89–92

Geraci C, Neri P, Paterno C, Rocco C, Tringali C (2000) An unusual nitrogenous terphenyl derivative from fruiting bodies of the basidiomycete *Sarcodon leucopus*. J Nat Prod 63:347–351

Gilardoni G, Clericuzio M, Tosi S, Zanoni G, Vidari G (2007) Antifungal acylcyclopentenediones from fruiting bodies of *Hygrophorus chrysodon*. J Nat Prod 70:137–139

Gill M (1999) Pigments of fungi (Macromycetes). Nat Prod Rep 16:301–317

Gill M, Steglich W (1987) Pigments of fungi (Macromycetes). In: Zechmeister (ed) Progress in the chemistry of organic natural products, vol 51. Springer, Heidelberg, pp 1–317

Gruhn N, Schöttler S, Sterner O, Anke T (2007) Biologically active metabolites from *Limacella illinita*(Fr.) Murr. Z Naturforsch 62c:808–8122

Hadvary P, Sidler W, Meister W, Vetter W, Wolfer H (1991) The lipase inhibitor tetrahydrolipstatin binds covalently to the putative active site serine of pancreatic lipase. J Biol Chem 266: 2021–2027

Harttig U, Anke T, Scherer A, Steglich W (1990) Antibiotics from basidiomycetes XXXVI. Leaianafulvene, a sesquiterpenoid fulvene derivative from cultures of *Mycena leaiana* (Agaricales). Phytochemistry 29:3942–3944

Hellwig V, Dasenbrock J, Graf C, Kaher L, Schumann S, Steglich W (2002) Calopins and cyclocalopins – bitter principles from *Boletus calopus* and related mushrooms. Eur J Org Chem 17:2895–2904

Hellwig V, Dasenbrock J, Klostermeyer D, Kroiß S, Sindlinger T, Spiteller P, Steffan B, Steglich W, Engler-Lohr M, Semar S, Anke T (1999) New benzodioxepin-type strobilurins from basidiomycetes. Structural revision and determination of the absolute configuration of strobilurin D and related β-methoxyacylate antibiotics. Tetrahedron 55:10101–10118

Hellwig V, Dasenbrock J, Schumann S, Steglich W, Leonhardt K, Anke T (1998) New triquinane-type sesquiterpenoids from *Macrocystidia cucumis* (Basidiomycetes). Eur J Org Chem 1:73–79

Hirota M, Morimura K, Shibata H (2002) Anti-inflammatory compounds from the bitter mushroom, *Sarcodon scabrosus*. Biosci Biotechnol Biochem 66:179–184

Högenauer G (1979) Tiamulin and pleuromutilin. In: Hahn FE (ed) Antibiotics, vol V-1. Springer, Heidelberg, pp 340–360

Hopmann C, Kurz M, Mueller G, Toti L (2001) Fermentative preparation of lipase-inhibiting percyquinnin using the fungus *Stereum complicatum* ST 001837 (DSM 13303). Patent EP 1142886

Huang Z, Dan Y, Huang Y, Lin L, Li T, Ye W, Wei X (2004) Sesquiterpenes from the mycelial cultures of *Dichomitus squalens*. J Nat Prod 67:2121–2123

Ito C, Abe N, Hirota A (1999) Isolation and structure of a novel indophenol-reducing and 1,1-diphenyl-2-picrylhydrazyl radical-scavenging compound from fungus. Biosci Biotechnol Biochem 63:1993–1996

Jiang M-Y, Wang F, Yang X-L, Fang L-Z, Dong Z-J, Zhu H-J, Liu J-K (2008) Derivatives of vibralactone from cultures of the basidiomycete *Boreostereum vibrans*. Chem Pharm Bull 56:1286–1288

Johansson M, Sterner O, Labischinski H, Anke T (2001) Coprinol a new antibiotic cuparane from a *Coprinus* species. Z Naturforsch 56c:31–34

Jülich W (1984) Die Nichtblätterpilze, Gallertpilze und Bauchpilze. Kleine Kryptogamenflora, Band IIb/1. Fischer, Stuttgart

Kamo T, Imura Y, Hagio T, Makabe H, Shibata H, Hirota M (2004) Antiinflammatory cyathane diterpenoids from *Sarcodon scabrosus*. Biosci Biotechnol Biochem 68:1362–1365

Kavanagh F, Hervey A, Robbins WJ (1951) Antibiotic substances from basidiomycetes. VIII. *Pleurotus mutilus*(Fr.) Sacc. and *Pleurotus passeckerianus* Pilat. Proc Natl Acad Sci USA 37:570–574

Kawai T, Mizutani S, Enoki T, Sagawa H, Sakai T, Shimanaka K, Kato I (2005) Antitumor agent. Patent PCT Int Appl WO 2005115364

Kettering M, Sterner O, Anke T (2004) Antibiotics in the chemical communication of fungi. Z Naturforsch 59c:816–823

Kita T, Takaya Y, Oshima Y, Ohta T, Aizawa K, Hirano T, Inakuma T (1998) Scabronines B, C, D, E and F, novel diterpenopids showing stimulating activity of nerve growth factor-synthesis, from the mushroom *Sarcodon scabrosus*. Tetrahedron 54:11877–11886

Kokubun T, Irwin D, Legg M, Veitch NC, Simmonds MSJ (2007) Serialynic acid a new phenol with an isopentenyne side chain from *Antrodia serialis*. J Antibiot 60: 285–288

Kozubek A, Tyman JHP (1995) Cereal grain resorcinolic lipids: mono and dienoic homologues are present in rye grains. Chem Phys Lip 78:29–35

Kupka J, Anke T, Steglich W, Zechlin L (1981) Antibiotics from basidiomycetes. XI. The biological activity of siccayne, isolated from the marine fungus *Halocyphina villosa* J. & E. Kohlmeyer. J Antibiot 34:298–304

Lang G, Cole ALJ, Blunt JW, Munro MHG (2006) An unusual oxalylated tetramic acid from the New Zealand basidiomycete *Chamonixia pachydermis*. J Nat Prod 69:151–153

León F, Brouard I, Rivera A, Torres F, Rubio S, Quintana J, Estévez F, Bermelo J (2006) Isolation, structure elucidation, total synthesis, and evaluation of new natural and synthetic ceramides on human SK-MEL-1 melanoma cells. J Med Chem 49:5830–5839

Léon F, Brouard I, Torres F, Quintana J, Rivera A, Estévez A, Bermejo J (2008) A new ceramide from *Suillus luteus* and its cytotoxic activity against human melanoma cells Chem Biodiv 5:120–125

Levy LM, Cabrera GM, Wright JE, Seldes AM (2003) 5H-Furan-2-ones from fungal cultures of *Aporpium caryae*. Phytochemistry 62:239–243

Levy LM, Cabrera GM, Wright JE, Seldes AM (2000) Indole alkaloids from a culture of the fungus *Aporpium caryae*. Phytochemistry 54:941–943

Lingham RB, Silverman KC, Jayasuriya H, Moon Kim B, Amo SE, Wilson FR, Rew DJ, Schaber MD, Bergstrom JD, Koblan KS, Graham SL, Kohl NE, Gibbs JB, Singh SB (1998) Clavaric acid and steroidal analogues as ras- and FPP-directed inhibitors of human farnesyl-protein transferase. J Med Chem 41:4492–4501

Liu J-K (2005) N-Containing compounds of macromycetes. Chem Rev 105:2723–2744

Liu D-Z, Wang F, Liao T-G, Tang J-G, Steglich W, Zhu H-J, Liu J-K (2006) Vibralactone: a lipase inhibitor with an unusual lactone produced by cultures of the basidiomycete *Boreostereum vibrans*. Org Lett 8:5749–5752

Liu D-Z, Wang F, Liu J-K (2007) Sesquiterpenes from cultures of the basidiomycete *Conocybe siliginea*. J Nat Prod 70:1503–1506

Liu J-K (2006) Natural terphenyls: developments since 1877. Chem Rev 106:2209–2223

Lorenzen K, Anke T (1998) Biologically active metabolites from basidiomycetes. Curr Org Chem 2:329–364

Lübken T, Schmidt J, Porzel A, Arnold N, Wessjohann L (2004) Hygrophorones A–G: fungicidal cyclopentenones from *Hygrophorus* species (Basidiomycetes). Phytochemistry 65:1061–1071

Lübken T (2006) Hygrophorone, neue antifungische Cyclopentenonderivate aus *Hygrophorus*-Arten (Basidiomycetes). Dissertation, University of Halle

Luebken T, Arnold N, Wessjohann L, Boettcher C, Schmidt J (2006) Analysis of fungal cyclopentenone derivatives from *Hygrophorus* spp. by liquid chromatography/electrospray-tandem mass spectrometry. J Mass Spectrom 41:361–371

Ma BJ, Liu JK (2005) An unusual nitrogenous terphenyl derivative from fruiting bodies of the basidiomycete *Sarcodon scabrosus*. Z Naturforsch 60b:565–568

Ma B-Ji A, Zhu H-J, Liu J-K (2004) Isolation and characterization of new bitter diterpenoids from the basidiomycete *Sarcodon scabrosus*. Helv Chim Acta 87:2877–2881

McMorris TC, Lira R, Gantzel PK, Kleiner MJ, Dawe R (2000) Sesquiterpenes from the basidiomycete *Omphalotus illudens*. J Nat Prod 63:1557–1559

McMorris TC, Staake MD, Kelner MJ (2004) Synthesis and biological activity of enantiomers of antitumor irofulven. J Org Chem 69:619–623

Mierau V, Anke T, Sterner O (2003) Dacrymenone and VM 3298-2 – new antibiotics with antibacterial and antifungal activity. Z Naturforsch 58c:541–546

Moulé Y, Moreau S, Bousquet JF (1977) Relationships between the chemical structure and the biological properties of some eremophilane compounds related to PR Toxin. Chem Biol Interact 17:185–192

Nicholas GM, Blunt JW, Munro MHG (2001) Cortamidine oxide a novel disulfide metabolite from the New Zealand basidiomycete (mushroom) *Cortinarius* species. J Nat Prod 64:341–344

Nicholas GM, Blunt JW, Cole ALJ, Munro MHG (1997) Investigation of the New Zealand basidiomycete *Favolaschia calocera*: Revision of the structures of 9-methoxystrobilurins K and L, strobilurin D, and hydroxystrobilurin D. Tetrahedron Lett 38:7465–7468

Obara Y, Nakahata N, Kita T, Takaya Y, Kobayashi H, Hosoi S, Kiuchi F, Ohta T, Oshima Y, Ohizumi Y (1999) Stimulation of neurotrophic factor secretion from 1321N1 human astrocytoma cells by novel diterpenoids, scabronines A and G. Eur J Pharmacol 370:79–84

Obara Y, Kobayashi H, Ohta T, Ohizumi Y, Nakahata N (2001) Scabronine G-methylester enhances secretion of neurotrophic factors mediated by an activation of protein kinase C-ζ. Mol Pharmacol 59:1287–1297

Ohashi Y, Kan Y, Watanabe T, Honda Y, Watanabe T (2007) Redox silencing of the Fenton reaction system by an alkylitaconic acid, ceriporic acid B produced by a selective lignin-degrading fungus, *Ceriporiopsis subvermispora*. Org Biomol Chem 5:840–847

Ohta T, Kita T, Kobayashi N, Obara Y, Nakahata N, Ohizumi Y, Takaya Y, Oshima Y (1998) Scabronine A, a novel diterpenoid having a potent inductive activity of nerve growth factor synthesis isolated from the mushroom *Sarcodon scabrosus*. Tetrahedron Lett 39:6229–6232

Opatz T, Kolshorn H, Birnbacher J, Schüffler A, Deininger F, Anke T (2007) The creolophins: a family of linear triquinanes from *Creolophus cirrhatus* (Basidiomycete). Eur J Org Chem 33:5546–5550

Paci A, Rezai K, Deroussent A, De Valeriola D, Re M, Weill S, Cvitkovic E, Kahatt C, Sha A, Waters S, Weems G, Vassal G, Lokiec F (2006) Pharmakoikinetics, metabolism, and routes of excretion of intravenous irofulven in patients with advanced solid tumors. Drug Metabol Dispos 34:1918–1926

Quang DN, Hashimoto T, Hitaka Y, Tanaka M, Nukada M, Yamamoto I, Asakawa Y (2003a) Thelephantins D–H: five p-terphenyl derivatives from the inedible mushroom *Thelephora aurantiotincta*. Phytochemistry 63:919–924

Quang DN, Hashimoto T, Nukada M, Yamamoto I, Hitaka Y, Tanaka M, Asakawa Y (2003b) Thelephantins A, B and C: three benzoyl p-terphenyl derivatives from the inedible mushroom *Thelephora aurantiotincta*. Phytochemistry 62:109–113

Quang DN, Hashimoto T, Hitaka Y, Tanaka M, Nukada M, Yamamoto I, Asakawa Y (2004) Thelephantins I–N; p-terphenyl derivatives from the inedible mushroom *Hydnellum caeruleum*. Phytochemistry 65:1179–1184

Radulović N, Quang DN, Hashimoto T, Nukada M, Asakawa Y (2005) Terrestrins A–G: p-terphenyl derivatives from the inedible mushroom *Thelephora terrestris*. Phytochemistry 66:1052–1059

Rahmawati N, Ohashi Y, Watanabe T, Honda Y, Watanabe T (2005) Ceriporic acid B an extracellular metabolite of *Ceriporiopsis subvermispora* suppresses the depolymerization of cellulose by the Fenton reaction. Biomacromolecules 6:2851–2856

Rasser F, Anke T, Sterner O (2002) Terpenoids from *Bovista* sp. 96042. Tetrahedron 58:7785–7789

Rasser F, Anke T, Sterner O (2000) New secondary metabolites from a *Gloeophyllum* species. Phytochemistry 54:511–516

Schultes RE, Hofmann A (1980) Pflanzen der Götter. Hallwag, Bern

Shibata H, Irie A, Morita Y (1998) New antibacterial diterpenoids from *Sarcodon scabrosus* fungus. Biosci Biotechnol Biochem 62:2450–2452

Silberborth S, Erkel G, Anke T, Sterner O (2000) The irpexans, a new group of biologically active metabolites produced by the basidiomycete *Irpex* sp. 93028. J Antibiot 53:1137–1144

Silberborth S, Stumpf A, Erkel G, Anke T, Sterner O (2002) Gerronemins A-F, cytotoxic biscatechols from a *Gerronema* species. Phytochemistry 59:643–648

Son KH, Kwon JY, Jeong HW, Kim HK, Kim CJ, Chang YH, Chio JD, Kwon BM (2002) 5-Demethylovalicin, as a methionine aminopeptidase-2 inhibitor produced by *Chrysoporium*. Bioorg Med Chem 10:185–189

Sontag B, Steglich W, Anke T (1999) Montadial a cytotoxic metabolite from *Bondarzewia montana* (Aphyllophorales). J Nat Prod 62:1425–1426

Sugawara F, Hallock YF, Bunkers GD, Kenfield DS, Strobel G, Yoshida S (1993) Phytoactive eremophilanes produced by the weed pathogen *Drechslera gigantea*. Biosci Biotech Biochem 57:236–239

Suzuki Y, Esumi Y, Hyakutake H, Kono Y Sakurai A (1996) Isolation of 5-(8-Z-heptadecenyl)-resorcinol from etiolated rice seedlings as an antifungal agent. Phytochemistry 41:1485–1489

Tan J, Dong Z, Hu L, Liu J-K (2003) Lepidamine the 1st aristolane-type sesquiterpene alkaloid from the basidiomycete *Russula lepida*. Helv Chim Acta:86:307–309

Tan J-W, Dong Z-J, Ding Z-H, Liu J-K (2002) Lepidolide a novel seco-ring-A cucurbitane triterpenoid from *Russula lepida* (Basidiomycetes). Z Naturforsch 57c:963–965

Tanaka N, Kitamura A, Mizushina Y, Sugawara F, Sakaguchi K (1998) Fomitellic acids triterpenoid inhibitors of eukaryotic DNA polymerases from a basidiomycete *Fomitella fraxinea*. J Nat Prod 61:193–197 (erratum: J Nat Prod 61:1180)

Tateishi K, Hoshi H, Matsunaga K (2005) Thelephoric acid as antiallergic agent. Patent PCT Int Appl WO 2005095412

Weber D, Erosa G, Sterner O, Anke T (2006) New bioactive sesquiterpenes from *Ripartites metrodii* and *R tricholoma*. Z Naturforsch 61c:663–669

Weibel E, Hadvary P, Hochuli E, Kupfer E, Lengsfeld H (1987) Lipstatin, an inhibitor of pancreatic lipase, produced by *Streptomyces toxytricini*. I. Producing organism, fermentation, isolation and biological activity. J Antibiot 40:1081–1105

Wilkins WH, Harris GCM (1944) Investigation into the production of bacteriostatic substances by fungi. VI. Examination of the larger Basidiomycetes. Ann Appl Biol 31:261–270

Wood KA, Kau DA, Wrigley SK, Beneyto R, Renno DV, Ainsworth AM, Penn J, Hill D, Killacky J, Depledge P (1996) Novel β-methoxyacrylates of the 9-methoxystrobilurin and oudemansin classes produced by the basidiomycete *Favolaschia pustulosa*. J Nat Prod 59:646–649

Yun B-S, Lee I-K, Kim J-P, Yoo I-D (2000) Curtisians A-D new free radical scavengers from the mushroom *Paxillus curtisii*. J Antibiot 53:114–122

Yun B-S, Cho Y, Lee I-K, Cho S-M, Lee TH, Yoo I-D (2002) Sterins A and B new antioxidative compounds from *Stereum hirsutum*. J Antibiot 55:208–210

Zähner H, Anke H, Anke T (1983) Evolution of secondary pathways. In: Bennett JW, Ciegler A (eds) Secondary metabolites and differentiation in fungi. Dekker, New York, pp 153–171

Zapf S, Werle A, Anke T, Klostermeyer D, Steffan B, Steglich W (1995) 9-Methoxystrobilurine – Bindeglieder zwischen Strobilurinen und Oudemansinen. Angew Chem 107:255–257

Zapf S, Wunder A, Anke T, Klostermeyer D, Steglich W, Shan R, Sterner O, Scheuer W (1996) (+)-10-a-Hydroxy-4-muurolen-3-one, a new inhibitor of leukotriene biosynthesis from a *Favolaschia* species. Comparison with other sesquiterpenes. Z Naturforsch 51c:487–492

Zhang Y, Mills GL, Nair MG (2002) Cyclooxygenase inhibitory and antioxidant compounds from the mycelia of the edible mushroom *Grifola frondosa*. J Agric Food Chem 50:7581–7585

11 Identification of Fungicide Targets in Pathogenic Fungi

Andrew J. Foster[1], Eckhard Thines[1]

CONTENTS

I. Introduction

Fungicides have become an extremely important means to combat problems caused by fungal pathogens both within the agricultural and medicinal fields. Although a number of very effective fungicides have been developed, the emergence of resistance has brought a constant need for the identification of new fungicide targets and for novel agents which act against these targets. This chapter discusses the current situation with regard to fungicide targets and available fungicides and the outlook for the development of novel fungicides, with special focus on the application of the 'omics' technologies in the drug development process. The availability of these genome-wide technologies represents a major new opportunity to identify novel fungicide targets within pathogenic fungi. Even within the best studied pathogenic fungi there is still a great deal to learn about how disease is established and maintained. A major challenge for drug development today is how best to employ the new genome-wide technologies to indentify which metabolic pathways and gene products are critical for disease establishment and progression and thus to increase the probability of finding novel fungicide targets. This chapter therefore reviews genome-wide approaches that have emerged in the post-genomics era, e.g. comparative genomics and gene expression profiling experiments, including technologies such as microarray-based transcriptional profiling, SAGE, and MPSS. It also reviews results of the application of such projects to date in the field.

II. Currently Deployed Fungicides

The frequent occurrence of resistance to commonly deployed anti-fungals necessitates a continued search for novel fungicides. Ideally new compounds should have a novel mode of action, have a minimal risk in terms of resistance development, and show a high degree of specificity towards the target organism. These compounds must also satisfy the current stringent regulations regarding ecotoxicity and user safety prior to registration.

The most successful fungicides on the market today act relatively broadly by targeting fungal vegetative growth and thus the entire fungal life cycle. Examples of a successful mode of action classes interfering with biochemical processes essential in fungi include compounds targeting

[1]Institut für Biotechnologie und Wirkstoff-Forschung e.V./ Institute for Biotechnology and Drug Research, Erwin-Schrödinger-Strasse 56, 67663 Kaiserslautern, Germany; e-mail: foster@ibwf.de, thines@ibwf.de

Physiology and Genetics, 1st Edition
The Mycota XV
T. Anke and D. Weber (Eds.)
© Springer-Verlag Berlin Heidelberg 2009

respiration and sterol biosynthesis. An example of the former mode of action class is given by compounds which target mitochondrial electron transport within the respiration chain. Fungicides of the strobilurin class have this mode of action. These compounds are structurally based on natural products and interfere with the ubiquinol-cytochrom C oxidoreductase of the cytochrome bc1 in complex III (Becker et al. 1981; Anke and Steglich 1999). A number of fungicides whose mode of action is the inhibition of sterol biosynthesis have been developed. The sterol biosynthetic pathway includes several drugable targets, such as a 3-keto reductase of an enzyme complex of the sterol C-4 demethylation, the squalene epoxidase, Δ^{14}-reductase and/or $\Delta^8 \rightarrow \Delta^7$-isomerase, and sterol 14α-demethylation, which is inhibited by azole fungicides (Köller et al. 1992; Debieu et al. 2001). Although both respiration and sterol biosynthesis have proved good targets for which very successful fungicides have been produced, resistance against such fungicides develops rapidly resulting in a high demand for new targets and fungicides. Furthermore, several fungicides with a broad range of activity have side-effects on benign fungal species, animals, or plants which cannot be excluded. For this reason there is a need for new types of fungicides which exhibit novel modes of action.

III. What Are the Ideal Attributes of the Fungicides of the Future?

Novel targets can be located either in biochemical or signalling pathways essential for vegetative growth or for pathogenic development of the fungus. Targets with high specificity can thereby be expected in pathways involved in adhesion, host-recognition/pre-penetration processes, host colonisation, and the final reproductive differentiation processes during pathogenic development. An attractive proposition is the development of fungicides interfering with pathogenic development but not with vegetative growth. Such a strategy could prevent or cure infections by fungal pathogens without affecting neutral or benign species.

Within the melanin biosynthetic pathway enzymes such as tetra- or trihydroxynaphthalene reductase and sytalone dehydratase have been identified as targets for non-fungitoxic fungicides.

Fungicides interfering with these targets, such as tricyclazole, phthalide, pyrochilon, and capropamid have been successfully used in plant protection (Nakasako et al. 1998; Thompson et al., 2000). Such inhibitors of melanin biosynthesis prove effective against the rice blast fungus *Magnaporthe grisea* where penetration of this species is prevented by blocking the formation of the appressorial melanin layer required for building turgor pressure within this cell (Howard and Ferrari 1989). A good degree of specificity for the target organism is achieved in this case because these inhibitors specifically target enzymes involved in the production of melanin from a pentaketide precursor while melanin in other organisms uses alternative biosynthetic pathways not affected by such agents (Thines et al. 2004).

The differentiation process of appressorium formation itself may offer novel targets for modern plant protection strategies. Inhibitors interfering with the formation of infection structures might be found which do not interfere with mycelial growth. One example for such a protective fungicide is quinoxyfen, a compound which prevents appressorium formation in *Blumeria graminis*. The mode of action of quinoxyfen is still unclear; however it was proposed recently that quinoxyfen interferes with signalling (Wheeler et al. 2003) and studies of quinoxyfen-resistant mutants recently implicated inhibition of a serine esterase (Lee et al. 2008).

The disadvantage of fungicides with high specificity is their low range of application, which makes them an unattractive commercial proposition for companies. Considering the high costs incurred during the development and registration of a novel fungicide, companies must have to opportunity to recover the enormous costs of development; fungicides active only against a narrow range of species are obviously much less likely to bring a profit to the developer.

IV. The Role of Traditional Screening Approaches Used in Fungicide Development

In recent decades traditional screening approaches with compounds derived from chemical synthesis or natural products in greenhouses were very successful and led to a generation of very effective fungicides with toxicologically favourable profiles. Therefore screening on living organisms/plants will play an important role in the future of plant protection research. However, the drawback of this traditional approach is that large numbers of compounds must be tested in order to get to a hit. To overcome this, target-based screening platforms and virtual molecular modelling systems have been developed in which large numbers of molecules can be tested.

V. Determining the Mode of Action of an Anti-Fungal Compound

Determining whether a new compound has a novel mode of action is a critical step in the drug development process and this must be achieved at an early stage. To this end in vitro screening systems targeting distinct biochemical reactions/pathways can also be used to identify the mode of action of compounds showing good antifungal activity during in vivo screening experiments. New compounds for which the mode of action is not known can initially be assessed in simple biochemical tests, such as inhibition of respiration or sterol biosynthesis. Furthermore fungicide-resistant mutants have successfully been used to identify which protein is likely targeted by the agent. When the mode of action of fungicidal compounds cannot be identified by such approaches, novel methods such as metabolome, proteome, or transcriptome analysis can provide signature patterns which can be compared to patterns obtained from experiments with known compounds and may indicate whether or not it is likely that the drug has a novel mode of action.

VI. Genome-Wide Approaches

Genome-wide technologies, e.g. gene expression profiling and comparative genomics, lead to the identification of a large number of potential candidate genes encoding factors essential either for invasive growth and development or for vegetative growth. However, these potential targets have to be validated by gene deletion or gene-silencing experiments. Even if deletion of the gene is lethal or the mutant shows a non-pathogenic phenotype,

it must be further established whether the target is drugable in vivo.

Once a target has been validated, target-based ultra-high throughput biochemical screening systems can be established in which hundreds of thousands of compounds can be tested daily. The challenge for modern agrochemical research is to convert the in vitro activity of compounds identified in such screening systems to products active in vivo.

The post-genomics era is now truly upon us. More than 50 different fungal genome sequences are now publicly available and more fungal sequencing projects are either planned or in progress. More than half of the finished genomes are of pathogenic species and cover the most important fungal pathogens of humans as well as many of the most devastating fungal pathogens of commercially important crops (Table 11.1). These resources open the door to genome-wide analysis within these species. Recently developed technologies allow the analysis of mRNA (transcriptomics), protein (proteomics), and metabolite (metabolomics) profiles. We focus here largely on transcriptomics and proteomics.

A. Transcriptomics: Microarrays

Microarrays were first employed within fungi more than ten years ago (DeRisi et al. 1997) and microarray technology has since been widely used to study a broad array of processes within fungi (reviewed by Breakspear and Momany 2007). The technology itself is a logical development from Southern/Northern blotting and, very basically, the microarray itself is a series of spots of nucleic acids (cDNAs or oligonucleotides) which are known as features or probes and which are attached to a solid support. cDNA or cRNA samples (somewhat confusingly these are known as targets in this context) can then be hybridised to the microarray and, following high stringency washes, the degree of probe to target annealing at each spot can be quantified and, by comparison of the signal under different conditions, the relative abundance of the target within two different samples can be assessed. Microarray technology is described in detail elsewhere (e.g. see Heller 2002). Several applications of microarray technology have emerged in recent years; however here we confine our discussion largely to the

Table 11.1. Fungal species whose genome sequence has been determined or for which a sequence release is imminent. The latest status for ongoing projects and links to finished genomes are available at http://fungalgenomes.org/wiki/Fungal_Genome_Links

Species	Division	Important features of the fungus	Responsible body
Agaricus bisporus	Basidiomycota	Common edible button or table mushroom	Joint Genome Institute
Alternaria brassicicola	Ascomycota	Plant pathogen (*Brassica* dark leaf spot of *Brassica* species)	WUSTL (access also via Broad Institute)
Amanita bisporigera	Basidiomycota	Common deadly mushroom (Destroying Angel)	Michigan State University
Ashbya gossypii	Ascomycota	Plant pathogen (cotton)	Broad Institute
Aspergillus fumigatus	Ascomycota	Human pathogen, especially in the immunocompromised	Sanger Centre/TIGR
A. oryzae	Ascomycota	A species widely used in biotechnology	National Institute of Technology and Evaluation
A. nidulans	Ascomycota	A model filamentous fungus; widely studied	Broad Institute
A. terreus	Ascomycota	Opportunistic human pathogen	Broad Institute
A. niger	Ascomycota	Common contaminant of food	Joint Genome Institute
A. clavatus	Ascomycota	A rare opportunistic human pathogen	TIGR
Batrachochytrium dendrobatidis	Chytridiomycota	Causes the amphibian disease chytridiomycosis	Broad Institute/Joint Genome Institute (two different strains sequenced)
Botrytis cinerea	Ascomycota	Plant pathogen with broad host range	Broad Institute
Candida albicans	Ascomycota	Opportunistic human pathogen; dimorphic yeast	Broad Institute
C. dubliniensis	Ascomycota	Opportunistic human pathogen; close relative of *C. albicans*	Sanger Centre
C. glabrata	Ascomycota	Opportunistic pathogen (non-dimorphic)	Sanger (finishing), Génolevures
C. guilliermondii	Ascomycota	A xylitol producer	Broad/Génolevures
C. lusitaniae	Ascomycota	Opportunistic pathogen (amphotericin B-resistant)	Broad Institute
C. parapsilosis	Ascomycota	Opportunistic human pathogen; forms biofilms	Sanger
C. tropicalis	Ascomycota	Opportunistic human pathogen	Broad Institute
Chaetomium globosum	Ascomycota	Causes infections of human nails and skin; allergen	Broad Institute
Coccidioides immitis	Ascomycota	Human pathogen; causes Valley fever	Broad Institute (four different strains sequenced)
C. posadsii	Ascomycota	Human pathogen	Broad Institute
Colletotrichum graminicola	Ascomycota	Plant pathogen (causes anthracnose diseases of cereals)	Broad Institute (still in progress)
Coprinus cinereus	Basidiomycota	Model basidiomycete 'ink cap' fungus	Broad Institute/Duke University
Cryptococcus neoformans[a]	Basidiomycota	Human pathogen; serotypes A and B have been sequenced	Broad Institute
C. gattii	Basidiomycota	Human pathogen (cryptococcosis)	Broad/British Columbia Genome Sequencing Centre
Debaryomyces hansenii	Ascomycota	A cryotolerant, marine yeast	Génolevures
Fusarium graminearum	Ascomycota	Plant pathogen (wheat head blight)	Broad Institute
F. oxysporum	Ascomycota	Plant pathogen (wilt diseases of many species of plants)	Broad Institute
F. verticillioides	Ascomycota	Plant pathogen ('foolish seedling' disease of rice seedlings)	Broad Institute

(*continued*)

Table 11.1. (continued)

Species	Division	Important features of the fungus	Responsible body
Heterobasidion annosum	Basidiomycete	Plant pathogen; causes root and butt rot to conifers	Joint Genome Institute
Histoplasma capsulatum	Ascomycota	Causes histoplasmosis; NAm1 and G186AR strains	Broad Institute
Kluyveromyces lactis	Ascomycota	Industrial and model yeast species	Génolevures
K. marxianus	Ascomycota	Yeast species/commercial producer of lactase	Génolevures
K. thermotolerans	Ascomycota	Yeast species associated with fruits	Génolevures
K. waltii	Ascomycota	Yeast species	Génolevures
Lacazia loboi	Ascomycota	Causal agent of lobomycosis (human skin disease)	Broad Institute
Laccaria bicolor	Basidiomycota	Ectomycorrhizal fungus	Joint Genome Institute
Lodderomyces elongisporus	Ascomycota	Closely related to *Candida parapsilosis*	Broad Institute
Magnaporthe grisea	Ascomycota	Plant pathogen (rice blast); model cereal pathogen	Broad Institute
Malassezia globosa	Basidiomycota	Associated with dandruff	Commercial (available through NCBI)
M. restricta	Basidiomycota	Associated with dandruff	Commercial (available through NCBI)
Microsporum gypseum	Ascomycota	Human pathogen (Dermatophytes)	Broad Institute
Moniliophthora perniciosa	Basidiomycota	Plant pathogen (causes cacao disease)	Unicamp
Mycosphaerella fijiensis	Ascomycota	Plant pathogen (banana Black Sigatoka)	Joint Genome Institute
M. graminicola	Ascomycota	Plant pathogen (*Septoria tritici* blotch of wheat)	Joint Genome Institute
Nectria haematococca	Ascomycota	Human and plant pathogen	Joint Genome Institute
Neosartorya fischeri	Ascomycota	*Aspergillus* species causes food spoilage/pathogenic (rare)	TIGR
Neurospora crassa	Ascomycota	Model filamentous fungus for basic research	Broad Institute
Paracoccidioides brasiliensis	Ascomycota	Dimorphic; causal agent of paracoccidioidomycosis	Broad Institute (strains Pb01, Pb03, Pb18)
Paxillus involutus	Basidiomycota	Common ectomycorrhizal fungus	Joint Genome Institute
Penicillium marneffei	Ascomycota	Human pathogen/thermally dimorphic species	TIGR
P. chrysogenum	Ascomycota	Industrially important source of several β-lactam antibiotics	Commercial (available through NCBI)
Phanerochaete chrysosporium	Basidiomycota	White rot fungus, used for bioremediation	Joint Genome Institute
P. carnosa	Basidiomycota	White rot fungus	Joint Genome Institute
Pichia angusta	Ascomycota	Methylotrophic yeast species with biotechnological uses	Génolevures
P. stipitis	Ascomycota	Yeast species; used for xylose fermentation	Joint Genome Institute/Génolevures
P. farinosa (sorbitophila)	Ascomycota	Yeast species; emerging human pathogen	Génolevures
Phycomyces blakesleeanus	Zygomycota	Model zygomycete; widely studied	Joint Genome Institute
Phytophthora infestans	Oomycete	Plant pathogen; potato blight	Sanger Centre
Postia placenta	Basidiomycota	Wood-rooting fungus; causes brown rot	Joint Genome Institute
Puccinia graminis	Basidiomycota	Plant pathogen; rusts of cereals	Broad Institute
Pyrenophora tritici-repentis	Ascomycota	Plant pathogen; yellow leaf spot of cereals and grasses	Broad Institute
Rhizopus oryzae	Zygomycota	Human pathogen	Broad Institute

(*continued*)

Table 11.1. (continued)

Species	Division	Important features of the fungus	Responsible body
Sclerotinia sclerotiorum	Ascomycota	Plant pathogen (root, crown, stem rots; varied hosts)	Broad Institute
Saccharomyces cerevisiae[b]	Ascomycota	Baker's yeast; a leading model organism in research	Broad Institute and others (several strains sequenced)
S. bayanus	Ascomycota	Yeast species; used in wine making	Broad Institute
S. castellii	Ascomycota	Yeast species	WUSTL
S. kluyveri	Ascomycota	Yeast species	WUSTL
S. kudriavzevii	Ascomycota	Yeast species	WUSTL/Stanford
S. mikatae	Ascomycota	Yeast species	Broad Institute/Stanford
S. paradoxus	Ascomycota	Yeast species	Broad Institute/Stanford
Schizophyllum commune	Basidiomycota	Common fungus; causes white rot	Joint Genome Institute I (near completion)
Schizosaccharomyces octosporus	Ascomycota	Fission yeast species; grows filamentously	Broad Institute
S. japonicus	Ascomycota	Fission yeast species	Broad Institute
S. pombe	Ascomycota	Fission yeast, very widely studied model fungus	Sanger Centre
Stagonospora nodorum	Ascomycota	Plant pathogen, blotch disease of wheat	Broad Institute
Trichoderma reesei	Ascomycota	Industrially important fungus	Joint Genome Institute
T. virens	Ascomycota	Used in biocontrol against fungal pathogens	Joint Genome Institute
Uncinocarpus reesii	Ascomycota	A close relative of *Coccidioides*	Broad Institute
Ustilago maydis	Basidiomycota	Plant pathogen (corn smut disease); model organism	Broad Institute
Verticillium dahliae	Ascomycota	Cause of wilt diseases in important crop species	Broad Institute
V. albo-atrum	Ascomycota	Cause of wilt diseases in important crop species	Broad Institute

[a]Three different strains of *Cryptococcus neoformans* have been sequenced: *C. neoformans* var. *neoformans* JEC21 (TIGR), *C. neoformans* var. *neoformans* B-3501A (Stanford University) and *C. neoformans* var. *grubbii* H99 (Duke/FGI)

[b]Several strains of *Saccharomyces cerevisiae* have been sequenced or are being sequenced. For further details, see http://www.sanger.ac.uk/Teams/Team71/durbin/sgrp/index.shtml

application of microarrays to the study of transcript abundance and how microarray analysis might be exploited to assist in the identification of novel fungicide targets. A number of different examples to illustrate the different potential applications of microarrays are given below.

B. Studying Genome-Wide Transcriptional Changes Which Accompany Differentiation

Within plant pathogenic fungi, early studies used sub-genome-wide technologies, often derived from the results of EST sequencing projects, to study development (See Breakspear and Momany 2007 and references therein). In investigations of the rice blast fungus *Magnaporthe grisea*, for example, the first microarray experiments used cDNA-based technology to investigate the relative transcript abundance of 3500 features (cDNAs with some redundancy) during infection-related morphogenesis to investigate appressorium formation (Takano et al. 2003).

A whole genome microarray of *M. grisea* is now commercially available from Agilent Biotechnologies and has been used to identify genes important for appressorium formation (Oh et al. 2008). Genes responding to nitrogen starvation, a situation believed to mimic the nutritional state of the fungus post-penetration and which is purported to induce infection-related development, have also been studied using these commercially available microarrays (Donofrio et al. 2006). One gene which responds to nitrogen starvation, *SPM1*, is predicted to encode a vacuolar serine protease. It was deleted and shown to be required for normal levels of virulence (Donofrio et al. 2006). Novel gene products which are critical for appressorium formation have been uncovered by microarray-based analyses, including a subtilisin-like protease and a NAD-specific glutamate dehydrogenase (Oh et al. 2008). These studies clearly illustrate the potential of genome-wide technologies to potentiate the identification of potential new fungicide targets by narrowing the search for possible pathogenicity factors.

In *Fusarium graminearum*, a whole genome microarray has been used to investigate which pathways are induced during the germination of conidia, an important first step in the establishment of disease (Seong et al. 2008). Clearly some of these differentially regulated genes might prove attractive as targets for drug intervention, although obviously with such an analysis significant follow-up experiments would be required in order to identify which genes are essential for the process studied, as these are likely only a small subset of the total number identified.

Further commercially produced whole-genome microarrays are in development for other plant pathogenic fungi, including *Botrytis cinerea* and *Mycosphaerella graminicola*. It can be anticipated that the range of microarrays available and their application within plant pathology will expand greatly in the future. Whether the application of these microarrays will eventually reveal fungicide targets which are conserved across a broad range of species is open to question; however there is no doubt about the potential of this technology to generate new experimental hypothesis which could greatly facilitate the identification of novel targets for drug control within individual species.

Extensive microarray-based analysis can also proceed in the absence of a whole-genome sequence. An example is the study of Both and co-workers (2005), who looked at more than 2000 *Blumeria graminis* genes from an infection time-course and revealed how several genes previously implicated in disease progression as well as many new unknown genes were expressed during infection-related development. Some of the genes identified during this study might be potential candidates for fungicide targets, although proving this is not straightforward in an obligate pathogen such as *Blumeria graminis*.

The application of microarrays to the study of fungal pathogens of humans is also becoming increasingly commonplace. For example, *Histoplasma capsulatum* is an ascomycete dimorphic fungus responsible for the human disease histoplasmosis. Microarray analysis using genomic DNA-derived targets has identified genes specifically expressed in the pathogenic yeast form (Hwang et al. 2003). Again significant follow-up experiments are required to test whether any of the products of these genes are essential for virulence but, in terms of identifying potential novel targets, such studies offer valuable new hypotheses which will no doubt lead to the identification of novel drug targets which could prove valuable in future years.

C. Cross-Species Comparisons: Comparative Genomic

Where significant similarity exists between closely related fungal species there is a possibility to conduct cross-species comparisons using microarrays. A good example of such an application is provided by the study of Moran and co-workers (2004) who exploited *Candida albicans*-based genomic microarrays to compare the genome of this species to the very closely related but less virulent species *C. dubliniensis*. Although the vast majority of genes were found to be highly conserved, several hundred genes were identified which were either absent in *C. dubliniensis* or whose sequence was poorly conserved between these two species. This study not only generated several testable hypotheses which might shed light on the difference in virulence between these two species but also provided a great deal of information about the *C. dubliniensis* gene repertoire in the absence of a genome sequence. Interestingly several of the genes poorly conserved between the two species studied are predicted to encode transcription factors and it would therefore be of significant interest to extend these analyses by comparing the transcriptome of the two species.

D. Identifying Possible Drug Targets by Comparison of the Transcriptome of Mutant and Wild-Type Cells

Using whole-genome microarrays to study a major developmental switch such as germination or appressorium formation, several hundred to thousands of genes have been found to be differentially expressed (e.g. Oh et al. 2008; Seong et al. 2008). An alternative route to identify the genes involved in a particular process is to examine a mutant blocked at that developmental stage. With a view to drug development, this may be a more attractive route, as the number of target genes identified by this approach is more likely in the tens rather than in the hundreds. Such an application of microarrays is nicely illustrated by a recent study of the human pathogen *Cryptococcus neoformans* (Cramer et al. 2006). These workers identified genes which are transcriptionally down-regulated in a mutant which lacks a component of the cAMP signalling pathway which is required for capsule formation, a critical determinant of virulence. One of these genes was found to encode a transcription factor with similarity to yeast transcription factor Nrg1p. A *C. neoformans* strain lacking this transcription factor was generated and was also found to be deficient in capsule formation. Microarrays were then used to identify genes transcriptionally dependent on the product of *NRG1* and among the 71 genes identified was *UGD1*, a gene whose product was previously implicated in capsule formation in this fungus (Moyrand and Janbon 2004). Although this gene had previously been identified, this study illustrates the potential of the combination of microarray analysis and specific mutations to uncover potentially drugable targets. In a similar manner, Odenbach and co-workers (2007) used microarrays to identify approximately 100 *Magnaporthe grisea* genes whose transcription depends on the Con7p transcription factor during germination of spores of this fungus. These included a chitin synthase encoding gene *CHS7* which, like the *con7⁻* mutant, is affected in the formation of the appressoria (Odenbach et al. 2007, 2009). Again chitin synthase is already considered a potential target for drug intervention in human disease, so although this is not a novel fungicide target, this study nicely illustrates the potential of using microarrays in combination with specific mutants to identify potential drug targets. In a similar manner using a human pathogenic fungus, Ngugen and Sil (2008) identified a *Histoplasma capsulatum* regulator Ryp1p which is required for yeast phase gene expression and identified several genes controlled by this transcription factor by comparing transcript profiles of mutant and wild-type cells.

Identification of novel fungicide targets is ultimately a process of elimination. Eliminating which factors are not good targets from the likely comparatively small number which are useful (drugable) targets is an important step in target identification. Mutant generation will continue to be a decisive experiment in this process for years to come. Any means by which it is possible to narrow the search for new targets is obviously attractive. Much research in the past few decades in the academic world has focused on signal transduction and, although it is questionable whether any of the proteins identified by such studies are themselves good targets, the mutants generated by such studies could be exploited to identify downstream components of which a small subset are likely to be essential for the process and which may include useful targets for drug intervention.

E. Exploring Drug Resistance Using Transcriptome Analysis: *Candida albicans*

A further application for microarrays within the field of drug development is in the study of resistance to fungicides. Such an approach has already been adopted by a number of groups attempting to understand the transcriptional response to fungicide treatment. In this respect by far the best studied species within fungi is the opportunistic fungal pathogen of humans *Candida albicans*. Two of the most commonly deployed drugs in the treatment of diseases caused by *C. albicans* are the polyene fungicide amphotericin B (AMB) and fungistatic azole compounds such as fluconazole. Although the mode of action of these drugs differs, resistance to both drugs within clinical isolates is often associated with induction of drug efflux transporter activity and/or alterations to the ergosterol biosynthetic enzymes (reviewed by Akins 2005). This is not surprising, as both drugs are thought to ultimately impact on membrane integrity. A broader understanding of the resistance mechanisms to these and other drugs has come from several microarray-based studies which have examined transcript profiles within amphotericin B-resistant isolates or drug-treated wild-type strains. These studies both confirmed previous reports which suggested induction of drug efflux transporter activity and/or alterations to the ergosterol biosynthetic enzymes and also led to the identification of further genes which are induced specifically on treatment with drugs of a specific class or which seem to be part of a more general response to drug challenge. Several genes whose transcription is altered by drug treatment are otherwise uncharacterised and their participation in drug response may therefore be the first clue to their function. Furthermore the specific transcript profile induced by treatment with a particular drug might also be used as a 'fingerprint' for drugs of that type and might assist in assigning a tentative mode of action to a drug whose targets is unknown or unclear. *C. albicans* is also known to commonly form biofilms which show increased resistance to commonly deployed antifugals (Chandra et al. 2001; Ramage et al. 2001). Farnesol, a quorum-sensing molecule, was shown to prevent biofilm at high concentrations and is considered a possible novel *Candida* control agent (Ramage et al. 2002). Microarray analysis subsequently shed some light on the mode of action which was previously not known, suggesting that the compound might in part prevent filamentous growth by inducing the transcription of the *TUP* gene which encodes a repressor of hyphal development (Cao et al. 2005). Again this study highlights the utility of microarray experiments in understanding drug response and mechanisms of resistance.

The response to azole fungicides has also been studied in *Aspergillus fumigatus* where, as is the case for *C. albicans*, resistance to clinically deployed drugs of this class is increasingly problematic. Several thousand genes were found to be differentially regulated in response to voriconazole (da Silva Ferreira et al. 2006). Up-regulated genes included several transporters of ABC and MFS type which might play a role in drug efflux.

The use of microarrays for the study of resistance to drug treatment can be expected to lead to valuable insights into what are the critical determinants of resistance and increasing application in this field as well as in determination of mode of action is anticipated in the future.

F. MPSS and SAGE

Serial analysis of gene expression (SAGE; Velculescu et al. 1995) and massively parallel signature sequencing (MPSS; Brenner et al. 2000) are related technologies which yield short sequence signatures derived from mRNAs. By comparison of the relative abundance of these signatures within libraries derived from different mRNA populations, a semi-quantitative measure of transcript abundance in a sample is obtained.

Examples of the application of these technologies are studies conducted using the plant pathogens *Magnaporthe grisea* (Irie at al. 2003) and *Blumeria graminis* (Thomas et al. 2002). Several genes dependent on cAMP were revealed using SAGE in *Magnaporthe*, including a number which have already been shown to be essential for pathogenicity (Irie et al. 2003). In *B. graminis*, an obligate pathogen, several thousand SAGE tags were sequenced and more than 100 genes differential transcribed during infection-related development were highlighted (Thomas et al. 2002). This study highlights one great advantage of these technologies: they can be applied to any organism irrespective of whether a genome sequence is available or not. A further application for such technologies is in determining which of the predicted genes within an annotated genome is really transcribed. Thus in the case of *M. grisea*, a large number of the genes predicted to exist in the genome by automated annotation were subsequently shown to be really transcribed using SAGE/MPSS (Gowda et al. 2006). In fact the same study indicted the existence of several thousand additional transcripts which do not correspond to any annotated gene and has therefore expanded the *M. grisea* gene repertoire.

SAGE has also been used to study protein kinase A (PKA)-dependent gene expression in the corn smut pathogen *Ustilago maydis* (Larraya et al. 2005). By analysis of transcript profiles based on the sequences of at least 40 000 tags for the wild type and two mutants in the PKA pathway, the authors were able demonstrate novel functions for PKA signalling in this organism, including a link to phosphate metabolism. Because PKA signalling is critical for virulence in this species, such analyses pinpoint downstream targets of PKA signalling which may well include potential new targets for fungicide control of corn smut.

Within fungi which are pathogenic to humans, SAGE has been employed for transcriptome analysis most extensively in *Cryptococcus neoformans*. One comparison used 49 224 SAGE tags from each of two protein kinase A (PKA) pathway mutants and a wild-type control under conditions known to induce capsule formation in this species (Hu et al. 2007). Among the 599 tags exhibiting altered abundance between these SAGE libraries were several corresponding to genes whose products might play a role in secretion. Furthermore, using inhibitors of secretion, the authors were able to demonstrate that capsule formation is inhibited at drug concentrations which do not affect growth in culture. This indicates that, as one might anticipate, secretion plays an important role in capsule formation. These results not only expand the understanding of the targets of PKA signalling but also highlight the potential of targeting components of the secretion machinery in order to control this species.

In conclusion SAGE and MPSS are powerful means to obtain a semi-quantitative assessment of relative transcript abundance under different conditions. There is no doubting the value of these technologies in providing information which could suggest which pathways or enzymes are likely to be activated under specific conditions. Additionally these technologies have been successfully employed in fungal species which have not yet been completely sequenced. As with microarrays, these technologies could readily be employed to suggest possible new targets for drug intervention; however, in comparison to microarrays, MPSS and SAGE have been much less broadly applied experimentally and it is therefore probable that their application in future drug development will also be peripheral.

G. Transcriptomics: Outlook

Microarrays and related technologies have unquestionably provided the researcher aiming to study plant–pathogen interactions with powerful tools which is are now very broadly accepted and increasingly widely used. It is likely that with ever improving technologies and decreasing prices such forms of analysis will become available to most researchers in the coming years. Nevertheless it is important to be aware that transcript abundance does not always correlate well with protein abundance. It has been common to extrapolate from the results of such analyses and suggest that certain pathways are activated in response to certain developmental or environmental conditions or as a result of certain mutations. Although this is likely to be true in the majority of cases, these technologies should at best be viewed as a very elegant means to generate hypotheses in the absence of any preconceptions of the experimental outcome. Although it is easy to be blinded by the power of this technology, any study which reports only microarray data is likely to provide very many

new questions but no real answers. The real power of genome wide analysis of transcript abundance will only be realised in combination with other approaches which address the response of the organism at other levels and by genetic manipulations to address the question as to whether the observed alteration in transcript abundance has any meaningful consequences for the test organism. Genetic manipulation of most fungal pathogens remains the major bottleneck in moving from the results of genome-wide studies of transcript abundance to establishing that the product of a particular gene is really important or essential in establishing disease (and therefore a potential drug target).

H. Other 'omics'

The ultimate targets of most known fungicides are proteins and there is some logic to working directly at the level of the protein. Additionally there are many examples where the major level of control which determines protein abundance is post-transcriptional. For these reasons proteome profiling represents an attractive approach to identify potential new fungicide targets. Proteome profiling studies in *Aspergillus* species have been used to identify proteins expressed during various physiological states of the cell (for a review, see Kim et al. 2008). Especially in the opportunistic human pathogenic fungus *Aspergillus fumigatus*, comparative proteome studies have been conducted in order to identify proteins involved in virulence or essential for mycelial growth that may serve as targets for antimycotic therapies. Bruneau and co-workers focussed on surface proteins thereby identifying glycosylphosphatidylinositol-anchored membrane proteins which had previously been shown to be involved in cell wall biosynthesis (Mouyna et al. 2000). It was shown that, out of five proteins with unknown function, the Ecm33 protein influences the conidial cell wall biosynthesis/cell wall morphogenesis (Chabane et al. 2006).

A proteome study conducted by Asif and co-workers (2006) led to the identification of 26 conidial cell surface proteins as potential vaccine candidates. Several of the proteins identified had no known function and may therefore be novel targets for therapeutic approaches.

In proteome studies using mutants it was recently shown that enhanced PKA activity results in the activation of stress-associated proteins, enzymes involved in protein biosynthesis, and glucose catabolism. Enzymes involved in nucleotide and amino acid biosynthesis and enzymes involved in catabolism other than glucose were down-regulated by PKA signalling (Große et al. 2008). It was furthermore found that septin and β-tubulin were down-regulated by enhanced PKA signalling, indicating that cAMP/PKA signalling is involved in the regulation of fungal morphogenesis and may therefore be of relevance for virulence. These findings are backed by mutant studies showing the relevance of PKA/cAMP signalling in *Aspergillus fumigatus* (Liebmann et al. 2004).

Proteomic analysis has also been carried out in the fungal human pathogen *Candida glabrata*, where a recent study using mutants lacking the Ace2 transcription factor which have an increased ability to cause disease. The types of proteins identified suggested that this factor controls the transcription of genes involved in cell wall biogenesis in this species (Stead et al. 2005).

In conclusion proteome analysis must been regarded as powerful tool to identify proteins required for structural integrity of the cells or for pathogenicity and it is likely that this approach might find application in the search for suitable targets for therapeutic approaches.

VII. Conclusions/Outlook

Genome-wide studies typically yield large numbers of candidate genes of which it is likely that only a small subset will be essential for the process studied. So for example, of several genes identified as differentially regulated during infection-related morphogenesis in *Magnaporthe grisea* only three out of 16 selected for mutation were found to be essential for pathogenicity. In order to fully exploit the results of genome-wide analysis there is a clear need for streamlined strategies for high-throughput analysis of gene function. Given that a subset of genes may be essential for viability, an attractive approach to investigate whether a particular gene represent a useful fungicide target would be gene silencing because inducible gene silencing might then allow the analysis of essential genes. Again taking *Magnaporthe* as an example, gene silencing has been shown to be effective in this fungus (Kadotani et al. 2003) and tools for

high-throughput analysis using this technology show great promise for the future (Nguyen et al. 2008). A collection of mutants deleted for all of the predicted genes, as has been developed for *Saccharomyces cerevisiae*, is an attractive resource; however the generation of such a resource is a major undertaking and it is likely that the intelligent use of transcription profiling combined with mutation of a subset of likely candidates is a more economical way to identify novel fungicide targets. Although the targets of most of the currently deployed fungicides were discovered by more traditional approaches, it is important to remember that genome-wide technologies are relatively new and that it may be several years before they yield their first fruits in terms of identification of new drug targets.

References

Akins RA (2005) An update on antifungal targets and mechanisms of resistance in *Candida albicans*. Med Mycol 43:285–318

Anke T, Steglich W (1999) Strobilurins and oudemansins. In: Grabley S, Thiericke R (eds) Drug discovery from nature. Springer, Heidelberg, pp 320–334

Asif AR, Oellerich M, Amstrong VW, Riemenschneider B, Monod M, Reichard U (2006) Proteome of conidial surface associated proteins of *Aspergillus fumigatus* reflecting potential vaccine candidates and allergens. J Proteome Res 5:954–962

Barker KS, Rogers PD (2005) Application of deoxyribonucleic acid microarray analysis to the study of azole antifungal resistance in *Candida albicans*. Methods Mol Med 118:45–56

Becker WF, von Jagow G, Anke T, Steglich W (1981) Oudemansin, strobilurin A, strobilurin B and myxothiazol: New inhibitors of the bc₁ segment of the respiratory chain with a E-ß-methoxyacrylate system as common structural element. FEBS Lett 132:329–333

Boddu J, Cho S, Kruger WM, Muehlbauer GJ (2006) Transcriptome analysis of the barley–*Fusarium graminearum* interaction. Mol Plant Microbe Interact 19:407–417

Both M, Eckert SE, Csukai M, Müller E, Dimopoulos G, Spanu PD (2005) Transcript profiles of *Blumeria graminis* development during infection reveal a cluster of genes that are potential virulence determinants. Mol Plant Microbe Interact 18:125–133

Breakspear A, Momany M (2007) The first fifty microarray studies in filamentous fungi. Microbiology 153:7–15

Brenner S, Johnson M, Bridgham J, Golda G, Lloyd DH, Johnson D, Luo S, McCurdy S, Foy M, Ewan M, Roth R, George D, Eletr S, Albrecht G, Vermaas E, Williams SR, Moon K, Burcham T, Pallas M, DuBridge RB, Kirchner J, Fearon K, Mao J, Corcoran Ka (2000) Gene expression analysis by massively parallel signature sequencing (MPSS) on microbead arrays. Nat Biotechnol 18:630–634

Bruneau JM, Magnin T, Tagat E, Legrand R, Bernard M, Diaquin M, Fudali C, Latgé JP (2001) Proteome analysis of *Aspergillus fumigatus* identifies glycosylphosphatidyl-inositol-anchored proteins associated to the cell wall biosynthesis. Electrophoresis 22:2812–2823

Cao YY, Cao YB, Xu Z, Ying K, Li Y, Xie Y, Zhu ZY, Chen WS, Jiang YY (2005) cDNA microarray analysis of differential gene expression in *Candida albicans* biofilm exposed to farnesol. Antimicrob Agents Chemother 49:584–589

Chabane S, Sarfati J, Ibrahim-Granet O, Du C, Schmidt C, Mouyna I, Prevost MC, Calderone R, Latgé JP (2006) Glycosylphosphatidylinositol-anchored Ecm33p influences conidial cell wall biosynthesis in *Aspergillus fumigatus*. Appl Environ Biol 72:3259–3267

Chandra J, Kuhn DM, Mukherjee PK, Hoyer LL, McCormick T, Ghannoum MA (2001) Biofilm formation by the fungal pathogen *Candida albicans*: development, architecture, and drug resistance. J Bacteriol 183:5385–5394

Cramer KL, Gerrald QD, Nichols CB, Price MS, Alspaugh JA (2006) Transcription factor Nrg1 mediates capsule formation, stress response, and pathogenesis in *Cryptococcus neoformans*. Eukaryot Cell 5:1147–1156

da Silva Ferreira ME, Malavazi I, Savoldi M, Brakhage AA, Goldman MH, Kim HS, Nierman WC, Goldman GH (2006) Transcriptome analysis of *Aspergillus fumigatus* exposed to voriconazole. Curr Genet 50:32–44

De Backer MD, Ilyina T, Ma XJ, Vandoninck S, Luyten WH, Vanden Bossche H (2001) Genomic profiling of the response of *Candida albicans* to itraconazole treatment using a DNA microarray. Antimicrob Agents Chemother 45:1660–1670

Debieu D, Bach J, Hugon M, Malosse C, Leroux P (2001) The hydroxyanilide fenhexamid, a new sterol biosynthesis inhibitor fungicide efficient against the plant pathogenic fungus *Botryotinia fuckeliana* (*Botrytis cinerea*). Pest Manag Sci 57:1060–1067

DeRisi JL, Iyer VR, Brown PO (1997) Exploring the metabolic and genetic control of gene expression on a genomic scale. Science 278:680–686

Donofrio NM, Oh Y, Lundy R, Pan H, Brown DE, Jeong JS, Coughlan S, Mitchell TK, Dean RA (2006) Global gene expression during nitrogen starvation in the rice blast fungus, *Magnaporthe grisea*. Fungal Genet Biol 43:605–617

Gioti A, Simon A, Le Pêcheur P, Giraud C, Pradier JM, Viaud M, Levis C (2006) Expression profiling of *Botrytis cinerea* genes identifies three patterns of up-regulation in planta and an FKBP12 protein affecting pathogenicity. J Mol Biol 358:372–386 Erratum in: J Mol Biol 364:550

Gowda M, Venu RC, Raghupathy MB, Nobuta K, Li H, Wing R, Stahlberg E, Couglan S, Haudenschild CD, Dean R, Nahm BH, Meyers BC, Wang GL (2006) Deep and comparative analysis of the mycelium and appressorium transcriptomes of *Magnaporthe grisea* using MPSS, RL-SAGE, and oligoarray methods. BMC Genomics 8:310

Grosse C, Heinekamp T, Kniemeyer O, Gehrke A, Brakhage AA (2008) Protein kinase A regulates growth,

sporulation, and pigment formation in *Aspergillus fumigatus*. Appl Environ Microbiol 74:4923–4933

Güldener U, Seong KY, Boddu J, Cho S, Trail F, Xu JR, Adam G, Mewes HW, Muehlbauer GJ, Kistler HC (2006) Development of a *Fusarium graminearum* Affymetrix GeneChip for profiling fungal gene expression in vitro and in planta. Fungal Genet Biol 43:316–325

Heller MJ (2002) DNA microarray technology: devices, systems, and applications. Annu Rev Biomed Eng 4:129–153

Howard RJ, Ferrari MA (1989) Role of melanin in appressorium function. Exp Mycol 13:403–418

Hu G, Steen BR, Lian T, Sham AP, Tam N, Tangen KL, Kronstad JW (2007) Transcriptional regulation by protein kinase A in *Cryptococcus neoformans*. PLoS Pathog 3:e42

Hwang L, Hocking-Murray D, Bahrami AK, Andersson M, Rine J, Sil A (2003) Identifying phase-specific genes in the fungal pathogen *Histoplasma capsulatum* using a genomic shotgun microarray. Mol Biol Cell 14:2314–2326

Irie T, Matsumura H, Terauchi R, Saitoh H (2003) Serial Analysis of Gene Expression (SAGE) of *Magnaporthe grisea*: genes involved in appressorium formation. Mol Genet Genomics 270:181–189

Kadotani N, Nakayashiki H, Tosa Y, Mayama S (2003) RNA silencing in the phytopathogenic fungus *Magnaporthe oryzae*. Mol Plant Microbe Interact 16:769–776

Kim Y, Nandakumar MP, Marten MR (2008) The state of proteome profiling in the fungal genus *Aspergillus*. Brief Funct Genomics Proteomics 7:87–94

Larraya LM, Boyce KJ, So A, Steen BR, Jones S, Marra M, Kronstad JW (2005) Serial analysis of gene expression reveals conserved links between protein kinase A, ribosome biogenesis, and phosphate metabolism in *Ustilago maydis*. Eukaryot Cell 4:2029–2043

Liebmann B, Müller M, Braun A, Brakhage AA (2004) The cyclic AMP-dependent protein kinase a network regulates development and virulence in *Aspergillus fumigatus*. Infect Immun 72:5193–5203

Köller W (1992) Antifungal agents with target sites in sterol functions and biosynthesis. In: Köller W (ed) Target sites of fungicide action CRC, Boca Raton, pp 119–206

Lee S, Gustafson G, Skamnioti P, Baloch R, Gurr S (2008) Host perception and signal transduction studies in wild-type *Blumeria graminis* f. sp. *hordei* and a quinoxyfen-resistant mutant implicate quinoxyfen in the inhibition of serine esterase activity. Pest Manag Sci 64:544–555

Lees ND, Bard M, Kirsch DR (1999) Biochemistry and molecular biology of sterol synthesis in *Saccharomyces cerevisiae*. Crit Rev Biochem Mol Biol 34:33–47

Lian T, Simmer MI, D'Souza CA, Steen BR, Zuyderduyn SD, Jones SJ, Marra MA, Kronstad JW (2005) Iron-regulated transcription and capsule formation in the fungal pathogen *Cryptococcus neoformans*. Mol Microbiol 55:1452–1472

Liebmann B, Müller M, Braun A, Brakhage AA (2004) The cyclic AMP-dependent protein kinase a network regulates development and virulence in *Aspergillus fumigatus*. Infect Immun 72:5193–203

Linksda Silva Ferreira ME, Malavazi I, Savoldi M, Brakhage AA, Goldman MH, Kim HS, Nierman WC,

Goldman GH (2006) Transcriptome analysis of *Aspergillus fumigatus* exposed to voriconazole. Curr Genet 50:32–44

Moran G, Stokes C, Thewes S, Hube B, Coleman DC, Sullivan D (2004) Comparative genomics using *Candida albicans* DNA microarrays reveals absence and divergence of virulence-associated genes in *Candida dubliniensis*. Microbiology 150:3363–3382

Mouyna I, Fontaine T, Vai M, Monod M, Fonzi WA, Diaquin M, Popolo L, Hartland RP, Latgé JP (2000) Glycosylphosphatidylinositol-anchored glucanosyl-transferases play an active role in the biosynthesis of the fungal cell wall. J Biol Chem 275:14882–14889

Moyrand F, Janbon G (2004) UGD1, encoding the *Cryptococcus neoformans* UDP-glucose dehydrogenase, is essential for growth at 37 degrees C and for capsule biosynthesis. Eukaryot Cell 3:1601–1608

Nakasako M, Motoyama T, Kurahashi Y, Yamaguchi I (1998) Cryogenic X-ray crystal structure analysis for the complex of scytalone dehydratase of a rice blast fungus and its tight-binding inhibitor, carpropamid: the structural basis of tight-binding inhibition. Biochemistry 37:9931–9939

Nguyen QB, Kadotani N, Kasahara S, Tosa Y, Mayama S, Nakayashiki H (2008) Systematic functional analysis of calcium-signalling proteins in the genome of the rice-blast fungus, *Magnaporthe oryzae*, using a high-throughput RNA-silencing system. Mol Microbiol 68:1348–1365

Nguyen VQ, Sil A (2008) Temperature-induced switch to the pathogenic yeast form of *Histoplasma capsulatum* requires Ryp1, a conserved transcriptional regulator. Proc Natl Acad Sci USA 105:4880–4885

Odenbach D, Breth B, Thines E, Weber RWS, Anke H, Foster AJ (2007) The transcription factor Con7p is a central regulator of infection-related morphogenesis in the rice blast fungus *Magnaporthe grisea*. Mol Microbiol 64:293–307

Odenbach D, Thines E, Anke H, Foster AJ (2009) The *Magnaporthe grisea* class VII chitin synthase is required for normal appressorial development and function. Mol Plant Pathol 10:81–94

Oh Y, Donofrio N, Pan H, Coughlan S, Brown DE, Meng S, Mitchell T, Dean RA (2008) Transcriptome analysis reveals new insight into appressorium formation and function in the rice blast fungus *Magnaporthe oryzae*. Genome Biol 9:R85

Ramage G, Wickes BL, Lopez-Ribot JL (2001) Biofilms of *Candida albicans* and their associated resistance to antifungal agents. Am Clin Lab 20:42–44

Ramage G, Saville SP, Wickes BL, López-Ribot JL (2002) Inhibition of *Candida albicans* biofilm formation by farnesol, a quorum-sensing molecule. Appl Environ Microbiol 68:5459–5463

Rogers PD, Barker KS (2002) Evaluation of differential gene expression in fluconazole-susceptible and -resistant isolates of *Candida albicans* by cDNA microarray analysis. Antimicrob Agents Chemother 46:3412–3417

Seong KY, Zhao X, Xu JR, Güldener U, Kistler HC (2008) Conidial germination in the filamentous fungus *Fusarium graminearum*. Fungal Genet Biol 45:389–399

Stead D, Findon H, Yin Z, Walker J, Selway L, Cash P, Dujon BA, Hennequin C, Brown AJ, Haynes K (2005) Proteomic changes associated with inactivation of the *Candida glabrata* ACE2 virulence-moderating gene. Proteomics 5:1838–1848

Takano Y, Choi WB, Mitchell TK, Okuno T, Dean RA (2003) Large scale parallel analysis of gene expression during infection-related morphogenesis of *Magnaporthe grisea*. Mol Plant Pathol 4:337–346

Thines E, Anke H, Weber RWS (2004) Fungal secondary metabolites as inhibitors of infection-related morphogenesis in phytopathogenic fungi. Mycol Res 108:14–25

Thomas SW, Glaring MA, Rasmussen SW, Kinane JT, Oliver RP (2002) Transcript profiling in the barley mildew pathogen *Blumeria graminis* by serial analysis of gene expression (SAGE). Mol Plant Microbe Interact 15:847–856

Thompson JE, Fahnestock S, Farrall L, Liao DI, Valent B, Jordan DB (2000) The second naphthol reductase of fungal melanin biosynthesis in *Magnaporthe grisea*. J Biol Chem 275:34867–34872

Velculescu VE, Zhang L, Vogelstein B, Kinzler KW (1995) Serial analysis of gene expression. Science 270:484–487

Wheeler IE, Hollomon DW, Gustafson G, Mitchell JC, Longhurst C, Zhang Z, Gurr S (2003) Quinoxyfen perturbs signal transduction in barley powdery mildew (*Blumeria graminis* f. sp. *hordei*). Mol Plant Pathol 4:177–186

12 Helminth Electron Transport Inhibitors Produced by Fungi

Rokuro Masuma[1], Kazuro Shiomi[1], Satoshi Ōmura[1]

CONTENTS

I. Introduction

The electron transport chain is present in mitochondria (eukaryotes) or plasma membrane (prokaryotes) and is linked with oxidative phosphorylation to produce ATP. The chain consists of complex I (NADH-ubiquinone reductase), complex II (succinate-ubiquinone reductase), complex III (ubiquinol–cytochrome-c reductase, cytochrome bc_1 complex), complex IV (cytochrome-c oxidase) and complex V (ATP synthase, F_OF_1-ATPase) which culminates in ATP via oxidative phosphorylation (Saraste 1999). Electrons generated from NADH and $FADH_2$ pass through complexes I–IV to produce a proton gradient, which is harnessed by complex V (Fig. 12.1).

Inhibitors of electron transport and oxidative phosphorylation enzymes are used to study the mechanism of energy conversion. Some of them have been developed into antifungal, insecticidal, antiparasite and other anti-infectious agents. For example, carboxin (1, complex II inhibitor) and azoxystrobin (2, complex III inhibitor) are used against plant-pathogenic fungi, fenpyroximate (3, complex I inhibitor) and chlorfenapyr (4, uncoupler) are used to combat insects or acari, and bithionol (5, complex II inhibitor) and atovaquone (6, complex III inhibitor) are used against various parasites (Fig. 12.2). Though they are synthetic compounds, several are derived from natural compounds. Compound 2 is an analog of strobilurin A (Fig. 12.3, 7) produced by a basidiomycete (Sauter et al. 1999), and the origin of 6 is dioxapyrrolomycin (Fig. 12.3, 8), produced by *Streptomyces* spp. (Addor et al. 1992).

Many electron transport inhibitors and oxidative phosphorylation enzyme inhibitors have been isolated from natural origins (Lardy 1980; Degli Esposti 1998; Ueki et al. 2000). A famous complex I inhibitor, rotenone (Fig. 12.4, 9), is a plant metabolite and used effectively as an insecticide. Piericidin A (Fig. 12.4, 10) and antimycin A_{3a} (Fig. 12.4, 11) are produced by *Streptomyces* spp. and widely used as complex I and complex III inhibitors respectively. Siccanin (Fig. 12.4, 12), also produced by a fungus, inhibits complex II and is used against surface mycosis.

During the course of screening for anthelmintic antibiotics, some new compounds isolated from the culture broths of fungi were found to be NADH-fumarate reductase (NFRD) inhibitors (Shiomi and Ōmura 2004; Kita et al. 2007; Ōmura and Shiomi 2007). Among them, some compounds show specific inhibition against

[1]The Kitasato Institute for Life Sciences and Graduate School of Infection Control Sciences, Kitasato University, 5-9-1 Shirokane, Minato-ku, Tokyo 108-8641, Japan; e-mail: masuma@lisci.kitasato-u.ac.jp, shiomi@lisci.kitasato-u.ac.jp

Physiology and Genetics, 1st Edition
The Mycota XV
T. Anke and D. Weber (Eds.)
© Springer-Verlag Berlin Heidelberg 2009

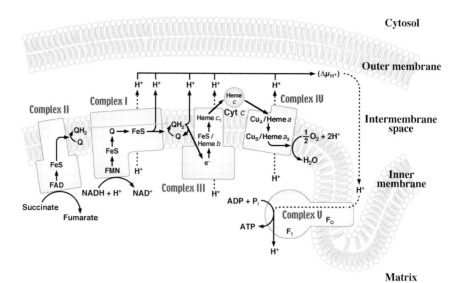

Fig. 12.1. Electron transport and oxidative phosphorylation in mitochondria

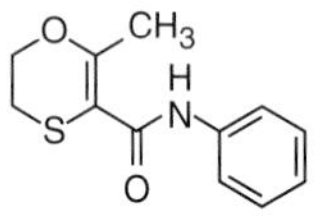

Carboxin **(1)** 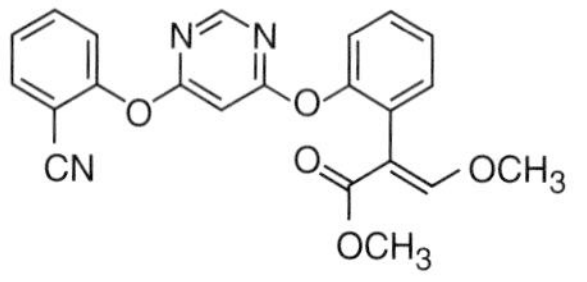   Azoxystrobin **(2)** 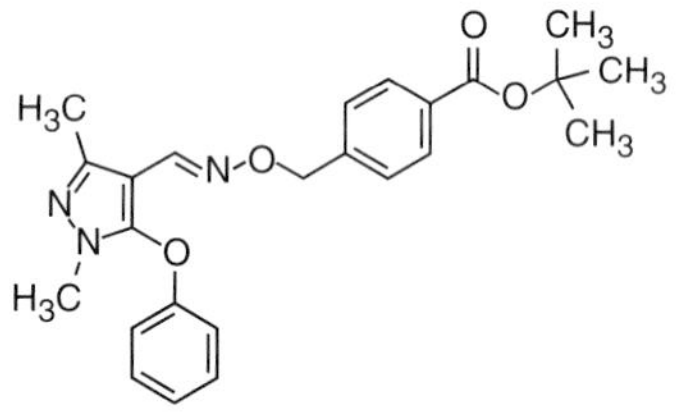   Fenpyroximate **(3)**

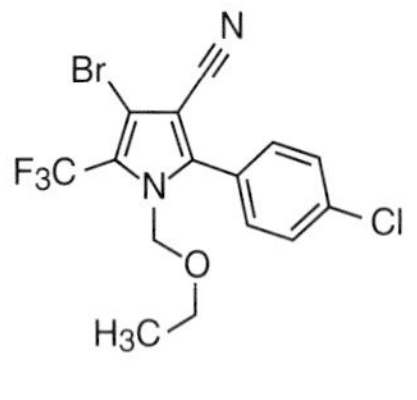

Chlorfenapyr **(4)** 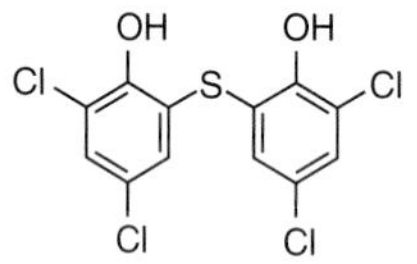   Bithionol **(5)** 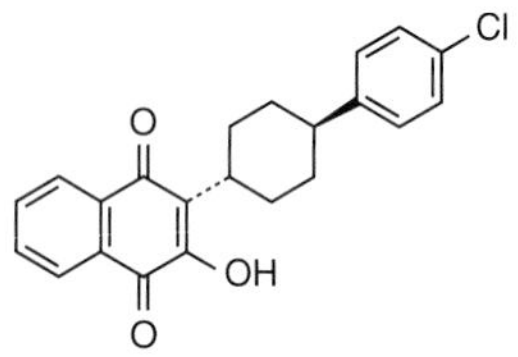   Atovaquone **(6)**

Fig. 12.2. Structures of practically used electron transport inhibitors

Fig. 12.3. Structures of strobilurin A and dioxapyrrolomycin

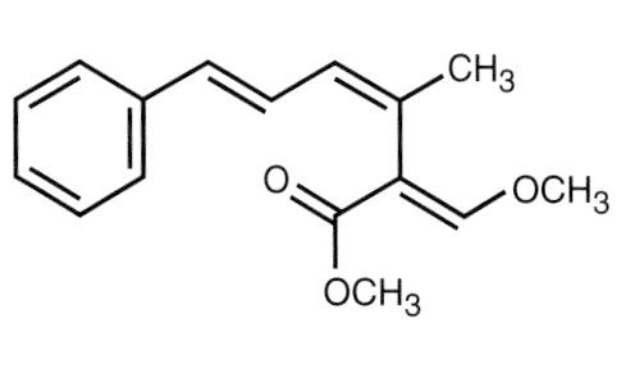

Strobilurin A **(7)** 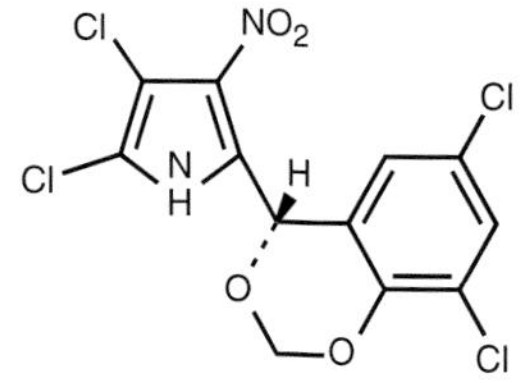   Dioxapyrrolomycin **(8)**

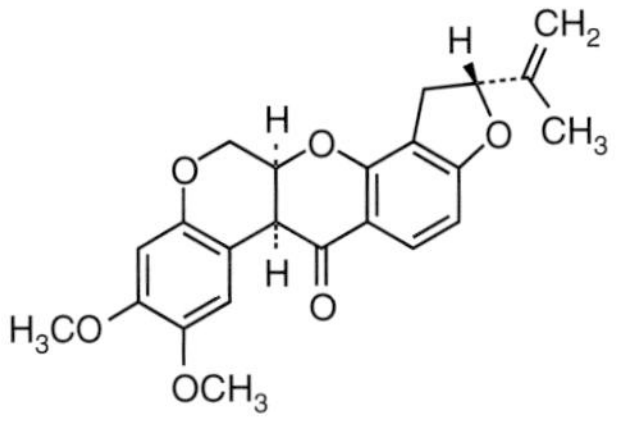

Rotenone (**9**)

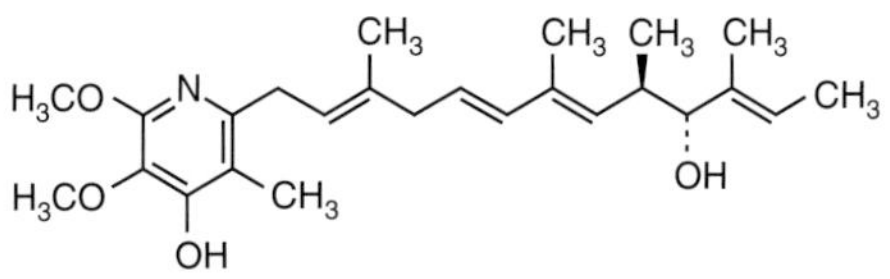

Piericidin A (**10**)

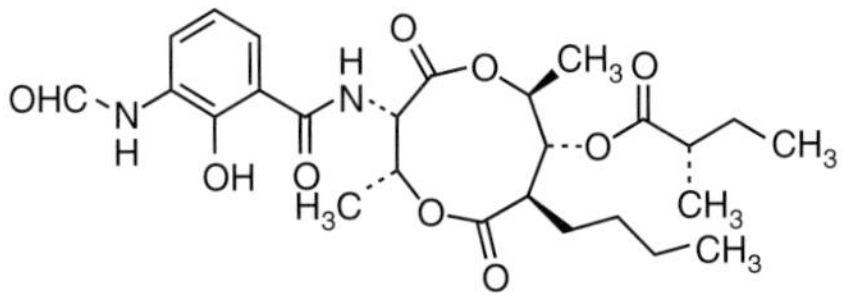

Antimycin A_{3a}(**11**)

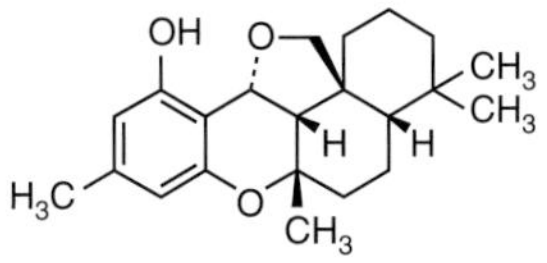

Siccanin (**12**)

Fig. 12.4. Popular natural electron transport inhibitors

Table 12.1. Electron transport and oxidative phosphorylation inhibitors produced by fungi

Compound	Producer	Biological activity
Complex I inhibitors		
Cochlioquinones A (**16**) and B (**17**)	*Cochliobolus miyabeanus*	Anthelmintic, phytotoxic, etc.
Isocochlioquinone A (**18**)	*Bipolaris bicolor*	Phytotoxic
Stemphone A (**19**)	*Stemphylium sarcinaeforme*	Phytotoxic, antibacterial
Pterulone (**20**) and pterulinic acid (**21**)	*Pterula* sp.	Antifungal
Nafuredin (**22**)	*Aspergillus niger*	Anthelmintic
Paecilaminol (**33**)	*Paecilomyces* sp.	Anthelmintic, insecticidal
Verticipyrone (**37**)	*Verticillium* sp.	Anthelmintic, insecticidal
Ukulactones A (**52**) and B (**53**)	*Penicillium* sp.	Anthelmintic
Complex II inhibitors		
Harzianopyridone (**58**)	*Trichoderma harzianum*	Antifungal
Atpenins A4 (**59**) and A5 (**60**)	*Penicillium* sp.	Antifungal
Siccanin (**12**)	*Helminthosporium siccans*	Antifungal
Anhydrofulvic acid (**71**)	*Penicillium* spp.	Antifungal
Complex III inhibitors		
Strobilurin A (**7**)	*Strobilurus tenacellus*	Antifungal
Ilicicolin H (**64**)	*Cylindrocladium ilicicola*	Antifungal, cytotoxic
Funiculosin (**65**)	*Penicillium funiculosum*	Antifungal, antiviral
Sambutoxin (**66**)	*Fusarium sambucinum*	Platelet aggregation inhibitor
Oudemansin A (**74**)	*Oudemansiella mucida*	Antifungal
Complex V inhibitors		
Aurovertin B (**77**)	*Calcarisporium arbuscula*	Apoptosis inducer
Citreoviridin (**78**)	*Penicillium citreoviride*	Neurotoxic
Asteltoxin (**79**)	*Aspergillus stellatus*	Toxic
Efrapeptin D (**80**)	*Tolypocladium niveum*	Antimalarial, antifungal, insecticidal
Tentoxin (**81**)	*Alternaria tenuis*	Chloroplast F_1-ATPase Inhibitor, phytotoxic
Uncouplers		
ACR-toxin I (**82**)	*Alternaria citri*	Phytotoxic
Leucinostatin A (**83**)	*Penicillium lilacinum*	Antifungal, antimalarial

complex I of anaerobic helminths. One compound, atpenin A5, inhibited complex II potently. Here, we review electron transport and oxidative phosphorylation inhibitors of fungal origin, including the NFRD inhibitors. These inhibitors are summarized in Table 12.1.

II. Inhibitors of Complex I

Complex I (NADH-ubiquinone reductase) is a relatively massive enzyme. In mammals, complex I consists of 45 different subunits with a combined molecular mass approaching 1 MDa, together with noncovalently bound FMN and eight iron–sulfur clusters (Carroll et al. 2006). Electrons from NADH are accepted by ubiquinone through complex I, ubiquinone being reduced to ubiquinonol. In this process, protons are transferred from the mitochondrial matrix to the intermembrane space, thus producing the electrochemical proton gradient. This is the first step of the electron transport system.

All known high-affinity inhibitors of complex I act at the terminal electron transfer step (quinone-binding site). They are classified into two or three groups (Friedrich et al. 1994; Degli Esposti and Ghelli 1999; Okun et al. 1999). Type A (class I) inhibitors, including piericidin A (**10**) and fenpyroximate (**3**), are quinone antagonists and inhibit in a partially-competitive manner with regard to ubiquinone. Type B (class II) inhibitors, including rotenone (**9**), aureothin (Fig. 12.5, **13**) and phenoxan (Fig. 12.5, **14**), are semiquinone antagonists and inhibit in a noncompetitive fashion (Washizu et al. 1954; Kunze et al. 1992). Aureothin is produced by *Streptomyces* sp. and **14** is produced by myxobacteria. Type C inhibitors are quinol antagonists, such as capsaicin (Fig. 12.5, **15**), a pungent principle of chili peppers (Shimomura et al. 1989; Yagi 1990). However, this classification does not mean the existence of two or three distinct inhibitors and quinone-binding sites. Okun et al. (1999) proposed the existence of only one large

inhibitor binding pocket in the hydrophobic part of complex I.

Many natural complex I inhibitors have been reported, especially from plants and myxobacteria, whereas the number of reported fungal complex I inhibitors is very small.

Cochlioquinones (Fig. 12.6) are isolated from *Cochliobolus miyabeanus* and some other fungi (Carruthers et al. 1971). They have a structure of benzoquinone joined to a sesquiterpene and they show various biological activities, such as nematocidal (against *Caenorhabditis elegans*), phytotoxic and anti-angiogenic properties and competitive inhibition of specific [³H]ivermectin-binding (Schaeffer et al. 1990). Inhibitory activities (IC_{50} values) of cochlioquinones A (**16**) and B (**17**), isocochlioquinone A (**18**) and stemphone A (**19**) against bovine heart NADH oxidase (complexes I+III+IV) were 115, 83, 56 and 160 nmol/mg of protein, respectively (Lim et al. 1996). Compound **17** inhibited complex I at an IC_{50} value of 370 nmol/mg of protein. It did not inhibit the other complexes, which indicated **17** was a specific complex I inhibitor. The inhibition against NADH is uncompetitive but inhibition against quinone changes from noncompetitive to competitive when the exogenous quinone concentration increases. A similar complicated inhibition against complex I was reported for capsaicin (**15**), a type C inhibitor of complex I (Yagi 1990).

Pterulone (Fig. 12.6, **20**) and pterulinic acid (Fig. 12.6, **21**) are produced by the submerged culture of the basidiomycete *Pterula* sp. (Engler et al. 1997). They have a 1-benzoxepin ring with chloromethylidene. Though **20** is a single *E*

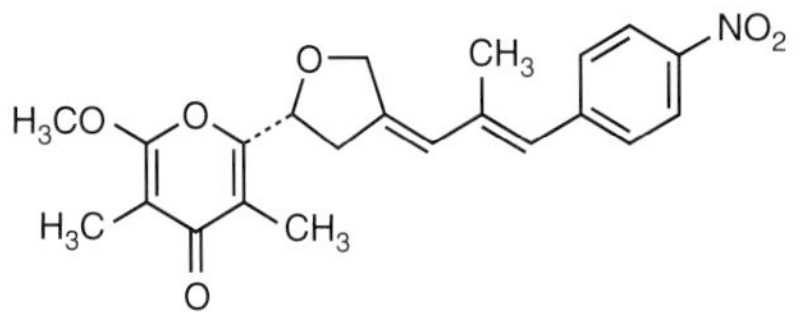

Aureothin (**13**)

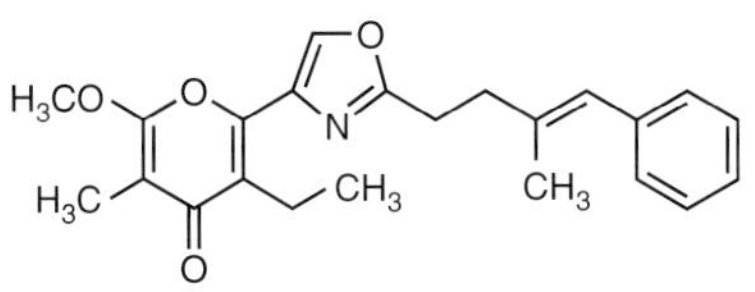

Phenoxan (**14**)

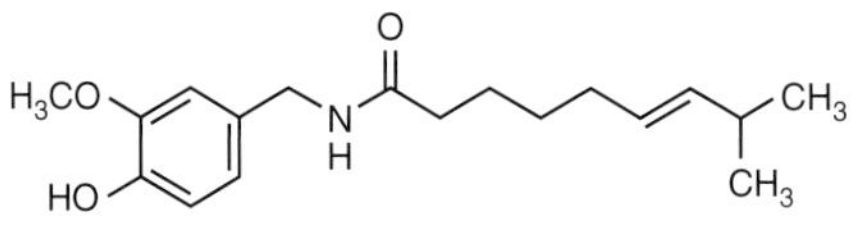

Capsaicin (**15**)

Fig. 12.5. Structures of aureothin, phenoxan, and capsaicin

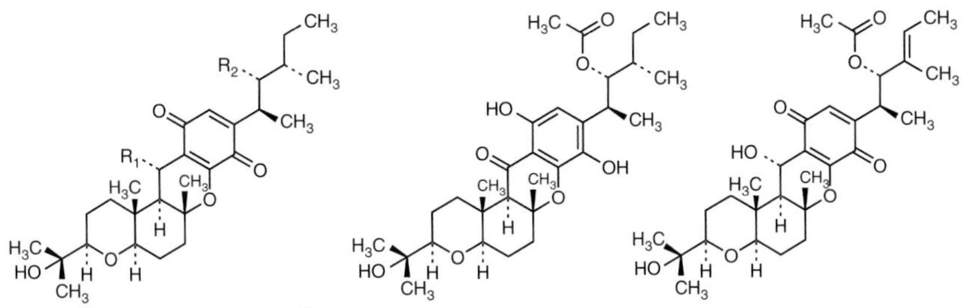

R1 R2
Cochlioquinone A (**16**) OH $OCOCH_3$
Cochlioquinone B (**17**) H =O

Isocochlioquinone A (**18**)

Stemphone A (**19**)

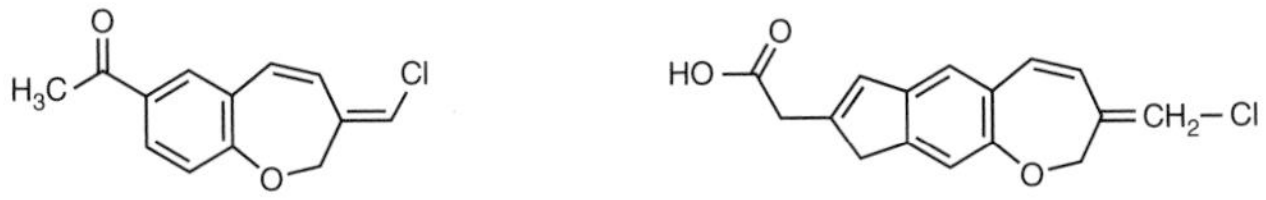

Pterulone (**20**)

Pterulinic acid (**21**)

Fig. 12.6. Structures of complex I inhibitors produced by fungi

isomer, **21** is a 5:1 mixture of *E*:*Z* that cannot be separated from each other. The compounds inhibit the growth of fungi. The IC_{50} values of **20** and **21** against bovine heart NADH oxidase were 36 and 450 μM, respectively, while they did not inhibit succinate oxidase (complexes II+III+IV). Therefore, they may be complex I inhibitors.

Since complex I reduces ubiquinone and complex III oxidizes ubiquinol, both complexes have quinone-binding sites. Therefore, some inhibitors targeting quinone-binding sites of complex I inhibit complex III and vice versa (Degli Esposti et al. 1993). Strobilurin A (**7**) inhibits complex III potently (see Sect. V.A) and reportedly inhibits bovine heart complex I weakly (15.5% inhibition at 3.3 μM; Degli Esposti et al. 1993).

III. Inhibitors of Helminth Complex I

A. NADH-Fumarate Reductase

Energy metabolism in many adult helminths differs from that in larvae and host (Kita et al. 2001; Komuniecki and Tielens 2003). They produce ATP in low oxygen concentration using a special respiratory system (Fig. 12.7). Phosphoenolpyruvate produced via an anaerobic glycolytic pathway is converted to oxalacetate by phosphoenolpyruvate carboxykinase, and oxalacetate is metabolized to malate and then to fumarate. Electrons from NADH are accepted by rhodoquinone through complex I (NADH-rhodoquinone reductase) and then transferred to fumarate through complex II (rhodoquinol-fumarate reductase). Therefore, the anaerobic complex II catalyzes the reverse reaction of aerobic complex II (succinate-ubiquinone reductase). The end-products of this glucose catabolism are volatile fatty acids, such as 2-methylpentanoate. This anaerobic electron transport system can provide ATP in the absence of oxygen. Rhodoquinone is used for this anaerobic respiration. Having a lower redox potential than ubiquinone enables rhodoquinone to be used for the reverse reaction of aerobic complex II. NFRD, composed of the above complex I and complex II, plays an important role in the anaerobic respiratory system. Therefore, NFRD inhibitors have been sought from microbial origin to produce anthelmintics and some helminth-specific complex I inhibitors have been discovered.

B. Nafuredin

1. Producing Strain and Fermentation

Nafuredin (Fig. 12.8, **22**) was isolated as an NFRD inhibitor (Ōmura et al. 2001; Ui et al. 2001). The producing fungal strain, FT-0554, was isolated from a marine sponge collected in the Palau Islands, Republic of Palau. The strain FT-0554 was identified as *Aspergillus niger* from morphological characteristics. *A. niger* is well known as a terrestrial fungus, so the effect of seawater concentration in a culture medium on fungal growth and nafuredin production was studied (Masuma et al. 2001).

The mycelial growth and the production of **22** were evaluated in different natural seawater

Fig. 12.7. Difference of energy metabolism between aerobic mammals and anaerobic helminths

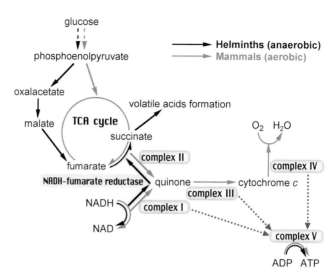

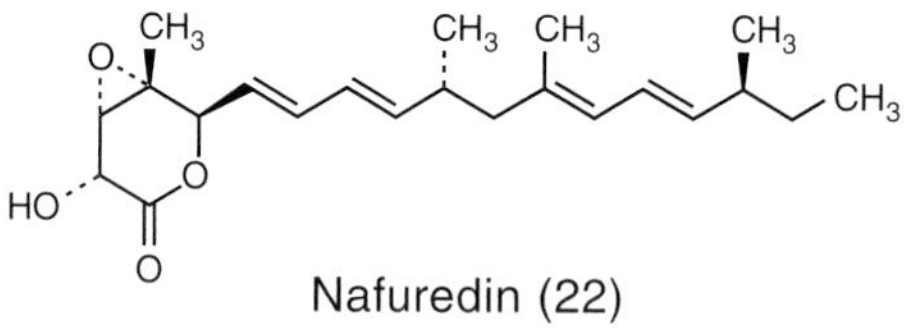

Fig. 12.8. Structure of nafuredin

concentrations (0–100%) after incubation at 25°C for 7 days on potato-dextrose agar or broth. The mycelial growth of strain FT-0544 increased with natural seawater concentration, and the addition of natural seawater (25–100%) enhanced the production of nafuredin. The strain FT-0554 also grew abundantly on medium containing 4% of sodium chloride. Generally, the growth of terrestrial fungi tends to be suppressed in the presence of natural seawater. Therefore, *A. niger* FT-0554 is suggested to be adapted to the marine environment.

After the isolation of **22**, more than five fungal strains have been found as producers of **22** during the screening of NFRD inhibitors. It is interesting that all strains were terrestrial fungi, and moreover, all were *Trichoderma* spp.

2. Structure

The structure of **22** was elucidated by NMR and mass spectra analysis (Ui et al. 2001). The total synthesis of **22** revealed its absolute configuration (Takano et al. 2001a, b). It has a β,γ-epoxy-δ-lactone ring with an alkenyl side chain. The biosynthesis study by Ui et al. (2001) suggested that

22 is composed of nine acetates and four methionines (branched four methyl carbons).

Many natural δ-lactones have been reported. δ-Decalactone (Fig. 12.9, **23**) is found in several kinds of foods and it is also produced by fungi (Nago et al. 1993). It is used for flavoring and fragrance. With the exception of the simple lactones, natural alkenyl-δ-lactones are usually very cytotoxic. Leptomycin B (Fig. 12.9, **24**) and kazusamycin A (Fig. 12.9, **25**), produced by *Streptomyces* spp., are antitumor and antifungal compounds (Hamamoto et al. 1983; Umezawa et al. 1984). They inhibit nuclear export signal-dependent nuclear export of proteins (Wolff et al. 1997). Fostriecin (Fig. 12.9, **26**) and pironetin (Fig. 12.9, **27**) are also antitumor compounds produced by *Streptomyces* spp. (Tunac et al. 1983; Kobayashi et al. 1994). The former inhibits protein phosphatase 2A (Roberge et al. 1994) and the latter causes microtubule disassembly (Kondoh et al. 1999). Aurovertins, citreoviridin and asteltoxin are fungal metabolites possessing alkenyl-δ-lactones. They are mycotoxins and inhibit F_1-ATPase of complex V (see Sect. V.B). ACR-toxin I is a fungal phytotoxin produced by *Alternaria* spp. It is an uncoupler (as shown in Sect. V.C).

Natural compounds having a β,γ-epoxy-δ-lactone moiety have only been found in clerodane furanoditerpene group, such as palmarin (Fig. 12.9, **28**), isolated as a bitter component of Calumba root (Wessely et al. 1936), but their δ-lactones are parts of fused rings (Yonemitsu et al. 1989). Therefore, **22** is the first natural compound having nonfused β,γ-epoxy-δ-lactone.

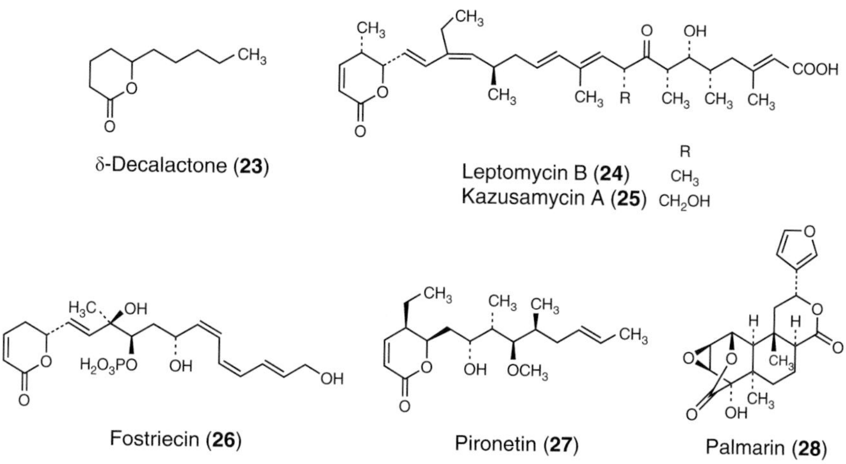

Fig. 12.9. Structures of δ-lactones produced by microorganisms

3. Enzyme Inhibition and Biological Activity

The screening for NFRD inhibitors was carried out using *Ascaris suum* (roundworm) mitochondria. Compound **22** inhibited NFRD at an IC_{50} value of 12 nM (Table 12.2; Ōmura et al. 2001). Since NFRD consists of NADH-rhodoquinone reductase (complex I) and rhodoquinol-fumarate reductase (complex II), inhibitory activity against each enzyme of *A. suum* was evaluated. As shown in Table 12.2, **21** potently inhibited complex I (IC_{50} = 24 nM) and the inhibition against complex II was very weak (IC_{50} = 80 μM). It is interesting that **22** inhibited not only anaerobic adult complex I of *A. suum* (NADH-rhodoquinone reductase) but also larval complex I (NADH-ubiquinone reductase), which has an aerobic energy metabolism like that of mammals. The IC_{50} value against larval complex I was 8.9 nM. Therefore, **22** is a complex I inhibitor. However, **22** only inhibited rat liver complex I at very high concentration (IC_{50} = 10 μM). Therefore, **22** seems to be a selective inhibitor of helminth

complex I. From a kinetic study against NADH-rhodoquinone reductase of adult *A. suum*, **22** revealed competitive inhibition with rhodoquinone and noncompetitive inhibition with NADH, which indicated that the site of inhibition of **22** is the quinone-binding domain in complex I.

Haemonchus contortus (barberpole worm) is reported to have an NFRD system (Van Hellemond et al. 1995). Therefore, in vivo anthelmintic activity of **22** was evaluated using *H. contortus*-infected sheep (Ōmura et al. 2001). As shown in Table 12.3, **22** (2 mg/kg p.o.) significantly reduced the number of fecal eggs of infected sheep. This anthelmintic activity may be due to the inhibition of complex I, because **22** also inhibits the enzyme of *H. contortus* (Table 12.2). Moreover, **22** was effective in mice infected with dwarf tapeworm, *Hymenolepis nana*. There were no signs of any side-effects and no loss of body weight during tests in either sheep (2 mg/kg p.o.) or mice (50 mg/kg p.o. and i.p.).

Table 12.2. Effects of nafuredin on electron transport enzymes of nematodes and rat. *NT* Not tested

Enzyme	Complex	IC_{50} (nM)			
		Ascaris suum (adult)	*Ascaris suum* (L2)	*Haemonchus contortus* (adult)	Rat liver
NADH-fumarate reductase	I+II	12	NT	NT	1000
NADH-ubiquinone reductase	I	8	8.9	86	10 000
NADH-rhodoquinone reductase	I	24	9.0	195	>100 000
Rhodoquinol-fumarate reductase	II	80 000	NT	NT	NT
Succinate-ubiquinone reductase	II	>100 000	NT	NT	>100 000

Table 12.3. Effects of treatment with nafuredin on fecal egg counts in sheep infected with *Haemonchus contortus*. Nafuredin (2 mg/kg) was given orally at day 0. Values are means of three experiments (treated animals) or two experiments (controls) $\pm$ SD

	Number of eggs per gram of feces	
	Treated	Control
Day −1	5689 ± 1120	5200 ± 754
Day 4	1489 ± 655	5100 ± 424
Day 11	289 ± 214	4034 ± 801

4. Nafuredin-γ and its Analogs

During total synthesis of **22**, it was found to convert into a novel γ-lactone derivative, named nafuredin-γ (Fig. 12.10, **29**), under mild basic conditions (Nagamitsu et al. 2003; Shiomi et al. 2005). The epoxide of **22** opens and the δ-lactone recyclizes to form γ-lactone with keto–enol tautomerism. Compound **29** inhibited NADH-rhodoquinone reductase of adult *A. suum* with an IC_{50} value of 2.3 nM, and did not inhibit bovine liver NADH-ubiquinone reductase at 10 µM. It showed anthelmintic activity against *H. contortus* using treatment with two oral doses each of 2 mg/kg. Thus, **29** has similar enzyme inhibitory and anthelmintic activities as those of **22**. Though **22** was converted into **29** in basic condition, only a part of **22** was converted in neutral buffer. Therefore, it is not likely that **21** was converted into **29** and showed inhibitory activity. Both **22** and **29** may inhibit complex I directly.

Since the lactone moiety synthesis of **29** is simpler than that of **22**, **29** is useful for studying structure-activity relationships. The total synthesis of **29** has been achieved (Nagamitsu et al. 2003), and Nagamitsu et al. (2008) prepared

some analogs (Fig. 12.11). The inhibitory activities of the 2-deoxy derivative (**30**) and C4 epimer (**31**) against NFRD of adult *A. suum* were similar to that of **29**. Therefore, the enol (or 2-ketone) functionality and the C4 stereochemistry of **29** are not responsible for the inhibitory activity. However, the NFRD inhibitory activity of the C5 epimer (**32**) was weakened 20-fold, which suggests that the stereochemistry of the C5 hydroxy group is important for the inhibitory activity of **29**.

C. Paecilaminol

Paecilaminol (Fig. 12.12, **33**) was produced by a fungus, *Paecilomyces* sp. FKI-0550, isolated from a soil sample collected on Miyakojima Island, Okinawa Prefecture, Japan (Ui et al. 2006a). The structure of **33** was elucidated by NMR and mass spectra as 2-amino-14,16-dimethyl-3-octadecanol. Compound **32** inhibited the growth of free-living nematode *C. elegans* and brine shrimp *Artemia salina* at the MIC values of 20 µg/ml and 5 µg/ml, respectively.

The adult *Ascaris suum* NFRD inhibitory activity of **33** was moderate (IC_{50} = 5.1 µM), but the inhibition was about four times more potent than that of bovine heart NADH oxidase (complexes I+III+IV, IC_{50} = 19.8 µM). The IC_{50} values of **33** against NADH-rhodoquinone reductase (complex I) and rhodoquinol-fumarate reductase (complex II) of *A. suum* were 23 µM and 35 µM, respectively. The values against NADH-ubiquinone reductase (complex I) and ubiquinol–cytochrome-*c* reductase (complex III) of bovine heart were 16 µM and 20 µM, respectively. However, **33** does not inhibit bovine succinate-ubiquinone reductase

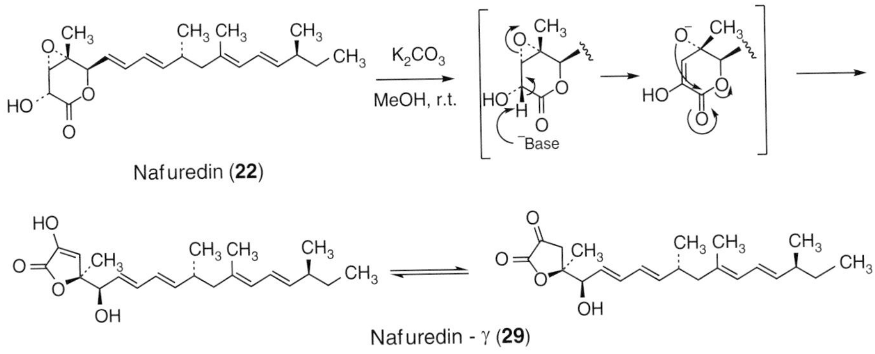

Fig. 12.10. Proposed mechanism of the conversion of nafuredin to nafuredin-γ

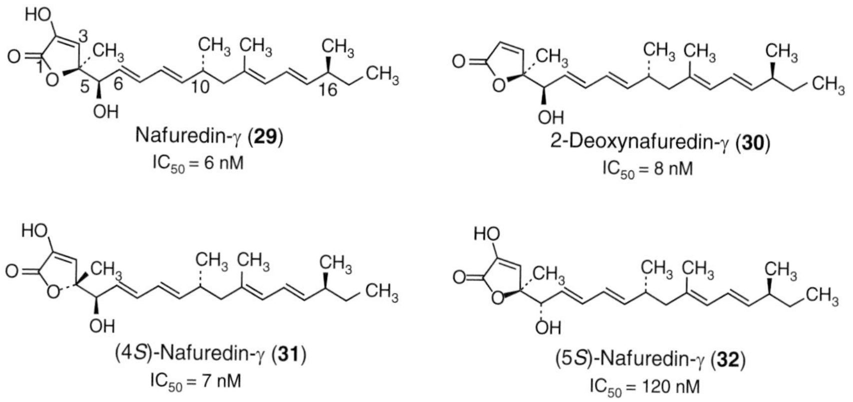

Fig. 12.11. NADH-fumarate reductase inhibitory activities of nafuredin-γ analogs

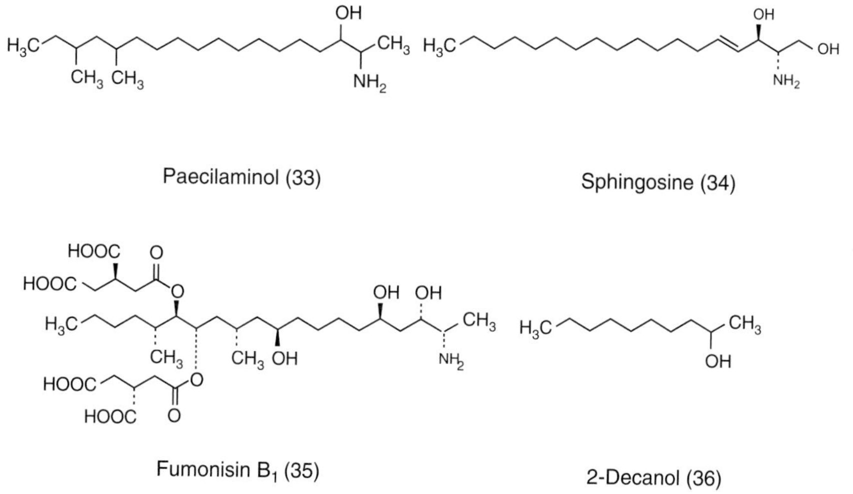

Fig. 12.12. Structures of paecilaminol and its related compounds

(complex II) at 100 μM. Therefore, **33** shows similar inhibitory activities against complexes I, II and III of *A. suum* and bovine heart, except bovine complex II. It is not common for electron transport inhibitors to show such low selectivity. The only group that show such wide inhibitions are 2-alkyl-4,6-dinitrophenols, with their inhibitory activities against complex II also being weaker than that for complexes I and III (Tan et al. 1993). The low selectivity of **33** may be due to its linear structure, because both amino and hydroxy groups can freely rotate and be attached to enzymes.

Compound **33** is an amino alcohol. The NFRD inhibitory activity of a similar amino alcohol sphingosine (Fig. 12.12, **34**), a long-chain base of sphingolipids, was weaker ($IC_{50} = 28$ μM) than **33**. However, Fumonisin B$_1$ (Fig. 12.12, **35**), an amino

alcohol produced by the fungus *Fusarium moniliforme* (Gelderblom et al. 1988), did not inhibit NFRD at 100 μM. Desai et al. (2002) reported that fumonisins inhibited ceramide synthase (sphingosine *N*-acyltransferase). A simple alcohol 2-decanol (Fig. 12.12, **36**) also showed no inhibition against NFRD at 100 μM.

D. Verticipyrone

Verticipyrone (Fig. 12.13, **37**) was produced by a fungus, *Verticillium* sp. FKI-1083, isolated from a soil sample collected on Yakushima Island, Kagoshima Prefecture, Japan (Ui et al. 2006b). The structure was shown to be (*E*)-2-methoxy-3, 5-dimethyl-6-(3-methyl-2-undecenyl)

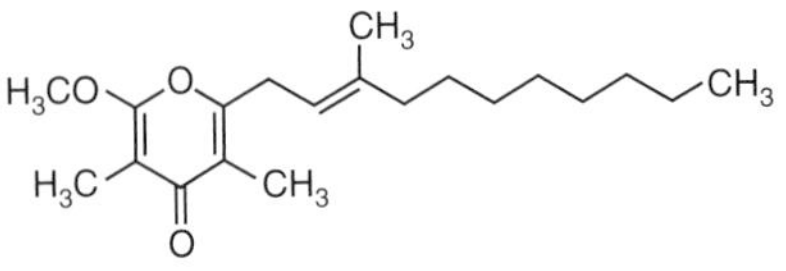

Verticipyrone (**37**): IC_{50} = 4.1 nM

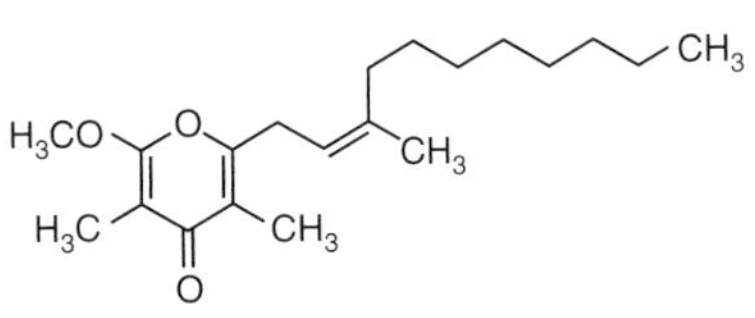

38: IC_{50} = 2.0 nM

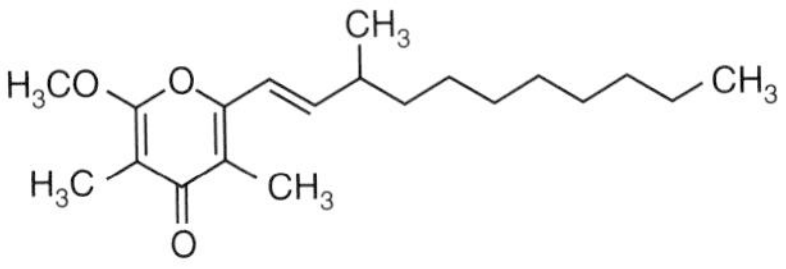

39: IC_{50} = 4.7 nM

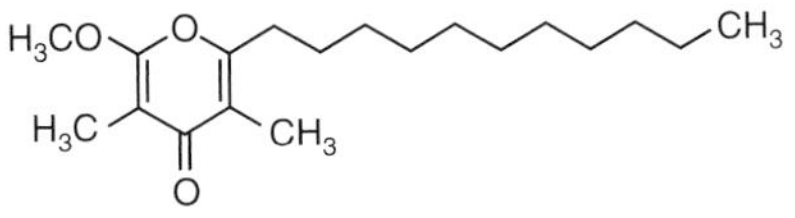

40: IC_{50} = 4.8 nM

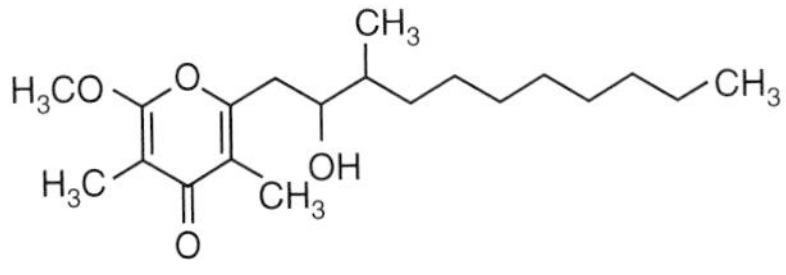

41: IC_{50} = 0.65 nM

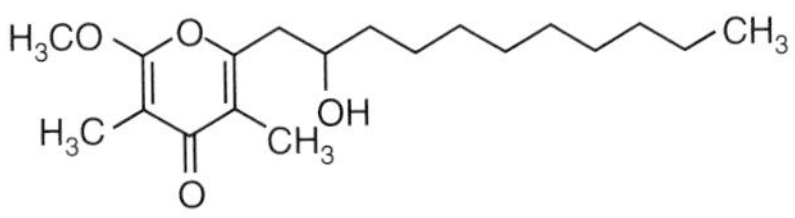

42: IC_{50} = 0.30 nM

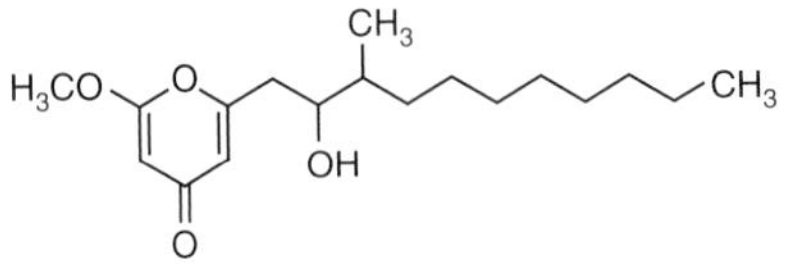

43: IC_{50} > 100 nM

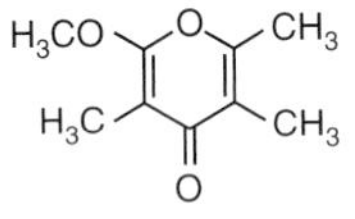

44: IC_{50} > 100 nM

Fig. 12.13. NADH-fumarate reductase inhibitory activities of verticipyrone and its analogs

-4*H*-pyran-4-one by NMR and mass spectra studies. MIC values of **37** against *Caenorhabditis elegans* and *A. salina* were 20 μg/ml and 2.0 μg/ml, respectively.

Compound **37** inhibited NFRD from *A. suum* with an IC_{50} value of 4.1 nM and the inhibition was specific to NADH-rhodoquinone reductase (complex I) as shown in Table 12.4. However, its inhibitory activity against NADH-ubiquinone reductase (complex I) from bovine heart was similar to that of *A. suum* complex I.

The total synthesis of **37** has been accomplished and some analogs of **37** were prepared (Shimamura et al. 2007), as shown in Fig. 12.13. Olefin isomers **38** and **39** and alkyl side chain analog **40** showed *Ascaris* NFRD inhibitory activities similar to **37**, which suggests that the olefin in the side chain is not important. The two alcohols **41** and **42** showed much potent NFRD inhibitory activities than **37** suggesting the newly introduced hydroxy group on the side chain may contribute to the inhibition. However, **43** (didemethyl **41**) showed no NFRD inhibition. Therefore, 3,5-dimethyl groups on the γ-pyrone moiety may be essential for NFRD inhibition. The γ-pyrone **44** showed no NFRD inhibition, which suggests that the long side chain is also important for the inhibition.

Among the analogs, **41** showed >20-fold more potent inhibition against *Ascaris* complex I compared with that of **37**, whereas inhibition of **41** against bovine heart complex I was four times less potent than that of **37** (Table 12.4). Therefore **41** is a selective inhibitor of *Ascaris* complex I.

Alveolar echinococcosis caused by larval *Echinococcus multilocularis* is a life-threatening parasitic zoonosis. Recently, *E. multilocularis* was found to use an anaerobic NADH-fumarate reductase system for its energy metabolic pathways (Matsumoto et al. 2008). Therefore, the efficacy of **42** (the most potent NFRD inhibitor among the analogs of **37**) against larval *E. multilocularis* was evaluated. The viability of the *E. multilocularis* protoscolex was progressively reduced during in vitro treatment of the

Table 12.4. Effects of verticipyrone and its analog **41** on electron transport enzymes

	Enzyme	Complex	IC$_{50}$ (nM)	
			Verticipyrone	**41**
Ascaris suum	NADH-fumarate reductase	I+II	4.1	0.65
	NADH-rhodoquinone reductase	I	49	2.0
	Rhodoquinol-fumarate reductase	II	>100 000	>100 000
Bovine heart	NADH oxidase	I+III+IV	1.3	20
	NADH-ubiquinone reductase	I	46	200
	Succinate-ubiquinone reductase	II	>100 000	>100 000
	Ubiquinol–cytochrome-*c* reductase	III	26 000	80 000

parasites with **42**, and more than 90% of the parasites were eliminated on days 5 and 18 when **42** was used at 50 µM and 5 µM, respectively (Matsumoto et al., unpublished data).

Compound **37** has the structure of 2-methoxy-3,5-dimethyl-γ-pyrone with a side chain at C6. Aureothin (**13**) produced by *Streptomyces thioluteus* has the same skeleton and inhibits complex I as shown in Sect. II. Neoaureothin (spectinabilin; Fig. 12.14, **45**), an analog of **13** produced by *Streptomyces* spp. (Cassinelli et al. 1966), was found to inhibit *Ascaris* NFRD with an IC$_{50}$ value of 15 nM (Ui et al. 2006b).

Some γ-pyrones with a side chain at C6 have been also isolated from the culture broths of fungi. Himeic acid A (Fig. 12.14, **46**), produced by *Aspergillus* sp. (Tsukamoto et al. 2005) inhibited ubiquitin-activating enzyme (E1). Carbonarone A (Fig. 12.14, **47**) produced by *Aspergillus carbonarius* (Zhang et al. 2007) showed moderate cytotoxicity. The producing strains of **46** and **47** were both marine-derived fungi. Funicone (**48**), deoxyfunicone (**49**) and vermistatin (**50**) have (*E*)-propenyl side chains at C6 of the γ-pyrone (Fig. 12.14). They were produced by *Penicillium funiculosum*, *P. vermiculatum* and some other fungi (Merlini et al. 1970; Fuska et al. 1979; Sassa et al. 1991). Compounds **48** and **49** showed antifungal activities and **50** showed cytotoxicity. Compounds **49** and **50** also potentiated the antifungal activity of miconazole against *Candida albicans* (Arai et al. 2002). 2,3,5-Trimethyl-6-(3-oxobutan-2-yl)-4*H*-pyran-4-one (Fig. 12.14, **51**), produced by *Aspergillus sydowi* isolated from the deep sea, was not cytotoxic (Li et al. 2007).

E. Ukulactones

Ukulactones A and B (Fig. 12.15, **52** and **53**) were produced by a fungus, *Penicillium* sp. FKI-3389, isolated from a soil sample collected on Hawaii

Island, Hawaii, United States (Ōmura et al. 2007). They have the same planar structure, with 2-oxabicyclo[2.2.1]heptane-3,5-dione and 5,6-dihydro-2*H*-pyran linked by pentaene. The configurations at C2 of the pyran (junction of the pyran and the pentaene) are opposite between **52** and **53**. The absolute configurations of **52** and **53** have not yet been elucidated. Compound **52** inhibited the growth of parasite nematode *Nippostrongylus brasiliensis* at 1 µg/ml in vitro.

Compound **52** inhibited *Ascaris* NFRD (Table 12.5) and is an NADH-rhodoquinone reductase (complex I)-specific inhibitor (IC$_{50}$ = 55 nM). Its inhibition against bovine heart NADH-ubiquinone reductase (complex I) was about 500 times weaker (IC$_{50}$ = 28 µM) than that of *Ascaris* complex I. The inhibitions against bovine complexes II and III were also weak. Therefore, **52** is a selective inhibitor of *Ascaris* complex I, comparable to nafuredin (**22**). The difference between **52** and **53** is the configuration of only one carbon. However, the inhibition of **53** against *Ascaris* NFRD was about 200 times weaker than that of **52** (Table 12.5). Thus, the configuration at C2 of the pyran may be very important for the inhibitory activity.

Recently, demethyl **52** compound, prugosene A1 (Fig. 12.15, **54**), was isolated from the surface-cultured mycelium of a marine sponge-derived *Penicillium rugulosum* by Lang et al. (2007). Compound **54** had no antimicrobial activity. Dong et al. isolated 195-A (Fig. 12.15, **55**) from the culture broth of *Talaromyces wortmannii*. The pentaene moiety of **54** is altered to tetraene at **55**. 2-Oxabicyclo[2.2.1]heptane-3,5-dione is also found in the structure of shimalactone A (Fig. 12.15, **56**) produced by the marine fungus *Emericella variecolor* (Wei et al. 2005). Compound **56** induced neuritogenesis against neuroblastoma cells. Compounds **52** and **53** have an (all-*E*)-2,10-dimethyldodeca-2,4,6,8,10-pentaene moiety, which is also found in

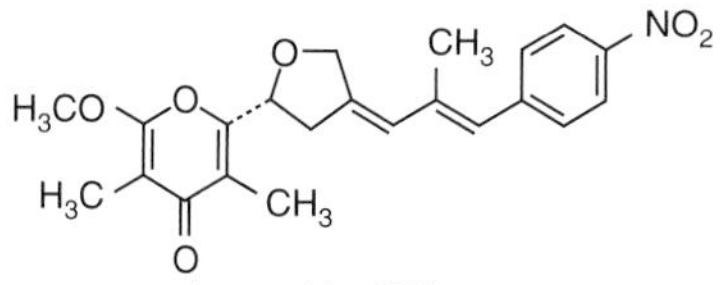

Aureothin (**13**)

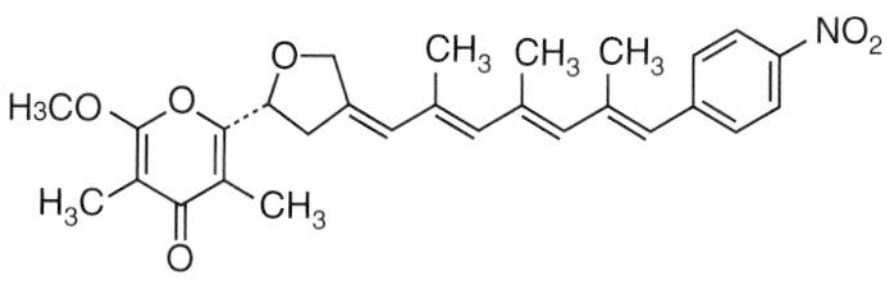

Neoaureothin (Spectinabilin, **45**)

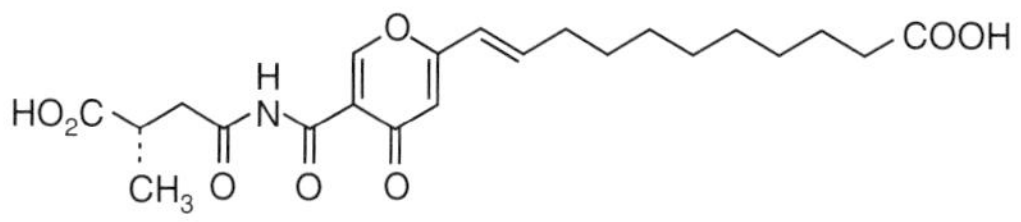

Himeic acid A (**46**)

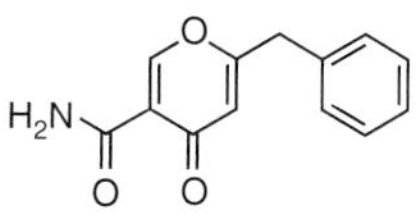

Carbonarone A (**47**)

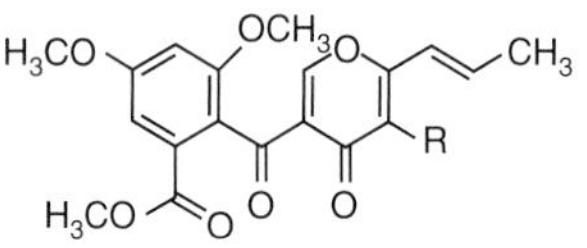

Funicone (**48**)   R

Deoxyfunicone (**49**)   H

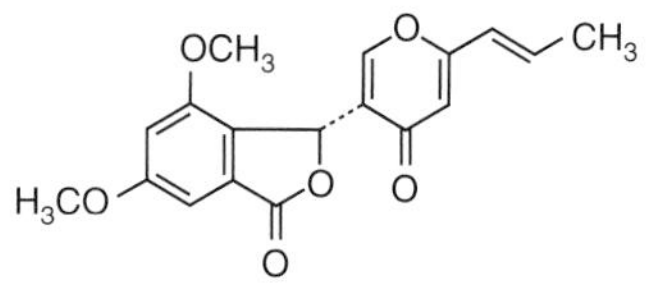

Vermistatin (**50**)

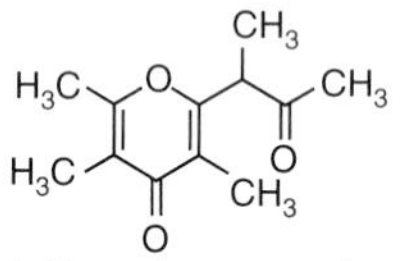

2,3,5-Trimethyl-6-(3-oxobutan-2-yl)-4*H*-pyran-4-one (51)

Fig. 12.14. Structures of verticipyrone related γ-pyrones

phenalamide A$_2$ (**57**) produced by the myxobacterium *Myxococcus stipitatus* (Trowitzsch-Kienast et al. 1992). It suppressed HIV-1 replication in cell cultures and inhibited NADH-ubiquinone reductase (complex I) potently (Friedrich et al. 1994). It is interesting that mammalian complex I was inhibited potently by **57**, but not by **52** and **53**.

IV. Inhibitors of Complex II

A. Atpenins and Harzianopyridone

1. Structures

Harzianopyridone (Fig. 12.16, **58**), produced by the fungus *Trichoderma* sp. was isolated during screening for NFRD inhibitors (Miyadera et al. 2003). It was originally isolated from *T. harzianum* and showed antifungal, antibacterial and herbicidal activities, but its mode of action had

not been studied (Dickinson et al.1989; Cutler and Jacyno 1991). Structurally similar antifungal antibiotics, atpenins A4, A5 and B (Fig. 12.16, **59–61**), have been isolated from the culture broth of the fungus *Penicillium* sp. (Ōmura et al. 1988; Kumagai et al. 1990) before the report of **58**. Since **61** was suggested to inhibit the ATP-generating system (Oshino et al. 1990), the effects of **58** together with atpenins on the inhibitory activities of electron transport enzymes were examined and were revealed to be potent and selective complex II (succinate-ubiquinone reductase) inhibitors, as shown below.

Though the structure of **58** is depicted as 2-pyridone and the structures of **59-61** are depicted as 2-pyridinol in Fig. 12.16, structures in the figure only reflect those of the original papers. 2-Pyridone and 2-pyridinol are tautomers, and they may be equivalent. The other compounds having the same chromophore are WF-16775 A$_1$ and A$_2$ (Fig. 12.16, **62**, **63**), isolated from the fungus

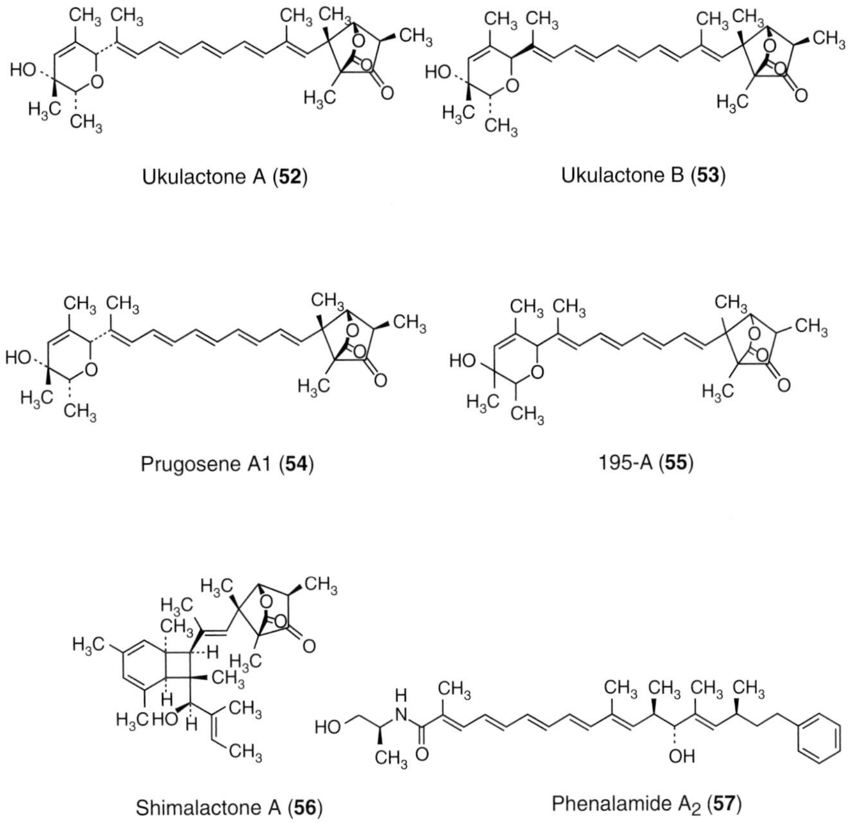

Ukulactone A (**52**)

Ukulactone B (**53**)

Prugosene A1 (**54**)

195-A (**55**)

Shimalactone A (**56**)

Phenalamide A$_2$ (**57**)

Fig. 12.15. Structures of ukulactones and their related compounds

Table 12.5. Effects of ukulactones on electron transport enzymes. *NT* Not tested

	Enzyme	Complex	IC$_{50}$ (nM)	
			Ukulactone A	Ukulactone B
Ascaris suum	NADH-fumarate reductase	I+II	2.4	470
	NADH-rhodoquinone reductase	I	55	NT
	Rhodoquinol-fumarate reductase	II	>100 000	NT
Bovine heart	NADH oxidase	I+III+IV	9000	16 000
	Succinate–cytochrome-*c* reductase	II+III	68 000	30 000
	NADH-ubiquinone reductase	I	28 000	NT
	Succinate-ubiquinone reductase	II	>100 000	NT
	Ubiquinol–cytochrome-*c* reductase	III	32 000	NT

Chaetasbolisia erysiophoides (Otsuka et al. 1992), and they display potent angiogenetic activity.

Fungi also produce some 2-pyridone compounds. Ilicicolin H (Fig. 12.17, **64**) was isolated by Hayakawa et al. (1971). Funiculosin (Fig. 12.17, **65**) and sambutoxin (Fig. 12.17, **66**) are *N*-methyl-2-pyridones (Ando et al.1969; Kim et al. 1995). Compounds **64–66** inhibit complex III (as shown in Sect. V.A). Flavipucine

(Fig. 12.17, **67**), produced by *Aspergillus flavipes* (Findlay and Radics 1972), shows antibacterial and fungicidal activities, and apiosporamide (Fig. 12.17, **68**), produced by *Apiospora montagnei* (Alfatafta et al. 1994), possesses antifungal activity. Militarinone D (Fig. 12.17, **69**), isolated from the culture broth of *Paecilomyces militaris* (Schmidt et al. 2003), exhibits cytotoxicity.

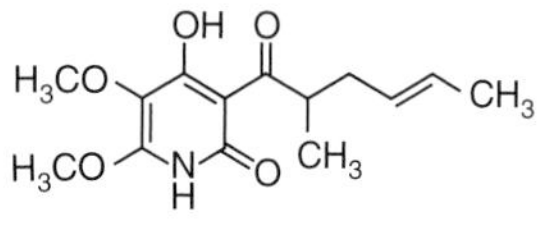

Harzianopyridone (**58**)

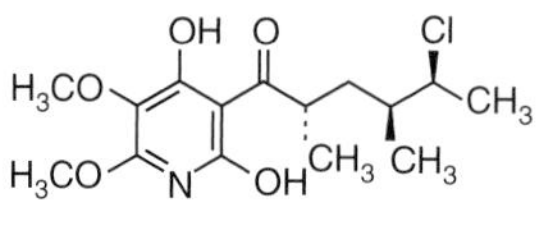

Atpenin A4 (**59**)

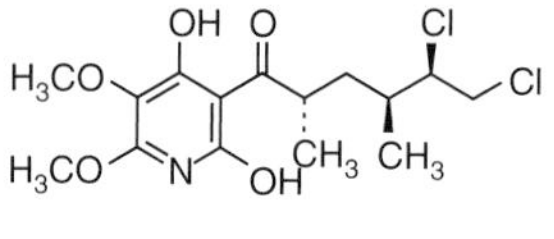

Atpenin A5 (**60**)

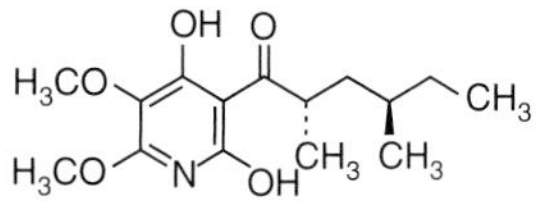

Atpenin B (**61**)

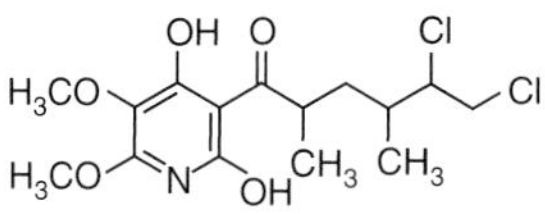

WF-16775 A_1 (**62**)

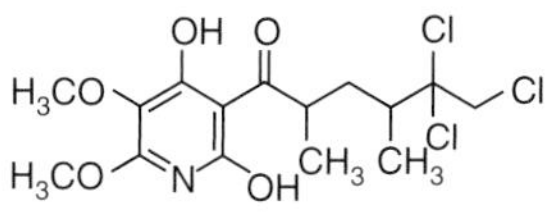

WF-16775 A_2 (**63**)

Fig. 12.16. Structures of atpenins and their related compounds

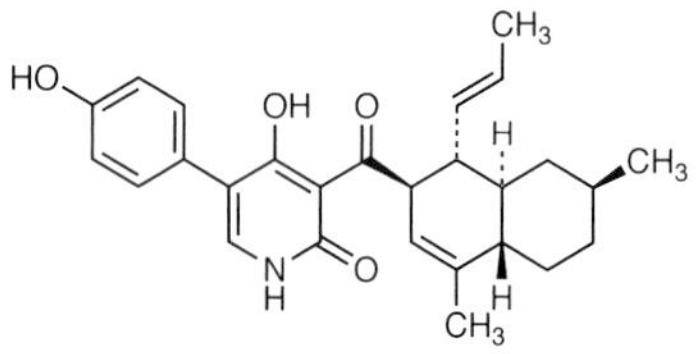

Ilicicolin H (**64**)

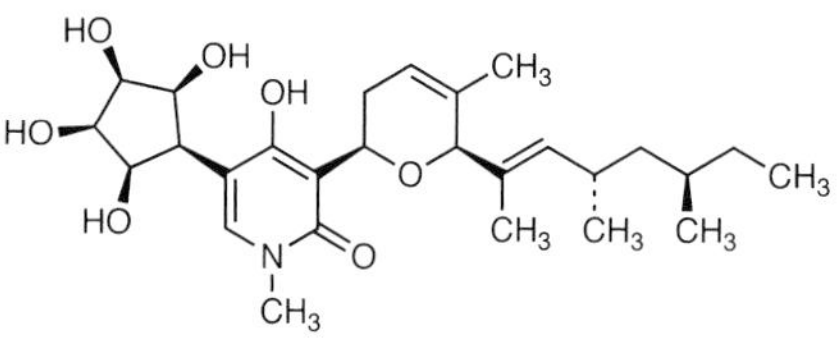

Funiculosin (**65**)

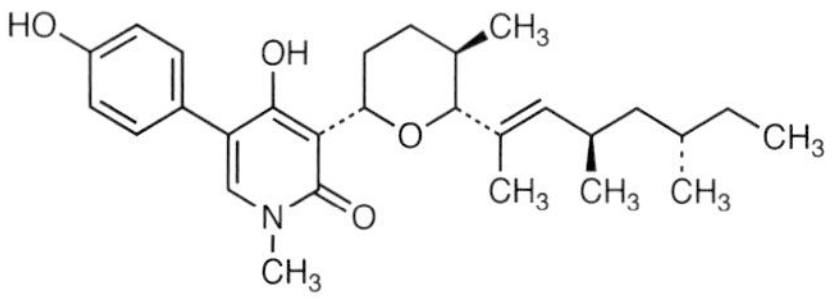

Sambutoxin (**66**)

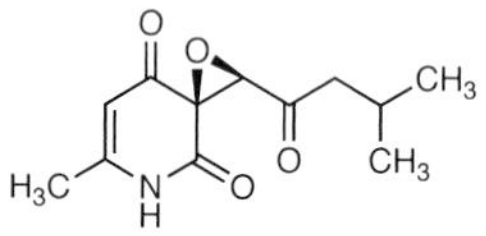

Flavipucine (**67**)

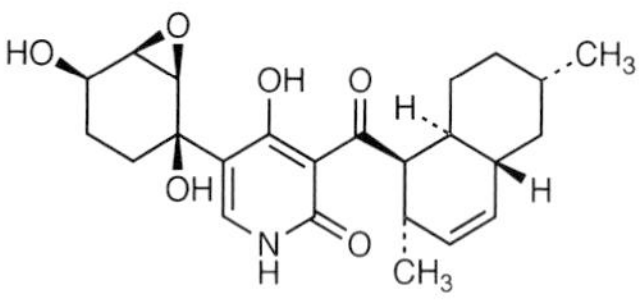

Apiosporamide (**68**)

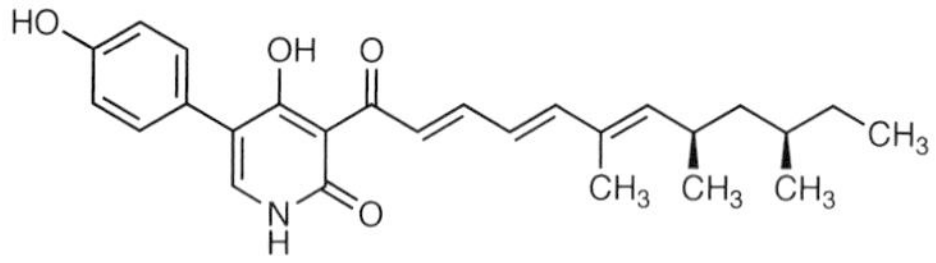

Militarinone D (**69**)

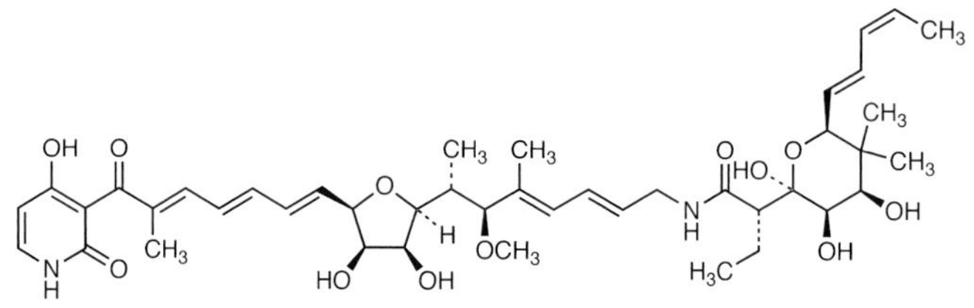

Kirromycin (**70**)

Fig. 12.17. Structures of 2-pyridones produced by microorganisms

As for Actinomycetes, most 2-pyridone-type products belong to the kirromycin (Fig. 12.17, **70**) group. Kirromycin (mocimycin), produced by *Streptomyces collinus* (Wolf and Zähner 1972), is an antibacterial antibiotic and inhibits protein synthesis interfering with transpeptidation by acting on elongation factor Tu (Parmeggiani and Nissen 2006).

2. Enzyme Inhibition and Biological Activity

The effects of 58–60 on electron transport enzymes are shown in Table 12.6 (Miyadera et al. 2003). Though they inhibited NFRD as potently as nafuredin (22), verticipyrone (37) and ukulactone A (52), they did not inhibit complex I. However, they inhibited *Ascaris* rhodoquinol-fumarate reductase and bovine heart succinate-ubiquinone reductase. Therefore, they are specific complex II inhibitors and not just selective for helminths. The inhibition against mammalian complex II is more potent than that of helminth complex II. Atpenins inhibited complex II more potently than 58, and 60 exhibited the most potent inhibition. Carboxin (1) is known as a potent complex II inhibitor (Mowery et al. 1977). However, the IC_{50} value against bovine heart succinate-ubiquinone reductase was 1.1 μM, which is 300 times weaker than that for 60. Therefore, atpenins may be useful tools for clarifying the biochemical and structural properties of complex II (Martens et al. 2005; Adebiyi et al. 2008).

X-ray crystallographic analyses of *Escherichia coli* and porcine heart complex II have been reported (Yankovskaya et al. 2003; Sun et al. 2005). Each complex is composed of four subunits, FAD-containing flavoprotein (Fp or SdhA), iron-sulfur protein (Ip or SdhB) and two membrane anchor subunits, CybL (SdhC) and CybS (SdhD). The succinate dehydrogenase catalytic portion is formed by Fp and Ip. The membrane anchor subunits are required for electron transfer to ubiquinone. Ubiquinone binds to complex II at the interface between Ip and the membrane anchor subunits. Kinetic analyses of atpenins revealed that they exhibited mixed inhibition with ubiquinone (K_i = 1.0 nM, K_i' = 5.9 nM for 60; Miyadera et al. 2003). This indicated that atpenins may block electron transfer between the enzyme and ubiquinone by binding to a region that partly overlaps with the physiological ubiquinone-binding site (Q-site). This can explain the observation that atpenins also affected the succinate dehydrogenase activity of bovine heart complex II (Table 12.6).

Compound 60 was co-crystallized with *E. coli* succinate-ubiquinone reductase and analyzed in detail by X-ray crystallography (Horsefield et al. 2006). Yankovskaya et al. (2003) observed that the interaction between ubiquinone at the Q-site appeared to be mediated solely by hydrogen bonding between the carbonyl oxygen (O1) of ubiquinone and the hydroxy group of tyrosine 83 in SdhD (Q_1-site). The co-crystallization study of complex II and 60 showed that 60 existed in the same hydrophobic pocket as ubiquinone but deeper within the pocket (Q_2-site). The protein–ligand docking model of complex II and ubiquinone was analyzed in silico, and it revealed that ubiquinone docked at the Q_2-site of complex II. At the Q_2-site, ubiquinone can interact with complex II via additional hydrogen bonds between carbonyl oxygen (O4) and the hydroxy group of serine 27 in SdhC and between 4-methoxy group and the imidazole of histidine 207 in SdhB. These interactions were observed in the co-crystallization result of complex II and 60. The above results support the proposition that the Q_1-site may be the initial binding site and the Q_2-site may be the catalytic site.

B. Other Complex II Inhibitors

Only a few complex II inhibitors have been found in microbial metabolites. Siccanin (Fig. 12.18, 12), produced by the fungus *Helminthosporium siccans*, showed antifungal activity, and it is used clinically for dermatophytosis as an ointment (Ishibashi 1962; Ishibashi et al. 1970). The structure can be regarded as derived from a *cis*-fused drimane condensed with orcinol. It showed 66% inhibition

Table 12.6. Effects of harzianopyridone and atpenins on electron transport enzymes

	Enzyme	Complex	IC_{50} (nM)		
			Harzianopyridone	Atpenin A4	Atpenin A5
Ascaris suum	NADH-fumarate reductase	I + II	1600	110	14
	NADH-quinone reductase	I	>100 000	>100 000	>100 000
	Rhodoquinol-fumarate reductase	II	360	220	12
Bovine heart	NADH–cytochrome-*c* reductase	I + III	420 000	140 000	82 000
	Succinate-ubiquinone reductase	II	17	11	3.6
	Succinate dehydrogenase	II	80	9.2	5.5

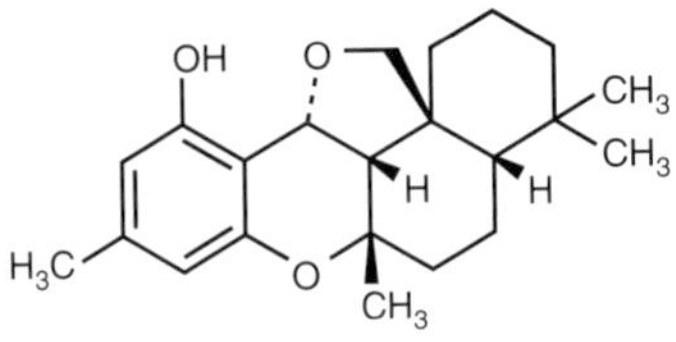

Siccanin (**12**) 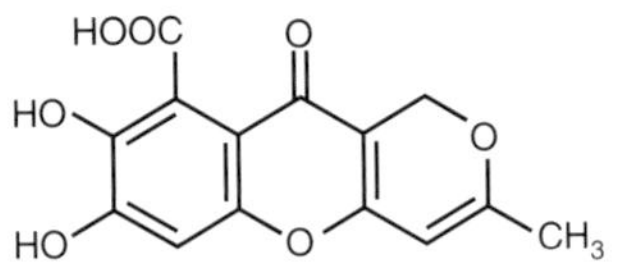   Anhydrofulvic acid (**71**)

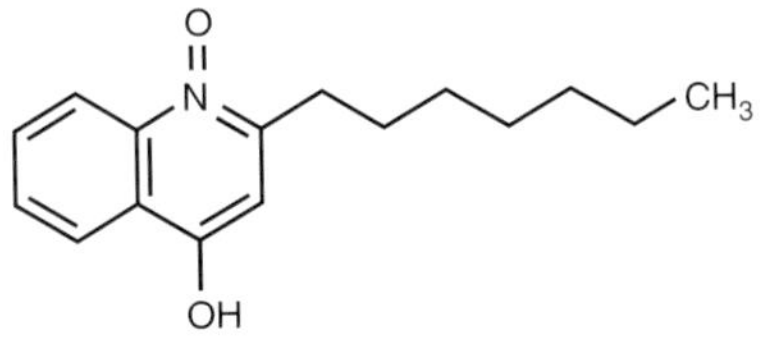

2-*n*-Heptyl-4-hydroxyquinoline *N*-oxide (**72**)

Fig. 12.18. Structures of complex II inhibitors produced by microorganisms

against the succinate dehydrogenase of a fungus, *Trichophyton mentagrophytes*, at 0.09 μM (Ishibashi et al. 1970; Nose and Endo 1971). Siccanin is a species-selective succinate dehydrogenase inhibitor, and it was effective against succinate dehydrogenases of *P. aeruginosa*, *P. putida*, rat and mouse mitochondria but ineffective or less effective against those of *E. coli*, *Corynebacterium glutamicum* and porcine mitochondria (Mogi et al. 2009). Anhydrofulvic acid (Fig. 12.18, **71**) was produced by *Penicillium* spp. and showed antifungal activity (Wrigley et al. 1994; Fujita et al. 1999). It exhibited 67% inhibition against succinate oxidase (complexes II +III+IV) of a fungus, *Candida utilis*, at 1.3 μM, but inhibited NADH oxidase (complexes I+III+IV) only weakly (18%) at 86 μM, which suggests **71** may be a complex II inhibitor (Fujita et al. 1999).

As for bacteria, 2-*n*-heptyl-4-hydroxyquinoline *N*-oxide (HQNO; Fig. 12.18, **72**) was isolated from the culture broth of *Pseudomonas aeruginosa* (Hays et al. 1945) and reported to inhibit bacterial complex II. It inhibited succinate-menaquinone reductase from *Bacillus subtilis* at the K_i value of 0.2 μM (Smirnova et al. 1995). Compound **72** is well known as a complex III inhibitor (Izzo et al. 1978).

V. Other Electron Transport Inhibitors

A. Inhibitors of Complex III

Complex III (ubiquinol–cytochrome-*c* reductase, cytochrome bc_1 complex) accepts electrons from

ubiquinol and passes them to cytochrome *c*, thereby transferring protons from mitochondrial matrix to intermembrane space. Mammalian complex III is a dimer with a molecular mass of 490 kDa, and each monomer consists of 11 subunits containing cytochrome *b*, cytochrome c_1 and an iron–sulfur protein. Cytochrome *b* has two quinone-binding sites, an ubiquinol oxidation site, Q_o site (Q_P site), and a ubiquinone reduction site, Q_i site (Q_N site). Most complex III inhibitors bind to either site.

The most famous Q_i site inhibitor is antimycin A_{3a} (**11**) produced by *Streptomyces* spp. As for fungal complex III inhibitors, funiculosin (Fig. 12.19, **65**) produced by *Penicillium funiculosum* (Ando et al. 1969) and ilicicolin H (Fig. 12.19, **64**) produced by *Cylindrocladium ilicicola* (Hayakawa et al. 1971) are Q_i site inhibitors (Rotsaert et al. 2008) and possess antifungal activities. While **65** inhibits both yeast and bovine heart complex III at $IC_{50} \sim 10$ nM (Rotsaert et al. 2008), the IC_{50} values of **64** against yeast and bovine heart complex III were 3–5 nM and 200–250 nM, respectively (Gutierrez-Cirlos et al. 2004). These results suggest a high degree of specificity in the determinants of ligand-binding at the Q_i site. Sambutoxin (Fig. 12.19, **66**) is structurally related to **64** and **65** and produced by *Fusarium sambucinum* (Kim et al. 1995). It was also reported to inhibit complex III (Kawai et al. 1997) and may be a Q_i site inhibitor. As shown in Sect. IV. B, HQNO (**72**) inhibits both complexes II and III. It is a Q_i site inhibitor, but the K_d against bovine heart mitochondria was about 3 orders of magnitude higher than that of antimycin (von Jagow and Link 1986).

The common Q_o site inhibitor, myxothiazol (Fig. 12.19, 73), is produced by the myxobacterium *Myxococcus fulvus* (Gerth et al. 1980). It has an *E*-β-methoxyacrylamide moiety. Similar *E*-β-methoxyacrylate moieties are found in some fungal metabolites, such as strobilurin A (mucidin, 7), and oudemansin A (Fig. 12.19, 74) produced by basidiomycetes. Compound 7 was originally isolated from culture broths of *Oudemansiella mucida* (Musilek et al. 1969) and *Strobilurus tenacellus* (Anke et al. 1977) and 74 from *Oudemansiella mucida* (Anke et al. 1979). All proved to be antifungal antibiotics. Their production and properties have been reviewed by Anke and Erkel (2002) in *The Mycota, Vol. X*. Compounds 7 and 74 inhibited bovine heart complex III at IC_{50} values of 65 nM and 290 nM, respectively (Brandt et al. 1988), and their targets are also Q_o site (von Jagow and Link 1986). While 7 inhibits both mammalian and fungal complex III, its toxicity against mammals is very weak. Therefore, many analogs of 7 have been synthesized, and some of them,

such as azoxystrobin (2), are commercially used for crop protection against phytopathogenic fungi (Sauter et al. 1999). Compounds having the same α-substituted methyl (*E*)-β-methoxyacrylate moiety as 7 and 74 were isolated from the myxobacteria *Cystobacter armeniaca* and *Archangium gephyra* in 2003 (Sasse et al. 2003). They are named cyrmenins, and cyrmenin B_1 (Fig. 12.19, 75) was shown to inhibit complex III.

As described in Sect. II, some inhibitors targeting quinone-binding sites of complex I inhibit complex III. Compound 7 inhibited bovine heart complex I weakly (15.5% inhibition at 3.3 μM) although the inhibition of 73 was more potent (84.0% inhibition at 3.0 μM). However, 2.9 μM of 65 showed no inhibition against complex I, and Q_o site inhibitors may affect complex I more potently than Q_i site inhibitors (Degli Esposti et al. 1993).

B. Inhibitors of Complex V

The electrochemical proton gradients produced by the electron transport chain drive complex V

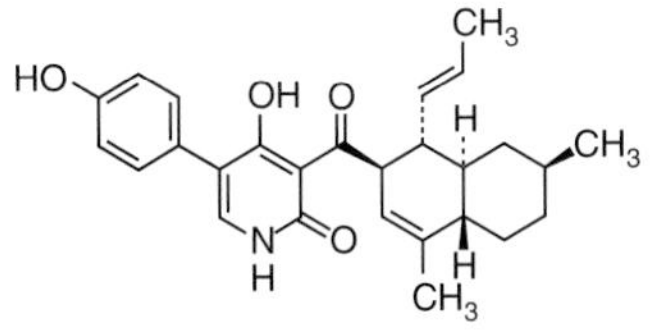

Ilicicolin H (**64**)

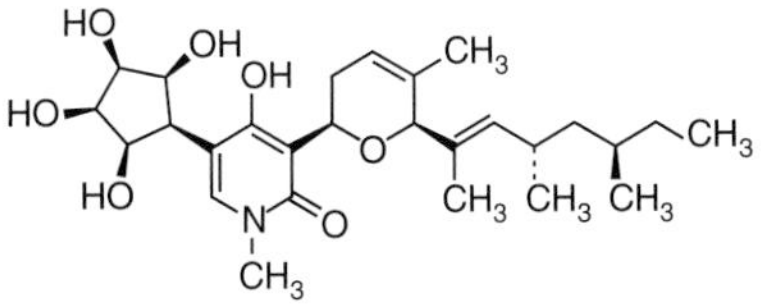

Funiculosin (**65**)

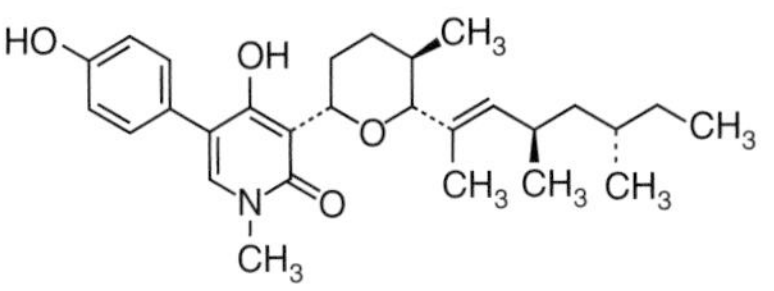

Sambutoxin (**66**)

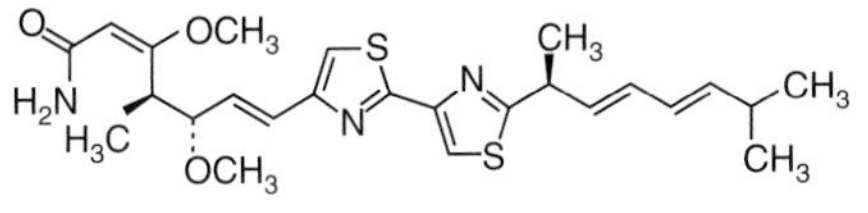

Myxothiazol A (**73**)

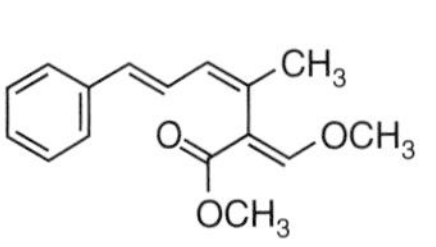

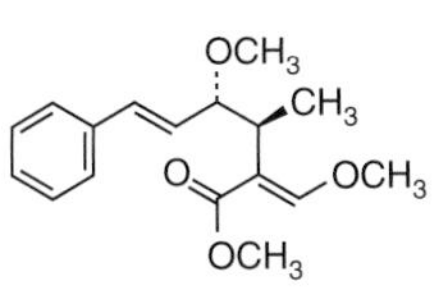

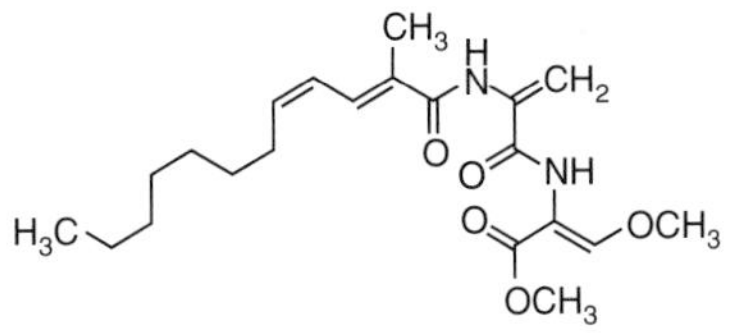

Strobilurin A (**7**)   Oudemansin A (**74**)   Cyrmenin B$_1$ (**75**)

Fig. 12.19. Structures of complex III inhibitors produced by microorganisms

(ATP synthase, $F_O F_1$-ATPase) to produce ATP in the critical process of oxidative phosphorylation. Mammalian complex V is suggested to be a dimeric protein, with each 600-kDa monomer consisting of 15 different protein subunits (Wittig and Schägger 2008). The monomer can be separated into the F_O domain and F_1 domain. The F_O domain is inserted in the membrane and translocates protons, while the F_1 domain protrudes into the matrix and synthesizes ATP. The F_1 domain is also called F_1-ATPase because it can hydrolyze ATP to ADP as the reverse reaction of ATP synthesis.

The name of F_O is derived from the oligomycin-sensitive factor (Racker 1963). Oligomycin A (Fig. 12.20, 76), produced by *Streptomyces diastatochromogenes*, inhibits proton transport through the F_O domain and is suggested to bind subunits *a* and *c* of the F_O domain (Devenish et al. 2000).

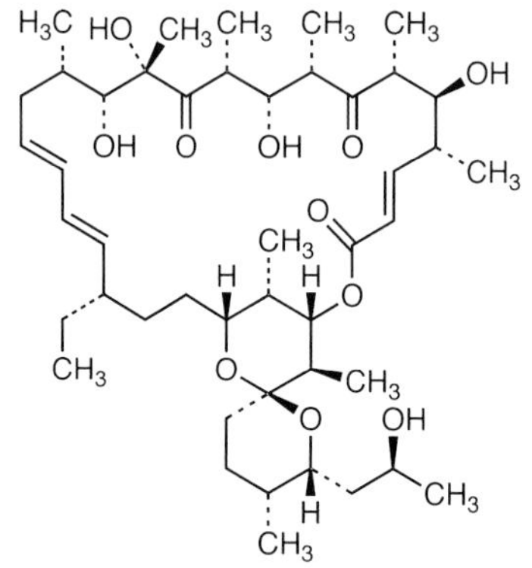

Oligomycin A (**76**)

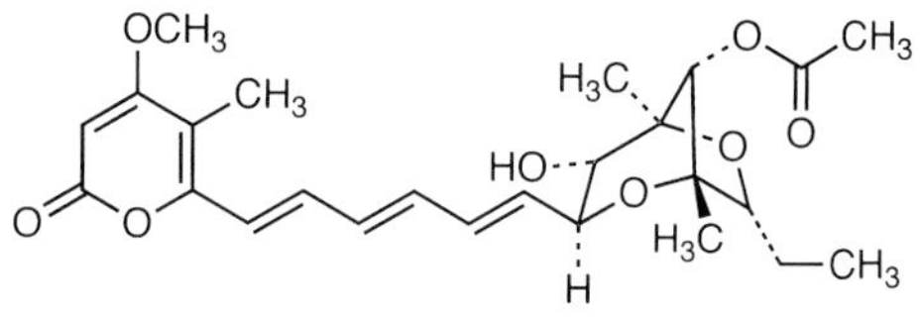

Aurovertin B (**77**)

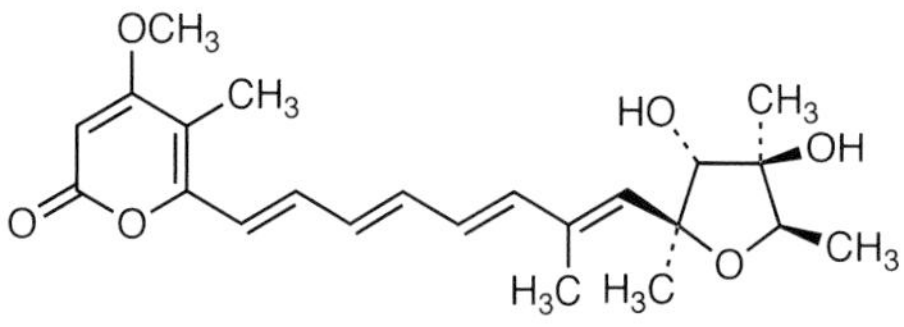

Citreoviridin (**78**)

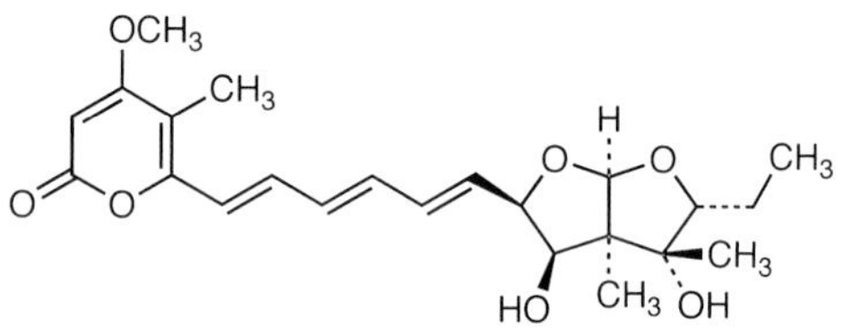

Asteltoxin (**79**)

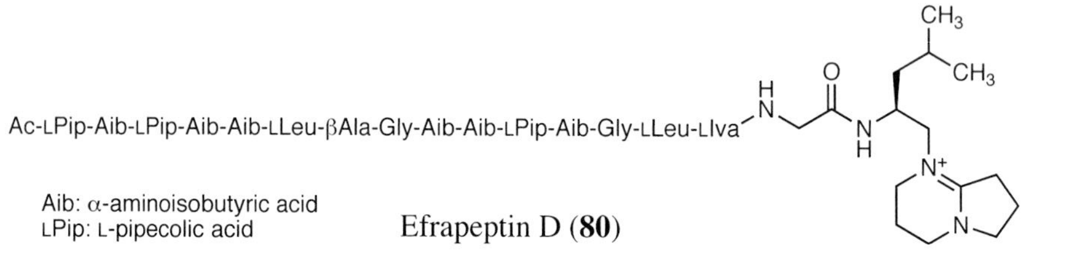

Ac-LPip-Aib-LPip-Aib-Aib-LLeu-βAla-Gly-Aib-Aib-LPip-Aib-Gly-LLeu-LIva

Aib: α-aminoisobutyric acid
LPip: L-pipecolic acid

Efrapeptin D (**80**)

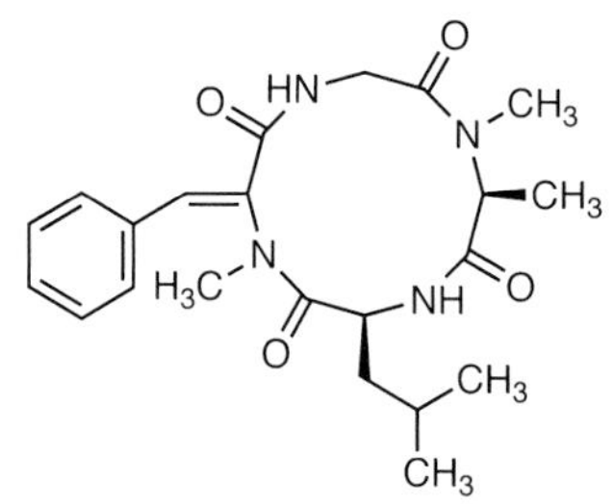

Tentoxin (**81**)

Fig. 12.20. Structures of complex V inhibitors produced by microorganisms

The subunit composition of the F_1 domain is $\alpha_3\beta_3\gamma_1\delta_1\epsilon_1$, and X-ray analysis of the bovine heart mitochondrial F_1 domain revealed its structure (Abrahams et al. 1994). Three β-subunits are catalytic. An ADP binds to the first subunit (β_{DP} subunit), and an ATP binds to the second subunit (β_{TP} subunit). No nucleotide binds to the third subunit (β_E subunit), and the three catalytic subunits interconvert through the cycle of conformations.

As for F_1-ATPase inhibitors, at least five inhibitory sites have been identified: nonhydrolysable NTP analog-binding site (catalytic site), aurovertin B-binding site, efrapeptin-binding site, natural inhibitor protein IF1-binding site and rhodamine 6G-binding site (Gledhill and Walker 2005). A mycotoxin, aurovertin B (Fig. 12.20, **77**), has been isolated from culture mycelia of the ascomycete *Calcarisporium arbuscula* (Osselton et al. 1974) and the basidiomycete *Albatrellus confluens* (Wang et al. 2005). The K_D value of **77** against bovine heart-soluble ATPase was 0.10 μM (Linnett and Beechey 1979). The co-crystallization study of F_1-ATPase and **77** revealed that **77** binds to bovine F_1 at two equivalent sites in the β_{TP} and β_E subunits in a cleft between the nucleotide binding and C-terminal domains (van Raaij et al.1996).

Compound **77** has a 4-methoxy-5-methyl-2-pyrone with a triene side chain at C-6. The same moiety is found in two other mycotoxins: citreoviridin (Fig. 12.20, **78**), produced by the fungus *Penicillium citreoviride* (Sakabe et al. 1964), and asteltoxin (Fig. 12.20, **79**), produced by the fungus *Aspergillus stellatus* (Kruger et al. 1979). The K_D value of **78** against bovine heart soluble ATPase was 3.1 μM (Linnett and Beechey 1979). The IC_{50} value of **79** against rat liver F_1-ATPase was about 0.5 μM (Kawai et al. 1997).

Efrapeptin D (Fig. 12.20, **80**) is produced by the fungus *Tolypocladium inflatum* (Jackson et al. 1979). Efrapeptins are α-aminoisobutyric acid (α-Aib)-rich peptides with acetylated N-terminus (Gupta et al. 1992). They are similar to fungal peptaibols (Whitmore and Wallace 2004) but their C-terminus is different. Instead of the amino alcohol C-terminus of peptaibols, efrapeptins have pyrrolo[1,2-*a*]pyrimidine moiety at the C-terminus. Efrapeptins are mycotoxins and their antifungal, insecticidal and antimalarial activities have been reported (Krasnoff et al. 1991; Nagaraj et al. 2001). Efrapeptins are potent F_1-ATPase

inhibitors. The K_D value of efrapeptins (major components were **80** and efrapeptins E, F and G) against bovine heart ATPase was 0.014 μM (Cross and Kohlbrenner 1978). The co-crystallization study of F_1-ATPase and efrapeptins revealed that efrapeptins made hydrophobic contact with the α-helical structure in the γ-subunit, which traversed the cavity, and with the β_E subunit and the two adjacent α subunits (Abrahams et al. 1996).

Tentoxin (Fig. 12.20, **81**) was isolated from a still culture broth of the fungus *Alternaria tenuis* as a chlorosis-inducing toxin (Saad et al. 1970). It is a cyclic tetrapeptide composed of *N*-methyl-L-alanine, L-leucine, (αZ)-α,β-didehydro-*N*-methyl-phenylalanine and glycine. It inhibits chloroplast F_1-ATPases (CF_1s) in sensitive species (e.g. lettuce) but not in insensitive species, such as radish (Linnett and Beechey (1979). It has no effect on bacterial and mitochondrial F_1-ATPases. The K_D value of **81** against lettuce CF_1 was 3–5 nM. The co-crystallization study of spinach CF_1 and **81** revealed that **81** bound to the αβ interface of the CF_1 in a cleft (Groth 2002). Single molecule studies of the $\alpha_3\beta_3\gamma$ complex of a cyanobacterium, *Thermosynechococcus elongatus*, with beads attached on the γ subunit suggested that **81** inhibited ATPase reaction after substrate binding to the β_E subunit by keeping the catalytic site in a closed conformation with bound **81** (Meiss et al. 2008).

C. Uncouplers

Respiration-dependent ATP synthesis is abolished by 2,4-dinitrophenol without inhibiting respiration itself. Compounds with this ability are called "uncouplers", and they selectively prevent utilization of electrochemical proton gradients derived from respiratory electron transport for net phosphorylation of ADP to ATP (Heytler 1979).

To date, some fungal uncouplers have been reported. ACR-toxin I (ACRL toxin I, Fig. 12.21, **82**) is a phytotoxin produced by *Alternaria citri* (Gardner et al. 1985) and *A. alternata* (Kohmoto et al. 1985). It is a δ-lactone, like nafuredin (**22**), and 1 μg/ml (2.8 μM) of **82** causes uncoupling of oxidative phosphorylation and changes in membrane potential in mitochondria from leaves of the susceptible rough lemon (*Citrus jambhiri* Lush., Akimitsu et al. 1989). Ohtani et al. (2002) discovered the ACR-toxin sensitivity gene (ACRS) in rough lemon mitochondrial DNA. Though ACRS was present in the genome of both toxin-sensitive and–insensitive citrus, the ACRS transcripts of insensitive plants were shorter than those of sensitive plants. It is suggested that the gene product of ACRS may be a pore-forming transmembrane receptor of **82** which leads to uncoupling.

Leucinostatin A (Fig. 12.21, **83**) is a nonapeptide produced by *Penicillium lilacinum* (Arai et al. 1973) and some other fungi. It showed an uncoupling effect against rat liver mitochondria at concentrations above 0.3 μM (Shima et al. 1990). However, **83** also inhibits ATPase activity at a lower concentration (0.2 μM). It facilitates the transport of monovalent and divalent cations with a half-maximal effect concentration in the range of 0.2–0.8 μM (Csermely et al. 1994). This ionophoric property of **83** may cause its uncoupling effect.

VI. Conclusions

Many inhibitors of electron transport and oxidative phosphorylation enzymes have been isolated from fungal cultures. Some of them, or their analogs, are used as medicines or agrochemicals. The inhibitors are also highly valuable for elucidating the mechanism of electron transport and oxidative phosphorylation systems. Specific inhibitors are used as important tools to study the systems. Recently, the X-ray crystallography structures of mitochondrial complexes II, III and IV and the F_1 domain of complex V have been reported. As for complex I, Sazanov and Hinchliffe (2006) reported

the crystal structure of the hydrophilic domain from a bacterium, *Thermus thermophilus*. Co-crystallization studies of the complexes and inhibitors clarified the mechanisms of both inhibition and electron transport.

Over the past few decades, mitochondria have attracted interest due to their relationship with various diseases; Leigh syndrome, paraganglioma, Parkinson's disease, Huntington's disease, Alzheimer disease and so on (Eng et al. 2003; Wallace 2005). Oxidative phosphorylation generates reactive oxygen species as toxic byproducts and they are suggested to cause a wide range of age-related disorders and various forms of cancer. The electron transport and oxidative phosphorylation inhibitors are extremely valuable tools for use in studies of mitochondria-associated diseases.

Recent research on the respiratory chain of the helminth *Ascaris suum* has shown that the mitochondrial NADH-fumarate reductase system has an important role in the anaerobic energy metabolism of adult parasites. Nafuredin (**22**) is a potent and selective inhibitor of complex I in this system, and it showed anthelmintic activity against *Haemonchus contortus* in an in vivo study. A verticipyrone analog (**42**) was effective against *Echinococcus multilocularis* in vitro. Such helminth-specific complex I inhibitors may be good lead compounds for anthelmintic drugs. A crystal of adult *A. suum* complex II (fumarate reductase), another component of the NADH-fumarate reductase system, was obtained recently (Shimizu et al. 2007), and analysis of parasite-specific factors in the enzyme is now in progress. This

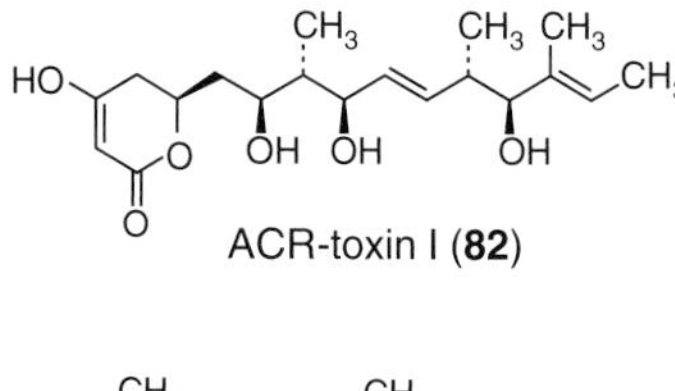

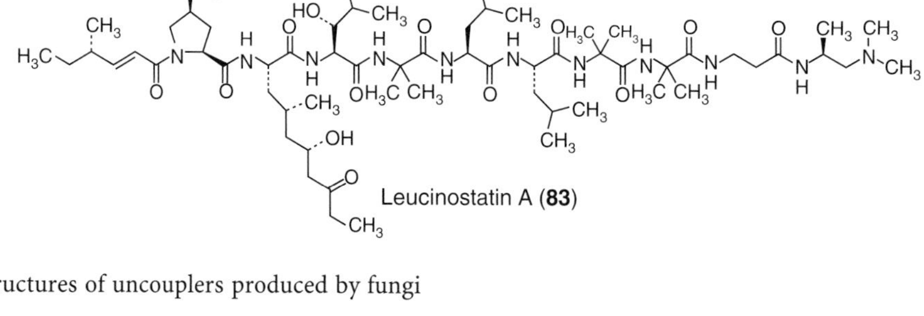

Fig. 12.21. Structures of uncouplers produced by fungi

information and the co-crystallization study of the enzyme with atpenin A5 (**60**) may clarify the interaction of **60** and helminth complex II and give a clue to the design of helminth-specific analogs of **60**. The first enantioselective total synthesis of atpenin A5 (**60**) has been achieved recently (Ohtawa et al. 2009). Thus helminth-specific inhibitors of elctron transport enzymes are good candidates for the treatment of heminthic diseases.

References

Abrahams JP, Leslie AG, Lutter R, Walker JE (1994) Structure at 2.8 Å resolution of F_1-ATPase from bovine heart mitochondria. Nature 370:621–628

Abrahams JP, Buchanan SK, van Raaij MJ, Fearnley IM, Leslie AGW, Walker JE (1996) The structure of bovine F_1-ATPase complexed with the peptide antibiotic efrapeptin. Proc Natl Acad Sci USA 93:9420–9424

Addor RW, Babcock TJ, Black BC, Brown DG, Diehl RE, Furch JA, Kameswaran V, Kamhi VM, Kremer KA, Kuhn DG, Lovell JB, Lowen GT, Miller TP, Peevey RM, Siddens JK, Treacy MF, Trotto SH, Wright DP (1992) Insecticidal pyrroles. Discovery and overview. ACS Symposium Ser 504:283–297

Adebiyi A, McNally EM, Jaggar JH (2008) Sulfonylurea receptor-dependent and -independent pathways mediate vasodilation induced by ATP-sensitive K^+ channel openers. Mol Pharmacol 74:736–743

Akimitsu K, Kohmoto K, Otani H, Nishimura S (1989) Host-specific effects of toxin from the rough lemon pathotype of *Alternaria alternata* on mitochondria. Plant Physiol 89:925–931

Alfatafta AA, Gloer JB, Scott JA, Malloch D (1994) Apiosporamide, a new antifungal agent from the coprophilous fungus *Apiospora montagnei*. J Nat Prod 57:1696–1702

Ando K, Suzuki S, Saeki T, Tamura G, Arima K (1969) Funiculosin, a new antibiotic. I. Isolation, biological and chemical properties. J Antibiot 22:189–194

Anke T, Erkel G (2002) Non-β-lactam antibiotics. In: Osiewacz HD (ed) Industrial applications. The Mycota X. Springer, Heidelberg, pp 93–108

Anke T, Oberwinkler F, Steglich W, Schramm G (1977) The strobilurins – new antifungal antibiotics from the basidiomycete *Strobilurus tenacellus* (Pers. ex Fr.) Sing. J Antibiot 30:806–810

Anke T, Hecht HJ, Schramm G, Steglich W (1979) Antibiotics from basidiomycetes. IX. Oudemansin, an antifungal antibiotic from *Oudemansiella mucida* (Schrader ex Fr.) Hoehnel (Agaricales). J Antibiot 32:1112–1117

Arai M, Tomoda H, Okuda T, Wang H, Tabata N, Masuma R, Yamaguchi Y, Omura S (2002) Funicone-related compounds, potentiators of antifungal miconazole activity, produced by *Talaromyces flavus* FKI-0076. J Antibiot 55:172–180

Arai T, Mikami Y, Fukushima K, Utsumi T, Yazawa K (1973) New antibiotic, leucinostatin, derived from *Penicillium lilacinum*. J Antibiot 26:157–161

Brandt U, Schägger H, von Jagow G (1988) Characterization of binding of the methoxyacrylate inhibitors to mitochondrial cytochrome *c* reductase. Eur J Biochem 173:499–506

Carroll J, Fearnley IM, Skehel JM, Shannon RJ, Hirst J, Walker JE (2006) Bovine complex I is a complex of 45 different subunits. J Biol Chem 281:32724–32727

Carruthers JR, Cerrini S, Fedeli W, Casinovi CG, Galeffi C, Vaccaro AMT, Scala A (1971) Structures of cochlioquinones A and B, new metabolites of *Cochliobolus miyabeanus*: chemical and X-ray crystallographic determination. J Chem Soc Chem Commun 164-166

Cassinelli G, Grein A, Orezzi P, Pennella P, Sanfilippo A (1966) New antibiotics produced by *Streptoverticillium orinoci*. Arch Mikrobiol 55:358–368

Cross RL, Kohlbrenner WE (1978) The mode of inhibition of oxidative phosphorylation by efrapeptin (A23871). Evidence for an alternating site mechanism for ATP synthesis. J Biol Chem 253:4865–4873

Csermely P, Radics L, Rossi C, Szamel M, Ricci M, Mihály K, Somogyi J (1994) The nonapeptide leucinostatin A acts as a weak ionophore and as an immunosuppressant on T lymphocytes. Biochim Biophys Acta 1221:125–132

Cutler HG, Jacyno JM (1991) Biological activity of (–)-harzianopyridone isolated from *Trichoderma harzianum*. Agric Biol Chem 55:2629–2631

Degli Esposti M (1998) Inhibitors of NADH-ubiquinone reductase: an overview. Biochim Biophys Acta 1364:222–235

Degli Esposti M, Ghelli A (1999) Ubiquinone and inhibitor sites in complex I: one, two or three?. Biochem Soc Trans 27:606–609

Degli Esposti M, Ghelli A, Crimi M, Estornell E, Fato R, Lenaz G (1993) Complex I and complex III of mitochondria have common inhibitors acting as ubiquinone antagonists. Biochem Biophys Res Commun 190:1090–1096

Desai K, Sullards MC, Allegood J, Wang E, Schmelz EM, Hartl M, Humpf H-U, Liotta DC, Peng Q, Merrill AH Jr (2002) Fumonisins and fumonisin analogs as inhibitors of ceramide synthase and inducers of apoptosis. Biochim Biophys Acta 1585:188–192

Devenish RJ, Prescott M, Boyle GM, Nagley P (2000) The oligomycin axis of mitochondrial ATP synthase: OSCP and the proton channel. J Bioenerg Biomembr 32:507–515

Dickinson JM, Hanson JR, Hitchcock PB, Claydon N (1989) Structure and biosynthesis of harzianopyridone, an antifungal metabolite of *Trichoderma harzianum*. J Chem Soc Perkin Trans 1:1885–1887

Dong Y, Lin J, Lu X, Liu M, Li Y, Ren X, Cui X, Shi Y, Zheng Z, Zhu J, Zhang H, He J (2007) Novel polyene compound with anti-tumor and antivirus effects, and preparation method and application thereof. China Patent Appl CN101081848

Eng C, Kiuru M, Fernandez MJ, Aaltonen LA (2003) A role for mitochondrial enzymes in inherited neoplasia and beyond. Nat Rev Cancer 3:193–202

Engler M, Anke T, Sterner O, Brandt U (1997) Pterulinic acid and pterulone, two novel inhibitors of NADH:

ubiquinone oxidoreductase (complex I) produced by a *Pterula* species. I. Production, isolation and biological activities. J Antibiot 50:325–329

Findlay JA, Radics L (1972) Flavipucine [3′-isovaleryl-6-methylpyridine-3-spiro-2′-oxirane-2(1*H*),4(3*H*)-dione], an antibiotic from *Aspergillus flavipes*. J Chem Soc Perkin Trans 1:2071–2074

Friedrich T, van Heek P, Leif H, Ohnishi T, Forche E, Kunze B, Jansen R, Trowitzsch-Kienast W, Höfle G, Reichenbach H, Weiss H (1994) Two binding sites of inhibitors in NADH:ubiquinone oxidoreductase (complex I). Relationship of one site with the ubiquinone-binding site of bacterial glucose:ubiquinone oxidoreductase. Eur J Biochem 219:691–698

Fujita K, Nagamine Y, Ping X, Taniguchi M (1999) Mode of action of anhydrofulvic acid against *Candida utilis* ATCC 42402 under acidic condition. J Antibiot 52:628–634

Fuska J, Fuskova A, Nemec P (1979) Vermistatin, an antibiotic with cytotoxic effects, produced by *Penicillium vermiculatum*. Biologia 34:735–739

Gardner JM, Kono Y, Tatum JH, Suzuki Y, Takeuchi S (1985) Structure of the major component of ACRL-toxins, host-specific pathotoxic compounds produced by *Alternaria citri*. Agric Biol Chem 49:1235–1238

Gelderblom WCA, Jaskiewicz K, Marasas WFO, Thiel PG, Horak RM, Vleggaar R, Kriek NPJ (1988) Fumonisins-novel mycotoxins with cancer-promoting activity produced by *Fusarium moniliforme*. Appl Envir Microbiol 54:1806–1811

Gerth K, Irschik H, Reichenbach H, Trowitzsch W (1980) Myxothiazol, an antibiotic from *Myxococcus fulvus* (Myxobacterales). I. Cultivation, isolation, physicochemical and biological properties. J Antibiot 33:1474–1479

Gledhill JR, Walker JE (2005) Inhibition sites in F_1-ATPase from bovine heart mitochondria. Biochem J 386:591–598

Groth G (2002) Structure of spinach chloroplast F_1-ATPase complexed with the phytopathogenic inhibitor tentoxin. Proc Natl Acad Sci USA 99:3464–3468

Gupta S, Krasnoff SB, Roberts DW, Renwick JAA, Brinen LS, Clardy J (1992) Structure of efrapeptins from the fungus *Tolypocladium niveum*: peptide inhibitors of mitochondrial ATPase. J Org Chem 57:2306–2313

Gutierrez-Cirlos EB, Merbitz-Zahradnik T, Trumpower BL (2004) Inhibition of the yeast cytochrome bc_1 complex by ilicicolin H, a novel inhibitor that acts at the Qn site of the bc_1 complex. J Biol Chem 279:8708–8714

Hamamoto T, Gunji S, Tsuji H, Beppu T (1983) Leptomycins A and B, new antifungal antibiotics. I. Taxonomy of the producing strain and their fermentation, purification and characterization. J Antibiot 36:639–645

Hayakawa S, Minato H, Katagiri K (1971) Ilicicolins, antibiotics from *Cylindrocladium ilicicola*. J Antibiot 24:653–654

Hays EE, Wells IC, Katzman PA, Cain CK, Jacobs FA, Thayer SA, Doisy EA, Gaby WL, Roberts EC, Muir RD, Carroll CJ, Jones LR, Wade NJ (1945) Antibiotic substances produced by *Pseudomonas aeruginosa*. J Biol Chem 159:725–750

Heytler PG (1979) Uncouplers of oxydative phosphorylation. In: Fleischer S, Packer L (eds) Methods in enzymology. Biomembranes part F: bioenergetics – oxidative phosphorylation, vol 55. Academic, New York, pp 462–472

Horsefield R, Yankovskaya V, Sexton G, Whittingham W, Shiomi K, Ōmura S, Byrne B, Cecchini G, Iwata S (2006) Structural and computational analysis of the quinone-binding site of complex II (succinate-ubiquinone oxidoreductase): A mechanism of electron transfer and proton conduction during ubiquinone reduction. J Biol Chem 281:7309–7316

Ishibashi K (1962) Studies on antibiotics from *Helminthosporium* sp. fungi. VII. Siccanin, a new antifungal antibiotic produced by *Helminthosporium siccans*. J Antibiot Ser A 15:161–167

Ishibashi K, Hirai K, Arai M, Sugawara S, Endo A, Yasumura A, Masuda H, Muramatsu T (1970) Siccanin, a new antifungal antibiotic. Ann Sankyo Res Lab 22:1–33

Izzo G, Guerrieri F, Papa S (1978) On the mechanism of inhibition of the respiratory chain by 2-heptyl-4-hydroxyquinoline-*N*-oxide. FEBS Lett 93:320–322

Jackson CG, Linnett PE, Beechey RB, Henderson PJF (1979) Purification and preliminary structure analysis of the efrapeptins, a group of antibiotics that inhibit the mitochondrial adenosine triphosphatase. Biochem Soc Trans 7:224–226

Kawai K, Fukushima H, Nozawa Y (1985) Inhibition of mitochondrial respiration by asteltoxin, a respiratory toxin from *Emericella variecolor*. Toxicol Lett 28:73–77

Kawai K, Suzuki T, Kitagawa A, Kim J-C, Lee Y-W (1997) A novel respiratory chain inhibitor, sambutoxin from *Fusarium sambucinum*. Cereal Res Commun 25:325–326

Kim J-C, Lee Y-W, Tamura H, Yoshizawa T (1995) Sambutoxin: a new mycotoxin isolated from *Fusarium sambucinum*. Tetrahedron Lett 36:1047–1050

Kita K, Miyadera H, Saruta F, Miyoshi H (2001) Parasite mitochondria as a target for chemotherapy. J Health Sci 47:219–239

Kita K, Shiomi K, Ōmura S (2007) Parasitology in Japan. Advances in drug discovery and biochemical studies. Trends Parasitol 23:223–229

Kobayashi S, Tsuchiya K, Harada T, Nishide M, Kurokawa T, Nakagawa T, Shimada N, Kobayashi K (1994) Pironetin, a novel plant growth regulator produced by *Streptomyces* sp. NK 10958. I. Taxonomy, production, isolation and preliminary characterization. J Antibiot 47:697–702

Kohmoto K, Kohguchi T, Kondoh Y, Otani H, Nishimura S, Nakatsuka S, Goto T (1985) The mitochondrion: the prime site for a host-selective toxin (ACR-toxin I) produced by *Alternaria alternata* pathogenic to rough lemon. Proc Jpn Acad, Ser B 61:269–272

Komuniecki R, Tielens AGM (2003) Carbohydrate and energy metabolism in parasitic helminths. In: Marr JJ, Nielsen TW, Komuniecki RW (eds) Molecular medical parasitology. Academic, London, pp 339–358

Kondoh M, Usui T, Nishikiori T, Mayumi T, Osada H (1999) Apoptosis induction via microtubule disassembly by an antitumor compound, pironetin. Biochem J 340:411–416

Krasnoff SB, Gupta S, St. Leger RJ, Renwick JAA, Roberts DW (1991) Antifungal and insecticidal properties of the efrapeptins: metabolites of the fungus *Tolypocladium niveum*. J Invertebr Pathol 58:180–188

Kruger GJ, Steyn PS, Vleggaar R, Rabie CJ (1979) X-ray crystal structure of asteltoxin, a novel mycotoxin from *Aspergillus stellatus* Curzi. J Chem Soc Chem Commun 1979:441–442

Kumagai H, Nishida H, Imamura N, Tomoda H, Ōmura S, Bordner J (1990) The structures of atpenins A4, A5 and B, new antifungal antibiotics produced by *Penicillium* sp. J Antibiot 43:1553–1558

Kunze B, Jansen R, Pridzun L, Jurkiewicz E, Hunsmann G, Höfle G, Reichenbach H (1992) Phenoxan, a new oxazole-pyrone from myxobacteria: production, antimicrobial activity and its inhibition of the electron transport in complex I of the respiratory chain. J Antibiot 45:1549–1552

Lang G, Wiese J, Schmaljohann R, Imhoff JF (2007) New pentaenes from the sponge-derived marine fungus *Penicillium rugulosum*: structure determination and biosynthetic studies. Tetrahedron 63:11844–11849

Lardy HA (1980) Antibiotic inhibitors of mitochondrial energy transfer. Pharmacol Ther 11:649–660

Li D-H, Cai S-X, Tian L, Lin Z-J, Zhu T-J, Fang Y-C, Liu P-P, Gu Q-Q, Zhu W-M (2007) Two new metabolites with cytotoxicities from deep-sea fungus, *Aspergillus sydowi* YH11-2. Arch Pharm Res 30:1051–1054

Lim C-H, Ueno H, Miyoshi H, Miyagawa H, Iwamura H, Ueno T (1996) Phytotoxic compounds cochlioquinones are inhibitors of mitochondrial NADH-ubiquinone reductase. J Pestic Sci 21:213–215

Linnett PE, Beechey RB (1979) Inhibitors of the ATP synthetase system. In: Fleischer S, Packer L (eds) Methods in enzymology. Biomembranes part F: bioenergetics – oxidative phosphorylation, vol. 55. Academic, New York, pp 472–518

Martens GA, Cai Y, Hinke S, Stange G, Van de Casteele M, Pipeleers D (2005) Glucose suppresses superoxide generation in metabolically responsive pancreatic β cells. J Biol Chem 280:20389–20396

Masuma R, Yamaguchi Y, Noumi M, Ōmura S, Namikoshi M (2001) Effect of sea water on hyphal growth and antimicrobial metabolite production in a marine fungi. Mycoscience 42:455–459

Matsumoto J, Sakamoto K, Shinjyo N, Kido Y, Yamamoto N, Yagi K, Miyoshi H, Nonaka N, Katakura K, Kita K, Oku Y (2008) Anaerobic NADH-fumarate reductase system is predominant in the respiratory chain of *Echinococcus multilocularis*, providing a novel target for the chemotherapy of alveolar echinococcosis. Antimicrob Agents Chemother 52:164–170

Meiss E, Konno H, Groth G, Hisabori T (2008) Molecular processes of inhibition and stimulation of ATP synthase caused by the phytotoxin tentoxin. J Biol Chem 283:24594–24599

Merlini L, Nasini G, Selva A (1970) Structure of funicone, a new metabolite from *Penicillium funiculosum*. Tetrahedron 26:2739–2749

Miyadera H, Shiomi K, Ui H, Yamaguchi Y, Masuma R, Tomoda H, Miyoshi H, Osanai A, Kita K, Ōmura S (2003) Atpenins, potent and specific inhibitors of mitochondrial complex II (succinate-ubiquinone oxidoreductase). Proc Natl Acad Sci USA 100:473–477

Mogi T, Kawakami T, Arai H, Igarashi Y, Matsushita K, Mori M, Shiomi K, Ōmura S, Harada S, Kita K (2009) Siccanin rediscovered as a species-selective succinate dehydrogenase inhibitor. J Biochem 15

Mowery PC, Steenkamp DJ, Ackrell AC, Singer TP, White GA (1977) Inhibition of mammalian succinate dehydrogenase by carboxins. Arch Biochem Biophys 178:495–506

Musilek V, Cerna J, Sasek V, Semerdzieva M, Vondracek M (1969) Antifungal antibiotic of the basidiomycete, *Oudemansiella mucida*. I. Isolation and cultivation of a producing strain. Folia Microbiol 14:377–387

Nagamitsu T, Takano D, Shiomi K, Ui H, Yamaguchi Y, Masuma R, Harigaya Y, Kuwajima I, Ōmura S (2003) Total synthesis of nafuredin-γ, a γ-lactone related to nafuredin with selective inhibitory activity against NADH-fumarate reductase. Tetrahedron Lett 44: 6441–6444

Nagamitsu T, Takano D, Seki M, Arima S, Ohtawa M, Shiomi K, Harigaya Y, Ōmura S (2008) The total synthesis and biological evaluation of nafuredin-γ and its analogues. Tetrahedron 64:8117–8127

Nagaraj G, Uma MV, Shivayogi MS, Balaram H (2001) Antimalarial activities of peptide antibiotics isolated from fungi. Antimicrob Agents Chemother 45:145–149

Nago H, Matsumoto M, Nakai S (1993) Degradation pathway of 2-deceno-δ-lactone by the lactone-producing fungus, *Fusarium solani*. Biosci Biotechnol Biochem 57:2111–2115

Nose K, Endo A (1971) Mode of action of the antibiotic siccanin on intact cells and mitochondria of *Trichophyton mentagrophytes*. J Bacteriol 105:176–184

Ohtani K, Yamamoto H, Akimitsu K (2002) Sensitivity to *Alternaria alternata* toxin in citrus because of altered mitochondrial RNA processing. Proc Natl Acad Sci USA 99:2439–2444

Ohtawa M, Ogihara S, Sugiyama K, Shiomi K, Harigaya Y, Nagamitsu T, Ōmura S (2009) Enantioselective total synthesis of atpenin A5. J Antibiot 62 (in press)

Okun JG, Lummen P, Brandt U (1999) Three classes of inhibitors share a common binding domain in mitochondrial complex I (NADH:ubiquinone oxidoreductase). J Biol Chem 274:2625–2630

Ōmura S, Shiomi K (2007) Discovery, chemistry, and chemical biology of microbial products. Pure Appl Chem 79:581–591

Ōmura S, Tomoda H, Kimura K, Zhen DZ, Kumagai H, Igarashi K, Imamura N, Takahashi Y, Tanaka Y, Iwai Y (1988) Atpenins, new antifungal antibiotics produced by *Penicillium* sp. Production, isolation, physicochemical and biological properties. J Antibiot 41:1769–1773

Ōmura S, Miyadera H, Ui H, Shiomi K, Yamaguchi Y, Masuma R, Nagamitsu T, Takano D, Sunazuka T, Harder A, Kölbl H, Namikoshi M, Miyoshi H, Sakamoto K, Kita K (2001) An anthelmintic compound, nafuredin, shows selective inhibition of complex I in helminth mitochondria. Proc Natl Acad Sci USA 98:60–62

Ōmura S, Shiomi K, Masuma R (2007) Novel FKI-3389 substance and method for producing the same. Patent Appl WO2007/148412

Oshino K, Kumagai H, Tomoda H, Ōmura S (1990) Mechanism of action of atpenin B on Raji cells. J Antibiot 43:1064–1068

Osselton MD, Baum H, Beechey RB (1974) Isolation, purification, and characterization of aurovertin B. Biochem Soc Trans 2:200–202

Otsuka T, Takase S, Terano H, Okuhara M (1992) New angiogenesis inhibitors, WF-16775 A_1 and A_2. J Antibiot 45:1970–1973

Parmeggiani A, Nissen P (2006) Elongation factor Tu-targeted antibiotics: Four different structures, two mechanisms of action. FEBS Lett 580:4576–4581

Racker E (1963) A mitochondrial factor conferring oligomycin sensitivity on soluble mitochondrial ATPase. Biochem Biophys Res Commun 10:435–439

Roberge M, Tudan C, Hung SMF, Harder KW, Jirik FR, Anderson H (1994) Antitumor drug fostriecin inhibits the mitotic entry checkpoint and protein phosphatases 1 and 2A. Cancer Res 54:6115–6121

Rotsaert FAJ, Ding MG, Trumpower BL (2008) Differential efficacy of inhibition of mitochondrial and bacterial cytochrome bc_1 complexes by center N inhibitors antimycin, ilicicolin H and funiculosin. Biochim Biophys Acta 1777:211–219

Saad SM, Halloin JM, Hagedorn DJ (1970) Production, purification, and bioassay of tentoxin. Phytopathology 60:415–418

Sakabe N, Goto T, Hirata Y (1964) Structure of citreoviridin, a toxic compound produced by *Penicillium citreoviride* on rice. Tetrahedron Lett 1825–1830

Saraste M (1999) Oxidative phosphorylation at the *fin de siècle*. Science 283:1488–1493

Sassa T, Nukina M, Suzuki Y (1991) Deoxyfunicone, a new γ-pyrone metabolite from a resorcylide-producing fungus (*Penicillium* sp.). Agric Biol Chem 55:2415–2416

Sasse F, Leibold T, Kunze B, Höfle G, Reichenbach H (2003) Cyrmenins, new β-methoxyacrylate inhibitors of the electron transport. Production, isolation, physico-chemical and biological properties. J Antibiot 56:827–831

Sauter H, Steglich W, Anke T (1999) Strobilurins: evolution of a new class of active substances. Angew Chem Int Ed 38:1329–1349

Sazanov LA, Hinchliffe P (2006) Structure of the hydrophilic domain of respiratory complex I from *Thermus thermophilus*. Science 311:1430–1436

Schaeffer JM, Frazier EG, Bergstrom AR, Williamson JM, Liesch JM, Goetz MA (1990) Cochlioquinone A, a nematocidal agent which competes for specific [^{3}H] ivermectin binding sites. J Antibiot 43:1179–1182

Shima A, Fukushima K, Arai T, Terada H (1990) Dual inhibitory effects of the peptide antibiotics leucinostatins on oxidative phosphorylation in mitochondria. Cell Struct Funct 15:53–58

Shimamura H, Sunazuka T, Izuhara T, Hirose T, Shiomi K, Ōmura S (2007) Total synthesis and biological evaluation of verticipyrone and analogues. Org Lett 9:65–67

Shimizu H, Harada S, Osanai A, Inaoka DK, Otani H, Sakamoto K, Kita K (2007) Crystal structure analysis of *Ascaris suum* adult mitochondrial complex II. Int Symp Diffract Struct Biol 2:P015

Shimomura Y, Kawada T, Suzuki M (1989) Capsaicin and its analogs inhibit the activity of NADH-coenzyme Q oxidoreductase of the mitochondrial respiratory chain. Arch Biochem Biophys 270:573–577

Shiomi K, Ōmura S (2004) Antiparasitic agents produced by microorganisms. Proc Jpn Acad Ser B 80B:245–258

Shiomi K, Ui H, Suzuki H, Hatano H, Nagamitsu T, Takano D, Miyadera H, Yamashita T, Kita K, Miyoshi H, Harder A, Tomoda H, Ōmura S (2005) A γ-lactone form nafuredin, nafuredin-γ, also inhibits helminth complex I. J Antibiot 58:50–55

Smirnova IA, Hägerhäll C, Konstantinov AA, Hederstedt L (1995) HOQNO interaction with cytochrome *b* in succinate:menaquinone oxidoreductase from *Bacillus subtilis*. FEBS Lett 359:23–26

Sun F, Huo X, Zhai Y, Wang A, Xu J, Su D, Bartlam M, Rao Z (2005) Crystal structure of mitochondrial respiratory membrane protein complex II. Cell 121:1043–1057

Takano D, Nagamitsu T, Ui H, Shiomi K, Yamaguchi Y, Masuma R, Kuwajima I, Ōmura S (2001a) Absolute configuration of nafuredin, a new specific NADH-fumarate reductase inhibitor. Tetrahedron Lett 42:3017–3020

Takano D, Nagamitsu T, Ui H, Shiomi K, Yamaguchi Y, Masuma R, Kuwajima I, Ōmura S (2001b) Total synthesis of nafuredin, a selective NADH-fumarate reductase inhibitor. Org Lett 3:2289–2291

Tan AK, Ramsay RR, Singer TP, Miyoshi H (1993) Comparison of the structures of the quinone-binding sites in beef heart mitochondria. J Biol Chem 268:19328–19333

Trowitzsch-Kienast W, Forche E, Wray V, Reichenbach H, Jurkiewicz E, Hunsmann G, Höfle G (1992) Antibiotika aus Gleitenden Bakterien, 45. Phenalamide, neue HIV-1-Inhibitoren aus *Myxococcus stipitatus* Mx s40. Liebigs Ann Chem 659–664

Tsukamoto S, Hirota H, Imachi M, Fujimuro M, Onuki H, Ohta T, Yokosawa H (2005) Himeic acid A: A new ubiquitin-activating enzyme inhibitor isolated from a marine-derived fungus, *Aspergillus* sp. Bioorg Med Chem Lett (2005) 15:191–194

Tunac JB, Graham BD, Dobson WE (1983) Novel antitumor agents CI-920, PD 113,270 and PD 113,271. I. Taxonomy, fermentation and biological properties. J Antibiot 36:1595–1600

Ueki M, Machida K, Taniguchi M (2000) Antifungal inhibitors of mitochondrial respiration: discovery and prospects for development. Curr Opin Anti-Infect Invest Drugs 2:387–398

Ui H, Shiomi K, Yamaguchi Y, Masuma R, Nagamitsu T, Takano D, Sunazuka T, Namikoshi M, Ōmura S (2001) Nafuredin, a novel inhibitor of NADH-fumarate reductase, produced by *Aspergillus niger* FT-0554. J Antibiot 54:234–238

Ui H, Shiomi K, Suzuki H, Hatano H, Morimoto H, Yamaguchi Y, Masuma R, Sakamoto K, Kita K, Miyoshi H,

Tomoda H, Tanaka H, Ōmura S (2006a) Paecilaminol, a new NADH-fumarate reductase inhibitor, produced by *Paecilomyces* sp. FKI-0550. J Antibiot 59:591–596

Ui H, Shiomi K, Suzuki H, Hatano H, Morimoto H, Yamaguchi Y, Masuma R, Sunazuka T, Shimamura H, Sakamoto K, Kita K, Miyoshi H, Tomoda H, Ōmura S (2006b) Verticipyrone, a new NADH-fumarate reductase inhibitor, produced by *Verticillium* sp. FKI-1083. J Antibiot 59:785–790

Umezawa I, Komiyama K, Oka H, Okada K, Tomisaka S, Miyano T, Takano S (1984) A new antitumor antibiotic, kazusamycin. J Antibiot 37:706–711

Van Hellemond JJ, Klockiewicz M, Gaasenbeek CPH, Roos MH, Tielens AGM (1995) Rhodoquinone and complex II of the electron transport chain in anaerobically functioning eukaryotes. J Biol Chem 270:31065–31070

van Raaij MJ, Abrahams JP, Leslie AGW, Walker JE (1996) The structure of bovine F_1-ATPase complexed with the antibiotic inhibitor aurovertin B. Proc Natl Acad Sci USA 93:6913–6917

von Jagow G, Link TA (1986) Use of specific inhibitors on the mitochondrial bc_1 complex. In: Fleischer S, Fleischer B (eds) Methods in enzymology, vol 126. Biomembranes, part N. Transport in bacteria, mitochondria, and chloroplasts: protonmotive force. Academic, London, pp 253–271

Wallace DC (2005) A mitochondrial paradigm of metabolic and degenerative diseases, aging, and cancer: a dawn for evolutionary medicine. Annu Rev Genet 39:359–407

Wang F, Luo D-Q, Liu J-K (2005) Aurovertin E, a new polyene pyrone from the basidiomycete *Albatrellus confluens*. J Antibiot 58:412–415

Washizu F, Umezawa H, Sugiyama N (1954) Chemical studies on a toxic product of *Streptomyces thioluteus*, aureothin. J Antibiot Ser A 7:60

Wei H, Itoh T, Kinoshita M, Kotoku N, Aoki S, Kobayashi M (2005) Shimalactone A, a novel polyketide, from marine-derived fungus *Emericella variecolor* GF10. Tetrahedron 61:8054–8058

Wessely F, Schönol K, Isemann W (1936) Zur Kenntnis der Bitterstoffe der Colombowurzel III. Über das Palmarin. Monats Chem 68:21–28

Whitmore L, Wallace BA (2004) The peptaibol database: a database for sequences and structures of naturally occurring peptaibols. Nucleic Acids Res 32:D593–D594

Wittig I, Schägger H (2008) Structural organization of mitochondrial ATP synthase. Biochim Biophys Acta 1777:592–598

Wolf H, Zähner H (1972) Metabolic products of microorganisms. 99. Kirromycin. Arch Mikrobiol 83:147–154

Wolff B, Sangliere J-J, Wang Y (1997) Leptomycin B is an inhibitor of nuclear export: inhibition of nucleocytoplasmic translocation of the human immunodeficiency virus type 1 (HIV-1) Rev protein and Rev-dependent mRNA. Chem Biol 4:139–147

Wrigley SK, Latif MA, Gibson TM, Chicarelli-Robinson MI, Williams DH (1994) Structure elucidation of xanthone derivatives with CD4-binding activity from *Penicillium glabrum* (Wehmer) Westling. Pure Appl Chem 66:2383–2386

Yagi T (1990) Inhibition by capsaicin of NADH-quinone oxidoreductases is correlated with the presence of energy-coupling site 1 in various organisms. Arch Biochem Biophys 281:305–311

Yankovskaya V, Horsefield R, Toernroth S, Luna-Chavez C, Miyoshi H, Leger C, Byrne B, Cecchini G, Iwata S (2003) Architecture of succinate dehydrogenase and reactive oxygen species generation. Science 299:700–704

Yonemitsu M, Fukuda N, Kimura T, Komori T, Lindner HJ, Habermehl G (1989) Crystal structure and NMR spectrometric analysis of palmarin. Liebigs Ann Chem 485–487

Zhang Y, Zhu T, Fang Y, Liu H, Gu Q, Zhu W (2007) Carbonarones A and B, new bioactive γ-pyrone and α-pyridone derivatives from the marine-derived fungus *Aspergillus carbonarius*. J Antibiot 60:153–157

13 Cyclic Peptides and Depsipeptides from Fungi

Heidrun Anke[1], Luis Antelo[1]

CONTENTS

I. Introduction

Cyclic peptides and depsipeptides are widely distributed in nature. They are found in plants (Gournelis et al. 1998; Tan and Zhou 2006), sponges and other lower sea animals (Bertram and Pattenden 2007), cyanobacteria (Welker and von Döhren 2006), bacteria and fungi alike and their bioactivities range from antimicrobial, insecticidal, nematicidal, antiviral, hepatotoxic, cytotoxic/cytostatic to immunosuppressive and other pharmacological properties (Kleinkauf and von Döhren 1997; Pomilio et al. 2006).

Some of the peptides and depsipeptides produced by fungi have gained entrance into the pharmaceutical market, like cyclosporins (Kürnsteiner et al. 2002), ergopeptides (Keller and Tudzynski 2002), penicillins (Demain and Elander 1999) and cephalosporins (Schmidt 2002), or are currently undergoing clinical trials, like the can-

dines, promising antifungal drugs against aspergillosis and candidiasis (Denning 2002; Johnson and Perfect 2003; Pasqualotto and Denning 2008). Caspofungin derived from pneumocandin and micafungin derived from FR901379 are examples of those novel drugs targeting fungal cell wall synthesis, e.g. biosynthesis of 1,3-β-glucan (Odds et al. 2003; Butler 2004). For a recent review, see Hashimoto (2009). Emodepsin, a semi-synthetic depsipeptide, is used in veterinary medicine against helminths (von Samson-Himmelstjerna et al. 2005). The drug is derived from PF1022A, a metabolite of an endophytic fungus from *Camellia japonica* (Sasaki et al. 1992; Scherkenbeck et al. 2002). As these groups of compounds are well covered in the literature, they are not addressed here in detail.

The biosynthesis of cyclic peptides and depsipeptides has attracted the interest of biochemists since the mid1960s (Gevers et al. 1968). Today, the focus has shifted from enzymology to genetics, e.g. the biosynthetic genes and their regulation. Therefore Chap. 15 is dedicated to this topic, to which the reader is referred.

A special group of cyclopeptides are the diketopiperazines, which consist of two amino acids linked by two peptide bonds. In the related epipolythiodioxopiperazines the 6-ring is bridged by one to four sulfur atoms. The structural diversity of diketopiperazines (more than 100 different compounds are known from fungi; Buckingham 2008) is matched by their biological activities. Recently published reviews are available (Cole and Schweikert 2003; Gardiner et al. 2005). Interestingly, functions in the producing organisms have been detected for some of these compounds, e.g. gliotoxin and related compounds play a role as virulence factors in invasive aspergillosis (Sugui et al. 2007) and coprogens in host invasion of plant-pathogenic fungi (Oide et al. 2006; Hof et al. 2007). The reported biological activities of gliotoxin are very broad and diverse. Antibacterial, antifungal, antiviral, amoebicidal and

[1]Institute for Biotechnology and Drug Research, IBWF e.V., Erwin-Schroedinger-Strasse 56, 67663 Kaiserslautern, Germany; e-mail: anke@ibwf.de, antelo@ibwf.de

Physiology and Genetics, 1st Edition
The Mycota XV
T. Anke and D. Weber (Eds.)
© Springer-Verlag Berlin Heidelberg 2009

immunosuppressive properties have been described (see below). Most of these activities are based on interactions with essential thiol groups in proteins (Waring and Beaver 1996). Iron chelators like dimerumic acid, rhodotorulic acid, coprogen and its derivatives are involved in iron uptake (Winkelmann and Drechsler 1997; Renshaw et al. 2002; Antelo et al. 2006), while other siderophores, e.g. the hexapeptides ferrichrome or ferricrocin, in addition to iron transport or storage functions act as virulence factors in some human and plant pathogens similar to coprogens (Howard 1999; Haas et al. 2008).

The group of peptaibiotics, a constantly growing family of linear α-aminobutyric acid (Aib)-containing linear peptides has been enlarged by a small group of cyclic peptides also containing Aib, now called cyclopeptaibiotics. Whereas the linear group comprises more than 800 compounds, only nine cyclic compounds have been reported to date. These are seven tetrapeptides structurally related to chlamydocin (Degenkolb et al. 2008) and the scytalidamides, two heptapeptides containing Aib residues (Tan et al. 2003).

II. Occurrence of Cyclic Peptides and Depsipeptides Within the Kingdom Eumycota (True Fungi)

A. Siderophores

The occurrence and distribution of siderophores among the taxonomic groups of fungi is very well covered by the reviews of Renshaw et al. (2002) and Haas et al. (2008). Zygomycetes very rarely produce cyclic peptide or depsipeptide siderophores. Up to now the hexapeptide ferrichrysin seems to be the only example. It is produced by *Cunninghamella blakesleeana* (Patil et al. 1995). The production of diketopiperazine and hexapeptide siderophores is common among asco- and basidiomycetes (Renshaw et al. 2002). The fact that members of some orders have not yet been reported to produce siderophores reflects a lack of investigation rather than presence. There are a few fungi, however, which do not produce siderophores: the ascomycetous yeasts *Saccharomyces cerevisiae* and *Candida albicans* or *Geotrichum candidum* and the basidiomycete

Cryptococcus neoformans (teleomorph *Filobasidiella*; Howard 1999; Haas et al. 2008). The investigation of basidiomycetes is difficult because iron-free media, which upregulate the biosynthesis of siderophores, often hardly support mycelial growth, requiring incubation times of eight to ten weeks (Welzel et al. 2005). In contrast, modern analytical techniques like HPLC-MS^n are sensitive enough to allow the detection and characterization of very small amounts (µg/l of culture). In addition, as more fungal genomes and NRPS genes and products become available, it is clear that siderophores and iron metabolism are important virulence determinants (Eichhorn et al. 2006; Oide et al. 2006; Haas et al. 2008).

It is remarkable that extracellular and intracellular siderophores are not identical and that the synthesis of intracellular siderophores is often not iron-dependent.

As an example, most *Trichoderma* species excrete coprogen-type siderophores and ferricrocin for the capture and transport of iron and use palmitoylcoprogen located within the mycelia as storage compound. In *T. pseudokoningii* and *T. longibrachiatum* however, palmitoylcoprogen was not detected, but these two species excrete fusigen-type siderophores in addition to coprogen and fericrocin (Anke et al. 1991). *Magnaporthe grisea* uses intracellular ferricrocin for iron storage and under iron deprivation excretes four coprogen derivatives (Hof et al. 2007). In other plant-pathogenic fungi like *Fusarium graminearum*, *F. culmorum*, *F. pseudograminearum*, *Cochliobolus heterostrophus* and *Gibberella zeae* ferricrocin has also been reported as intracellular siderophore (Oide et al. 2007; Tobiasen et al. 2007). The situation in the human pathogen *Aspergillus fumigatus* is similar. Ferricrocin is located in the mycelia, a hydroxylated derivative in the conidia and triacetylfusigen is excreted (Schrettl et al. 2007).

The structures of several iron-free siderophores, e.g. rhodotorulic acid, 2-*N*-methylcoprogen, palmitoylcoprogen, ferricrocin and ferrichrome are given in Fig. 13.1.

B. Diketopiperazines

Simple diketopiperazines may be detected in fermentations of many fungi. Sometimes it is difficult to decide whether these are degradation products of proteins and peptides or synthesized de novo (Prasad 1995). In the future, this problem might be solved by molecular genetics, since the presence of the relevant biosynthetic genes can be

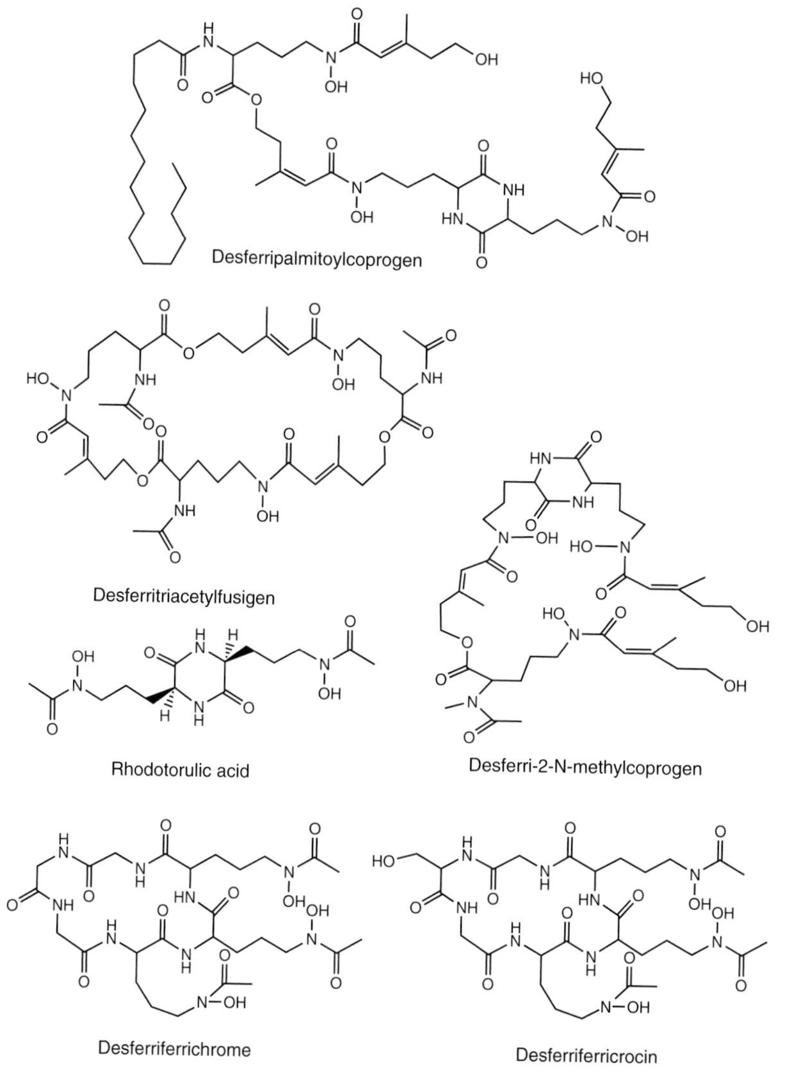

Fig. 13.1. Structures of some intracellular and extracellular siderophores produced by fungi

Desferripalmitoylcoprogen

Desferritriacetylfusigen

Rhodotorulic acid

Desferri-2-N-methylcoprogen

Desferriferrichrome

Desferriferricrocin

proof of de novo synthesis (Chap. 15). The recently demonstrated behavioral effects and occurrence in humans of cyclo(His-Pro) stimulated research on such compounds which are easily accessible by chemical synthesis. However, cyclo(His-Pro) has not yet been reported from fungi. This may be due to the fact that its bioactivities, e.g. inhibition of food intake and inhibition of prolactin secretion or modulation of pain perception (Prasad 1995) are not suited for a screening of microbial cultures. Usually these compounds are detected during the isolation of other metabolites and described as side-products. A recent example is L-alanyl-L-tryptophan anhydride isolated together with golmaenone, a radical scavenger compound,

and neoechinulin from an marine *Aspergillus* species (Li et al. 2004). As in many other cases, the simple alkaloid is the biogenetic precursor of the other two compounds. With antimicrobial, cytotoxic, phytotoxic, insecticidal and other test systems which have been extensively used in screenings for bioactive natural products, simple diketopierazines are less frequently detected. One example is the fungistatic mactanamide from a marine *Aspergillus* species (Lorenz et al. 1998). Simple diketopiperazines have been described from hetero- and homobasidiomycetes, for example *Ustilago cynodontis*, *Entoloma haastii* and *Stereum hirsutum* (Turner and Aldridge 1983), ascomycetes like *Rosellinia necatrix*,

Claviceps species, *Eurotium* and *Emericella* species, *Leptosphaeria* species including their anamorphs, *Aspergillus*, *Phoma* and *Coniothyrium* species (Turner and Aldridge 1983; Cole and Schweikert 2003; Blunt et al. 2006).

Aspergillus and *Penicillium* species are very prolific producers of cyclic dipeptide-derived mycotoxins like fumitremorgins, verruculogens or roquefortine C, while sporidesmins, mycotoxins that cause facial eczema in grazing sheep, are produced by *Pithomyces chartarum* (Betina 1989). From several *Penicillium* species, mycelianamide, one of the very "old" diketopiperazines, has been known since 1931. This compound was detected during early screenings after the discovery of penicillin G. The

recently described sulfur-containing gliovictin was obtained from an endophytic *Penicillium janczewskii* (Gunatilaka 2006) and diketopiperazine-derived rostratins from a marine *Exserohilium rostratum* (Tan et al. 2004). To the long list of *Penicillium* species producing diketopiperazines, *P. dipodomyis*, *P. nalgiovense*, *P. fellutanum* and *P. simplicissimum* were recently added (Lewis 2002).

Examples for structures of simple and complex diketopiperazines are found in Fig. 13.2.

From cultures of a number of fungi producing cyclic depsipeptides, e.g. *Beauveria bassiana*, dipeptides composed of the amino acids occurring in the depsipeptides have been

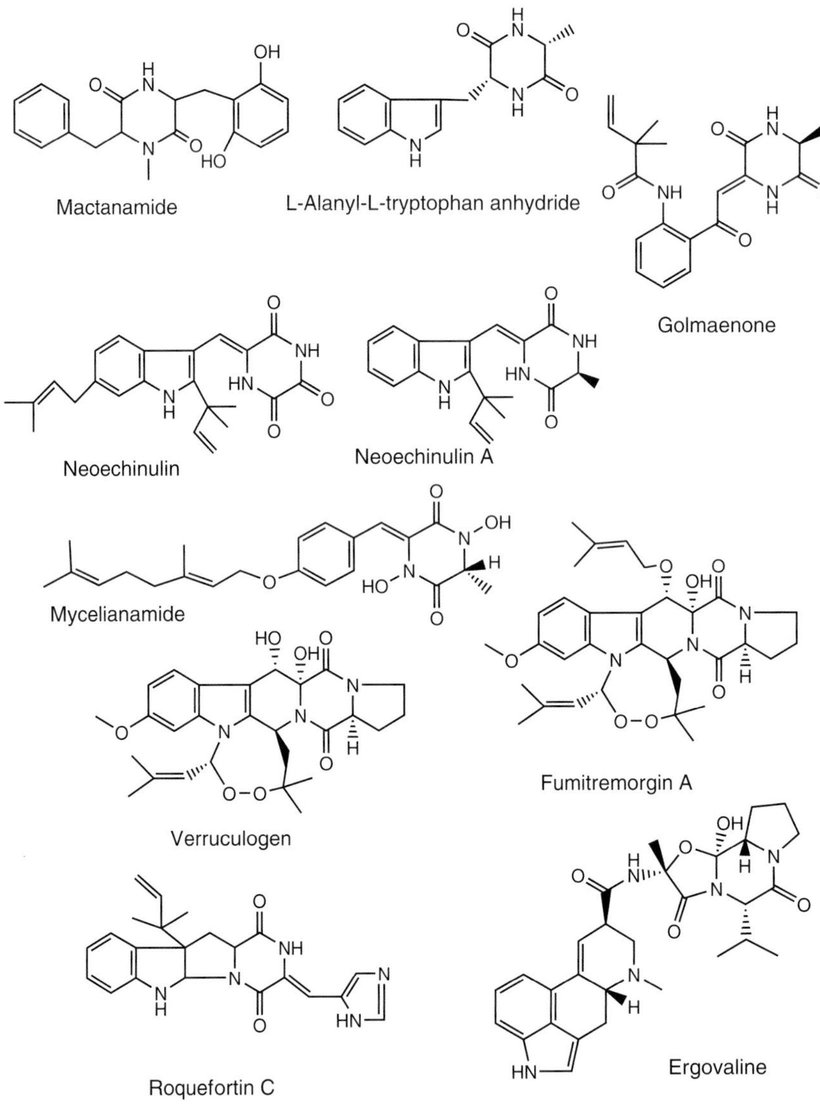

Fig. 13.2. Structures of some diketopiperazines

isolated. Other insect pathogens like *Verticillium* species and *Metarhizium anisopliae* as well as plant-pathogenic fungi, e.g. *Colletotrichum gloeosporioides, Exserohilum holmi, Gliocladium deliquenscens, Alternaria* and *Trichoderma* produce dipeptides. An unidentified endophyte from mangrove leaf produces two cyclic depsipeptides and three diketopiperazines (Huang et al. 2007). The role of the compounds, dipeptides and depsipeptides, in insect and plant-pathogenicity has not yet been completely elucidated. As molecular tools become more easily available, this question might be addressed or even answered in the near future, especially since the elucidation of the ecological function of secondary metabolites for the producers becomes more interesting (see below).

Epipolythiopiperazines with more than 60 members, gliotoxin being the most prominent, are widely distributed in nature. Their producers are mainly found among the ascomycete genera *Aspergillus, Penicillium, Gliocladium, Verticillium, Chaetominum, Emericella, Acrostalagmus* (syn. *Verticillium*), *Pithomyces, Bionectria, Leptosphaeria, Hyalodendron, Trichoderma, Sirodesmium* (syn. *Coniosporium), Epicoccum, Arachniotus* and *Pseudallescheria* (Turner and Aldridge 1983; Betina 1989; Takahashi et al. 1994; Gardiner at al. 2005; Li et al. 2006; Zheng et al. 2007). There is one report on the occurrence of an epipolythiopiperazine in lichens, e.g. *Xanthoparmelia scabrosa* (Ernst-Russell et al. 1999). As is true for many lichen metabolites, it may be also in this case the ascomycetous fungal

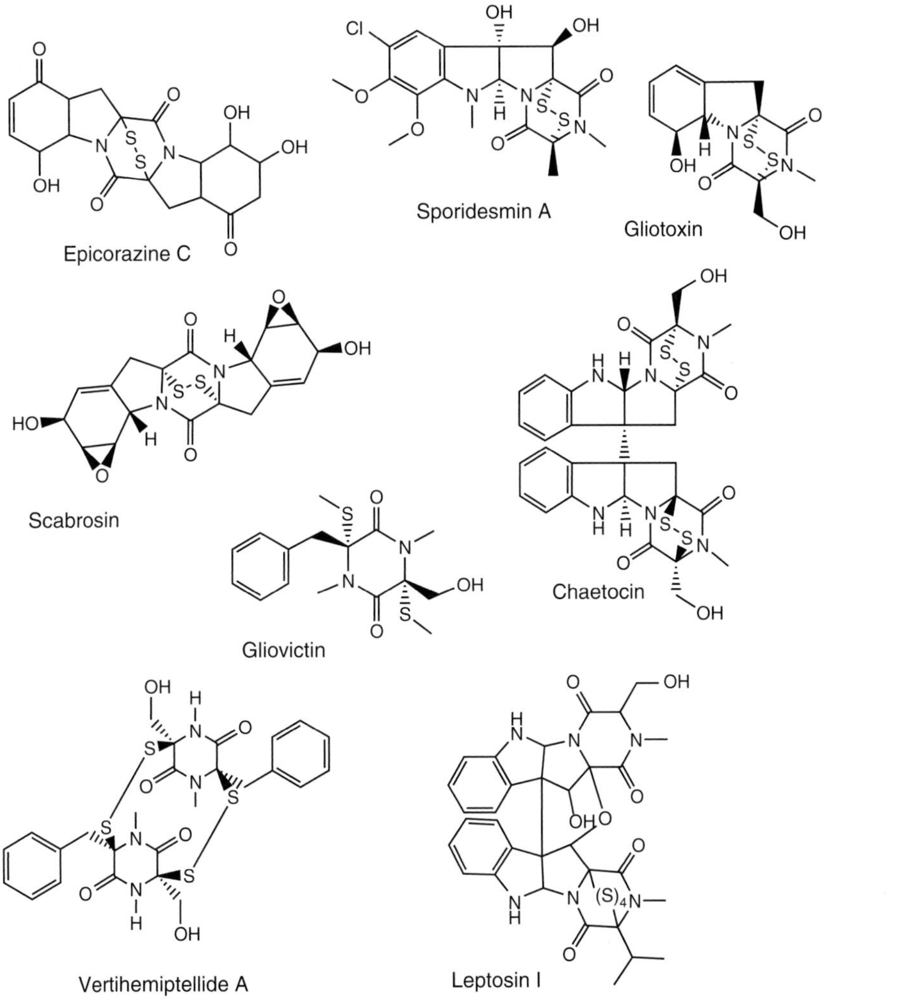

Epicorazine C

Sporidesmin A

Gliotoxin

Scabrosin

Gliovictin

Chaetocin

Vertihemiptellide A

Leptosin I

Fig. 13.3. Structures of some epipolythiopiperazines

partner which is responsible for the production of scabrosin. The production of epicorazine C by *Stereum hirsutum*, a basidiomycete, seems a bit questionable since related epicorazines are produced by *Epicoccum nigrum* and *E. purpurascens* (Kleinwachter et al. 2001). Overlaps between metabolites from basidiomycetes and ascomycetes are fairly rare but do occur occasionally. Other examples may be beauvericin and chlamydocin (see below). The structures of gliotoxin, epicorazines, scabrosin, vertihemiptellide A and other epipolythiopiperazines are given in Fig. 13.3.

C. Cyclic Peptides

Cyclic peptides are mainly produced by ascomycetes and their anamorphs. Among cyclic peptides, the immunomodulating cyclosporins constitute the largest group with 46 members. The producing organisms are found mainly in the ascomycetous families *Hypocreaceae* and *Clavivipitaceae* and their anamorphs *Tolypocladium inflatum*, *T. tundrense* and *T. terricola*. In addition, three soil-borne insect pathogens, *Neocosmospora vasinfecta*, *Acremonium luzulae*, a *Cyclindrotrichum* species, *Stachybotrys chartarum*,

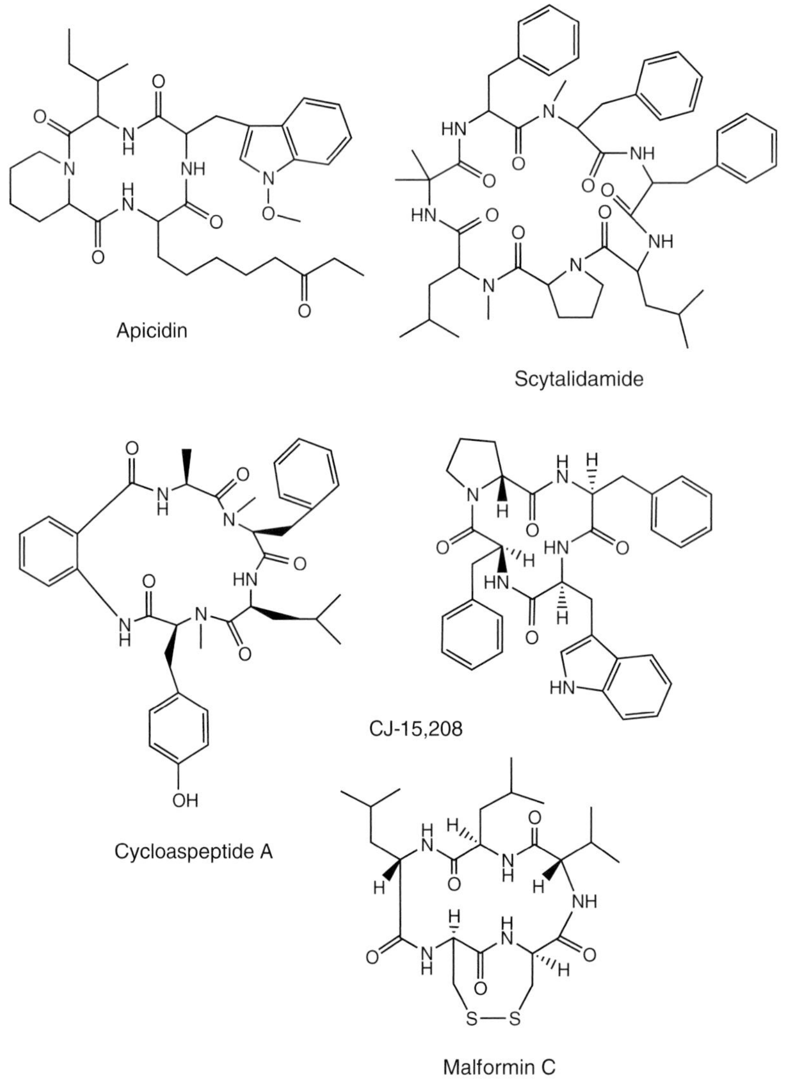

Apicidin

Scytalidamide

Cycloaspeptide A

CJ-15,208

Malformin C

Fig. 13.4A. Structures of some simple cyclopeptides

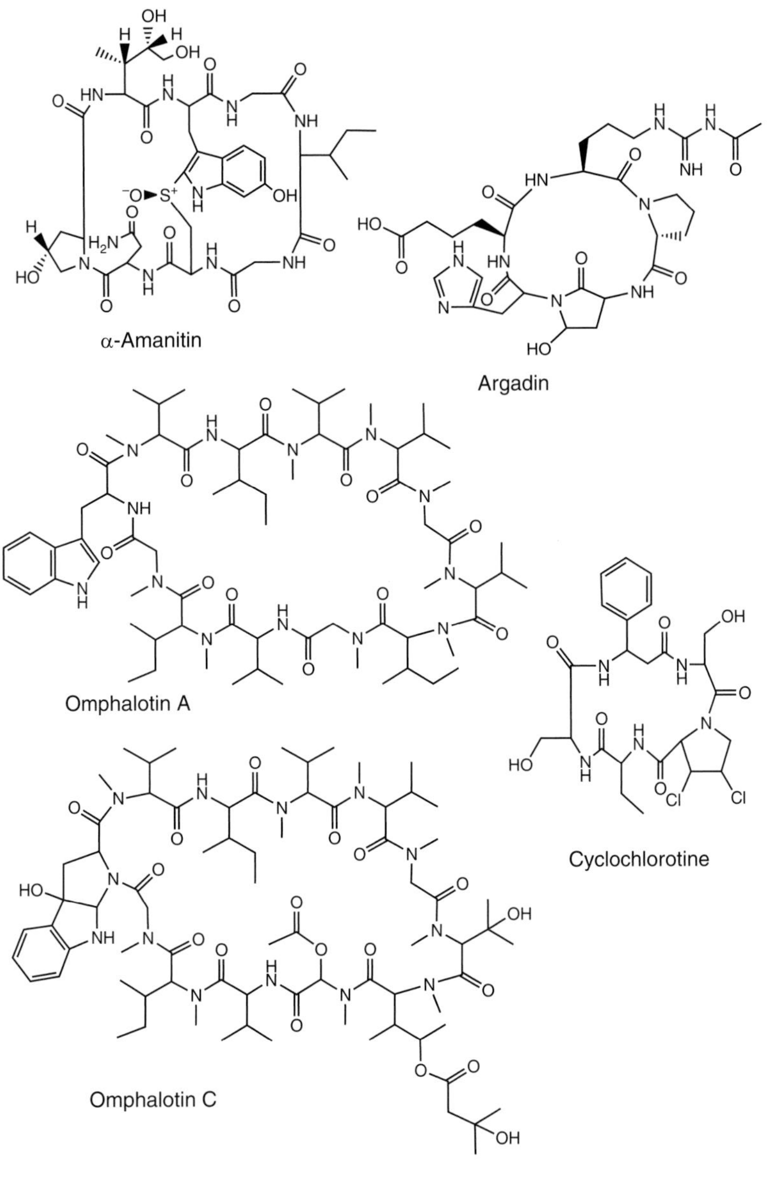

Fig. 13.4B. Structures of some complex cyclopeptides.

Trichoderma viride, a *Leptostroma* anamorph of *Hypoderma eucalyptii*, *Chaunopycnis alba* and an unidentified mycelium sterilium have been reported to produce cyclosporins (Matha et al. 1992; Traber and Dreyfuss 1996). The structure of cyclosporin A is found in Fig. 13.4C. Figure. 13.4A shows examples of simpler cyclospeptides.

The malformins, a group of nine phytotoxic compounds, are only found within the *Aspergillus niger* group (Kobbe et al. 1977). Some authors classify the compounds as mycotoxins even so they are rarely found in food or feed stuff.

The antifungal echinocandins comprising different compounds (aculeacin A, echinocandin B, pneumocandins, mulundocandins, FR901379, WF11899A, B, C, FR227673, FR190293, etc.) have been reported from several Aspergilli, *Coleophoma empetri*, *C. crateriformis*, *Chalara* species, *Tolypocladium parasiticum* and *Zalerion arboricola* (Iwamoto et al. 1994b; Anke and Erkel 2002; Denning 2002; Kanasaki et al. 2006a, b, c,).

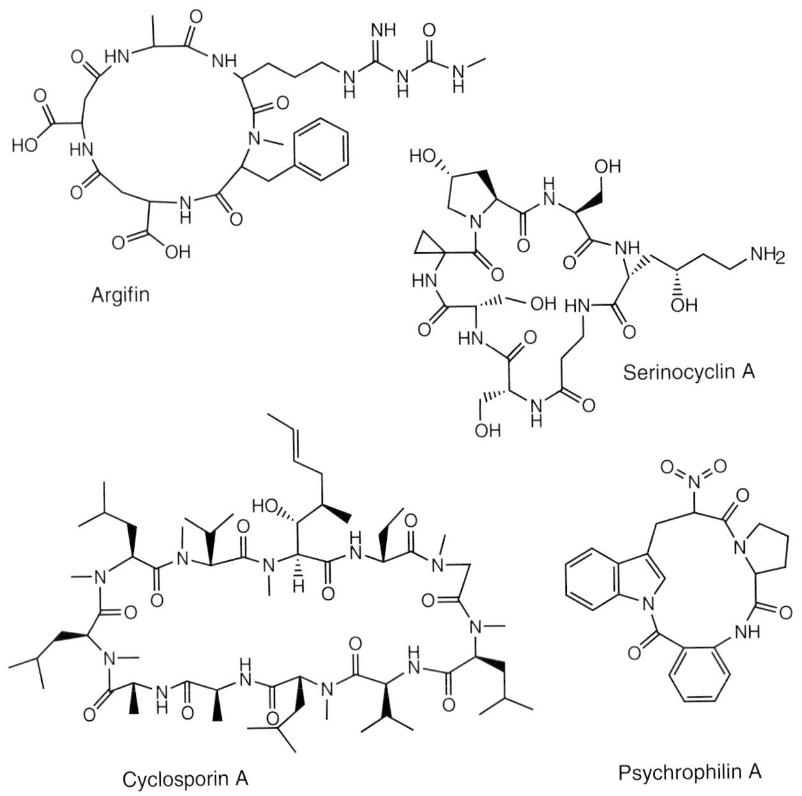

Fig. 13.4C. Structures of other complex cyclopeptides

The *Zalerion* strain producing echinocandin B was later reclassified as *Glarea lozoyensis*, a new anamorph genus and species within the Leotiales (Bills et al. 1999). The fungus producing arborcandins (Ohyama et al. 2000), has not been identified.

The structures of some of these compounds can be found in Fig. 13.5.

Producers of various cyclic peptides are found in many other families and genera, for example *Diheterospora*, *Gliocladium*, *Cylindrocarpon*, *Clonostachys*, *Cochliobolus* and *Fusarium* (Lewis 2002; Adachi et al. 2005; Weber et al. 2006; Degenkolb et al. 2008).

As endophytic fungi have recently come into focus as producers of bioactive natural compounds, it is not astonishing that also novel cyclic peptides have been reported from these fungi.

A pentapeptide was isolated from an unidentified endophyte from the seed of *Avicinnia marina* (Gunatilaka 2006), other cyclopeptides from endophytic *Fusarium* species (Shiono et al. 2007), *Epichloe typhina* (Seto et al. 2007) or endophyte

"2221" from *Castaniopsis fissa* (Yin et al. 2005). More than 450 cyclic peptides are known from plants (Tan and Zhou 2006); some of these actually may be produced by endophytic fungi in planta.

In recent years, marine habitats have drawn much interest as ecological niches for producers of novel bioactive metabolites. The unguisins were isolated from a marine-derived strain of *Emericella unguis* (Malstrom 2002). Among cyclic peptides from obligate marine ascomycetes are the highly cytotoxic trapoxin A produced by *Corollospora intermedia* (Daferner 2000) or scytalidamides from a *Scytalidium* species from a marine alga (Tan et al. 2003). JM47, structurally related to HC-toxins and trapoxin, was isolated together with enniatin from a marine-derived *Fusarium* species (Jiang et al. 2002). Trapoxins are also known from terrestrial fungi, e.g. *Helicoma ambiens*, the anamorph of *Thaxteriella pezicula* (Itazaki et al. 1990) and structurally related metabolites have been described from the phytopathogenic *Cyclindrocladium scorparium* (teleomorph *Calonectria morganii*) and *Cochliobolus carbonum* (Degenkolb et al. 2008).

For structures see Fig. 13.6.

The only cyclopeptides, besides the siderophores, known from *submerged cultures* of basi-

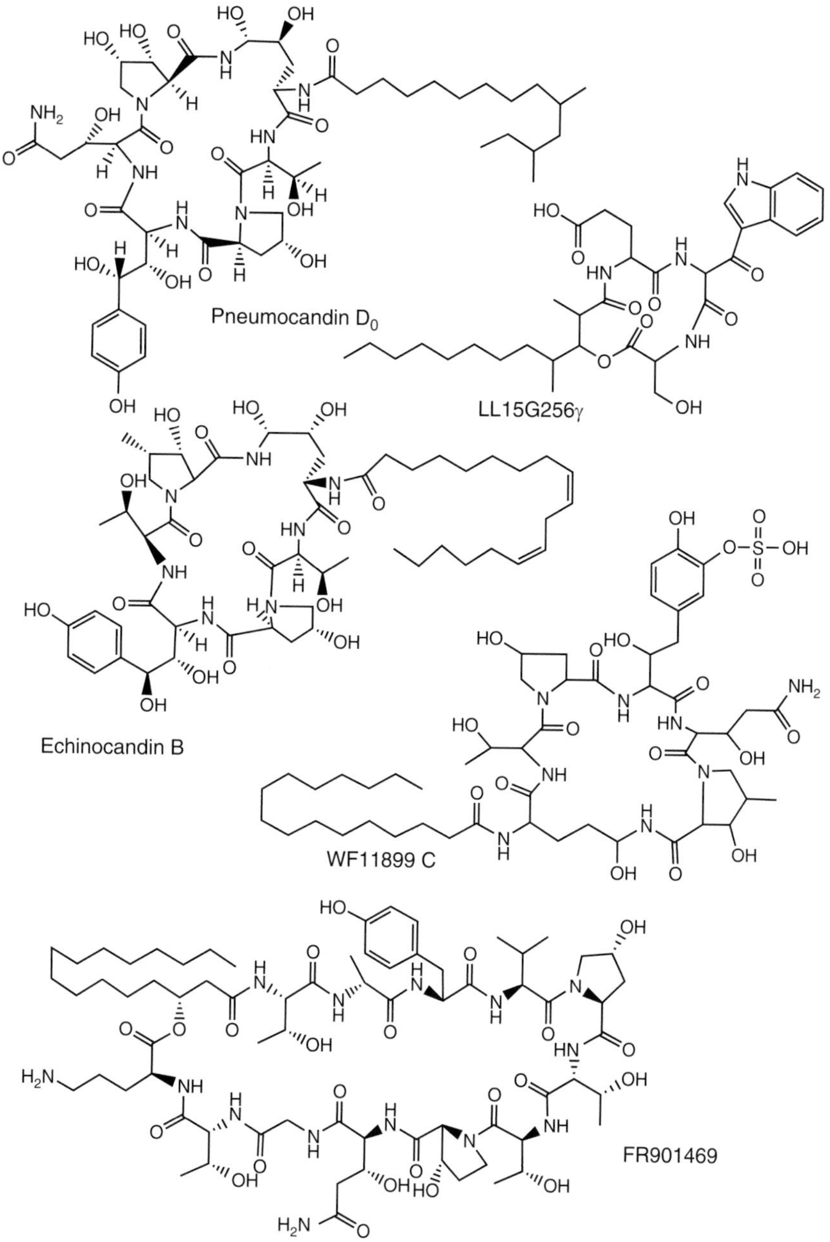

Fig. 13.5. Structures of some β-1,3-glucan synthase inhibitors

diomycetes are the omphalotins from *Omphalotus olearius* (Büchel et al. 1998a,b), amanitins from *Amanita exitialis* (Zhang et al. 2005) and chlamy-docins from a *Peniophora* strain isolated from soil (Tani et al. 2001). The chlamydocins are tetrapeptides with Aib and an unusual amino acid. Most of these are produced by ascomycetes, e.g. *Diheterospora chlamyydosporia* (Closse and Huguenin 1974) and *V. coccosporum* (Gupta et al. 1994). Interestingly, the omphalotins produced by a monokaryotic strain differ from those found in the dikaryotic parental strain (Liermann et al. 2009). However, all *O. olearius* strains irrespective of their geographical origin produce omphalotin derivatives (Anke et al., unpublished data). In fruiting bodies omphalotins could not

Fig. 13.6. Structures of some histone deacetylase inhibitors

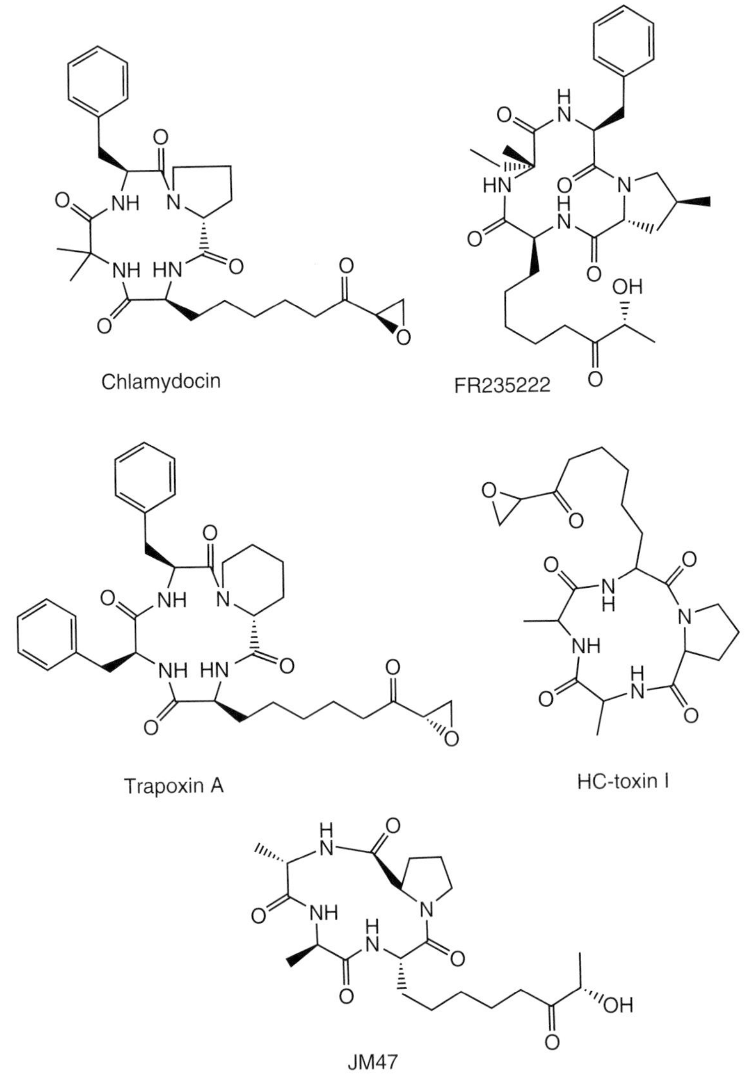

Chlamydocin

FR235222

Trapoxin A

HC-toxin I

JM47

be detected, contrary to *Amanita exitialis* carpophores which contained tenfold more α- and β-amanitin as compared to the slow growing mycelial cultures (Zhang et al. 2005). For recent surveys of *Amanita* toxins from fruiting bodies see Li and Oberlies (2005), Liu (2005) and Pomilio et al. (2006). Figure 13.4B shows the structures of omphalotins and α-amanitin.

D. Cyclic Depsipeptides

Most depsipeptides are metabolites from ascomycetes and their anamorphs. They are widespread in phytopathogens (e.g. *Cochliobolus* with

anamorphs *Helminthosporium* and *Bipolaris*, *Calonectria* and its anamorph *Cyclindrocladium*, as well as *Fusarium* and *Alternaria*), insect pathogens (*Aschersonia, Beauveria, Cordyceps, Diheterospora, Fusarium, Hirsutella, Isaria, Metharizium, Paecilomyces, Verticillium*) and others (Zimmermann 2007a, b; Buckingham 2008). For a compilation of beauvericins and enniatins produced by *Cordyceps* species and their anamorphs as well as other insect pathogens see Isaka et al. (2005a, b). Figure 13.7 gives the structures of some cyclodepsipeptides.

Up to now the pteratides (Fig. 13.7C) are the only depsipeptides reported from basidiomycetes, namely from the fruiting bodies of a

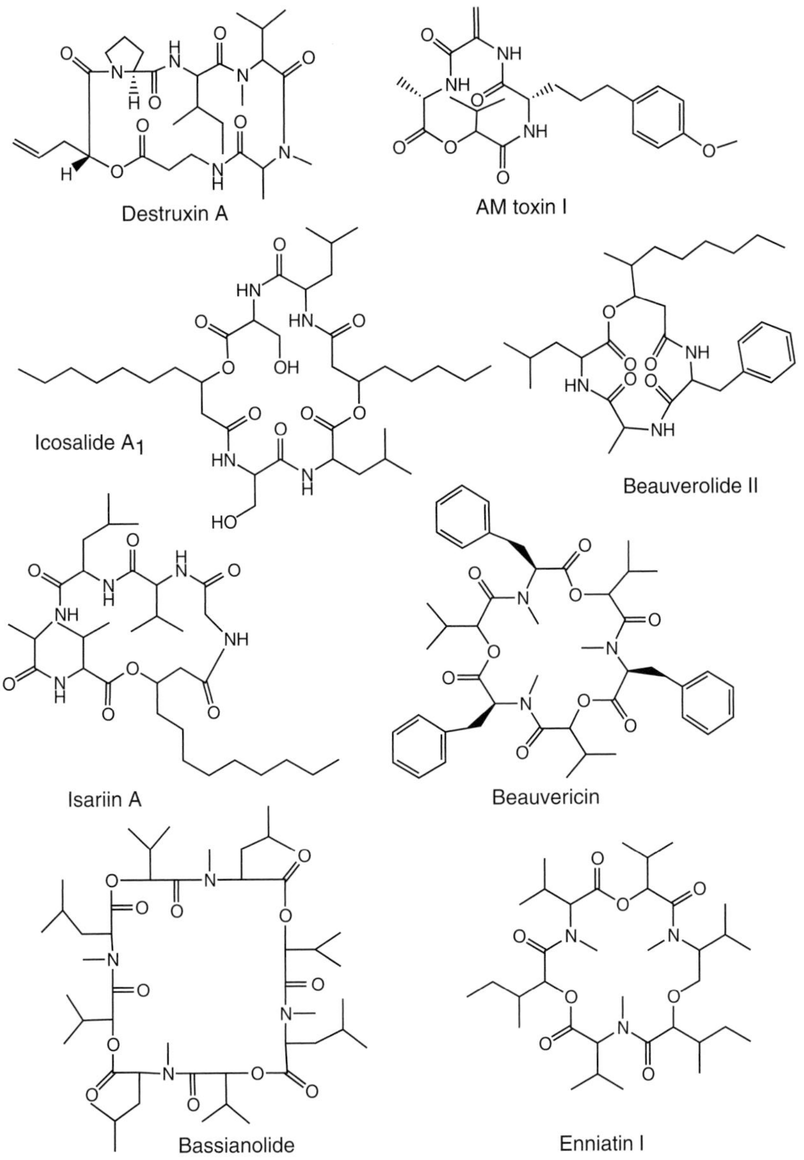

Fig. 13.7A. Structures of some simple cyclodepsipeptide

Pterula species (Chen et al. 2006). From zygomycetes none have been described. One report on the production of beauvericin by *Laetiporus sulfureus* (Badan et al. 1978) could not be confirmed by other groups. In our cultures from *L. sulfureus* from different locations we could only detect laetiporic acid and its derivatives (Davoli et al. 2005).

Since the review of Anke and Sterner (2002), additional producers of bioactive depsipeptides have been reported, for example marine-derived strains of *Beauveria fellina* (Lira et al. 2006), *Verticillium* sp. FKI-1033 (Monma et al. 2006), *Aspergillus carneus* (Capon et al. 2003), *Torrubiella luteorostrata* and its anamorph *Paecilomyces cinnamomeus* (both isolated from a scale insect; Isaka et al. 2007), *Verticillium hemipterigenum* (Supothina et al. 2004), an *Aureobasidium* species from the tropical rain forest (Boros et al. 2006), an unidentified endophytic fungus (Huang et al. 2007) and a soil-borne *Phoma* species (Aoyagi et al. 2007). Pseudodestruxins have been

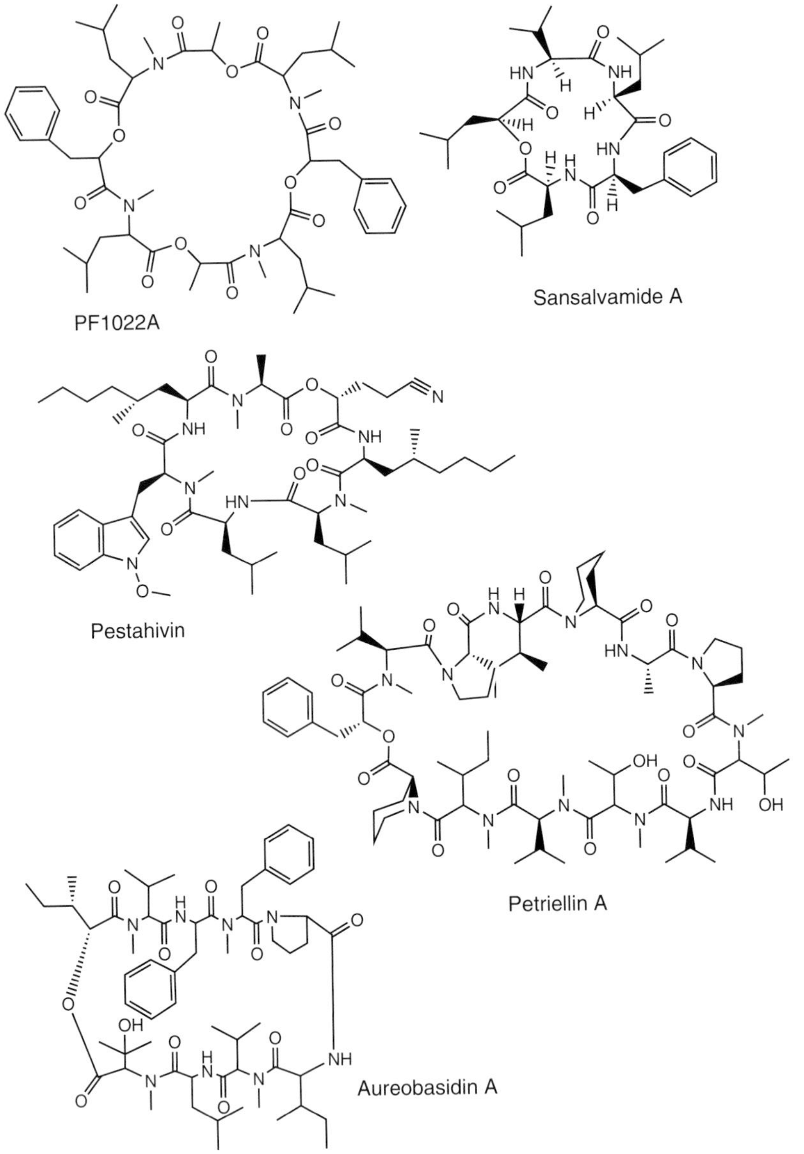

PF1022A

Sansalvamide A

Pestahivin

Petriellin A

Aureobasidin A

Fig. 13.7B. Structures of some complex cyclodepsipeptides

reported from *Nigrosabulum globosum* (Che et al. 2001) and reviews on destruxins and the producing organisms have been published by (Pedras et al. 2002) and Zimmermann (2007b).

The endophyte-producing PF1022A (and related anthelmintic cyclooctadepsipeptides) isolated from leaves of a camellia has been identified based on its 18S rRNA gene sequence as a member of the Xylariaceae close to *Xylaria polymorpha* and *Rosellinia necatrix* (Miyado et al. 2000).

One of the few aquatic fungi investigated for secondary metabolite production is *Clavariopsis aquatica* from which the antifungal clavariopsins A and B were isolated (Kaida et al. 2001).

Analogues of the lipopeptides with 1,3-β-glucan synthase inhibitory activity are the lipodepsipeptides FR901469 or LL15G256γ (see Fig. 13.5). The former is produced by an unidentified fungus, the latter (identical to arthrichitin from *Arthrinum*

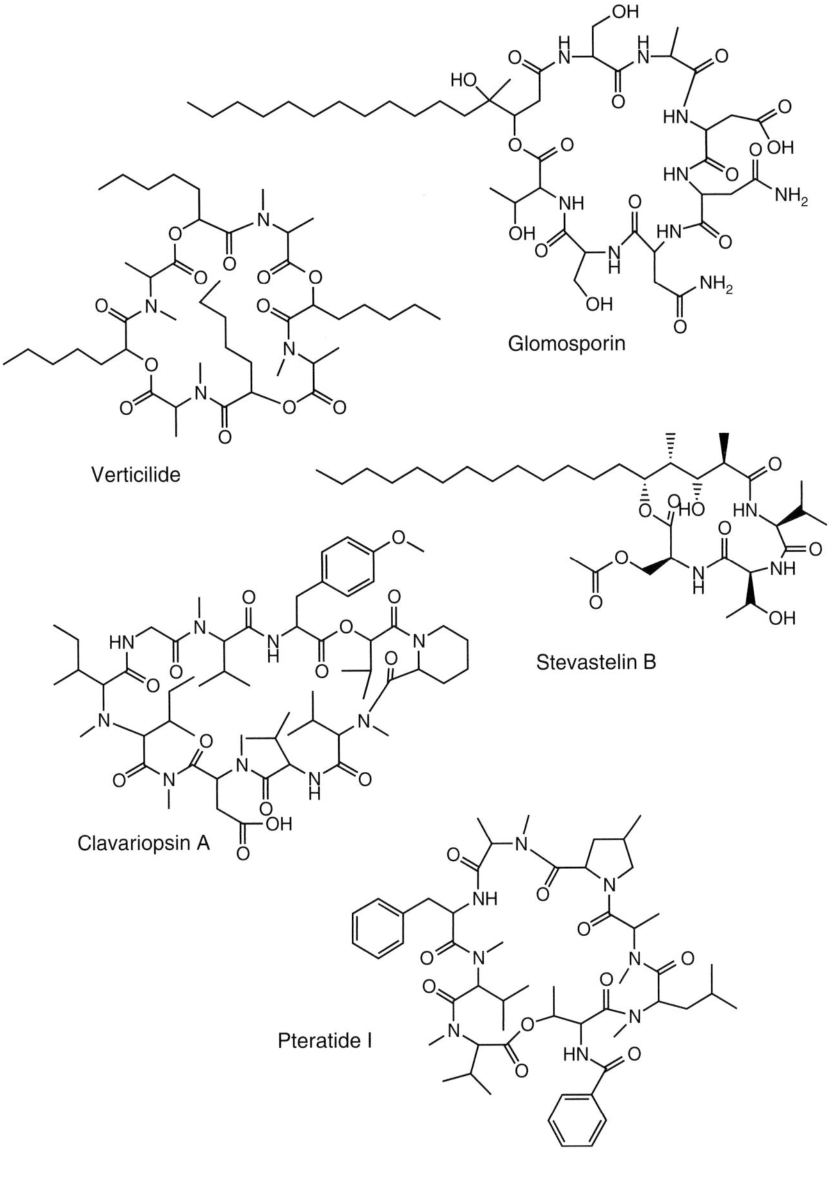

Fig. 13.7C. Structures of other complex cyclodepsipeptides

spaeospermum) by *Hypoxylon oceanicum* (Abbanat et al. 1998; Fujie et al. 2000).

III. Chemical and Biological Diversity of Cyclic Peptides and Depsipeptides

A. Diversity of Building Blocks

Cyclic peptides and depsipeptides constitute a class of natural compounds with an enormous structural diversity. This diversity is brought upon by the different building blocks in the ring: proteinogenic amino acids including their D-isomers, nonproteinogenic amino acids, branched or unbranched lipoamino acids and hydroxylated short-, medium- and long-chain fatty acids. The diversity of the building blocks can be deduced from Tables 13.1–13.3, which give a compilation of unusual building blocks (Table 13.1 unusual amino acids, Table 13.2 unusual fatty acids) and various modifications (Table 13.3).

Table 13.1. Diversity of amino acid building blocks in cyclic peptides and depsipeptides

Amino acid	Example	Figure	Reference
α-Aminoadipic acid	Argadin	13.4B	Arai et al. (2000)
Prolyl-homoserine	Argadin	13.4B	Arai et al. (2000)
β-Keto tryptophan	LL15G256γ	13.5	Abbanat et al. (1996)
Propylleucine	Pestahivin	13.7B	Hommel et al. (1996)
Dehydroalanine	AM-toxin I	13.7A	Ueno et al. (1975)
α-Amino-p-methoxyphenylvaleric acid	AM-toxin I	13.7A	Ueno et al. (1975)
N^5-Hydroxyornithine	Siderophores	13.1	Renshaw et al. (2002)
4-Methylproline	FR-235222	13.6	Mori et al. (2003)
2-Butenyl-4-methylthreonine	Cyclosporin A	13.4C	Rüegger et al. (1975)
α-Aminobutyric acid	Cyclosporin A	13.4C	Rüegger et al. (1975)
3-Hydroxyhomotyrosine	WF-11899C	13.5	Iwamoto et al. (1994b)
5-Hydroxyornithine	WF-11899C	13.5	Iwamoto et al. (1994b)
Dichloro-proline	Cyclochlorotine	13.4B	Yoshioka et al. (1973)
β-Phenyl-β-aminopropionic acid	Cyclochlorotine	13.4B	Yoshioka et al. (1973)
β-Alanine	Destruxin A	13.7A	Rees et al. (1996)
β-Aspartic acid	Argifin	13.4C	Arai et al. (2000)
α-Aminoisobutyric acid	Chlamydocin	13.6	Closse and Huguenin (1974)
Isovaline	FR-235222	13.6	Mori et al. (2003)
1-Aminocyclopropane-1-carboxylic acid	Serinocyclin A	13.4C	Krasnoff et al. (2007)
Pipecolinic acid	Trapoxin A	13.6	Itazaki et al. (1990)
Anthranilic acid	Psychrophilin D	13.4C	Dalsgaard et al. (2005)
2-Amino-8-oxo-9-hydroxydecanoic acid	JM47	13.6	Jiang et al. (2002)
2-Amino-9,10-epoxy-8-oxodecanoic acid	HC-toxin I	13.6	Gross et al. (1982)

Table 13.2. Diversity of hydroxyacid building blocks in cyclic depsipeptides

Acid	Example	Figure	Reference
2-Hydroxyisovaleric acid	Clavariopsin A	13.7C	Kaida et al. (2001)
3,4-Dihydroxy-4-methylhexadecanoic acid	Glomosporin	13.7C	Ishiyama et al. (2000)
2-Hydroxy-3-methylpentanoic acid	Enniatin I	13.7A	Nilanonta et al. (2003)
2-Hydroxyheptanoic acid	Verticilide	13.7C	Monma et al. (2006)
2-Hydroxy-4-metylpentanoic acid	Sansalvamide A	13.7B	Belofsky et al. (1999)
Phenyllactic acid	PF1022A	13.7B	Sasaki et al. (1992)
Lactic acid	PF1022A	13.7B	Sasaki et al. (1992)
3-Hydroxydodecanoic acid	Isariin A	13.7A	Wolstenholme and Vining (1966)
3-Hydroxy-4-methyldecanoic acid	Beauverolide II	13.7A	Mochizuki et al. (1993)
3-Hydroxydecanoic acid	Icosalide A_1	13.7A	Boros et al. (2006)
3,5-Dihydroxy-2,4-dimethylstearic acid	Stevastelin B	13.7C	Morino et al. (1994)
2-Hydroxy-4-cyanobutyric acid	Pestahivin	13.7B	Hommel et al. (1996)
2-Hydroxy-4-enoylpentanoic acid	Destruxin A	13.7A	Rees et al. (1996)
2,4-Dimethyl-3-hydroxydodecanoic acid	LL15G256γ	13.5	Abbanat et al. (1996)

B. Diversity of Structures

Additional variations are due to the different numbers of building blocks, their arrangement (e.g. sequence in the ring) and their linkage (e.g. amide and ester bonds). Some depsipetides like the enniatins, beauvericins, bassianolide or verticilide show a symmetric arrangement in the ring. The majority however are asymmetric, like the destruxins, beauverolides, isariins or *Alternaria* toxins (Fig. 13.7A,C).

Cyclic peptides including the cyclosporins are asymmetric, as are the echinocandins. The number of building blocks in cyclic peptides varies from two in the diketopiperazines, some of which are symmetric if composed of two residues of the same amino acid, to 12 in the omphalotins, which are at present the largest cyclopeptides known from fungi. In addition, the omphalotins are an example of modifications after ring closure. Omphalotins B, C and D are derived from omphalotin A by hydroxylation followed by acylation to

Table 13.3. Modifications in cyclic peptide and depsipeptides

Modification/substitution	Example	Figure	Reference
O-Methyl	Clavariopsin A	13.7C	Kaida et al. (2001)
N-Methyl	Omphalotin A	13.4B	Sterner et al. (1997)
Methoxy	Pestahivin	13.7B	Hommel et al. (1996)
Acetyl	Omphalotin C	13.4B	Büchel et al. (1998a)
3-Hydroxy-methylbutanoyl	Omphalotin C	13.4B	Büchel et al. (1998a)
Palmitic acid	WF-11899C	13.5	Iwamoto et al. (1994b)
3-Hydroxypalmitic acid	FR 901469	13.5	Fujie et al. (2000)
Linoleic acid	Echinocandin B	13.5	Keller-Juslen et al. (1976)
Sulfate	WF-11899C	13.5	Iwamoto et al. (1994b)
Nitro	Psychrophilin A	13.4C	Dalsgaard et al. (2005)
Halogenation	Sporidesmin A	13.3	Fridrichsons and Mathieson (1962)
Isoprenyl	Roquefortine C	13.2	Scott et al. (1979)
Prenyl	Fumitremorgin A	13.2	Eickman et al. (1975)
Geranyl	Mycelianamide	13.2	Birch et al. (1956)
N-Methylcarbamoyl	Argifin	13.4C	Arai et al. (2000)
Hydroxylation			
3-Hydroxyvaline	Omphalotin C	13.4B	Büchel et al. (1998a)
4,5-Dihydroxyornithine	Echinocandin B	13.5	Keller-Juslen et al. (1976)
3,4-Dihydroxyhomotyrosine	Echinocandin B	13.5	Keller-Juslen et al. (1976)
2,6-Dihydroxyphenylalanine	Mactanamide	13.2	Lorenz et al. (1998)
3,4-Dihydroxyproline	Pneumocandin D_0	13.5	Morris et al. (1994)

the corresponding esters and formation of additional ring structures (Büchel et al. 1998).

Novel omphalotins were recently isolated from a monokaryotic strain. The elucidation of their structures was greatly hampered by their instability (Liermann et al. 2009). These omphalotins bear additional hydroxyl groups, thus bringing the number of known cyclic peptides from *O. olearius* to 11. A second hydroxylation at the tryptophan leads to a novel ring system (Fig. 13.4B). HPLC-MS spectra of enriched extracts indicate the presence of additional members of the group. The psychrophilins are nitropeptides with unusual structures (Fig. 13.4A). The compounds are produced by several psychrotolerant *Penicillium* species (Dalsgaard et al. 2004a, b; 2005). Cyclochlorotine, a mycotoxin from *P. islandicum* contains a dichloroprolyl residue (Betina 1989).

The depsipeptides start with four building blocks (angolide, beauverolides) up to 12 in the antibiotic FR901469 (a member of the 1,3-β-glucan synthase inhibitors; Fujie et al. 2001) and 13 in petriellin A (Lee et al. 1995). The latter contains β-phenyllactic acid, a building block not often found in cyclopeptides and -depsipeptides. Further modifications of cyclic peptides and depsipeptides include N-methylation, hydroxylations, acylation, isoprenylation and the introduction of sulfate-, nitrochloro- or cyano- groups. These modifications can occur at the beginning of biosynthesis, like N-methylations, or after cyclization, e.g. C- and

N-hydroxylations followed by an acylation (Glinski et al. 2001; Chap. 15). In many cases however it is not clear at which step the modifications occur. The low substrate specificity of the NRPS enzymes allows the incorporation of modified ring components. In fact, Zocher and his group have made use of this to produce novel enniatin derivatives in vitro (Feifel et al. 2007).

C. Diversity of Biological Activities

The structural diversity of diketopiperazines, cyclopeptides and -depsipeptides is matched by the diversity of their biological activities. To list all activities and compounds would be beyond of the scope of this chapter. An overview on biological activities of diketopiperazines is given by Martins and Carvalho (2007), cyclic depsipeptides and their biological activities are reviewed by (Sarabia et al. 2004), while insecticidal and other biological activities of destruxins, isariins, enniatins, and beauverolides are reviewed by Anke and Sterner (2002) and by Zimmermann (2007a, b). Some of the compounds exhibit rather selective activities like the antifungal, 1,3-β-glucan synthesis inhibitors (see below) whereas others like gliotoxin show a broad spectrum of activities. While the former (due to fewer side-effects) generally have a higher potential to be developed into drugs or pesticides, the latter might be of interest as biochemical tools or chemical building blocks. In the following, we attempt to give an overview on the different biological activities exhibited by fungal cyclopeptides and -depsipeptides.

Gliotoxin, already isolated in 1932, recently regained interest not only due to its immunosuppressive and apoptosis-inducing activities (Waring et al. 1988) but moreover due to its occurrence in the blood of aspergillosis patients and its effects on various human cells among them an inhibition of cell adherence in macrophages (Amitani et al. 1995; Kamei and Watanabe 2005). The plethora of biological activities is evident from the number of papers published on gliotoxin and related epipolythiodioxopiperazines (Waring and Beaver 1996; Hume et al. 2002; Gardiner et al. 2005).

The vertihemiptellides A and B and their S-methylated monomers exhibit antimycobacterial and cytotoxic effects (Isaka et al. 2005b). Sirodesmin PL produced by *Leptosphaeria maculans* has phytotoxic, antibacterial and insecticidal properties (Rouxel et al. 1988; Boudart 1989) and the leptosins inhibited the proliferation of P388 lymphocytic leukemia cells with an ED_{50} of 1.1–1.3 µg/ml (Takahashi et al. 1994).

The HC-toxins, host-specific toxins from *Cochliobolus carbonum* (anamorph *Helminthosporium carbonum*), are cyto- and phytotoxic and inhibitors of histone deacetylase (Taunton et al. 1996).

Structurally related tetrapeptides (Fig. 13.6) like apicidin from a *Fusarium* species (Darkin-Rattray et al. 1996; Singh et al. 2002), JM47 from a marine *Fusarium* species (Jiang et al. 2002), FR235222 from an *Acremonium* species (Mori et al. 2003) or the chlamydocins from *Diheterospora chlamydosporia* (Closse and Huguenin 1974) and *Peniophora* sp. (Tani et al. 2001) have been reported to exhibit antiprotozoal activity, to induce apoptosis, to have immunosuppressive effects or to retard plant growth (de Schepper et al. 2003).

Due to their toxic effects in animal and humans and their occurrence in food and feedstuff, fumitremorgins, verruculogens, roquefortins C and D, sporidesmins, chaetocin, cyclochlorotine and malformins were classified as mycotoxins (Betina 1989). For their different biological activities the reader is referred to the vast online literature on this group of fungal products.

Malformin C (Fig. 13.4), despite its antibacterial, plant-deforming and fibrinolytic activities, recently aroused some interest due to its inhibitory effects on bleomycin-induced G2 arrest, thus potentiating its DNA-damaging action, a mode of action that might be useful for the treatment of cancer (Hagimori et al. 2007).

Cyclosporins are not the only immunomodulating fungal metabolites. Many epipolythiodioxopiperazines, in addition to other biological activities, are immunosuppressants.

Sevastelins, cyclodepsipeptides with a lipophilic side-chain, from a *Penicillium* species blocked human T cell activation in vitro and showed low acute toxicity in mice (Morino et al. 1994). HUN-7293 acts as inhibitor of cytokine-induced expression of vascular cell adhesion molecule-1 on human endothelial cells (Hommel et al. 1996). It is structurally identical to pestahivin.

The depsipeptide aureobasidin A has an interesting mode of action, the inositol phosphoceramide synthase (IPS). The fungal enzyme is considered to be an attractive target for novel fungicides. Further development of aureobasidin A was hampered by its inhibitory effects on ABC transporters in yeasts and humans (Fostel and Lartey 2000). The pleofungins from a *Phoma* species showed antifungal activity towards *Candida albicans*, *Cryptococcus neoformans* and *A. fumigatus* with minimal inhibitory concentrations in the range of 1 µg/ml or lower (Yano et al. 2007). The compounds inhibited the *A. fumigatus* IPS with IC_{50} values of 1 ng/ml (Aoyagi et al. 2007).

Neoechinulin A has protective activity in PC12 cells against lethal effects of peroxynitrite and against 1-methyl-4-phenylpyridine, a neurotoxin capable of inducing neurodegeneration in humans (Kajimura et al. 2008). The cyclic tetrapeptide CJ-15,208 is a kappa opinoid receptor antagonist (Saito et al. 2002) and four depsipeptides were reported to be selective and competitive human tachykinin receptor antagonsits (Hedge et al. 2001).

Among nine beauverolides tested for acyl-CoA: cholesterol acyltransferase (ACAT) inhibitory activity in CHO-cells expressing ACAT1 or ACAT2, beauverolides I and III inhibited ACAT1 rather selectively, no antimicrobial or cytotoxic activities were detected and beauvericin was cytotoxic (Matsuda et al. 2004; Ohshiro et al. 2007). ACAT is discussed as a target for new antiatherosclerotic agents (Roth 1998; Namatame et al. 2004).

The outstanding anthelmintic activity of PF1022A combined with its mode of action, e.g. binding to the latrophilin-like receptor of

Haemonchus contortus (Conder et al. 1995; Saeger et al. 2001) and low toxicity led to the development of emodepsin, a novel drug used in animal health.

Antiparasitic properties have been reported for cycloaspeptides A and D (Dalsgaard et al. 2004b). Verticilide, a cyclic depsipeptide isolated from the culture broth of *Verticillium* sp. FKI-1033, inhibits the binding of ryanodine to the receptor (RyR) and has insecticidal activity (Monma et al. 2006). Serinocyclin A isolated from *M. anisopliae* condia produced a sublethal locomotory defect in mosquito larvae (Krasnoff et al. 2007). Argifin and argadin, two cyclopentapeptides from a *Gliocladium* and a *Clonostachys* species, are potent inhibitors of chitinase B from *Serratia marcescens* (Houston et al. 2002). When injected into cockroach larvae, the moult was arrested. Besides cyclopeptides and -depsipeptides fungi also produce other peptides with insecticidal activities, recent examples are the neofrapeptins from *Geotrichum candidum* (Fredenhagen et al. 2006). Selective nematicidal properties have been reported only for the omphalotins with high inhibitory activity towards *Meloidogyne incognita* and low activity towards *Caenorhabditis elegans* (Mayer et al. 1997, 1999; Sterner et al. 1997). The nematicidal properties of the hydroxylated omphalotins are higher than those of the parent compound, but unfortunately they are not stable (Büchel et al. 1998a, Liermann et al. 2009).

Antiviral properties have been reported for sansalvamide A, a cyclodepsipeptide from a marine *Fusarium*, which inhibits viral topoisomerase-catalyzed DNA relaxation (Hwang et al. 1999).

The clavariopsins, cyclic depsipeptides from *Clavariopsis aquatica*, show selective antifungal activity, bacteria are not affected and mice tolerate 100 mg/kg of clavariopsin A. As mode of action, an inhibition of cell components was proposed (Kaida et al. 2001). Glomosporin from a *Glomospora* species is a lipophilic depsipeptide with antifungal activity (Sato et al. 2000). Whether this compound also inhibits cell wall synthesis was not reported. Antifungal and cytotoxic activities were reported for petriellin A (Lee et al. 1995). Cytotoxic activities are exhibited by many cyclopeptides and -depesipeptides. The destruxins have been intensively investigated (Vey et al. 2002; Skrobek and Butt 2005). Psychrophilin D is weakly cytotoxic towards P388 mouse leukaemia cells with an IC_{50} value of 10 µg/ml (Dalsgaard et al. 2005), while the icosalides inhibit the replication of MDCK cells with LD_{50} of 5–10 µg/ml (Boros et al. 2006). The aspergillicins are weakly cytotoxic with LD_{99} of 25–50 µg/ml (Capon et al. 2003).

As inhibitors of 1,3-β-glucan synthesis have high potential as antimycotic drugs (Fostel and Lartey 2000), fungi have been intensively screened for the production of inhibitors of cell wall synthesis and cyclic peptides as well as cyclic depsipeptides have been found.

The antimycotic drugs already on the market (caspofungin, micafungin, anidulafungin) are derived from lipopeptides (Butler 2004; Morrison 2006). Their spectrum of activity is mainly restricted to *Candida* and *Aspergillus* species. *Cryptococcus neoformans*, *Trichosporon* and *Fusarium* species or Zygomycetes are not affected (Denning 2003), although the glucan synthase from *C. neoformans* is sensitive to echinocandins (Maligie and Selitrennikoff 2005).

IV. Ecological Role of Cyclic Peptides and Depsipeptides

Many secondary metabolites play a crucial role for fungi in their natural habitats. Endophytic fungi of grasses belonging to the genera *Neotyphodium/ Epichloe* confer protection from mammalian and insect herbivores, or enhanced resistance against nematodes and phytopathogenic fungi (Schardl et al. 2004; Panaccione et al. 2006). Some of these beneficial effects are due to NRPS products. Ergovaline has been identified among the fungal metabolites in the plant host. Malformins have been detected in onion scales after infection with *A. niger* (Curtis et al. 1974).

The role of siderophores in plant and human pathogens is currently elucidated by many research groups (for a review see Haas et al. 2008). Additional functions of siderophores for the producing organism are acquisition and storage of iron as well as regulation of asexual and sexual development and protection against oxidative stress (Einsendle et al. 2006; Hof et al. 2009). Nonproducing organisms like *Saccharomyces cerevisiae* are able to use, e.g. transport iron-siderophore complexes, thus the compounds might also play a role in fungus–fungus interactions.

In plant-pathogenic fungi cyclic peptides like HC-toxins in *Cochliobolus carbonum*, AM toxins in *Alternaria alternata*, sirodesmin PL in *Leptosphaeria maculans* (anamorph *Phoma lingam*) or enniatins in *Fusarium* species act as putative virulence factors. In some cases this has already been proven, when gene deletions result in apathogenic strains or strains with reduced virulence (Ahn and Walton 1998; Pedley and Walton 2001; Elliott et al. 2007). Likewise the insecticidal depsipetides of insect pathogens have the same function. Investigation on the role of destruxins in the pathogenicity of *Metarhizium anisopliae* against three species of insects revealed a direct relationship between the titer of destruxins produced by the strains in vitro and their destructive action (Kershaw et al. 1999). In the plant-pathogenic *Alternaria brassicae*, destruxin B is a host-specific toxin. In three *Brassica* species the degree of their sensitivity to destruxin B positively correlated with their degree of susceptibility (Pedras et al. 2002).

The function of shearamide A, an insecticidal cyclopeptide isolated from the ascostromata of *Eupenicillium shearii* (Belofsky et al. 1998) may be in protecting the fungus against insects, similar to ergopeptides in the sklerotia of *Claviceps* species (Chap. 9).

V. Conclusions

The capability to produce secondary metabolites derived from amino acids by NRPS is widespread among the higher fungi and not dependent on the ecological niches inhabited by them. There are no special habitats from which highly prolific secondary metabolite producers are isolated.

Cyclic peptides and -depsipeptides constitute an interesting class of secondary metabolites with great potential not only in medicine but also in agriculture. This can easily be grasped from the wide array of biological activities exhibited by these compounds. Their chemical diversity is enhanced by the possibility of producing an array of related compounds by precursor-supplemented fermentations of the correspondent fungus. This readily facilitates investigations on structure–activity relationships.

In agriculture, fungally derived pesticides offer ecological advantages and strains with enhanced production of bioactive compounds might be developed as biopesticides. For both agriculture and pharmacology bioactive natural compounds may lead to novel targets and serve as lead structures.

Acknowledgements. Work in our Institute was supported by the State of Rhineland–Palatinate, BASF SE, Bayer AG, BMBF and the DFG.

References

Abbanat D, Leighton M, Maiese W, Jones EBG, Pearce C, Greenstein M (1998) Cell wall active compounds produced by the marine fungus *Hypoxylon oceanicum* LL-15G56. J Antibiot 51:296–302

Adachi K, Kanoh K, Wisespong P, Nishijima M, Shizuri Y (2005) Clonostachysins A and B, new antidinoflagellate cyclic peptides from a marine-derived fungus. J Antibiot 58:145–150

Ahn JH, Walton JD (1998) Regulation of cyclic peptide biosynthesis and pathogenicity in *Cochliobolus carbonum* by TOXEP, a novel protein with a bZIP basic DNA-binding motif and four ankyrin repeats. Mol Gen Genet 260:462–469

Amitani R, Taylor G, Elezis EN, Liewellyn-Jones C, Mitchell J, Kuze F, Cole PJ, Wilson R (1995) Purification and characterization of factors produced by *Aspergillus fumigatus* which affect human ciliated respiratory epithelium. Infect Immun 63:3266–3271

Anke H, Sterner O (2002) Insecticidal and nematicidal metabolites from fungi. In: Osiewacz HD (ed) Industrial applications. Mycota X. Springer, Heidelberg, pp 109–127

Anke H, Kinn J, Bergquist KE, Sterner O (1991) Production of siderophores by strains of the genus *Trichoderma*: Isolation and characterization of the new lipophilic coprogen derivative, palmitoylcoprogen. Biol Metals 4:176–180

Anke T, Erkel G (2002) Non β-lactam antibiotics. In: Osiewacz HD (ed) Industrial applications. Mycota X. Springer, Heidelberg, pp 93–108

Antelo L, Hof C, Eisfeld K, Sterner O, Anke H (2006) Siderophores produced by *Magnaporthe grisea* in the presence and absence of iron. Z Naturforsch. 61c:461–464

Aoyagi A, Yano T, Kozuma S, Takatsu T (2007) Pleofungins, novel inositol phosphorylceramide synthase inhibitors, from *Phoma* sp. SANK 13899. J Antibiot 60:143–152

Arai N, Shiomi K, Iwai Y, Omura S (2000) Argifin, a new chitinase inhibitor, produced by *Gliocladium* sp. FTD-0668. II. Isolation, physico-chemical properties, and structure elucidation. J Antibiot 53:609–614

Badan SD, Ridley DD, Singh P (1978) Isolation of cyclodepsipeptides from plant pathogenic fungi. Aust J Chem 31:1397–1399

Belofsky GN, Gloer JB, Wicklow DT, Dowd PF (1998) Shearamide A: a new cyclic peptide from the ascostromata of *Eupenicillium shearii*. Tetrahedron Lett 39:5497–5500

Belofsky GN, Jensen PR, Fenical W (1999) Sansalvamide: a new cytotoxic cyclic depsipeptide produced by a marine fungus of the genus *Fusarium*. Tetrahedron Lett 40:2913–2916

Bertram A, Pattenden G (2007) Marine metabolites: metal binding and metal complexes of azole-based cyclic peptides of marine origin. Nat Prod Rep 24:18–30

Betina V (1989) Epipolythiopiperazine-3,6-diones. In: Mycotoxins, chemical, biological and evironmental aspects. Elsevier, Amsterdam, pp 388–405

Bills GF, Platas G, Peláez F, Masurekar P (1999) Reclassification of a pneumocandin-producing anamorph, *Glarea lozoyensis* gen. et sp. nov., previously identified as *Zalerion arboricola*. Mycol Res 103:179–192

Birch AJ, Massy-Westropp RA, Rickards RW (1956) Studies in relation to biosynthesis. Part VIII. The structure of mycelianamide. J Chem Soc 3717-3721

Blunt JW, Copp BR, Munro MHG, Northcote PT, Prinsep MR (2006) Marine natural products. Nat Prod Rep 23:26–78

Boros C, Smith CJ, Vasina Y, Che Y, Dix AB, Darveaux B, Pearce C (2006) Isolation and identification of the icosalides – cyclic peptolides with selective antibiotic and cytotoxic activities. J Antibiot 59:486–494

Boudart G (1989) Antibacterial activity of sirodesmin PL phytotoxin: application to the selection of phytotoxin-deficient mutants. Appl Enivron Microbiol 55:1555–1559

Büchel E, Martini U, Mayer A, Anke H, Sterner O (1998a) Omphalotins B, C, and D, nematicidal cyclopeptides from *Omphalotus olearius*. Absolute configuaration of omphalotin A. Tetrahedron 54:5345–5352

Büchel E, Mayer A, Martini U, Anke H, Sterner O (1998b) Structure elucidation of omphalotin, a cyclic dodecapeptide with potent nematicidal activity from *Omphalotus olearius*. Pest Sci 54:309–311

Buckingham J (2008) (Ed) Dictionary of natural products on DVD, version 17.1. Chapman and Hall/CRC, Boca Raton

Butler MS (2004) The role of natural product chemistry in drug discovery. J Nat Prod 67:2141–2154

Capon RJ, Skene C, Stewart M, Ford J, O'Hair RAJ, Williams L, Lacey E, Gill JH, Heiland K, Friedel T (2003) Aspergillicins A–E: five novel depsipeptides from the marine-derived fungus *Aspergillus carneus*. Org Biomol Chem 1:1856–1862

Che Y, Swenson DC, Gloer JB, Koster B, Malloch D (2001) Pseudodestruxins A and B: new cycllic depsipeptides from the coprophilous fungus *Nigrosabulum globosum*. J Nat Prod 64:555–558

Chen CH, Lang G, Mitova MI, Murphy AC, Cole ALJ, Din LB, Blunt JW, Munro MHG (2006) Pteratides I–IV, new cytotoxic cyclodepsipeptides from the Malaysian basidiomycete *Pterula* sp. J Org Chem 71:7947–7951

Closse A, Huguenin R (1974) Isolierung und Strukturaufklärung von Chlamydocin. Helv Chim Acta 57:533–545

Cole RJ, Schweikert MA (2003) Diketopiperazines. In: Handbook of secondary fungal metabolites, vol 1. Academic, Amsterdam, pp 145–244

Conder GA, Johnson SS, Nowakowski DS, Blake TE, Dutton FE, Nelson SJ, Thomas EM, Davis JP, Thompson DP (1995) Anthelmintic profile of the cyclodepsipeptide PF1022A in in vitro and in vivo models. J Antibiot 48:820–823

Curtis RW, Stevenson WR, Tuite J (1974) Malformin in *Aspergillus niger*-infected onion bulbs (*Allium cepa*). Appl Environ Microbiol 28:362–365

Daferner M (2000) Antibiotisch aktive Sekundärstoffe aus höheren marinen Pilzen. Dissertation, University of Kaiserslautern

Dalsgaard PW, Blunt JW, Munro MHG, Larsen TO, Christophersen C (2004a) Psychrophilin B and C: cyclic nitropeptides from the psychrotolerant fungus *Penicillium rivulum*. J Nat Prod 67:1950–1952

Dalsgaard PW, Larsen TO, Frydenvang K, Christophersen C (2004b) Psychrophilin A and cycloaspeptide D, novel cyclic peptides from the psychotolerant fungus *Penicillium ribeum*. J Nat Prod 67:878–881

Dalsgaard PW, Larsen TO, Christophersen C (2005) Bioactive cyclic peptides from the psychrotolerant fungus *Penicillium algidum*. J Antibiot 58:141–144

Darkin-Rattray SJ, Gurnett AM, Myers RW, Dulski PM, Crumley TM, Allocco JJ, Cannova C, Meinke PT, Colletti SL, Bednarel MA, Singh SB, Goetz MA, Dombrowski AW, Polishook ED, Schmatz DM (1996) Apicidin, a novel antiprotozoal agent that inhibits parasite histone deacetylase Proc Natl Acad Sci USA 93:13143–31147

Davoli P, Mucci A, Schenetti L, Weber RWS (2005) Laetiporic acids, a family of non-carotenoid polyene pigments from fruit-bodies and liquid cultures of *Laetiporus sulphureus* (Polyporales, Fungi). Phytochemistry 66:817–823

!de Schepper S, Bruwiere H, Verhulst T, Steller U, Andries L, Wouters W, Janicot M, Arts J, van Heusden J (2003) Inhibition of histone deacylases by chlamydocin induces apoptosis and proteasome-mediated degradation of survivin. J Pharmacol Exp Ther 304:881–888

Degenkolb T, Gams W, Brückner H (2008) Natural cyclopeptaibols and related cyclic tetrapeptides: structural diversity and future prospects. Chem Biodiver 5:693–706

Demain AL, Elander RP (1999) The beta-lactam antibiotics: past, present, and future. Antonie Van Leeuwenhoek 75:5–19

Denning DW (2002) Echinocandins: a new class of antifungals. J Antimicrob Chemother 49:889–891

Denning DW (2003) Echinocandin antifungal drugs. Lancet 362:1142–1151

Eichhorn H, Lessing F, Winterberg B, Schirawski J, Kamper J, Mueller P, Kahmann R (2006) A ferroxidation/permeation iron uptake system is required for virulence in *Ustilago maydis*. Plant Cell 18:3332–3345

Eickman N, Clardy J, Cole RJ, Kirksey JW (1975) The structure of fumitremorgin A. Tetrahedron Lett 16:1051–1054

Eisendle M, Schrettl M, Kragl C, Müller D, Illmer P, Haas H (2006) The intracellular siderophore ferricrocin is involved in iron storage, oxidative-stress resistance, germination, and sexual development in *Aspergillus nidulans*. Eukaryot Cell 5:1596–603

Elliott CE, Gardiner DM, Thoma G, Cozijnsen A, van de Wouw A, Howlett BJ (2007) Production of the toxin sirodesmin PL by *Leptosphaeria maculans* during infection of *Brassica napus*. Mol Plant Pathol 8:791–802

Ernst-Russell M, Chai CL, Hurne AM, Waring P, Hockless DCR, Elix JA (1999) Structure revision and cytotoxic activity ot the scabrosin esters, epithiopiperazine-diones from the lichen *Xanthoparmelia scabrosa*. Aust J Chem 52:279–283

Feifel SC, Schmiederer T, Hornbogen T, Berg H, Süssmuth RD, Zocher R (2007) In vitro synthesis of new ennia-tins: probing the α-d-hydroxy carboxylic acid bind-ing pocket of the multienzyme enniatin synthetase. ChemBioChem 8:1767–1770

Fostel JM, Lartey PA (2000) Emerging novel antifungal agents. Drug Discov Today 5:25–32

Fredenhagen A, Molleyres LP, Böhlendorf B, Laue G (2006) Structure determination of neofrapeptins A to N: peptides with insecticidal activity produced by the fungus *Geotrichum candidum*. J Antibiot 59:267–280

Fridrichsons J, Mathieson AMCL (1962) The structure of sporidesmin: causative agent of facial eczema in sheep. Tetrahedron Lett 3:1265–1268

Fujie A, Iwamoto T, Muramatsu H, Okudaira T, Nitta K, Nakanishi T, Sakamoto K, Hori Y, Hino M, Hashi-moto S, Okuhara M (2000) FR901469, a novel anti-fungal antibiotic from an unidentified fungus No 11243. I. Taxonomy, fermentation, isolation, phy-sico-chemical properties and biological properties. J Antibiot 53:912–919

Fujie A, Muramatsu H, Yoshimura S, Hashimoto M, Shigematsu N, Takase S (2001) FR901469, a novel antifungal antibiotic from an unidentified fungus No 112434. III. Structure determination. J Antibiot 54:588–594

Gardiner DM, Waring P, Howlett BJ (2005) The epipo-lythiodioxopiperazine (ETP) class of fungal toxins: distribution, mode of action, functions and biosyn-thesis. Microbiology 151:1021–1032

Gevers W, Kleinkauf H, Lipmann F (1968) The activation of amino acids for biosynthesis of gramicidin S. Proc Natl Acad Sci USA 63:1335–1342

Glinski M, Hornbogen T, Zocher R (2001) Enzymatic syn-thesis of fungal N-methylated cyclopeptides and dep-sipeptides. In: Kirst H, Yeh WK, Zmijewski M (eds) Enzyme technologies for pharmaceutical and bio-technological applications. Dekker, New York, pp 471–497

Gournelis DC, Laskaris GG, Verpoorte R (1998) Cyclopep-tide alkaloids In: Herz W, Falk H, Kirby GW, Moore RE, Tamm Ch (eds) Fortschritte der Chemie orga-nischer Naturstoffe, vol 75. Springer, Heidelberg, pp 1–179

Gross ML, McCrery D, Crow F, Tomer KB, Pope MR, Ciuffetti LM, Knoche HW, Daly JM, Dunkle DL (1982) The structure of the toxin from *Helminthos-porium carbonum*. Tetrahedron Lett 51:5381–5384

Gunatilaka AAL (2006) Natural products from plant-associated microorganisms: distribution, structural

diversity, bioactivity, and implications of their occurrence. J Nat Prod 69:509–526

Gupta S, Peiser G, Nakajima T, Hwang Y-S (1994) Charac-terization of a phytotoxic cyclotetrapeptide, a novel chlamydocin analogue, from *Verticillium coccos-porum*. Tetrahedron Lett 35:6009–6012

Haas H, Eisendle M, Turgeon BG (2008) Siderophores in fungal physiology and virulence. Annu Rev Phyto-pathol 46:149–187

Hagimori K, Fukuda T, Hasegawa Y, Omura S, Tomoda H (2007) Fungal malformins inhibit bleomycin-induced G2 checkpoint in Jurkat cells. Biol Pharm Bull 30:1379–1383

Hashimoto S (2009) Micafungin: a sulfated echinocandin. J Antibiot 62:27–35

Hedge VR, Puar MS, Dai P, Pu H, Patel M, Anthes JC, Richard C, Terracciano J, Das PR, Gullo V (2001) A family of depsipeptide fungal metabolites, as selec-tive and competitive human tachykinin receptor (NK2) antagonists: fermentation, isolation, physico-chemical properties, and biological activity. J Anti-biot 54:125–135

Hof C, Eisfeld K, Welzel K, Antelo L, Foster AJ, Anke H (2007) Ferricrocin synthesis in *Magnaporthe grisea* and its role in pathogenicity. Mol Plant Pathol 8:163–172

Hof C, Eisfeld K, Antelo L, Foster AJ, Anke H (2009) Side-rophore synthesis in *Magnaporthe grisea* is essential for vegetative growth, conidiation and resistance to oxidative stress. Fungal Genet Biol 46:321–332

Hommel U, Weber H-P, Oberer L, Naegeli HU, Oberhauser B, Foster CA (1996) The 3D-structure of a natural inhibitor of cell adhesion molecule expression. FEBS Lett 379:69–73

Houston DR, Shiomi K, Arai N, Omura S, Peter MG, Tur-berg A, Synstad B, Eijsink VG, van Aalten DMF (2002) High-resolution structures of a chitinase complex with natural product cyclopentapeptide inhibitors: Mimicry of carbohydrate substrate Proc Natl Acad Sci USA 99:9127–9132

Howard DH (1999) Acquisition, transport, and storage of iron by pathogenic fungi. Clin Microbiol Rev 12:394–404

Huang H, She Z, Lin Y, Vrijmoed LLP, Lin W (2007) Cyclic peptides from an endophytic fungus obtained from a Mangrove leaf (*Kandelia candel*). J Nat Prod 70:1696–1699

Hume AM, Chai CLL, Moermann K, Waring P (2002) Influx of calcium through a redox-sensitive plasma membrane channel in thymocytes causes early necrotic cell death induced by the epipolythiodioxo-piperazine toxins. J Biol Chem 35:31631–31638

Hwang Y, Rowley D, Rhodes D, Gertsch J, Fenical W, Bushman F (1999) Mechanism of inhibition of a pox-virus topoisomerase by the marine natural product sansalvamide A. Mol Pharmacol 55:1049–1053

Isaka M, Kittakoop P, Kirtikara K, Hywel-Jones NI, Thebtaranonth Y (2005a) Bioactive substances from insect pathogenic fungi. Acc Chem Res 38:813–823

Isaka M, Palasarn S, Rachtawee P, Vimuttipong S, Kongsaeree P (2005b) Unique diketopiperazine dimers from the insect pathogenic fungus *Verticillium hemipterigenum* BCC 1449. Org Lett 7:2257–2260

Isaka M, Palasarn S, Kocharin K, Hywel-Jones NI (2007) Comparison of the bioactive secondary metabolites from the scale insect pathogens, anamorph *Paecilomyces cinnamomeus*, and teleomorph *Torrubiella luteorostrata*. J Antibiotics 60:577–581

Ishiyama D, Sato T, Honda R, Senda H, Konno H, Kanazawa S (2000) Glomosporin, a novel antifungal cyclic depsipeptide from *Glomospora* sp. II. Structure elucidation. J Antibiot 53:525–631

Itazaki H, Nagashima K, Sugita K, Yoshida H, Kawamura Y, Yashuda Y, Matsumoto K, Ishii K, Uotani N, Nakai H, Terui A, Yoshimatsu S, Ikenishi Y, Nakagawa Y (1990) Isolation and structural elucidation of new cyclotetrapeptides, trapoxins A and B, having detransformation. J Antibiot 43:1524–1532

Iwamoto T, Fujie A, Nitta K, Hashimoto S, Okuhara M, Kohsaka M (1994a) WF11899A, B and C, novel antifungal lipopeptides II. Biological properties. J Antibiot 45:1092–1097

Iwamoto T, Fujie A, Sakamota K, Tsurumi Y, Shigematsu N, Yamashita M, Hashimoto S, Okuhara M, Kohsaka M (1994b) WF11899A, B and C, novel antifungal lipopeptides I. Taxonomy, fermentation, isolation and physico-chemical properties. J Antibiot 47:1084–1091

Jiang Z, Barret MO, Boyd KG, Adams DR, Boid ASF, Burgess JG (2002) JM47, a cyclic tetrapeptide HC-toxin analogue from a marine *Fusarium* species. Phytochemistry 60:33–38

Johnson MD, Perfect JR (2003) Caspofungin: first approved agent in a new class of antifungals. Expert Opin Pharmacother 4:807–823

Kaida K, Fudou R, Kameyama T, Tubaki K, Suzuki Y, Ojika M, Sakagami Y (2001) New cyclic depsipeptide antibiotics, clavariopsins A and B, produced by an aquantic hyphomycete, *Clavariopsis aquatica*. J Antibiot 54:17–21

Kajimura Y, Aoki T, Kuramochi K, Kobayashi S, Sugawara F, Watanabe N, Arai T (2008) Neoechinulin A protects PC12 cells against MPP+-induced cytotoxicity. J Antibiot 61:330–333

Kamei K, Watanabe A (2005) *Aspergillus* mycotoxins and their effect on the host. Med Mycol 43[Suppl 1]:95–99

Kanasaki R, Abe F, Kobayashi M, Katsuoka M, Hashimoto M, Takase S, Tsurumi Y, Fujie A, Hino M, Hashimoto S, Hori Y (2006a) FR220897 and FR220899, novel antifungal lipopeptides from *Coleophoma empetri* No. 14573. J Antibiot 59:149–157

Kanasaki R, Kobayashi M, Fujine K, Sato I, Hashimoto M, Takase S, Tsurumi Y, Fujie A, Hino M, Hashimoto S (2006b) FR227673 and FR190293, novel antifungal lipopeptides from *Chalara* sp. No22210 and *Tolypocladium parasiticum* No 16616. J Antibiot 59:158–167

Kanasaki R, Sakamota K, Hashimoto M, Takase S, Tsurumi Y, Fujie A, Hino M, Hashimoto S, Hori Y (2006c) FR209602 and related compounds, novel antifungal lipopeptides from *Coleophoma crateriformis* No. 738. J Antibiot 59:137–144

Keller U, Tudzynski P (2002) Ergot alkaloids. In: Osiewacz HD (ed) Industrial applications. Mycota X. Springer, Heidelberg, pp 157–181

Keller-Juslen C, Kuhn M, Loosli HR, Petcher TJ, Weber HP, von Wartburg A (1976) Struktur des Cyclopeptid-Antibiotikums SL 7810 (= Echinocandin B) Tetrahedron Lett 17:4147–4150

Kershaw M, Moorhouse ER, Bateman R, Reynolds SE, Charnley AK (1999) The role of destruxins in the pathogenicity of *Metarhizium anisopliae* for three species of insect. J Invert Pathol 74:213–223

Kleinkauf H, von Döhren H (1997) Peptide antibiotics. In: Kleinkauf H, von Döhren H (eds) Products of secondary metabolism. Biotechnology, vol 7. VCH, Weinheim, pp 277–322

Kleinwachter P, Dahse HM, Luhmann U, Schlegel B, Dornberger K (2001) Epicorazine C, an antimicrobial metabolite from *Stereum hirsutum* HKI 0195. J Antibiot 54:521–525

Krasnoff SB, Keresztes I, Gillilan RE, Szebenyi DME, Donzelli BGG, Vhurchill ACL, Gibson DM (2007) Serinocyclins A and B, cyclic heptapeptides from *Metarhizium anisopliae*. J Nat Prod 70:1919–1924

Kobbe B, Cushman M, Wogan GN, Demain AL (1977) Production and antibacterial activity of malformin C, a toxic metabolite of *Aspergillus niger*. Appl Environ Microbiol 33:996–997

Kürnsteiner H, Zinner M, Kück U (2002) Immunosuppressants. In: Osiewacz HD (ed) Industrial applications. Mycota X. Springer, Heidelberg, pp 129–155

Lee KK, Gloer JB Scott JA, Malloch D (1995) Petriellin A: a novel antifungal depsipeptide from the coprophilous fungus *Petriella sordida*. J Org Chem 60:5384–5385

Lewis JR (2002) Amaryllidaceae, *Sceletium*, imidazole, oxazole, thiazole, peptide and miscellaneous alkaloids. Nat Prod Rep 19:223–258

Li C, Oberlies NH (2005) The most widely recognized mushroom: chemistry of the genus *Amanita*. Life Sci 78:532–538

Li X, Kim S-K, Nam KW, Kang JS, Choi HD, Son BW (2006) A new antibacterial dioxopiperazine alkaloid related to gliotoxin from a marine isolate of the fungus *Pseudallescheria*. J Antibiot 59:248–250

Li Y, Li X, Kim S-K, Kang JS, Choi HD, Rho JR, Son BW (2004) Golmaenone, a new diketopiperazine alkaloid from the marine-derived fungus *Aspergillus* sp. Chem Pharm Bull 52:375–376

Liermann JC, Kolshorn H, Antelo L, Hof C, Anke H, Opatz T (2009) Omphalotins E-I, oxidatively modified nematicidal cyclopeptides from *Omphalotus olearius*. Eur J Org Chem 2009:1256–1262

Lira SP, Vita-Marques AM, Seleghim MHR, Bugni TS, LaBarbera DV, Sette LD, Sponchiado SRP, Ireland CM, Berlinck RGS (2006) New destruxins from the marine-derived fungus *Beauveria felina*. J Antibiot 59:553–563

Liu J-K (2005) N-containing compounds of macromycetes. Chem Rev 105:2723–2744

Lorenz P, Jensen PR, Fenical W (1998) Mactanamide, a new fungistatic diketopiperazine produced by a marine *Aspergillus* sp. Nat Prod Lett 12:55–60

Maligie MA, Selitrennikoff CP (2005) *Cryptococcus neoformans* resistance to echinocandins: (1,3) β-glucan synthase activity is senitive to echinocandins. Antimicrob Agents Chemother 49:2851–2856

Malmstrom J, Ryager A, Anthoni U, Nielsen PH (2002) Unguisin C, a GABA-containing cyclic peptide from the fungus *Emericella unguis*. Phytochemistry 60:869–887

Martins MB, Carvalho I (2007) Diketopiperazines: biological activity and synthesis. Tetrahedron 64:9923–9932

Matha V, Jegorov A, Weiser J, Pillai JS (1992) The mosquitocidal activity of conidia of *Tolypocladium tundrense* and *Tolypocladium terricola*. Cytobios 69:163–170

Matsuda D, Namatame I, Tomoda H, Kobayashi S, Zocher R, Kleinkauf H, Omura S (2004) New beauverolides produced by amino acid-supplemented fermentation of *Beauveria* sp. FO-6979. J Antibiot 57:1–9

Mayer A, Sterner O, Anke H (1997) Omphalotin, a new cyclic peptide with potent nematicidal activity from *Omphalotus olearius*. 1. Fermentation and biological activity. Nat Prod Lett 10:25–33

Mayer A, Kilian M, Hoster B, Sterner O, Anke H (1999) In vitro and in vivo nematicidal activities of the cyclic dodecapeptide omphalotin A. Pest Sci 55:27–30

Miyado S, Kawasaki H, Aoyagi K, Yaguchi T, Okada T, Sugiyama J (2000) Taxonomic position of the fungus producing the anthelmintic PF1022 based on the 18S rRNA gene base sequence. Nippon Kinzoku Gakkai Kaiho 41:183–188

Mochizuki K, Ohmori K, Tamura H, Shizuri Y, Nishiyama S, Miyoshi E, Yamamura S (1993) The structures of bioactive cyclodepsipeptides, beauveriolides I and II, metabolites of entomopathogenic fungi *Beauveria* sp. Bull Chem Soc Jpn 66:3041–3046

Monma S, Sunazuka T, Nagai K, Arai T, Shiomi K, Matsui R, Mura S (2006) Verticilide: elucidation of absolute configuration and total synthesis. Org Lett 8:5601–5604

Mori H, Urano Y, Abe F, Furukawa S, Tsurumi Y, Sakamoto K, Hashimoto M, Takase S, Hino M, Fujii T (2003) FR235222, a fungal metabolite, is a novel immunosuppressant that inhibits mammalian histone deacetylase (HDAC) 1. Taxonomy, fermentation, isolation, and biological activities. J Antibiot 56:72–79

Morino T, Masuda A, Yamada M, Nishimoto Y, Nishikiori T, Saito S, Shimada (1994) Stevastelins, novel immunosuppresssants produced by *Penicillium*. J Antibiot 47:1341–1343

Morris SA, Schwartz RE, Sesin DF, Masurekar P, Hallada TC, Schmatz DM, Bartizal K, Hensens OD, Zink DL (1994) Pneumocandin D$_0$, a new antifungal agent and potent inhibitor of *Pneumocystis carinii*. J Antibiot 47:755–764

Morrison VA (2006) Echinocandin antifungals: review and update. Expert Rev Anti Infect Ther 4:325–342

Namatame I, Zomoda H, Ishibashi S, Omura S (2004) Antiatherogenic activity of fungal beauverolides, inhibitors of lipid droplet accumulation in macrophages. Proc Natl Acad Sci USA 101:737–742

Nilanonta C, Isaka M, Chanphen R, Thong-orn N, Tanticharoen M, Thebtaranonth Y (2003) Unusual enniatins produced by the insect pathogenic fungus *Verticillium hemipterigenum*: isolation and studies on precursor-directed biosynthesis. Tetrahedron 59:1015–1020

Odds FC, Brown AJ, Gow NA (2003) Antifungal agents: mechanisms of action. Trends Microbiol 11:272–279

Oide S, Moeder W, Krasnoff S, Gibson D, Haas H, Yoshioka K, Turgeon BG (2006) *NPS6*, encoding a nonribosomal peptide synthetase involved in siderophore-mediated iron metabolism, is a conserved virulence determinant of plant pathogenic ascomycetes. Plant Cell 18:2836–2853

Oide S, Krasnoff SB, Gibson DM, Turgeon BG (2007) Intracellular siderophores are essential for ascomycete sexual development in heterothallic *Cochliobolus heterostrophus* and homothallic *Gibberella zeae*. Eukaryot Cell 6:1339–1353

Ohshiro T, Rudel LL, Omura S, Tomoda H (2007) Selectivity of microbial acyl-CoA:cholesterol acyltransferase inhibitors towards isoenzymes. J Antibiot 60:43–51

Ohyama T, Kurihara Y, Ono Y, Ishikawa T, Miyakoshi S, Hamano K, Arai M, Suzuki T, Igari H, Suzuki Y, Inukai M (2000) Arborcandins A, B, C, D, E, and F, novel 1,3-beta-glucan synthase inhibitors: production and biological activities. J Antibiot 53:1108–1116

Panaccione DC, Cipoletti JR, Sedlock AB, Blemings KP, Schradl CL, Machado C, Seidel GE (2006) Effects of ergot alkaloids on food preference and satiety in rabbits, as assessed with gene-knockout endophytes in perennial ryegrass (*Lolium perenne*). J. Agric Food Chem 54:4582–4587

Pasqualotto AC, Denning DW (2008) New and emerging treatments for fungal infections. J Antimicrob Chemother 61[Suppl 1]:i19–i30

Patil BB, Wakharkar RD, Chincholkar SB (1995) Siderophores of *Cunninghamella blakesleeana* NCIM 687. World J Microbiol Biotechnol 15:265–268

Pedley KF, Walton JD (2001) Regulation of cyclic peptide biosynthesis in a plant pathogenic fungus by a novel transcription factor. Proc Natl Acad Sci USA 98:14174–14179

Pedras MSC, Zaharia LI, Ward DE (2002) The destruxins: synthesis, biosynthesis, biotransformation, and biological activity. Phytochemistry 59:579–596

Pomilio AB, Battista ME, Vitale AA (2006) Naturally-occurring cycopeptides: structures and bioactivity. Curr Org Chem 10:2075–2121

Prasad C (1995) Bioactive cyclic peptides. Peptides 16:151–164

Rees NH, Penfold DJ, Rowe ME, Chowdhry BZ, Cole SCJ, Samuels RI, Turner DL (1996) NMR studies of the conformation of destruxin A in water and in acetonitrile. Magn Reson Chem 34:237–241

Renshaw JC, Robson GD, Trinci APJ, Wiebe MG, Livens FR, Collison DC, Taylor RJ (2002) Fungal siderophores: structures, functions and applications. Mycol Res 106:1123–1142

Roth BD (1998) ACAT inhibitors: evolution from cholesterol-absorption inhibitors to antiatherosclerotic agents. Drug Discov Today 3:19–25

Rouxel T, Chupeau Y, Fritz R, Kollmann A, Bousquet J-F (1988) Biological effects of sirodesmin PL, a phytotoxin produced by *Leptosphaeria maculans*. Plant Sci 57:45–53

Rüegger A, Kuhn M, Lichti H, Loosli HR, Huguenin R, Quiquerez C, von Wartburg A (1975) Cyclosporin A, ein immunsuppressiv wirksamer Peptidmetabolit aus *Trichoderma polysporum* (Link ex Pers.) *Rifai*. Helv Chim Acta 59:1075–1092

Saeger B, Schmitt-Wrede HP, Dehnhardt M, Benten WP, Krucken J, Harder A, Samson-Himmelstjerna von G, Wiegand H, Wunderlich F (2001) Latrophilin-like receptor from the parasitic nematode *Haemonchus contortus* as target for the anthelmintic depsipeptide PF1022A. FASEB J 15:1332–1334

Saito T, Hirai H, Kim Y-J, Kojima Y, Matsunaga Y, Nishida H, Sakakibara T, Suga O, Sujaku T, Kojima N (2002) CJ 15208, a novel kappa opioid receptor antagonist from a fungus, *Ctenomyces serratus* ATCC15502. J Antibiot 55:847–854

Samson-Himmelstjerna von G, Harder A, Sangster NC, Coles GC (2005) Efficacy of two cyclooctadepsipeptides, PF022A and emodepside, against anthelmintic-resistant nematodes in sheep and cattle. Parasitology 130:343–347

Sarabia F, Chammaa S, Sánchez Ruiz A, Martín Ortiz L, López Herrera FJ (2004) Chemistry and biology of cyclic depsipeptides of medicinal and biological interest. Curr Med Chem 11:1309–1332

Sasaki T, Takagi M, Yaguchi T, Miyado S, Okada T, Koyama M (1992) A new anthelmintic cyclodepsipeptide, PF1022. J Antibiot 45:692–697

Sato T, Ishiyama D, Honda R, Senda H, Konno H, Tokumasu S, Kanazawa S (2000) Glomosporin, a novel cyclic depsipeptide from *Glomospora* sp. I. Production, isolation, physico-chemical properties, and biological activities. J Antibiot 53:597–602

Schardl CL, Leuchtmann A, Spiering MJ (2004) Symbioses of grasses with seedborne fungal endophytes. Annu Rev Plant Biol 55:315–340

Scherkenbeck J, Jeschke P, Harder A (2002) PF1022A and related cyclodepsipeptides - a novel class of anthelmintics. Curr Topics Med Chem 7:759–777

Schmidt FR (2002) Beta-lactam antibiotics: aspects of manufacture and therapy. In: Osiewacz HD (ed) Industrial applications. Mycota X. Springer, Heidelberg, pp 69–91

Schrettl M, Bignell E, Kragl C, Sabiha Y, Loss O, Eisendle M, Wallner A, Arst HN Jr, Haynes K, Haas H (2007) Distinct roles for intra- and extracellular siderophores during *Aspergillus fumigatus* infection. PLoS Pathog 3:1195–1207

Scott PM, Polonsky J, Merrien MA (1979) Configuration of the 3,12 double bond of roquefortine. J Agric Food Chem 27:201–202

Seto Y, Takahasi K, Matsuura H, Kogami Y, Yada H, Yoshihara T, Nabeta K (2007) Novel cyclic peptide, epichlicin, from the endophytic fungus, *Epichloe typhina*. Biosci Biotechnol Biochem 71:1470–1475

Shiono Y, Tschuchinari M, Shimanuki K, Miyajima T, Murayama T, Koseki T, Laatsch H, Funakoshi T, Takanami K, Suzuki K (2007) Fusaristatins A and B, two new cyclic lipopeptides from an endophytic *Fusarium* sp. J Antibiot 60:309–316

Singh SB, Zink DL, Liesch JM, Mosley RT, Dombrowski AW, Bills GF, Darkin-Rattray SJ, Schmatz DM, Goetz MA (2002) Structure and chemistry of apicidins, a class of novel cyclic tetrapeptides without a terminal α-keto epoxide as inhibitors of histone deacetylase with potent antiprotozoal activities. J Org Chem 67:815–825

Skrobek A, Butt TM (2005) Toxicity testing of destruxins and crude extracts from the insect-pathogenic fungus *Metarhizium anisopliae*. FEMS Microbiol Lett 251:23–28

Sterner O, Etzel W, Mayer A, Anke H (1997) Omphalotin, a new cyclic peptide with potent nematicidal activity from *Omphalotus olearius*. II. Isolation and structure determination. Nat Prod Lett 10: 33–38

Sugui JA, Pardo J, Chang YC, Zarember KA, Nardone G, Galvez EM, Müllbacher A, Gallin JI, Simon MM, Kwon-Chung KJ (2007) Gliotoxin is a virulence factor of *Aspergillus fumigatus*: gliP deletion attenuates virulence in mice immunosuppressed with hydrocortisone. Eukaryot Cell 6:1562–1569

Supothina S, Isaka M, Kirtikara K, Tanticharoen M, Thebtaranonth Y (2004) Enniatin production by the entomopathogenic fungus *Verticillium hemipterigenum* BCC 1449. J Antibiot 57:732–738

Takahashi C, Numata A, Matsumura E, Minoura K, Eto H, Shingu T, Ito T, Hasgawa T (1994) Leptosins I and J, cytotoxic substances produced by a *Leptosphaeria* sp. physico-chemical properties and structures. J Antibiot 47:1242–1249

Tan LT, Cheng XC, Jensen PR, Fenical W (2003) Scytalidamides A and B, new cytotoxic cyclic heptapeptides from a marine fungus of the genus *Scytalidium*. J Org Chem 68:8767–8773

Tan NH, Zhou J (2006) Plant cyclopeptides. Chem Rev 106:840–895

Tan RX, Jensen PR, Williams PG, Fenical W (2004) Isolation and structure assignments of rostratins A–D, cytotoxic disulfides produced by the marine-derived fungus *Exserohilum rostratum*. J Nat Prod 67:1374–1382

Tani H, Fujii Y, Nakajima H (2001) Chlamydocin analogues from the soil fungus *Peniophora* sp.: structures and plant growth-retardant activity. Phytochemistry 58:305–310

Taunton J, Hassig CA, Schreiber SL (1996) A mammalian histone deacetylase related to the yeast transcriptional regulator Rpd3p. Science 272:408–411

Tobiasen C, Aahman J, Ravnholt KS, Bjerrum MJ, Grell MN, Giese H (2007) Nonribosomal peptide synthetase (NPS) genes in *Fusarium graminearum*, *F. culmorum* and *F. pseudograminearum* and identification of NPS2 as the producer of ferricrocin. Curr Genet 51:43–58

Traber R, Dreyfuss MM (1996) Occurrence of cyclosporins and cyclosporin-like peptolides in fungi. J Indust Microbiol 17:397–401

Turner WB, Aldridge DC (1983) Diketopiperazines and related compounds. In: Fungal metabolites II. Academic, London, pp 405–423

Ueno T, Nakashima T, Hayashi Y, Fukami H (1975) Structures of AM-toxin I and II, host-specific phytotoxic metabolites produced by *Alternaria mali*. Agric Biol Chem 39:1115–1122

Vey A, Matha V, Dumas C (2002) Effects of the peptide mycotoxin destruxin E on insect haemocytes and on dynamics and efficiency of the multicellular immune reaction. J Invert Pathol 80:177–187

Waring P, Beaver J (1996) Gliotoxin and related epipoly-thiodioxopiperazines. Gen Pharmacol 27:1311–1316

Waring P, Eichner RD, Müllbacher A (1988) The chemistry and biology of the immunomodulating agent glio-toxin and related epipolythiodioxopiperazines. Med Res Rev 8:499–524

Weber D, Erosa G, Sterner O, Anke T (2006) Cyclindrocy-clin A, a new cytotoxic cyclopeptide from *Cylindro-carpon* sp. J Antibiot 59:495–499

Welker M, von Döhren H (2006) Cyanobacterial peptides – nature's own combinatorial biosynthesis. FEMS Microbiol Rev 30:530–563

Welzel K, Eisfeld K, Antelo L, Anke T, Anke H (2005) Characterization of the ferrichrome A biosynthetic gene cluster in the homobasidiomycete *Omphalotus olearius*. FEMS Microbiol Lett 249:157–163

Winkelmann W, Drechsel H (1997) Microbial sidero-phores. In: Kleinkauf H, von Döhren H (eds) Products of secondary metabolism. Biotechnology, vol 7. VCH, Weinheim, pp 199–246

Wolstenholme WA, Vining LC (1966) Determination of amino acid sequences in oligopeptides by mass spectrometry VIII. The structure of isariin. Tetrahe-dron Lett 7:2785–2791

Yano T, Aoyagi A, Kozuma S, Kawamura Y, Tanaka I, Suzuki Y, Takamatsu Y, Takatsu T, Inukai M (2007) Pleofungins, novel inositol phosphorylceramide synthase inhibitors, from *Phoma* sp. SANK 13899. J Antibiot 60:136–142

Yin WQ, Zou JM, She ZG, Vrijmoed LLP, Jones EBG, Lin YC (2005) Two cyclic peptides produced by the endophytic fungus 2221 from *Castaniopsis fissa* on the South China sea coast. Chin Chem Lett 16:219–222

Yoshioka H, Nakatsu K, Sato M, Tatsuno T (1973) The molecular structure of cyclochlorotine, a toxic chlo-rine-containing peptapentide. Chem Lett 12:1319–1322

Zhang P, Chen Z, Hu J, Wei B, Zhang Z, Hu W (2005) Production and characterization of amanitin toxins from a pure culture of *Amanita exitialis*. FEMS Microbiol Lett 252:223–228

Zheng CJ, Oark SH, Koshino H, Kim YH, Kim WG (2007) Verticillin G, a new antibacterial compound from *Bionectria byssicola*. J Antibiot 60:61–64

Zimmermann G (2007a) Review on safety of the entomo-pathogenic fungi *Beauveria bassiana* and *Beauveria brongniartii*. Biocontrol Sci Technol 17:553–596

Zimmermann G (2007b) Review on safety of the entomo-pathogenic fungus *Metarhizium anisopliae*. Biocon-trol Sci Technol 17:879–920

14 Fungal Genome Mining and Activation of Silent Gene Clusters

Axel A. Brakhage[1], Sebastian Bergmann[1], Julia Schuemann[2], Kirstin Scherlach[2], Volker Schroeckh[1], Christian Hertweck[2]

CONTENTS

I. Introduction

The metabolism of all organisms can be divided into two parts: primary metabolism provides the cells with energy and chemical precursors that are essential for growth and reproduction of the organisms; secondary metabolism seems to possess no obvious function for cell growth in the laboratory (Bennett and Bentley 1989). However, these compounds remain a constant source of drug leads with more than 40% of new chemical entities reported since 1981 from microbial sources (Clardy and Walsh 2004; Khosla 1997; Sieber and Marahiel 2005). Even more remarkable is the fact that more than 60% of anticancer agents and 70% of anti-infectives currently in clinical use are natural products or natural product-based compounds (Yin et al. 2007). Secondary metabolites from microorganisms have been, and contin-ue to be, a leading source of molecules for drug discovery, but new technologies are required to increase the probability of identifying new entities (McAlpine et al. 2005).

II. Fungal Genome Mining and Activation of Silent Gene Clusters

A. Genetic Potential for Secondary Metabolism Biosynthesis of Fungi

A literature survey covering more than 23 000 bioactive microbial products, i.e. antifungal, anti-bacterial, antiviral, cytotoxic and immunosuppressive agents, shows that the producing strains are mainly from the fungal kingdom (ca. 42%), followed by strains belonging to the genus *Streptomyces* (32.1%; Lazzarini et al. 2000). Hence, fungi represent one of the most promising sources of bioactive compounds (Bhatnagar et al. 2002).

Despite the large number of known fungal metabolites, the biosynthetic potential of these organisms is greatly underestimated by an order of magnitude. Analyzing the increasing number of sequenced genomes indicates that fungi encode the genetic information for the biosynthesis of many more yet unknown compounds (e.g. Bergmann et al. 2007; Chiang et al. 2008; Peric-Concha and Long 2003). This information can be gained bioinformatically by scanning the genome for the presence of characteristic biosynthesis genes. Many biologically active compounds are produced by multifunctional enzymes, namely polyketide synthases (PKSs) and nonribosomal peptide synthetases (NRPSs). PKSs and NRPSs utilize simple malonyl (PKS) or amino acid (NRPS) building blocks, respectively. They follow very similar strategies for their assembly.

[1]Molecular and Applied Microbiology, Leibniz Institute for Natural Product Research and Infection Biology (HKI), and Friedrich Schiller University, Beutenbergstrasse 11a, 07745 Jena, Germany; e-mail: axel.brakhage@hki-jena.de
[2]Biomolecular Chemistry, Leibniz Institute for Natural Product Research and Infection Biology (HKI), and Friedrich Schiller University, Beutenbergstrasse 11a, 07745 Jena, Germany

Physiology and Genetics, 1st Edition
The Mycota XV
T. Anke and D. Weber (Eds)
© Springer-Verlag Berlin Heidelberg 2009

Although they utilize different classes of substrates, both PKSs and NRPSs show striking similarities in the architecture and mechanism of the modularized assembly lines. Each module is responsible for one or more chain-elongation steps and can be subdivided into domains controlling the choice of the extender unit and connection with the growing peptide chain. A typical NRPS module minimally consists of an adenylation (A) domain responsible for amino acid activation, a condensation (c) domain, connecting the activated amino acid with the growing chain, and a thiolation (T) domain – also known as peptidyl carrier protein (PCP) that serves as an anchor for the growing peptide chain. Additionally, a variety of optional [e.g. methyltransferase (MT) or epimerisation (E)] domains may be present (Walsh et al. 2001). Similarly, three domains are the basic equipment of most PKS elongation modules: an acyltransferase (AT) domain for extender unit selection and transfer, an acyl carrier protein (ACP) for extender unit loading and a ketoacyl synthase (KS) domain for decarboxylative condensation of the extender unit (usually malonyl-CoA) with an acyl thioester. The resulting β-keto thioester may subsequently be processed by β-ketoacyl reductase (KR) domains, dehydratase (DH) domains, enoyl reductase (ER) domains and MT domains (Cane et al. 1997). Modular PKS and NRPS systems can also closely cooperate to form so-called hybrid products (Cane and Walsh 1999; Paitan et al. 1999; Silakowski et al. 1999; Wenzel et al. 2005). Typically, the nascent polyketide or peptide backbone is further modified by tailoring enzymes in post-PKS or post-NRPS steps of the pathway, such as oxygenases and glycosyltransferases or other transferases, in order to imbue additional structural functionalities to the final natural product.

B. Genome Mining

The conserved features within these PKS and NRPS machineries have been cornerstones of the genomic-guided discovery of natural products (Ikeda et al. 2003; Liu et al. 2002; Menzella et al. 2005; Yin et al. 2007). In all cases known to date, genes codings for PKS and NRPS are part of a cluster containing additional genes for modifying enzymes. With the genome sequence at hand, it is possible to estimate the biosynthetic potential for a given organism by in silico screening for

typical secondary metabolite biosynthesis genes. Systematic mining of the genomes has uncovered the occurrence of cryptic biosynthesis genes in actinomycetes (McAlpine et al. 2005; Lautru et al. 2005; Zazopoulos et al. 2003; Challis 2007), myxobacteria (Wenzel et al. 2005; Bode and Muller 2005), *Pseudomonas* (Brendel et al. 2007; Gross et al. 2007) and *Burkholderia* (Nguyen et al. 2008). Recently the genomes of various filamentous fungi, including *Aspergillus nidulans*, *A. fumigatus* and *A. oryzae*, have been sequenced (http://www.broad.mit.edu/annotation/genome/aspergillus_group/MultiHome.html).

The analysis of deduced gene products suggests that *A. nidulans* has the potential to generate up to 32 polyketides, 14 nonribosomal peptides and two indole alkaloids; similar predictions can be made from the *A. fumigatus* and *A. oryzae* genome data (Table 14.1). Interestingly, there appear to be almost no orthologs among these genes across the three species, thus representing a loss of synteny to a degree not seen in other regions of the genomes (Bok et al. 2006). This high number of putative metabolites is greater than the known metabolites ascribed to these species, and it may be a reflection of incomplete natural product analysis in these species or failure of many clusters to be expressed, at least under the culture conditions commonly used in laboratories (Scherlach and Hertweck 2006). For example, the aflatoxin gene cluster is not expressed in *A. oryzae* (van den Broek et al. 2001; Zhang et al. 2004; Bok et al. 2006). It is apparent that this vast number of predicted fungal biosynthesis genes is not reflected by the metabolic profile observed under standard fermentation conditions. Apparently, in the absence of a particular trigger these gene loci remain silent.

One of the major challenges is to understand the physiological meaning of the many secondary metabolites for the producing microorganism

Table 14.1 Putative secondary metabolism gene types. *NRPS* Nonribosomal peptide synthetase, *PKS* polyketide synthase, *DMAT* dimethylallyl tryptophan synthase (Brakhage et al. 2008)

Putative gene	*Aspergillus nidulans*	*A. fumigatus*	*A. oryzae*
NRPS	14[a]	14[a]	17
PKS	32[a]	14[a]	28
DMAT	2	7	2

[a]Includes one hybrid PKS/NRPS.

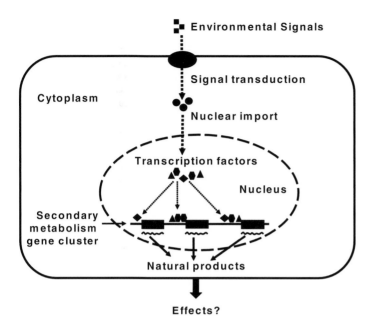

Fig. 14.1. Scheme of the processes involved in the activation of fungal secondary metabolism gene clusters in a fungal cell

(Fig. 14.1). As long as this is not understood it is often not possible to directly activate gene clusters under natural conditions and to predict the regulatory circuits involved in the regulation of their biosyntheses. Notably, in many fungal PKS gene clusters genes encoding transcription factors can be found, while such genes are scarce in NRPS gene clusters.

C. Activation of Silent Gene Clusters

For the functional analysis of cryptic or orphan genes, gene inactivation would only provide an option if the metabolite was constitutively produced by the strain. This method is clearly not applicable for gene clusters that are silent under standard fermentation conditions, and thus tools are needed to assign their biosynthetic role. In general, several strategies would be conceivable: gene expression could be induced, for example (i) by heterologous expression under the control of defined promoters (Fig. 14.2a) or (ii) by promoter exchange within the genome (homologous expression; Fig. 14.2b). The latter method would be feasible in particular for bacterial gene clusters with polycistronic gene organization. In the case of fungal secondary metabolism gene clusters, promoter exchange can be cumbersome and is usually employed for individual genes only (Kennedy and Turner 1996). Obviously the latter approach has

numerous disadvantages. The method requires the availability of genetic techniques for the particular organism. Furthermore, many fungi only show a very low frequency of homologous recombination, which hampers promoter exchange. In addition to the cumbersome handling of large gene constructs, another drawback is that overexpression of a single gene of a cluster often leads to limitation of another gene product of the same cluster. Finally, gene expression and, as in the lovastatin pathway, the functioning of enzymes may be context-dependent (Bergmann et al. 2007).

One solution of this problem is: (iii) the expression of pathway-specific regulatory genes, which are present in many secondary metabolite gene clusters (Fig. 14.2c). This approach is rendered feasible by the fact that all of the genes encoding the large number of enzymes required for the synthesis of a typical secondary metabolite are clustered and that in some cases a single regulator controls the expression of all members of a gene cluster to a certain extent (McAlpine et al. 2005). Consequently, only a small gene needs to be handled. Furthermore, ectopic integration of simple gene cassettes would be sufficient, bypassing all limitations of homologous recombination. Most conveniently, this strategy would allow for the concerted expression of all pathway genes.

Recently, we developed the latter strategy for the successful induction of a silent metabolic pathway in the fungal model organism *A. nidulans*.

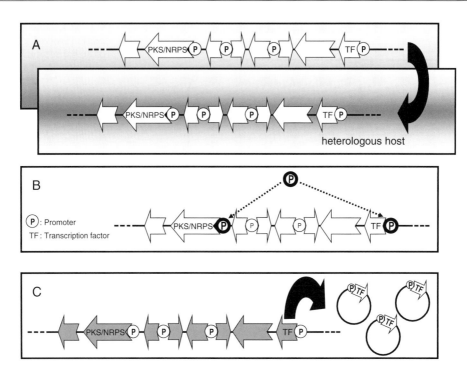

Fig. 14.2. Molecular methods to induce silent gene clusters. (**a**) Expression of gene clusters in a heterologous host. (**b**) Promoter exchange of individual genes of a cluster using strong, regulatable promoters. (**c**) Overexpression of transcription factors using strong inducible promoters

By mining the *A. nidulans* genome for cryptic (or orphan) gene clusters potentially coding for the biosynthesis of polyketides or polypeptides, we noted the presence of a putative hybrid PKS-NRPS gene (*apdA*) (Table 14.1, Fig. 14.3). The deduced gene product shows typical motifs of PKS and NRPS domains, as well as an additional C-terminal reductase domain. It obviously represents a rare fungal PKS-NRPS hybrid. Despite the large number of metabolic data available for *A. nidulans* the role of this gene was obscure. To date, only a few fungal PKS-NRPS hybrid synthases have been attributed to metabolic functions. Gene inactivation experiments revealed that two related ones are involved in the biosynthesis of the tetramic acid derivatives equisetin and fusarin in *Fusarium heterosporum* and *F. fujikuroi*, respectively (Sims et al. 2005; Song et al. 2004). The cryptic PKS-NRPS-encoding gene in the *A. nidulans* genome is flanked by upstream genes coding for several putative oxidoreductases: two cytochrome P450-monooxygenases (ApdB, ApdE), an FAD-dependent monooxygenase (ApdD) and an enoyl reductase (ApdC). Downstream of *apdA* the putative exporter *apdF* and activator *apdR* genes are located. The deduced gene product of *apdR* is related to a putative C6 transcription factor of *A. fumigatus* and to a putative regulator with a GAL4-type Zn_2Cys_6 binuclear cluster DNA-binding domain from *A. flavus*. To prove the concept that the homologous overexpression of a regulatory gene can lead to activation of a silent gene cluster the putative activator gene *apdR* was amplified from genomic DNA and cloned into an expression vector carrying the promoter of the alcohol dehydrogenase of *A. nidulans* (Fig. 14.3). This promoter can be induced by the addition of cyclopentanone to the medium and repressed by the use of glucose as the carbon source (Waring et al. 1989).

Transformants of *A. nidulans* carrying the *alcAp-apdR* gene fusion ectopically integrated into the genome were checked for transcription of genes of the cluster by Northern blot analysis. Whereas the transcripts were completely absent in the wild type under both non-inducing and inducing conditions, in the transformant strain under inducing conditions strong mRNA signals were detected for the genes of the cluster including the PKS-NRPS hybrid (Fig. 14.4). This observation further confirmed that under the conditions

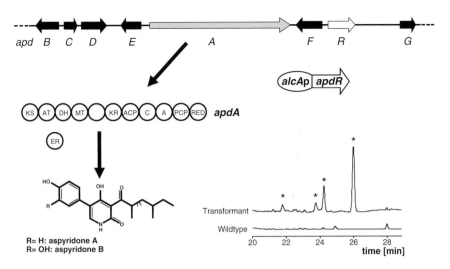

Fig. 14.3. Activation of a silent gene cluster in *Aspergillus nidulans* by overexpression of the transcription factor gene *apdR* using the inducible promoter of the alcohol dehydrogenase gene *alcA* of *A. nidulans*. This leads to the production of a mixed PKS-NRPS encoded by the *apdA* gene and the production of the novel natural products aspyridone A and B under inducing conditions which is shown by the HPLC profile (Bergmann et al. 2007). New metabolites are labeled by asterisks

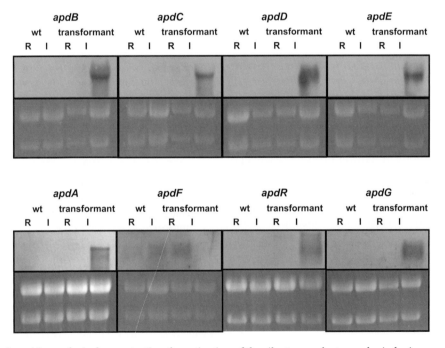

Fig. 14.4. Northern blot analysis demonstrating the activation of the silent gene cluster under inducing conditions in the transformant strain encoding the extra copies of the inducible transcription factor *apdR*. *R* repressing conditions; *I* inducing conditions with cyclopentanone in the medium; *apdR* transcription factor gene (Bergmann et al. 2007)

applied this gene cluster was silent without induction. Furthermore, it also helped to confine the borders of the gene cluster (Bergmann et al. 2007).

The culture extracts were analysed by HPLC coupled to DAD and MS detectors. Under inducing conditions, the transformant strain produced novel compounds called aspyridones (Fig. 14.3). They are similar, but not identical to a variety of other pyridones isolated from fungi, such as militarinone D and tenellin

(Eley et al. 2007; Schmidt et al. 2003). In a broad bioactivity screening they exhibited moderate cytotoxic activities. The results cited here provide the proof of principle for a strategy that may be generally applicable to the activation of silent biosynthesis gene clusters, in particular in eukaryotes.

Recently, rather global regulatory strategies were explored for the activation of silent fungal biosynthesis genes. It is known that some secondary metabolite gene clusters are located in chromatin regions that are controlled by epigenetic regulation such as histone deacetylation and DNA methylation. Keller and colleagues have demonstrated that the disruption of histone deacetylase activity (Δhda) in *A. nidulans* led to transcriptional activation of secondary metabolite gene clusters (Perrin et al. 2007; Shwab et al. 2007). This observation stimulated the use of epigenetic modifiers for inducing silent natural product biosynthetic pathways. Cichewicz and colleagues successfully applied this concept for remodeling the metabolome in various filamentous fungi (Williams et al. 2008).

III. Conclusions

The rapidly growing number of sequenced fungal genomes reveals that the biosynthetic potential of fungi has been greatly under-explored by traditional methods of natural product discovery. Obviously, a multitude of potentially useful metabolites still awaits discovery. Molecular methods indicate that many gene clusters are silent under standard laboratory cultivation conditions. However, if these genes are still intact, their expression by genetic engineering may lead to the discovery of so far unknown natural products and therefore potential drug candidates.

References

Bennett JW, Bentley R (1989) What's in a name? – microbial secondary metabolism. Adv Appl Microbiol 34:1–28

Bergmann S, Schumann J, Scherlach K, Lange C, Brakhage AA, Hertweck C (2007) Genomics-driven discovery of PKS-NRPS hybrid metabolites from *Aspergillus nidulans*. Nat Chem Biol 3:213–217

Bhatnagar D, Yu C, Ehrlich KC (2002) Toxins of filamentous fungi. Chem Immunol 81:167–206

Bode HB, Muller R (2005) The impact of bacterial genomics on natural product research. Angew Chem Int Ed Engl 44:6828–6846

Bode HB, Bethe B, Hofs R, Zeeck A (2002) Big effects from small changes: possible ways to explore nature's chemical diversity. Chembiochem 3:619–627

Bok JW, Hoffmeister D, Maggio-Hall LA, Murillo R, Glasner JD, Keller NP (2006) Genomic mining for *Aspergillus* natural products. Chem Biol 13:31–37

Brakhage AA, Schuemann J, Bergmann S, Scherlach K, Schroeckh V, Hertweck C (2008) Activation of fungal silent gene clusters: a new avenue to drug discovery. Prog Drug Res 66:3–12

Brendel N, Partida-Martinez LP, Scherlach K, Hertweck C (2007) A cryptic PKS-NRPS gene locus in the plant commensal *Pseudomonas fluorescens* Pf-5 codes for the biosynthesis of an antimitotic rhizoxin complex. Org Biomol Chem 5:2211–2213

Cane DE (1997) Introduction: Polyketide and nonribosomal polypeptide biosynthesis. From collie to coli. Chem Rev 97:2463–2464

Cane DE, Walsh CT (1999) The parallel and convergent universes of polyketide synthases and nonribosomal peptide synthetases. Chem Biol 6:R319–R325

Challis GL (2007) A widely distributed bacterial pathway for siderophore biosynthesis independent of nonribosomal peptide synthetases. Chembiochem 8:1477

Chiang YM, Szewczyk E, Nayak T, Davidson AD, Sanchez JF, Lo HC, Ho WY, Simityan H, Kuo E, Praseuth A, Watanabe K, Oakley BR, Wang CC (2008) Molecular genetic mining of the *Aspergillus* secondary metabolome: discovery of the emericellamide biosynthetic pathway. Chem Biol 15:527–532

Clardy J, Walsh C (2004) Lessons from natural molecules. Nature 432:829–837

Eley KL, Halo LM, Song Z, Powles H, Cox RJ, Bailey AM, Lazarus CM, Simpson TJ (2007) Biosynthesis of the 2-pyridone tenellin in the insect pathogenic fungus *Beauveria bassiana*. Chembiochem 8:289–297

Gross H, Stockwell VO, Henkels MD, Nowak-Thompson B, Loper JE, Gerwick WH (2007) The genomisotopic approach: a systematic method to isolate products of orphan biosynthetic gene clusters. Chem Biol 14:53–63

Ikeda H, Ishikawa J, Hanamoto A, Shinose M, Kikuchi H, Shiba T, Sakaki Y, Hattori M, Omura S (2003) Complete genome sequence and comparative analysis of the industrial microorganism *Streptomyces avermitilis*. Nat Biotechnol 21:526–531

Kennedy J, Turner G (1996) delta-(L-alpha-aminoadipyl)-L-cysteinyl-D-valine synthetase is a rate limiting enzyme for penicillin production in *Aspergillus nidulans*. Mol Gen Genet 253:189–197

Khosla C (1997) Harnessing the biosynthetic potential of modular polyketide synthases. Chem Rev 97:2577–2590

Lautru S, Deeth RJ, Bailey LM, Challis GL (2005) Discovery of a new peptide natural product by *Streptomyces coelicolor* genome mining. Nat Chem Biol 1:265–269

Lazzarini A, Cavaletti L, Toppo G, Marinelli F (2000) Rare genera of actinomycetes as potential producers of new antibiotics. Antonie Van Leeuwenhoek 78:399–405

Liu W, Christenson SD, Standage S, Shen B (2002) Biosynthesis of the enediyne antitumor antibiotic C-1027. Science 297:1170–1173

McAlpine JB, Bachmann BO, Piraee M, Tremblay S, Alarco AM, Zazopoulos E, Farnet CM (2005) Microbial genomics as a guide to drug discovery and structural elucidation: ECO-02301, a novel antifungal agent, as an example. J Nat Prod 68:493–496

Menzella HG, Reid R, Carney JR, Chandran SS, Reisinger SJ, Patel KG, Hopwood DA, Santi DV (2005) Combinatorial polyketide biosynthesis by de novo design and rearrangement of modular polyketide synthase genes. Nat Biotechnol 23:1171–1176

Nguyen T, Ishida K, Jenke-Kodama H, Dittmann E, Gurgui C, Hochmuth T, Taudien S, Platzer M, Hertweck C, Piel J (2008) Exploiting the mosaic structure of trans-acyl-transferase polyketide synthases for natural product discovery and pathway dissection. Nat Biotechnol 26:225–233

Paitan Y, Alon G, Orr E, Ron EZ, Rosenberg E (1999) The first gene in the biosynthesis of the polyketide antibiotic TA of *Myxococcus xanthus* codes for a unique PKS module coupled to a peptide synthetase. J Mol Biol 286:465–474

Peric-Concha N, Long PF (2003) Mining the microbial metabolome: a new frontier for natural product lead discovery. Drug Discov Today 8:1078–1084

Perrin RM, Fedorova ND, Bok JW, Cramer RA, Wortman JR, Kim HS, Nierman WC, Keller NP (2007) Transcriptional regulation of chemical diversity in *Aspergillus fumigatus* by LaeA. PLoS Pathog 3:e50

Scherlach K, Hertweck C (2006) Discovery of aspoquinolones A-D, prenylated quinoline-2-one alkaloids from *Aspergillus nidulans*, motivated by genome mining. Org Biomol Chem 4:3517–3520

Schmidt K, Riese U, Li Z, Hamburger M (2003) Novel tetramic acids and pyridone alkaloids, militarinones B, C and D, from the insect pathogenic fungus *Paecilomyces militaris*. J Nat Prod 66:378–383

Shwab EK, Bok JW, Tribus M, Galehr J, Graessle S, Keller NP (2007) Histone deacetylase activity regulates chemical diversity in *Aspergillus*. Eukaryot Cell 6:1656–1664

Sieber SA, Marahiel MA (2005) Molecular mechanisms underlying nonribosomal peptide synthesis: approaches to new antibiotics. Chem Rev 105:715–738

Silakowski B, Schairer HU, Ehret H, Kunze B, Weinig S, Nordsiek G, Brandt P, Blocker H, Hofle G, Beyer S, Muller R (1999) New lessons for combinatorial biosynthesis from myxobacteria. The myxothiazol biosynthetic gene cluster of *Stigmatella aurantiaca* DW4/3-1. J Biol Chem 274:37391–37399

Sims JW, Fillmore JP, Warner DD, Schmidt EW (2005) Equisetin biosynthesis in *Fusarium heterosporum*. Chem Commun (Camb) 2:186–188

Song Z, Cox RJ, Lazarus CM, Simpson TT (2004) Fusarin C biosynthesis in *Fusarium moniliforme* and *Fusarium venenatum*. Chembiochem 5:1196–1203

van den Broek P, Pittet A, Hajjaj H (2001) Aflatoxin genes and the aflatoxigenic potential of Koji moulds. Appl Microbiol Biotechnol 57:192–199

Walsh CT, Chen H, Keating TA, Hubbard BK, Losey HC, Luo L, Marshall CG, Miller DA, Patel HM (2001) Tailoring enzymes that modify nonribosomal peptides during and after chain elongation on NRPS assembly lines. Curr Opin Chem Biol 5:525–534

Waring RB, May GS, Morris NR (1989) Characterization of an inducible expression system in *Aspergillus nidulans* using *alcA* and tubulin-coding genes. Gene 79:119–130

Wenzel S, Kunze CB, Hofle G, Silakowski B, Scharfe M, Blocker H, Muller R (2005) Structure and biosynthesis of myxochromides S1-3 in *Stigmatella aurantiaca*: evidence for an iterative bacterial type I polyketide synthase and for module skipping in nonribosomal peptide biosynthesis. Chembiochem 6:375–385

Williams RB, Henrikson JC, Hoover AR, Lee AE, Cichewicz RH (2008) Epigenetic remodeling of the fungal secondary metabolome. Org Biomol Chem 6:1895–1897

Yin J, Straight PD, Hrvatin S, Dorrestein PC, Bumpus SB, Jao C, Kelleher NL, Kolter R, Walsh CT (2007) Genome-wide high-throughput mining of natural-product biosynthetic gene clusters by phage display. Chem Biol 14:303–312

Zazopoulos E, Huang K, Staffa A, Liu W, Bachmann BO, Nonaka K, Ahlert J, Thorson S, Shen B, Farnet CM (2003) A genomics-guided approach for discovering and expressing cryptic metabolic pathways. Nat Biotechnol 21:187–190

Zhang Y-Q, Wilkinson H, Keller NP, Tsitsigiannis DI (2004) Secondary metabolite gene clusters. In: An, Z (ed) Handbook of industrial microbiology. Dekker, New York, pp 355–386

15 Non-Ribosomal Peptide Synthetases of Fungi

Katrin Eisfeld[1]

CONTENTS

I. Introduction

Non-ribosomal peptides (NRPs) are a class of secondary metabolites produced by bacteria (*Bacilli*, *Streptomyces*, some Pseudomonades and Myxobacteria) and filamentous fungi. Secondary metabolites are compounds that are not directly involved in processes indispensable to life but often have potent physiological activities. Although chemically diverse, all secondary metabolites are produced by a few common biosynthetic pathways, including non-ribosomal peptide synthesis (Keller et al. 2005).

Fungal and bacterial NRPs are a structurally very diverse family of natural products which display a wide variety of biological activities. The most famous non-ribosomally produced peptide is penicillin which was discovered by Alexander Fleming's observation on the inhibition of bacterial growth by *Penicillium notatum* (Fleming 1929). This antibiotic gained importance during World War II when penicillin was used as a "miracle drug" and saved countless lives. Hence many NRPs (e.g. penicillin, cyclosporin, ergotamine) have tremendous importance as pharmaceutically relevant agents. Many peptides with structures unique to fungi are known, including siderophores of the ferrichrome and coprogen type, cyclodepsipeptides and peptaibols. Biosynthesis of these compounds is independent of ribosomes and mRNA but is catalyzed by large multifunctional enzymes called non-ribosomal peptide synthetases (NRPSs) which are discussed here. NRPSs are the largest enzymes known in nature. Peptides produced by NRPSs show peculiar features compared to traditional proteins. They contain not only standard amino acids but also non-proteinogenic amino acids. Peptides can be linear, but there also exist cyclic and branched NRPS configurations. The amino acids incorporated can be modified by N-methylation, epimerization or can be reduced. Sizes of the non-ribosomally produced peptides range from two to 40 amino acids (for reviews see for example Stachelhaus and Marahiel 1995; Kleinkauf and von Döhren 1996; Schwarzer and Marahiel 2001; Schwarzer et al. 2003; Finking and Marahiel 2004; von Döhren 2004).

[1] Institute of Biotechnology and Drug Research, Erwin-Schrödinger-Strasse 56, 67663 Kaiserslautern, Germany;
e-mail: eisfeld@ibwf.de

Physiology and Genetics, Ist Edition
The Mycota XV
T. Anke and D. Weber (Eds.)
© Springer-Verlag Berlin Heidelberg 2009

Protein data show that NRPSs exhibit a modular structure, with one module being a semi-autonomous unit that recognizes, activates and modifies a single residue of the final peptide. The modules themselves are organized in domains with specific catalytic functions, thereby defining the amino acid sequence and structure of the products (Smith et al. 1990; Zocher and Keller 1997; Konz and Marahiel 1999). Some non-ribosomal peptide synthetases (including enniatin synthetase, PF1022 synthetase, beavericin synthetase) have been purified and enzymatically characterized (Zocher et al. 1982; Peeters et al. 1988; Weckwerth et al. 2000). Biochemical investigations reveal that these enzymes consist of one single polypeptide chain which comprises all catalytic functions necessary for peptide synthesis, i.e. ATP-dependent substrate activation, thiolation reactions, N-methylation and condensation of the covalently bound substrates. To date, far more peptides which are supposed to be produced non-ribosomally are known than the genes that encode their corresponding NRPSs. Table 15.1 gives a summary of NRPs from ascomycetes whose biosynthetic pathways have been fully characterized (fungal siderophore synthetases are not included; they were reviewed very recently by Johnson 2008).

In genome sequencing projects a great number of new NRPS genes have been identified while the products of the NRPSs they encode have not been detected. For example, in *Aspergillus fumigatus* a total of at least 14 NRPS genes have been discovered, but only four of them can be associated with their NRPS products: the gliotoxin synthetase gene (*gliP*), the ferricrocin synthetase gene (*sidC*), the gene encoding fusarinine synthetase (*sidD*) and the brevanamide synthetase gene *ftmA* (Nierman et al. 2005; Reiber et al., 2005; Balibar and Walsh 2006; Cramer et al. 2006; Maiya et al. 2006; Schrettl et al. 2007; Stack et al. 2007). Other fungi of the genus *Aspergillus* also produce a considerable variety of non-ribosomally produced peptides. However, although conserved orthologous NRPS genes exist, the products of most of the NRPSs remain unknown. In *A. terreus* for example, only four of a total of 20 NRPS genes can be assigned to a hypothetical function by means of homology comparison (Cramer et al. 2006). Cross-genome comparisons indicate that most filamentous ascomycetes carry many genes encoding NRPSs. Some of them appear to be discontinuously distributed, while most are not conserved from one species to another (Turgeon et al. 2008). This implies that a lot of unknown NRPS products exhibiting potentially interesting properties still await discovery.

II. Non-Ribosomal Peptide Synthetases

A. Structure of NRPSs

NRPSs have been proposed to operate via a thiotemplate mechanism with a pantotheine "swinging arm" transferring activated amino acids to the condensation domain of the enzyme (Stein et al. 1996). A multimodular arrangement of NRPSs was confirmed by the isolation and characterization of genes encoding NRPSs from bacterial and fungal origin. One module is responsible for the introduction of one amino acid. Each module consists of several domains with defined functions (Stachelhaus and Marahiel 1995; Marahiel 1997; Schwarzer and Marahiel 2001; Weber and Marahiel 2001; Mootz et al. 2002a, b). Usually the order of modules and domains of a fungal NRPS is collinear to the sequence of the synthesized product. Unlike in bacteria, fungal non-ribosomal peptide synthesis is mostly catalyzed by one enzyme. These enzymes can reach an enormous size; for example, the *Trichoderma virens PES* gene which is responsible for peptaibol production comprises a 62.8-kb continuous open reading frame and the peptide synthetase consists of 18 modules (Wiest et al. 2002).

A minimal module for incorporation of one amino acid into the growing peptide chain consists of an adenylation (A) domain, a thiolation (T) or peptidyl carrier protein (PCP) domain and a condensation (C) domain. The A domain is responsible for substrate recognition and activation, the T/PCP domain binds a 4′-phosphopantetheine cofactor to which the activated amino acid is tethered and the C domain is required for peptide bond formation. Besides the domains being essential for the elongation of the peptide chain there are optional domains including domains for N-methylation (N-MTase domain), epimerization (E domain) and heterocyclic ring formation which are inserted at specific locations within the module. Furthermore, besides proteinogenic amino acids often unusual amino acids or carboxylic acids are incorporated. Further variability is achieved through cyclization of the

Table 15.1. Summary of characterized NRPSs, their products and properties

Producer	NRPS/Structure	Product	Properties
Penicillium chrysogenum and others	ACV synthetase (ACVS) $(ATC)_2ATETe$	Tripeptide	Precursor of penicillin and cephalosporin
Fusarium oxysporum and others	Enniatin synthetase CATCAMTTC	Cyclo-hexadepsipeptide	Ionophore; inhibitor of acyl-CoA-cholesterol acyltransferase and of phosphodiesterase; potential anticancer compound.
Beauveria nivea, *Tolypocladium niveum*	Cyclosporin synthetase $CAT(CAMT)_4CAT$ $(CAMT)_2CATCAMTCATC$	11-Cyclopeptide	Immunosuppressive: antifungal; anti-inflammatory
Claviceps purpurea	Ergotamin synthetases ATC and ATCATCATCyc	Acyl-tripeptides	Aeurotoxic; vasoconstrictor
Cochliobolus carbonum	HC-toxin synthetase (HTS1) $ATEC(ATC)_3$	Cyclo-tetrapeptide	Inhibitor of histone-deacetylases: maize-pathogen
Alternaria alternata	AM-toxin synthetase$(ATC)_4$	4-Peptidolactone	Phytotoxic
Trichoderma virens	Peptaibol synthetase KS^a-AC^b-$C(ATC)_{17}$ AT-DH^c	Acylpeptides (18, 14 or 11 amino acids)	fungicidal
Leptosphaeria maculans	Sirodesmin synthetase SirP $(ATC)_2$	Diketopiperazine epipolythiodioxopiperazin	Phytotoxin mycotoxin; antibacterial activity
Aspergillus fumigatus and others	Gliotoxin synthetase$(ATC)_2T$	Cyclodipeptide	Induces apoptosis; immunosuppressive
A. fumigatus	Brevianamide F synthetase $(ATC)_2$	Diketopiperazine	Precursor of fumitremorgins
Epichloe festucae	Peramine synthetase perA ATCAMTR	Pyrrolopyrazine	Repellent to insects
A. nidulans	TdiA ATTe	Bisindolebenzoquinone	Precursor of terrequinone
Xylaria sp. BCC1067	Bassianolide synthetase CATCAMTTCR	Cyclooctadepsipeptide	Pathogenic to insects

[a]KS = Ketoacyl synthase.
[b]AC = Acyltransferase.
[c]DH = β-Hydroxysteroid dehydrogenase.

peptide or modifications that take place after the assembly of the peptide chain (Stachelhaus and Marahiel 1995; Marahiel 1997; Schwarzer and Marahiel 2001; Weber and Marahiel 2001; Mootz et al. 2002a, b; Samel and Marahiel 2008).

Two types of modules can be distinguished: the initiation or starting module lacks the C domain and therefore is A–T. Elongation or extending modules are –C–A–T–. Methylation domains are located within A-domains, while epimerization domains are inserted between T– and C-domains. The products are released at thioesterase-like domains, reductive domains or specialized C-domains. Fig. 15.1 shows the domain composition within one module.

Comparison of bacterial and fungal NRPSs led to the identification of conserved sequence motifs ("core motifs") characteristic for each domain type (Stachelhaus and Marahiel 1995; Konz and Marahiel 1999). These motifs are similar in bacteria and fungi but differ to some extend. Table 15.2 displays a comparison of bacterial and fungal NRPS core motifs.

B. Module Arrangement

Originally, many linearly organized NRPS enzymes were discovered. This led to the development of the co-linearity rule: the order in which the amino or hydroxyl acids are combined into the peptide product is normally determined by the arrangement of the A domains within the multi-domain NRPS enzyme. However, in the meantime many NRPSs with alternative synthesis patterns

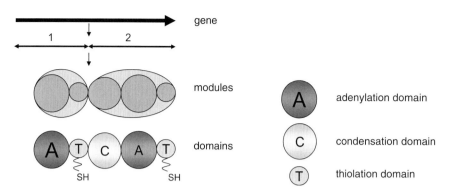

Fig. 15.1. The amino acid sequence of NRPSs, deriving from the gene, allows identification of the modules responsible for amino acid activation. Modules can be subdivided into domains. Adenylation (*A*) domains are responsible for substrate recognition and activation, thiolation (*T*) domains covalently bind the activated substrate and the condensation (*C*) domains are the site for peptide bond formation. Modules lacking a C domain are used for initiation of peptide synthesis, while those harboring a C domain are elongation modules (from Finking and Marahiel 2004)

Table 15.2. Conserved sequence motifs of NRPS domains

Domain	Motif	Sequence in bacteria (Stachelhaus and Marahiel 1995)	Sequence in *C. heterostrophus* (ascomycete; Lee et al. 2005)	Sequence in *O. olearius* (basidiomycete; Eisfeld, unpublished results)
A	A1	L(TS)YxEL	LTYxEL	xTYxxL
	A2	LKAGxAYL(VL)P(LI)D	LKAGA(AG)F(VY)P(LI)DP	LKAGxAxxPxx
	A3	LAYxxYTSG(ST)TGxPGK	LAY(VI)(IL)FTSGTGxPKGV	xAYxxxTSG(TS)TGxPKxV
	A4	FDxS	FDxSxxE	FDxxxx(ED)
	A5	NxYGPTE	N(GA)YGP(TA)E	NxYGPxE
	A6	GELxIxGxG(VL)ARGYL	GELxIxxxGxG(VL)ARGYL	GELx(IVL)GGxxx(AG)xGYxx
	A7	Y(RK)TGDL	RxY(RK)TGDLxR	X(YF)xTGDxxR
	A8	GRxDxQVKIRGxRIELGEIE	GRxDxQVK(IL)RGxR(IV)ELGEIE	GRxDxxxVKxxGxR(VI)xLxE (IV)x
	A9	LPxYM(IV)P	LPxYM(IV)P	LPxxMxP
	A10	NGK(VL)DR	SGKxDR	xGKxDR
C	C1	Not listed	PxxPxQ	PxxPxQ
	C2	RHExLRTxF	xxxxLRTxx	xxxxLR(TS)xF
	C3	MHHxISDG(WV)S	xHHxxxDGxS	xHHA(IL)YDGxS
	C4	YxD)FY)AVW	xxxxAAWA	/
	C5	(IV)GxFVNT(QL)xR	xGPxxxTxPxR	xGPxxxTxPVRV
E	E1	xxxPIQxWF	PLxPIQxxF	PLxxxQxxx
	E2	HHxISDG(WV)Sx	HHLxVDxVSW	HHLxxDxVSx
	E3	DxLLxAxG	Not listed	xxLxxxG
	E4	EGHGRE	Not listed	ExHGRx
	E5	RTVGWFTxxYP(YV)PF	xTVGWFTMxPxxx	xTVGWFTxxxPx
	E6	PxxGxGYG	Not listed	PxxGxxYx
	E7	FNYLG(QR)	Not listed	FNxLGx

and hence different strategies of biosynthesis have been identified. Mootz et al. (2002a, b) proposed classifying the known NRPS systems into three groups according to their biosynthetic logic: linear (type A), iterative (type B) and non-linear NRPSs (type C). Examples for the three groups can be found not only in bacterial systems but also in fungi.

A typical *linear* NRPS template has the domain organization A–T–(C–A–T)$_{n-1}$–TE/C with *n* being the number of amino acids that form the peptide. Modification domains can be inserted into the respective modules. In linear NRPSs, predictions can be made about the products being synthesized. Examples for fungal linear NRPS are

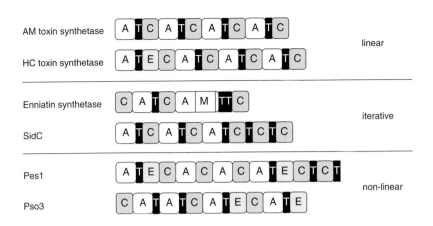

Fig. 15.2. Examples for linear, iterative and non-linear NRPSs in fungi

ACV synthetase, peptaibol synthetases, and cyclosporin synthetases (Fig. 15.2).

Iterative NRPSs use their modules or domains repeatedly during the assembly of a product. For example, in enniatin synthesis, the cyclo-hexadepsipeptide is synthesized by repeated use of two modules which activate D-hydroxy-isovaleric acid and N-methylvaline (Glinski et al. 2002). PF1022A synthetase most likely exhibits the same domain architecture as enniatin synthetase and also produces a cyclo-hexadepsipeptide which is assembled by successive condensations of dipeptidol building blocks (Weckwerth et al. 2000). Fungal siderophore synthetases of the ferrichrome type combine three small amino acids (Gly, Ala or Ser) and three acylated L-5-hydroxy-ornithines but contain only three A domains. These hexapeptides are also expected to be produced iteratively, mediated by additional truncated modules lacking the adenylation domains (Haas et al. 2003; Schwecke et al. 2006). This architecture is also found at the C-terminus of the enniatin synthetase Glinski et al. 2002). Mootz et al. (2002a, b) suggested that iterative NRPSs can be regarded as linear NRPS with an iterative activity at the C-terminus (Fig. 15.2).

The architecture of *non-linear* NRPSs differs from that of type A NRPSs and the domain interaction is more complex. Their possible products can not be predicted. An unusual domain arrangement is found at a large number of bacterial NRPSs while most fungal NRPSs described so far are of the linear or iterative type.

For example, in syringomycin and yersiniabactin biosynthesis occurs in *trans* aminoacylation. In yersiniabactin synthesis one A domain loads three different PCPs (Zhang et al. 1995; Guenzi et al. 1998; Stachelhaus et al. 1999). Another example for a type C NRPS is myxochelin synthesis where one C domain catalyses the formation of two amide bonds (Silakowski et al. 2000). Non-linear NRPSs are also often found to incorporate free soluble molecules without prior binding to a thioester, as in vibriobactin or bleomycin synthesis (Butterton et al. 2000; Keating et al. 2000a, b; Wyckoff et al. 2001). By now, in Ascomycetes only three NRPSs belonging to type C are described: the *A. fumigatus* NRPS Afpes1 has the domain organization ATC-CA-ATC-A (Neville et al. 2005). The *Alternaria brassicae* NRPS Abrepsy1 consists of five complete modules and an initiation lacking the A domain (TEC-ATC-ATEC-ATC-ATEC-TC), similar to yersiniabactin synthetase. Therefore Guillemette et al. (2004) suggested the presence of a second NRPS with at least one A domain. Stand-alone adenylation domains can be found performing Blast searches in many fungal genomes (Eisfeld, unpublished data). Also, the basidiomycete *O. olearius* NRPS Pso3 has an unusual domain arrangement (CAT-ATC-ATEC-ATE; Eisfeld et al., unpublished data). Unfortunately, the products of these fungal non-linear NRPSs have not been identified so far. Thus, no statement about the strategy of synthesis can be made.

C. Domain Types of Non-Ribosomal Peptide Synthetases

1. Adenylation Domain

a) Functions of Adenylation Domains

The core element of NRPS modules is the adenylation domain (A-domain). This domain is about 550 amino acids in size and fulfils two functions: the selection and the activation of the cognate substrate. The substrate to be activated is an amino acid, an imino acid, a hydroxyl acid or a carboxy acid that will subsequently be added to the growing peptide chain (von Döhren 2004). The activation step is ATP dependent, consuming one

ATP and generating one PPi (Dieckmann et al. 1995; Stachelhaus and Marahiel 1995; Mootz and Marahiel 1997). It has been studied by the ATP-PPi exchange assay which was developed to measure the substrate specificity for an A domain (Santi et al. 1974; Konz et al. 1999; Mootz and Marahiel 2002).

Adenylation domains catalyze a two-step reaction. The first step is the adenylation of the substrate resulting in the formation of an acyl-AMP. The second step is the formation of a covalent binding via a thioester bond to the 4' phosphopantetheine (Ppant) cofactor at the PCP domain (Fig. 15.3).

A set of conserved sequence motifs for recognition and binding of the substrates is characteristic for A domains (Fig. 15.4). These motifs are highly conserved and almost all of them are located around the active sites where the substrates are bound (Marahiel et al. 1997).

The activation reaction is analogous to the first reaction performed by aminoacyl-tRNA synthetases on ribosomes (Arnez and Moras 1997). Crystal structures exist for two bacterial adenylation domains, PheA (Conti et al. 1997) and DhbE (May et al. 2002). PheA is the L-phenylalanine-activating domain of the *Bacillus brevis* gramicidin S synthetase GrsA. Conti et al. (1997) found that this A-domain shares no homology with tRNA synthetases but shows low (16%) sequence homology with acyl-CoA ligases and firefly luciferases, which also belong to the superfamily of peptide synthetases and adenylate-forming enzymes and these enzymes exhibit an almost identical fold. May et al. (2002) solved the structure of the stand-alone 2,3-dihy-

droxybenzoic acid activating domain DhbE which supported the findings that A domains share a very similar fold. Both adenylation domains consist of two major subdomains, a smaller C-terminal subdomain of about 100 residues and a larger N-terminal subdomain of about 400 residues.

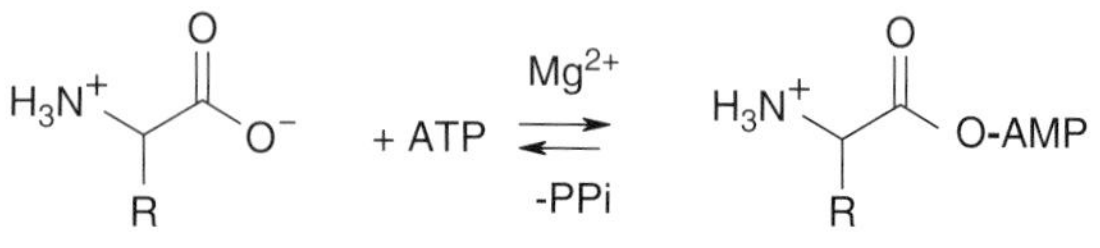

b) Substrate Specificity

The study of the specificity of A domains was greatly facilitated by the determination of the crystal structures of PheA and DhbE (Conti et al. 1997; May et al. 2002). Based on the crystal structure of PheA, Conti et al. (1997) determined the amino acid residues that are crucial for substrate binding. Adenylation domain sequence alignments identified ten sequence positions which lie in a 100 amino acid stretch between core motifs A4 and A5 and are lining the binding pocket (Fig. 15.4). This allowed the correlation with known substrates and the proposal of a non ribosomal specificity-conferring code. The crystal structure of DhbE allowed the adaptation of a specificity conferring code for aryl acid activating domains (Stachelhaus et al. 1999). Challis et al. (2000) presented a more detailed analysis allowing the prediction of the specificity of adenylation domains of unknown function. The predictive structure-based models permit the prediction of amino acid specificity in NRPS of unknown function but derive largely by systematic comparative analysis of bacterial A domains. However, these models are not fully applicable for all fungal A domains. For example, while the three adenylation domains of ACV synthetases have identical amino acid residues at the positions defining the non-ribosomal code, no other valine specificity domains correlate. The D-alanine-specific domains of cyclosporin synthetase and HC-toxin synthetase have identical residues, but the predicted specificity-conferring residues for alanine and proline show wide variations in fungi (von

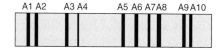

Fig. 15.3. The adenylation domain catalyzes the activation of a carboxy acid by hydrolysis of ATP

```
A1 A2    A3 A4       A5 A6 A7A8    A9 A10
```

A3 interaction with pyrophosphate leaving group
A4 interaction with amino group
A7 interaction via hydrogen bonds with oxygen atoms of nucleoside ribose moiety
A8 essential for adenylation
A10 binds to substrate and ribose oxygens

Fig. 15.4. Position and function of conserved motifs of the adenylation domain

Döhren 2004). Siderophore synthetases of the ferrichrome type are restricted to fungi and Schwecke et al. (2006) analyzed the codons derived from code positions in detail, as they did not match available specificities. The authors found three different codes for glycine activating domains and three major groups of adenylation domains for ornithine-derived substrates. Rausch et al. (2005) invented a new support vector machine-based approach to predict substrate specificity using 1230 NRPS A domains and were able to predict the specificity for 18% more sequences than Stachelhaus et al. (1999) and Challis et al. (2000). For 2.4% of the sequences no prognosis was possible, a large ratio of them of fungal origin. Therefore, eukaryotic A domains might have developed a differing domain architecture of the substrate binding sites or an alternative selection of the residues lining the binding pocket might define slightly different contacts (Rausch et al. 2005; Schwecke et al. 2006).

Furthermore, in contrast to aminoacyl-tRNA synthetases, A domains of peptide synthetases often do not display unique substrate specificity. Many non-ribosomally formed peptides display both highly conserved and variable amino acid positions. Wiest et al. (2002) suggested that the multiplicity of products of one NRPS is a result of the ability of the A domain to bind multiple substrates. Both in phylogenetic analysis and using the signature sequences proposed by Challis et al. (2000) the A domains of peptaibol synthetases from Trichoderma/Hypocrea display a high diversity and prediction of amino acids bound by these modules often was impossible (Komon-Zelazowska et al. 2007). These variations lead to a wide range of peptide families. For example, more than 30 analogues of cyclosporin A have been described (Billich and Zocher 1987; Lawen et al. 1989; Traber et al. 1989; Kleinkauf and von Döhren 1995; Eckstein and Fung 2003; Wei et al. 2004). Cyclosporin A is a cyclic undecapeptide with various biological properties including antifungal, antiparasitic, anti-inflammatory and immunosuppressive activities (Borel 1986) and is widely used in human transplantation surgery (Kahan 1984) and in the treatment of autoimmune diseases (Talal 1988). The cyclosporin synthetase of Beauveria nivea exhibits a broad spectrum of activities. Incorporation of amino acids ranges from highly specific sites (e.g. sarcosine at position 3, methylleucine at positions 4, 6, 9) to very unspecific ones (e.g. butenylmethythreonine at position 1, L-α-aminobutyric acid at position 2, alanine at position 8). Interestingly, the specificity of the cyclosporin synthetase system differs in vivo and in vitro (Lawen and Traber 1993). Most sites of the related peptolide SDZ 214-103 synthetase appear to be more specific than those of cyclosporine synthetase (Lawen and Traber 1993).

Isolated A domains of ACV synthetase which synthesizes the penicillin precursor do not discriminate between L-α-amino-adipic acid (Aad) and valine. In didomain constructs with the cognate carrier proteins however, both substrates were activated at similar levels but only Aad was detected as an acid-stable thioester, which suggests a role in selection of the carrier domain (von Döhren et al. 1999). There also exist various natural enniatins caused by a lowered substrate specificity of the A domain of the second module (EB) of the enniatin synthetase which activates the branched chain N-methyl-L-amino acid (Madry et al. 1984). The first module of the enniatin synthetase (EA; activation of α-D-hydroxyisovaleric acid) also has a relaxed substrate specificity (Krause et al. 2001; Feifel et al. 2007), but a highly specific dehydrogenase supplies the enzyme exclusively with D-Hiv (Lee et al. 1992) and is found in most enniatin synthesizing strains. Therefore a high conservation in this position is preserved. Feifel et al. (2007) demonstrated, that the enniatin synthetase has substrate tolerance towards a variety of different α-D-hydroxy acids and furthermore permits the synthesis of various new enniatins with versatile functional groups.

Peptaibols are synthesized by various fungi, including fungi of the genera Acremonium, Paecilomyces, Emericellopsis and several species of Trichoderma (reviewed by Daniel and Filho 2007). Peptaibols are linear oligopeptides which self-assemble into oligomers forming pores or channels across the plasma membrane of their target organisms (Chugh and Wallace 2001). They typically occur as mixtures of isoforms and more than 300 peptaibols have been found to date (Whitmore and Wallace 2004; http://www.cryst.bbk.ac.uk/peptaibol). A special characteristic of peptaibols is that they are produced as families of similar peptides, often differing in one or a few amino acids and displaying similar biological activities. Particularly noticeable, however, is that most positions are conserved while others are less stringently controlled. Two examples are shown in Fig. 15.5.

2. Thiolation Domains and 4′-Phosphopantetheine Transferases

Following selection and activation, the substrate is covalently attached to the 80-amino-acid thiolation (T) or peptidyl carrier protein (PCP) domain

Antiamoebins (Lehr et al. 2006)

I   Ac-Phe-Aib-Aib-Aib-D-Iva-Gly-Leu-Aib-Aib-Hyp-Gln-D-Iva-Hyp-Aib-Pro-Phe-OH
II   Ac-Phe-Aib-Aib-Aib-Aib-Gly-Leu-Aib-Aib-Hyp-Gln-D-Iva-Hyp-Aib-Pro-Phe-OH
XVI   Ac-Leu-Aib-Aib-Aib-D-Iva-Gly-Leu-Aib-Aib-Hyp-Gln-D-Iva-Hyp-Aib-Pro-Phe-OH

Ampullosporin A and analogues (Kronen et al. 2001)

A   Ac-Trp-Ala-Aib-Aib-Leu-Aib-Gln-Aib-Aib-Aib-Gln-Leu-Aib-Gln-Leu-OH
B   Ac-Trp-Ala-Aib-Aib-Leu-Aib-Gln-Aib-Ala-Aib-Gln-Leu-Aib-Gln-Leu-OH
C   Ac-Trp-Ala-Aib-Aib-Leu-Aib-Gln-Aib-Aib-Ala-Gln-Leu-Aib-Gln-Leu-OH
D   Ac-Trp-Ala-Aib-Aib-Leu-Aib-Gln-Aib-Aib-Aib-Gln-Leu-Ala-Gln-Leu-OH
E1   Ac-Trp-Ala-Aib-Aib-Leu-Aib-Gln-Aib-Ala-Ala-Gln-Leu-Aib-Gln-Leu-OH
E2   Ac-Trp-Ala-Aib-Aib-Leu-Aib-Gln-Aib-Ala-Ala-Gln-Leu-Ala-Gln-Leu-OH
E3   Ac-Trp-Ala-Aib-Aib-Leu-Aib-Gln-Aib-Aib-Aib-Gln-Leu-Ala-Gln-Leu-OH
E4   Ac-Trp-Ala-Aib-Aib-Leu-Aib-Gln-Ala-Ala-Aib-Gln-Leu-Aib-Gln-Leu-OH

Fig. 15.5. Examples for similar peptides of the peptaibol family. Less conserved positions are shaded in *gray*

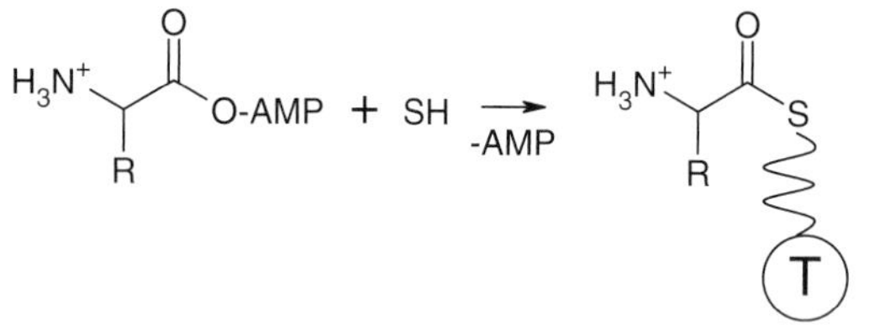

Fig. 15.6. Aminoacyl adenylate substrates of the elongating peptide chain are covalently attached to the 4'Ppant cofactor of the T domain

located at the C terminus of an A domain. This domain accepts the activated amino acid and acts as a transport vehicle (Stachelhaus et al. 1996; Weber et al. 2000; Fig. 15.6). Before this occurs, the T domain has to be post-translationally converted to the active form by the covalent attachment of a 4' Ppant cofactor derived from coenzyme A (Lambalot et al. 1996; Reuter et al. 1999). This modification is catalyzed by a member of a special class of phosphopantetheinyl transferases (the carrier protein superfamily) which also activate acyl carrier proteins of fatty acid synthases and polyketide synthases.

Bacterial Ppant transferases (PPTases) are subdivided into three groups according to their primary sequences and their substrate specificities (Lambalot et al. 1996). The bacterial AcpS-type PPTases are usually associated with primary metabolism while a second group of eukaryotic PPTases are integral domains of fatty acid synthases. Ppant transferases of the third class (sfp type; *surfactin phosphotransferase*) often are associated with NRPS gene clusters (Nakano et al. 1992; Lambalot 1996). These enzymes modify a variety of carrier proteins and domains including PCPs of NRPSs and acyl-carrier proteins (ACPs) of fatty acid synthases and polyketide synthases in bacteria.

In *Aspergillus fumigatus* and in *A. nidulans*, the npgA/cwfA PPTases have been characterized and shown to contribute to penicillin biosynthesis, pigmentation, lysine and siderophore biosynthesis and biosynthesis of other polyketides and peptides (Kim et al. 2001; Mootz et al. 2002a, b; Keszenman-Pereyra et al. 2003; Oberegger et al. 2003; Márquez-Fernández et al. 2007). It was suggested that a single PPTase is responsible for the activation of all PKSs and NRPSs in *A. nidulans* (Márquez-Fernández et al. 2007).

T domains exhibit a conserved structure consisting of a four-helix bundle fold (Weber et al. 2000). Helix 2 contains the site of recognition for other protein partners (including AcpS, A-domains, C-domains, E-domains; reviewed by Finking and Marahiel 2004).

In T domains of NRPSs the core motif is LGG (HD)**S**(LI). The motif encloses a highly conserved serine residue whose hydroxyl group functions as attachment site for Ppant. The activated amino acid is transferred onto the thiol moiety of Ppant and a thioester bond is formed. The cofactor acts according to the swinging arm mechanism described for fatty acid synthesis to allow the bound substrate to travel between different catalytic centers (Finking and Marahiel, 2004; Stachelhaus et al. 1996; Stein et al. 1996).

Alignments of carrier domains of NRPSs led to clustering that correlates with function (i.e. interaction with their protein partners and accepted amino acid; Dieckmann and von Döhren 1997) and localization of the T domain within a module (Linne et al. 2001). In bacteria, thiolation domains interacting with epimerization domains contain

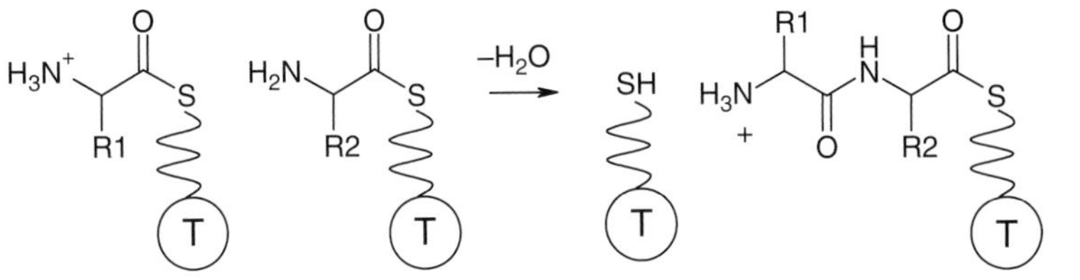

Fig. 15.7. The condensation domain catalyzes peptide bond formation between the acyl groups of activated substrates attached to the T domains of adjacent modules

the conserved motif LGGDS*L*, while interaction with condensation domains requires the motif LGGDS*I*. Again, the assumption that an asparagine residue is essential for the epimerization reaction is not fully applicable to fungi, as fungal T domains contain most frequently an isoleucine residue in the presence and in the absence of an E domain (von Döhren 2004).

3. Condensation Domains

The condensation (C) domain consists of ~450 amino acids and is the third domain type, being essential for elongation of the peptide chain in non-ribosomal peptide synthesis. The C domain catalyzes the peptide bond formation between the aminoacyl substrates attached to T domains of adjacent modules (Stachelhaus et al. 1998). According to the thiotemplate model (Stein et al. 1996), C domains comprise a donor and an acceptor site. The latter binds the aminoacylated cofactor which acts as nucleophile while the donor harbors the cofactor of the upstream module which is loaded with an aminoacyl or peptidyl group. The C domain catalyzes the nucleophilic attack which results in the formation of a peptide bond between the activated acyl group and the free amino (or imino or hydroxyl) group of a second amino acid (de Crécy-Lagard et al. 1995; von Döhren et al. 1997; Fig. 15.7). During assembly of the peptide chain in an N- to C-terminal direction the intermediates remain bound to the enzyme (Roskoski et al. 1970).

The C domains appear to be enzymes to be found exclusively in NRPSs. Solely the highly conserved core motif HHxxxDG (C3; see Table 15.2) resembles the active site motif of acyl transferases (de Crécy-Lagard et al. 1995; Marahiel et al. 1997). Mutation of the second histidine residue in this motif abolishes the condensation reaction (Stachelhaus et al. 1998).

As described above, the specificity of an NRPS module is principally controlled at the adenyla-

tion/pantetheinylation step. However, by means of mischarged donor and acceptor aa-S-Ppant T domains Belshaw et al. (1999) demonstrated that there is also a selectivity of the C domain at the acceptor site. Both donor and acceptor site exhibit strong stereoselectivity while the acceptor site also shows significant side-chain selectivity (Lautru and Challis 2004). Rausch et al. (2007) recently reviewed the current knowledge of subtypes of the C domain. LDL-catalysts catalyze the condensation of two L-amino acids. DCL-catalysts are C-domains that are located immediately downstream of epimerization domains and are D-specific for the upstream donor and L-specific for the downstream acceptor, and therefore form a peptide bond between a D- and an L-residue (Clugston et al. 2003). Dual epimerization/condensation (E/C) domains are DCL catalysts with epimerase activity, which catalyze the epimerization of an L-residue and subsequently promote the condensation of this amino acid with the amino acid attached to the upstream module. E/C-domains exhibit a second His-motif (HH(I/L)xxxGD) which is located near the N-terminus (Balibar et al. 2005). Heterocyclization (Cyc) domains catalyze peptide bond formation and subsequent cyclization of cysteine, serine and threonine which results in formation of a five-membered heterocyclic ring (Sieber and Marahiel 2005). Cyc domains are related to C-domains but the active site motif is replaced by the motif DxxxxD in which the aspartate residues are critical for both reactions performed by this domain (Keating et al. 2002).

4. Modifying Domains

Typically NRPs contain unusual or modified amino acids. Fungal non-ribosomal peptides often contain modifications such as N-acylation and C-terminal reduction (peptaibols), N-methylations (cyclosporin, enniatin) and site-specific epimerizations. N-methylations and epimerizations are conducted by particular domains while the latter can also be conducted by distinct enzymes. Here, N-methylation (N-MTase) domains and epimerization (E) domains are described.

a) N-Methylation Domain

N-Methylated peptides produced by fungi include pharmacological interesting peptides such as cyclosporin, members of the enniatin family and aureobasidin A. The genes encoding cyclosporin synthetase, enniatin synthetase and peramine synthetase have been studied (see Table 15.1), but N-methylated amino acids are also found in PF1022, beauvericin, tentoxin, cyclopeptin and omphalotins and there are predicted N-methylation domains in genomes of e.g. *Magnaporthe grisea*, *Aspergillus flavus*, *A. nidulans* and *Phanerochaete chrysosporium* (Peeters et al. 1988; Liebermann and Ramm 1991; Sterner et al. 1997; Weckwerth et al. 2000; Eisfeld, unpublished data). N-Methylation of peptides contributes to their biological activity and to peptide bond stabilization against proteolytic cleavage (Marahiel et al. 1997).

N-Methylation of amino acids during nonribosomal peptide synthesis occurs at the thioester stage prior to peptide bond formation (Zocher et al. 1983, 1986; Schauwecker et al. 2000; Weckwerth et al. 2000). The N-methyltransferase activity is determined by N-MTase domains (~420 amino acids) which are inserted between core motifs A8 and A9 of the corresponding A domains (Fig. 15.4, Table 15.2). The 4'-phosphopantetheine-bound amino acids are substrates for the corresponding methyltransferase. S-Adenosyl-methionine (SAM) serves as methyl donor. The N-MTase domains of NRPSs catalyze the transfer of the methyl residue onto the covalently bound aminoacyl (or peptidyl) group, releasing S-adenosyl-L-homocysteine (AdoHcy) as a reaction product (Lawen and Zocher 1990; Dittmann et al. 1994; Zocher and Keller 1997; Glinski et al. 2002). Recently, Hornbogen et al. (2007) demonstrated that AdoHyc acts as a regulator of enniatin synthesis by inhibiting the formation of desmethyl enniatin. However, in the absence of SAM, non-methylated products form at a reduced rate (Glinski et al. 2001), indicating that N-methylation is not obligatory for the following condensation reaction.

N-MTase domains contain four signature core motifs including a glycine-rich sequence (motif I;

I	VLEIGTGSGMIL
II/Y	SYVGLDPS
IV	DLVVFNSVVQVFTPPEYL
V	ATNGHFLAARA

AdoMet binding activity

Fig. 15.8. Position and function of conserved motifs of the methylation domain

VL(D/E)GxGxG), which is similar to the SAM-binding site of the heterologous class of cosubstrate-dependent methyltransferases (Hacker et al. 2000; Fig. 15.8). Aside from structural features typical of other AdoMet-dependent methyltransferases, N-MTase domains of NRPS share only weak sequence similarity to other methyltransferases (Billich and Zocher 1987; Burmester et al. 1995; Hacker et al. 2000).

b) Epimerization Domains and Amino Acid Racemases

Besides N-methylation, incorporation of D-amino acids is also a common feature of non-ribosomal peptide syntheses in fungi. The corresponding NRPSs mostly contain an epimerization (E) domain. These domains are about 420 amino acids in size. E domains strongly resemble condensation domains but can be distinguished by the presence of seven signature sequences characteristic for these domains (Table 15.2). In a linear NRPS, E domains are located at the C-terminus of the respective module's T domain. Two types of E domains can be distinguished: E domains originating from elongation modules (peptidyl-E domains) and E domains of initiation modules (aminoacyl-E domains). Epimerization of aminoacyl and peptidyl intermediates takes place at the thioester stage. The reaction is reversible, thus an equilibrium of both isomers is established (Stachelhaus and Walsh 2000). The condensation reaction is involved in the control of stereospecifity by catalyzing solely the elongation of the D-isomer (Belshaw et al. 1999; Clugston et al. 2003). However, the active site of the E domain selectively stabilizes the D-isomer in the E domain of gramicidin S synthetase. Therefore, stereospecifity is not only controlled in the subsequent reaction but also at the epimerization stage (Stachelhaus and Walsh 2000).

Incorporation of a D-amino acid may also be determined by the specificity of the A domain. In this case, an external racemase provides the D-amino acid. Well known examples in fungal peptide synthesis are alanine residues in cyclosporin and HC toxin. Hoffmann et al. (1994) characterized the racemase involved in cyclosporin formation. The enzyme catalyzes the reversible racemization of alanine and requires pyridoxal phosphate as the exclusive cofactor. In *Cochliobolus carbonum*, the *TOXG* gene encodes the alanine racemase required for synthesis of D-Ala for incorporation into HC toxin (Cheng and Walton 2000).

5. Termination Domains

The final step in non-ribosomal peptide synthesis is the release of the covalently bound peptide from the enzyme. This step is catalyzed by specialized C-terminal domains which on the one hand must accept the mature product and on the other hand must recognize the full-length peptide chain to prevent hydrolysis of incomplete peptides (Schneider and Marahiel 1998). In bacteria, C-terminal thioesterase (TE) domains commonly release the final product. However, termination of non-ribosomal peptide synthesis may also be catalyzed by reductase (R) domains or condenation domains. The latter are found mainly in fungal NRPSs (Keating et al. 2001).

a) Thioesterase Domain
In bacterial NRPSs, the release of the nascent peptide is mostly catalyzed by a thioesterase (TE) domain which comprises about 250 amino acids. This domain has homology to thioesterases found in the fatty acid metabolism (Pazirandeh et al. 1991). After transfer of the peptidyl intermediate onto a catalytic serine or cysteine the TE domain leads to the release of the nascent peptide by hydrolysis, cyclization or oligomerization (Samel et al. 2006). Most fungal NRPSs, however, lack TE domains (Walton 2004). One exception is the ACV synthetase which is highly conserved between bacteria and fungi. There is also a terminal TE domain in an uncharacterized *Phanerochaete chrysosporium* NRPS and in *A. nidulans* TdiA (see Table 15.1; Balibar et al. 2007).

b) Reductase Domain
A terminal aldehyde or alcohol group has been found in various non-ribosomally produced peptides, for example peptaibols. These result in liberation of the nascent peptide from its covalent linkage to the enzyme by a reductase (R) domain. R domains (~400 amino acids in size) share homology with NAD(P)H-dependent reductases, e.g. in the yeast biosynthetic pathway for lysine synthesis (Guo et al. 2001; von Döhren 2004).

c) Condensation Domain
In fungi, most NRPSs carry a C-terminal condensation domain which results in cyclization of the peptide product. Chain termination in this case is accomplished by amide bond formation. For example, in cyclosporin synthesis the terminal C domain has been assumed to couple the free amino group of D-Ala1 to the carboxyl group of L-Ala11. However, there is neither direct evidence for this reaction nor a lack of evidence for an acyl-O-C domain intermediate (Keating et al. 2001). Enniatin synthesis, HC-toxin synthesis and the synthesis of PF1022A are predicted to follow another mechanism. In analogy to acyl transferase 1, a direct nucleophilic attack on the thioester bond was suggested (Scott-Craig et al. 1992; Haese et al. 1993; Pieper et al. 1995; Shaw-Reid et al. 1999; Weckwerth et al. 2000). The enniatin synthetase carries a terminal didomain T-C module replacing the TE domain and allowing the successive accumulation of oligomeres on the T domain prior to the cyclization mediated by the C-domain (Feifel et al. 2007).

In LPS1, the ergotamine synthetase of *Claviceps pupurea*, a domain with limited similarity to both, C and cyclization domains of NRPS is located at the carboxy terminal side. This domain differs in the highly conserved motif C3, suggesting a special mechanism in the final step of the D-lysergyl-peptide lactam synthesis where a diketopiperazine is formed from the Phe and Pro moieties (Walzel et al. 1997).

III. Distinctions Between Fungal and Bacterial NRPSs

Although fungal and bacterial NRPSs share the domain architecture typical for these enzymes, there are some characteristics of fungal peptide synthetases that differ from those of bacterial origin. As described above, conserved sequence motifs strongly resemble each other in bacteria and fungi, but also differ to some extend and the "non-ribosomal code" established with sequence data from bacterial NRPSs cannot be (fully) applied to fungal systems (von Döhren 2004; Schwecke et al. 2006).

The most obvious difference between fungal and bacterial NRPSs is that most fungal peptide synthetases consist of one polypeptide chain while bacterial NRPSs in most cases consist of more than one protein (Weber et al. 1994). Initially, this was thought to be characteristic for fungal NRPSs and still applies to most of them. The ergot peptide system, however, operates with two interacting multienzymes, lysergyl peptide synthetase 1 (Lps1) and lysergyl peptide synthetase 2 (Lps2; Riederer et al. 1996). A model of two interacting NRPSs has also been proposed for ampullosporin synthesis in *Sepedonium ampullosporum* (Reiber et al. 2003).

Most bacterial NRPSs carry a TE domain at the C-terminus (Keating et al. 2001). In fungi, TE domains are rare and mostly a C-terminal C domain is found. A TE domain is found in the ACV synthetase, but this enzyme is an exception as it shares a very high homology to bacterial ACV synthetases. It has been suggested that the penicillin biosynthetic gene cluster derived from prokaryotic origin and has been transmitted by horizontal gene transfer (Aharonowitz et al. 1992). Furthermore, typical for fungal NRPSs is that they often contain D-amino acids, but the corresponding NRPSs contain no epimerization domain (Doekel and Marahiel 2001). In these cases, external epimerases catalyze the formation of the D-amino acid which is then incorporated by an adenylation domain with substrate specificity for the D-isomer (see above). External epimerases have not yet been described for bacterial NRPSs where this reaction is typically carried out in *cis*. A lot of modifications carried out subsequent to the assembly of the peptides have been described for bacterial NRPs. These reactions include halogenations, hydroxylations and glycosylations. The chemical modifications of the peptides are catalyzed by enzymes which are associated with the corresponding gene cluster (reviewed by Samel and Marahiel 2008). Such enzymes associated with fungal NRPSs have not yet been described.

Non-ribosomal peptide synthesis has been studied mainly in bacteria and at lot of information about bacterial NRPSs is available. Apart from that, these enzymes have been described mostly for filamentous ascomycete fungi. As described above, many ascomycetes produce NRPs which are useful in medicine (penicillin, cyclosporine), or are involved in pathogenesis (enniatin, HC toxin, AM toxin, gliotoxin), act as siderophores for iron acquisition or in iron storage or possess yet unknown activities (Stachelhaus and Marahiel 1995; Kleinkauf and von Döhren 1996; Schwarzer and Marahiel 2001; Haas 2003; von Döhren 2004). The first fungal NRPS genes examined (e.g. ACV synthetase, HC toxin synthetase, cyclosporin synthetase and others) were large genes which contained no introns (Smith et al. 1990; Gutierrez et al. 1991; Scott-Craig et al. 1992; Weber et al. 1994). Therefore, it was suggested, that these genes derived from horizontal gene transfer from bacteria. The fact, that fungal NRPS genes are intron-less is still true for many NRPS genes but meanwhile also many NRPS genes have been discovered that bear introns. Some examples are shown in Table 15.3. The first NRPS gene identified which is interrupted by at least two introns was a peptide synthetase gene in *Metarhizium anisopliae* (Bailey et al. 1996). However, the number of introns in ascomycetous NRPS genes ranges from one to seven which still is a very low intron content, considering the enormous size of the genes and compared to other genes. However, in *A. fumigatus*, ten out of the 14 NRPS sequences identified contain introns with NRPS1 having the most with six (Cramer et al. 2006).

Table 15.3. Selected NRPS genes of ascomycetes and the number of introns contained

Producer	NRPS gene	Number of introns
Claviceps purpurea	CPPS2	2
Neotyphodium lolii	lpsA	2
Metarhizium anisopliae	PESA	2
Alternaria brassicae	AbrePsy1	7
Aspergillus nidulans	SIDC	1
A. fumigatus	SIDC	1
Schizosaccharomyces pombe	SIP1	1
Magnaporthe grisea	SSM1	1

IV. Non-Ribosomal Peptide Synthesis in Basidiomycetes

Most fungal NRPSs described so far derive from ascomycetes. Basidiomycetes are also well known as producers of interesting natural compounds and possess a manifold secondary metabolism (Lorenzen and Anke 1998). Peptides produced by basidiomycetes include for example the toxic amanitines and phalloidines (*Amanita* sp.; Vetter 1998), the immunosuppressive cycloamanides (*Amanita phalloides*; Chiang et al. 1982) and the nematicidal omphalotins (*Omphalotus olearius*; Sterner et al. 1997). However, there is not much information about non-ribosomal peptide synthesis in basidiomycetes.

The bicyclic amatoxins and phallotoxins are octapeptides and heptapeptides, respectively. These peptides were thought to be produced non-ribosomally. However, sequencing of *Amanita bisporogena* revealed that this fungus contains no NRPS genes but two genes, *AMA*1 and *PHA*1, which directly encode α-amanitin and phallacidin. Both peptides are synthesized as proproteins from which they are predicted to be cleaved by a prolyl oligopeptidase (Hallen et al. 2007). Hallen et al. suggested that fungi of the *Amanita* sp. have a broad capacity to synthesize cyclic peptides ribosomally as AMA1 and PHA1 are also present in other toxic species. *Amanita virosa* produces virotoxins, toxic peptides which are solely found in this fungus. The structure of viroisin, the main virotoxin, is in part the same as in phallotoxins but contains D-serine instead of L-cysteine and two unnatural amino acids (2,3-trans-3,4-dihydroxy-L-proline and 2'-(methylsulfonyl)-L-tryptophan). It remains to be elucidated whether this toxin also is synthesized ribosomally.

However, NRPS genes have been found in other basidiomycetes. Siderophore biosynthesis has been investigated in the ustilaginomycete *Ustilago maydis* and in the homobasidiomycete *Omphalotus olearius*. *U. maydis* produces ferrichrome and ferrichrome A (Budde and Leong 1989). The corresponding siderophore synthetases have been identified. *SID2* and *FER3* encode the ferrichrome synthetase and the ferrichrome A synthetase, respectively (Yuan et al. 2001; Eichhorn et al. 2006). A putative ferrichrome A synthetase gene has also been found in *O. olearius*. The NRPS gene encoding the ferrichrome A synthetase in this fungus (*FSO1*) was the first NRPS gene described in a higher basidiomycete. The basidiomycete ferrichrome-type siderophore synthetase possesses high levels of similarity to siderophore synthetases of ascomycetes.

Interestingly, *FSO1* is interrupted by a total of 48 introns (Welzel et al. 2005). Although a high number of introns is very unusual for NRPS genes, basidiomycete genes generally are interrupted by many introns (Larrondo et al. 2004; Martinez et al. 2004). Two more NRPS genes have been identified in *O. olearius*: *PSO3* and *PSO4* (Welzel 2005; Eisfeld, unpublished data). Both NRPSs have unusual domain architectures and contain a relatively high number of introns. *PSO4* consists of one minimal module and an additional C domain at its N-terminus (CATC). *PSO3* is displayed in Fig. 15.2. The 15 656-bp gene is disrupted by 25 introns. Noticeably, introns often interrupt conserved sequence motifs. Attempts to identify the products of these NRPSs have been unsuccessful so far.

The recent annotation of several basidiomycete genomes facilitates the search for NRPS genes in these fungi. In *U. maydis*, besides the siderophore synthetases a partial open reading frame that contains the N-terminal fragment of an NRPS was identified (um10543) which has the predicted domain architecture ATC-ATC-ATC and contains at least 12 introns (Bölker et al. 2008). The *Coprinus cinerea* genome seems to comprise one NRPS gene, CC1G_04210, with a length of 7782 bp, which is interrupted by 14 introns (Blast searches at the Broad Institute). The truncated modules (TC) suggest that the deduced peptide synthetase belongs to the iterative NRPSs (Fig. 15.9). CC1G_04210 is clustered with an L-ornithine monooxygenase gene, therefore this NRPS is likely to encode a siderophore synthetase. L-Ornithine monooxygenases comprise the first step in synthesis of siderophores of the hydroxamate type. The domain architecture resembles more that of NRPSs of the coprogen type than that of the ferrichrome type which in ascomycetes is mostly ATC-TC or ATC-TTC (Oide et al. 2006). In *O. olearius* and in several ascomycetes the siderophore synthetase genes are also clustered with the L-ornithine monooxygenase-encoding gene (Haas 2003; Welzel et al. 2005).

Phanerochaete chrysosporium was predicted to contain seven NRPS genes (Martinez et al. 2004). However, Blast searches at NCBI or JGI could only identify one NRPS. This gene encodes a dimodular NRPS with the first module comprising an N-MTase domain. Interestingly, a terminal Te domain is found in this NRPS (Fig. 15.9), which is rare in fungal NRPSs. In silico analyses lead to the assumption that the *P. chrysosporium* NRPS gene is interrupted by at least five introns (Eisfeld, unpublished data). In the genome of the stem rust fungus, *Puccinia graminis*, one monomodular NRPS (ATC) can be identified performing Blast searches at NCBI and this gene is interrupted by three predicted introns. In the *Laccaria bicolor* genome no NRPS-encoding gene could be identified so far.

Comparison of conserved sequence motifs in basidiomycete NRPSs reveals, that sequences show

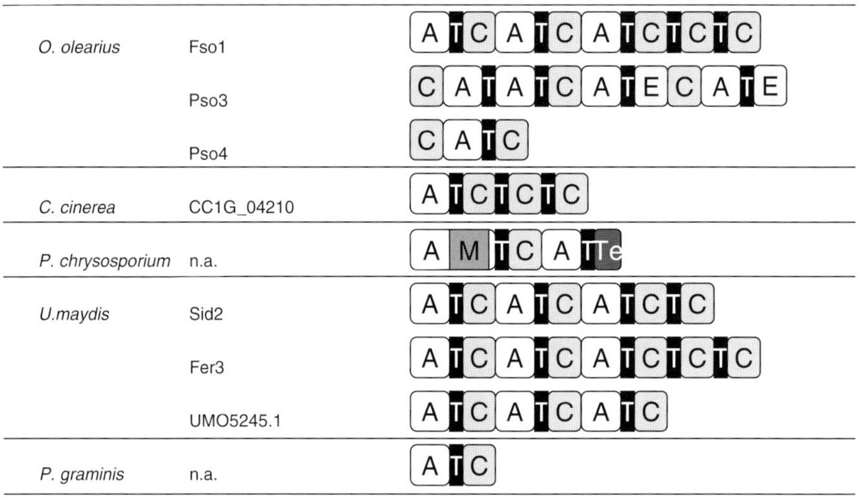

Fig. 15.9. NRPSs in basidiomycetes

Table 15.4. Selected NRPS genes of basidiomycetes and the number of introns contained

Producer	NRPS gene	Number of introns
Puccinia graminis	Not annotated	2 (in silico analysis)
Ustilago maydis	SID2	0
	FER3	1
	UMO5245.1	8
Coprinus cinerea	CC1G_04210	14 (predicted)
Phanerochaete chrysosporium	Not annotated	
Omphalotus olearius	FSO1	48
	PSO3	25
	PSO4	3

much more variations than those of bacterial origin, even within one species (Eisfeld, unpublished data). While amplification of NRPS gene fragments using degenerate primers deduced from core motifs often is successful in bacteria and in ascomycetes, sequence variations may hamper or even impede the amplification of yet unknown NRPSs in basidiomycetes using this approach.

It is particularly noticeable that the homobasidiomycete NRPS genes are interrupted by a higher number of introns than those of *U. maydis* and *P. graminis*. The *P. graminis* NRPS-encoding gene seems to be interrupted by two introns (Table 15.4). The divergence of ustilaginomycetes (including *U. maydis* and *P. graminis*) from hymenomycetes, to which the homobasidiomycetes belong, occurred at an early stage in evolution at least 500×10^6 years ago (Taylor and Berbee 2006). According to the "intron early theory" a near intron-less state of diverse eukaryotes seems to be due to the massive loss of ancestral introns, also bolstering the idea of complete intron loss in prokaryotes (reviewed by Roy and Gilbert 2006). This might be also reflected in the intron content of fungal NRPS genes.

V. Physiological Significance of Peptides

The activities of peptide products with respect to interactions with other organisms have been extensively studied. For example, the mode of action of cyclosporin and penicillin and their derivatives is very well documented. However, interorganismal activities mostly are not the primary function of NRPs and the physiological significance of most of the peptides is unknown (Turgeon et al. 2008). Among the few well studied functions of small peptides is the function as siderophores in complexing Fe^{3+}. Siderophores of the ferrichrome and of the coprogen type have been described in many fungi as for example *A. fumigatus*, *A. nidulans*, *U. maydis*, *O. olearius*, *S. pombe*, and various phytopathogenic fungi (*M. grisea*, *C. heterostrophus*, *F. graminearum*, *Alternaria brassicicola*, etc.) and an additional role as pathogenicity factor has been demonstrated for some of them (Eisendle et al. 2003, 2006; Hof et al. 2005; Welzel et al. 2005; Oide et al. 2006, 2007; Schwecke et al. 2006). The functions of siderophores were reviewed very recently by Johnson (2008) and Haas et al. (2008), and Schwecke et al. (2006) compared peptide synthetases of the ferrichrome type. In brief, siderophores play very diverse roles. Their primary function is the acquisition of iron under iron limited conditions and the storage of iron. Some siderophores are also required for resistance to oxidative stress, asexual/sexual development or virulence.

There are more non-ribosomally synthesized peptides which act as virulence factors during plant pathogenesis. These peptides include AM toxin from *Alternaria alternata* (Panaccione et al. 1992; Johnson et al. 2000) which causes the apple pathotype, and the maize pathogenic HC toxin from *Cochliobolus carbonum* (Scott-Craig et al. 1992; Jones and Dunkle 1995), and also enniatin from *Fusarium* spp. (Herrmann et al. 1996). A possible function for petaibols as virulence factor has also been described. Peptaibol-producing fungi of the genus *Sepedonium* infect fruiting bodies of basidiomycetes of the order *Boletales*. The peptaibols are thought to facilitate the invasion process by damaging host tissues in the basidiomycetes and also in insect invasion processes (Matha et al. 1992; Engelberth et al. 2001).

Endophyte toxins may play a protective role for their hosts. Peramine for example is a potent insect feeding deterrent. This pyrrolopyrazine is the product of a two-module NRPS and is synthesized by *Epichloë/Neophytodium* mutualistic endophytes (Tanaka et al. 2005). Some perennial ryegrass/*Neotyphodium* sp. symbiota accumulate ergovaline which reduces the appeal to rabbits. Rabbits prefer plants that are endophyte-infected but free of ergot alkaloids over endophyte-free plants. Therefore, the accumulation of ergot alkaloids seems to counteract the added appeal of endophyte-infected plants (Panaccione et al. 2006).

In order to characterize the function of all non-ribosomally produced peptides in one fungus, Turgeon et al. (2008) deleted each of the 12 NRPS genes in *C. heterosprophus*. Four of the genes are conserved in ascomycete genomes, and these genes were also the only ones yielding phenotypes. Two of the genes (*NPS*6, *NPS*2) are siderophore synthetases. Deletion of *NPS*10 led to an alteration in the morphology of the colonies and hypersensitivity

to various stresses. Deletion of *NPS*4 led to loss of hydrophobicity, a phenotype also described for the homologous genes in *A. brassicicola* (AbNPS2) and *G. zeae* (Kim et al. 2007; Turgeon et al. 2008). Turgeon et al. (2008) speculated that the corresponding NRPS produces a metabolite similar to acuminatum, a cyclodepsipeptide which has surfactant properties or a peptide that regulates the production of hydrophobins (Tobiasen et al. 2007). In *A. brassicicola*, deletion of the homologous gene led not only to a decreased hydrophobicity phenotype but also to an abnormal spore cell wall morphology and decreased spore germination rates, indicating that AbNPS2 plays an important role in development and virulence (Kim et al. 2007). Consequently, these studies showed that NRPs may play a role in basic biological processes such as nutrient acquisition and sexual and asexual development. However, the physiological function of many of them still remains to be elucidated.

VI. Examples for NRPS Gene Clusters and Cluster Evolution

In contrast to genes involved in primary metabolism, in fungi secondary metabolite genes are often organized in gene clusters (Keller and Hohn 1996). These genes are located adjacent to one another and are co-regulated. This is also the case for those pathways involved in non-ribosomal peptide synthesis (Walton 2000; Gardiner et al. 2004; Haarmann et al. 2005; Cramer et al. 2006; Zaleta-Rivera et al. 2006; Johnson et al. 2007). Models to explain this phenomenon include the suggestion that these genes derived from horizontal gene transfer from prokaryotes. More recent evidence suggests regulatory mechanisms as evolutionary driving force. Also, the clustering increases the probability of co-mobilization. Thus, there is a selective advantage to the gene cluster itself (Walton 2000; Hoffmeister and Keller 2006). There are some examples of horizontal transfer of fungal nuclear genes, and horizontal gene transfer has often been suggested as an explanation for the discontinuity in distribution of some fungal secondary metabolite genes when they are present only in some isolates of a species and are absent from others (Walton 2000). The horizontal transfer of an entire chromosome between two vegetative incompatible isolates of the same species has been demonstrated by He et al. (1998). This may explain the origin of supernumerary chromosomes in fungi which has been observed in several species. For example, Wang et al. (2003) demonstrated that a dispensable (CD) chromosome exists in the insect pathogenic fungus, *M. anisopliae*. The genes conferring the ability to produce destruxin are most likely located on this chromosome (Wang et al. 2003). The cyclic peptide synthetase gene encoding the AM toxin synthetase in *Alternaria alternata* also resides on a CD chromosome. Loss of this chromosome leads to loss of AM toxin production and pathogenicity (Johnson et al. 2001). The *O. olearius* peptide synthetase gene *PSO3* also is located on a CD chromosome (Eisfeld, unpublished data).

Some NRPS gene clusters are widely distributed in different ascomycete lineages. For example, putative epipolythiodioxopiperazine (ETPs) gene clusters are present in 14 ascomycete taxa. The gene content of the homologous clusters is not identical but common genes were identified in these 14 species. The ETP gene clusters appear to have a single origin and have been inherited relatively intact rather than assembling independently in the different ascomycete lineages (Patron et al. 2007). In *M. grisea*, the avirulence gene ACE1 which codes for a hybrid PKS-NRPS is organized in a cluster of 15 genes. Related clusters were found in several other ascomycetes, e.g. *Chaetomium globosum* and *A. clavatus*, which appear to have obtained the cluster by horizontal gene transfer form a donor closely related to *M. grisea* (Khaldi et al. 2008).

A. Penicillin and Cephalosporin Biosynthesis

Penicillins and cephalosporins are β-lactam containing antibiotics produced both of prokaryotes and eukaryotes. Penicillin G production is described solely for fungi (e.g. *P. chrysogenum*, *P. nalgiovenese*, *P. griseofulvum*, *P. dipodomys*, *P. flavigenum*, *A. nidulans*). Cephalosporin is produced by both bacteria and fungi, for example the Ascomycetes *Acremonium chrysogenum* and *Kallichroma thetys* and the Actinomycetes *Streptomyces clavuligerus*, *Streptomyces lipmanii* and *Nocardia lactamdurans*. Biosynthesis and gene clusters for β-lactam antibiotics and the control of their expression were recently reviewed by Brakhage et al. (2005) and Liras

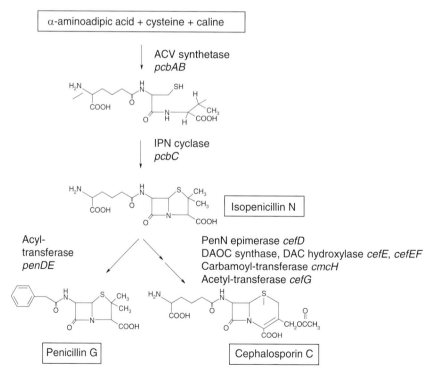

Fig. 15.10. Biosynthetic pathways of β-lactam antibiotics (modified from Liras and Martin 2006)

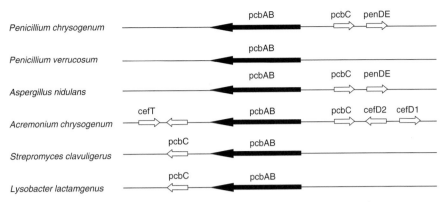

Fig. 15.11. Excerpt of penicillin/cephalosporin gene cluster of several fungi and bacteria. The NRPS gene is indicated as a *black arrow*. The *arrows* indicate the orientation of transcription (modified from Liras and Martín 2006)

and Martín (2006). In brief, the first two enzymatic steps are common to all β-lactam producers. First, a NRPS (ACV synthetase, encoded by *pcbAB*) catalyzes the formation of the tripeptide δ-(L-α-aminoadipyl)-L-cysteinyl-D-valine which in the second step is converted to the bicyclic isopenicillin N by the action of isopenicillin N (IPN) synthase (also named ACV cyclase), encoded by the *pcbC* gene (Fig. 15.10). This is the branch point for various penicillin and cephalosporin pathways. Fungal penicillin producers contain a third gene, *penDE*, which encodes an isopenicillin N acyltransferase that leads to the formation of penicillin G by hydrolyzation of the α-aminoadipic sidechain of isopenicillin N and introduction of phenylacetyl-CoA. For cephalosporin C synthesis, a set of genes is needed which encode the enzymes responsible for isomerization of isopenicillin N (*cefD*), conversion of the thiazolidine ring of penicillin N to a dihydrothiazine ring and hydroxylation at C-3 (*cefEF* in fungi, *cefE* and *cefF* in bacteria) and acetylation (*cefG*).

The genes responsible for β-lactam biosynthesis are clustered in all producers (pro- and eukaryotes). Figure 15.11 shows the structure and orientation of some of the genes. The *pcbC* and *pcbAB* genes are always located next to each other. In penicillin-producing fungi, the *penDE* gene is directly adjacent. The epimerization reaction to form the D-isomer penicillin N in *Acremonium* is encoded by two genes, *cefD1* and *cefD2* (while in bacteria only one gene, *cefD*, is necessary for this reaction). *cefEF* and *cefG*, whose gene products are involved in the final steps of cephalosporin synthesis, are linked and located on a different chromosome.

It was suggested that the β-lactam gene cluster has bacterial origin and was transmitted by horizontal gene transfer (Weigel et al. 1988; Aharonowitz et al. 1992). This theory is supported by the fact that the structure and orientation of the genes in the penicillin gene cluster is the same in both *Aspergillus* (MacCabe et al. 1990) and *Penicillium* (Laich et al. 1999). Furthermore, high homologies exist between the bacterial and fungal genes that are shared in both groups (ACV synthetase, IPN synthase, CefE, CefF) and which exceed those of genes belonging to primary metabolism. In addition, introns are lacking in *pcbAB*, pcbC and *cefEF*, while genes absent in bacterial clusters do contain introns (Liras and Martín 2006). Therefore, horizontal gene transfer of *pcbAB* and *pcbC* either took place about 370×10^6 years ago, before the split between *Aspergillus* and *Penicillium* occurred or multiple gene transfer events took place (Weigel et al. 1988; Aharonowitz et al. 1992). Laich et al. (2002) found a cluster of genes almost identical to that in *P. chrysogenum* in several other *Penicillium* spp. (*P. nalgiovense*, *P. griseofulvum*) while a truncated cluster (consisting of *pcbAB* and lacking *pcbC* and *penDE*) is present in *P. verrucosum*. Therefore, the penicillin gene cluster or just parts of the cluster as in *P. verrucosum* may have been lost throughout the evolution from a common ancestor (Fig. 15.11).

B. Ergot Alkaloid Biosynthetic Gene Clusters

The ergot alkaloids and their synthesis have been extensively studied and were recently reviewed, for

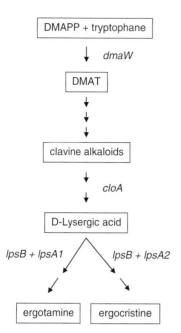

DMAPP + tryptophane

↓ *dmaW*

DMAT

↓
↓
↓

clavine alkaloids

↓ *cloA*
↓

D-Lysergic acid

lpsB + lpsA1   *lpsB + lpsA2*

ergotamine   ergocristine

Fig. 15.12. Biosynthetic pathway of ergot alkaloids in *Claviceps purpurea* (modified from Haarmann et al. 2005)

example by Lorenz et al. (2007) and Schardl et al. (2006). Ergot alkaloids share a four-membered ergoline ring but differ in the number, type and position of the side-chains. These pharmacological interesting mycotoxins are among others produced by fungi of the relatively distant taxons *Clavicipitaceae* and *Epichloë* endophytes of grasses and *A. fumigatus.*

As ergoline alkaloids which are synthesized by epibiotic fungi present on different plant species is the topic of Chap. 9, this chapter focuses on the peptide synthetases involved in ergot alkaloid synthesis and the evolution of the gene cluster. The first ergot alkaloid gene cluster analyzed was that in *Claviceps purpurea*. The first enzyme identified was dimethylallyl-tryptophan synthase (DMATS) which catalyzes the first step in the pathway (prenylation of tryptophan with formation of dimethylallyltryprophan; DMAT; see Figs. 15.12, 15.13; Gebler et al. 1992). The gene encoding this protein, *dmaW*, was identified in *C. purpurea* (Tsai et al. 1995; Tudzynski et al. 1999) and also characterized in *A. fumigatus* (Coyle and Panaccione 2005), *Claviceps fusiformis* (Tsai et al. 1995), and *Neotyphodium sp.* strain Lp1 (*Epichloë typhina* x *N. lolii*; Wang et al. 2004). Using this gene as a starting point, Tud-

zynski et al. (1999) cloned additional genes, providing evidence for an ergot alkaloid gene cluster which now comprises 14 genes in a genomic region of 68.5 kb (Correia et al. 2003; Haarmann et al. 2005; Lorenz et al. 2007). Besides *dmaW*, the function of *cloA* (conversion of elymoclavine to paspalic acid) and others has been functionally analyzed by heterologous expression or gene replacement approaches (Tsai et al. 1995; Correia et al. 2003; Haarmann et al. 2006). The pathway leading to two different ergot alkaloids is depicted (simplified) in Fig. 15.12.

The ergot alkaloid gene cluster in *C. purpurea* comprises four NRPS genes. The lysergyl peptide synthetase is a complex consisting of two NRPSs, LpsB and one of two different versions of LpsA which are encoded by *lpsA1* and *lpsA2*, respectively. LpsA from an ergocristine-producing isolate of *C. purpurea* differs from that of a previously characterized ergotamine-producing isolate concerning the signature sequences in the first module leading to the recognition of different amino acid substrates by the adenylation domains. The function of the fourth NRPS in the ergot alkaloid gene cluster of *C. purpurea*, *lpsC*, is still unknown (Haarmann et al. 2005).

The LpsA-encoding gene of *Neotyphodium lolii* was also analyzed and compared to that of *C. purpurea*. Both fungi produce different ergopeptines as most abundant ergot alkaloids. In *N. lolii* ergovaline is produced. Damrongkool et al. (2005) found that a copy of the NRPS gene was present which at the 5' end encodes a partial condensation domain in *N. lolii*. This truncated C domain is missing in *C. purpurea* LpsA. Both NRPSs have similar signature sequences in modules one and three but divergent signature sequences in module two. This finding is consistent with the amino acid found at the corresponding position in the major alkaloids of these fungi. The LpsA-encoding genes in *N. lolii* and *C. purpurea* both are interrupted by two introns. The position of the introns is conserved and, furthermore, located at the same point preceding the first and third module of the NRPSs. The latter finding supports the hypothesis that multimodular NRPS genes arose by duplication of an ancestral monomodular NRPS-encoding gene (Damrongkool et al. 2005).

Claviceps fusiformis and *A. fumigatus* are relatively distant taxons but both produce clavines.

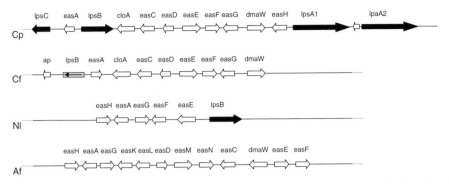

Fig. 15.13. Schematic comparison of the ergot alkaloid gene clusters of *C. purpurea* (*Cp*), *C. fusiformis* (*Cf*), *Neotyphodium lolii* (*Nl*) and *A. fumigatus* (*Af*). NRPS genes in the species are indicated by *black arrows*. The *arrows* indicate the orientation of transcription (modified from Fleetwood et al. 2007; Lorenz et al. 2007)

Therefore, the NRPSs and *cloA* are dispensable in these organisms. Coyle and Panaccione (2005) investigated whether the ergot alkaloids of *A. fumigatus* have a common biosynthetic origin with those of the clavicipitaceous family. *dmaW* was identified and could be complemented with the corresponding *C. purpurea* gene after it had been deleted. Besides this, five further genes which are proposed to encode steps in the ergot alkaloid pathway were detected, but orientation and positions of several genes differed. This is also the case for *C. fusiformis*. In this fungus, nine homologues of genes identified in the *C. purpurea* gene cluster were identified. Interestingly, *C. fusiformis* contains homologues of *cloA* and *lpsB*, but both are pseudogenes which contain frameshifts and stop codons. These findings support the proposal that the ergot alkaloid gene cluster evolved from a more complex cluster and that during evolution some genes were lost or inactivated due to mutations and rearrangements. In an ergovaline-producing *Neotyphodium lolii* strain, the ergot alkaloid gene cluster was analyzed and homologues to all genes found in the *C. purpurea* cluster were identified except for two genes (Fleetwood et al. 2007). Despite conservation of gene sequences, gene order is substantially different between the *N. lolii*, *A. fumigatus* and *C. purpurea* gene clusters (Fig. 15.13). *In N. lolii*, several long terminal repeat retrotransposons and non-autonomous transposable elements were identified in intergenic regions of the cluster which could be the reason for rearrangement events in the overall structure of the cluster (Fleetwood et al. 2007).

VII. Conclusions

Identification of new NRPS products with potentially interesting biological effects has been facilitated by genome sequencing projects. Knowledge of regulators of secondary metabolism help to identify transcriptionally active clusters and also to specifically manipulate their expression. Furthermore, silent or cryptic gene clusters may be activated leading to new compounds. Genetic engineering using combinatorial biosynthesis may also lead to novel NRPs.

A. Approaches to Identify New NRPSs

Identification of pharmaceutically interesting natural products has been merely achieved by screening technologies. This strategy has been even more successful with the advent of high-throughput screening (Wilkinson and Micklefield 2007). However, evidence is emerging that the genomes of fungi are richer in genes that are likely to be involved in the synthesis of secondary metabolites and many of these metabolites have not been discovered yet. The capacity to produce NRPs in some fungi is very high, as information about fungal genomes revealed. The conservation of DNA and protein sequences greatly facilitates identification of NRPS genes in the completed genomes. For example, *Aspergilli* genomes comprise 14 or more NRPS genes (Cramer et al. 2006), *C. heterosptrophus* contains 11 NRPS genes (Lee et al. 2005) and *Fusarium graminearum* 15 NRPS genes (Varga et al. 2005; Tobiasen 2007). The genes were identified by in silico analyses and the ongoing growth of DNA sequence data will lead to the discovery of more natural-product biosynthesis pathway genes.

The number of putative NRPSs is much greater than the known peptides described for these species. One reason for this might be that these compounds are not produced under laboratory conditions (Bok et al. 2006; Brakhage et al. 2008). These "silent" or cryptic biosynthetic genes may encode the information for potentially useful metabolites that still await discovery. Silent gene clusters can be activated for example by heterologous expression under the control of a defined promoter or promoters of biosynthesis genes within the genome can be exchanged by defined promoters (Brakhage et al. 2008). Bergmann et al. (2006) identified a silent gene cluster in *A. nidulans* which contained a gene encoding a hybrid PKS-NRPS and an activator gene. This activator gene was retransformed into *A. nidulans* and expressed under the control of an inducible promoter. Under inducing conditions, the gene encoding the PKS-NRPS hybrid synthetase was transcribed and the metabolite could be identified. This most promising approach takes into account that all pathway genes are co-regulated.

Another interesting approach was enabled by the identification of a global regulator of secondary metabolism in *Aspergillus* spp., LaeA, by Bok and Keller (2004). Deletion of the gene blocks the expression of metabolic gene clusters including that responsible for penicillin production in *A. nidulans* while overexpression of *laeA* triggers an increased gene expression. Bok et al. (2006) performed microarrays and demonstrated that transcribed secondary metabolite gene clusters can be revealed by profiling *laeA* deletion and overexpression mutants. Using this method, the authors were able not only to identify transcriptionally active clusters but also to specifically manipulate their expression and characterize a novel natural product, terrequinone, and its gene cluster, in *A. nidulans*.

B. Combinatorial Biosynthesis of New NRPs

A current challenge in NRPS research is to re-engineer natural products in order to increase or alter their biological activities. This can be achieved by the construction and combination of domains and modules at the genetic level, for example by exchanging A-T units, artificial fusion of modules, module swaps and deletions. These applications have been successful in bacterial systems especially by work with *Streptomyces* spp., reviewed by Sieber and Marahiel (2005). A recent chemoenzymatic approach combines synthetic peptide synthesis with TE domain cyclization (reviewed by Kopp and Marahiel 2007). However, combinatorial biosynthesis has not yet been applied to fungal systems to a large extend. The establishment of fungal systems for domain exchange requires a more profound knowledge of the domain interactions, A domain specificity and linker regions between the domains, as identified in bacterial NRPSs (Mootz et al. 2000; Keating et al. 2002; von Döhren 2004).

References

Aharonowitz Y, Cohen G, Martin JF (1992) Penicillin and cephalosporin biosynthetic genes: structure, organization, regulation, and evolution. Annu Rev Microbiol 46:461–495

Arnez JG, Augustine JG, Moras D, Francklyn CS (1997) The first step of aminoacylation at the atomic level in histidyl-tRNA synthetase. Proc Natl Acad Sci USA 94:7144–7149

Bailey AM, Kershaw MJ, Hunt BA, Paterson IC, Charnley AK, Reynolds SE, Clarkson JM (1996) Cloning and sequence analysis of an intron-containing domain from a peptide synthetase-encoding gene of the entomopathogenic fungus *Metarhizium anisopliae*. Gene 173:195–197

Balibar CJ, Walsh CT (2006) GliP, a multimodular nonribosomal peptide synthetase in *Aspergillus fumigatus*, makes the diketopiperazine scaffold of gliotoxin. Biochemistry 45:15029–15038

Balibar CJ, Vaillancourt FH, Walsh CT (2005) Generation of D amino acid residues in assembly of arthrofactin by dual condensation/epimerization domains. Chem Biol 12:1189–1200

Balibar CJ, Howard-Jones AR, Walsh CT (2007) Terrequinone A biosynthesis through L-tryptophan oxidation, dimerization and bisprenylation. Nat Chem Biol 3:584–592

Belshaw PJ, Walsh CT, Stachelhaus T (1999) Aminoacyl-CoAs as probes of condensation domain selectivity in nonribosomal peptide synthesis. Science 284:486–489

Bergmann S, Schumann J, Scherlach K, Lange C, Brakhage AA, Hertweck C (2007) Genomics-driven discovery of PKS-NRPS hybrid metabolites from *Aspergillus nidulans*. Nat Chem Biol 3:213–217

Billich A, Zocher R (1987) Enzymatic synthesis of cyclosporine. A. J Biol Chem 262:17258–17259

Bok JW, Keller NP (2004) LaeA, a regulator of secondary metabolism in *Aspergillus* spp. Eukaryot Cell 3:527–535

Bok JW, Hoffmeister D, Maggio-Hall LA, Murillo R, Glasner JD, Keller NP (2006) Genomic mining for *Aspergillus* natural products. Chem Biol 13:31–37

Bölker M, Basse CW, Schirawski J (2008) *Ustilago maydis* secondary metabolism – From genomics to biochemistry. Fungal Genet Biol 45:88–93

Borel JF (1986) Cyclosporine forever? Transplant Proc 18:271–272

Brakhage AA, Al-Abdallah Q, Tuncher A, Sprote P (2005) Evolution of beta-lactam biosynthesis genes and recruitment of trans-acting factors. Phytochemistry 66:1200–1210

Brakhage AA, Schuemann J, Bergmann S, Scherlach K, Schroeckh V, Hertweck C (2008) Activation of fungal silent gene clusters: a new avenue to drug discovery. Prog Drug Res 66:3–12

Budde AD, Leong SA (1989) Characterization of siderophores from *Ustilago maydis*. Mycopathologia 108:125–133

Burmester J, Haese A, Zocher R (1995) Highly conserved N-methyltransferases as an integral part of peptide synthetases. Biochem Mol Biol Int 37:201–207

Butterton JR, Choi MH, Watnick PI, Carroll PA, Calderwood SB (2000) *Vibrio cholerae* VibF is required for vibriobactin synthesis and is a member of the family of nonribosomal peptide synthetases. J Bacteriol 182:1731–1738

Challis GL, Ravel J, Townsend CA (2000) Predictive, structure-based model of amino acid recognition by nonribosomal peptide synthetase adenylation domains. Chem Biol 7:211–224

Cheng YQ, Walton JD (2000) A eukaryotic alanine racemase gene involved in cyclic peptide biosynthesis. J Biol Chem 275:4906–4911

Chiang C, Karle IL, T W (1982) Unusual intramolecular hydrogen bonding in cycloamanide A, cyclic (LPro-LVal-LPhe-LPhe-LAla-Gly). A crystal structure analysis. Int J Pept Protein Res 20:414–412

Chugh JK, Wallace BA (2001) Peptaibols: models for ion channels. Biochem Soc Trans 29:565–570

Clugston SL, Sieber SA, Marahiel MA, Walsh CT (2003) Chirality of peptide bond-forming condensation domains in nonribosomal peptide synthetases: the C5 domain of tyrocidine synthetase is a (D)C(L) catalyst. Biochemistry 42:12095–12104

Conti E, Stachelhaus T, Marahiel MA, Brick P (1997) Structural basis for the activation of phenylalanine in the non-ribosomal biosynthesis of gramicidin S. EMBO J 16:4174–4183

Correia T, Grammel N, Ortel I, Keller U, Tudzynski P (2003) Molecular cloning and analysis of the ergopeptine assembly system in the ergot fungus *Claviceps purpurea*. Chem Biol 10:1281–1292

Coyle CM, Panaccione DG (2005) An ergot alkaloid biosynthesis gene and clustered hypothetical genes from *Aspergillus fumigatus*. Appl Environ Microbiol 71:3112–3118

Cramer RA Jr, Stajich JE, Yamanaka Y, Dietrich FS, Steinbach WJ, Perfect JR (2006) Phylogenomic analysis of non-ribosomal peptide synthetases in the genus *Aspergillus*. Gene 383:24–32

Damrongkool P, Sedlock AB, Young CA, Johnson RD, Goetz KE, Scott B, Schardl CL, Panaccione DG (2005) Structural analysis of a peptide synthetase gene required for ergopeptine production in the endophytic fungus *Neotyphodium lolii*. DNA Seq 16:379–385

Daniel JF, Filho ER (2007) Peptaibols of *Trichoderma*. Nat Prod Rep 24:1128–1141

De Crécy-Lagard V, Marlière P, Saurin W (1995) Multienzymatic non ribosomal peptide biosynthesis: identification of the functional domains catalysing peptide elongation and epimerisation. C R Acad Sci Ser C 318:927–936

Dieckmann R, von Döhren H (1997) Structural model of acyl carrier domains in integrated biosynthetic systems forming peptides, polyketides and fatty acids based on analogy to the *E. coli* acyl carrier protein. In: Baltz RH, Hegeman GD and Skatrud PL (eds) Developments in industrial microbiology. Society for Industrial Microbiology, Fairfax, pp 79–87

Dieckmann R, Lee YO, van Liempt H, von Dohren H, Kleinkauf H (1995) Expression of an active adenylate-forming domain of peptide synthetases corresponding to acyl-CoA-synthetases. FEBS Lett 357:212–216

Dittmann J, Wenger RM, Kleinkauf H, Lawen A (1994) Mechanism of cyclosporin A biosynthesis. Evidence for synthesis via a single linear undecapeptide precursor. J Biol Chem 269:2841–2846

Doekel S, Marahiel MA (2001) Biosynthesis of natural products on modular peptide synthetases. Metab Eng 3:64–77

Eichhorn H, Lessing F, Winterberg B, Schirawski J, Kamper J, Muller P, Kahmann R (2006) A ferroxidation/permeation iron uptake system is required for virulence in *Ustilago maydis*. Plant Cell 18:3332–3345

Eisendle M, Oberegger H, Zadra I, Haas H (2003) The siderophore system is essential for viability of *Aspergillus nidulans*: functional analysis of two genes encoding l-ornithine N 5-monooxygenase (*sidA*) and a non-ribosomal peptide synthetase (*sidC*). Mol Microbiol 49:359–375

Eisendle M, Schrettl M, Kragl C, Muller D, Illmer P, Haas H (2006) The intracellular siderophore ferricrocin is involved in iron storage, oxidative-stress resistance, germination, and sexual development in *Aspergillus nidulans*. Eukaryot Cell 5:1596–1603

Feifel SC, Schmiederer T, Hornbogen T, Berg H, Sussmuth RD, Zocher R (2007) In vitro synthesis of new enniatins: probing the alpha-D-hydroxy carboxylic acid binding pocket of the multienzyme enniatin synthetase. Chembiochem 8:1767–1770

Finking R, Marahiel MA (2004) Biosynthesis of nonribosomal peptides. Annu Rev Microbiol 58:453–488

Fleetwood DJ, Scott B, Lane GA, Tanaka A, Johnson RD (2007) A complex ergovaline gene cluster in epichloe endophytes of grasses. Appl Environ Microbiol 73:2571–2579

Fleming A (1929) On the antibacterial action of cultures of a *Penicillium*, with special reference to their use in the isolation of *B. influenzae*. Br J Exp Pathol 10:226–236

Gardiner DM, Cozijnsen AJ, Wilson LM, Pedras MS, Howlett BJ (2004) The sirodesmin biosynthetic gene cluster of the plant pathogenic fungus *Leptosphaeria maculans*. Mol Microbiol 53:1307–1318

Gebler JC, Poulter CD (1992) Purification and characterization of dimethylallyl tryptophan synthase from *Claviceps purpurea*. Arch Biochem Biophys 296:308–313

Glinski M, Hornbogen T, Zocher R (2001) Enzymatic synthesis of fungal N-methylated cyclopeptides and depsipeptides. In: Kirst H, Yeh W-K, Zmijewski M (eds) Enzyme technologies for pharmaceutical and biotechnological applications. Dekker, New York, pp 471–497

Glinski M, Urbanke C, Hornbogen T, Zocher R (2002) Enniatin synthetase is a monomer with extended structure: evidence for an intramolecular reaction mechanism. Arch Microbiol 178:267–273

Guenzi E, Galli G, Grgurina I, Gross DC, Grandi G (1998) Characterization of the syringomycin synthetase gene cluster. A link between prokaryotic and eukaryotic peptide synthetases. J Biol Chem 273:32857–32863

Guillemette T, Sellam A, Simoneau P (2004) Analysis of a nonribosomal peptide synthetase gene from Alternaria brassicae and flanking genomic sequences. Curr Genet 45:214–224

Guo S, Evans SA, Wilkes MB, Bhattacharjee JK (2001) Novel posttranslational activation of the *LYS2*-encoded alpha-aminoadipate reductase for biosynthesis of lysine and site-directed mutational analysis of conserved amino acid residues in the activation domain of *Candida albicans*. J Bacteriol 183:7120–7125

Gutierrez S, Diez B, Montenegro E, Martin JF (1991) Characterization of the *Cephalosporium acremonium* pcbAB gene encoding alpha-aminoadipyl-cysteinyl-valine synthetase, a large multidomain peptide synthetase: linkage to the pcbC gene as a cluster of early cephalosporin biosynthetic genes and evidence of multiple functional domains. J Bacteriol 173:2354–2365

Haarmann T, Machado C, Lubbe Y, Correia T, Schardl CL, Panaccione DG, Tudzynski P (2005) The ergot alkaloid gene cluster in *Claviceps purpurea*: extension of the cluster sequence and intra species evolution. Phytochemistry 66:1312–1320

Haarmann T, Ortel I, Tudzynski P, Keller U (2006) Identification of the cytochrome P450 monooxygenase that bridges the clavine and ergoline alkaloid pathways. Chembiochem 7:645–652

Haas H (2003) Molecular genetics of fungal siderophore biosynthesis and uptake: the role of siderophores in iron uptake and storage. Appl Microbiol Biotechnol 62:316–330

Haas H, Eisendle M, Turgeon BG (2008) Siderophores in fungal physiology and virulence. Annu Rev Phytopathol 46:149–187

Hacker C, Glinski M, Hornbogen T, Doller A, Zocher R (2000) Mutational analysis of the N-methyltransferase domain of the multifunctional enzyme enniatin synthetase. J Biol Chem 275:30826–30832

Haese A, Schubert M, Herrmann M, Zocher R (1993) Molecular characterization of the enniatin synthetase gene encoding a multifunctional enzyme catalysing

N-methyldepsipeptide formation in *Fusarium scirpi*. Mol Microbiol 7:905–914

Hallen HE, Luo H, Scott-Craig JS, Walton JD (2007) Gene family encoding the major toxins of lethal *Amanita* mushrooms. Proc Natl Acad Sci USA 104:19097–19101

He C, Rusu AG, Poplawski AM, Irwin JA, Manners JM (1998) Transfer of a supernumerary chromosome between vegetatively incompatible biotypes of the fungus *Colletotrichum gloeosporioides*. Genetics 150:1459–1466

Herrmann M, Zocher R, Haese A (1996) Enniatin Production by *Fusarium* Strains and Its Effect on Potato Tuber Tissue. Appl Environ Microbiol 62:393–398

Hof C, Eisfeld K, Welzel K, Antelo L, Foster AJ, Anke H (2007) Ferricrocin synthesis in *Magnaporthe grisea* and its role in pathogenicity in rice. Mol Plant Pathol 8:163–172

Hoffmann K, Schneider-Scherzer E, Kleinkauf H, Zocher R (1994) Purification and characterization of eucaryotic alanine racemase acting as key enzyme in cyclosporin biosynthesis. J Biol Chem 269:12710–12714

Hoffmeister D, Keller NP (2006) Natural products of filamentous fungi: enzymes, genes, and their regulation. Nat Prod Rep 24:393–416

Hornbogen T, Riechers SP, Prinz B, Schultchen J, Lang C, Schmidt S, Mugge C, Turkanovic S, Sussmuth RD, Tauberger E, Zocher R (2007) Functional characterization of the recombinant N-methyltransferase domain from the multienzyme enniatin synthetase. Chembiochem 8:1048–1054

Johnson LJ, Johnson RD, Akamatsu H, Salamiah A, Otani H, Kohmoto K, Kodama M (2001) Spontaneous loss of a conditionally dispensable chromosome from the *Alternaria alternata* apple pathotype leads to loss of toxin production and pathogenicity. Curr Genet 40:65–72

Johnson R, Johnson L, Itoh Y, Kodama M, Otani H, Kohmoto K (2000) Cloning and characterization of a cyclic peptide synthetase gene from *Alternaria alternata* apple pathotype whose product is involved in AM-toxin synthesis and pathogenicity. Mol Plant Microbe Interact 13:742–753

Johnson R, Voisey C, Johnson L, Pratt J, Fleetwood D, Khan A, Bryan G (2007) Distribution of NRPS gene families within the *Neotyphodium/Epichloe* complex. Fungal Genet Biol 44:1180–1190

Jones MJ, Dunkle LD (1995) Virulence gene expression during conidial germination in *Cochliobolus carbonum*. Mol Plant Microbe Interact 8:476–479

Kahan BD (1984) Cyclosporine: a powerful addition to the immunosuppressive armamentarium. Am J Kidney Dis 3:444–455

Keating TA, Marshall CG, Walsh CT (2000a) Reconstitution and characterization of the *Vibrio cholerae* vibriobactin synthetase from VibB, VibE, VibF, and VibH. Biochemistry 39:15522–15530

Keating TA, Marshall CG, Walsh CT (2000b) Vibriobactin biosynthesis in *Vibrio cholerae*: VibH is an amide synthase homologous to nonribosomal peptide synthetase condensation domains. Biochemistry 39:15513–15521

Keating TA, Ehmann DE, Kohli RM, Marshall CG, Trauger JW, Walsh CT (2001) Chain termination steps in nonribosomal peptide synthetase assembly lines: directed acyl-S-enzyme breakdown in antibiotic and siderophore biosynthesis. Chembiochem 2:99–107

Keating TA, Marshall CG, Walsh CT, Keating AE (2002) The structure of VibH represents nonribosomal peptide synthetase condensation, cyclization and epimerization domains. Nat Struct Biol 9:522–526

Keller NP, Hohn TM (1996) Metabolic pathway gene clusters in filamentous fungi. Fungal Genet Biol 21:17–29

Keller NP, Turner G, Bennett JW (2005) Fungal secondary metabolism – from biochemistry to genomics. Nat Rev Microbiol 3:937–947

Keszenman-Pereyra D, Lawrence S, Twfieg ME, Price J, Turner G (2003) The npgA/ cfwA gene encodes a putative 4'-phosphopantetheinyl transferase which is essential for penicillin biosynthesis in Aspergillus nidulans. Curr Genet 43:186–190

Khaldi N, Collemare J, Lebrun MH, Wolfe KH (2008) Evidence for horizontal transfer of a secondary metabolite gene cluster between fungi. Genome Biol 9:R18

Kim KH, Cho Y, La Rota M, Cramer RA, Lawrence CB (2007) Functional analysis of the Alternaria brassicicola non-ribosomal peptide synthetase gene AbNPS2 reveals a role in conidial cell wall construction. Molecular Plant Microbe Interactions 13:23–39

Kim M, Han D-M, Chae K-S, Chae K-S, K-Y J (2001) Isolation and characterization of the npgA gene involved in pigment formation in Aspergillus nidulans. Fungal Genet News 1:48–52

Kleinkauf H, von Dohren H (1995) The nonribosomal peptide biosynthetic system-on the origins of structural diversity of peptides, cyclopeptides and related compounds. Antonie Van Leeuwenhoek 67:229–242

Kleinkauf H, Von Dohren H (1996) A nonribosomal system of peptide biosynthesis. Eur J Biochem 236: 335–351

Komon-Zelazowska M, Neuhof T, Dieckmann R, von Dohren H, Herrera-Estrella A, Kubicek CP, Druzhinina IS (2007) Formation of atroviridin by Hypocrea atroviridis is conidiation associated and positively regulated by blue light and the G protein GNA3. Eukaryot Cell 6:2332–2342

Konz D, Marahiel MA (1999) How do peptide synthetases generate structural diversity? Chem Biol 6:R39–R48

Kopp F, Marahiel MA (2007) Where chemistry meets biology: the chemoenzymatic synthesis of nonribosomal peptides and polyketides. Curr Opin Biotechnol 18:513–520

Krause M, Lindemann A, Glinski M, Hornbogen T, Bonse G, Jeschke P, Thielking G, Gau W, Kleinkauf H, Zocher R (2001) Directed biosynthesis of new enniatins. J Antibiot (Tokyo) 54:797–804

Kronen M, Kleinwachter P, Schlegel B, Hartl A, Grafe U (2001) Ampullosporines B,C,D,E1,E2,E3 and E4 from Sepedonium ampullosporum HKI-0053: structures and biological activities. J Antibiot (Tokyo) 54:175–178

Laich F, Fierro F, Cardoza RE, Martin JF (1999) Organization of the gene cluster for biosynthesis of penicillin in Penicillium nalgiovense and antibiotic production in cured dry sausages. Appl Environ Microbiol 65:1236–1240

Laich F, Fierro F, Martin JF (2002) Production of penicillin by fungi growing on food products: identification of a complete penicillin gene cluster in Penicillium griseofulvum and a truncated cluster in Penicillium verrucosum. Appl Environ Microbiol 68:1211–1219

Lambalot RH, Gehring AM, Flugel RS, Zuber P, LaCelle M, Marahiel MA, Reid R, Khosla C, Walsh CT (1996) A new enzyme superfamily – the phosphopantetheinyl transferases. Chem Biol 3:923–936

Larrondo LF, Gonzalez B, Cullen D, Vicuna R (2004) Characterization of a multicopper oxidase gene cluster in Phanerochaete chrysosporium and evidence of altered splicing of the mco transcripts. Microbiology 150:2775–2783

Lautru S, Challis GL (2004) Substrate recognition by nonribosomal peptide synthetase multi-enzymes. Microbiology 150:1629–1636

Lawen A, Traber R (1993) Substrate specificities of cyclosporin synthetase and peptolide SDZ 214-103 synthetase. Comparison of the substrate specificities of the related multifunctional polypeptides. J Biol Chem 268:20452–20465

Lawen A, Zocher R (1990) Cyclosporin synthetase. The most complex peptide synthesizing multienzyme polypeptide so far described. J Biol Chem 265:11355–11360

Lawen A, Traber R, Geyl D, Zocher R, Kleinkauf H (1989) Cell-free biosynthesis of new cyclosporins. J Antibiot (Tokyo) 42:1283–1289

Lee BN, Kroken S, Chou DY, Robbertse B, Yoder OC, Turgeon BG (2005) Functional analysis of all nonribosomal peptide synthetases in Cochliobolus heterostrophus reveals a factor, NPS6, involved in virulence and resistance to oxidative stress. Eukaryot Cell 4:545–555

Lee C, Gorisch H, Kleinkauf H, Zocher R (1992) A highly specific D-hydroxyisovalerate dehydrogenase from the enniatin producer Fusarium sambucinum. J Biol Chem 267:11741–11744

Lehr NA, Meffert A, Antelo L, Sterner O, Anke H, Weber RW (2006) Antiamoebins, myrocin B and the basis of antifungal antibiosis in the coprophilous fungus Stilbella erythrocephala (syn. S. fimetaria). FEMS Microbiol Ecol 55:105–112

Liebermann B, Ramm K (1991) N-methylation in the biosynthesis of the phytotoxin tentoxin. Phytochemistry 30:1815–1817

Linne U, Doekel S, Marahiel MA (2001) Portability of epimerization domain and role of peptidyl carrier protein on epimerization activity in nonribosomal peptide synthetases. Biochemistry 40:15824–15834

Liras P, Martin JF (2006) Gene clusters for beta-lactam antibiotics and control of their expression: why have clusters evolved, and from where did they originate? Int Microbiol 9:9–19

Lorenz N, Wilson EV, Machado C, Schardl CL, Tudzynski P (2007) Comparison of ergot alkaloid biosynthesis gene clusters in Claviceps species indicates loss of late pathway steps in evolution of C. fusiformis. Appl Environ Microbiol 73:7185–7191

Lorenzen K, Anke T (1998) Basidiomycetes as a source for new bioactive natural products. Curr Org Chem 2:329–364

MacCabe AP, Riach MB, Unkles SE, Kinghorn JR (1990) The *Aspergillus nidulans npeA* locus consists of three contiguous genes required for penicillin biosynthesis. EMBO J 9:279–287

Madry N, Zocher R, Grodzki K, Kleinkauf H (1984) Selective synthesis of depsipeptides by the immobilized multienzyme enniatin synthetase. Appl Microbiol Biotechnol 20:83–86

Maiya S, Grundmann A, Li SM, Turner G (2006) The fumitremorgin gene cluster of *Aspergillus fumigatus*: identification of a gene encoding brevianamide F synthetase. Chembiochem 7:1062–1069

Marahiel MA (1997) Protein templates for the biosynthesis of peptide antibiotics. Chem Biol 4:561–567

Marahiel MA, Stachelhaus T, Mootz HD (1997) Modular peptide synthetases involved in nonribosomal peptide synthesis. Chem Rev 97:2651–2674

Marquez-Fernandez O, Trigos A, Ramos-Balderas JL, Viniegra-Gonzalez G, Deising HB, Aguirre J (2007) Phosphopantetheinyl transferase CfwA/NpgA is required for *Aspergillus nidulans* secondary metabolism and asexual development. Eukaryot Cell 6:710–720

Martinez D, Larrondo LF, Putnam N, Gelpke MD, Huang K, Chapman J, Helfenbein KG, Ramaiya P, Detter JC, Larimer F, Coutinho PM, Henrissat B, Berka R, Cullen D, Rokhsar D (2004) Genome sequence of the lignocellulose degrading fungus *Phanerochaete chrysosporium* strain RP78. Nat Biotechnol 22:695–700

May JJ, Kessler N, Marahiel MA, Stubbs MT (2002) Crystal structure of DhbE, an archetype for aryl acid activating domains of modular nonribosomal peptide synthetases. Proc Natl Acad Sci USA 99:12120–12125

Mootz HD, Marahiel MA (1997) Biosynthetic systems for nonribosomal peptide antibiotic assembly. Curr Opin Chem Biol 1:543–551

Mootz HD, Schwarzer D, Marahiel MA (2000) Construction of hybrid peptide synthetases by module and domain fusions. Proc Natl Acad Sci USA 97: 5848–5853

Mootz HD, Schorgendorfer K, Marahiel MA (2002a) Functional characterization of 4'-phosphopantetheinyl transferase genes of bacterial and fungal origin by complementation of *Saccharomyces cerevisiae* lys5. FEMS Microbiol Lett 213:51–57

Mootz HD, Schwarzer D, Marahiel MA (2002b) Ways of assembling complex natural products on modular nonribosomal peptide synthetases. Chembiochem 3:490–504

Nakano MM, Xia LA, Zuber P (1991) Transcription initiation region of the srfA operon, which is controlled by the comP-comA signal transduction system in *Bacillus subtilis*. J Bacteriol 173:5487–5493

Nierman WC, Pain A, Anderson MJ, Wortman JR, Kim HS, Arroyo J, Berriman M, Abe K, Archer DB, Bermejo C, Bennett J, Bowyer P, Chen D, Collins M, Coulsen R, Davies R, Dyer PS, Farman M, Fedorova N, Feldblyum TV, Fischer R, Fosker N, Fraser A, Garcia JL, Garcia MJ, Goble A, Goldman GH, Gomi K, Griffith-Jones S, Gwilliam R, Haas B, Haas H, Harris D, Horiuchi H, Huang J, Humphray S, Jimenez J, Keller N, Khouri H, Kitamoto K, Kobayashi T, Konzack S, Kulkarni R, Kumagai T, Lafon A, Latge JP, Li W, Lord A, Lu C, Majoros WH, May GS, Miller BL, Mohamoud Y, Molina M, Monod M, Mouyna I, Mulligan S, Murphy L, O'Neil S, Paulsen I, Penalva MA, Pertea M, Price C, Pritchard BL, Quail MA, Rabbinowitsch E, Rawlins N, Rajandream MA, Reichard U, Renauld H, Robson GD, Rodriguez de Cordoba S, Rodriguez-Pena JM, Ronning CM, Rutter S, Salzberg SL, Sanchez M, Sanchez-Ferrero JC, Saunders D, Seeger K, Squares R, Squares S, Takeuchi M, Tekaia F, Turner G, Vazquez de Aldana CR, Weidman J, White O, Woodward J, Yu JH, Fraser C, Galagan JE, Asai K, Machida M, Hall N, Barrell B, Denning DW (2005) Genomic sequence of the pathogenic and allergenic filamentous fungus *Aspergillus fumigatus*. Nature 438:1151–1156

Oberegger H, Eisendle M, Schrettl M, Graessle S, Haas H (2003) 4'-phosphopantetheinyl transferase-encoding npgA is essential for siderophore biosynthesis in *Aspergillus nidulans*. Curr Genet 44:211–215

Oide S, Moeder W, Krasnoff S, Gibson D, Haas H, Yoshioka K, Turgeon BG (2006) NPS6, encoding a nonribosomal peptide synthetase involved in siderophore-mediated iron metabolism, is a conserved virulence determinant of plant pathogenic ascomycetes. Plant Cell 18:2836–2853

Oide S, Krasnoff SB, Gibson DM, Turgeon BG (2007) Intracellular siderophores are essential for ascomycete sexual development in heterothallic *Cochliobolus heterostrophus* and homothallic *Gibberella zeae*. Eukaryot Cell 6:1339–1353

Panaccione DG, Scott-Craig JS, Pocard JA, Walton JD (1992) A cyclic peptide synthetase gene required for pathogenicity of the fungus *Cochliobolus carbonum* on maize. Proc Natl Acad Sci USA 89:6590–6594

Panaccione DG, Cipoletti JR, Sedlock AB, Blemings KP, Schardl CL, Machado C, Seidel GE (2006) Effects of ergot alkaloids on food preference and satiety in rabbits, as assessed with gene-knockout endophytes in perennial ryegrass (*Lolium perenne*). J Agric Food Chem 54:4582–4587

Patron NJ, Waller RF, Cozijnsen AJ, Straney DC, Gardiner DM, Nierman WC, Howlett BJ (2007) Origin and distribution of epipolythiodioxopiperazine (ETP) gene clusters in filamentous ascomycetes. BMC Evol Biol 7:174–189

Pazirandeh M, Chirala SS, Wakil SJ (1991) Site-directed mutagenesis studies on the recombinant thioesterase domain of chicken fatty acid synthase expressed in *Escherichia coli*. J Biol Chem 266:20946–20952

Peeters H, Zocher R, Kleinkauf H (1988) Synthesis of beauvericin by a multifunctional enzyme. J Antibiot (Tokyo) 41:352–359

Pieper R, Haese A, Schroder W, Zocher R (1995) Arrangement of catalytic sites in the multifunctional enzyme enniatin synthetase. Eur J Biochem 230:119–126

Rausch C, Hoof I, Weber T, Wohlleben W, Huson DH (2007) Phylogenetic analysis of condensation domains in NRPS sheds light on their functional evolution. BMC Evol Biol 7:78

Reiber K, Reeves EP, Neville CM, Winkler R, Gebhardt P, Kavanagh K, Doyle S (2005) The expression of selected non-ribosomal peptide synthetases in *Aspergillus fumigatus* is controlled by the availability of free iron. FEMS Microbiol Lett 248:83–91

Reuter K, Mofid MR, Marahiel MA, Ficner R (1999) Crystal structure of the surfactin synthetase-activating enzyme sfp: a prototype of the 4′-phosphopantetheinyl transferase superfamily. EMBO J 18:6823–6831

Riederer B, Han M, Keller U (1996) D-Lysergyl peptide synthetase from the ergot fungus *Claviceps purpurea*. J Biol Chem 271:27524–27530

Roskoski R Jr, Kleinkauf H, Gevers W, Lipmann F (1970) Isolation of enzyme-bound peptide intermediates in tyrocidine biosynthesis. Biochemistry 9:4846–4851

Roy SW, Gilbert W (2006) The evolution of spliceosomal introns: patterns, puzzles and progress. Nat Rev Genet 7:211–221

Samel SA, Wagner B, Marahiel MA, Essen LO (2006) The thioesterase domain of the fengycin biosynthesis cluster: a structural base for the macrocyclization of a non-ribosomal lipopeptide. J Mol Biol 359:876–889

Santi DV, Webster RW Jr, Cleland WW (1974) Kinetics of aminoacyl-tRNA synthetases catalyzed ATP-PPi exchange. Methods Enzymol 29:620–627

Schardl CL, Panaccione DG, Tudzynski P (2006) Ergot alkaloids – biology and molecular biology. Alkaloids Chem Biol 63:45–86

Schauwecker F, Pfennig F, Grammel N, Keller U (2000) Construction and in vitro analysis of a new bi-modular polypeptide synthetase for synthesis of N-methylated acyl peptides. Chem Biol 7:287–297

Schneider A, Marahiel MA (1998) Genetic evidence for a role of thioesterase domains, integrated in or associated with peptide synthetases, in non-ribosomal peptide biosynthesis in *Bacillus subtilis*. Arch Microbiol 169:404–410

Schrettl M, Bignell E, Kragl C, Sabiha Y, Loss O, Eisendle M, Wallner A, Arst HN Jr, Haynes K, Haas H (2007) Distinct roles for intra- and extracellular siderophores during *Aspergillus fumigatus* infection. PLoS Pathog 3:1195–1207

Schwarzer D, Marahiel MA (2001) Multimodular biocatalysts for natural product assembly. Naturwissenschaften 88:93–101

Schwarzer D, Finking R, Marahiel MA (2003) Nonribosomal peptides: from genes to products. Nat Prod Rep 20:275–287

Schwecke T, Gottling K, Durek P, Duenas I, Kaufer NF, Zock-Emmenthal S, Staub E, Neuhof T, Dieckmann R, von Döhren H (2006) Nonribosomal peptide synthesis in *Schizosaccharomyces pombe* and the architectures of ferrichrome-type siderophore synthetases in fungi. Chembiochem 7:612–622

Scott-Craig JS, Panaccione DG, Pocard JA, Walton JD (1992) The cyclic peptide synthetase catalyzing HC-toxin production in the filamentous fungus *Cochliobolus carbonum* is encoded by a 15.7-kilobase open reading frame. J Biol Chem 267:26044–26049

Shaw-Reid CA, Kelleher NL, Losey HC, Gehring AM, Berg C, Walsh CT (1999) Assembly line enzymology by multimodular nonribosomal peptide synthetases: the thioesterase domain of *E. coli* EntF catalyzes both elongation and cyclolactonization. Chem Biol 6:385–400

Sieber SA, Marahiel MA (2005) Molecular mechanisms underlying nonribosomal peptide synthesis: approaches to new antibiotics. Chem Rev 105:715–738

Silakowski B, Kunze B, Nordsiek G, Blocker H, Hofle G, Muller R (2000) The myxochelin iron transport regulon of the myxobacterium *Stigmatella aurantiaca* Sg a15. Eur J Biochem 267:6476–6485

Smith DJ, Earl AJ, Turner G (1990) The multifunctional peptide synthetase performing the first step of penicillin biosynthesis in *Penicillium chrysogenum* is a 421,073 dalton protein similar to *Bacillus brevis* peptide antibiotic synthetases. Embo J 9:2743–2750

Stachelhaus T, Marahiel MA (1995) Modular structure of genes encoding multifunctional peptide synthetases required for non-ribosomal peptide synthesis. FEMS Microbiol Lett 125:3–14

Stachelhaus T, Walsh CT (2000) Mutational analysis of the epimerization domain in the initiation module Phe-ATE of gramicidin S synthetase. Biochemistry 39:5775–5787

Stachelhaus T, Huser A, Marahiel MA (1996) Biochemical characterization of peptidyl carrier protein (PCP), the thiolation domain of multifunctional peptide synthetases. Chem Biol 3:913–921

Stachelhaus T, Mootz HD, Bergendahl V, Marahiel MA (1998) Peptide bond formation in nonribosomal peptide biosynthesis. Catalytic role of the condensation domain. J Biol Chem 273:22773–22781

Stachelhaus T, Mootz HD, Marahiel MA (1999) The specificity-conferring code of adenylation domains in nonribosomal peptide synthetases. Chem Biol 6:493–505

Stack D, Neville C, Doyle S (2007) Nonribosomal peptide synthesis in *Aspergillus fumigatus* and other fungi. Microbiology 153:1297–1306

Stein T, Vater J, Kruft V, Otto A, Wittmann-Liebold B, Franke P, Panico M, McDowell R, Morris HR (1996) The multiple carrier model of nonribosomal peptide biosynthesis at modular multienzymatic templates. J Biol Chem 271:15428–15435

Sterner O, Etzel W, A M, Anke H (1997) Omphalotin, a new cyclic peptide with potent nematicidal activity from *Omphalotus olearius*. II. Isolation and structure determination. Nat Prod Lett 10:33–38

Talal N (1988) Cyclosporine as an immunosuppressive agent for autoimmune disease: theoretical concepts and therapeutic strategies. Transplant Proc 20:11–15

Tanaka A, Tapper BA, Popay A, Parker EJ, Scott B (2005) A symbiosis expressed non-ribosomal peptide synthetase from a mutalistic fungal endophyte of perennial ryegrass confers protection to the symbiotum from insect herbivory. Mol Microbiol 57:1036–1050

Taylor JW, Berbee ML (2006) Dating divergences in the Fungal Tree of Life: review and new analyses. Mycologia 98:838–849

Tobiasen C, Aahman J, Ravnholt KS, Bjerrum MJ, Grell MN, Giese H (2007) Nonribosomal peptide synthetase (NPS) genes in *Fusarium graminearum*, *F. culmorum* and *F. pseudograminearium* and identification of NPS2 as the producer of ferricrocin. Curr Genet 51:43–58

Traber R, Hofmann H, Kobel H (1989) Cyclosporins-new analogues by precursor directed biosynthesis. J Antibiot (Tokyo) 42:591–597

Tsai HF, Wang H, Gebler JC, Poulter CD, Schardl CL (1995) The *Claviceps purpurea* gene encoding dimethylallyltryptophan synthase, the committed step for ergot alkaloid biosynthesis. Biochem Biophys Res Commun 216:119–125

Tudzynski P, Holter K, Correia T, Arntz C, Grammel N, Keller U (1999) Evidence for an ergot alkaloid gene cluster in *Claviceps purpurea*. Mol Gen Genet 261:133–141

Turgeon BG, Oide S, Bushley K (2008) Creating and screening *Cochliobolus heterostrophus* non-ribosomal peptide synthetase mutants. Mycol Res 112:200–206

Varga J, Kocsube S, Toth B, Mesterhazy A (2005) Nonribosomal peptide synthetase genes in the genome of *Fusarium graminearum*, causative agent of wheat head blight. Acta Biol Hung 56:375–388

Vetter J (1998) Toxins of *Amanita phalloides*. Toxicon 36:13–24

von Döhren H (2004) Biochemistry and general genetics of nonribosomal peptide synthetases in fungi. Adv Biochem Eng Biotechnol 88:217–264

von Döhren H, Keller U, Vater J, Zocher R (1997) Multifunctional peptide synthetases. Chem Rev 97:2675–2706

von Döhren H, Dieckmann R, Pavela-Vrancic M (1999) The nonribosomal code. Chem Biol 6:273–279

Walton JD (2000) Horizontal gene transfer and the evolution of secondary metabolite gene clusters in fungi: an hypothesis. Fungal Genet Biol 30:167–171

Walton JD, Panaccione DG, Hallen HE (2004) Peptide synthesis without ribosomes. Adv Fungal Biotechnol Ind Agric Med 1:127–162

Walzel B, Riederer B, Keller U (1997) Mechanism of alkaloid cyclopeptide synthesis in the ergot fungus *Claviceps purpurea*. Chem Biol 4:223–230

Wang C, Skrobek A, Butt TM (2003) Concurrence of losing a chromosome and the ability to produce destruxins in a mutant of Metarhizium anisopliae. FEMS Microbiol Lett 226:373–378

Wang J, Machado C, Panaccione DG, Tsai HF, Schardl CL (2004) The determinant step in ergot alkaloid biosynthesis by an endophyte of perennial ryegrass. Fungal Genet Biol 41:189–198

Weber G, Schorgendorfer K, Schneider-Scherzer E, Leitner E (1994) The peptide synthetase catalyzing cyclosporine production in *Tolypocladium niveum* is encoded by a giant 45.8-kilobase open reading frame. Curr Genet 26:120–125

Weber T, Marahiel MA (2001) Exploring the domain structure of modular nonribosomal peptide synthetases. Structure 9:R3–R9

Weber T, Baumgartner R, Renner C, Marahiel MA, Holak TA (2000) Solution structure of PCP, a prototype for the peptidyl carrier domains of modular peptide synthetases. Structure 8:407–418

Weckwerth W, Miyamoto K, Iinuma K, Krause M, Glinski M, Storm T, Bonse G, Kleinkauf H, Zocher R (2000) Biosynthesis of PF1022A and related cyclooctadepsipeptides. J Biol Chem 275:17909–17915

Wei L, Steiner JP, Hamilton GS, Wu YQ (2004) Synthesis and neurotrophic activity of nonimmunosuppressant cyclosporin A derivatives. Bioorg Med Chem Lett 14:4549–4551

Weigel BJ, Burgett SG, Chen VJ, Skatrud PL, Frolik CA, Queener SW, Ingolia TD (1988) Cloning and expression in Escherichia coli of isopenicillin N synthetase genes from *Streptomyces lipmanii* and *Aspergillus nidulans*. J Bacteriol 170:3817–3826

Welzel K (2005) Molekularbiologische Untersuchungen zur nicht-ribosomalen Peptidsynthese in *Omphalotus olearius*. Dissertation, Technische Universität Kaiserslautern

Welzel K, Eisfeld K, Antelo L, Anke T, Anke H (2005) Characterization of the ferrichrome A biosynthetic gene cluster in the homobasidiomycete *Omphalotus olearius*. FEMS Microbiol Lett 249:157–163

Whitmore L, Wallace BA (2004) The Peptaibol Database: a database for sequences and structures of naturally occurring peptaibols. Nucleic Acids Res 32:D593–D594

Wiest A, Grzegorski D, Xu BW, Goulard C, Rebuffat S, Ebbole DJ, Bodo B, Kenerley C (2002) Identification of peptaibols from *Trichoderma virens* and cloning of a peptaibol synthetase. J Biol Chem 277:20862–20868

Wilkinson B, Micklefield J (2007) Mining and engineering natural-product biosynthetic pathways. Nat Chem Biol 3:379–386

Wyckoff EE, Smith SL, Payne SM (2001) VibD and VibH are required for late steps in vibriobactin biosynthesis in *Vibrio cholerae*. J Bacteriol 183:1830–1834

Yuan WM, Gentil GD, Budde AD, Leong SA (2001) Characterization of the *Ustilago maydis sid2* gene, encoding a multidomain peptide synthetase in the ferrichrome biosynthetic gene cluster. J Bacteriol 183:4040–4051

Zaleta-Rivera K, Xu C, Yu F, Butchko RA, Proctor RH, Hidalgo-Lara ME, Raza A, Dussault PH, Du L (2006) A bidomain nonribosomal peptide synthetase encoded by *FUM14* catalyzes the formation of tricarballylic esters in the biosynthesis of fumonisins. Biochemistry 45:2561–2569

Zhang JH, Quigley NB, Gross DC (1995) Analysis of the *syrB* and *syrC* genes of *Pseudomonas syringae pv. syringae* indicates that syringomycin is synthesized by a thiotemplate mechanism. J Bacteriol 177:4009–4020

Zocher R, Keller U (1997) Thiol template peptide synthesis systems in bacteria and fungi. Adv Microb Physiol 38:85–131

Zocher R, Keller U, Kleinkauf H (1982) Enniatin synthetase, a novel type of multifunctional enzyme catalyzing depsipeptide synthesis in *Fusarium oxysporum*. Biochemistry 21:43–48

Zocher R, Keller U, Kleinkauf H (1983) Mechanism of depsipeptide formation catalyzed by enniatin synthetase. Biochem Biophys Res Commun 110:292–299

Zocher R, Nihira T, Paul E, Madry N, Peeters H, Kleinkauf H, Keller U (1986) Biosynthesis of cyclosporin A: partial purification and properties of a multifunctional enzyme from *Tolypocladium inflatum*. Biochemistry 25:550–553

16 Biosynthesis of Fungal Polyketides

JULIA SCHUEMANN[1], CHRISTIAN HERTWECK[1]

CONTENTS

I. Introduction

Fungal secondary metabolism is a rich source of biologically active compounds. Besides non-ribosomal peptides and terpenoids, fungi are known as producers of structurally diverse polyketides. Many of these metabolites are infamous toxins, such as aflatoxin (Minto and Townsend 1997) and fumonisin (Chu 1991; Sweeney and Dobson 1999), pigments or virulence factors like melanin (Perpetua et al. 1996; Fig. 16.1). However, fungal polyketides may also be pharmaceutically relevant, like the cholesterol-lowering agent lovastatin (Hendrickson et al. 1999). While a large

body of knowledge exists on bacterial polyketide biosynthesis, the corresponding fungal systems are far less explored. Not until 1990 did scientists clone and sequence the first fungal polyketide synthase (PKS) gene, encoding 6-methylsalicylic acid synthase (6-MSAS; Beck et al. 1990). While the product (6-MSA) has a relatively simple structure, some highly complex polyketides are also produced by fungi, for example T-toxin (Yang et al. 1996) and the cytochalasins (Binder and Tamm 1973). The derivatisation or total synthesis of such compounds can be cumbersome or inefficient, making pathway engineering approaches attractive. In the era of whole-genome sequencing we have an insight into the biosynthetic potential of fungi. Even so, yet very few of those identified PKS genes have been linked to natural products. Thus, the improvement of fungal molecular and genetic techniques is leading to enormous progress in understanding and manipulating fungal polyketide biosynthesis. This review focuses on the ecological and medical aspects of polyketide biosynthesis in fungi, the mechanisms involved in polyketide assembly and approaches to manipulate the pathways.

II. Ecological Importance and Pharmaceutical Use of Fungal Polyketides

Polyketides are a class of secondary metabolites that exhibit a broad spectrum of relevant activities. Fungal polyketides may serve as precursors to toxins that have adverse effects on animals or against plants. Important examples are lovastatin from *Aspergillus terreus* (Hendrickson et al. 1999), citrinin from *Monascus purpureus* (Shimizu et al. 2005), fumonisin from *Gibberella fujikuroi* (Proctor et al. 1999), T-toxin from *Cochliobolus*

[1]Leibniz Institute for Natural Product Research and Infection Biology, HKI, Beutenbergstrasse 11a, 07745 Jena, Germany; e-mail: Christian.Hertweck@hki-jena.de

Physiology and Genetics, 1st Edition
The Mycota XV
T. Anke and D. Weber (Eds)
© Springer-Verlag Berlin Heidelberg 2009

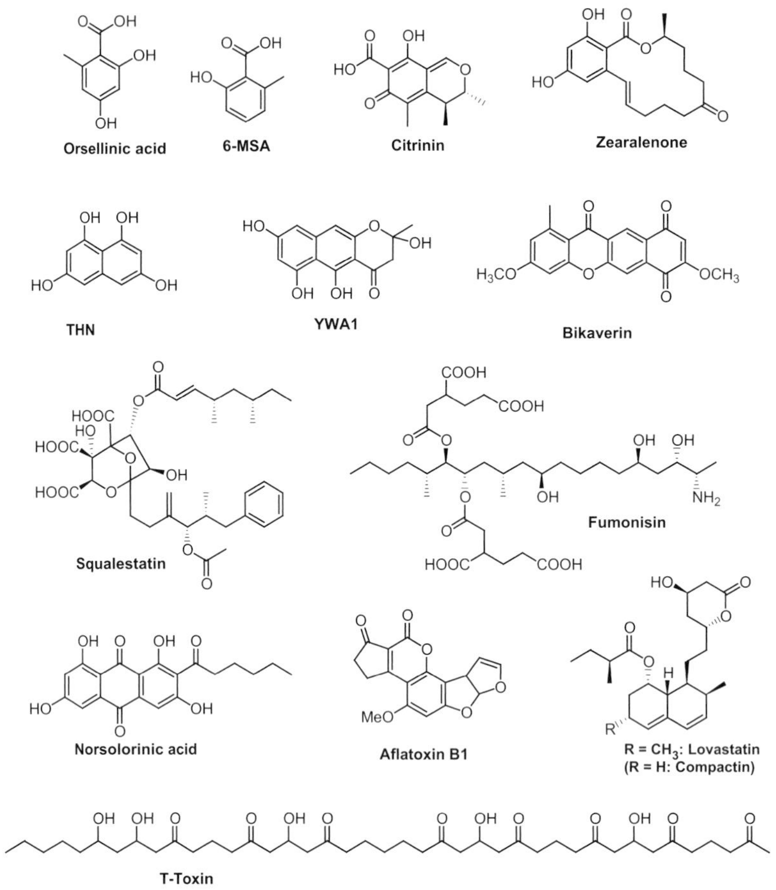

Fig. 16.1. Structures of selected fungal polyketide metabolites

heterostrophus (Yang et al. 1996) and aflatoxin produced by *Aspergillus parasiticus* (Feng and Leonard 1995). These compounds, among others, are infamous for causing severe damage in food industry, agriculture and human health. Aflatoxins are potent animal and human pathogens (Eaton and Groopman 1994). They show immunosuppressive, mutagenic, teratogenic and hepatocarcinogenic properties. Aflatoxins are produced by several species of *Aspergillus*, contaminating commodities, corn, peanuts, cotton seed and tree nuts. The production of aflatoxins probably protects the producer strain from feeding competitors.

A similar function in habitat defence but also in plant destruction might be due to fumonisins, produced by a number of cereal-pathogenic

G. fujikuroi. Fumonisins cause several mycotoxicoses, like leucoencephalomalacia in horses, pulmonary oedema in pigs and liver cancer in rats (Nelson et al. 1993). Their chemical structure resembles sphingosine. Fumonisins are competitive inhibitors of sphingosine *N*-acetyltransferase. They block complex sphingolipid biosynthesis and lead to the accumulation of sphingosine (Moss 1998). *G. fujikuroi* MP-A is one of the most common pathogens of maize. It is associated with diseases of the roots, stalk and ears but additionally the fungus is frequently present in apparently healthy maize tissues (Munkvold and Desjardins 1997).

Cytochalasins comprise a miscellaneous group of fungal polyketides with an impressing variety of biological activities. Beside cytotoxic activities various members have antibiotic, anti-inflammatory, antitumoral and antiviral activities. They can affect glucose transport, release of thyroid and growth hormones and inhibit cholesterol synthesis. Because of their defined binding of actin filaments cytochalasins are used as tools in cell and molecular biology (Binder and Tamm 1973; Steglich et al. 1997; Lodish et al. 2001). According to Deacon and others, cytochalasin production by insect pathogenic fungi, such as *Metarhizium anisopliae*, can lead to a lethal toxification of insects (Deacon 2001). Cytochalasins have also been implicated as virulence factors produced by the plant-pathogenic fungus *Phoma exigua* (Scott et al. 1975). Saprophytic fungi like *Penicillium expansum* or *Xylaria* spp. might produce cytochalasins in order to defend their habitat and secure their nutrial resources, e.g. wood, leafs, fruits or seeds.

Polyketides may also function as pigments like melanin (Perpetua et al. 1996) and bikaverin (Linnemannstons et al. 2002), or as pathogenicity factors, e.g. reported for melanin (Perpetua et al. 1996). The pentaketide 1,3,6,8-tetrahydroxy-naphthalene (THN) plays a key role in the melanisation of various fungi (Butler and Day 1998). THN is reduced to further intermediates such as 1,8-diydroxynaphthalene (DHN), which undergoes polymerisation to DHN-melanin (Fig. 16.2). For plant pathogens, e.g. *Colletotrichum lagenarium* or *Magnaporthe grisea*, melanin may support apressorium formation and penetration of plant cells (Howard et al. 1991). DHN-melanin likely plays a protective role in human pathogenic fungi like *Wangiella dermatitidis*, *Sporothrix schenckii* or *Aspergillus fumigatus* (Langfelder et al. 2003). There is strong evidence that melanin protects the fungus from the host immune system, e.g. through scavenging reactive oxygen species, and that it stabilises fungal hyphae for penetration of host tissue.

T-toxins are a family of C35 to C49 polyketides produced by *C. heterostrophus* (Kono et al.

1980, 1981; Kono et al. 1983). They are involved in virulence on maize (Yoder 1980; Yoder et al. 1997; Turgeon and Lu 2000).

Various fungal polyketides are medicinally important or have served as lead structure for the development of therapeutics. Lovastatin and the related compound compactin, also known as mevinolin and mevastatin, are used to lower serum cholesterol (Endo et al. 1976; Alberts et al. 1980). Compactin produced by *Penicillium* species differs from lovastatin only in the absence of the C-12 methyl group (Brown et al. 1976). Both compounds efficiently inhibit mammalian hydroxymethylglutaryl (HMG)-CoA reductase. Another potent pharmaceutical against elevated serum cholesterol is squalestatin from *Phoma* sp. and other filamentous fungi (Dawson et al. 1992). Squalestatin is an inhibitor of mammalian squalene synthase and shows curative activity against prion-infected neurons (Bate et al. 2004).

III. Biosynthesis of Polyketides

Polyketide biosynthesis is closely related to fatty acid biosynthesis (Birch and Donovan 1953). Isotope labelling experiments showed that repetitive Claisen condensations of an acyl-coenzyme A (CoA) starter unit with malonyl-CoA elongation units form the carbon backbone of fatty acids and polyketides (Fig. 16.3). Fatty acid synthases (FAS) and PKS contain an obligatory set of ketosynthase (KS), acyltransferase (AT) and acyl carrier protein (ACP) domains. A conserved serine of the *apo*-ACP needs to be post-translational modified with a CoA-derived 4-phosphopantetheine (PP) arm that is transferred by means of a PP transferase (Keszenman-Pereyra et al. 2003). The ACP domains serve as an anchor for both the malonyl extender units and transiently the growing acyl chain. The Claisen condensation is catalysed by the KS while the AT transfers acyl groups from

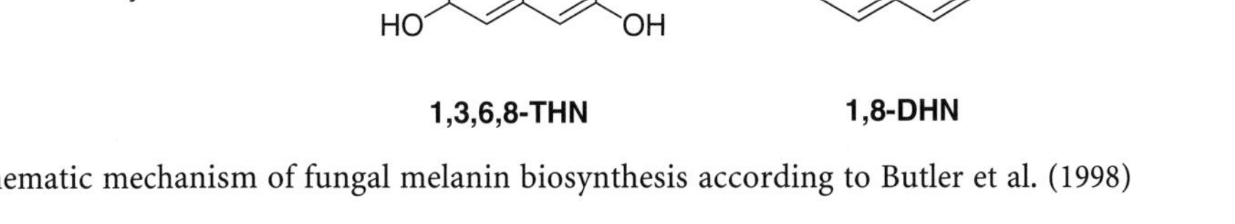

Fig. 16.2. Schematic mechanism of fungal melanin biosynthesis according to Butler et al. (1998)

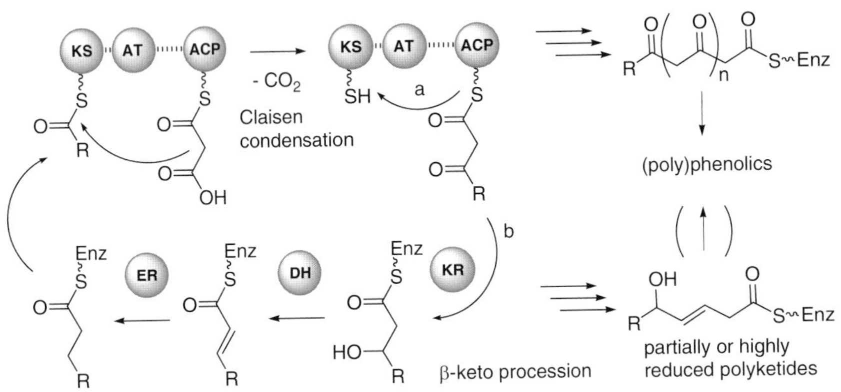

Fig. 16.3. Basic mechanisms of fungal polyketide biosynthesis

CoA onto the KS and ACP domains. While FAS reduce every newly formed β-ketothioester to a saturated chain while attached to the ACP, there is greater flexibility in the PKS systems. Consequently, fungal PKS can be classified according to the degree of reduction (or more exactly β-keto processing) of the products into non-reducing, partially or highly reducing PKS. Several additional domains are involved in β-keto processing: First a β-ketoreductase (KR) reduces the keto group to a secondary alcohol. Then elimination by a dehydratase (DH) leads to an unsaturated thioester before an enoyl reductase (ER) produces a saturated ester.

As soon as the polyketide chain has reached its final length, a thioesterase (TE) hydrolyses the thioester and releases the carbon chain from the enzyme. Notably, TE domains are not found in every PKS. Fungal PKS C-terminal domains can also have cyclisation activity and release products through intramolecular ring formation (Fujii et al. 2001).

Fungal PKS can possess another intrinsic domain for *C*-methyl transfer (CMeT). Unlike *O*- and *N*-methylations, which are performed by "post-PKS" enzymes after polyketide assembly, fungal methylation of the polyketide carbon backbone takes place during chain formation (Hutchinson et al. 2000; Nicholson et al. 2001).

Although the multidomain enzymes usually act in an iterative fashion, interestingly the degree of reduction can vary in each extended unit. β-Keto processing domains KR, DH, ER are optionally used in every extension round and therefore lead to a large variability in polyketides not only in chain length (Staunton and Weissman

2001; Cox 2007). The factors governing this variability are largely unknown.

IV. Fungal Polyketide Synthase Classes

Polyketide synthases are classified according to their protein architecture and way of action. In analogy to multidomain type I FAS, type I PKS are large enzymes that contain all necessary enzymatic activities on a single protein. In contrast, type II PKS form a complex of single proteins with separated enzyme activities. While bacteria possess both non-iterative type I PKS containing individual modules with all necessary domains for every single elongation step and iterative type II PKS, in fungi mainly iterative type I PKS were found. They repeatedly use their enzymatic activities to elongate the growing acyl ester. An exception are lovastatin diketide synthase (LDKS) from *A. terreus* (Kennedy et al. 1999) and compactin diketide synthase (CDKS) from *Penicillium citrinum* (Abe et al. 2002). These type I PKS catalyse only a single elongation to synthesise a reduced diketide product. Type III PKS were found in plants, bacteria and fungi. They are simple KS proteins producing chalcones, stilbenes and small aromatic compounds (Staunton and Weissman 2001; Seshime et al. 2005).

As mentioned above, fungal type I PKS are further classified according to the degree of reduction of their products into non-reducing (NR) PKS, partially reducing (PR) PKS and highly reducing (HR) PKS (Bingle et al. 1999; Nicholson et al. 2001). A particular type of HR PKS is a hybrid synthetase composed of a HR-PKS that is fused to a non-ribosomal peptide synthetase (NRPS) module. These PKS-NRPS systems produce polyketides that are fused to an amino acid by an amide bond (Song et al. 2004).

Phylogenetic studies have shown that the programming of polyketide assembly and degree of reduction is reflected in the KS gene and deduced amino acid sequences (Kroken et al. 2003). Bingle et al. (1999) designed degenerated primers based on conserved DNA sequences of KS domains to selectively amplify fragments of the fungal NR type (WA type) and PR PKS (MSAS type). In a subsequent study Nicolson et al. (2001) complemented this set with selective degenerated primers targeting KS domains of HR PKS (lovastatin type), KR domains and CMeT domains, respectively. These genetic tools were successfully used to identify various fungal PKS genes like citrinin synthase from *M. purpureus* (Shimizu et al. 2005), squalestatin synthase from *Phoma* sp. C2932 (Cox et al. 2004) and chaetoglobosin synthetase from *P. expansum* (Schuemann and Hertweck 2004).

A. Non-Reducing PKS

Orsellinic acid is the simplest known fungal polyketide because it is formed through the cyclocondensation of a non-reduced tetraketide. Four decades ago it was shown that a cell-free extract from *Penicillium madriti* is capable of producing orsellinic acid in vitro, thus representing the first observed PKS activity (Gaucher and Shepherd 1968). However, the nature of the PKS is still unknown. NR PKS generally lack domains for β-keto processing. They possess N-terminal starter unit acyl transferase (SAT) domains, responsible for loading a starter unit, followed by the typical KS, AT and ACP domains necessary for chain elongation (Crawford et al. 2006). Another conserved region has been identified between AT and ACP, named product template (PT) domain. Crawford et al. (2008) showed that this domain is involved in correct cyclisation chemistry of ACP-bound polyketides and increased product release from the enzyme. Some known NR PKS feature additional processing activities like cyclase (Watanabe et al. 1998), methyl transferase (Shimizu et al. 2005), reductase domains (Cox 2007) or additional ACPs (Fujii et al. 2001).

Several NR PKS genes have been characterised that are involved in the biosynthesis of norsolorinic acid in *A. parasiticus* (Minto and Townsend 1997), *A. nidulans* (Minto and Townsend 1997) and *Dothistroma septosporum* (Bradshaw et al. 2006) and in the formation of pigment precursors like YWA1 from *A. nidulans* (Watanabe et al. 1999) and *A. fumigatus* (Watanabe et al. 2000). Another important example is tetrahydroxyl naphthalene (THN) synthase from *C. lagenarium* (Takano et al. 1995) and *W. dermatitidis* (Feng et al. 2001).

Labelling experiments revealed that some fungal NR PKS can use short-chain fatty acids as starter units instead of acetyl-CoA (Minto and Townsend 1997; Kroken et al. 2003). Norsoloronic acid, a precursor of aflatoxin, is biosynthesised from a hexanoate starter unit. Two genes of the aflatoxin biosynthesis gene cluster from *A. parasiticus*, *fas-1* and *fas-2*, are related to fungal FAS genes and were shown to be involved in hexanoate synthesis (Hitchman et al. 2001; Watanabe and Townsend 2002). Zearalenone biosynthesis in *Gibberella zeae* requires two PKS genes that code for one HR and one NR PKS (Gaffoor and Trail 2006). The HR PKS (ZEA2p) likely generates a hexaketide starter unit that is loaded onto the NR PKS (ZEA1p) SAT domain and then undergoes three further elongations (Fig. 16.4).

The PT domain is a yet little understood region of about 350 residues. Sequences of known PT domain regions group according to the chain length of their products. This suggests that they have a similar function in the determination of extension number like the KS_β (or chain length factor, CLF) component of bacterial type II PKS (Keatinge-Clay et al. 2004; Cox 2007). Deconstruction experiments by domain dissection and reassembly of the *A. parasiticus* PksA revealed the function of PT in product formation (Crawford et al. 2008). During aflatoxin biosynthesis PT was shown to drive aromatisation chemistry to irreversibly form rings A and B and to increase the flux of the norsoloronic acid precursor from the PKS enzyme.

TE domains are the most common N-terminal processing domains found in NR PKS. They can operate as thioesterases or as Claisen cyclases (CYC) as shown by Ebizuka and others (Watanabe et al. 1998; Watanabe et al. 1999; Fig. 16.5). Heterologous expression of the *A. nidulans* YWA1 synthase (WAS) gene *wA* in full-length and truncated variants thereof showed that the CYC/TE is not necessarily required for the release of the

Fig. 16.4. Biosynthesis of zearalenone by NR-PKS ZEA1p and
HR-PKS ZEA2p

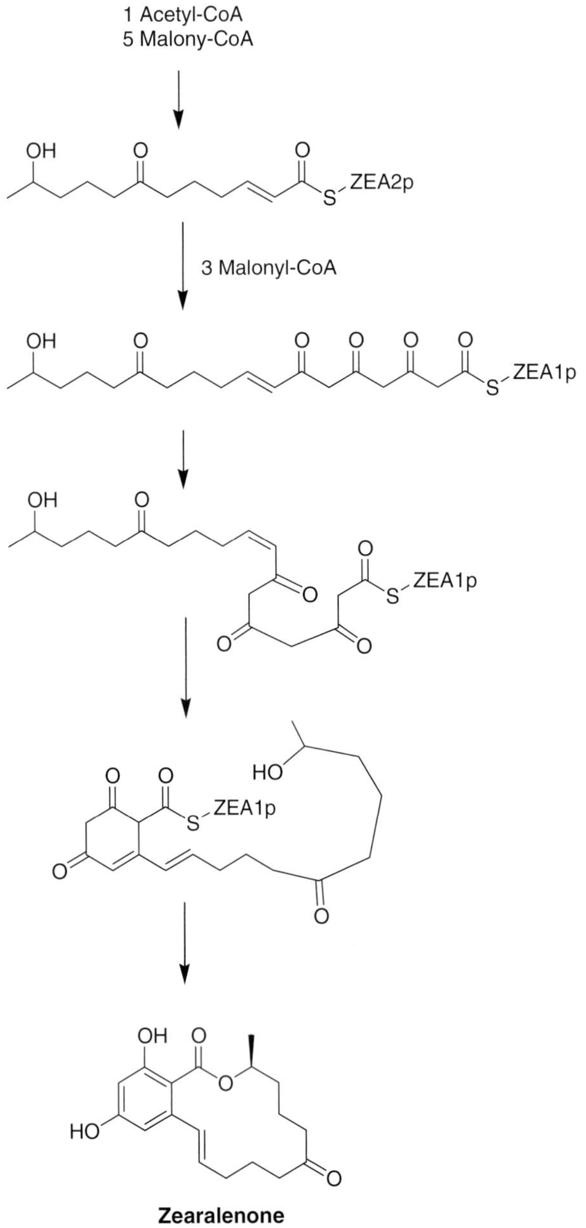

1 Acetyl-CoA
5 Malony-CoA

3 Malonyl-CoA

Zearalenone

polyketide product. Nonetheless, a conserved histidine and a serine residue are crucial for CYC/TE
activity (Fujii et al. 2001). A similar CYC/TE domain occurs in *A. nidulans* sterigmatocystin
synthase PKSst (Yu and Leonard 1995).

Another function of C-terminal TE domains is
the determination of chain length. *C. aegenarium*
PKS1 naturally produces the pentaketide THN
(Takano et al. 1995). Mutants of the PKS1 TE
domain resulted in production of mainly a

hexaketide product showing that the deletion of
TE or mutation of conserved residues influences
the polyketide chain length (Takano et al. 1995;
Watanabe and Ebizuka 2004).

Some fungal NR PKS contain multiple ACP
domains. *A. nidulans* WAS (Watanabe et al.
1999), NSAS (Minto and Townsend 1997) and
C. lagenarium THNS (Takano et al. 1995) contain
two ACPs and *D. septosporum* NSAS even three
copies (Bradshaw et al. 2006). The additional

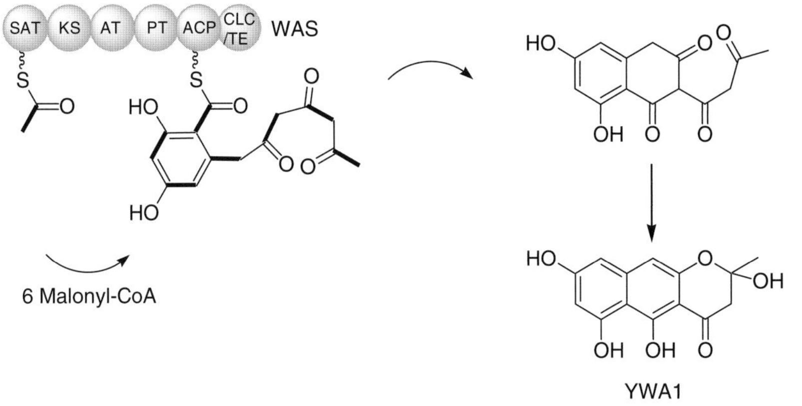

Fig. 16.5. Structure of WAS and biosynthesis model

ACPs are not required for biosynthesis, but all ACP domains are functional and exchangeable even among different species (Fujii et al. 2001; Watanabe and Ebizuka 2002).

Kroken et al. (2003) identified a group of NR PKS containing a CMeT domain downstream of ACP. The only known sequence that is correlated with a secondary metabolite is the citrinin synthase (CitS) gene from *M. pupureus*. CitS likely elongates a reduced diketide starter unit to citrinin (Shimizu et al. 2005). In this case CMeT must act twice but it is not clear how the point in time for methylation is programmed (Hajjaj et al. 1999).

Sequence analyses show that *M. pupureus* CitS possesses a C-terminal thioester reductase. In a mechanism similar to NRPS systems, the CitS reductase domain might be involved in the release of the product as an aldehyde or primary alcohol (Shimizu et al. 2005; Cox 2007).

B. Partially Reducing PKS

Fungal PR PKS differ from NR PKS in that they contain an N-terminal KS, followed by AT, DH, a core domain (core), KR and a C-terminal ACP. There are no SAT, PT or TE domains. Only three fungal genes coding for PR PKS have been matched to secondary metabolites, although many more fungal PR PKS genes are known. These three studied genes code for fungal 6-MSAS (Fig. 16.6). One was isolated from *P. patulum* (*MSAS*; Beck et al. 1990), one from *A. terreus* (*atX*; Fujii et al. 1996) and another

gene (*pks2*) was discovered in *Glarea lozoyensis* (Lu et al. 2005).

P. patulum and *A. terreus* MSAS are both supposed to form homotetramers. Fujii and coworkers studied the recombination of functional MSAS complexes from deletion mutants of the *A. terreus atX* gene (Moriguchi et al. 2006). They expressed two copies of *atX* in *S. cerevisiae* simultaneously. This allowed them to mutate independently and to cross-complement mutants. Almost all combinations of truncated MSAS proteins were functional, showing that all truncations can be complemented by another functional copy of the protein. Only one short 122-amino-acid region of the core domain was necessary for complementation. Probably this region codes for a motif required for subunit–subunit interaction. A similar core sequence was found in other fungal PR PKS and in bacterial PKS. Also, mammalian FAS possess a core region with similar function in enzyme assembly (Witkowski et al. 2004). However, no sequence similarities were found between fungal MSAS and mammalian FAS core regions.

C. Highly Reducing PKS

HR PKS differ from NR and PR PKS by an extended set of domains. The N-terminal KS domain is followed by AT, DH, ER, KR domains, and a C-terminal ACP domain. In some cases, however, the protein lacks a conserved ER domain, in lieu harbouring a region without known function. These PKS apparently utilize a free-standing ER in *trans*. In addition to the full set of β-keto

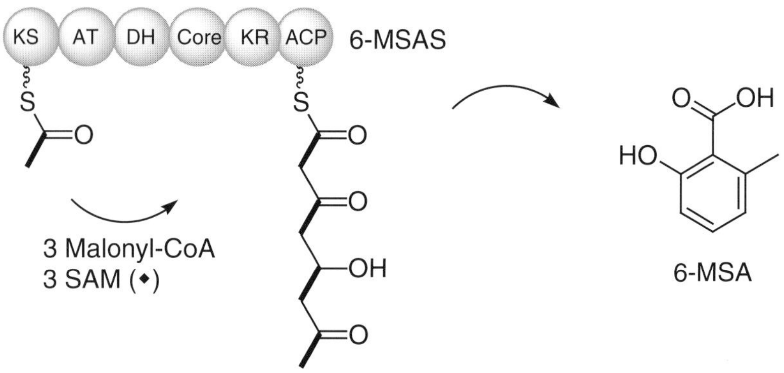

Fig. 16.6. Architecture of the MSAS and model for the biosynthesis of 6-MSA

processing domains, many HR PKS possess a CMeT domain following a DH.

Genes linked to known highly reduced metabolites are involved in the biosynthesis of lovastatin produced by *A. terreus* (Kennedy et al. 1999), compactin from *P. citrinum* (Abe et al. 2002), zearalenone from *G. zeae* (Gaffoor and Trail 2006), squalestatin produced by *Phoma* sp. (Cox et al. 2004), T-toxin from *C. heterostrophus* (Yang et al. 1996; Baker et al. 2006) and fumonisin from *G. fujikuroi* (Proctor et al. 1999).

Lovastatin and compactin are both biosynthesised by two separate PKS. Genes *lovB* and *lovF* (Hendrickson et al. 1999; Kennedy et al. 1999) encode lovastatin nonaketide synthase (LNKS) and lovastatin diketide synthase (LDKS), respectively, that are involved in lovastatin biosynthesis (Fig. 16.7). Compactin nonaketide synthase (CNKS; encoded by *mlcA*) and compactin diketide synthase (CDKS; encoded by *mlcB*) assemble compactin (Abe et al. 2002). Isotope feeding experiments suggested the synthesis of a reduced nonaketide chain that would undergo a Diels–Alder reaction to form dihydromonacolin L (Moore et al. 1985; Witter and Vederas 1996). LNKS has an inactive ER domain and requires the assistance of the separately encoded ER LovC for correct function. The involvement of a *trans*-acting ER was supposed for CNKS as well. Surprisingly, LNKS possesses a C-terminal truncated condensation (C) domain typical for NRPS. It is not clear whether this domain has a function in offloading or if it is just a relict of truncation.

LDKS is a HR PKS with KS, AT, CMeT, ER, KR and ACP domains. Notably, this is not an iterative PKS; in order to synthesise a diketide only a single elongation step is required. Acyl transferase LovD

transfers the diketide onto C-10 hydroxyl of monacolin J to assemble lovastatin (Kennedy et al. 1999). Xie et al. (2006) showed that LovD, produced in *E. coli*, can transfer various acyl-CoA residues onto monacolin J.

Squalestatin tetraketide synthase (SQTKS) was cloned from *Phoma* sp. (Cox et al. 2004). It shows high similarity to LDKS but produces a tetraketide chain. Squalestatin is assembled from two methylated polyketide chains, a hexaketide and a tetraketide chain. Notably, the hexaketide chain is synthesised from an unusual benzoate starter unit (Jones et al. 1992).

T-toxin is a very long, linear C_{41} polyketide produced by the maize pathogen *C. heterostrophus* (Kono and Daly 1979). It was one of the first compounds linked to a defined PKS gene (Yang et al. 1996). Later it turned out that two, not clustered genes are involved in T-toxin biosynthesis (Baker et al. 2006). Genes *pks1* and *pks2* code for T-toxin synthases TTS1 and TTS2, which both represent typical HR PKS. Interestingly, TTS1 possesses a CMeT domain although T-toxin does not show any methyl branches.

The mycotoxin fumonisin B1 causes several human and animal diseases (Nelson et al. 1993). It contains a methylated nonaketide chain, esterified to two unusual C_7 tricarboxylic acids. Fumonisin synthase (FUMS) appears to be a typical HR PKS with a functional ER domain. It is encoded by *fum1* in *G. fujikuroi* (Proctor et al. 1999).

The plant pathogen *Alternaria solani* produces numerous polyketides, for example solanopyrone A and alternaric acid (Fig. 16.8). PCR screening with primers for conserved PKS sequences led to the identification of two PKS genes (Fujii et al. 2005). One gene (*alt5*) codes

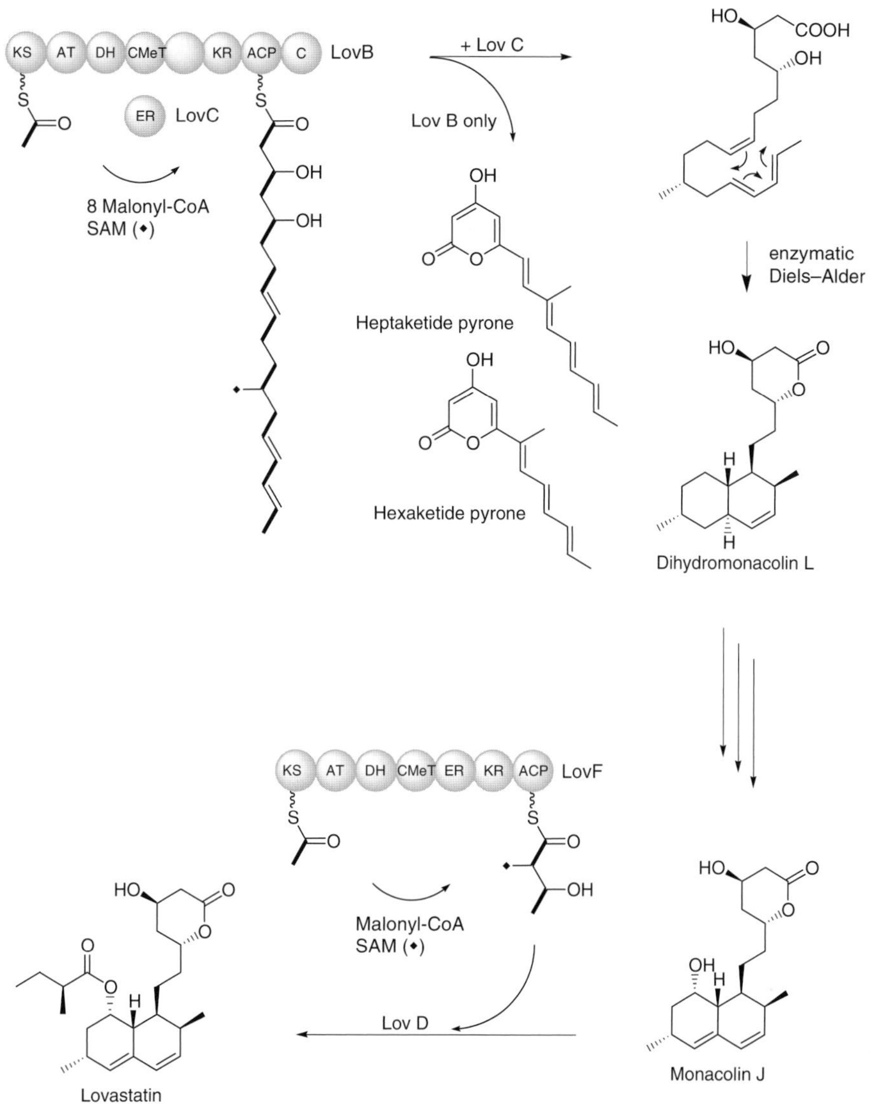

Fig. 16.7. Lovastatin biosynthesis involving LovB and LovF

for a HR PKS (PksN) and another (*pksA*) for a NR PKS. Heterologous production of PksN in *Aspergillus oryzae* led to the production of the methylated decaketide alternapyrone. Two more HR PKS sequences (*pksF*, *pksK*) were identified through gene probing of the *Aspergillus solani* genome. Both PKS genes lack a CMeT region (Kasahara et al. 2006). PksK was inactive when expressed in *A. oryzae*. However, the heterologous expression of PksF produced two mayor compounds, aslanipyrone and aslaniol. Both are polyunsaturated suggesting that the ER domain is inactive.

D. Polyketide Synthase–Non-Ribosomal Peptide Synthetase Hybrids

Many structurally intriguing fungal metabolites with remarkable biological activities feature a polyketide backbone fused to an amino acid. The biosynthesis of such hybrid metabolites is well studied in bacteria, e.g. biosynthesis of the siderophore and virulence factor yersiniabactin from *Yersinia pestis* (Miller et al. 2002). Fungal PKS-NRPS hybrids have a highly reducing PKS region containing an N-terminal KS, AT, DH, CMeT, inactive ER, KR and ACP domains. The ACP is flanked by domains of a

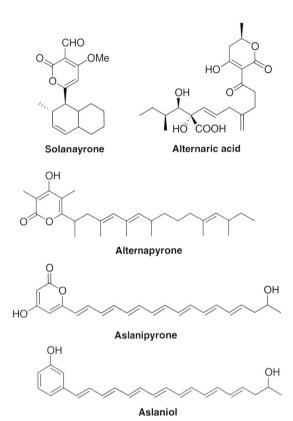

Fig. 16.8. Compounds produced by *Alternaria solani* PKS

typical NRPS module: condensation (C), adenylation (A) and peptidyl carrier protein (PCP) domains. Furthermore, the hybrid systems feature an additional C-terminal domain that might either function as a thioester reductase (R) or cyclase domain (Song et al. 2004). Like in lovastatin biosynthesis, a *trans* ER may be required for the correct PKS function (Halo et al. 2008).

PKS-NRPS hybrid synthetases are involved in the biosynthesis of fusarin C in *Fusarium moniliforme* (Song et al. 2004), equisetin from *F. heterosporum* (Sims et al. 2005), tenellin produced in *Beauveria bassiana* (Eley et al. 2007), aspyridone from *A. nidulans* (Bergmann et al. 2007), chaetoglobosin produced by *P. expansum* (Schuemann and Hertweck 2004) and pseurotin A from *A. fumigatus* (Maiya et al. 2007) (Fig. 16.9).

Biosynthesis of the mycotoxin fusarin C starts with the assembly of a tetramethylated heptaketide that is fused to homoserin (Song et al. 2004). The reductive release of the hybrid product and subsequent cyclisation putatively forms

prefusarin, which is then transformed into fusarin C by various tailoring enzymes.

Biosynthesis of the HIV-1 integrase inhibitor equisetin has been examined by Schmidt and colleagues (Sims et al. 2005). Equisetin synthetase (EqiS) has the same overall domain architecture as fusarin synthetase (FUSS). It differs from FUSS in the formation of a methylated oktaketide chain that is linked to serine. Another gene cluster containing a PKS-NRPS gene was identified within the genome of the plant pathogen *M. grisea* (Böhnert et al. 2004). The product of the encoded hybrid enzyme is yet unknown. It was shown that the gene cluster is expressed only during appressorium formation during penetration of the rice plant. Böhnert et al. (2004) could further show that production of the yet unknown fungal compound leads to resistance of the rice plant to further fungal penetration.

The yellow pigment tenellin from the insect pathogen *B. bassiana* is produced by the PKS-NRPS TENS (Eley et al. 2007). A dimethylated pentaketide fused to tyrosine is synthesised to build pre-tenellin after reductive release. Tailoring reactions performed by two putative P450 monooxygenases and rearrangements probably convert the precursor into tenellin. The hypothesis that tenellin might be involved in the pathogenicity of *B. bassiana* was ruled out by the fact that a direct knockout of TENS did not significantly reduce pathogenicity towards wax moth larvae.

A genome mining approach identified the only gene cluster of the *A. nidulans* genome containing a silent PKS-NRPS hybrid gene (Bergmann et al. 2007). Promoter exchange of the only clustered regulator gene led to the production of novel polyketide-amino acid hybrid compounds (Fig. 16.10). Aspyridone synthetase ApdA likely produces a dimethylated tetraketide and fuses it to a tyrosine moiety. A reductive downloading mechanism might release an amino aldehyde that could undergo cyclisation to a pyrrolinone. The involvement of oxidative tailoring reactions towards tetramic acid intermediates and ring rearrangements probably form the moderately cytotoxic aspyridones A and B. Aspyridone biosynthesis seems to be closely related to tenellin biosynthesis reported by Eley et al. (2007).

Cytochalasin (chaetoglobosin) biosynthesis was investigated in the fungus *P. expansum* (Schuemann and Hertweck 2004). Screening a genomic library of *P. expansum* led to the

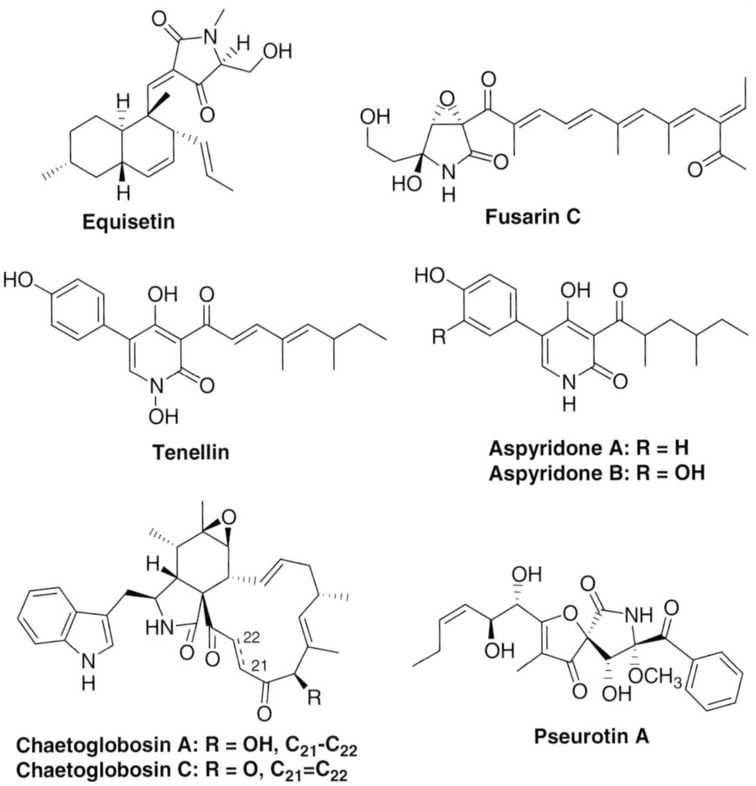

Equisetin

Fusarin C

Tenellin

Aspyridone A: R = H
Aspyridone B: R = OH

Chaetoglobosin A: R = OH, C_{21}-C_{22}
Chaetoglobosin C: R = O, C_{21}=C_{22}

Pseurotin A

Fig. 16.9. PKS-NRPS hybrid metabolites

identification of a gene cluster encoding a PKS-NRPS (CheA; Fig. 16.11). A RNA-silencing method proved that CheA is involved in chaetoglobosin production in *P. expansum*. It was suggested that CheA produces a trimethylated nonaketide that is then fused to a tryptophan moiety (Fig. 16.12). After reductive release of the hybrid product, catalysed by the C-terminal R domain, cyclisation to a pyrrolinone product likely occurs. This would act as a dienophile and undergo a Diels–Alder reaction with the diene moiety of the polyketide chain. Tailoring reactions seem to be carried out by the putative cytochrome P450 monooxygenases CheD and CheG and by the encoded FAD-dependent monooxygenase CheE to yield chaetoglobosin A and C. A Diels–Alder-like cyclisation mechanism was also proposed for lovastatin and equisetin biosynthesis (Oikawa and Tokiwano 2004; Sims et al. 2005). Oikawa and Tokiwano proposed the involvement of enzymatic Diels–Alderases for cyclisation of natural product precursors, e.g. polyketides. It is conceivable that PKS–NRPS synthetases enzymatically catalyse this reaction under physiological conditions or that they act as an entropic trap.

A single PKS-NRPS hybrid gene was identified within the sequenced genome of *A. fumigatus*. Deletion of the hybrid gene named *psoA* in

A. fumigatus led to the absence of pseurotin A, while overexpression led to accumulation of the hybrid metabolite (Maiya et al. 2007). Pseurotin A is a compound with unusual heterospirocyclic gamma-lactame structure and is a competitive inhibitor of chitin synthase and inducer of nerve-cell proliferation (Bloch et al. 1976; Komagata et al. 1996).

V. Post-PKS Modifications

The large structural diversity of polyketides results not only from varying chain lengths, degree of β-keto processing and cyclisation, but also from post-PKS modifications of the carbon chain by alkylation, acylation or oxygenation.

An impressive set of post-PKS modifications is involved in aflatoxin biosynthesis (Fig. 16.13). Aflatoxins, produced by some *Aspergillus* species, are a group of structurally related polyketide-derived furanocoumarins. In a 70-kb gene cluster at least 25 genes are involved in a complex biosynthetic pathway. Beside the PKS gene (*pksA*) and two FAS genes (*fas-1*, *fas-2*) they code for cytochrome P450 monooxygenases, dehydrogenases, methyltransferases, pathway regulators and a

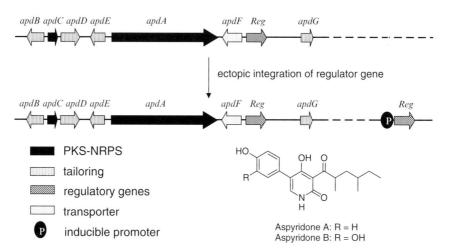

Fig. 16.10. Activation of the silent aspyridone gene cluster via ectopic integration of inducible regulator copy

transporter. Fifteen of those genes catalyse oxidative reactions. Most of the clustered genes are regulated by a single Zn2Cys6-type transcription factor AflR and by the co-regulator AfkJ. In total about 21 enzymatic steps are required for aflatoxin biosynthesis (Minto and Townsend 1997; Bhatnagar et al. 2003).

Treatment of fungal cultures with inhibitors of post-PKS enzymes can significantly affect the product spectrum. For example, *Chaetomium subaffine*, producer of chaetoglobosin A, was incubated with the cytochrome P450 inhibitor metyrapone (Oikawa et al. 1991). Chaetoglobosin A production was reduced while accumulation of less oxidised precursor compounds was observed, thus providing an insight into the final biosynthetic steps.

In post-PKS reactions polyketides can also be fused to further polyketide chains. Acyl transferase Lov D transfers a methylbuturyl side chain onto the lovastatin precursor monacolin J in order to fully assemble lovastatin (Hendrickson et al. 1999; Kennedy et al. 1999).

VI. Phylogeny

In an extensive phylogenetic study Kroken et al. (2003) compared the amino acid sequences of mainly fungal KS domains. Fungal type I PKS sequences form a major clade together with bacterial type I PKS. This clade is clearly sister to animal FAS and also bacterial type II PKS. Type I PKS fall into three main clades; two of these contain exclusively fungal sequences and correlate with their products reduction: reducing PKS and non-reducing PKS. They are mainly characterised by the presence or absence of β-keto processing domains (DH, ER, KR). Each main fungal clade is further divided into four groups with characteristic domain sets resulting from loss of ancestral domains or from gain of novel ones (e.g. NRPS domains). The ancestral type I PKS domain structure is supposed to be: KS, AT, DH, CMeT, ER, KR, ACP. The first main fungal clade contains PKS that are involved in the production of reduced, mainly linear polyketides like precursors of toxins (e.g. lovastatin, fumonisin, T-toxin). The second main fungal clade includes PKS that synthesise unreduced, cyclic polyketides, such as toxins (e.g. aflatoxin) and pigments (e.g. melanin). Two more fungal clades were found both nested into the bacterial type I PKS clade. One contains 6-MSA-type PKS. The other consists of a single sequence, the *C. heterostrophus* PKS24 probably a PKS-NRPS hybrid. Orthologous PKS genes, producing nearly identical polyketides, were found among closely related but also distant species. Due to the diversity and distribution of fungal PKS it was speculated that gene duplication, divergence and differential gene loss are responsible for generation and maintenance of this diversity. Horizontal gene transfer is not necessarily involved (Kroken et al. 2003).

In contrast, horizontal gene transfer was proposed as one way for the evolution of virulence

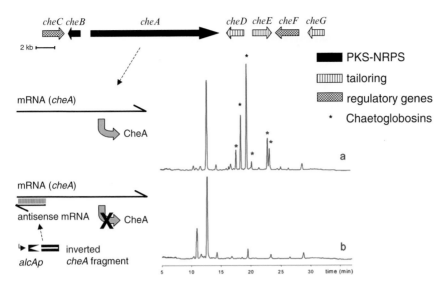

Fig. 16.11. Organisation of the chaetoglobosins gene cluster. (a) *P. expansum* wild type producing chaetoglobosins (*) and (b) silencing mutant with reduced chaetoglobosin synthesis

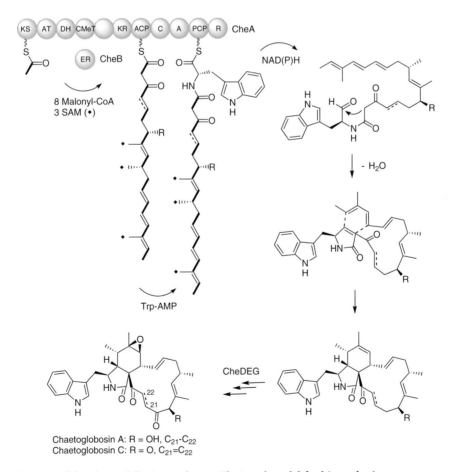

Fig. 16.12. Architecture of the chaetoglobosin synthetase CheA and model for biosynthesis

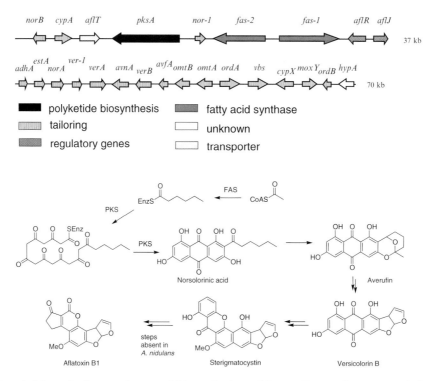

Fig. 16.13. Aflatoxin biosynthetic gene cluster and biosynthesis model according to Bhatnagar et al. (2003)

genes in filamentous fungi, e.g. the PKS-NRPS encoding gene *ACEI* from *M. grisea* (Böhnert et al. 2004). While phylogenetic analyses reveal complex duplication events for the evolution of *ACEI*-like clusters in *Chaetomium globosum* and *M. grisea*, horizontal gene transfer is proposed from a relative of *M. grisea* into an ancestor of *Aspergillus clavatus* (Khaldi et al. 2008).

Additional phylogenetic analysis of the hybrid synthetase EqiS indicate that fungal PKS-NRPS have likely diversified by point mutation and deletion (Sims et al. 2005). This is in contrast to bacterial PKS systems, which use module and domain swaps to generate structural diversity (Crosa and Walsh 2002).

Phylogenetic studies were also carried out on KS sequences of NR PKS from Pertusariales (lichenised fungi) and other NR PKS sequences (Schmitt et al. 2005). In alignments the sequences grouped in 18 clades containing only lichenised taxa, only non-lichenised taxa or sequences from both. Interestingly, clades could not be associated with secondary metabolism classes. Another study concerned sequence comparison of mainly *Leconora* spp. PKS sequences (Grube and Blaha 2003).

They clustered in two clades of Schmitt's alignment. Amino acid sequence of KS domains from 15 *Leconora* species, three other lichen species and other non-lichenised fungi were compared. *Leconora* PKS sequences clustered mainly with PKS producing complex aromatic compounds, with non-lichen-derived PKS producing precursors of DHN-melanins and in another clade without relationship to known fungal PKS.

VII. Functional Analysis and Engineering of Fungal Polyketide Biosynthesis

Engineering fungal polyketide biosynthesis genes or gene clusters could be an efficient way to diversify metabolite structures and their biological activities (Burkart 2003; Reeves 2003; Schuemann and Hertweck 2006).

From the gene loci that have been investigated it appears that nearly all genes necessary for fungal polyketide biosynthesis, post-PKS processing, regulation and resistance are clustered, as exemplified for the lovastatin (Kennedy et al. 1999) and aflatoxin biosynthesis gene clusters (Yu et al. 1995; Bhatnagar et al. 2003; Fig. 16.13). This organisation is an advantage for identification and handling of gene cassettes or even whole gene clusters.

Pathway engineering towards novel compounds has been intensely explored in bacteria (Cane et al. 1998; Walsh 2002; Weber et al. 2003), but in fungi such combinatorial biosynthesis approaches have been far more challenging from a biotechnical point of view. The incidence of introns in fungal genes can be problematic for heterologous expressions and also the directed integration of DNA into fungal genomes is not guaranteed and may be cumbersome.

A. Inactivation of PKS Genes

1. Knockout

The disruption or replacement of PKS genes is a common approach for their functional analyses. However, the efficiency of targeting a genetic locus can vary significantly between fungal species. Apart from challenges regarding selection markers and introduction of DNA into the fungus, the integration of transformed DNA can occur in a homologous or non-homologous manner.

The lovastatin biosynthesis genes were identified by screening mutants of the producer strain *A. terreus* in 1999 (Hendrickson et al. 1999; Kennedy et al. 1999). Several mutants were investigated that were unable to produce lovastatin or that yielded polyketide shunt products only, which

led to the discovery of the lovastatin biosynthesis gene cluster including both PKS genes *lovB* and *lovF* (Fig. 16.7). Further directed gene disruptions proved the involvement of *lovC* and *lovD* in lovastatin biosynthesis (Kennedy et al. 1999). Similar knockout experiments in *P. citrinum* correlated gene *mlcA* with compactin production (Abe et al. 2002).

Targeted gene knockout experiments also revealed the function of PksA from *A. parasiticus* (Chang et al. 1995). The corresponding gene *pksA* is similar to the *A. nidulans wA* gene and converts hexanoyl-CoA into norsolonic acid, an intermediate in sterigmatocystin and aflatoxin biosynthesis (Trail et al. 1995; Brown et al. 1996; Bennett et al. 1997; Minto and Townsend 1997; Bhatnagar et al. 2003).

Directed gene replacement proved the function of *pks1*, a gene coding for a PKS involved in DHN-melanin synthesis in *G. lozoyensis*. Knockout mutants were achieved by double crossover recombination via *Agrobacterium*-mediated transformation (Zhang et al. 2003).

Directed knockout experiments in *F. moniliforme* and *F. venenatum* correlated a PKS-NRPS hybrid gene with fusarin C biosynthesis (Song et al. 2004; Fig. 16.14). Likewise, a gene knockout by homologous recombination in *F. heterosporum* revealed that equisetin biosynthesis requires the gene product of *pks2* (Sims et al. 2005).

In 2007 the PKS-NRPS hybrid gene *psoA* was associated with pseurotin A production in *A. fumigatus* by deletion and overexpression (Maiya et al. 2007).

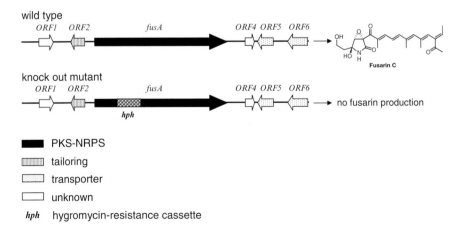

Fig. 16.14. Gene knockout of the *F. moniliforme* gene *fusA*

The generation of random mutations through restriction enzyme-mediated integration (REMI; Brown et al. 1998), UV-mutagenesis (Jahn et al. 1997) or transposon mutagenesis (Kempken et al. 1998; Kempken and Kuck 2000; Brakhage and Langfelder 2002) has been successfully used to target fungal genes. The REMI procedure was used to identify a PKS gene involved in T-toxin production in the fungal pathogen *C. heterostrophus* (Yang et al. 1996).

2. Knockdown

Another genetic tool to study PKS gene functions in fungi is RNA-mediated gene silencing. In this post-transcriptional down-regulation method double-stranded RNA triggers the degradation of a sequence homologous mRNA. With this method Nakayashiki et al. (2005) successfully silenced specific genes in *Magnaporthe oryzae* and *C. lagenarium*.

RNA silencing is an attractive method to analyse lethal knockouts or cases when directed gene disruptions failed due to a low efficacy of homologous integration events in the fungal genome. This method was successfully applied in the investigation of chaetoglobosin biosynthesis in *P. expansum* (Schuemann and Hertweck 2004). All efforts to disrupt the proposed PKS-NRPS gene *cheA* of the sequenced gene cluster failed. Finally, *cheA* RNA-silencing resulted in a significant decrease in chaetoglobosin production, which proved the involvement of the targeted gene in metabolite biosynthesis (Fig. 16.11).

Sequence similar genes can be silenced simultaneously. A disadvantage is that silencing not necessarily leads to a null phenotype. This may be crucial for the down-regulation of non-major activity enzymes (like tailoring enzymes), which might have no detectable effect on metabolite production level (Thierry and Vaucheret 1996; Baulcombe 1999).

B. Heterologous Expression of PKS Genes

A general expression system for fungal PKS genes is preferable in order to engineer metabolite biosynthesis and improve yields. The use of native polyketide producers is often hampered by a high background of metabolites, sub-optimal growth conditions or low production of the desired polyketide.

In contrast to bacterial genes, fungal genes mostly contain introns that require a strain-specific splicing mechanism. Introns are hard to remove and impair gene expression in non-eukaryotic systems. In addition, apo-ACP domains need to be post-translationally 4-phosphopantetheinylated, thus making co-expression of a specific PPTase essential. Beside fungal hosts (e.g. *A. nidulans*), yeast (*Saccharomyces cerevisiae*), bacteria (Actinomycetes) and plants (tobacco) have proven suitable for PKS gene expression (Pfeifer and Khosla 2001).

The first successful expression of a fungal PKS gene in bacteria was reported in 1995 (Bedford et al. 1995). Bedford et al. (1995) expressed the 6-MSAS gene from *P. patulum* in *Streptomyces coelicolor* CH999. Extensive modification of the expressed gene was necessary. The intron was removed, a Shine–Dalgarno sequence was cloned in the 3' region and four codons were adapted into more frequently used ones in Streptomycetes (Wright and Bibb 1992). The 6-MSAS gene was expressed under control of the *S. coelicolor* actinorhodin *actI* promoter.

One year later the 6-MSAS encoding gene *atX* from *A. terreus* was identified and expressed in the fungal host *A. nidulans* (Fujii et al. 1996). For successful expression the *atX* gene was cloned downstream the *amyB* promoter of the Takaamylase A gene.

In 1999 the expression of the LNKS gene in *A. nidulans* using the *alcA* promoter for expression control was reported by Kennedy et al. (1999). Apparently LNKS alone was malfunctioning and synthesised polyunsaturated pyrones only (Fig. 16.7). Enoyl reduction and ketoreduction had not occurred and the polyketide chain was released prematurely. Sequence analysis revealed that the NADPH binding pocket and this for ER activity of LNKS was dysfunctional. Co-expression of the separately encoded and *trans*-acting ER LovC complemented ER activity and led to the production of the expected dihydromonacolin L.

Squalestatin tetraketide synthase from *Phoma* sp. was introduced into *A. oryzae*, and the transformant produced detectable amounts of the tetraketide side-chain of sqalestatin S1 (Cox et al. 2004).

The only successful expression of a fungal PKS in *Escherichia coli* and additionally in *S. cerevisiae* was reported by Kealey et al. (1998). They co-expressed the gene coding for 6-MSAS together with the gene encoding *Bacillus subtilis* PPTase Sfp in both hosts. Using suitable promoters, alcohol dehydrogenase 2 (ADH2) promoter and T5 RNA polymerase promoter, respectively, high production levels were achieved in yeast (twofold higher than in the wild type), while 6-MSA production was just detectable in *E. coli* (Kealey et al. 1998).

The 6-MSAS gene from *P. patulum* was also expressed in tobacco, showing that plants could also serve for the production of fungal PKS. Interestingly, the plant PPTase seemed to convert fungal apo-ACP into the holo enzyme (Yalpani et al. 2001).

Ebizuka and co-workers demonstrated the expression of three HR PKS genes from *A. solani* in *A. oryzae* (Kasahara, Fujii et al. 2006). While PksK was non-functional, production of PksN and PksF led to the production of the octamethylated decaketide alternapyrone and aslanipyrone/aslaniol, respectively (Fujii et al. 2005; Fig. 16.8).

Halo et al. (2008) succeeded in expressing the tenellin synthetase gene *tenS* in *A. oryzae*. Heterologous expression of the PKS-NRPS hybrid synthetase-encoding gene *tenS* without the separately encoded ER gene led to the production of acyl tetramic acids, albeit their structure elucidation revealed errors in polyketide assembly. A correctly built polyketide could be achieved by co-expression of *tenS* with *orf3*, coding for the *trans*-acting ER of the tenellin biosynthesis gene cluster.

In 2007 the iterative, non-reducing bikaverin PKS, PKS4 from *G. fujikuroi*, was expressed at high levels in *E. coli*, purified and reconstituted to in vitro activity (Ma et al. 2007). In the presence of malonyl-CoA PKS4 synthesised the precursor of bikaverin as main product, demonstrating the functionality of the bacterial expressed fungal PKS. This work provides a basis for enzymatic in vitro synthesis of fungal polyketide compounds. Mutations of the *G. fujikuroi* PKS4 C-terminal thioesterase/Claisen cyclase (TE/CYC) domain were performed by Ma et al. (2008). The mutant PKS were disabled in correct cyclisation of the nascent polyketide, proving the involvement of TE/CYC in the folding of bikaverin. Complementation of the mutant PKS4 with a standalone TE/CLC domain restored the regioselective cyclisation steps. In further in vitro experiments dissociated bacterial tailoring enzymes were added to the reconstituted PKS4. Addition of the *S. coelicolor* actinothodin KR led to the C9 reduced and shortened oktaketide mutactin. The addition of first ring aromatase/cyclase (GRIS ARO/CYC)

from the griseusin PKS (Zawada and Khosla 1999) and second ring cyclase OxyN (Zhang et al. 2007) resulted in anthraquinone products. These experiments impressively show how fungal PKS can be complemented with *trans*-acting domains and tailoring enzymes even from distinct families of PKS.

C. PKS Pathway Regulation

For several reasons it might be attractive to regulate the activity of a whole biosynthesis gene cluster in order to increase, decrease or temporally control metabolite production. Methodologies to alter fungal secondary metabolic synthesis involve varying environmental factors such as nutrients or temperature. The marine fungus *Phomopsis asparagi* was treated with the F-actin inhibitor jasplakinolide in order to turn on pathways (Christian et al. 2005). The *Phomopsis* genus is a rich source for secondary metabolites. Incubation of *P. asparagi* with jasplakinolide, first isolated from the sponge *Jaspis splendens*, resulted in the production of three new biological active chaetoglobosins.

Mining the genome of *A. nidulans*, an orphan PKS gene cluster was identified (Bergmann et al. 2007). Analyses revealed the presence of one PKS-NRPS gene, one gene encoding an ER, four putative tailoring genes, one gene coding for an exporter and one single regulator gene named *apdR*. All efforts to cultivate this fungus under conditions that lead to the production of a polyketide-non-ribosomal peptide related product failed. The gene cluster remained silent under test conditions. A novel activation method was established in order to identify the encoded natural product biosynthesis pathway. A copy of the regulator gene *apdR*, coding for a C6 transcription factor, was cloned and expressed under control of the inducible *alcA* promoter from *A. nidulans* (Fig. 16.10). Induction of the regulator induced the transcription of all biosynthesis genes and finally led to the production of new polyketide-non-ribosomal peptide products in *A. nidulans*. New members of pyridone alkaloids were isolated, the aspyridones.

VIII. Conclusions

Fungal PKS are highly specialised programmed nanomachines. With a simple set of active domains some of the most complex natural products are synthesised. Fungal PKS use their enzymatic domains iteratively, and it is not yet thoroughly understood how chain length, degree of reduction and C-methylation pattern are programmed. During the past years remarkable progress was made in the genetic engineering of fungal PKS. Modifications of PKS genes, tailoring genes or whole biosynthesis pathways were demonstrated. Expression in heterologous hosts was realised and pilot experiments in fungal PKS domain swaps were made. An enormous biosynthetic potential holds the combination of polyketide biosynthesis with other pathways, for example nonribosomal peptide biosynthesis. Nature combines those pathways in mixed biosynthetic enzymes, such as PKS-NRPS hybrids.

Still we are far away from the status quo of understanding and handling bacterial PKS. To take advantage of the biosynthetic potential of fungal PKS we have to learn much more. Future insights in the structure and action of fungal PKS will set the basis to identify and synthesise new pharmaceutical active products.

References

Abe Y, Suzuki T, et al (2002) Molecular cloning and characterization of an ML-236B (compactin) biosynthetic gene cluster in *Penicillium citrinum*. Mol Genet Genomics 267:636–646

Alberts AW, Chen J, et al (1980) Mevinolin: a highly potent competitive inhibitor of hydroxymethylglutaryl-coenzyme A reductase and a cholesterol-lowering agent. Proc Natl Acad Sci USA 77:3957–3961

Baker SE, Kroken S, et al (2006) Two polyketide synthase-encoding genes are required for biosynthesis of the polyketide virulence factor, T-toxin, by *Cochliobolus heterostrophus*. Mol Plant Microbe Interact 19:139–149

Bate C, Salmona M, et al (2004) Squalestatin cures prion-infected neurons and protects against prion neurotoxicity. J Biol Chem 279:14983–14990

Baulcombe DC (1999) Gene silencing: RNA makes RNA makes no protein. Curr Biol 22:1559–1566

Beck J, Ripka S, et al (1990) The multifunctional 6-methylsalicylic acid synthase gene of *Penicillium patulum*. Its gene structure relative to that of other polyketide synthases. Eur J Biochem 192:487–498

Bedford DJ, Schweizer E, et al (1995) Expression of a functional fungal polyketide synthase in the bacterium *Streptomyces coelicolor* A3(2). J Bacteriol 177:4544–4548

Bennett JW, Chang PK, et al (1997) One gene to whole pathway: the role of norsolorinic acid in aflatoxin research. Adv Appl Microbiol 45:1–15

Bergmann S, Schuemann J, et al (2007) Genomics-driven discovery of PKS-NRPS hybrid metabolites from *Aspergillus nidulans*. Nat Chem Biol 3:213–217

Bhatnagar D, Ehrlich KC, et al (2003) Molecular genetic analysis and regulation of aflatoxin biosynthesis. Appl Microbiol Biotechnol 61:83–93

Binder M, Tamm C (1973) Die Cytochalasane, eine neue Klasse biologisch aktiver Metabolite von Mikroorganismen. Angew Chem Int Ed Engl 9:369–420

Bingle LE, Simpson TJ, et al (1999) Ketosynthase domain probes identify two subclasses of fungal polyketide synthase genes. Fungal Genet Biol 26:209–223

Birch AJ, Donovan FW (1953) Studies in relation to biosynthesis. I. Some possible routes to derivatives of orcinol and phloroglucinol. Aust J Chem 6:360–368

Bloch P, Tamm C, et al (1976) Pseurotin, a new metabolite of Pseudeurotium ovalis Stolk having an unusual hetero-spirocyclic system. Helv Chim Acta 59:133–137

Böhnert HU, Fudal I, et al (2004) A putative polyketide synthase/peptide synthetase from *Magnaporthe grisea* signals pathogen attack to resistant rice. Plant Cell 16:2499–2513

Bradshaw RE, Jin H, et al (2006) A polyketide synthase gene required for biosynthesis of the aflatoxin-like toxin, dothistromin. Mycopathologia 161:283–294

Brakhage AA, Langfelder K (2002) Menacing mold: the molecular biology of *Aspergillus fumigatus*. Annu Rev Microbiol 56:433–455

Brown AG, Smale TC, et al (1976) Crystal and molecular structure of compactin, a new antifungal metabolite from *Penicillium brevicompactum*. J Chem Soc [Perkin 1]:1165–1170

Brown DW, Yu JH, et al (1996) Twenty-five coregulated transcripts define a sterigmatocystin gene cluster in *Aspergillus nidulans*. Proc Natl Acad Sci USA 93:1418–1422

Brown JS, Aufauvre-Brown A, et al (1998) Insertional mutagenesis of *Aspergillus fumigatus*. Mol Gen Genet 259:327–335

Burkart MD (2003) Metabolic engineering – a genetic toolbox for small molecule organic synthesis. Org Biomol Chem 1:1–4

Butler MJ, Day AW (1998) Fungal melanins: a review. Can J Microbiol 44:1115–1136

Cane DE, Walsh CT, et al (1998) Harnessing the biosynthetic code: combinations, permutations, and mutations. Science 282:63–68

Chang PK, Cary JW, et al (1995) The *Aspergillus parasiticus* polyketide synthase gene pksA, a homolog of *Aspergillus nidulans* wA, is required for aflatoxin B1 biosynthesis. Mol Gen Genet 248:270–277

Christian OE, Compton J, et al (2005) Using jasplakinolide to turn on pathways that enable the isolation of new chaetoglobosins from *Phomopsis asparagi*. J Nat Prod 68:1592–1597

Chu FS (1991) Mycotoxins: food contamination, mechanism, carcinogenic potential and preventive measures. Mutat Res 259:291–306

Cox RJ (2007) Polyketides, proteins and genes in fungi: programmed nano-machines begin to reveal their secrets. Org Biomol Chem 5:2010–2026

Cox RJ, Glod F, et al (2004) Rapid cloning and expression of a fungal polyketide synthase gene involved in squalestatin biosynthesis. Chem Commun (Camb) :2260–2261

Crawford JM, Dancy BC, et al (2006) Identification of a starter unit acyl-carrier protein transacylase domain in an iterative type I polyketide synthase. Proc Natl Acad Sci USA 103:16728–16733

Crawford JM, Thomas PM, et al (2008) Deconstruction of iterative multidomain polyketide synthase function. Science 320:243–246

Crosa JH, Walsh CT (2002) Genetics and assembly line enzymology of siderophore biosynthesis in bacteria. Microbiol Mol Biol Rev 66:223–249

Dawson MJ, Farthing JE, et al (1992) The squalestatins, novel inhibitors of squalene synthase produced by a species of Phoma. I. Taxonomy, fermentation, isolation, physico-chemical properties and biological activity. J Antibiot 45:639–647

Deacon JW (2001) Modern mykology. Blackwell, London

Eaton DL, Groopman JD (1994) The toxicology of aflatoxins: human health, veterinary, and agricultural significance. Academic, San Diego

Eley KL, Halo LM, et al (2007) Biosynthesis of the 2-pyridone tenellin in the insect pathogenic fungus Beauveria bassiana. Chembiochem 8:289–297

Endo A, Kuroda M, et al (1976) ML-236A, ML-236B, and ML-236C, new inhibitors of cholesterogenesis produced by Penicillium citrinium. J Antibiot (Tokyo) 29:1346–1348

Feng B, Wang X, et al (2001) Molecular cloning and characterization of WdPKS1, a gene involved in dihydroxynaphthalene melanin biosynthesis and virulence in Wangiella (Exophiala) dermatitidis. Infect Immun 69:1781–1794

Feng GH, Leonard TJ (1995) Characterization of the polyketide synthase gene (pksL1) required for aflatoxin biosynthesis in Aspergillus parasiticus. J Bacteriol 177:6246–6254

Fujii I, Ono Y, et al (1996) Cloning of the polyketide synthase gene atX from Aspergillus terreus and its identification as the 6-methylsalicylic acid synthase gene by heterologous expression. Mol Gen Genet 253:1–10

Fujii I, Watanabe A, et al (2001) Identification of Claisen cyclase domain in fungal polyketide synthase WA, a naphthopyrone synthase of Aspergillus nidulans. Chem Biol 8:189–197

Fujii I, Yoshida N, et al (2005) An iterative type I polyketide synthase PKSN catalyzes synthesis of the decaketide alternapyrone with regio-specific octa-methylation. Chem Biol 12:1301–1309

Gaffoor I, Trail F (2006) Characterization of two polyketide synthase genes involved in zearalenone biosynthesis in Gibberella zeae. Appl Environ Microbiol 72:1793–1799

Gaucher GM, Shepherd MG (1968) Isolation of orsellinic acid synthase. Biochem Biophys Res Commun 32:664–671

Grube M, Blaha J (2003) On the phylogeny of some polyketide synthase genes in the lichenized genus Lecanora. Mycol Res 107:1419–1426

Hajjaj H, Klaebe A, et al (1999) Biosynthetic pathway of citrinin in the filamentous fungus Monascus ruber as revealed by 13C nuclear magnetic resonance. Appl Environ Microbiol 65:311–314

Halo LM, Marshall JW, et al (2008) Authentic heterologous expression of the tenellin iterative polyketide synthase nonribosomal peptide synthetase requires coexpression with an enoyl reductase. Chembiochem 9:585–594

Hendrickson L, Davis CR, et al (1999) Lovastatin biosynthesis in Aspergillus terreus: characterization of blocked mutants, enzyme activities and a multifunctional polyketide synthase gene. Chem Biol 6:429–439

Hitchman TS, Schmidt EW, et al (2001) Hexanoate synthase, a specialized type I fatty acid synthase in aflatoxin B1 biosynthesis. Bioorg Chem 29:293–307

Howard RJ, Ferrari MA, et al (1991) Penetration of hard substrates by a fungus employing enormous turgor pressures. Proc Natl Acad Sci USA 88:11281–11284

Hutchinson RC, Kennedy J, et al (2000) Aspects of the biosynthesis of non-aromatic fungal polyketides by iterative polyketide synthases. Antonie Van Leeuwenhoek 78:287–295

Jahn B, Koch A, et al (1997) Isolation and characterization of a pigmentless-conidium mutant of Aspergillus fumigatus with altered conidial surface and reduced virulence. Infect Immun 65:5110–5117

Jones CA, Sidebottom PJ, et al (1992) The squalestatins, novel inhibitors of squalene synthase produced by a species of Phoma. III. Biosynthesis. J Antibiot 45:1492–1498

Kasahara K, Fujii I, et al (2006) Expression of Alternaria solani PKSF generates a set of complex reduced-type polyketides with different carbon-lengths and cyclization. Chembiochem 7:920–924

Kealey JT, Liu L, et al (1998) Production of a poyketide natural product in nonpolyketide-producing procaryotic and eukaryotic hosts. Proc Natl Acad Sci USA 95:505–509

Keatinge-Clay AT, Maltby DA, et al (2004) An antibiotic factory caught in action. Nat Struct Mol Biol 11:888–893

Kempken F, Kuck U (2000) Tagging of a nitrogen pathway-specific regulator gene in Tolypocladium inflatum by the transposon Restless. Mol Gen Genet 263:302–308

Kempken F, Jacobsen S, et al (1998) Distribution of the fungal transposon Restless: full-length and truncated copies in closely related strains. Fungal Genet Biol 25:110–118

Kennedy J, Auclair K, et al (1999) Modulation of polyketide synthase activity by accessory proteins during lovastatin biosynthesis. Science 284:1368–1372

Keszenman-Pereyra D, Lawrence S, et al (2003) The npgA/cfwA gene encodes a putative 4′-phosphopantetheinyl transferase which is essential for penicillin biosynthesis in Aspergillus nidulans. Curr Genet 43:186–190

Khaldi N, Collemare J, et al (2008) Evidence for horizontal transfer of a secondary metabolite gene cluster between fungi. Genome Biol 9:R18

Komagata D, Fujita S, et al (1996) Novel neuritogenic activities of pseurotin A and penicillic acid. J Antibiot (Tokyo) 49:958–959

Kono Y, Daly JM (1979) Characterization of the host-specific pathotoxin produced by *Helminthosporium maydis*, race T, affecting corn with Texas male sterile cytoplasm. Bioorg Chem 8:391–397

Kono Y, Takeuchi S, et al (1980) Studies on the host-specific pathotoxins produced by *Helminthosporium maydis* race T. Agric Biol Chem 44:2613–2622

Kono Y, Takeuchi S, et al (1981) Studies on the host-specific pathotoxins produced in minor amounts by *Helminthosporium maydis*, race T. Bioorg Chem 10:206–218

Kono Y, Danko SJ, et al (1983) Structure of the host-specific pathotoxins produced by *Phyllosticta maydis*. Tetrahedron Lett 24:3803–3806

Kroken S, Glass LN, et al (2003) Phylogenetic analysis of type I polyketide synthase genes in pathogenic and saprobic ascomycetes. Proc Natl Acad Sci USA 100:15670–15675

Langfelder K, Streibel M, et al (2003) Biosynthesis of fungal melanins and their importance for human pathogenic fungi. Fungal Genet Biol 38:143–158

Linnemannstons P, Schulte J, et al (2002) The polyketide synthase gene pks4 from *Gibberella fujikuroi* encodes a key enzyme in the biosynthesis of the red pigment bikaverin. Fungal Genet Biol 37:134–148

Lodish H, Berk A, et al (2001) Molekulare Zellbiologie. Spektrum, Vienna

Lu P, Zhang A, et al (2005) A gene (pks2) encoding a putative 6-methylsalicylic acid synthase from *Glarea lozoyensis*. Mol Genet Genomics 273:207–216

Ma SM, Zhan J, et al (2007) Enzymatic synthesis of aromatic polyketides using PKS4 from *Gibberella fujikuroi*. J Am Chem Soc 129:10642–10643

Ma SM, Zhan J, et al (2008) Redirecting the cyclization steps of fungal polyketide synthase. J Am Chem Soc 130:38–39

Maiya S, Grundmann A, et al (2007) Identification of a hybrid PKS/NRPS required for pseurotin A biosynthesis in the human pathogen *Aspergillus fumigatus*. Chembiochem 8:1736–1743

Miller DA, Lunsong L, et al (2002) Yersiniabactin synthetase: A four-protein assembly line producing the nonribosomal peptide/polyketide hybrid siderophore of *Yersinia pestis*. Chem Biol 9:333–344

Minto RE, Townsend CA (1997) Enzymology and molecular biology of aflatoxin biosynthesis. Chem Rev 97:2537–2556

Moore RN, Gigam G, et al (1985) Biosynthesis of the hypercholesterolemic agent mevinolin by *Aspergillus terreus*: determination of the origin of carbon, hydrogen and oxygen atoms by carbon-13 NMR and mass spectrometry. J Am Chem Soc 107:3694–3701

Moriguchi T, Ebizuka Y, et al (2006) Analysis of subunit interactions in the iterative type I polyketide synthase ATX from *Aspergillus terreus*. Chembiochem 7:1869–1874

Moss MO (1998) Recent studies of mycotoxins. Symp Ser Soc Appl Microbiol 27:62S–76S

Munkvold GP, Desjardins AE (1997) Fumonisins in maize: can we reduce their occurrence?. Plant Dis 81:556–565

Nakayashiki H, Hanada S, et al (2005) RNA silencing as a tool for exploring gene function in ascomycete fungi. Fungal Genet Biol 42:275–283

Nelson PE, Desjardins AE, et al (1993) Fumonisins, mycotoxins produced by fusarium species: biology, chemistry, and significance. Annu Rev Phytopathol 31:233–252

Nicholson TP, Rudd BA, et al (2001) Design and utility of oligonucleotide gene probes for fungal polyketide synthases. Chem Biol 8:157–178

Oikawa H, Tokiwano T (2004) Enzymatic catalysis of the Diels–Alder reaction in the biosynthesis of natural products. Nat Prod Rep 21:321–352

Oikawa H, Murakami Y, et al (1991) New plausible precursors of chaetoglobosin A accumulated by treatment of *Chaetomium subaffine* with cytochrome P-450 inhibitors. Tetrahedron Lett 32:4533–4536

Perpetua NS, Kubo Y, et al (1996) Cloning and characterization of a melanin biosynthetic THR1 reductase gene essential for appressorial penetration of *Colletotrichum lagenarium*. Mol Plant Microbe Interact 9:323–329

Pfeifer BA, Khosla C (2001) Biosynthesis of polyketides in heterologous hosts. Microbiol Mol Biol Rev 65:106–118

Proctor RH, Desjardins AE, et al (1999) A polyketide synthase gene required for biosynthesis of fumonisin mycotoxins in *Gibberella fujikuroi* mating population A. Fungal Genet Biol 27:100–112

Reeves CD (2003) The enzymology of combinatorial biosynthesis. Crit Rev Biotechnol 23:95–147

Schmitt I, Martin MP, et al (2005) Diversity of non-reducing polyketide synthase genes in the Pertusariales (lichenized Ascomycota): a phylogenetic perspective. Phytochemistry 66:1241–1253

Schuemann J, Hertweck C (2006) Advances in cloning, functional analysis and heterologous expression of fungal polyketide synthase genes. J Biotechnol 124:690–703

Schuemann J, Hertweck C (2007) Molecular basis of cytochalasan biosynthesis in fungi: gene cluster analysis and evidence for the involvement of a PKS-NRPS hybrid synthase by RNA silencing.

Scott PM, Harwig J, et al (1975) Cytochalasins A and B from strains of *Phoma exigua* var. *exigua* and formation of cytochalasin B in potato gangrene. J Gen Microbiol 87:177–180

Seshime Y, Juvvadi PR, et al (2005) Discovery of a novel superfamily of type III polyketide synthases in Aspergillus oryzae. Biochem Biophys Res Commun 331:253–260

Shimizu T, Kinoshita H, et al (2005) Polyketide synthase gene responsible for citrinin biosynthesis in *Monascus purpureus*. Appl Environ Microbiol 71:3453–3457

Sims JW, Fillmore JP, et al (2005) Equisetin biosynthesis in *Fusarium heterosporum*. Chem Commun (Camb) 2005:186–188

Song Z, Cox RJ, et al (2004) Fusarin C biosynthesis in *Fusarium moniliforme* and *Fusarium venenatum*. Chembiochem 5:1196–1203

Staunton J, Weissman KJ (2001) Polyketide biosynthesis: a millenium review. Nat Prod Rep 18:380–416

Steglich W, Fugmann B, et al (1997). Römpp Lexikon Naturstoffe, Thieme

Sweeney MJ, Dobson AD (1999) Molecular biology of mycotoxin biosynthesis. FEMS Microbiol Lett 175:149–163

Takano Y, Kubo Y, et al (1995) Structural analysis of PKS1, a polyketide synthase gene involved in melanin biosynthesis in *Colletotrichum lagenarium*. Mol Gen Genet 249:162–167

Thierry D, Vaucheret H (1996) Sequence homology requirements for transcriptional silencing of 35S transgenes and post-transcriptional silencing of nitrite reductase (trans) genes by the tobacco 271 locus. Plant Mol Biol 32:1075–1083

Trail F, Mahanti N, et al (1995) Molecular biology of aflatoxin biosynthesis. Microbiology 141:755–765

Turgeon BG, Lu S-W (2000) Evolution of host specific virulence in *Cochliobolus heterostrophus*. Kronstad JW (ed) Fungal pathology. Kluwer, Dordrecht, pp 93–126

Walsh CT (2002) Combinatorial biosynthesis of antibiotics: challenges and opportunities. Chembiochem 3:125–134

Watanabe A, Ebizuka Y (2002) A novel hexaketide naphthalene synthesized by a chimeric polyketide synthase composed of fungal pentaketide and heptaketide synthases. Tetrahedron Lett 43:843–846

Watanabe A, Ebizuka Y (2004) Unprecedented mechanism of chain length determination in fungal aromatic polyketide synthases. Chem Biol 11:1101–1106

Watanabe A, Ono Y, et al (1998) Product identification of polyketide synthase coded by *Aspergillus nidulans* wA gene. Tetrahedron Lett 39:7733–7736

Watanabe A, Fujii I, et al (1999) Re-identification of *Aspergillus nidulans* wA gene to code for a polyketide synthase of naphthopyrone. Tetrahedron Lett 40:91–94

Watanabe A, Fujii I, et al (2000) *Aspergillus fumigatus* alb1 encodes naphthopyrone synthase when expressed in *Aspergillus oryzae*. FEMS Microbiol Lett 192:39–44

Watanabe CM, Townsend CA (2002) Initial characterization of a type I fatty acid synthase and polyketide synthase multienzyme complex NorS in the biosynthesis of aflatoxin B(1). Chem Biol 9:981–988

Weber T, Welzel K, et al (2003) Exploiting the genetic potential of polyketide producing streptomycetes. J Biotechnol 106:221–232

Witkowski A, Joshi AK, et al (2004) Characterization of the beta-carbon processing reactions of the mammalian cytosolic fatty acid synthase: role of the central core. Biochemistry 43:10458–10466

Witter DJ, Vederas JC (1996) Putative Diels–Alder-catalyzed cyclization during the biosynthesis of lovastatin. J Org Chem 61:2613–2623

Wright F, Bibb MJ (1992) Codon usage in the G+C-rich *Streptomyces genome*. Gene 113:55–65

Xie X, Watanabe K, et al (2006) Biosynthesis of lovastatin analogs with a broadly specific acyltransferase. Chem Biol 13:1161–1169

Yalpani N, Altier DJ, et al (2001) Production of 6-methylsalicylic acid by expression of a fungal polyketide synthase activates disease resistance in tobacco. Plant Cell 13:1401–1409

Yang G, Rose MS, et al (1996) A polyketide synthase is required for fungal virulence and production of the polyketide T-toxin. Plant Cell 8:2139–2150

Yoder OC (1980) Toxins in pathogenesis. Annu Rev Phytopathol 18:103–129

Yoder OC, Macko V, et al (1997) *Cochliobolus* spp. and their host-specific toxins. In: Carroll G, Tudzynski P (eds) The Mycota, vol V. Springer, Heidelberg, pp 145–166

Yu J, Leonard TJ (1995) Sterigmatocystin biosynthesis in *Aspergillus nidulans* requires a novel type I polyketide synthase. J Bacteriol 177:4792–800

Yu J, Chang PK, et al (1995) Comparative mapping of aflatoxin pathway gene clusters in *Aspergillus parasiticus* and *Aspergillus flavus*. Appl Environ Microbiol 61:2365–2371

Zawada RJ, Khosla C (1999) Heterologous expression, purification, reconstitution and kinetic analysis of an extended type II polyketide synthase. Chem Biol 6:607–615

Zhang A, Lu P, et al (2003) Efficient disruption of a polyketide synthase gene (pks1) required for melanin synthesis through Agrobacterium-mediated transformation of *Glarea lozoyensis*. Mol Genet Genomics 268:645–655

Zhang W, Watanabe K, et al (2007) Investigation of early tailoring reactions in the oxytetracycline biosynthetic pathway. J Biol Chem 282:25717–25725

17 Physiological and Molecular Aspects of Ochratoxin A Biosynthesis

ROLF GEISEN[1], MARKUS SCHMIDT-HEYDT[1]

CONTENTS

I. Introduction

Ochratoxin A was first described as a toxic metabolite produced by *Aspergillus ochraceus*, giving this secondary metabolite its name (van der Merwe et al. 1968). Secondary metabolites of filamentous fungi can be more or less artificially divided in antibiotics (toxic for microorganisms), phytotoxins (toxic for plants) and mycotoxins (toxic for humans and animals). About 300–400 secondary metabolites, which are categorized as mycotoxins, are known; however most of them do not have an impact on the health of human and animal beings, simply because they do not occur in substantial amounts in foods and feeds. The most important mycotoxins from that respect are the aflatoxins, which were the first mycotoxins discovered, because of the death of 100 000 turkeys in Great Britain, described by Asao et al. (1963). Other important mycotoxins are trichothecenes produced by several

plant-pathogenic *Fusarium* species, fumonisins also produced by *Fusarium* species, patulin produced mainly by *Penicillium* species and ochratoxin A produced by *Aspergillus* and *Penicillium* species. Especially ochratoxin A attracted much attention in recent time which led to a considerable progress in the knowledge about this secondary metabolite. Ochratoxin is mainly a nephrotoxic mycotoxin but has also hepatotoxic, immunotoxic and teratogenic activities (Pfohl-Leszkowicz and Manderville 2007). Ochratoxin A is considered to be a human carcinogen of the 2B class (Petzinger and Weidenbach 2002). It is discussed that ochratoxin A is implicated in an endemic disease in certain areas of the Balkan, the so-called Balkan endemic nephropathy. The symptoms which develop during that disease very much resemble the symptoms of ochratoxin A toxicity (Pfohl-Leszkowicz et al. 2002). For that reason statutory limits were recently set by the European Union. The biological background of aflatoxin, trichothecene and fumonisin biosynthesis is well elucidated; however despite of the importance of ochratoxin A, surprisingly not that much is known about this secondary metabolite. This chapter tries to summarize the current knowledge about the biology of ochratoxin biosynthesis.

II. Ochratoxin A-Producing Fungal Species

Ochratoxin A was first discovered from a culture of *A. ochraceus* (van der Merwe et al. 1965). These authors analyzed the toxicity of five strains of *A. ochraceus*. This species is a saprophyte which often occurs on decaying plant material. However *A. ochraceus* can also grow on stored wheat, as long as the moisture content is above 16%. It was also reported as a contaminant of "katsuobushi" a

[1]Max Rubner Institute, Federal Research Institute of Nutrition and Food, Haid-und-Neu-Strasse 9, 76131 Karlsruhe, Germany; e-mail: Rolf.Geisen@MRI.Bund.de, Markus.Schmidt-Heydt@MRI.Bund.de

Physiology and Genetics, 1st Edition
The Mycota XV
T. Anke and D. Weber (Eds.)
© Springer-Verlag Berlin Heidelberg 2009

traditional Asian fermented fish. Of the analyzed five strains, three caused the death of test animals after feeding; and an LD_{50} of 0.5 µg/g was observed. This work described the formula of ochratoxin for the first time. *A. ochraceus* has long been regarded as a major ochratoxin A-producing organism on plant commodities like coffee beans, cocoa, grapes and spices. This species belongs to *Aspergillus*, Section Circumdati. However, after detailed examination of the taxonomic relationships, new ochratoxin A-producing species within this section were recently described (Frisvad et al. 2004). According to these authors eight species of this section consistently produce large amounts of ochratoxin A, in particular *A. cretensis, A. flocculosus, A. pseudoelegans, A. roseoglobulosus, A. westerdijkiae, A. steynii, A sulphorosus* and *Neopetromyces muricatus*. Two other species produce moderate amounts of ochratoxin A, but less consistently, e.g. *A. ochraceus*, the original ochratoxin-producing species, and *A. sclerotiorum*. Four other species produce only very inconsistently ochratoxin A to a very low amount, in particular *A. melleus, A. ostianus, A. petrakii* and *A. persii*. In the *Aspergillus* Nigri section two very important ochratoxin-producing species are well known: (i) *A. carbonarius* is a strong and consistent ochratoxin A producer and (ii) *A. niger* produces inconsistently lower amounts of ochratoxin. *A. niger* is a GRAS organism (generally *recognized as* safe) which is used extensively for biotechnological processes like the production of citric acid, hydrolytic enzymes like pectinases, proteases and amylases. However in 1994 it was shown for the first time that this species is also able to produce ochratoxin A (Abarca et al. 1994). These authors analyzed 19 strains of *A. niger* isolated from feed and identified two strains which were able to produce ochratoxin A. Both strains produced low amounts of ochratoxin A, in particular 0.21 µg/g and 0.36 µg/g. Generally only about 6% of the naturally isolated strains of *A. niger* are able to produce ochratoxin A (Schuster et al. 2002). These facts must be taken into account when strains of *A. niger* are selected for biotechnological purposes. However, because of the low incidence of ochratoxinogenic strains within this species, it should not be a problem to select suitable technological strains. The situation with *A. carbonarius* is completely different. According to many reports in the literature >90% (up to 97%) of the

A. carbonarius strains isolated from the natural environment are able to produce ochratoxin A (Abarca et al. 2003), which is therefore a very consistent feature in this species. Recently Samson et al. (2004) updated the taxonomic situation of the ochratoxin A-producing species within Section Nigri and described new ochratoxigenic species, two of them (*A. lacticoffeatus, A. sclerotioniger*) produces high amounts of ochratoxin A. According to these authors, four of the 15 accepted species within this section are ochratoxin producers, namely *A. carbonarius, A. niger, A. lacticoffeatus* and *A. sclerotioniger*. The taxonomy with section Nigri is not yet completely solved (Abarca et al. 2004). For this reason occasionally members of this group, like *A. japonicus* and *A. foetidus* (which are usually not regarded as ochratoxigenic species), have been reported to produce ochratoxin A. Recently two new members of this section were described, *A. uvarum* (Perrone et al. 2008) and *A. ibericus* (Serra et al. 2006), but neither produces ochratoxin A. Accensi et al. (1999) described an RFLP of the ITS 1 region which can distinguish between potential ochratoxin A-producing *A. niger* strains (N-type pattern) and the highly related, but non-ochratoxigenic *A. tubingensis* strains (T-type pattern). Other described ochratoxin A-producing *Aspergillus* species are *A. alliaceus, A. sulphurus, A. albertensis, A. auricomus* and *A. wendtii* (Varga et al. 1996).

In contrast only two species of *Penicillium* which produce ochratoxin A are known: *P. verrucosum* and *P. nordicum*. Before the clear clarification of the taxonomic status of ochratoxin A-producing *Penicillium* species by Pitt (1987), ochratoxin A-producing Penicillia were assigned to the species *P. viridicatum*. Ciegler et al. (1973) first described three subgroups within *P. viridicatum*, based on the production of ochratoxin A and other metabolites; two of these subgroups which were able to produce ochratoxin A now form *P. verrucosum* (Pitt 1987) or *P. nordicum*. *P. verrucosum* has long been regarded as the sole *Penicillium* species able to produce ochratoxin A. However it was recently demonstrated by chemotaxonomical and molecular methods that *P. nordicum* is clearly a separate species with characteristics distinct from *P. verrucosum* (Larsen et al. 2001; Castella et al. 2002). *P. verrucosum* and *P. nordicum* are morphologically very similar. That is the reason why most *P. nordicum* strains were formerly identified as *P. verrucosum* by morphological means. However nowadays rapid and objective molecular diagnostic methods exist to differentiate between the two species (Bogs et al. 2006). Also both species exhibit a different chemotype after analysis of the produced secondary metabolites. *P. nordicum* for example is able to produce anacines, sclerotigenin and ochratoxin A, whereas *P. verrucosum* synthesizes ochratoxin A, verrucines and citrinin. In addition both species differ considerably in their genome organization, because they showed consistent clearly distinguishable RAPD and AFLP profiles (Castella et al. 2002). Generally most strains of *P. nordicum* produce high amounts of ochratoxin, whereas the incidence of ochratoxin A production in *P. verrucosum* is much less pronounced. From a natural population of *P. verrucosum* only 66% were able to produce ochratoxin A (Frisvad et al. 2005). Moreover most strains produce only moderate amounts of the toxin. However, because of their frequent occurrence in certain plant commodities, *P. verrucosum* plays an important role as an ochratoxin A-producing contaminant.

The Aspergilli and Penicillia able to produce ochratoxin A are generally not plant pathogens, but rather saprophytes or maximally opportunistic pathogens which can colonize wounds or plant material already disturbed by more serious pathogens or insects. This is especially true for the Aspergilli, which can for example colonize bunches of grapes and produce ochratoxin A in that environment (Battilani et al. 2003; Varga and Kozakiewicz 2006). The ochratoxigenic Penicillia in contrast are not regarded as pathogens, but as so-called storage fungi, colonizing the product only after harvest. That is especially the case for cereals. This plant product is frequently colonized by *P. verrucosum*. In fact *P. verrucosum* is the only confirmed ochratoxin A-producing species in this habitat. Occasionally found Aspergilli do not contribute to ochratoxin A biosynthesis in this environment (Frisvad et al. 2005). Generally the geographical habitats differ between ochratoxin A-producing Aspergilli and Penicillia. Aspergilli are more adapted to warmer temperatures. Some of them have their growth optimum between 30°C and 37°C. For this reason they mainly occur on plants growing in regions with tropical and subtropical climate like coffee,

Table 17.1. Occurrence of ochratoxin A-producing fungi

Habitat/product	Species
Cereals	*Penicillium verrucosum*
Salt-rich, dry-cured meat, cheese	*P. nordicum*
Grapes, wine	*Aspergillus niger, A. carbonarius, A. ochraceus*
Coffee, spices	*A. ochraceus, A. carbonarius, A. niger, A. westerdijkiae, A. steynii*

grapes spices or cocoa. Penicillia grow better at moderate temperatures, e.g. between 20°C and 25°C. Therefore they can be isolated mainly from sources of the northern hemisphere. Especial problems with this fungus have the north European countries like Denmark, Sweden or the United Kingdom (Frisvad 2005). Until now only a few isolates have been found in locations with higher temperature, like Italy, Spain, Portugal, Nicaragua and Bulgaria (Frisvad et al. 2005). *P. nordicum* is a special case because it rarely occurs on plant material. It is rather adapted to man made environment. That is the production plants of salt-rich, dry-cured foods like cheeses and meats. It is a moderate xerophilic species and likes a certain concentration of salt. It even shows increasing growth rates at concentrations of more than 40g/L NaCl per liter growth medium. Beside the temperature preferences also the substrate directs the species which can grow. Table 17.1 shows the typical occurrence of ochratoxin A-producing Aspergilli and Penicillia.

III. Structure and Biosynthesis of Ochratoxin A

Ochratoxin is a composite mycotoxin consisting of a polyketide part and the amino acid phenylalanine. Chemically ochratoxin A is the pentaketide 3,4-dihydro-3-R-methyl-isocoumarin linked via its 12-carboxyl group by an peptide bond to the amino acid phenylalanine (Steyn 1993) The full nomenclatural name is L-phenylalanine-N-[5-chloro-3,4-dihydro-8-hydroxy-3-methyl-1-oxo-1H2-benzopyran-7-yl]carbonyl-*R*-isocoumarin (Chemical Abstracts, No. 303-47-9). Ochratoxin A is chlorinated at position 5 of the polyketide. Various other forms have been identified, including ochratoxin B, which carries a

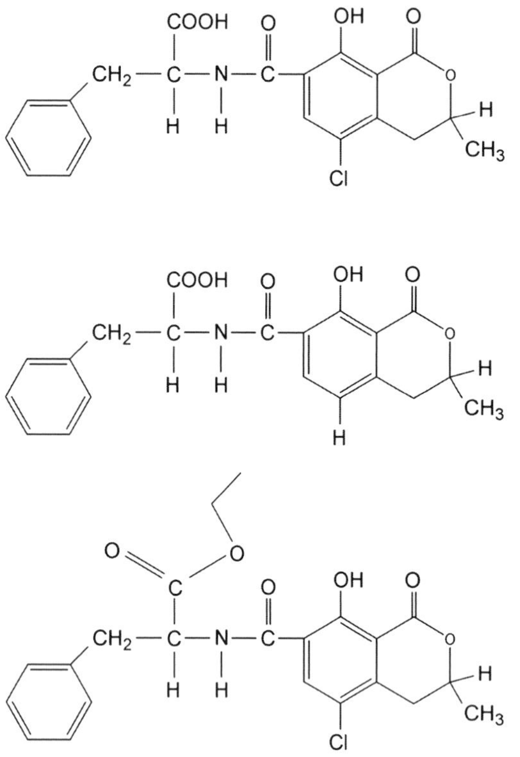

Fig. 17.1. Structural formulas of ochratoxin A, B and C (from *top* to *bottom*)

hydrogen atom instead of chlorine at position 5. Ochratoxin C is the ethylester of ochratoxin A (Fig. 17.1). Ochratoxins α and β respectively also differ in the chlorination at position 5, however in contrast to ochratoxin A and B, they are not linked to phenylalanine.

Ochratoxin A is the most abundant end-product synthesized by strains isolated from nature, followed by ochratoxin B. The α and β forms are obviously precursors during the biosynthesis of ochratoxin A (Harris and Mantle 2001); however they are not produced as end-products. Until today the biosynthesis of ochratoxin A has not been completely described in detail. Three major steps must be involved in ochratoxin A biosynthesis: i.e. (i) biosynthesis of the polyketide, (ii) peptide formation between the polyketide and the amino acid phenylalanine, which is derived from the shikimic acid pathway and (iii) chlorination to ochratoxin A. Today the order of this reactions is still not completely clear. Huff and Hamilton (1979) proposed a pathway based on feeding studies with labeled precursors. The dihydroisocoumarin polyketide moiety is synthesized by a single

multifunctional enzyme, the ochratoxin A polyketide synthase. Acetate or malonate precursors are added in a stepwise head to tail addition until the pentaketide is formed. Intermediates of this reaction have not been detected. Huff and Hamilton (1979) proposed mellein as an end-product of the polyketide pathway.

Harris and Mantle (2001) identified the transitory production of mellein during the biosynthesis of ochratoxin A in liquid medium, but not during biosynthesis of ochratoxin A on solid medium. For this reason the authors were not sure whether mellein plays an important role during the biosynthesis of ochratoxin A. According to the scheme of Huff and Hamilton (1970) 7-carboxymellein is further methylated via S-adenosyl-methionine at position 7 to synthesize 7-carboxymellein or ochratoxin β. This reaction might be carried out through the action of a methyltransferase domain from the polyketide synthase as it is the case with the related mycotoxin citrinin (Shimizu et al. 2005). In fact the polyketide moiety of citrinin is very similar to that of ochratoxin β. Both polyketides differ only in the presence (citrinin) or absence (ochratoxin A) of the methyl groups. As a next step, Huff and Hamilton (1979) suggested a chloroperoxidase reaction for the addition of the chlorine to position 5 of the isocoumarin ring system. The chlorine is part of the structure of ochratoxin A and ochratoxin C. According to Harris and Mantle (2001), either ochratoxin β (7-carboxymellein) or ochratoxin B can be chlorinated to give rise to ochratoxin A. These authors suggest a major pathway from ochratoxin β to ochratoxin α to ochratoxin A with a possible branch from β to B. This would explain the frequent occurrence of ochratoxin B which can be found in fungal cultures. However ochratoxin B can also be generated from ochratoxin A non-specifically. No clear evidence was found if this conversion is true to a biotransformation process. (Harris and Mantle 2001). However in this case ochratoxin α is not needed for the biosynthesis, but it can be identified naturally. According to the results of these authors mellein per se seems not to be involved in the biosynthesis pathway. Wei et al. (1971) showed that chlorine-36 is most effectively incorporated into ochratoxin A, when added to the cultures at day two or three. According to this experiment only ochratoxin A is labelled by the precursor. In a similar experiment Searcy et al. (1969) showed that radiolabeled phenylalanine-1-^{14}C was readily incorporated into the end-product ochratoxin A and also sodium acetate-2-^{14}C. Both are the expected main precursors of the pathway.

In the scheme of Huff and Hamilton (1979), the next step is the joining reaction between the amino acid phenylalanine and the polyketide dihydroisocoumarin. Huff and Hamilton (1979) described an ochratoxin A synthetase as the enzymatic function performing these step. This ochratoxin synthetases is probably a non-ribosomal peptide synthetase which ligates the amino acid to the polyketide and forms the amid bond between the amino group of phenylalanine and the carboxy group of 7-carboxymellein. However the ethylester of phenylalanine seems to be involved in this reaction. The esterification of the carboxy group of phenylalanine may be important for protection purposes (Harris and Mantle 2001). This reaction would result in ochratoxin C in which the carboxy group is still esterified. In a later step this ester can be hydrolyzed to form ochratoxin A. Ochratoxin C has been isolated from *A. ochraceus* cultures.

Taken the published facts together, the pentaketide ochratoxin β and its chlorinated derivative ochratoxin α are methylated via the polyketide synthase reaction. Feeding experiments have shown that ochratoxin α can be converted into ochratoxin A, but ochratoxin β can be converted to ochratoxin A and B. That means that the chlorinating step is intermediate and prior to the ligation of the polyketide to the amino acid phenylalanine. This is described for *A. ochraceus*, however new molecular evidence obtained in the case of *Penicillium* suggests a slightly different order of reactions, with the chlorination as the last reaction before excretion. This is described below (Fig. 17.2).

IV. Physiological Conditions for Production and Occurrence of Ochratoxin A

Like all mycotoxins, ochratoxin A is only produced under certain growth conditions and is influenced by growth phase, substrate, temperature, water activity and pH. Moreover the production profiles of mycotoxins are species-specific and may even vary from strain to strain. Despite these variability, data about the influence of environmental parameters on the biosynthesis of ochratoxin A are important for estimating the activation of mycotoxin biosynthesis under a given situation. Because of this aspect, a wealth of literature data is available describing ochratoxin A biosynthesis under a given set of conditions.

One of the most important influences is the growth substrate. Skrinjar and Dimic (1992) showed that the nutrient rich YES medium was an adequate medium for the biosynthesis of ochratoxin A for *A. ochraceus, A. sclerotiorum,*

A. sulphurus and *P. verrucosum.* More natural model media (e.g. rice meal, oat meal, corn grits) supported ochratoxin A biosynthesis to a much lesser extent. The same could be shown for *P. nordicum.* Also with this species YES was the most supportive medium for ochratoxin A biosynthesis (Geisen 2004). Two other media with strong regulatory influence on ochratoxin A biosynthesis gene expression have been developed (Geisen et al. 2004). Both are minimal media with the same composition. Only differences in the carbon and nitrogen sources exist. One medium (MMp) contains NH_4^+ as nitrogen source and glycerol as carbon source and is fully supportive for ochratoxin A biosynthesis, whereas the other medium (MMr) contains the combination NO_3^-/ glucose and nearly completely represses the toxin biosynthesis, albeit growth is still possible. Expression analysis of ochratoxin A biosynthesis genes revealed that, after growth in the supportive medium with glycerol and NH_4^+, ochratoxin transcript levels are several-fold higher compared to the non-supportive medium (Geisen et al. 2006). These results indicate that glucose represses and NH_4^+ activates ochratoxin A biosynthesis in *Penicillium*. The effect of glucose on the expression of secondary metabolite genes is well known. For example the biosynthesis of the antibiotic actinomycin is tightly regulated by catabolite repression (Gallo and Katz 1972; Yu and Keller 2005). Catabolite repression is mediated via the *cre*A zinc finger DNA binding factor (Dowzer and Kelly 1991) which binds to a consensus sequence located in front of regulated promoters, thereby down-regulating transcription of the respective gene. According to the results described above, this seems to be the case for the biosynthesis of ochratoxin A. In media containing glucose as sole carbon source, ochratoxin A biosynthesis is strongly inhibited. Binding sites for the CREA protein have been described in *Penicillium* (Diez et al. 2005).

In contrast to the situation of ochratoxin A, the biosynthesis of aflatoxin seems not to be regulated by carbon catabolite repression. According to Bhatnagar et al. (2006) aflatoxin biosynthesis by *A. flavus* and *A. parasiticus* is activated by simple carbon sources like glucose and sucrose, but not by peptone or complex carbohydrates like starch. Interestingly in case of aflatoxin a hexose utilization gene cluster is located just adjacent to the aflatoxin biosynthesis gene cluster, indicating a possible relationship. Generally the regulation of aflatoxin biosynthesis by simple carbon sources is not well understood. In contrast

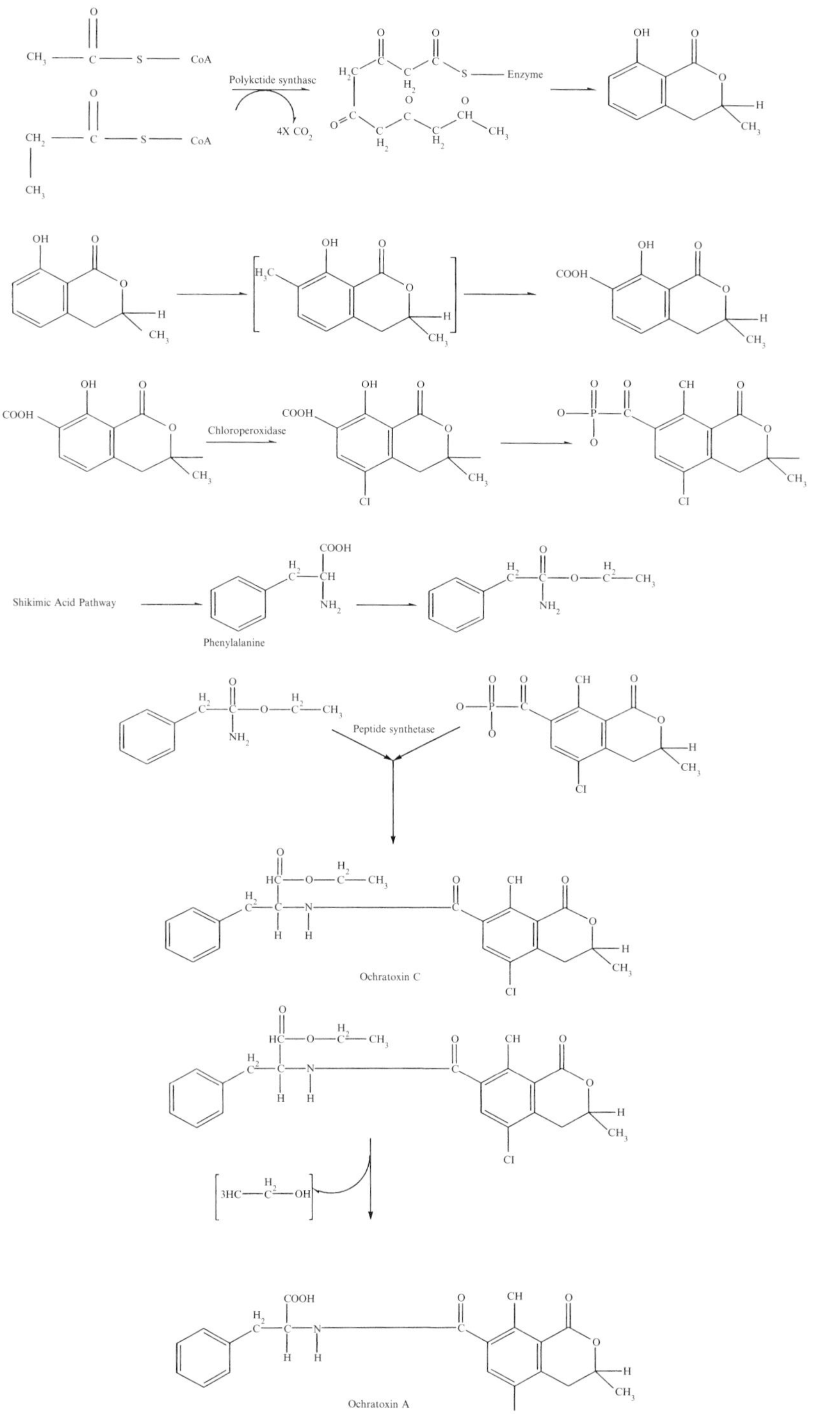

Fig. 17.2. Proposed biosynthesis pathway according to the literature data mentioned above. The molecular data of *Penicillium* imply that the chlorine atom is incorporated during the very last reaction, shortly before the excretion of the molecule into the environment

to the regulation of many other secondary metabolites, aflatoxin biosynthesis is activated by these simple carbohydrates. Most of the aflatoxin biosynthesis genes studied so far do not have CREA binding sites in front of their promoters, which explains the independence from carbon catabolite repression. However CREA may play a role in the regulation of the expression of an antisense transcript of the regulatory factor *afl*R (Bhatnagar et al. 2003).

Ehrlich et al. (2002) reported a few putative CREA sites identified in the intergenic region between the *nor-1*–*pks*A genes of *A. nomius*, *A. bombycis*, *A. pseudotamarii* and *A. flavus* strain S$_{BG}$. These are all less important and "atypical" aflatoxin-producing species. According to present evidence, CREA regulation seems not to play an important role during aflatoxin biosynthesis, which is in contrast to the results found for ochratoxin A biosynthesis in *Penicillium*, in which CREA seems to play a role. Tudzynski et al. (2000) characterized the creA genes from two plant-pathogenic fungi, in particular from *Gibberella fujikuroi* and *Botrytis cinerea*, the former producing the secondary metabolite gibberellin (Tudzynski et al. 2000).

Also the nitrogen source plays a prominent role in regulating mycotoxin biosynthesis. In the case of *Penicillium*, NH_4^+ increases and NO_3^- decreases the biosynthesis of ochratoxin A. This is also the case for aflatoxin biosynthesis. Ehrlich and Cotty (2002) described that a strain of *A. flavus* S$_{BG}$ produced 20-fold lower amounts of aflatoxin in buffered nitrate medium than in ammonium salts medium. This effect was not true for a changed expression of the regulatory gene *aflR*, but the expression of the coregulatory gene *aflJ* was reduced by a factor of two. The promoter region of the *aflJ* genes carries some potential AREA binding sites. AREA is a global regulatory protein which can regulate expression of certain genes in response to the presence (repressed) or absence (induced) of NH_4^+ (Marzluf 1997).

According to Price et al. (2005), organic nitrogen sources induce, while inorganic sources retard aflatoxin biosynthesis. According to these authors, nitrate repression may occur due to an altered redox state in the cell. By microarray analysis during aflatoxin production conditions, a potential regulon was identified, which was upregulated under growth on nitrate medium. One gene of this regulon may be involved in heme biosynthesis. Heme is a structural component of nitrate reductase, which is the first enzyme in the catabolic pathway leading from nitrate to ammonia. This observation might constitute a link to the nitrate repression of many mycotoxin biosynthesis pathways. Other genes of the regulon code for proteins involved in the electron transport system, which might play a role during reduction of nitrate to ammonia. Also, the genes in this regulon carry AREA binding sites in their promoters, indicating their coregulation. Bhatnagar et al. (2006) also described AREA binding sites in the promoter elements of *aflJ* and *aflR*. The recognition site has the GATA consensus motive. Certain strains of *A. flavus* which respond differentially to nitrate had differences in the number of GATA sites within the promoter region of *aflJ*. According to Bhatnagar et al. (2006), the direct cause of nitrogen regulation is still not clear, but it could be mediated by AREA.

Orvehed et al. (1988) described the inhibiting effect of $NaNO_3$ on alternariol biosynthesis by *Alternaria alternata*. The biosynthesis of the secondary metabolite gibberellin by *G. fujikuroi* is strongly suppressed by high amounts of nitrogen (e.g. ammonium, glutamate, glutamine, nitrate). For this fungus it has been shown that AREA controls expression of the gibberellin biosynthesis genes. The expression of the gibberellin biosynthesis genes in mutants of the *areA*-Gf gene is strongly reduced, indicating its involvement in regulation (Tudzynski 2005).

These examples pinpoint the importance of AREA regulation on the transcription of secondary metabolites including mycotoxins. Recently an *areA*/*nmr*C homologue was identified in *P. nordicum*, which is upregulated under NO_3^- and downregulated under NH_4^+ growth condition. Therefore all experimental evidence suggest that a similar mechanism is responsible for nitrate suppression of ochratoxin A biosynthesis in these fungi.

Another external factor which plays an important role on ochratoxin A biosynthesis is the pH. Arroyo et al. (2003) reported that the biosynthesis of ochratoxin A increases in *A. ochraceus* when the pH drops to a range of 6.0–4.5. O'Callaghan et al. (2006) also found at the phenotypical and molecular level that biosynthesis gene expression and ochratoxin A biosynthesis was higher at low pH values (down to pH 4.0) than at neutral or slightly alkaline conditions. This is in contrast to the situation in *Penicillium*. *P. nordicum* shows highest production at neutral to slightly alkaline conditions, e.g. between pH 6.0 and pH 8.0 (Geisen 2004), whereas production is reduced under acidic conditions (<pH 5.0). In case of aflatoxin biosynthesis also acidic conditions favor toxin biosynthesis (Bhatnagar et al. 2006). Aflatoxin biosynthesis by *A. flavus* is highest at a pH range between 3.4 and 5.5. Keller et al. (1997) found a five- to tenfold increase of aflatoxin biosynthesis in *A. parasiticus* and sterigmatocystin synthesis in *A. nidulans*, when cultures were grown at pH 4.0–5.0 versus pH 8.0. These authors also analyzed the expression of certain key genes of the biosynthesis pathways of these secondary metabolites and showed that *stc*U (a gene of the sterigmatocystin pathway) and *ver*-1 (a gene of the aflatoxin pathway) were upregulated under acidic conditions.

Interestingly the influence of pH on the biosynthesis of ochratoxin A is clearly detectable but much less pronounced than the influence of temperature and water activity. This might be true because of the presence of a pH controlled system which in turn regulates downstream genes, e.g. the pacC/palA, B, C, F, H, I system as elucidated in *A. nidulans*. The pacC gene codes for a transcription factor, which acts downstream of the six *pal* genes. The products of these genes are components of a signal transduction pathway. PALI and PALH are potential plasma membrane pH sensors. PACC can either activate or repress downstream genes (Arst and Penalva 2003). In a microarray expression analysis of genes upregulated under aflatoxin-producing conditions in *A. flavus*, Price et al. (2005) found genes with creA, areA and pacC transcription factor binding sites in their promoter regions, indicating the involvement of this mechanism in the biosynthesis of the mycotoxin aflatoxin.

Ehrlich et al. (2002) analyzed the expression of *pks*A an acid-expressed gene, involved in the biosynthesis of aflatoxin in *A. flavus*. Regulation of acid- or alkaline-regulated genes is a process involving the proteolytic degradation of the PACC protein, giving rise to the active form that supports the expression of alkaline-activated genes and represses the expression of acid-activated genes. Since the biosynthesis of aflatoxin is high under acidic conditions, but repressed under neutral and alkaline conditions, this mechanism seems to act negatively on genes of the pathway containing PACC binding sites. This was confirmed by the fact that a deletion of one or more of the PACC binding sites in the promotor region of *pks*A mimics acid conditions. Albeit no molecular evidence is available yet for ochratoxin A biosynthesis in Penicillia, just the opposite seems to be true for that fungal species. The ochratoxin A biosynthesis genes seems to be regulated positively by PACC under alkaline conditions, indicating that these genes are alkaline-activated. *Fusarium oxysporum*, another mycotoxin-producing fungus, also contains a characterized homologue of the pacC gene (Caracuel et al. 2003) which is involved in virulence regulation.

V. Molecular Biology of Ochratoxin A Biosynthesis

Despite the importance of ochratoxin A as a mycotoxin for human health, not much is known about the genetical background of ochratoxin A biosynthesis, especially in *Aspergillus*, compared to the other important mycotoxins like aflatoxin, trichothecenes or the fumonisins. In contrast the situation in *Penicillium* is somewhat further elucidated.

Generally the genes for the biosynthesis of mycotoxins are organized in clusters (Proctor et al. 2003; Yu et al. 2004; Brown et al. 2005). One or two regulatory genes (regulated by global regulatory factors as mentioned above) control the expression of the mycotoxin biosynthesis structural genes and activate these genes under permissive conditions (Fig. 17.3).

Recently a polyketide synthase responsible for the biosynthesis of ochratoxin A in *A. ochraceus* was partly characterized (O'Callaghan et al. 2003). These authors identified a DNA fragment of 1.4 kb which shared 28–35% identity to acyl transferase regions from fungal polyketide synthases. Expression analysis of this gene revealed that it is only activated under conditions which are supportive for ochratoxin A biosynthesis by *A. ochraceus*. The expression could also only be found at the early phases of ochratoxin A biosynthesis. Moreover a transformant in which the *pks* gene was inactivated by integration was no longer able to produce ochratoxin A, giving evidence for the involvement of this gene in ochratoxin A biosynthesis. The authors used an SSH-PCR approach (suppressive substractive hybridization PCR) to identify the *pks* gene, by using permissive cDNA as tester and restrictive cDNA as driver. Before identifying the correct *pks* gene, the authors cloned three other *pks* genes which were actively transcribed, but were not involved in ochratoxin biosynthesis. This indicates that *A. ochraceus* contains several polyketide synthases, as was also found in other fungi like for example *G. zeae* (Gaffor et al. 2005). The work by Atoui et al. (2006) clearly shows that the two ochratoxin A-producing species *A. ochraceus* and *A. carbonarius* carries several *pks* genes. These authors used several primer sequences specific for ketosynthase (KS) domains from *pks* genes and could identify nine different KS domain sequences in *A. ochraceus* and five in *A. carbonarius*. In a phylogenetic tree based on the similarity of the DNA sequences of the domains, the identified gene fragments were distributed over five different similarity clusters, indicating different functions of these enzymes. No hint for the involvement of any of these genes in ochratoxin A biosynthesis was given in this report. Based on one of these *pks* genes, a

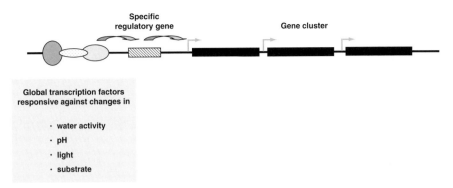

Fig. 17.3. Organization of a typical secondary metabolite gene cluster. The arrows indicate the hierarchical regulation of global versus specific transcription factors. Global factors are responsive to environmental changes

molecular detection system to specifically identify *A. carbonarius* on infected grapes was described (Atoui et al. 2007). Beside the polyketide synthase of *A. ochraceus*, whose expression is correlated with the production of ochratoxin A (O'Callaghan et al. 2003), the same authors identified two p450 monooxygenase genes, in particular p450-H11 and p450-B03 which had a very similar expression profile compared to the polyketide synthase gene (O'Callaghan et al. 2006). For this reason the authors conclude that both gene products might be involved in ochratoxin A biosynthesis. These three genes have been identified using an SSH approach, with which 152 genes, upregulated under ochratoxin A-producing conditions have been found. In a later analysis the same authors compared the sequences of the ochratoxin A *pks* gene of *A. ochraceus*, with that of other known *pks* genes (O'Callaghan 2006b). They found that the ochratoxin A *pks* gene has a high homology, especially in the ketosynthase domain, with *pks* genes of the MSAS type (methyl salicylic acid synthase). That is in nice agreement with the *otapks*PN gene identified in *P. nordicum* (see below), which also shows high homology to MSAS type *pks* genes (Karolewiez and Geisen 2005).

Another report of the cloning of an adjacent fragment of the same gene was given by Dao et al. (2005). They used the identified DNA fragment for the development of primers specific for ochratoxin A-producing fungi. These primers were used in diagnostic PCR reactions to identify potential ochratoxin A-producing organisms in environmental samples.

In the case of *P. nordicum* the gene cluster responsible for ochratoxin A biosynthesis is partly elucidated (Karolewiez and Geisen 2005; Geisen

et al. 2006). Figure 17.4 shows the information currently available about the gene cluster in *P. nordicum*.

Interestingly the presence of homologues of the genes has also been demonstrated for *P. verrucosum*, the closely related species. All the genes seem to be very similar, except the polyketide synthase which differs considerably between the two species. Part of a polyketide synthase from *P. verrucosum* CBS 302.48, obviously involved in ochratoxin A biosynthesis, was deposited in GeneBank under accession number EF527420. This fragment covers the acyltransferase domain of the polyketide synthase of *P. verrucosum* and has only limited homology to the respective domain of *P. nordicum* (similarity index 13.7, after the Martinez/Needleman_Wunsch DNA alignment). If this dissimilarity between the two polyketide synthases of the two related species is due to the fact that *P. verrucosum* is also able to produce citrinin, another polyketide mycotoxin is not yet clear. It is speculated that the *P. verrucosum* polyketide synthase might be involved in the biosynthesis of both toxins. The identified genes of *P. nordicum* involved in ochratoxin A biosynthesis are a polyketide synthase (*otapks*PN), a nonribosomal peptide synthetase (*otanps*PN), a gene with some homology to chlorinating enzymes (*otachl*PN; which is thought to be involved in the chlorination step) and a gene with high homology to the organic anion transporter (oat) from rat kidneys (*otatra*PN). The gene product of the rat gene is described to be involved in the export of ochratoxin A from kidney cells. This gives a strong indication that the *P. nordicum* gene might be involved in the transport of ochratoxin A out of the fungal cell. Until now no gene or gene

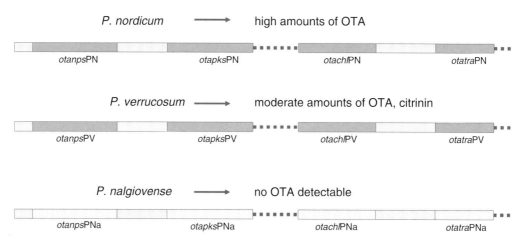

Fig. 17.4. Schematic drawing of the ochratoxin A gene clusters in *P. nordicum, P. verrucosum* and *P. nalgiovense*. For the *dotted line* no sequence information is available yet

product involved in the regulation of ochratoxin A has been described. The regulation of activation of the identified genes very congruently fits to the phenotypic production of ochratoxin A; and gene inactivation approaches demonstrate the involvement of three of the identified genes in ochratoxin A biosynthesis. Transformation of *P. nordicum* with either internal fragments of the *otapks*PN, the *otanps*PN or the *otachl*PN genes revealed strains unable to produce ochratoxin A, in contrast to the wild type. Interestingly, in the case of inactivation of the *otachl*PN gene, the respective transformant produced mainly ochratoxin B instead of ochratoxin A. This indicates the possible involvement of the gene product in the chlorination step.

Very unexpectedly homologues of these genes have not yet been identified in the major ochratoxin A-producing Aspergilli, indicating a convergent evolution of the non-homologous gene clusters, rather than separation from a common ancestor. Convergent evolution of secondary metabolite pathways is also suggested by Frisvad et al. (2008), trying to explain the presence of similar secondary metabolites in phylogenetically unrelated species. The *P. nordicum otapks*PN is a polyketide synthase of the MSAS type (*m*ethyl *s*alicylic *a*cid *s*ynthetases), leading to partially reduced polyketides, which fits very well with the structure of ochratoxin A. Atoui et al. (2006) found a *pks* gene in *A. carbonarius*, whose acyltransferase domain has about 47% identity to the PKS responsible for the biosynthesis of methyl salicylic acid (MSA) by *A. terreus* and *P. patulum*

and 44% identity to the *otapks*PN gene from *P. nordicum*. However until now it has not been ultimately shown that this *pks* gene from *A. carbonarius* is involved in ochratoxin A biosynthesis. The original report about the identification of the *otapks*PN gene also described highest homology to the MSAS-like *pks* genes from *A. terreus, Byssochlamys nivea* and *P. griseofulvum* (Karolewiez and Geisen 2005). As can be seen from Fig. 17.4, another species could be found which obviously carries homologues of the ochratoxin A biosynthesis genes. This species is *P. nalgiovense*, which is known as a biotechnological production culture for dry-cured, salt-rich foods, especially meats. This is the same habitat in which *P. nordicum* nearly exclusively occurs. Until now, *P. nalgiovense* has never been described to produce ochratoxin A. Comparison of part of the *pks* gene homologue in *P. nalgiovense* revealed high homology with *P. nordicum*. Only two nucleotide exchanges were found over a stretch of about 460 nucleotides for each of the two analyzed *P. nalgiovense* sequences (Bogs et al. 2006). However a subsequent expression analysis revealed silent *otapks*PNa genes (at least under the conditions applied) in the two analyzed *P. nalgiovense* strains, compared to a very active gene in case of *P. nordicum*.

This looks like a situation similar to the *A. flavus/A. oryzea* or *A. parasiticus/A. sojae* species pairs. In these cases the two former species are potent aflatoxin producers, whereas the two latter species are regarded as domesticated forms of the two former ones. Interestingly the latter species are used for food fermentations in Asian countries and have never been reported to produce aflatoxin. The reason for its non-aflatoxigenicity has been well elucidated in the case of *A. sojae* (Chang et al. 2007). In this species the interaction between the two regulatory proteins AFLR and AFLJ is disturbed, leading to a non-aflatoxigenic phenotype. This effect is caused by an early termination of the translation of the AFLR protein. In this case an AFLR variant is formed with a reduced ability for transcription activation. In addition this variant form is no longer able to interact with AFLJ. Other mutations are responsible for the inactivation of aflatoxin biosynthesis in *A. sojae*. For example the *pks*A gene, a polyketide synthase involved in aflatoxin biosynthesis, is defective because it contains a pretermination stop codon which leads to a truncated product of 1847 amino acids. In the case of *A. oryzea* this effect has not been shown yet. However Tominaga et al. (2006) recently compared the aflatoxin gene cluster of *A. flavus* with that of *A. oryzea*. Generally they found a high homology of 97–99% between both clusters. However three genes showed more pronounced differences. The *aflT* gene of *A. oryzea* had a deletion of 257 bp and there was a frame shift mutation in *nor*-A and a base pair substitution in *ver*A. In addition two substitutions in the promotor region of the *aflR* gene could be identified: one in an AREA binding site and one in a FACB binding site. Obviously these mutations are responsible for the aflatoxin negative phenotype of *A. oryzea*.

Until now nothing has been known about the reason for the non-ochratoxin A phenotype of *P. nalgiovense*. Similar mutations could be responsible for this feature, but it is not absolutely ruled out that *P. nalgiovense* may be able to produce ochratoxin under certain growth conditions.

Recently a systematic analysis of the occurrence of *P. nordicum* on salt-rich, dry-cured products isolated a set of strains which were partly able to produce low amounts of ochratoxin A and which carried homologues of the ochratoxin A biosynthesis genes of *P. nordicum*. Also the *pks* genes were similar to those of *P. nordicum*. However the morphology and RAPD types differed from those of *P. nordicum*, *P. nalgiovense* and *P. verrucosum*. In comparison, *P. nordicum* produced high amounts of ochratoxin, whereas this group of strains produced very low amounts of ochratoxin A, if at all. These preliminary results indicate that possibly a third species (as yet undescribed) occurs in this limited habitat of salt-fermented foods. That is a very interesting fact which means that two (plus one unidentified) of four species now known to carry the ochratoxin A biosynthesis genes occur in this limited habitat of salt-rich, dry-cured foods. This points to the intriguing idea that ochratoxin A might be involved in salt stress adaptation by Penicillia.

VI. Regulation of Expression of Ochratoxin A Biosynthesis Genes in *Penicillium* in Relation to Environmental Factors

Ochratoxin A is produced in the secondary metabolism of the fungus. The distinction between primary and secondary metabolism is more or less historical; however from the viewpoint of expression kinetics secondary metabolism is activated at later growth phases. Historically transmission of the culture from the actively growing trophophase state to the more static stationary phase is regarded as a signal for induction of secondary metabolite biosynthesis. However recent detailed analysis showed that the production of mycotoxins, although it does not take place from the very beginning of growth, can start very early and at any case before reaching the idiophase. This means that signals other than the shift from tropho- to idiophase must trigger activation of mycotoxin biosynthesis. The exhaustion of a nutrient, very often nitrogen, is regarded as one signal for secondary metabolite activation (Bennett and Christensen 1983). Shortage of a nutrient can be regarded as some kind of nutritional stress caused by unfavorable nutrient composition of the substrate. This may be occur quite fast after growth on solid agar medium, when the diffusion of nutrients plays a role. Likewise on natural substrates the overall nutrient amount is high, but nevertheless nutrient limitations in the microenvironment may also occur due to diffusion problems.

As shown above, several external parameters like substrate, temperature, water activity, pH, light, preservatives and fungicides do have an important influence on the biosynthesis of mycotoxins. Mycotoxin biosynthesis genes are not constitutively expressed, but are induced under favorable conditions and growth phases. This situation implies that these external effects have an influence on gene transcription. This was demonstrated for ochratoxin A biosynthesis by O'Callaghan et al. (2006), who showed that the composition of the growth medium has a strong effect on the expression of the ochratoxin A *pks* gene of *A. ochraceus*. After growth on potato dextrose medium, and on potato dextrose medium supplemented with casamino acids, they observed a very low expression of the *pks* gene and two additional cytochrome monooxygenase genes, thought to be involved in ochratoxin A biosynthesis. In contrast, after growth on yeast extract medium, expression of these genes were 39-fold higher. The same authors showed that the effect of external pH on ochratoxin A biosynthesis is also exerted at the transcriptional level. At low pH values (pH 3–4) when the biosynthesis of ochratoxin A by *A. ochraceus* was high, the transcript levels also were high, albeit only minor influences on the produced biomass could be observed. The transcript level of the *pks* gene was about twofold higher at low pH values (pH 3) compared to high pH values (pH 10). In contrast, the transcript accumulation of the glyceraldehyde-3-phosphate dehydrogenase gene was similar in all cultures, with about the same copy number at all conditions.

By microarray O'Brian et al. (2007) analyzed the expression pattern of an aflatoxin-producing strain of *A. flavus*. During this approach the authors analyzed 5000 genes, from which 144 where differentially expressed between the two analyzed temperatures (28°C, 37°C). It was already known from phenotypical analysis that the production optimum for aflatoxin is 28°C, whereas 37°C is the growth optimum. Very low amounts of aflatoxin are produced at the growth optimum. From the 144 differentially expressed genes 103 were more highly expressed at 28°C. About 25% of theses genes were related to secondary metabolism. As expected, the authors demonstrated that all aflatoxin biosynthesis genes are much more expressed at 28°C compared to 37°C. This decrease in expression at elevated temperatures explains the very low to undetectable amounts of aflatoxin produced by *A. flavus* at optimal growth conditions of 37°C. However on other substrates (e.g. ground nuts) the temperature profile of aflatoxin biosynthesis may be completely different. Interestingly some pathway genes (*aflR*, *aflJ*, *aflS*) showed no differential expression between the two temperatures. For this reason the authors concluded that the temperature per se does not have an influence on gene expression, but that less product of the *aflR* gene is produced under elevated temperatures. The *aflR* gene is the regulatory gene of the aflatoxin biosynthesis pathway. According to another suggestion by the authors, the AFLR protein might also be non-functional at higher temperatures. In another analysis, Price et al. (2005) used a whole genome microarray to study the influence of ammonia and pH on aflatoxin gene expression. They showed that beside the aflatoxin biosynthesis cluster other regulons are coregulated, indicating accessory involvement in aflatoxin biosynthesis. The authors stated for example that, for a carbon source to support both growth and aflatoxin production, it must be accessible to both the hexose monophosphate and glycolytic pathways. In the current analysis genes upregulated in response to sucrose and aflatoxin biosynthesis could be identified.

This situation indicated the sometimes complex mechanisms controlling regulation of mycotoxin biosynthesis genes by environmental factors. This microarray study could also reveal a possible influence of changes in pH on aflatoxin biosynthesis. The authors identified a set of 27 genes coordinately up-regulated with increasing pH. This group of genes also includes *pkaA*, a phosphokinase involved in regulation of aflatoxin biosynthesis. The activity of *pkaA* inhibits aflatoxin biosynthesis.

For the regulation of ochratoxin A biosynthesis in *Penicillium* such an in depth description of the molecular mechanisms behind the phenotypic influences cannot be given yet, nevertheless detailed and systematical analysis of ochratoxin A gene expression have been performed. It can be shown as expected that the polyketide synthase gene from *P. nordicum* is regulated and not constitutively expressed (Geisen et al. 2004). The first induction of the *otapksPN* gene can be observed after 3–4 days of growth on YES medium. The maximum of expression is reached between day 5 and day 8. After that time-point expression declines sharply. In contrast, expression of a house-keeping gene remains constant over the whole observation period of 10 days. The phenotypic production of ochratoxin A by *P. nordicum* completely fits to the expression pattern of the *otapksPN* gene. Low amounts of ochratoxin A can be detected by HPLC after 4–5 days of incubation, i.e. more than 24 h after induction of the gene took place. As long as the *otapksPN* gene is expressed, e.g. until day 8, the concentration of ochratoxin A increased. After the decline of *otapksPN* gene expression, the ochratoxin A concentration remains constant over the observation period (oscillates around a certain level, as discussed later). The expression behavior of the *otapksPN* gene very much resembles the pattern of the citrinin polyketide synthase of *Monascus purpureus* (Shimizu et al. 2005). Also this *pks* gene has an expression optimum after 3–6 days which is followed by citrinin production. The fact that the expression of both genes is well correlated with the production of the respective secondary metabolite demonstrates that both can be regarded as key enzymes of the pathway. The transcription activation of mycotoxin biosynthesis genes are not necessarily direct correlated to toxin biosynthesis. In a detailed analysis, Scherm et al. (2005) showed that the expression pattern of only certain genes of the aflatoxin biosynthesis pathway in *A. flavus* fits well the phenotypic

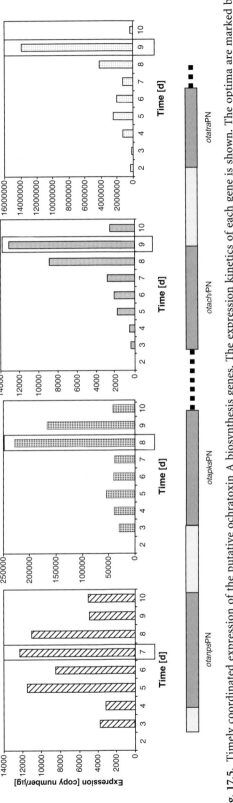

Fig. 17.5. Timely coordinated expression of the putative ochratoxin A biosynthesis genes. The expression kinetics of each gene is shown. The optima are marked by rectangles

aflatoxin production, in particular *nor*-1, *omt*B and *omt*-1. Also a detailed analysis of the known ochratoxin A-related genes of *P. nordicum* revealed different expression kinetics in correlation to ochratoxin A biosynthesis (Geisen et al. 2006). According to this analysis there seems to be a timely consecutive occurrence of expression optima in the ochratoxin A-related genes. The first gene, whose activation can be detected after 3–4 days and whose optimum is at 5–8 days, is the *otapks*PN gene, followed by the *otanps*PN (activation from day 5 to day 7; optimum days 8–9), the *otachl*PN (optimum days 8–9) and the *otatra*PN gene (optimum day 9; Fig. 17.5).

As mentioned above, the expression pattern of the *otapks*PN gene fits most to the phenotypical production of ochratoxin A. In addition the expression of the β-actin gene of *P. nordicum* remains constant over the whole observation period. It is intriguing to speculate that the consecutive occurrence of expression optima of the various genes is a hint for the consecutive enzymatic reactions during ochratoxin A biosynthesis. According to this order the *otapks*PN gene would be activated first, which is logical from the biosynthesis point of view, as its gene product first synthesizes the polyketide moiety dihydroisocoumarin. The second active gene, the *otanps*PN gene, after activation adds the amino acid phenylalanine, which apparently originates from the shikimic acid pathway. The gene product of the *otanps*PN gene is a non-ribosomal peptide synthetase, whose activity leads to the formation of a peptide bond between phenylalanine and the polyketide moiety. This reaction gives rise to ochratoxin B. The toxicity of ochratoxin B is by a factor of 10 to 100 less than the toxicity of ochratoxin A. At the molecular level ochratoxin A is a competitive inhibitor of the phenylalanine t-RNA synthetase. This might also be the case in the producer strains. A late chlorination and an immediate excretion might greatly reduce the toxic effect of ochratoxin A on the producer cell. Interestingly the *otachl*PN and the *otatra*PN genes have their expression optimum at the latest, which ideally fits with this hypothesis. This putative order of reactions would include a self-protection mechanism against the toxicity of ochratoxin A. The most toxic compounds are formed very late during biosynthesis and seem to be excreted just after formation.

A self-protection mechanism is also known in the biosynthesis of trichothecenes by *Fusaria*. The toxicity of trichothecenes is also exerted at the level of translation. In this case the gene *tri*101 codes for an acetyl transferase, which acetylates intracellularly the produced trichothecenes and thereby reducing their toxic activities (Kimura et al. 2007). This gene was originally identified as a clone in *Schizosacharomyces pombe* that conferred resistance to T-2 toxin, a type A trichothecene, when expressed in the heterologous host. Later however it was shown, that *F. sporotrichioides* strains in which this gene was inactivated were not lethal, but produced another set of trichothecene end-products (McCormick et al. 1999). For this reason the authors conclude that the gene product is involved in additional tasks within the biosynthesis of trichothecenes besides the self-protection effect. By transgenic expression of this gene in either tobacco (Muhitch et al. 2000) or wheat (Okubara et al. 2002), the resistance of these crops against the toxic action of the trichothecenes has been increased. A second self-protection mechanism in *Fusarium* is exerted by the *tri*12 gene product. This protein is an efflux pump which excretes the toxin out of the cell. This activation might be analogous to the activity of the *otatra*PN gene.

The influence of external abiotic factors like temperature, water activity and pH on the expression of the *otapks*PN gene of *P. nordicum* (Geisen 2004) or the *otapks*PV gene of *P. verrucosum* have been extensively studied. Generally after continuously changing a single parameter the growth rate has a clear optimum at a certain value and decreases when the parameter (for example temperature) approaches suboptimal conditions. This is true for all parameters analyzed (temperature, pH, water activity). In contrast to this behavior, expression of the *otapks*PV gene has several peaks of induction (Schmidt-Heydt et al. 2008): one major peak at or near the optimal growth conditions and one minor peak under conditions which impose mild stress to the fungus. This is indicated by the fact that, under these conditions, *P. verrucosum* shows a reduced growth. The phenotypic production of ochratoxin A by *P. verrucosum* completely follows this expression behavior (Fig. 17.6).

This typical expression profile was not only found in the case of water activity, but also at changing temperature or pH values. Moreover the trichothecene-producing *Fusarium culmorum* and aflatoxin-producing *Aspergillus flavus* strains also exhibited the same profile, demonstrating that this profile seems to be a general one. Also after microarray analysis of the whole known ochratoxin A cluster of genes the same expression profile was found in relation to changing

parameters, one major peak of expression at near optimal growth conditions and a smaller peak at conditions which lead to reduction, but not to a stop of growth. Obviously the influence of external abiotic factors, which leads to cellular stresses activate ochratoxin A biosynthesis. This microarray expression profile was found for all three species mentioned above, again indicating the general importance of this profile. The occurrence is interpreted as some kind of stress response triggered by environmental parameters reaching unfavorable conditions for growth. The major peaks at optimum growth conditions at a first glance do not look like stress adaptation. However *P. verrucosum* produces high amounts of ochratoxin A on solid agar medium. *P. verrucosum* grows very slowly and in a very compact form, producing quite high biomass per defined surface area. This growth morphology leads to the situation that high amounts of nutrients are needed during the trophophase. On a growth medium like agar plates, where diffusion plays a role, this situation might lead to a shortage of nutrients, for example nitrogen, whose exhaustion is known to induce mycotoxin biosynthesis (Bennet and Christensen 1988). It was shown that moderate growth conditions, e.g. at conditions where growth is not optimal (but not yet retarded by unfavorable conditions) produced the lowest amounts of ochratoxin A. This fact again demonstrates that certain kinds of stress, either exerted by abiotic factors or by nutritional constraints activate ochratoxin A biosynthesis. It is also known that suboptimal amounts of fungicides or preservatives, which retard but still allow growth, lead to an increase in mycotoxin biosynthesis. Ochiai et al. (2007) constructed a *Fusarium graminearum* reporter strain, with a GFP protein under the control of the *tri*5 promotor. TRI5, a trichodiene synthase, is the key enzyme of the trichothecene biosynthesis pathway. They used this strain to analyze the influence of various fungicides on the activation of trichothecene biosynthesis genes. According to their results, triazol fungicides transcriptionally activate the *tri*5 gene under sublethal conditions. Similar results were described by Schmidt-Heydt et al. (2007). During this work the influence of sublethal amounts of the preservatives sodium propionate and potassium sorbate on the expression of the *P. verrucosum* ochratoxin polyketide synthase gene was analyzed. At certain sublethal concen-

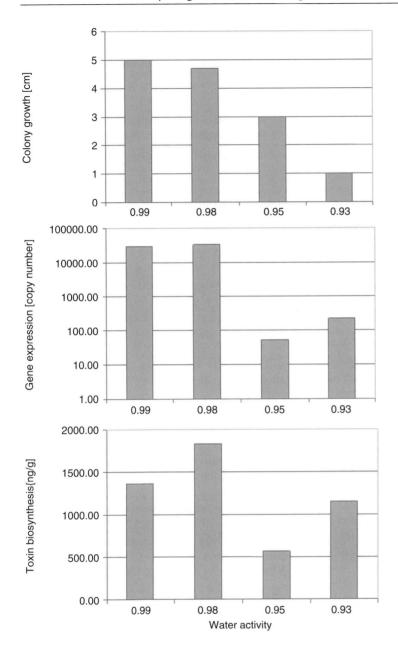

Fig. 17.6. Growth kinetics (*top*), *otapks*PV expression profile (*middle*) and ochratoxin A-production kinetics (*bottom*) of *P. verrucosum* at a changing single parameter (water activity). The major peak of ochratoxin biosynthesis arose at a water activity of 0.98, the minor peak at 0.93, which completely follows the respective gene expression

trations of both preservatives the gene was highly activated, which was paralleled by a higher production of ochratoxin A. This induction however was only observed if the concentration of the preservative did not exceed a certain level. When the concentration was too high, growth was still possible, but ochratoxin production was reduced. This means that only mild stress, obviously irrespective of the kind of stress, induces ochratoxin A biosynthesis, whereas more extreme conditions leads to an inactivation of the biosynthesis.

A similar behavior was described for patulin biosynthesis by *P. expansum* (Baert et al. 2007). Also, in the case of *P. expansum*, limited stress conditions (e.g. low temperature, low oxygen content) induced patulin biosynthesis, whereas the application of higher stress factors stopped patulin biosynthesis. An activation of mycotoxin biosynthesis genes by stress was also demonstrated for fumonisin production in *F. verticillioides* (Jurado et al. 2008). These authors found an activation of the *fum1* gene by reduction of the temperature or application of water stress. In contrast to the increase in fumonisin biosynthesis, these conditions lead to a reduction in the growth rate, indicating mild stress for the fungus. In the case of aflatoxin biosynthesis by *Aspergillus*, it was shown that oxidative stress induces toxin biosynthesis (Jayashree and Subramaniam 2000). The aflatoxin production by *A. flavus* on pistachio kernel agar can be increased by a factor of about two by applying oxidative stress by the addition of 100 μM tert-butyl hydroperoxide (Kim et al. 2006). The addition of a certain concentration of antioxidants nearly completely counteract the induction of aflatoxin biosynthesis because these substances reduce the external oxidative stress. The authors speculate, because of the fact that the precursors of aflatoxin are phenolic compounds with predictable antioxidative activities, that they contribute to the alleviation of the oxidative stress. This counter-action against the negative effects of oxidative stress by inducing aflatoxin biosynthesis might be an evolutionary advantage over species unable to produce the toxin. The same group (Kim et al. 2005) developed a model system in *S. cerevisiae* to analyze the stress response in relation to aflatoxin biosynthesis. This work identified 43 genes of *A. flavus* which are activated during stress responses and which are possibly linked to aflatoxin biosynthesis.

Because of the fact that obviously several myco-toxins (aflatoxin, ochratoxin A, patulin, trichothecenes, fumonisins) are activated under stress conditions, an involvement of these mycotoxins in stress adaptation seems evident. It seems also likely that the role of the different mycotoxins during stress adaptation is different. The involvement of aflatoxin in the inhibition of negative oxidative stress activities seems reasonable. The biosynthesis of ochratoxin A is activated under a set of different stresses (probably all stresses), which indicates a general role in stress adaptation.

Environmental stresses must somehow be recognized by the cell and the signal must be transduced to the nucleus to alter gene expression. Until now not much is known about the connection between signal transduction pathways, stress response and mycotoxin gene activation. Recently Ochiai et al. (2007b) connected the activity of an osmosensor histidine kinase and other osmotic stress-activated protein kinases to the regulation of trichothecene biosynthesis in *F. graminearum*. Deletion mutants in several of the kinase genes (MAPK) resulted in strains unable to produce trichothecenes, albeit other secondary metabolites like the red pigment aurofusarin were still produced. In parallel the expression of two genes of the trichothecene biosynthesis pathway were markedly reduced in particular *tri*4 (a p450-monooxygenase) and *tri*6 (a transcription activator). Interestingly salt stress completely inhibits toxin biosynthesis in the case of the trichothecenes. This is in sharp contrast to ochratoxin A biosynthesis in which higher salt concentrations

clearly activate toxin production. This again indicates that the role of various mycotoxins in stress adaptation seems to be different.

VII. Regulation of Ochratoxin A Biosynthesis in *Penicillium* in Relation to Light

It has long been recognized that light can have a profound effect on the development, morphology and secondary metabolite production of various fungi. Already in 1955 two German scientists described the influence of light on mycelium and conidium formation in *Alternaria brassicae* (Witsch and Wagner 1955). Since that time a wealth of information has been gleaned on this topic. An important biological feature with this respect is the molecular clock described in detail in *N. crassa* (Dunlap and Loros 2006). A well studied rhythm also exists in the Zygomycete *Pilobolus*. However most details about the molecular components of the circadian clock stems from *N. crassa*. Recently a circadian rhythm was detected in a mycotoxin-producing fungus, in particular in *A. flavus* (Greene et al. 2003). A typical expression behavior was described for certain genes which were under the control of a biological clock. However no circadian regulation of aflatoxin biosynthesis genes was described in this work, instead an oscillation of the glyceraldehyde-3-phosphate dehydrogenase was found. In addition the development of sclerotia was clock-regulated. During this work the typical molecular components of a fungal biological clock was also identified: these consist of the gene frequency (*frq*) which is the endogenous oscillator and entrains (synchronizes) the clock in relation to the light/dark period. Two other components are the white collar1 (WC-1) and white collar2 (WC-2) proteins. Both proteins can form a dimer which in turn can regulate downstream genes, the so-called clock-controlled genes or ccgs. These are often conidiation-related genes, but the mating-related genes, glyceraldehyde-3-phosphate dehydrogenase, copper methallothionine or trehalose synthase are also controlled by the clock. This set of known ccgs suggests that the biological clock of *Neurospora* regulates biological functions such as metabolism and stress responses not related to conidiation (Lakin-Thomas and Brody 2004).

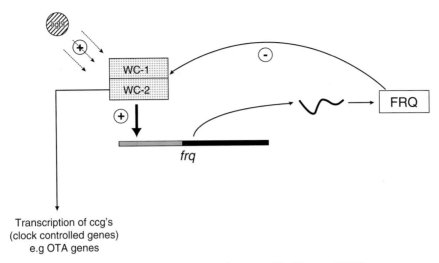

Fig. 17.7. Model of the circadian clock from *N. crassa* according to Lakin-Thomas (2000)

The regulation of ccgs often differs from gene to gene, so general predictions cannot be made. The *frq* gene is regulated by a feedback loop acting negatively on its own transcription. This negative feedback is supported by activities of the WC-1 and WC-2 proteins (Lakin-Thomas and Brody 2004). A complex of the two WC proteins (white collar complex; WCC) is required for full transcriptional regulation of *frq*. The FRQ protein itself positively regulates the levels of the WC proteins and negatively regulates the activity of the WCC. The *frq* gene has two light responsive elements (LREs) in its promoter region, suggesting that light regulation is mediated via transcription of the *frq* gene. Also one of the transcription factors, WC-1, can sense light at a specific wavelength, as it can act as a blue light receptor (Lakin-Thomas and Brody 2004). A simple model of the circadian clock of *Neurospora* is shown in Fig. 17.7.

Until now mycotoxin biosynthesis has not been described to be under circadian regulation. However first evidence for a circadian regulation of the ochratoxin A biosynthesis in *P. nordicum* and *P. verrucosum* came from variabilities in the expression data of ochratoxin A biosynthesis genes, when the samples were taken at variable day times. A systematical analysis of this phenomenon revealed that the expression kinetics of ochratoxin A biosynthesis genes (*otapks*) in *P. nordicum* and *P. verrucosum* exhibit a day/night oscillation of transcript levels, which was not the case for any analyzed housekeeping gene

(e.g. β-actin). This oscillation was much more symmetrical when a constant light/dark rhythm was ensured. After incubation under constant dark conditions, the oscillation became rapidly non-symmetrical and less regulated. This typical deregulation of the oscillation was also shown for *N. crassa*. In contrast to this, the expression of a housekeeping gene, e.g. the β-actin gene stayed more or less constant over the observation period. Figure 17.8 shows the oscillating expression of the *otapks* gene of *Penicillium*.

According to the profile, expression increases during the night and decreases during daytime. Similar observations were described by Stinnett et al. (2007). These authors described that also in *A. parasiticus* the production of versicolorin, a precursor of aflatoxin is higher in the dark than in the light (Calvo et al. 2004). This is due, beside the activity of the circadian clock, to a global regulatory protein velvet A (*ve*A) which is needed for the activation of secondary metabolite biosynthesis.

The product of *ve*A is required for expression of *afl*R and *afl*J two regulatory and activating proteins of the aflatoxin biosynthesis pathway in *A. parasiticus* and *A. flavus*. This gene is also involved in fungal development in that it affects cleistothecial development in *A. nidulans* or sclerotial development in *A. parasiticus* (Calvo et al. 2004). Light negatively influences the activity of this gene, which leads to the reduced formation of cleisthotecia and secondary metabolites. A deletion of this gene leads to a complete loss of transcription of genes involved in aflatoxin biosynthesis, like the two regulatory-activating

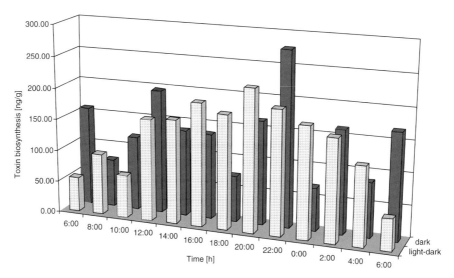

Fig. 17.8. Circadian regulated expression of the *otapks*PV gene of *P. verrucosum*. The expression kinetics is a smooth curve under alternating dark/light conditions, whereas the expression kinetics under constant dark conditions is rapidly deregulated. Highest expression and biosynthesis was found during the dark

genes *aflR* and *aflJ* and the two cluster genes *pks*L1 and *ver*1. This indicates the strong dependence of aflatoxin biosynthesis on a functional *veA* gene product. The *veA* product is imported into the nucleus by the action of importin α. This migration to the nucleus is light-dependent. In the dark VEA is located mainly in the nucleus, whereas in the light VEA can mainly be found in the cytoplasm (Stinnet et al. 2007). It could also be shown that various wavelengths of light have different effects. Blue light most prevents accumulation of VEA in the nucleus.

Recently homologoues of the *veA* gene were identified in *P. verrucosum* and *P. nordicum*, named *ve*PV and *ve*PN. These genes were identified by heterologous PCR with primers deduced from *A. parasiticus*. A PCR fragment from *P. verrucosum* or *P. nordicum* was sequenced and resulted in sequences homologous to the *veA* gene of *A. parasiticus*, indicating the conservation of this gene. After inactivation of the gene in *P. verrucosum* a phenotype arose which corresponds to the *veA*⁻ phenotype of *A. parasiticus* (Calvo et al. 2004) in that the *P. verrucosum* mutant was no longer able to produce ochratoxin A or citrinin. In addition the components of the molecular clock could also be identified by the same approach. Homologues of the *frq* and the *wc*-1 and *wc*-2 genes could be identified in *P. verrucosum*, demonstrating the molecular basis of a circadian clock in this fungus. Analysis of the expression kinetics of the *frq* and the *wc*-1 gene demonstrated that this clock is

functional. Both genes are conversely regulated. When the *frq* transcript level is high, the transcript level of *wc*-1 is low and vice versa, leading to a reversed oscillation of transcription of both genes, as also shown for *Neurospora* (Dunlap and Loros 2006). Unequivocal evidence that the circadian clock of *Penicillium* is involved in ochratoxin A biosynthesis comes from the fact that a gene-inactivation mutant of the *wc*-1 gene of *P. verrucosum* had a completely altered pattern of ochratoxin A regulation in response to light. Interestingly and unexpectedly the mutant produces more ochratoxin under light conditions compared to the wild type. That indicates that WC-1 has a negative activity on transcription of ochratoxin A biosynthesis genes under light conditions whereas *ve*PV has a positive activity and acts as an antagonist for the WC complex.

Purschwitz et al. (2008) recently analyzed the physical interactions between blue and red light sensors in *A. nidulans*. They found that a phytochrome component (FPHA) is part of a protein complex containing LREA (WC-1) and LREB (WC-2). It was shown for *A. nidulans* that FphA represses sexual development and mycotoxin (sterigmatocystin) formation, whereas LreA and LreB stimulate both. FPHA seems to interact with LREB and VeA, which was described already to be involved in light/dark-regulated mycotoxin biosynthesis. During this analysis it was shown that blue light repressed mycotoxin biosynthesis to a level similar to white light, which indicates that the blue light receptor (LreA) might in addition have a negative influence on mycotoxin biosynthesis under certain conditions. According to these authors LreA and LreB act positively and light inhibition on mycotoxin biosynthesis is mediated by negatively acting FphA.

In *Penicillium* this regulation seems not be completely identical, as an inactivation of a *wc-1* homologue increases ochratoxin A biosynthesis drastically. It might be possible that the inactivation of *wc-1* prevents interaction of a putative FphA protein with the WC complex and thereby ceases the negative action of the red light receptor, allowing increased mycotoxin biosynthesis.

Another important hint for differences in light regulation in *Penicillium* and *Aspergillus* comes from the fact that, although blue light has a negative impact on the biosynthesis of sterigmatocystin by *A. nidulans*, this effect is not stronger than the effect of white light (Purschwitz et al. 2008). In *Penicillium* white light has a negative effect on ochratoxin A biosynthesis too, but blue light completely blocks mycotoxin biosynthesis or even growth if applied at higher light intensities. A similar observation has been made by Häggblom and Unestam (1997). They described that blue light reduced the production of alternariol and alternariol-monomethylether by *Alternaria alternata* by 69% and 77% respectively. Red light had no effect on toxin levels. That is similar to the situation in *Penicillium*, where it also has been shown that red light had nearly no reducing effect on ochratoxin A biosynthesis.

There is another effect of light/dark alterations on the steady-state level of ochratoxin A produced by *Penicillium*. It has been stated already that gene activation and biosynthesis is increased under dark conditions and reduced during light. However, in addition to the drop in ochratoxin A biosynthesis under light conditions, the already produced ochratoxin A seems to be partly degraded. This results under stationary conditions also in an oscillation of the ochratoxin A concentration itself, beside the oscillation at the transcript level. The level of ochratoxin A oscillates around a fixed value. Ochratoxin A is produced during the night and partly degraded during the day. The reason for this oscillation is not completely clear yet. However Gillmann et al. (1998) showed that the toxin is susceptible to light and decomposes over time. Their experiments showed that especially blue light degrades ochratoxin A. The same authors however also showed

that ochratoxin A has a DNA photocleavage activity under certain light conditions. So the decrease in ochratoxin A concentration during the day may be due to an absorption of the blue light components of daylight, thereby protecting the fungus from these growth-inhibiting wavelengths, or alternatively the ochratoxin A might be actively degraded by peptidases, which cleave the peptide bond between phenylalanine and the polyketide moiety, until an ochratoxin A concentration is reached, which is no longer toxic for the fungus under light conditions due to the photocleavage effect. Stander et al. (2000) demonstrated that hydrolytic enzyme mixtures containing potential proteases and amidases are able to cleave ochratoxin A in ochratoxin α and phenylalanine. Varga et al. (2000) demonstrated that different Aspergilli, including *A. niger*, which itself is able to produce ochratoxin A, can degrade ochratoxin A to ochratoxin α and phenylalanine. The authors suggested a carboxypeptidase is responsible for the degradation. Figure 17.9 summarizes the influence of light/dark alterations on ochratoxin biosynthesis by *Penicillium*.

VIII. Conclusions

Current knowledge of the physiology, molecular biology and genetics of ochratoxin A biosynthesis in Penicillia has been described. The reported results imply that the ochratoxin A biosynthesis is activated under unfavorable growth conditions which suppose mild stress conditions to the fungus. Obviously various stress conditions exerted by different parameters (like temperature, water activity, pH, suboptimal amounts of preservatives or fungicides or unfavorable wavelengths of light) lead to the same biological answer, e.g. the production of ochratoxin A to elevated levels. This suggests that the activation of ochratoxin A biosynthesis is correlated to stress adaptation (Fig. 17.10).

The biosynthesis of ochratoxin A might simply be coregulated with the stress-adaptive response without having defined functions in stress adaptations. However, ochratoxin A might be a component of the stress adaptation. It is known that ochratoxin A binds to proteins. For this reason it may have some kind of protective or stabilizing functions which increase the activity

Fig. 17.9. Scheme of the influence of light/dark rhythms on ochratoxin A biosynthesis by *Penicillium*

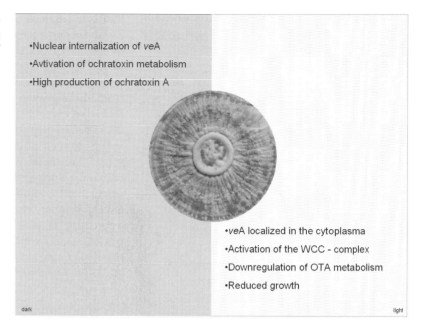

- Nuclear internalization of *veA*
- Avtivation of ochratoxin metabolism
- High production of ochratoxin A

- *veA* localized in the cytoplasma
- Activation of the WCC - complex
- Downregulation of OTA metabolism
- Reduced growth

dark   light

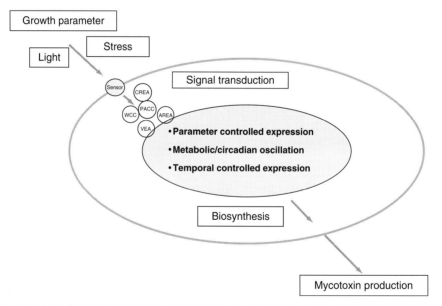

Fig. 17.10. Model of the influence between environment and the fungal cell leading to the activation of ochratoxin biosynthesis genes

window of proteins with respect to enzymatic or other activities. This of course should lead to an increased fitness under conditions which lead to an activation of the ochratoxin A biosynthesis genes. Indeed a very careful analysis of the growth rate of ochratoxin A-producing Penicillia under constantly changing environmental conditions (for example temperature) reveals two growth peaks, a major peak of the growth rate at the well known conditions of 20–25°C (depending on the strain) and a minor peak at the conditions which are exactly the mild stress conditions and which leads to the induction of ochratoxin A biosynthesis (for example 15°C). The minor peak can only be found after prolonged measurement of the growth rate. The rate at 15°C at the beginning of growth is below the rate at higher temperature (say 20°C) as expected; however after prolonged incubation the growth rate at 15°C increases and outruns the rate at higher temperature (20°C). Very interestingly the acceleration of the growth rate begins during the growth phase, when ochratoxin A biosynthesis starts at these mild stress conditions. Neither of the several non-ochratoxin A-producing *Penicillium* species tested showed this behavior, indicating that ochratoxin A biosynthesis and acceleration of growth rate is coupled. Similar increases in growth rate after onset of ochratoxin A biosynthesis were observed after changing other parameters like salt concentration, indicating that this might be a general phenomenon.

It is tempting to speculate according to these recent results, that the biosynthesis of ochratoxin A might increase the fitness of the fungus under mild stress conditions, like low temperature, a certain salt concentration or other parameters. What is clear at this point is the fact that under mild stress conditions ochratoxin A biosynthesis increases and at the same time the growth rate accelerates in a later growth phase when ochratoxin A is produced. However it is not clear at this stage whether the activation of ochratoxin A biosynthesis is the cause or consequence of growth acceleration. Further research in connecting stress responses and ochratoxin A biosynthesis at the molecular level is needed to understand this obvious link.

References

Abarca ML, Bragulat MR, Castella G, Cabanes FJ (1994) Ochratoxin A production by strains of *Aspergillus niger* ver. *niger*. Appl Environ Microbiol 60:2650–2652

Abarca ML, Accensi F, Bragulat MR, Castella G, Cabanes FJ (2003) *Aspergillus carbonarius* as the main source of ochratoxin A contamination in dried vine fruits from the Spanish market. J Food Protect 66:504–506

Abarca ML, Accensi F, Cano J, Cabanes FJ (2004) Taxonomy and significance of black aspergilli. Antonie Van Leeuwenhoek 86:33–49

Accensi F, Cano J, Figuera L, Abarca ML, Cabanes FJ (1999) New PCR method to differentiate species in the *Aspergillus niger* aggregate. FEMS Microbiol Lett 180:191–196

Arroyo M, Aldred D, Magan N (2003) Impact of environmental factors and preservatives on growth and ochratoxin A production by Aspergillus ochraceus in wheat-based media. Aspects Appl Biol 68:169–174

Arst HN, Penalva MA (2003) pH regulation in *Aspergillus* and parallels with higher eukaryotic regulatory systems. Trends Genet 19:224–231

Asao T, Wogan GN, Chang SB, Buchi G, Wick EL, Abdelkad MM (1963) Aflatoxin B and G. J Am Chem Soc 85:1706–1963

Atoui A, Dao HP, Mathieu F, Lebrihi A (2006) Amplification and diversity analysis of ketosynthase domains of putative polyketide synthase genes in *Aspergillus ochraceus* and *Aspergillus carbonarius* producers of ochratoxin A. Mol Nutr Food Res 50:488–493

Atoui A, Mathieu F, Lebrihi A (2007) Targeting a polyketide synthase gene for *Aspergillus carbonarius* quantification and ochratoxin A assessment in grapes using real-time PCR. Int J Food Microbiol 115:313–318

Battilani P, Giorni P, Pietri A (2003) Epidemiology of toxin-producing fungi and ochratoxin A occurrence in grape. Eur J Plant Pathol 109:715–722

Bennett JW, Christensen SB (1983) New perspectives on aflatoxin biosynthesis. Adv Appl Microbiol 29:53–92

Bhatnagar D, Ehrlich KC, Cleveland TE (2003) Molecular genetic analysis and regulation of aflatoxin biosynthesis. Appl Microbiol Biotechnol 61:83–93

Bhatnagar D, Carry JW, Ehrlich K, Yu J, Cleveland TE (2006) Understanding the genetics of regulation of aflatoxin production and *Aspergillus flavus* development. Mycopathologia 162:155–166

Bogs C, Battilani P, Geisen R (2006) Development of a molecular detection and differentiation system for ochratoxin A producing *Penicillium* species and its application to analyse the occurrence of *Penicillium nordicum* in cured meats. Int J Food Microbiol 107:39–47

Brown DW, Cheung F, Proctor RH, Butchko RAE, Zheng L, Lee Y, Utterback T, Smith S, Feldblyum T, Glenn AE, Plattner RD, Kendra DF, Town CD, Whitelaw CA (2005) Comparative analysis of 87,000 expressed sequence tags from the fumonisin-producing fungus *Fusarium verticillioides*. Fungal Genet Biol 42:848–861

Calvo AM, Bok J, Brooks W, Keller NP (2004) *veA* is required for toxin and sclerotial production in *Aspergillus parasiticus*. Appl Environ Microbiol 70:4733–4739

Caracuel Z, Roncero MIG, Espeso EA, González-Verdejo CI, García-Maceira FI, Di Pietro A (2003) The pH signalling transcription factor PacC controls virulence in the plant pathogen *Fusarium oxysporum*. Mol Microbiol 48:765–779

Chang PK, Matsushima K, Takahashi T, Yu J, Abe K, Bhatnagar D, Yuan GF, Koyama Y, Cleveland TE (2007) Understanding nonaflatoxigenicity of *Aspergillus sojae*: a windfall of aflatoxin biosynthesis research. Appl Microbiol Biotechnol 76:977–984

Ciegler A, Fennel DI, Sansing GA, Detroy RW, Bennet GA (1973) Mycotoxin producing strains of *Penicillium viridicatum*: classification into subgroups. Appl Microbiol 26:271–278

Comi G, Orlic S, Redzepovic S, Urso R, Iacumin L (2004) Moulds isolated from Istrian dried ham at the pre-ripening and ripening level. Int J Food Microbiol 96:29–34

Dao HP, Mathieu F, Lebrihi A (2005) Two primer pairs to detect OTA producers by PCR method. Int J Food Microbiol 104:61–67

Diez P, Rodriguez-Saiz M, del la Fuente JL, Moreno MA, Barredo JL (2005) The nagA gene of *Penicillium chrysogenum* encoding ß-N-acetylglucosaminidase. FEMS Microbiol Lett 242:257–264

Dowzer CEA, Kelly JM (1991) Analysis of the creA gene, a regulator of carbon catabolite repression in *Aspergillus nidulans*. Mol Cell Biol 11:5701–5709

Dunlap JC, Loros JJ (2006) How fungi keep time: circadian system in Neurospora and other fungi. Curr Opin Microbiol 9:579–587

Ehrlich KC, Cotty PJ (2002) Variability in nitrogen regulation of aflatoxin production by *Aspergillus flavus* strains. Appl Microbiol Biotechnol 60:174–178

Ehrlich KC, Montalbano BG, Cary JW, Cotty PJ (2002) Promoter elements in the aflatoxin pathway polyketide synthase gene. Biochim Biophys Acta 1576: 171–175

Frisvad JC, Elmholt S (2005) Ochratoxin A producing *Penicillium verrucosum* isolates from cereals reveal large AFLP fingerprinting variability. J Appl Microbiol 98:684–692

Frisvad J, Frank JM, Houbraken JAMP, Kuijpers AFA, Samson RA (2004) New ochratoxin A producing species of *Aspergillus* section *Circumdati*. Stud Mycol 50:23–43

Frisvad JC, Andersen B, Thrane U (2008) The use of secondary metabolite profiling in chemotaxonomy of filamentous fungi. Mycol Res 112:231–240

Gaffoor I, Brown DW, Plattner R, Proctor RH, Qi W, Trail F (2005) Functional analysis of the polyketide synthase genes in the filamentous fungus *Gibberella zeae* (anamorph *Fusarium graminearum*). Eukaryot Cell 4:1926–1933

Gallo M, Katz E (1972) Regulation of secondary metabolite biosynthesis: catabolite repression of phenoxazinone synthase and actinomycin formation by glucose. J Bacteriol 109:659–667

Geisen R (2004) Molecular monitoring of environmental conditions influencing the induction of ochratoxin A biosynthetic genes in *Penicillium nordicum*. Mol Nutr Food Res 48:532–540

Geisen R, Mayer Z, Karolewiez A, Färber P (2004) Development of a real time PCR system for detection of *Penicillium nordicum* and for monitoring ochratoxin A production in foods by targeting the ochratoxin polyketide synthase gene. Syst Appl Microbiol 27:501–507

Geisen R, Schmidt-Heydt M, Karolewiez A (2006) A gene cluster of the ochratoxin A biosynthetic genes in *Penicillium*. Mycotoxin Res 22:134–141

Gillman IG, Yezek JM, Manderville RA (1998) Ochratoxin A acts as a photoactivateable DNA cleaving agent. Chem Commun 647–648

Greene AV, Keller N, Haas H, Bell-Pedersen D (2003) A circadian oscillator in *Aspergillus* spp. regulates daily development and gene expression. Eukaryot Cell 2:231–237

Häggblom P, Unestam T (1997) Blue light inhibits mycotoxin production and increases total lipids and pigmentation in *Alternaria alternata*. Appl Environ Microbiol 38:1074–1077

Harris JP, Mantle PG (2001) Biosynthesis of ochratoxins by *Aspergillus ochraceus*. Phytochemistry 58:709–716

Huff WE, Hamilton PB (1979) Mycotoxins-their biosynthesis in fungi: Ochratoxins-metabolites of combined pathways. J Food Protect 42:815–820

Jayashree T, Subramanyam C (2000) Oxidative stress as a prerequisite for aflatoxin production by *Aspergillus parasiticus*. Free Radic Biol Med 29:981–985

Karolewiez A, Geisen R (2005) Cloning a part of the ochratoxin A biosynthetic gene cluster of *Penicillium nordicum* and characterization of the ochratoxin polyketide synthase gene. System Appl Microbiol 28:588–595

Keller NP, Nesbitt C, Sarr B, Phillips TD, Burow GB (1997) pH regulation of sterigmatocystin and aflatoxin biosynthesis in *Aspergillus* spp. Phytopathology 87:643–648

Kim JH, Campbell BC, Yu J, Mahoney N, Chan KL, Molyneux RJ, Bhatnagar D, Cleveland TE (2005) Examination of fungal stress response genes using *Saccharomyces cerevisiae* as a model system: targeting genes affecting aflatoxin biosynthesis by *Aspergillus flavus* Link. Appl Microbiol Biotechnol 67:807–815

Kim JH, Campbell BC, Molyneux R, Mahoney N, Chan KL, Yu J, Wilkinson J, Cary J, Bhatnagar D, Cleveland TE (2006) Gene targets for fungal and mycotoxin control. Mycotoxin Res 22:3–8

Kimura M, Tokai T, Takahashi-Ando N, Ohsato S, Fujimura M (2007) Molecular and genetic studies of *Fusarium* trichothecene biosynthesis: pathways, genes, and evolution. Biosci Biotechnol Biochem 71:2105–2123

Lakin-Thomas PL (2000) Circadian rhythms: new functions for old clock genes? Trends Genet 16:135–142

Lakin-Thomas PL, Brody S (2004) Circadian rhythms in microorganisms: new complexities. Annu Rev Microbiol 58:489–519

Larsen TO, Svendsen A, Smedsgaard J (2001) Biochemical characterization of ochratoxin A-producing strains of the genus *Penicillium*. Appl Environ Microbiol 67:3630–3635

Marzluf GA (1997) Genetic regulation of nitrogen metabolism in the fungi. Microbiol Mol Biol Rev 61:17–32

McCormick SP, Alexander NJ, Trapp SE, Hohn TM (1999) Disruption of TRI101, the gene encoding trichothecene 3-O-Acetyltransferase, from *Fusarium sporotrichioides*. Appl Environ Microbiol 65:5252–5256

Muhitch MJ, McCormick SP, Alexander NJ, Hohn TM (2000) Transgenic expression of the TRI101 or PDR5 gene increases resistance of tobacco to the phytotoxic effects of the trichothecene 4,15-diacetoxyscirpenol. Plant Sci 157:201–207

O'Callaghan J, Dobson ADW (2006) Phylogenetic analysis of polyketide synthase genes from *Aspergillus ochraceus*. Mycotoxin Res 22:125–133

O'Callaghan J, Caddick MX, Dobson ADW (2003) A polyketide synthase gene required for ochratoxin A biosynthesis in *Aspergillus ochraceus*. Microbiology 149:3485–3491

O'Callaghan J, Stapleton PC, Dobson ADW (2006) Ochratoxin A biosynthetic genes in *Aspergillus ochraceus* are differentially resulted by pH and nutritional stimuli. Fungal Genet Biol 43:213–221

Ochiai N, Tokai T, Nishiuchi T, Takahashi-Ando N, Fujimura M, Kimura M (2007a) Involvement of the osmosensor histidine kinase and osmotic stress-activated protein kinases in the regulation of secondary metabolism in *Fusarium graminearum*. Biochem Biophys Res Commun 363:639–644

Ochiai N, Tokai T, Takahashi-Ando N, Fujimura M, Kimura M (2007b) Genetically engineered *Fusarium* as a tool to evaluate the effects of environmental factors on initiation of trichothecene biosynthesis. FEMS Microbiol Lett 275:53–61

Okubara PA, Blechl AE, McCormick SP, Alexander NJ, Dill-Macky R, Hohn TM (2002) Engineering deoxynivalenol metabolism in wheat through the expression of a fungal trichothecene acetyltransferase gene. Theor Appl Genet 106:74–83

Orvehed M, Häggblom P, Söderhäll K (1988) Nitrogen Inhibition of Mycotoxin Production by *Alternaria alternata*. Appl Environ Microbiol 54:2361–2364

O'Brian GR, Georgianna DR, Wilkinson JR, Yu J, Abbas HK, Bhatnagar D, Cleveland TE, Nierman W, Payne GA (2007) The effect of elevated temperature on gene transcription and aflatoxin biosynthesis. Mycologia 99:232–239

Perrone G, Varga J, Susca A, Frisvad JC, Stea G, Kocsube S, Toth B, Kozakiewicz Z, Samson RA (2008) *Aspergillus uvarum sp.* nov., an uniseriate black *Aspergillus* species isolated from grapes in Europe. Int J Syst Evol Microbiol 58:1032–1039

Petzinger E, Weidenbach A (2002) Mycotoxins in the food chain: the role of ochratoxins. Livestock Prod Sci 76:245–250

Pfohl-Leszkowicz A, Petkova-Bocharova T, Chernozemsky IN, Castegnaro M (2002) Balkan endemic nephropathy and associated urinary tract tumours: a review on aetiological causes and the potential role of mycotoxins. Food Addit Contam 19:282–302

Pitt JI (1987) *Penicillium viridicatum, Penicillium verrucosum* and production of ochratoxin A. Appl Environ Microbiol 53:266–269

Pohl-Leszkowicz A, Manderville RA (2007) Ochratoxin A: An overview on toxicity and carcinogenicity in animals and humans. Mol Nutr Food Res 51:61–99

Price MS, Conners SB, Tachdjian S, Kelly RM, Payne GA (2005) Aflatoxin conducive and non-conducive growth conditions reveal new gene associations with aflatoxin production. Fungal Genet Biol 42:506–518

Proctor RH, Brown DW, Plattner RD, Desjardins AE (2003) Co-expression of 15 contiguous genes delineates a fumonisin biosynthetic gene cluster in *Gibberella moniliformis*. Fungal Genet Biol 38:237–249

Purschwitz J, Müller S, Kastner C, Schöser M, Haas H, Espeso EA, Atoui A, Calvo AM, Fischer R (2008) Functional and Physical Interaction of Blue- and Red Light Sensors in *Aspergillus nidulans*. Curr Biol 18:1–5

Samson RA, Houbraken JAMP, Kuijpers AFA, Frank JM, Frisvad JC (2004) New ochratoxin A or sclerotium producing species in *Aspergillus* section *Nigri*. Stud Mycol 50:45–61

Scherm B, Palomba M, Serra D, Marcello A, Migheli Q (2005) Detection of transcripts of the aflatoxin genes *aflD, aflO,* and *aflP* by reverse-transcription-polymerase chain reaction allows differentiation of aflatoxin-producing and non-producing isolates of *Aspergillus flavus* and *Aspergillus parasiticus*. Int J Food Microbiol 98:201–210

Schmidt-Heydt M, Baxter E, Geisen R, Magan N (2007) Physiological relationship between food preservatives, environmental factors, ochratoxin and *otapks*PV gene expression by *Penicillium verrucosum*. Int J Food Microbiol 119:277–283

Schmidt-Heydt M, Magan N, Geisen R (2008) Stress induction of mycotoxin biosynthesis genes by abiotic factors. FEMS Microbiol Lett 284:142–149

Schuster E, Dunn-Coleman N, Frisvad JC (2002) On the safety of *Aspergillus niger* – a review. Appl Microbiol Biotechnol 59:426–435

Searcy JW, Davis ND, Diener UL (1969) Biosynthesis of ochratoxin A. Appl Microbiol 18:622–627

Serra R, Cabanes FJ, Perrone G, Castellá G, Venancio A, Mulè G, Kozakiewicz Z (2006) *Aspergillus ibericus*: a new species of section Nigri isolated from grapes. Mycologia 98:295–306

Shimizu T, Kinoshita H, Ishihara S, Sakai K, Nagai S, Nihira T (2005) Polyketide synthase gene responsible for citrinin biosynthesis in *Monascus purpureus*. Appl Environ Microbiol 71:3453–3457

Skrinjar M, Dimic G (1992) Ochratoxigenicity of *Aspergillus ochraceus* group and *Penicillium verrucosum var. cyclopium* strains on various media. Acta Microbiol Hung 39:257–261

Stander MA, Bornscheuer UT, Henke E, Steyn PS (2000) Screening of Commercial Hydrolases for the Degradation of Ochratoxin A. J Agric Food Chem 48:5736–5739

Steyn PS (1993) Ochratoxin A: its chemistry, conformation and biosynthesis. In Creppy EE, Castegnaro M, Dirheimer G (eds) Human ochratoxicosis and its pathologies. Libbey Euro, Paris, pp 51–58

Stinnett SM, Espeso EA, Cobeno L, Araújo-Bazán L, Calvo AM (2007) *Aspergillus nidulans* veA subcellular localization is dependent on the importin a carrier and on light. Mol Microbiol 63:242–255

Tominaga M, Lee YH, Hayashi R, Suzuki Y, Yamada O, Sakamoto K, Gotoh K, Akita O (2006) Molecular analysis of an inactive aflatoxin biosynthesis gene cluster in *Aspergillus oryzae* RIB strains. Appl Environ Microbiol 72:484–490

Tudzynski B (2005) Gibberellin biosynthesis in fungi: genes, enzymes, evolution, and impact on biotechnology. Appl Microbiol Biotechnol 66:597–611

Tudzynski B, Liu S, Kelly JM (2000) Carbon catabolite repression in plant pathogenic fungi: isolation and characterization of the *Gibberella fujikuroi* and *Botrytis cinerea creA* genes. FEMS Microbiol Lett 184:9–15

van der Merwe KJ, Steyn PS, Fourie L (1965) Ochratoxin A, a toxic metabolite produced by *Aspergillus ochraceus* Wilh. Nature 205:1112–1113

Varga J, Kevei É, Rinyu E, Teren J, Kozakiewicz Z (1996) Ochratoxin production by *Aspergillus* species. Appl Environ Microbiol 62:4461–4464

Varga J, Kozakiewicz Z (2006) Ochratoxin A in grapes and grape-derived products. Trends Food Sci Technol 17:72–81

Varga J, Rigó K, Téren J (2000) Degradation of ochratoxin A by *Aspergillus species*. Int J Food Microbiol 59:1–7

von Witsch H, Wagner F (1955) Beobachtungen über den Einfluß des Lichtes auf Mycel- und Conidienbildung bei *Alternaria brassicae* var. *dauci*. Arch Microbiol 22:307–312

Wei R-D, Strong FM, Smalley EB (1971) Incorporation of Chlorine-36 into Ochratoxin A. Appl Microbiol 22:276–277

Yu J, Keller N (2005) Regulation of secondary metabolism in filamentous fungi. Annu Rev Phytopathol 43:437–458

Yu J, Bhatnagar D, Cleveland TE (2004) Completed sequence of aflatoxin pathway gene cluster in *Aspergillus parasiticus*. FEBS Lett 564:126–130

18 Genetic and Metabolic Engineering in Filamentous Fungi

Jochen Schmid[1], Ulf Stahl[1], Vera Meyer[1,2]

CONTENTS

I. Introduction

The groundwork for modern fungal biotechnology was laid in the beginning of the twentieth century, accompanied by advances in microbiology, biochemistry and fermentation technology. The pioneering works of Jokichi Takamine (production of amylase from koji mold *Aspergillus oryzae*, 1894), James Currie (development of fungal fermentation for citric acid production, 1917) and Alexander Fleming (discovery of penicillin production by *Penicillium notatum*, 1928) stimulated scientists to further explore fungal metabolic capacities and, moreover, prompted engineers to develop large-scale and controlled production processes for filamentous fungi. Improvements of fungal capacities to produce metabolites of interest were, however, mainly restricted to classic mutagenesis techniques. The development of recombinant DNA technologies for filamentous fungi, shown for the first time in 1979 for *Neurospora crassa*, was a milestone in obtaining insights into the molecular basis of product formation and to improve traditional fungal fermentations by genetic engineering.

The industrial relevance of filamentous fungi is based on their high capacity to produce primary and secondary metabolites as well as for secreting proteins, having a wide spectrum of activity such as hydrolases and proteases (Conesa 2001; Punt et al. 2002). Additionally, the fungal glycosylation machinery is capable of providing a more 'mammalian-like' glycosylation pattern to proteins compared to the commonly used yeast hosts (Karnaukhova et al. 2007; Nevalainen et al. 2005; Ward et al. 2004), making filamentous fungi very attractive for the production of proteins used in medical applications. Secretion is, in particular, related to fungal morphology as it is thought to take place at the growing fungal tip (Fischer et al. 2008; Khalaj et al. 2001; Torralba et al. 1998). One focus of current research is thus on fungal morphology to improve the secretory capacity of industrially used fungi (Grimm et al. 2005; Meyer et al. 2008; Papagianni et al. 2003).

The following three sections are devoted to three important areas of research of genetic engineering in filamentous fungi, all aiming at the improved application of these organisms in biotechnology.

Section II summarizes current genomic approaches for filamentous fungi and discusses their benefit for the identification of new commercially interesting products.

Section III highlights the progress made in the post-genomic era, concerning new omic techniques as well as related challenges and future perspectives.

Section IV deals with genetic and metabolic engineering tools applicable nowadays in

[1]University of Technology Berlin, Department of Microbiology and Genetics, Gustav-Meyer-Allee 25, 13355 Berlin, Germany; e-mail: Ulf.Stahl@lb.tu-berlin.de, j.schmid@lb.tu-berlin.de
[2]Leiden University, Institute of Biology, Research Group for Molecular Microbiology, Wassenaarseweg 64, 2333 AL Leiden, The Netherlands

Physiology and Genetics, 1st Edition
The Mycota XV
T. Anke and D. Weber (Eds.)
© Springer-Verlag Berlin Heidelberg 2009

filamentous fungi and highlights the progress made for different transformation techniques and gene knockout/knockdown approaches.

II. Fungal Genomics: Advances in Exploring Sequence Data

Identification of putative new and industrial applicable enzymes or secondary metabolites, e.g. of medical interest, requires evaluation of the genetic potential of a given organism. Sequence data has to be screened for genes coding for desired enzymes involved in the pathway of interest. To meet the demand for sequence information, more and more fungal genomes have been sequenced since the first genome of the yeast model organism *Saccharomyces cerevisiae* was published (Goffeau et al. 1996).

Generally, genome sequencing projects continue to unravel the genetic capabilities of many organisms, resulting in more than 800 fully sequenced genomes currently being available, 25 of which are fungal genomes. More than 2500 sequencing projects are ongoing, including more than 300 fungal projects (http://www.genomeson line.org/gold_statistics.htm), whereby industrial and medically interesting fungi are mainly in the focus of genome sequencing projects (Table 18.1).

These genomes have been partially or completely sequenced, or they are currently being re-sequenced. However, even fully sequenced genomes such as for *S. cerevisiae* are still undergoing correction with regard to sequence details, splicing or improved annotation (Jones 2007). Comparative genomics is a very useful tool to compare less well-studied species with better understood model organisms and to clarify hypothetical genes by computational annotation. Furthermore, they help to identify differences between closely related species as regards pathogenicity, secondary metabolism or other properties (reviewed by Jones 2007). For example, the genome of *Trichoderma reesei* is considered to be a promising candidate for biofuel production from cellulose fibers. It should eventually become a monosaccharide-producing factory through the use of rational metabolic engineering (Martinez et al. 2008).

'Gene mining' as a tool for unravelling new proteins masked in fungal genomes is an approach which focuses on the industrial exploitation of currently unknown proteins and enzymes. For example, over 200 new proteases have been identified during annotation of the *A. niger* genome, of which two have already been commercialized – a protease preventing chill-haze in beer (Lopez and Edens 2005) and a protease used for the production of sport drinks (Edens et al. 2005).

Detailed analysis of the *A. niger* genome combined with the reconstruction of the metabolic networks identified many gene duplications, transporters and intra- and extracellular enzymes, providing new insights into the efficient citric acid production by *A. niger* (Pel et al. 2007). The *A. oryzae* genome was found to be extremely rich in genes involved in biomass degradation and primary and secondary metabolism (Abe et al. 2006; Kobayashi et al. 2007). New α-amylases and α-glucosidases were found. The genome sequence revealed 134 peptidase genes, in contrast to 18 peptidases hitherto known (Kobayashi et al. 2007). Comparative genomics led to improved functional annotation of the genome of *A. nidulans* and deeper insights in pathway modelling (David et al. 2008).

Furthermore, the so-called 'genome-based strain reconstruction', based on a comparison of high-producing strains with the wild-type strain, led to reconstructed strains superior to former production strains (Adrio and Demain 2006; Ohnishi et al. 2002).

An absolute novelty decrypted by genome data and first metabolic approaches was the identification of (silent) gene clusters for mycotoxin production in fungal strains which have been regarded to be safe for many years. Species such as *A. oryzae*, *A. niger* and *A. sojae* were shown to produce or have the opportunity of mycotoxin synthesis dozing in the genome (Jones 2007; Pel et al. 2007). Dependent on sequencing results, these facts indicate that using these fungi (and other species) could be more risky than initially thought and might thus require new risk assessments (Abe et al. 2006).

'Metagenomics' or 'environmental sequencing' involves the sequencing of whole ecosystems, independent of specific organisms. This boosts the amount of sequence data available (Raes et al. 2007). Metagenomics can be used to identify and analyze fungal communities (O'Brien et al. 2005), explicitly for chosen proteins (mostly polysaccharide-modifying enzymes, proteases, nitrilases; Blackwood et al. 2007; Streit and Sehmitz 2004), or to sequence fungi which cannot yet be grown in culture and therefore cannot be analyzed by traditional means. Environmental sequencing will play a major role in the decryption of complex communities and synergisms among different organisms in their natural habitats.

'Functional sequencing' allows insight into the real natural behavior over and above the limiting

Table 18.1 Genome sequencing status of selected filamentous fungi

Organism	Strain	Remark	Relevance	URL
Aspergillus clavatus	NRRL 1	Annotated	Medical	http://www.tigr.org/tdb/e2k1/acla1/
A. flavus	NRRL 3357	Annotated	Medical	http://www.Aspergillusflavus.org/genomics/
A. fumigatus	Af293	Annotated	Medical	http://www.tigr.org/tdb/e2k1/afu1/intro.shtml
A. nidulans	FGSC A4	Annotated	Model organism	http://www.broad.mit.edu/annotation/genome/Aspergillus_nidulans/Home.html
A. niger	ATCC 1015	Annotated	Industrial	http://genome.jgi-psf.org/Aspni1/Aspni1.home.html
A. niger	CBS 513.88	Annotated	Industrial	http://www.dsm.com/en_US/html/dfs/genomics_aniger.htm
A. oryzae	RIB40	Annotated	Industrial	http://www.bio.nite.go.jp/dogan/MicroTop?GENOME_ID=ao
A. terreus	NIH 2624	Annotated	Medical	http://www.broad.mit.edu/annotation/genome/Aspergillus_terreus/Home.html
Coprinus cinereus	7#130	Annotated	Model organism	http://www.broad.mit.edu/annotation/genome/Coprinus_cinereus/Home.html
Fusarium oxysporum	FGSC 4286	Annotated	Plant pathogen	http://www.broad.mit.edu/annotation/genome/fusarium_group/MultiHome.html
Laccaria bicolor	S238N-H82	Annotated	Symbiotic	http://genome.jgi-psf.org/Lacbi1/Lacbi1.home.html
Magnaporthe grisea	70-15	Annotated	Plant pathogen	http://www.broad.mit.edu/annotation/genome/magnaporthe_grisea/Home.html
Neosatorya fisheri	NRRL 181	Annotated	Medical	http://www.tigr.org/tdb/e2k1/nfa1/
Neurospora crassa	OR74A	Annotated	Model organism	http://www.broad.mit.edu/annotation/genome/Neurospora/Home.html
Phanerochaete chrysosporium	RP-78	Draft assembly	Industrial	http://genome.jgi-psf.org/whiterot1/whiterot1.home.html
Sclerotinia sclerotiorum	1980	Annotated	Plant pathogen	http://www.broad.mit.edu/annotation/genome/sclerotinia_sclerotiorum/Home.html
Trichoderma reesei	QM6a	Annotated	Industrial	http://genome.jgi-psf.org/Trire2/Trire2.home.html
Ustilago maydis	521	Annotated	Plant pathogen	http://www.broad.mit.edu/annotation/genome/ustilago_maydis/Home.html

conditions of cultivating strains under laboratory conditions and provides excellent opportunities to identify new and as yet unknown proteins and enzymes.

III. Post-Genomic Approaches to Unravel the Metabolism of Filamentous Fungi

New 'omic' tools allow fast and easy decryption of fungal genomes and the metabolites produced.

These tools can be used to explore new commercially interesting products and to improve metabolic fluxes by rational design. The main characteristic of the so-called 'post-genomic' era is the need for methods and tools to analyze the huge amount of data produced. The suffix 'omics' was created in the 1990s in the field of bioinformatics and marks the realization of the importance of information processing in biology; omics is a general term for a broad discipline of science and engineering analyzing the interaction of biological information in various 'omes'. These include the genome, transcriptome, proteome, metabolome, expressome, interactome and many more defined fields, such as glycome, ionome, lipidome or even physiome and reactome.

In most cases, genomics, transcriptomics and proteomics together with metabolomics are used to identify theoretical knockout events or metabolic fluxes in pathways of interest. The main focus is on gathering information for engineering metabolic networks to manipulate the regulatory mechanisms of the entirety of biological processes.

The term 'systems biology' is often used in combination with omics and is the 'biology' that focuses on complex systems in life. It is a holistic approach which allows analysis of the topology of biochemical and signalling networks involved in cellular responses; in addition it is able to capture the dynamics of the response. While different omics deliver only a piece of the puzzle, it is hoped that systems biology will eventually map the complete picture (Siliang Zhang 2006).

A. Transcriptomics

RNA-based technologies have been developed enormously in the past few years, whereby two different strategies have emerged: (i) strategies based on the knowledge of DNA sequences (microarrays) and (ii) strategies which do not require any sequence information, such as suppression subtractive hybridization (SSH) and serial analysis of gene expression (SAGE Breakspear and Momany 2007). The first 50 fungal microarrays were reviewed by Breakspear and Momany (2007), describing the development of fungal array approaches. Interestingly, they highlight the lack of basidiomycete microarray experiments (only three have been performed to date).

Comparative transcriptional analysis of *A. oryzae* using DNA microarrays indicated the potential of new proteins identified; and it is as a tool which can be used to develop industrial systems (Abe et al. 2006). Anderson et al. (2008b) used a comparative transcriptomic approach where one Gene Chip was developed for transcriptome analysis of triplicate batch cultivations of *A. nidulans*, *A. niger* and *A. oryzae*.

They were able to identify 23 genes conserved across *Aspergillus* spp. (mainly sugar transporters and enzymes) and 365 genes which were differently expressed in only two of the *Aspergilli*. Thus, such a cross-species study based on transcriptome analysis can be used for the identification of new enzymes or even strains. In a former publication the authors described the use of transcriptome data to perform metabolic network modelling in *A. niger*, which is explained in detail in Sect. III.C (Andersen et al. 2008).

SSH and SAGE approaches also helped to understand the infection process of plant pathogen fungi and mycoparasitic fungi. (Carpenter et al. 2005; Gowda et al. 2006; Larraya et al. 2005; Matsumura et al. 2005; Salvianti et al. 2008; Schulze Gronover et al. 2004; Yu et al. 2008). Furthermore, transcriptomic analysis of different culture conditions will help to identify genes involved in product formation and may lead to overproducing strains which are 'debugged' with respect to bottlenecks or unnecessary genes (Foreman et al. 2003). Also, insights in energy and primary metabolism have been achieved by transcriptional analysis (Maeda et al. 2004). The method of transcriptional profiling has also been used to reveal previously unknown, but relevant pathways (Rautio et al. 2006; van der Werf 2005).

B. Proteomics

Transcriptomic approaches are mostly based on the straightforward development of microarray chips, whereas proteomic fields require analytical techniques such as mass spectrometry (MS) or nuclear magnetic field resonance. The latest review of *Aspergillus* proteomics provides a summary of all recent information on this technique (Kim et al. 2008).

In 10 different studies using *Aspergillus* proteomics, a total of 28 cell surface proteins, 102 secreted and 139 intracellular proteins were identified. A proteome map highlighting proteins identified in major metabolic pathway is summarized by Kim et al. (2008).

For *Phanerochaete chrysosporium*, more than 1100 intracellular and 300 mitochondrial proteins were resolved in 2D gels and the metabolic flux shift was determined by comparative proteomics for growth on vanillin. By using vanillin, which is the decomposition product of lignin, the lignin degradation by white rot fungi can be studied. Many more proteomic approaches have been performed using fungi to identify new products for industrial applications, as recently reviewed by Kim et al. (2007).

C. Metabolomics

It is assumed that around 10 000 secondary metabolites may be produced by fungi, examples of which are found in *Aspergillus* and *Penicillium*; however, less than 10% are known (Smedsgaard and Nielsen 2005). Much understanding of the central metabolism and regulation in less-studied filamentous fungi can be learned from comparative metabolite profiling and the metabolomics of yeast and filamentous fungi (Smedsgaard and Nielsen 2005). Metabolomics is considered to be more sensitive than transcriptomics or proteomics due to the measuring of metabolite concentrations caused by environmental perturbations that may not affect transcription or protein levels, an example being enzyme activity levels (Oliver et al. 1998). The use of metabolomics means that methods are available that can simultaneously determine over 1000 charged metabolites using capillary electrophoresis-MS (Soga 2007). Even techniques which simulate the whole cell metabolism have been developed (Ishii et al. 2004).

A. nidulans represents an important model organism for studies of cell biology and gene regulation. Kouskoumvekaki et al. (2008) initiated a metabolomics approach in recombinant *Aspergilli* for clustering and classification. More than 450 detected metabolites were analyzed and resulted in the identification of seven putative biomarkers by which classification into genotype was possible. Thus, metabolite profiling is a powerful tool for the classification of filamentous fungi as well as for the identification of targets for metabolic engineering (Kouskoumvekaki et al. 2008).

A metabolic model integration of the bibliome, genome, metabolome and reactome was constructed for *A. niger* (Andersen et al. 2008). A complete metabolic network was presented which shows great potential for expanding the use of *A. niger* as a high-yield production platform. By the multi-omic approach, the use of precursors was identified to increase productivity. In order to combine the knowledge of the underlying metabolic system and the analysis of the data, novel computational approaches and methods that go beyond commonly used statistical techniques are required (Van Dien and Schilling 2006).

Progress has been made using methods for analyzing quantitative metabolomics data in the context of the entire network. Methods based on Gibbs energies of formation, the second law of thermodynamics and on known

direction reactions within the cell allow dynamic analysis of metabolites in the cell (Kummel et al. 2006).

Mainly, metabolic flux analysis (MFA) has become a fundamental tool in metabolic engineering to elucidate the metabolic state of the cell and has been applied to various biotechnological processes (Spegel et al. 2007). 'Conventional MFA' is based on mass balances using stoichiometric constraints coupled to extracellular product formation rates and is a widely used technique (Vallino and Stephanopoulos 2000).

Applying the full range of omics technologies for strain improvement, it would be desirable if the 'concept of sample' was shared among these technologies (in particular, to focus on a biological sample that is prepared for use in a specific omics assay; Morrison et al. 2006). A common data file format should be found to enable fast and secure file sharing and to reduce redundant data and the potential for errors. The main issue as regards combining the analysis of metabolomics data with other omics results is the data integration problem (Mendes 2006).

There is also the call for standards in sample collection and processing to obtain reliable results and to avoid significant error in omics data. It is essential to use the same biological sample when confronted with diverse omic approaches (Martins et al. 2007; Weckwerth et al. 2004).

This has resulted in many omics standardization initiatives aiming at the development of new concepts to overcome the problems of data confusion and sample correlation (Jones et al. 2007; Morrison et al. 2006). Also claimed is the necessity to develop common databases for storage of metabolomics data (Kouskoumvekaki et al. 2008).

IV. Metabolic Engineering: Finding the Optimum Genetic Strategy

For successful optimization of a fungal metabolism, several issues of engineering strategy have to be taken into account. Every approach depends on the organism of choice, the target to be engineered and the aim of the intervention. From various genetic engineering tools available nowadays for filamentous fungi, the most suitable has to be chosen in order to specifically and efficiently improve the strain of interest.

A. Choosing the Right Transformation Technique

Highly sophisticated genetic methods are available nowadays for filamentous fungi; however, safe and suitable transformation techniques are fundamental prerequisites for genetic engineering approaches. Different transformation techniques enable scientists to design and develop rational metabolic engineering strategies for industrially important fungal species. Linear DNA or plasmids are transferred into the fungal cell, either by chemical treatment of protoplasts using Ca^{2+} and polyethylene glycol (PEG; Spegel et al. 2007), by physical treatment such as electroporation and biolistic procedures or by using *Agrobacterium tumefaciens*-mediated transformation (AMT; Casas-Flores et al. 2004; Meyer 2008; Michielse 2005; Ruiz-Diez 2002; reviewed by Fincham 1989).

Historically, PEG-mediated transformation of protoplasts was the first transformation technique established in 1978 for the budding yeast *S. cerevisiae* (Hinnen et al. 1978). This method was afterwards used in filamentous fungi (Fincham 1989; Punt et al. 1987; Ruiz-Diez 2002; Stahl 1987). Progress in the past few decades was achieved in establishing alternative transformation methods to overcome the limits related to protoplast formation, which is highly dependent on the quality of the cell wall-degrading enzyme preparation and often only leads to low transformation rates (Michielse 2005).

In most cases, asexual spores such as conidia or sporangiospores are used as they are considered to be the most favorable constituent for transformation. However, if these are not available, the whole mycelium has to be transformed, making the procedure more difficult and laborious. In addition, the multinuclear state of some conidia, mycelia and protoplasts hampers the selection of transformants and makes the selection process tedious (Deed 1989; Farina et al. 2004). When comparing the transformation rates of yeast and filamentous fungi, it becomes evident that transformation of yeast is usually much more efficient (Ruiz-Diez 2002).

New systems such as electroporation have been developed. This is actually a method often used for the transformation of filamentous fungi (reviewed by Ruiz-Diez 2002). There are three possible fungal structures that can be transformed by means of electroporation: protoplasts, conidia or young germlings (Chakraborty et al. 1991; Ruiz-Diez 2002).

Biolistic transformation was introduced in 1987 (Klein et al. 1992; Sanford 1987) and has been applied to a number of filamentous fungi (Ruiz-Diez 2002). This method is especially beneficial for those organisms which cannot be transformed by AMT. The use of *A. tumefaciens* as a mediator of foreign DNA into fungal cells is a tool which has been adapted for many different fungal organisms in the past few years (Michielse 2005). The transformation rate is up to 100–1000 times higher than with protoplastation and can be used for most fungi which cannot be transformed by other methods or where protoplasts do not regenerate sufficiently (Casas-Flores et al. 2004; MacKenzie 2004; Meyer et al. 2003). However, even this method is not suitable for all fungi. There are also reports on less successful attempts, for example in *A. niger* (Michielse 2005).

AMT and all other transformation techniques cannot be used for every fungal organism without first being adapted. This may be explained by major differences found in fungal genomes such as *A. nidulans* (Galagan et al. 2005), *A. oryzae* (Machida et al. 2005) and *A. fumigatus* (Nierman et al. 2005). Analysis of their genomes shows that there is only 68% of amino acids identity shared by all three species (Galagan et al. 2005), an evolutionary distance comparable to that between human and fish (Dujon et al. 2004; Fedorova et al. 2008). Roughly 70% of *A. nidulans* genes could be mapped to a syntenic block with either *A. fumigatus* or *A. oryzae*, with about 50% of *A. nidulans* in conserved synteny across all three species (Galagan et al. 2005), suggesting that this could be one reason for the observation that each transformation method has to be adapted and optimized for every single species and even strain.

Apart from transformation approaches, sexual crossing is a useful tool to improve the phenotype of filamentous fungi. Unfortunately, most medically or industrially interesting fungi lack a sexual cycle, precluding them from classic genetics. However, insights into the genome of *A. fumigatus* and *A. oryzae* revealed that both are potentially capable for sexual reproduction (Dyer and Paoletti 2005; Galagan et al. 2005; Paoletti et al. 2005), suggesting that classic genetic tools can also be established for these fungi. These findings challenge the whole taxon of *fungi imperfecti* and even highlight the power of genome analysis for taxonomical (re-)classification.

A very important and interesting question as regards the different transformation systems is the fate of the introduced DNA. DNA which has been introduced will be either maintained autonomously (occurs rarely) or integrated into the genome via homologous or heterologous recombination. Homologous recombination targets the foreign DNA to regions showing sufficient homology, whereby the DNA becomes integrated into the genome as a single or tandem copy. In contrast, heterologous recombination events occur randomly, resulting in single or multi-copy integrations (de Groot et al. 1998; Malonek and Meinhardt 2001; Mullins and Kang 2001).

In the case of AMT, mainly single-copy integration events occur, whereas mainly multi-copy integrations are observed after transformation by PMT. This observation can have an important impact on the choice of a suitable transformation system for the design of a metabolic engineering strategy. AMT can be the method of choice for targeted integration into the genome and PMT can be used for ectopic and multi-copy integration to improve protein expression in the strain of interest (Meyer 2008). The latter strategy has indeed been shown to be a powerful tool for protein overexpression in *Aspergillus* and *Trichoderma* (Askolin et al. 2001; Lee et al. 1998; Verdoes JC 1995).

In addition, it was found that AMT can increase homologous recombination frequency, for example in *A. awamori* (Michielse et al. 2005). This observation pointed to new application opportunities of AMT as a suitable method for directed and insertional mutagenesis (Betts et al. 2007; Lee and Bostock 2006; Michielse 2005; Sugui et al. 2005).

V. Enhancing Gene-Targeting Efficiency

Metabolic engineering can be used to delete genes of unwanted side-pathways, thereby redirecting the metabolic flux to the required product-forming pathway. In addition, targeted integration of genes to a genomic locus known to strengthen transcription is one strategy to enhance protein expression and thereby to improve productivity of an industrial process. Besides, gene targeting is also the method used for functional genomics. The mode of integration of foreign DNA is determined by two competing processes – homologous recombination (HR) and the non-homologous end-joining (NHEJ) pathway (Dudasova et al. 2004; Krogh and Symington 2004). The low rates of HR events usually observed can render some filamentous

fungi unattractive for many industrial applications.

However, advances in fungal gene targeting were recently achieved by suppressing the NHEJ pathway (for a review, see Meyer 2008). As summarized in Table 18.2, a dramatic increase in HR efficiency was reported when strains were used in which the NHEJ pathway was inactivated (Goins et al. 2006; Ishibashi et al. 2006; Kooistra et al. 2004; Krappmann et al. 2006; Meyer et al. 2007; Nayak et al. 2006; Ninomiya et al. 2004; Poggeler and Kuck 2006; Takahashi et al. 2006).

Phenotypic analysis of defective NHEJ strains revealed that these strains showed higher sensitivity to various toxins and irradiation (da Silva Ferreira et al. 2006; Meyer et al. 2007; Ninomiya et al. 2004). Therefore, some unexpected growth behavior could appear, and the more elegant solution would be a transient silencing of the NHEJ pathway, as recently reported for *Candida glabrata* and *A. nidulans* (Nielsen et al. 2008; Ueno et al. 2007).

A. RNA-Based Tools for Metabolic Engineering

RNA technology is an attractive alternative to DNA-based methods to silence gene expression post-transcriptionally and thereby to control unwanted metabolic pathways. Different RNA techniques such as antisense RNA, RNAi and hammerhead ribozymes have been shown to be valuable tools for filamentous fungi (Table 18.2). These methods provide the advantage of not deleting a gene, thereby bypassing the possibility of lethal or other unwanted effects on the organism.

RNA-based methods are especially valuable when: (i) gene-targeting approaches fail, (ii) multiple copies of a gene of interest are present in the genome or (iii) isogenes might compensate for the knockout of the deleted gene (Akashi et al. 2005). For example, silencing a whole gene family using only a single antisense RNA construct has been described for *A. oryzae* (Yamada et al. 2007). Another advantage of the RNAi mechanism is its locus independence due to its mediation by a mobile *trans*-acting signal in the cytoplasm. Consequently, this mechanisms can be used in fungi which have multi-nuclear hyphae or a low targeting efficiency, even in heterokaryotic fungal strains, RNA-based downregulation of genes is possible (de Jong et al. 2006; Nakayashiki 2005).

Table 18.2. Genetic tools applicable to filamentous fungi for gene targeting and silencing

Organism	References	Efficiency
Aspergillus awamori		
Antisense RNA	Moralejo et al. (2002), Lombrana et al. (2004)	80%
Aspergillus flavus		
RNAi	McDonald et al. (2005)	<~90%
Improved gene targeting	Chang (2008)	96%
Aspergillus fumigatus		
RNAi	Bromley et al. (2006), Henry et al. (2007)	~80%
Improved gene targeting	da Silva Ferreira et al. (2006), Krappmann et al. (2006)	75–96%
Aspergillus giganteus		
Ribozyme	Mueller et al. (2006)	20–100%
Aspergillus nidulans		
RNAi	Khatri and Rajam (2007)	~40%
Antisense RNA	Bautista et al. (2000), Hoffmann et al. (2000), Herrmann et al. (2006)	>50–90%
Improved gene targeting	Hammond and Keller (2005)	~90%
Aspergillus niger		
RNAi	Nayak et al. (2006), Barnes et al. (2008)	21–82%
Antisense RNA	Ngiam et al. (2000)	60–70%
Improved gene targeting	Meyer et al. (2007), Barnes et al. (2008)	~80%
Aspergillus oryzae		
RNAi	Yamada et al. (2007)	90%
Antisense RNA	Zheng et al. (1998), Kitamoto et al. (1999)	75–80%
Improved gene targeting		~63–87%
Aspergillus parasiticus		
RNAi	McDonald et al. (2005)	~90%
Improved gene targeting	Chang (2008)	96%
Aspergillus sojae		
Improved gene targeting	Kadotani et al. (2003), Nakayashiki (2005)	64–75%
Coprinopsis cinereus		
RNAi	Walti et al. (2006)	90%
Cryphonectria parasitica		
Improved gene targeting	Lan et al. (2008)	80–96%
Fusarium solani		
RNAi	Ha et al. (2006)	>98%
Magnaporthe oryzae		
RNAi	Kadotani et al. (2003), Nakayashiki et al. (2005), Caracuel-Rios and Talbot (2008)	70–90%
Antisense RNA	Kadotani et al. (2003), Nguyen et al. (2008)	
Improved gene targeting	Villalba et al. (2008)	>80%
Neurospora crassa		
RNAi	Cogoni et al. (1996), Goldoni et al. (2004), Choudhary et al. (2007)	11–49%
Antisense RNA	Tentler et al. (1997), Kramer et al. (2003), Fecke (2008)	50–80%
Improved gene targeting	Ishibashi et al. (2006)	91–100%
Penicillium expansum		
RNAi	Schumann and Hertweck (2007)	~95%
Antisense RNA	Garcia-Rico et al. (2007)	~95%
Podospora anserina		
RNAi	El-Khoury et al. (2008)	100%
Schizophyllum commune		
RNAi	de Jong et al. (2006)	80%
Trichoderma harzianum		
RNAi	Cardoza et al. (2007)	71–98%
Trichoderma reseei		
Antisense RNA	Wang et al. (2005)	65%

Antisense RNA gene silencing is performed by using single-stranded RNA which is complementary to an mRNA strand transcribed within the cell. Formation of a complementary mRNA hybrid physically blocks the translation machinery and thereby stops translation of the endogenous mRNA.

Successful gene silencing using artificial antisense constructs have been reported for different filamentous fungi (Bautista et al. 2000; Blanco and Judelson 2005; Kitamoto et al. 1999; Lombrana et al. 2004; Ngiam et al. 2000; Zheng et al. 1998). For example, antisense silencing of the protease aspergillopepsin B resulted in a reduction of 10–70% of protease levels and a 30% increase in heterologous thaumatin production in A. awamori (Moralejo et al. 2002).

However, antisense-mediated reduction of gene expression to zero levels has not been reported to date. Nevertheless, exactly this phenomenon can be used for knocking-down gene expression instead of knocking it out.

For instance, the wide-domain transcription factor CreA, the key component of carbon catabolite repression in Aspergillus (Dowzer and Kelly 1991; Ruijter and Visser 1997; Shroff et al. 1997), negatively regulates a number of industrially important enzymes. Bautista et al. (2000) reported partial suppression of creA expression in A. nidulans by its antisense molecule (about 50% reduced expression was estimated) yielding a partial alleviation of glucose repression and thereby a substantial increase of the productivities of intra- and extracellular glucose-repressible enzymes.

Another RNA-based technology for filamentous fungi is the use of catalytic RNA molecules, termed ribozymes. The spliceosome and ribosomes are two examples for naturally occurring ribozymes. The so-called hammerhead ribozyme is the smallest and best-studied class of catalytic RNAs (Akashi et al. 2005; Mueller et al. 2006). The hammerhead ribozyme and other ribozymes are antisense RNA molecules. They function by binding to the target RNA moiety through Watson–Crick base pairing and inactivate it by cleaving the phosphodiester backbone at a specific cutting site. The substrate-recognition arms of hammerhead ribozymes are engineered so that the arms are rendered complementary to any chosen RNA, enabling the ribozyme to bind to its target. The functionality of hammerhead ribozymes as a tool for RNA-based technology on gene expression has been shown for bacterial, yeast, plant and mammalian systems (Akashi et al. 2005; Bussiere et al.

2003; Isaacs et al. 2006). Mueller et al. (2006) recently provided a proof of principle for filamentous fungi.

Congruent to other systems, it has been shown that ribozymes, targeting the 5' region of a substrate mRNA can lead to complete gene silencing in A. giganteus, whereas ribozymes targeting the 3' region only lead to a partial reduction (about 20–50%).

RNA interference (RNAi) is a naturally occurring post-transcriptional gene-silencing phenomenon, first described in Caenorhabditis elegans and thereafter in other organisms such as N. crassa (quelling; Romano and Macino 1992), plants ('co-suppression'; Napoli et al. 1990) and animals ('RNA interference'; Elbashir et al. 2001a). The concept of RNAi can be used for artificial gene silencing in nearly all organisms, even in filamentous fungi. By this method, double-stranded RNA (dsRNA) trans-genetically delivered to the fungal interior, is cleaved by Dicer (type-III-ribonuclease) into 21–26 nt small interfering RNA (siRNA). Dicer is always associated with Argonaute proteins (which bind targeted si/mRNA) and acts by generating siRNA molecules which in turn target mRNAs to be silenced. Dicer cleaving products get incorporated into the ribonucleoprotein complex (RISC; Bernstein et al. 2001; Hammond et al. 2000). Homologous mRNAs are subsequently recognized and degraded via complementary base pairing by means of incorporated siRNA in the RISC (Elbashir et al. 2001b; Zamore et al. 2000). In some organisms, a RNA-dependent RNA polymerase (RdRP) can use the antisense siRNA to prime the conversion of endogenous mRNA into dsRNA amplifying the silencing signal (Forrest 2004).

The most effective way of post-transcriptional gene silencing in filamentous fungi can be achieved using ectopically integrated RNAi constructs which usually code for 'double-stranded RNA' molecules. These molecules are self-complementary hairpin RNAs, formed by an inverted repeat which is interrupted by a spacer sequence, and are identical to part of the endogenous sequence being targeted (Mouyna et al. 2004). Insertion of an intron in the spacer sequence greatly increases the silencing efficiency in N. crassa, possibly due to an enhanced export of the hairpin from the nucleus during splicing (Goldoni et al. 2004). Remarkably, some fungal strains, e.g. Ustilago maydis, Candida albicans and S. cerevisiae lack components of the RNAi-silencing machinery, indicating that this tool is not applicable for these organisms (Nakayashiki 2005). As summarized in Table 18.2, specific inhibition of gene expression by RNAi has been shown to be suitable for a multitude of filamentous fungi, such as *A. nidulans* (Hammond and Keller 2005; Khatri and Rajam 2007), *A. fumigatus* (Bromley et al. 2006; Henry et al. 2007), *A. oryzae* (Yamada et al. 2007), *Coprinus cinereus* (Namekawa et al. 2005; Walti et al. 2006), *Fusarium solani* (Ha et al. 2006), *Magnaportae oryzae* (Caracuel-Rios and Talbot 2008; Kadotani et al. 2003; Nakayashiki et al. 2005), *N. crassa* (Goldoni et al. 2004) and *Schizophyllum commune* (de Jong et al. 2006). Similar to antisense strategies, RNAi-induced silencing of fungal gene expression was most often found to be incomplete (maximal reduction up to 10% of wild-type level) and full knockout phenotypes were seldom observed.

A striking disadvantage of RNAi-based gene silencing is the instability of the silencing construct and the possibility of co-silencing unwanted genes ('off targets'), showing partial sequence homology to the target gene. Many transformants lose the RNAi construct after prolonged cultivation.

Chimeric constructs with two genes in tandem can result in very different silencing efficacies (Goldoni et al. 2004; Henry et al. 2007; Nakayashiki et al. 2005). In the case of *A. fumigatus*, it has been reported that approximately 50% of the transformants lack the complete RNAi construct or part of it after prolonged cultivation (Henry et al. 2007). One possible explanation for this phenomenon might be the loss of one of the inverted repeats after the first mitotic event (Henry et al. 2007).

Still, advances in using RNAi approaches have been achieved during recent years. For example, success has been reported by using inducible RNAi constructs in *A. fumigatus* (Khalaj et al. 2007) or by using new uptake methods of artificial siRNA constructs in *A. nidulans* (Khatri and Rajam 2007). Nevertheless, the possibility of off-target effects can form an obstacle and future work on the approach has to address the question what is the optimum sequence length and the minimum homology to avoid any unwanted co-silencing.

VI. Concluding Remarks and Prospects

The fungal post-genomic era is still in its infancy; however, the real power of the omic approaches is

the possibility to analyze cellular processes and responses on different levels, including DNA, RNA, protein and metabolite levels (Shulaev et al. 2008).

By integrated analysis of these levels, several important features of metabolic regulation has been and will be identified (Le Lay et al. 2006; Shulaev et al. 2008). Future strategies will combine all fields of omics and will allow the unravelling of the dynamics of cellular metabolic activities in various filamentous fungi. The identification of new proteins, enzymes, pathways and their involvement in metabolic networks as well as 'classic genetic' techniques and new metabolic engineering approaches will eventually enable scientists to develop optimum production strains. The use of the new and highly sophisticated omic methods developed in the fungal post-genomic era will open the floodgates towards high-throughput analysis and efficient rational metabolic engineering approaches. The knowledge gained leads to the improvement of industrial biotechnological processes and will help to meet the increasing need for sustainable fungal bio-products. In order to avoid bottlenecks in the post-genomic era, standardized methods and protocols have to be established for sampling, sample processing, sample collection, sample handling and integration into databanks.

References

Abe K, Gomi K, Hasegawa F, Machida M (2006) Impact of *Aspergillus oryzae* genomics on industrial production of metabolites. Mycopathologia 162:143–153

Adrio JL, Demain AL (2006) Genetic improvement of processes yielding microbial products. FEMS Microbiol Rev 30:187–214

Akashi H, Matsumoto S, Taira K (2005) Gene discovery by ribozyme and siRNA libraries. Nat Rev Mol Cell Biol 6:413–422

Andersen MR, Nielsen ML, Nielsen J (2008) Metabolic model integration of the bibliome, genome, metabolome and reactome of *Aspergillus niger*. Mol Syst Biol 4:178

Askolin S, Nakari-Setala T, Tenkanen M (2001) Over-production, purification, and characterization of the *Trichoderma reesei* hydrophobin HFBI. Appl Microbiol Biotechnol 57:124–130

Barnes SE, Alcocer MJ, Archer DB (2008) siRNA as a molecular tool for use in *Aspergillus niger*. Biotechnol Lett 30:885–890

Bautista LF, Aleksenko A, Hentzer M, Santerre-Henriksen A, Nielsen J (2000) Antisense silencing of the *creA* gene in *Aspergillus nidulans*. Appl Environ Microbiol 66:4579–4581

Bernstein E, Caudy AA, Hammond SM, Hannon GJ (2001) Role for a bidentate ribonuclease in the initiation step of RNA interference. Nature 409:363–366

Betts MF, Tucker SL, Galadima N, Meng Y, Patel G, Li L, Donofrio N, Floyd A, Nolin S, Brown D, Mandel MA, Mitchell TK, Xu JR, Dean RA, Farman ML, Orbach MJ (2007) Development of a high throughput transformation system for insertional mutagenesis in *Magnaporthe oryzae*. Fungal Genet Biol 44:1035–1049

Blackwood CB, Waldrop MP, Zak DR, Sinsabaugh RL (2007) Molecular analysis of fungal communities and laccase genes in decomposing litter reveals differences among forest types but no impact of nitrogen deposition. Environ Microbiol 9:1306–1316

Blanco FA, Judelson HS (2005) A bZIP transcription factor from Phytophthora interacts with a protein kinase and is required for zoospore motility and plant infection. Mol Microbiol 56:638–648

Breakspear A, Momany M (2007) The first fifty microarray studies in filamentous fungi. Microbiology 153:7–15

Bromley M, Gordon C, Rovira-Graells N, Oliver J (2006) The *Aspergillus fumigatus* cellobiohydrolase B (*cbhB*) promoter is tightly regulated and can be exploited for controlled protein expression and RNAi. FEMS Microbiol Lett 264:246–254

Bussiere F, Ledu S, Girard M, Heroux M, Perreault JP, Matton DP (2003) Development of an efficient cis-trans-cis ribozyme cassette to inactivate plant genes. Plant Biotechnol J 1:423–435

Caracuel-Rios Z, Talbot NJ (2008) Silencing the crowd: high-throughput functional genomics in *Magnaporthe oryzae*. Mol Microbiol Mol Microbiol 68:1341–1344

Cardoza RE, Hermosa MR, Vizcaino JA, Gonzalez F, Llobell A, Monte E, Gutierrez S (2007) Partial silencing of a hydroxy-methylglutaryl-CoA reductase-encoding gene in *Trichoderma harzianum* CECT 2413 results in a lower level of resistance to lovastatin and lower antifungal activity. Fungal Genet Biol 44:269–283

Carpenter MA, Stewart A, Ridgway HJ (2005) Identification of novel *Trichoderma hamatum* genes expressed during mycoparasitism using subtractive hybridisation. FEMS Microbiol Lett 251:105–112

Casas-Flores S, Rosales-Saavedra T, Herrera-Estrella A (2004) Three decades of fungal transformation: novel technologies. Methods Mol Biol 267:315–325

Chakraborty BN, Patterson NA, Kapoor M (1991) An electroporation-based system for high-efficiency transformation of germinated conidia of filamentous fungi. Can J Microbiol 37:858–863

Chang PK (2008) A highly efficient gene-targeting system for *Aspergillus parasiticus*. Lett Appl Microbiol 46:587–592

Choudhary S, Lee HC, Maiti M, He Q, Cheng P, Liu Q, Liu Y (2007) A double-stranded-RNA response program important for RNA interference efficiency. Mol Cell Biol 27:3995–4005

Cogoni C, Irelan JT, Schumacher M, Schmidhauser TJ, Selker EU, Macino G (1996) Transgene silencing of the al-1 gene in vegetative cells of *Neurospora* is mediated by a cytoplasmic effector and does not depend on DNA-DNA interactions or DNA methylation. EMBO J 15:3153–3163

Conesa AP, van Luijk PJ, van den Hondel CA (2001) The secretion pathway in filamentous fungi: a biotechnological view. Fungal Genet Biol 33:155–171

da Silva Ferreira ME, Kress MR, Savoldi M, Goldman MH, Hartl A, Heinekamp T, Brakhage AA, Goldman GH (2006) The *akuB* (KU80) mutant deficient for nonhomologous end joining is a powerful tool for analyzing pathogenicity in *Aspergillus fumigatus*. Eukaryot Cell 5:207–211

David H, Ozcelik IS, Hofmann G, Nielsen J (2008) Analysis of *Aspergillus nidulans* metabolism at the genome-scale. BMC Genomics 9:163

de Groot MJ, Bundock P, Hooykaas PJ, Beijersbergen AG (1998) *Agrobacterium tumefaciens*-mediated transformation of filamentous fungi. Nat Biotechnol 16:839–842

de Jong JF, Deelstra HJ, Wosten HA, Lugones LG (2006) RNA-mediated gene silencing in monokaryons and dikaryons of *Schizophyllum commune*. Appl Environ Microbiol 72:1267–1269

Deed ASR (1989) Formation and regeneration of protoplasts of *Sclerotium glucanicum*. Appl Microbiol Biotechnol 31:259–264

Dowzer CE, Kelly JM (1991) Analysis of the *creA* gene, a regulator of carbon catabolite repression in *Aspergillus nidulans*. Mol Cell Biol 11:5701–5709

Dudasova Z, Dudas A, Chovanec M (2004) Non-homologous end-joining factors of *Saccharomyces cerevisiae*. FEMS Microbiol Rev 28:581–601

Dujon B, Sherman D, Fischer G, Durrens P, Casaregola S, Lafontaine I, De Montigny J, Marck C, Neuveglise C, Talla E, Goffard N, Frangeul L, Aigle M, Anthouard V, Babour A, Barbe V, Barnay S, Blanchin S, Beckerich JM, Beyne E, Bleykasten C, Boisrame A, Boyer J, Cattolico L, Confanioleri F, De Daruvar A, Despons L, Fabre E, Fairhead C, Ferry-Dumazet H, Groppi A, Hantraye F, Hennequin C, Jauniaux N, Joyet P, Kachouri R, Kerrest A, Koszul R, Lemaire M, Lesur I, Ma L, Muller H, Nicaud JM, Nikolski M, Oztas S, Ozier-Kalogeropoulos O, Pellenz S, Potier S, Richard GF, Straub ML, Suleau A, Swennen D, Tekaia F, Wesolowski-Louvel M, Westhof E, Wirth B, Zeniou-Meyer M, Zivanovic I, Bolotin-Fukuhara M, Thierry A, Bouchier C, Caudron B, Scarpelli C, Gaillardin C, Weissenbach J, Wincker P, Souciet JL (2004) Genome evolution in yeasts. Nature 430:35–44

Dyer PS, Paoletti M (2005) Reproduction in *Aspergillus fumigatus*: sexuality in a supposedly asexual species? Med Mycol 43[Suppl 1]:7–14

Edens L, Dekker P, van der Hoeven R, Deen F, de Roos A, Floris R (2005) Extracellular prolyl endoprotease from *Aspergillus niger* and its use in the debittering of protein hydrolysates. J Agric Food Chem 53:7950–7957

El-Khoury R, Sellem CH, Coppin E, Boivin A, Maas MF, Debuchy R, Sainsard-Chanet A (2008) Gene deletion and allelic replacement in the filamentous fungus *Podospora anserina*. Curr Genet 53:249–258

Elbashir SM, Harborth J, Lendeckel W, Yalcin A, Weber K, Tuschl T (2001a) Duplexes of 21-nucleotide RNAs mediate RNA interference in cultured mammalian cells. Nature 411:494–498

Elbashir SM, Lendeckel W, Tuschl T (2001b) RNA interference is mediated by 21- and 22-nucleotide RNAs. Genes Dev 15:188–200

Farina JI, Molina OE, Figueroa LI (2004) Formation and regeneration of protoplasts in *Sclerotium rolfsii* ATCC 201126. J Appl Microbiol 96:254–262

Fecke W (2008) An antisense RNA expression vector for *Neurospora crassa*. ECFG 9 poster session 1

Fedorova ND, Khaldi N, Joardar VS, Maiti R, Amedeo P, Anderson MJ, Crabtree J, Silva JC, Badger JH, Albarraq A, Angiuoli S, Bussey H, Bowyer P, Cotty PJ, Dyer PS, Egan A, Galens K, Fraser-Liggett CM, Haas BJ, Inman JM, Kent R, Lemieux S, Malavazi I, Orvis J, Roemer T, Ronning CM, Sundaram JP, Sutton G, Turner G, Venter JC, White OR, Whitty BR, Youngman P, Wolfe KH, Goldman GH, Wortman JR, Jiang B, Denning DW, Nierman WC (2008) Genomic islands in the pathogenic filamentous fungus *Aspergillus fumigatus*. PLoS Genet 4:e1000046

Fincham JR (1989) Transformation in fungi. Microbiol Rev 53:148–170

Fischer R, Zekert N, Takeshita N (2008) Polarized growth in fungi–interplay between the cytoskeleton, positional markers and membrane domains. Mol Microbiol 68:813–826

Foreman PK, Brown D, Dankmeyer L, Dean R, Diener S, Dunn-Coleman NS, Goedegebuur F, Houfek TD, England GJ, Kelley AS, Meerman HJ, Mitchell T, Mitchinson C, Olivares HA, Teunissen PJ, Yao J, Ward M (2003) Transcriptional regulation of biomass-degrading enzymes in the filamentous fungus *Trichoderma reesei*. J Biol Chem 278:31988–31997

Forrest EC, Cogoni C, Macino G (2004) The RNA-dependent RNA polymerase, QDE-1, is a rate-limiting factor in post-transcriptional gene silencing in *Neurospora crassa*. Nucleic Acids Res 32:2123–2128

Galagan JE, Calvo SE, Cuomo C, Ma LJ, Wortman JR, Batzoglou S, Lee SI, Basturkmen M, Spevak CC, Clutterbuck J, Kapitonov V, Jurka J, Scazzocchio C, Farman M, Butler J, Purcell S, Harris S, Braus GH, Draht O, Busch S, D'Enfert C, Bouchier C, Goldman GH, Bell-Pedersen D, Griffiths-Jones S, Doonan JH, Yu J, Vienken K, Pain A, Freitag M, Selker EU, Archer DB, Penalva MA, Oakley BR, Momany M, Tanaka T, Kumagai T, Asai K, Machida M, Nierman WC, Denning DW, Caddick M, Hynes M, Paoletti M, Fischer R, Miller B, Dyer P, Sachs MS, Osmani SA, Birren BW (2005) Sequencing of *Aspergillus nidulans* and comparative analysis with *A. fumigatus* and *A. oryzae*. Nature 438:1105–1115

Garcia-Rico RO, Martin JF, Fierro F (2007) The pga1 gene of *Penicillium chrysogenum* NRRL 1951 encodes a heterotrimeric G protein alpha subunit that controls growth and development. Res Microbiol 158:437–446

Goffeau A, Barrell BG, Bussey H, Davis RW, Dujon B, Feldmann H, Galibert F, Hoheisel JD, Jacq C, Johnston M, Louis EJ, Mewes HW, Murakami Y, Philippsen P, Tettelin H, Oliver SG (1996) Life with 6000 genes. Science 274:563–567

Goins CL, Gerik KJ, Lodge JK (2006) Improvements to gene deletion in the fungal pathogen *Cryptococcus neoformans*: absence of Ku proteins increases homologous recombination, and co-transformation of independent DNA molecules allows rapid complementation of deletion phenotypes. Fungal Genet Biol 43:531–544

Goldoni M, Azzalin G, Macino G, Cogoni C (2004) Efficient gene silencing by expression of double stranded RNA in *Neurospora crassa*. Fungal Genet Biol 41:1016–1024

Gowda M, Venu RC, Raghupathy MB, Nobuta K, Li H, Wing R, Stahlberg E, Couglan S, Haudenschild CD, Dean R, Nahm BH, Meyers BC, Wang GL (2006) Deep and comparative analysis of the mycelium and appressorium transcriptomes of *Magnaporthe grisea* using MPSS, RL-SAGE, and oligoarray methods. BMC Genomics 7:310

Grimm LH, Kelly S, Krull R, Hempel DC (2005) Morphology and productivity of filamentous fungi. Appl Microbiol Biotechnol 69:375–384

Ha YS, Covert SF, Momany M (2006) FsFKS1, the 1,3-beta-glucan synthase from the caspofungin-resistant fungus *Fusarium solani*. Eukaryot Cell 5:1036–1042

Hammond SM, Bernstein E, Beach D, Hannon GJ (2000) An RNA-directed nuclease mediates post-transcriptional gene silencing in *Drosophila* cells. Nature 404:293–296

Hammond TM, Keller NP (2005) RNA silencing in *Aspergillus nidulans* is independent of RNA-dependent RNA polymerases. Genetics 169:607–617

Henry C, Mouyna I, Latge JP (2007) Testing the efficacy of RNA interference constructs in *Aspergillus fumigatus*. Curr Genet 51:277–284

Herrmann M, Sprote P, Brakhage AA (2006) Protein kinase C (PkcA) of *Aspergillus nidulans* is involved in penicillin production. Appl Environ Microbiol 72:2957–2970

Hinnen A, Hicks JB, Fink GR (1978) Transformation of yeast. Proc Natl Acad Sci USA 75:1929–1933

Hoffmann B, LaPaglia SK, Kubler E, Andermann M, Eckert SE, Braus GH (2000) Developmental and metabolic regulation of the phosphoglucomutase-encoding gene, pgmB, of *Aspergillus nidulans*. Mol Gen Genet 262:1001–1011

Isaacs FJ, Dwyer DJ, Collins JJ (2006) RNA synthetic biology. Nat Biotechnol 24:545–554

Ishibashi K, Suzuki KoY, Takakura C, Inoue H (2006) Nonhomologous chromosomal integration of foreign DNA is completely dependent on MUS-53 (human Lig4 homolog) in *Neurospora*. Proc Natl Acad Sci USA 103:14871–14876

Ishii N, Robert M, Nakayama Y, Kanai A, Tomita M (2004) Toward large-scale modeling of the microbial cell for computer simulation. J Biotechnol 113: 281–294

Jones AR, Miller M, Aebersold R, Apweiler R, Ball CA, Brazma A, Degreef J, Hardy N, Hermjakob H, Hubbard SJ, Hussey P, Igra M, Jenkins H, Julian RK Jr, Laursen K, Oliver SG, Paton NW, Sansone SA, Sarkans U, Stoeckert CJ Jr, Taylor CF, Whetzel PL, White JA, Spellman P, Pizarro A (2007) The Functional Genomics Experiment model (FuGE): an extensible framework for standards in functional genomics. Nat Biotechnol 25:1127–1133

Jones MG (2007) The first filamentous fungal genome sequences: *Aspergillus* leads the way for essential everyday resources or dusty museum specimens? Microbiology 153:1–6

Kadotani N, Nakayashiki H, Tosa Y, Mayama S (2003) RNA silencing in the phytopathogenic fungus *Magnaporthe oryzae*. Mol Plant Microbe Interact 16:769–776

Kadotani N, Nakayashiki H, Tosa Y, Mayama S (2004) One of the two Dicer-like proteins in the filamentous fungi *Magnaporthe oryzae* genome is responsible for hairpin RNA-triggered RNA silencing and related small interfering RNA accumulation. J Biol Chem 279:44467–44474

Karnaukhova E, Ophir Y, Trinh L, Dalal N, Punt PJ, Golding B, Shiloach J (2007) Expression of human alpha1-proteinase inhibitor in *Aspergillus niger*. Microb Cell Fact 6:34

Khalaj V, Brookman JL, Robson GD (2001) A study of the protein secretory pathway of *Aspergillus niger* using a glucoamylase-GFP fusion protein. Fungal Genet Biol 32:55–65

Khalaj V, Eslami H, Azizi M, Rovira-Graells N, Bromley M (2007) Efficient downregulation of alb1 gene using an AMA1-based episomal expression of RNAi construct in *Aspergillus fumigatus*. FEMS Microbiol Lett 270:250–254

Khatri M, Rajam MV (2007) Targeting polyamines of *Aspergillus nidulans* by siRNA specific to fungal ornithine decarboxylase gene. Med Mycol 45:211–220

Kim Y, Nandakumar MP, Marten MR (2007) Proteomics of filamentous fungi. Trends Biotechnol 25:395–400

Kim Y, Nandakumar MP, Marten MR (2008) The state of proteome profiling in the fungal genus *Aspergillus*. Brief Funct Genomic Proteomic 7:87–94

Kitamoto N, Yoshino S, Ohmiya K, Tsukagoshi N (1999) Sequence analysis, overexpression, and antisense inhibition of a beta-xylosidase gene, xylA, from *Aspergillus oryzae* KBN616. Appl Environ Microbiol 65:20–24

Klein RM, Wolf ED, Wu R, Sanford JC (1992) High-velocity microprojectiles for delivering nucleic acids into living cells. Biotechnology 24:384–386

Kobayashi T, Abe K, Asai K, Gomi K, Juvvadi PR, Kato M, Kitamoto K, Takeuchi M, Machida M (2007) Genomics of *Aspergillus oryzae*. Biosci Biotechnol Biochem 71:646–670

Kooistra R, Hooykaas PJ, Steensma HY (2004) Efficient gene targeting in *Kluyveromyces lactis*. Yeast 21:781–792

Kouskoumvekaki I, Yang Z, Jonsdottir SO, Olsson L, Panagiotou G (2008) Identification of biomarkers for genotyping Aspergilli using non-linear methods for clustering and classification. BMC Bioinformatics 9:59

Kramer C, Loros JJ, Dunlap JC, Crosthwaite SK (2003) Role for antisense RNA in regulating circadian clock function in *Neurospora crassa*. Nature 421:948–952

Krappmann S, Sasse C, Braus GH (2006) Gene targeting in *Aspergillus fumigatus* by homologous recombination is facilitated in a nonhomologous end-joining-deficient genetic background. Eukaryot Cell 5:212–215

Krogh BO, Symington LS (2004) Recombination proteins in yeast. Annu Rev Genet 38:233–271

Kummel A, Panke S, Heinemann M (2006) Systematic assignment of thermodynamic constraints in metabolic network models. BMC Bioinformatics 7:512

Lan X, Yao Z, Zhou Y, Shang J, Lin H, Nuss DL, Chen B (2008) Deletion of the *cpku80* gene in the chestnut blight fungus, *Cryphonectria parasitica*, enhances gene disruption efficiency. Curr Genet 53:59–66

Larraya LM, Boyce KJ, So A, Steen BR, Jones S, Marra M, Kronstad JW (2005) Serial analysis of gene expression reveals conserved links between protein kinase A, ribosome biogenesis, and phosphate metabolism in *Ustilago maydis*. Eukaryot Cell 4:2029–2043

Le Lay P, Isaure MP, Sarry JE, Kuhn L, Fayard B, Le Bail JL, Bastien O, Garin J, Roby C, Bourguignon J (2006) Metabolomic, proteomic and biophysical analyses of *Arabidopsis thaliana* cells exposed to a caesium stress. Influence of potassium supply. Biochimie 88:1533–1547

Lee DG, Nishimura-Masuda I, Nakamura A, Hidaka M, Masaki H, Uozumi T (1998) Overproduction of alpha-glucosidase in *Aspergillus niger* transformed with the cloned gene aglA. J Gen Appl Microbiol 44:177–181

Lee MH, Bostock RM (2006) Agrobacterium T-DNA-mediated integration and gene replacement in the brown rot pathogen *Monilinia fructicola*. Curr Genet 49:309–322

Lombrana M, Moralejo FJ, Pinto R, Martin JF (2004) Modulation of *Aspergillus awamori* thaumatin secretion by modification of *bipA* gene expression. Appl Environ Microbiol 70:5145–5152

Lopez M, Edens L (2005) Effective prevention of chill-haze in beer using an acid proline-specific endoprotease from *Aspergillus niger*. J Agric Food Chem 53:7944–7949

Machida M, Asai K, Sano M, Tanaka T, Kumagai T, Terai G, Kusumoto K, Arima T, Akita O, Kashiwagi Y, Abe K, Gomi K, Horiuchi H, Kitamoto K, Kobayashi T, Takeuchi M, Denning DW, Galagan JE, Nierman WC, Yu J, Archer DB, Bennett JW, Bhatnagar D, Cleveland TE, Fedorova ND, Gotoh O, Horikawa H, Hosoyama A, Ichinomiya M, Igarashi R, Iwashita K, Juvvadi PR, Kato M, Kato Y, Kin T, Kokubun A, Maeda H, Maeyama N, Maruyama J, Nagasaki H, Nakajima T, Oda K, Okada K, Paulsen I, Sakamoto K, Sawano T, Takahashi M, Takase K, Terabayashi Y, Wortman JR, Yamada O, Yamagata Y, Anazawa H, Hata Y, Koide Y, Komori T, Koyama Y, Minetoki T, Suharnan S, Tanaka A, Isono K, Kuhara S, Ogasawara N, Kikuchi H (2005) Genome sequencing and analysis of *Aspergillus oryzae*. Nature 438:1157–1161

MacKenzie DA, Archer DB (2004) Filamentous fungi as expression systems for heterologous proteins. Springer, Heidelberg The Mycota II:289–316

Maeda H, Sano M, Maruyama Y, Tanno T, Akao T, Totsuka Y, Endo M, Sakurada R, Yamagata Y, Machida M, Akita O, Hasegawa F, Abe K, Gomi K,

Nakajima T, Iguchi Y (2004) Transcriptional analysis of genes for energy catabolism and hydrolytic enzymes in the filamentous fungus *Aspergillus oryzae* using cDNA microarrays and expressed sequence tags. Appl Microbiol Biotechnol 65:74–83

Malonek S, Meinhardt F (2001) *Agrobacterium tumefaciens*-mediated genetic transformation of the phytopathogenic ascomycete *Calonectria morganii*. Curr Genet 40:152–155

Martinez D, Berka RM, Henrissat B, Saloheimo M, Arvas M, Baker SE, Chapman J, Chertkov O, Coutinho PM, Cullen D, Danchin EG, Grigoriev IV, Harris P, Jackson M, Kubicek CP, Han CS, Ho I, Larrondo LF, de Leon AL, Magnuson JK, Merino S, Misra M, Nelson B, Putnam N, Robbertse B, Salamov AA, Schmoll M, Terry A, Thayer N, Westerholm-Parvinen A, Schoch CL, Yao J, Barbote R, Nelson MA, Detter C, Bruce D, Kuske CR, Xie G, Richardson P, Rokhsar DS, Lucas SM, Rubin EM, Dunn-Coleman N, Ward M, Brettin TS (2008) Genome sequencing and analysis of the biomass-degrading fungus *Trichoderma reesei* (syn. *Hypocrea jecorina*). Nat Biotechnol 26:553–560

Martins AM, Sha W, Evans C, Martino-Catt S, Mendes P, Shulaev V (2007) Comparison of sampling techniques for parallel analysis of transcript and metabolite levels in *Saccharomyces cerevisiae*. Yeast 24:181–188

Matsumura H, Ito A, Saitoh H, Winter P, Kahl G, Kruger DH, Terauchi R (2005) SuperSAGE. Cell Microbiol 7:11–18

McDonald T, Brown D, Keller NP, Hammond TM (2005) RNA silencing of mycotoxin production in *Aspergillus* and *Fusarium* species. Mol Plant Microbe Interact 18:539–545

Mendes BMaP (2006) Bioinformatics approaches to integrate metabolomics and other systems biology data. Springer, Heidelberg Biotechnology in Agriculture and Forestry; 57:105–115

Meyer V (2008) Genetic engineering of filamentous fungi-progress, obstacles and future trends. Biotechnol Adv 26:177–185

Meyer V, Arentshorst M, El-Ghezal A, Drews AC, Kooistra R, van den Hondel CA, Ram AF (2007) Highly efficient gene targeting in the *Aspergillus niger kusA* mutant. J Biotechnol 128:770–775

Meyer V, Arentshorst M, van den Hondel CA, Ram AF (2008) The polarisome component SpaA localises to hyphal tips of *Aspergillus niger* and is important for polar growth. Fungal Genet Biol 45:152–164

Meyer V, Mueller D, Strowig T, Stahl U (2003) Comparison of different transformation methods for *Aspergillus giganteus*. Curr Genet 43:371–377

Michielse CB, Arentshorst M, Ram AF, van den Hondel CA (2005) *Agrobacterium*-mediated transformation leads to improved gene replacement efficiency in *Aspergillus awamori*. Fungal Genet Biol 42:9–19

Michielse CBH, van den Hondel CA, Ram AF (2005) *Agrobacterium*-mediated transformation as a tool for functional genomics in fungi. Curr Genet 48:1–17

Moralejo FJ, Cardoza RE, Gutierrez S, Lombrana M, Fierro F, Martin JF (2002) Silencing of the aspergillopepsin B (pepB) gene of *Aspergillus awamori* by antisense RNA expression or protease removal by gene disruption results in a large increase in thaumatin production. Appl Environ Microbiol 68:3550–3559

Morrison N, Cochrane G, Faruque N, Tatusova T, Tateno Y, Hancock D, Field D (2006) Concept of sample in OMICS technology. Omics 10:127–137

Mouyna I, Henry C, Doering TL, Latge JP (2004) Gene silencing with RNA interference in the human pathogenic fungus *Aspergillus fumigatus*. FEMS Microbiol Lett 237:317–324

Mueller D, Stahl U, Meyer V (2006) Application of hammerhead ribozymes in filamentous fungi. J Microbiol Methods 65:585–595

Mullins ED, Kang S (2001) Transformation: a tool for studying fungal pathogens of plants. Cell Mol Life Sci 58:2043–2052

Nakayashiki H (2005) RNA silencing in fungi: mechanisms and applications. FEBS Lett 579:5950–5957

Nakayashiki H, Hanada S, Nguyen BQ, Kadotani N, Tosa Y, Mayama S (2005) RNA silencing as a tool for exploring gene function in ascomycete fungi. Fungal Genet Biol 42:275–283

Namekawa SH, Iwabata K, Sugawara H, Hamada FN, Koshiyama A, Chiku H, Kamada T, Sakaguchi K (2005) Knockdown of LIM15/DMC1 in the mushroom *Coprinus cinereus* by double-stranded RNA-mediated gene silencing. Microbiology 151:3669–3678

Nayak T, Szewczyk E, Oakley CE, Osmani A, Ukil L, Murray SL, Hynes MJ, Osmani SA, Oakley BR (2006) A versatile and efficient gene-targeting system for *Aspergillus nidulans*. Genetics 172:1557–1566

Nevalainen KM, Te'o VS, Bergquist PL (2005) Heterologous protein expression in filamentous fungi. Trends Biotechnol 23:468–474

Ngiam C, Jeenes DJ, Punt PJ, Van Den Hondel CA, Archer DB (2000) Characterization of a foldase, protein disulfide isomerase A, in the protein secretory pathway of *Aspergillus niger*. Appl Environ Microbiol 66:775–782

Nguyen QB, Kadotani N, Kasahara S, Tosa Y, Mayama S, Nakayashiki H (2008) Systematic functional analysis of calcium-signalling proteins in the genome of the rice-blast fungus, *Magnaporthe oryzae*, using a high-throughput RNA-silencing system. Mol Microbiol 68:1348–1365

Nielsen JB, Nielsen ML, Mortensen UH (2008) Transient disruption of non-homologous end-joining facilitates targeted genome manipulations in the filamentous fungus *Aspergillus nidulans*. Fungal Genet Biol 45:165–170

Nierman WC, Pain A, Anderson MJ, Wortman JR, Kim HS, Arroyo J, Berriman M, Abe K, Archer DB, Bermejo C, Bennett J, Bowyer P, Chen D, Collins M, Coulsen R, Davies R, Dyer PS, Farman M, Fedorova N, Fedorova N, Feldblyum TV, Fischer R, Fosker N, Fraser A, Garcia JL, Garcia MJ, Goble A, Goldman GH, Gomi K, Griffith-Jones S, Gwilliam R, Haas B, Haas H, Harris D, Horiuchi H, Huang J, Humphray S, Jimenez J, Keller N, Khouri H, Kitamoto K, Kobayashi T, Konzack S, Kulkarni R, Kumagai T, Lafon A, Latge JP, Li W, Lord A, Lu C, Majoros WH, May GS, Miller BL, Mohamoud Y, Molina M, Monod M, Mouyna I,

Mulligan S, Murphy L, O'Neil S, Paulsen I, Penalva MA, Pertea M, Price C, Pritchard BL, Quail MA, Rabbinowitsch E, Rawlins N, Rajandream MA, Reichard U, Renauld H, Robson GD, Rodriguez de Cordoba S, Rodriguez-Pena JM, Ronning CM, Rutter S, Salzberg SL, Sanchez M, Sanchez-Ferrero JC, Saunders D, Seeger K, Squares R, Squares S, Takeuchi M, Tekaia F, Turner G, Vazquez de Aldana CR, Weidman J, White O, Woodward J, Yu JH, Fraser C, Galagan JE, Asai K, Machida M, Hall N, Barrell B, Denning DW (2005) Genomic sequence of the pathogenic and allergenic filamentous fungus *Aspergillus fumigatus*. Nature 438:1151–1156

Ninomiya Y, Suzuki K, Ishii C, Inoue H (2004) Highly efficient gene replacements in *Neurospora* strains deficient for nonhomologous end-joining. Proc Natl Acad Sci USA 101:12248–12253

O'Brien HE, Parrent JL, Jackson JA, Moncalvo JM, Vilgalys R (2005) Fungal community analysis by large-scale sequencing of environmental samples. Appl Environ Microbiol 71:5544–5550

Ohnishi J, Mitsuhashi S, Hayashi M, Ando S, Yokoi H, Ochiai K, Ikeda M (2002) A novel methodology employing *Corynebacterium glutamicum* genome information to generate a new L-lysine-producing mutant. Appl Microbiol Biotechnol 58:217–223

Oliver SG, Winson MK, Kell DB, Baganz F (1998) Systematic functional analysis of the yeast genome. Trends Biotechnol 16:373–378

Paoletti M, Rydholm C, Schwier EU, Anderson MJ, Szakacs G, Lutzoni F, Debeaupuis JP, Latge JP, Denning DW, Dyer PS (2005) Evidence for sexuality in the opportunistic fungal pathogen *Aspergillus fumigatus*. Curr Biol 15:1242–1248

Papagianni M, Mattey M, Kristiansen B (2003) Design of a tubular loop bioreactor for scale-up and scale-down of fermentation processes. Biotechnol Prog 19:1498–1504

Pel HJ, de Winde JH, Archer DB, Dyer PS, Hofmann G, Schaap PJ, Turner G, de Vries RP, Albang R, Albermann K, Kersen MR, Bendtsen JD, Benen JA, van den Berg M, Breestraat S, Caddick MX, Contreras R, Cornell M, Coutinho PM, Danchin EG, Debets AJ, Dekker P, van Dijck PW, van Dijk A, Dijkhuizen L, Driessen AJ, d'Enfert C, Geysens S, Goosen C, Groot GS, de Groot PW, Guillemette T, Henrissat B, Herweijer M, van den Hombergh JP, van den Hondel CA, van der Heijden RT, van der Kaaij RM, Klis FM, Kools HJ, Kubicek CP, van Kuyk PA, Lauber J, Lu X, van der Maarel MJ, Meulenberg R, Menke H, Mortimer MA, Nielsen J, Oliver SG, Olsthoorn M, Pal K, van Peij NN, Ram AF, Rinas U, Roubos JA, Sagt CM, Schmoll M, Sun J, Ussery D, Varga J, Vervecken W, van de Vondervoort PJ, Wedler H, Wosten HA, Zeng AP, van Ooyen AJ, Visser J, Stam H (2007) Genome sequencing and analysis of the versatile cell factory *Aspergillus niger* CBS 513.88. Nat Biotechnol 25:221–231

Poggeler S, Kuck U (2006) Highly efficient generation of signal transduction knockout mutants using a fungal strain deficient in the mammalian *ku70* ortholog. Gene 37:1–10

Punt PJ, Oliver RP, Dingemanse MA, Pouwels PH, van den Hondel CA (1987) Transformation of *Aspergillus* based on the hygromycin B resistance marker from *Escherichia coli*. Gene 56:117–124

Punt PJ, van Biezen N, Conesa A, Albers A, Mangnus J, van den Hondel C (2002) Filamentous fungi as cell factories for heterologous protein production. Trends Biotechnol 20:200–206

Raes J, Foerstner KU, Bork P (2007) Get the most out of your metagenome: computational analysis of environmental sequence data. Curr Opin Microbiol 10:490–498

Rautio JJ, Smit BA, Wiebe M, Penttila M, Saloheimo M (2006) Transcriptional monitoring of steady state and effects of anaerobic phases in chemostat cultures of the filamentous fungus *Trichoderma reesei*. BMC Genomics 7:247

Ruijter GJ, Visser J (1997) Carbon repression in *Aspergilli*. FEMS Microbiol Lett 151:103–114

Ruiz-Diez B (2002) Strategies for the transformation of filamentous fungi. J Appl Microbiol 92:189–195

Salvianti F, Bettini PP, Giordani E, Sacchetti P, Bellini E, Buiatti M (2008) Identification by suppression subtractive hybridization of genes expressed in pear (*Pyrus* spp.) upon infestation with *Cacopsylla pyri* (Homoptera:Psyllidae). J Plant Physiol 165:1808–1816

Sanford JCK, Wolf ED, Allen N (1987) Delivery of substances into cells and tissues using a microprojectile bombardment process. J Particle Sci Technol 5:27–37

Schulze Gronover C, Schorn C, Tudzynski B (2004) Identification of *Botrytis cinerea* genes up-regulated during infection and controlled by the Galpha subunit BCG1 using suppression subtractive hybridization (SSH). Mol Plant Microbe Interact 17:537–546

Schumann J, Hertweck C (2007) Molecular basis of cytochalasan biosynthesis in fungi: gene cluster analysis and evidence for the involvement of a PKS-NRPS hybrid synthase by RNA silencing. J Am Chem Soc 129:9564–9565

Shroff RA, O'Connor SM, Hynes MJ, Lockington RA, Kelly JM (1997) Null alleles of *creA*, the regulator of carbon catabolite repression in *Aspergillus nidulans*. Fungal Genet Biol 22:28–38

Shulaev V, Cortes D, Miller G, Mittler R (2008) Metabolomics for plant stress response. Physiol Plant 132:199–208

Siliang Zhang B-CY, Yingping Zhuang JC, Meijin G (2006) From multi-scale methodology to systems biology: to integrate strain improvement and fermentation optimization. J Chem Technol Biotechnol 81:734–745

Smedsgaard J, Nielsen J (2005) Metabolite profiling of fungi and yeast: from phenotype to metabolome by MS and informatics. J Exp Bot 56:273–286

Soga T (2007) Capillary electrophoresis-mass spectrometry for metabolomics. Methods Mol Biol 358:129–137

Spegel CF, Heiskanen AR, Kostesha N, Johanson TH, Gorwa-Grauslund MF, Koudelka-Hep M, Emneus J, Ruzgas T (2007) Amperometric response from the glycolytic versus the pentose phosphate pathway in *Saccharomyces cerevisiae* cells. Anal Chem 79:8919–8926

Stahl U, Leitner E, Esser K (1987) Transformation of *Penicillium chrysogenum* by a vector containing a mitochondrial origin of replication. Eur J Appl Microbiol Biotechnol 2:237–241

Streit WR, Schmitz RA (2004) Metagenomics-the key to the uncultured microbes. Curr Opin Microbiol 7: 492–498

Sugui JA, Chang YC, Kwon-Chung KJ (2005) *Agrobacterium tumefaciens*-mediated transformation of *Aspergillus fumigatus*: an efficient tool for insertional mutagenesis and targeted gene disruption. Appl Environ Microbiol 71:1798–1802

Takahashi T, Masuda T, Koyama Y (2006) Enhanced gene targeting frequency in *ku70* and *ku80* disruption mutants of *Aspergillus sojae* and *Aspergillus oryzae*. Mol Genet Genomics 275:460–470

Tentler S, Palas J, Enderlin C, Campbell J, Taft C, Miller TK, Wood RL, Selitrennikoff CP (1997) Inhibition of *Neurospora crassa* growth by a glucan synthase-1 antisense construct. Curr Microbiol 34:303–308

Torralba S, Raudaskoski M, Pedregosa AM, Laborda F (1998) Effect of cytochalasin A on apical growth, actin cytoskeleton organization and enzyme secretion in *Aspergillus nidulans*. Microbiology 144:45–53

Ueno K, Uno J, Nakayama H, Sasamoto K, Mikami Y, Chibana H (2007) Development of a highly efficient gene targeting system induced by transient repression of *YKU80* expression in *Candida glabrata*. Eukaryot Cell 6:1239–1247

Vallino JJ, Stephanopoulos G (2000) Metabolic flux distributions in *Corynebacterium glutamicum* during growth and lysine overproduction (reprint from Biotechnol Bioeng 41:633–646 1993). Biotechnol Bioeng 67:872–885

van der Werf MJ (2005) Towards replacing closed with open target selection strategies. Trends Biotechnol 23:11–16

Van Dien S, Schilling CH (2006) Bringing metabolomics data into the forefront of systems biology. Mol Syst Biol 2:0035

Verdoes JCPP, van der Hondel CAMJJ (1995) Molecular genetic strain improvement for the overproduction of fungal proteins by filamentous fungi. Appl Microbiol Biotechnol 43:195–205

Villalba F, Collemare J, Landraud P, Lambou K, Brozek V, Cirer B, Morin D, Bruel C, Beffa R, Lebrun MH (2008) Improved gene targeting in *Magnaporthe grisea* by inactivation of *MgKU80* required for non-homologous end joining. Fungal Genet Biol 45:68–75

Walti MA, Villalba C, Buser RM, Grunler A, Aebi M, Kunzler M (2006) Targeted gene silencing in the model mushroom *Coprinopsis cinerea* (*Coprinus cinereus*) by expression of homologous hairpin RNAs. Eukaryot Cell 5:732–744

Wang TH, Zhong YH, Huang W, Liu T, You YW (2005) Antisense inhibition of xylitol dehydrogenase gene, xdh1 from *Trichoderma reesei*. Lett Appl Microbiol 40:424–429

Ward M, Lin C, Victoria DC, Fox BP, Fox JA, Wong DL, Meerman HJ, Pucci JP, Fong RB, Heng MH, Tsurushita N, Gieswein C, Park M, Wang H (2004) Characterization of humanized antibodies secreted by *Aspergillus niger*. Appl Environ Microbiol 70:2567–2576

Weckwerth W, Wenzel K, Fiehn O (2004) Process for the integrated extraction, identification and quantification of metabolites, proteins and RNA to reveal their co-regulation in biochemical networks. Proteomics 4:78–83

Yamada O, Ikeda R, Ohkita Y, Hayashi R, Sakamoto K, Akita O (2007) Gene silencing by RNA interference in the *koji* mold *Aspergillus oryzae*. Biosci Biotechnol Biochem 71:138–144

Yu XM, Yu XD, Qu ZP, Huang XJ, Guo J, Han QM, Zhao J, Huang LL, Kang ZS (2008) Cloning of a putative hypersensitive induced reaction gene from wheat infected by stripe rust fungus. Gene 407:193–198

Zamore PD, Tuschl T, Sharp PA, Bartel DP (2000) RNAi: double-stranded RNA directs the ATP-dependent cleavage of mRNA at 21 to 23 nucleotide intervals. Cell 101:25–33

Zheng XF, Kobayashi Y, Takeuchi M (1998) Construction of a low-serine-type-carboxypeptidase-producing mutant of *Aspergillus oryzae* by the expression of antisense RNA and its use as a host for heterologous protein secretion. Appl Microbiol Biotechnol 49:39–44

Biosystematic Index

A

Subject Index